Advances in Intelligent Systems and Computing

Volume 527

Series editor

Janusz Kacprzyk, Polish Academy of Sciences, Warsaw, Poland
e-mail: kacprzyk@ibspan.waw.pl

About this Series

The series "Advances in Intelligent Systems and Computing" contains publications on theory, applications, and design methods of Intelligent Systems and Intelligent Computing. Virtually all disciplines such as engineering, natural sciences, computer and information science, ICT, economics, business, e-commerce, environment, healthcare, life science are covered. The list of topics spans all the areas of modern intelligent systems and computing.

The publications within "Advances in Intelligent Systems and Computing" are primarily textbooks and proceedings of important conferences, symposia and congresses. They cover significant recent developments in the field, both of a foundational and applicable character. An important characteristic feature of the series is the short publication time and world-wide distribution. This permits a rapid and broad dissemination of research results.

More information about this series at http://www.springer.com/series/11156

Manuel Graña · José Manuel López-Guede
Oier Etxaniz · Álvaro Herrero
Héctor Quintián · Emilio Corchado
Editors

International Joint Conference SOCO'16-CISIS'16-ICEUTE'16

San Sebastián, Spain, October 19th–21st, 2016
Proceedings

Editors
Manuel Graña
Computational Intelligence Group
University of the Basque Country (UPV/EHU)
San Sebastian
Spain

José Manuel López-Guede
University of the Basque Country
Vitoria
Spain

Oier Etxaniz
Computational Intelligence Group
University of the Basque Country (UPV/EHU)
San Sebastian
Spain

Álvaro Herrero
Department of Civil Engineering
University of Burgos
Burgos
Spain

Héctor Quintián
University of Salamanca
Salamanca
Spain

Emilio Corchado
University of Salamanca
Salamanca
Spain

ISSN 2194-5357 ISSN 2194-5365 (electronic)
Advances in Intelligent Systems and Computing
ISBN 978-3-319-47363-5 ISBN 978-3-319-47364-2 (eBook)
DOI 10.1007/978-3-319-47364-2

Library of Congress Control Number: 2016953011

Printed on acid-free paper

This Springer imprint is published by Springer Nature
The registered company is Springer International Publishing AG
The registered company address is: Gewerbestrasse 11, 6330 Cham, Switzerland

Preface

This volume of Advances in Intelligent and Soft Computing contains the accepted papers presented at SOCO 2016, CISIS 2016 and ICEUTE 2016, all conferences held in the beautiful and historic city of San Sebastián (Spain), in October 2016.

Soft computing represents a collection or a set of computational techniques in machine learning, computer science and some engineering disciplines, which investigate, simulate, and analyze very complex issues and phenomena.

After a thorough peer-review process, the 11th SOCO 2016 International Program Committee selected 45 papers which are published in these conference proceedings, and this represents an acceptance rate of 45%. In this relevant edition, a special emphasis was laid on the organization of special sessions. Two special sessions were organized related to relevant topics such as Optimization, Modeling and Control Systems by Soft Computing, and Soft Computing Methods in Manufacturing and Management Systems.

The aim of the 9th CISIS 2016 conference is to offer a meeting opportunity for academic and industry-related researchers belonging to the various, vast communities of computational intelligence, information security, and data mining. The need for intelligent, flexible behaviour by large, complex systems, especially in mission-critical domains, is intended to be the catalyst and the aggregation stimulus for the overall event.

After a thorough peer-review process, the CISIS 2016 International Program Committee selected 20 papers which are published in these conference proceedings achieving an acceptance rate of 40%.

In the case of 7th ICEUTE 2016, the International Program Committee selected 14 papers, which are published in these conference proceedings.

The selection of papers was extremely rigorous in order to maintain the high quality of the conference and we would like to thank the members of the program committees for their hard work in the reviewing process. This is a crucial process to the creation of a high-standard conference and the SOCO, CISIS and ICEUTE conferences would not exist without their help.

SOCO'16, CISIS'16, and ICEUTE'16 enjoyed outstanding keynote speeches by distinguished guest speakers: Prof. Ajith Abraham—MIR Labs (USA), Prof. Michal Wozniak—Wroclaw University of Technology (Poland), Carlos Toro—Vicomtech (Spain), and Sebastian Rios University of Chile (Chile).

SOCO'16 has teamed up with the Journal of Applied Logic (ELSEVIER) for a suite of special issue including selected papers from SOCO'16.

For this CISIS'16 special edition, as a follow-up of the conference, we anticipate further publication of selected papers in a special issue of the prestigious Logic Journal of the IGPL Published by Oxford Journals.

Particular thanks go as well to the conference main sponsors, COESI, IEEE Systems, Man and Cybernetics-Spanish Chapter, AEPIA, The International Federation for

Computational Logic, Python San Sebastián Society, University of Basque Country, who jointly contributed in an active and constructive manner to the success of this initiative. We want also to extend our warm gratitude to all the special sessions chairs for their continuing support to the SOCO, CISIS and ICEUTE series of conferences.

We would like to thank all the special session organizers, contributing authors, as well as the members of the program committees and the local organizing committee for their hard and highly valuable work. Their work has helped to contribute to the success of the SOCO 2016, CISIS 2016 and ICEUTE 2016 events.

The editors

October 2016

Manuel Graña
José Manuel López-Guede
Oier Etxaniz
Álvaro Herrero
Héctor Quintián
Emilio Corchado

SOCO 2016

Organization

General Chairs

Manuel Graña	University of Basque Country, Spain
Emilio Corchado	University of Salamanca, Spain

International Advisory Committee

Ashraf Saad	Armstrong Atlantic State University, USA
Amy Neustein	Linguistic Technology Systems, USA
Ajith Abraham	Machine Intelligence Research Labs -MIR Labs, Europe
Jon G. Hall	The Open University, UK
Paulo Novais	Universidade do Minho, Portugal
Amparo Alonso Betanzos	President Spanish Association for Artificial Intelligence (AEPIA), Spain
Michael Gabbay	Kings College London, UK
Aditya Ghose	University of Wollongong, Australia
Saeid Nahavandi	Deakin University, Australia
Henri Pierreval	LIMOS UMR CNRS 6158 IFMA, France

Program Committee Chairs

Emilio Corchado	University of Salamanca, Spain
Manuel Graña	University of Basque Country, Spain
Álvaro Herrero	University of Burgos, Spain

Program Committee

Ajith Abraham	Machine Intelligence Research Labs (MIR Labs), USA
Cesar Analide	University of Minho, Portugal
Ángel Arroyo	University of Burgos, Spain
Mehmet Emin Aydin	University of the West of England, UK
Antonio Bahamonde	University of Oviedo at Gijón, Spain
Marius Balas	Aurel Vlaicu University of Arad, Romania
Anna Bartkowiak	University of Wroclaw, Poland
Bruno Baruque	University of Burgos, Spain
Rosa Basagoiti	Mondragon University, Spain
Anna Burduk	Wrocław University of Technology, Poland
Robert Burduk	Wroclaw University of Technology, Poland

Humberto Bustince	Public University of Navarre, Spain
Jose Luis Calvo-Rolle	University of A Coruña, Spain
Andre Carvalho	University of Saõ Paulo, Brazil
José Luis Casteleiro-Roca	University of Coruña, Spain
M. Chadli	University of Picardie Jules Verne, France
Camelia Chira	Technical University of Cluj-Napoca, Romania
Sung-Bae Cho	Yonsei University, Korea, Republic of
Emilio Corchado	University of Salamanca, Spain
Jose Alfredo Ferreira Costa	Universidade Federal do Rio Grande do Norte, Brazil
José Valente De Oliveira	Universidade do Algarve, Portugal
Jorge Díez	University of Oviedo, Spain
Enrique Domínguez	University of Malaga, Spain
Abraham Duarte	University Rey Juan Carlos, Spain
Richard Duro	University of Coruna, Spain
Wilfried Elmenreich	Alpen-Adria-Universität Klagenfurt, Austria
Florentina-Magda Enescu	University of Pitesti, Romania
Florentino Fdez-Riverola	University of Vigo, Spain
Leocadio G. Casado	University of Almeria, Spain
José Gámez	University of Castilla-La Mancha, Spain
Poonthalir Ganesan	PSG College of Technology, India
Esteban García-Cuesta	Universidad Europea de Madrid, Spain
Jorge García-Gutiérrez	University of Seville, Spain
Jose García-Rodríguez	University of Alicante, Spain
George Georgoulas	TEI of Epiruw, Greece
Juan Gomez Romero	University of Granada, Spain
Manuel Graña	University of País Vasco, Spain
David Griol	Universidad Carlos III de Madrid, Spain
Maciej Grzenda	Warsaw University of Technology, Poland
Álvaro Herrero	University of Burgos, Spain
Cesar Hervas	University of Cordoba, Spain
Tzung-Pei Hong	National University of Kaohsiung, Taiwan
Wei-Chiang Hong	Nanjing Tech University, Taiwan
Valeriu-Manuel Ionescu	University of Pitesti, Romania
Krzysztof Kalinowski	Silesian University of Technology, Poland
Fanny Klett	German Workforce ADL Partnership Laboratory, Germany
Przemyslaw Korytkowski	West Pomeranian University of Technology in Szczecin, Poland
Damian Krenczyk	Silesian University of Technology, Poland
Mitiche Lahcene	University of Djelfa, Algeria
Carlos Laorden	University of Deusto, Spain
José Manuel López Guede	University of País Vasco, Spain
Martin Macas	Czech Technical University in Prague, Czech Republic
Eleni Mangina	University College Dublin, Ireland
Francesco Marcelloni	University of Pisa, Italy
María Martínez Ballesteros	University of Seville, Spain
Francisco Martínez-Álvarez	University Pablo de Olavide, Spain

Oliviu Matei	North University of Baia Mare, Romania
Manuel Mejia-Lavalle	Cenidet, Mexico
José M. Molina	University Carlos III de Madrid, Spain
María N. Moreno García	University of Salamanca, Spain
Mohamed Mostafa Fouad	Arab Academy for Science, Technology, and Maritime Transport, Egypt
Dimitris Mourtzis	University of Patras, Greece
Luís Nunes	Instituto Universitário de Lisboa, Portugal
Enrique Onieva	University of Deusto, Spain
Marcin Paprzycki	IBS PAN and WSM, Poland
Daniela Perdukova	Technical Univerzity of Kosice, Slovakia
Carlos Pereira	ISEC, Portugal
Henri Pierreval	LIMOS-IFMA, France
Stefano Pizzuti	Energy New technologies and sustainable Economic development Agency (ENEA), Italy
Jiri Pospichal	University of SS. Cyril and Methodius, Slovakia
Dilip Pratihar	Indian Institute of Technology Kharagpur, India
Héctor Quintián	University of Salamanca, Spain
Luis Paulo Reis	University of Minho, Portugal
Dragan Simic	University of Novi Sad, Serbia
Georgios Ch. Sirakoulis	Democritus University of Thrace, Greece
Aureli Soria-Frisch	Starlab Barcelona, S.L., Spain
Rui Sousa	University of Minho, Portugal
Alicia Troncoso	Universidad Pablo de Olavide, Spain
Zita Vale	GECAD - ISEP/IPP, Portugal
Mª Belén Vaquerizo	University of Burgos, Spain
Sebastián Ventura	University of Cordoba, Spain
José Ramón Villar	University of Oviedo, Spain
Eva Volna	Univerzity of Ostrava, Czech Republic
Michal Wozniak	Wroclaw University of Technology, Poland
Daniela Zaharie	West University of Timisoara, Romania
Cui Zhihua	Taiyuan University of Science and Technology, China
Urko Zurutuza	Mondragon University, Spain

Special Sessions

Optimization, Modeling and Control Systems by Soft Computing

Program Committee

Héctor Alaiz Moreton	University of Leon, Spain
José Luis Calvo Rolle	University of A Coruña, Spain
Jose Luis Casteleiro Roca	University of A Coruña, Spain
Luis Alfonso Fernández Serantes	FH-Joanneum University of Applied Sciences, Austria
Eloy Irigoyen	University of País Vasco, Spain

Esteban Jove Pérez	University of Salamanca, Spain
Maria Del Carmen Meizoso López	University of A Coruña, Spain
Andrés José Piñón Pazos	University of A Coruña, Spain
Matilde Santos	Complutense University of Madrid, Spain

Soft Computing Methods in Manufacturing and Management Systems

Program Committee

Maria Baron-Puda	University of Bielsko-Biala, Poland
Grzegorz Bocewicz	Koszalin University of Technology, Poland
Anna Burduk	Wroclaw University of Technology, Poland
Grzegorz Cwikla	Silesian University of Technology, Poland
Reggie Davidrajuh	University of Stavanger, Norway
Paul Eric Dossou	Icam, France
Laszlo Dudas	University of Miskolc, Hungary
Franjo Jović	Josip Juraj Strossmayer University of Osijek, Croatia
Krzysztof Kalinowski	Silesian University of Technology, Poland
Damian Krenczyk	Silesian University of Technology, Poland
Ivan Kuric	University of Žilina, Slovakia
Hongze Ma	KONE, Finland
Dariusz Plinta	University of Bielsko-Biala, Poland
Sebastian Saniuk	University of Zielona Gora, Poland
Bożena Skołud	Silesian University of Technology, Poland
Remus Zagan	Constanta Maritime University, Romania
Marcin Zemczak	University of Bielsko-Biala, Poland

Organising Committee

Manuel Graña	University of Basque Country, Spain
Oier Etxaniz	University of Basque Country, Spain
José Manuel López-Guede	University of Basque Country, Spain
Emilio Corchado	University of Salamanca, Spain
Álvaro Herrero	University of Burgos, Spain
Héctor Quintián	University of Salamanca, Spain

International Conference on
SOCO
Soft Computing Models in Industrial
and Environmental Applications

CISIS 2016

Organization

General Chairs

Manuel Graña	University of Basque Country, Spain
Emilio Corchado	University of Salamanca, Spain

International Advisory Committee

Ajith Abraham	Machine Intelligence Research Labs -MIR Labs, Europe
Michael Gabbay	Kings College London, UK
Antonio Bahamonde	University of Oviedo at Gijón

Program Committee Chairs

Emilio Corchado	University of Salamanca, Spain
Manuel Graña	University of Basque Country, Spain
Álvaro Herrero	University of Burgos, Spain
José López Guede	University of Basque Country, Spain

Program Committee

Isaac Agudo	University of Malaga, Spain
Cristina Alcaraz	University of Malaga, Spain
Salvador Alcaraz	Miguel Hernandez University, Spain
Rafael Alvarez	University of Alicante, Spain
Javier Areitio	University of Deusto, Spain
Angel Arroyo	University of Burgos, Spain
Juan Jesús Barbarán	University of Granada, Spain
Joan Borrell	Universitat Autònoma de Barcelona, Spain
José Daniel Britos	Universidad Nacional de Cordoba, Argentina
Robert Burduk	Wroclaw University of Technology, Poland
Pino Caballero-Gil	University of La Laguna, Spain
José Luis Calvo-Rolle	University of A Coruña, Spain
José Luis Casteleiro-Roca	University of Coruña, Spain
Michal Choras	ITTI Ltd., Poland
Ricardo Contreras	University of Concepción, Chile
Rafael Corchuelo	University of Seville, Spain

Enrique De La Hoz	University of Alcalá, Spain
Raúl Durán	University of Alcalá, Spain
Jesús Díaz-Verdejo	University of Granada, Spain
Josép Ferrer	Universitat de les Illes Balears, Spain
Amparo Fuster-Sabater	Institute of Applied Physics (CSIC), Spain
Leocadio G. Casado	University of Almeria, Spain
Debasis Giri	Haldia Institute of Technology, India
Ana I. González-Tablas	University Carlos III of Madrid, Spain
Manuel Grana	University of Basque Country, Spain
Juan Pedro Hecht	University of Buenos Aires, Argentina
Luis Hernandez Encinas	Institute of Physical and Information Tecnologies (ITEFI), Spain
José Luis Imana	Complutense University of Madrid, Spain
Gustavo Isaza	University of Caldas, Colombia
Tomasz Kajdanowicz	Wroclaw University of Technology, Poland
Wojciech Kmiecik	Wroclaw University of Technology, Poland
Zuzana Kominkova Oplatkova	Tomas Bata University in Zlin, Czech Republic
Bartosz Krawczyk	Wroclaw University Of Technology, Poland
Carlos Laorden	University of Deusto, Spain
Wenjian Luo	University of Science and Technology of China, China
José M. Molina	University Carlos III of Madrid, Spain
Guillermo Morales-Luna	CINVESTAV-IPN, Mexico
Enrique Onieva	University of Deusto, Spain
José A. Onieva	University of Malaga, Spain
Gianluca Papaleo	CNR-IEIIT, Italy
Alberto Peinado	University of Malaga, Spain
Carlos Pereira	ISEC, Portugal
Araceli Queiruga Dios	University of Salamanca, Spain
Héctor Quintián	University of Salamanca, Spain
Gerardo Rodriguez Sanchez	University of Salamanca, Spain
Luis Enrique Sanchez Crespo	University of the Armed Forces (ESPE-L), Ecuador
Gerald Schaefer	Loughborough University, UK
Hugo Scolnik	University of Buenos Aires, Argentina
Roman Senkerik	Tomas Bata University in Zlín, Czech Republic
Sorin Stratulat	Université de Lorraine, France
Javier Sánchez-Monedero	University Loyola Andalucía, Spain
Juan Tena	University of Valladolid, Spain
Antonio J. Tomeu-Hardasmal	University of Cádiz, Spain
Fernando Tricas	University of Zaragoza, Spain
Eva Volna	University of Ostrava, Czech Republic
Antonio Zamora Gómez	University of Alicante, Spain
Rodolfo Zunino	University of Genoa, Italy
Urko Zurutuza	Mondragon University, Spain

Special Sessions

Security in Wireless Networks: Mathematical Algorithms and Models

Program Committee

Raúl Durán Díaz	University of Alcalá, Spain
Luis Hernández Encinas	Institute of Physical and Information Tecnologies (ITEFI), Spain
Ángel Martín Del Rey	University of Salamanca, Spain
Agustín Martín Muñoz	Information Security Institute (CSIC), Spain
Alberto Peinado Domínguez	University of Malaga, Spain
Araceli Queiruga Dios	University of Salamanca, Spain
Gerardo Rodríguez Sánchez	University of Salamanca, Spain

Organising Committee

Manuel Graña	University of Basque Country, Spain
Oier Etxaniz	University of Basque Country, Spain
José Manuel López-Guede	University of Basque Country, Spain
Emilio Corchado	University of Salamanca, Spain
Álvaro Herrero	University of Burgos, Spain
Héctor Quintián	University of Salamanca, Spain

International Conference on

Computational Intelligence in
Security for Information Systems

ICEUTE 2016

Organization

General Chairs

Manuel Graña	University of Basque Country, Spain
Emilio Corchado	University of Salamanca, Spain

Program Committee Chairs

Emilio Corchado	University of Salamanca, Spain
Manuel Graña	University of Basque Country, Spain
Álvaro Herrero	University of Burgos, Spain
José López Guede	University of Basque Country, Spain

Program Committee

Estibaliz Apiñaniz-Fernandez de Larrinoa	University of Basque Country, Spain
Karmele Artano-Perez	University of Basque Country, Spain
Ruperta Delgado-Tercero	University of Basque Country, Spain
Manuel Graña	University of Basque Country, Spain
José Manuel López Guede	University of Basque Country, Spain
Pilar Martinez-Blanco	University of Basque Country, Spain
Amaia Mesanza-Moraza	University of Basque Country, Spain
Paulo Moura Oliveira	UTAD University, Portugal
Paulo Novais	University of Minho, Portugal
Iñaki Ochoade Eribe	University of Basque Country, Spain
Eduardo Solteiro Pires	UTAD University, Portugal
Inmaculada Tazo	University of Basque Country, Spain

Organising Committee

Manuel Graña	University of Basque Country, Spain
Oier Etxaniz	University of Basque Country, Spain
José Manuel López-Guede	University of Basque Country, Spain
Emilio Corchado	University of Salamanca, Spain
Álvaro Herrero	University of Burgos, Spain
Héctor Quintián	University of Salamanca, Spain

Contents

SOCO 2016: Soft Computing Applications

SOCO 2016: Genetic Algorithms

CISIS 2016: Applications of Intelligent Methods for Security

ICEUTE 2016

SOCO 2016: Classification

Predicting 30-Day Emergency Readmission Risk

Arkaitz Artetxe[1,2(✉)], Andoni Beristain[1,2], Manuel Graña[2], and Ariadna Besga[3]

[1] Vicomtech-IK4 Research Centre, Mikeletegi Pasealekua 57, 20009 San Sebastian, Spain
{aartetxe, aberistain}@vicomtech.org
[2] Computation Intelligence Group, Basque University (UPV/EHU), P. Manuel Lardizabal 1, 20018 San Sebastian, Spain
[3] Department of Internal Medicine, Hospital Universitario de Alava, Vitoria, Spain

Abstract. *Objective*: Predicting Emergency Department (ED) readmissions is of great importance since it helps identifying patients requiring further post-discharge attention as well as reducing healthcare costs. It is becoming standard procedure to evaluate the risk of ED readmission within 30 days after discharge. *Methods*. Our dataset is stratified into four groups according to the Kaiser Permanente Risk Stratification Model. We deal with imbalanced data using different approaches for resampling. Feature selection is also addressed by a wrapper method which evaluates feature set importance by the performance of various classifiers trained on them. *Results*. We trained a model for each scenario and subpopulation, namely case management (CM), heart failure (HF), chronic obstructive pulmonary disease (COPD) and diabetes mellitus (DM). Using the full dataset we found that the best sensitivity is achieved by SVM using over-sampling methods (40.62 % sensitivity, 78.71 % specificity and 71.94 accuracy). *Conclusions*. Imbalance correction techniques allow to achieve better sensitivity performance, however the dataset has not enough positive cases, hindering the achievement of better prediction ability. The arbitrary definition of a threshold-based discretization for measurements which are inherently is an important drawback for the exploitation of the data, therefore a regression approach is considered as future work.

Keywords: Readmission risk · Imbalanced datasets · SVM · Classification

1 Introduction

The number of people aged over 65 is projected to grow from an estimated 524 million in 2010 to nearly 1.5 billion in 2050 worldwide [1]. This trend has a direct impact on the sustainability of health systems, in maintaining both public policies and the required budgets.

This growing population group represents an unprecedented challenge for healthcare systems. In developed countries, older adults already account for 12 to 21 % of all ED visits and it is estimated that this will increase by around 34 % by 2030 [14].

M. Graña et al. (eds.), *International Joint Conference SOCO'16-CISIS'16-ICEUTE'16*,
Advances in Intelligent Systems and Computing 527, DOI 10.1007/978-3-319-47364-2_1

Older patients have increasingly complex medical conditions in terms of their number of morbidities and other conditions, such as the number of medications they use, existence of geriatric syndromes, their degree of physical or mental disability, and the interplay of social factors influencing their condition [9]. Recent studies have shown that adults above 75 years of age have the highest rates of ED readmission, and the longest stays, demanding around 50 % more ancillary tests [15]. Notwithstanding the intense use of resources, these patients often leave the ED unsatisfied, and with poorer clinical outcomes, and higher rates of misdiagnosis and medication errors [16] compared to younger patients. Additionally, once they are discharged from the hospital, they have a high risk of adverse outcomes, such as functional worsening, ED readmission, hospitalization, death and institutionalization [17].

In this paper we present our recent work on ED readmission risk prediction. We utilize historic patient information, including demographic data, clinical characteristics or drug treatment information among others. Our work focuses on high risk patients (two higher strata) according to the Kaiser Permanente Risk Stratification Model [11]. This includes patients with prominence of specific organ disease (heart failure, chronic obstructive pulmonary disease and diabetes mellitus) and patients with high multi-morbidity. Predictive models are built for each of the stratified groups using different classifiers such as Support Vector Machine (SVM) and Random Forest. In order to deal with class imbalance and high dimensional feature space, different filtering techniques have been proposed during experimental approach.

The main contributions of this work are:

- We extend the work by Besga et al. [2] applying well-known machine learning techniques such as class balancing and feature selection in order to obtain better sensitivity.
- We compare two well stablished supervised classification algorithms, Random Forests and SVM, and analyze their performance in different scenarios.
- We make use of a wrapper feature selection method that maximizes the prediction ability while minimizes models' complexity.

The paper is organized as follows. In Sect. 2 we present some related works on predictive modelling for readmission risk estimation. In Sect. 3 we present the dataset as well as the methodological approach followed in order to build our models. Next, we describe the evaluation methodology and the experimental results. In Sect. 5 we discuss the conclusions and future work.

2 Related Work

Readmission risk modelling is a research topic that has been extensively studied in recent years. The main objective is usually to reduce readmission costs by identifying those patients with higher risk of coming back soon. Patients with higher risk can be followed-up after discharge, checking their health status by means of interventions such as phone calls, home visits or online monitoring, which are resource intensive. Predictive systems generally try to model the probability of unplanned readmission (or death) of a patient within a given time period.

In a recent work, Kansagara et al. [9] presented a systematic review of risk prediction models for hospital readmission. Many of the analyzed models target certain subpopulation with specific conditions or diseases such as Acute Miocardial Infarction (AMI) or heart failure (HF) while others embrace general population.

One of the most popular models that focus on general populations is LACE [3]. The LACE index is based on a model that predicts the risk of death or urgent readmission (within 30 days) after leaving the hospital. The algorithm used to build the model is commonly used in the literature (logistic regression analysis) and, according to the published results, the model has a high discriminative ability. The model uses information of 48 variables collected from 4812 patients from several Canadian hospitals.

A variant called LACE + [4] is an extension of the previous model that makes use of variables drawn from administrative data.

A similar approach is followed by Health Quality Ontario (HQO) with their system called HARP (Hospital Admission Risk Prediction) [10]. The system aims to determine the risk of patients in short and long term future hospitalizations. HARP defines two periods of 30 days and 15 months for which the model infers the probability of hospitalization, relaying on several variables. From an initial set of variables of 4 different categories (demographic, feature community, disease and condition and meetings with the hospital system) the system identifies two sets of variables, a complex and a simpler one, with the most predictive variables. Using these sets of variables and a dataset containing approximately 382,000 episodes, two models for one month and 15 months are implemented. The models were developed using multivariate regression analysis. According to the committee of experts involved in the development of HARP, the most important metric was the sensitivity (i.e. the ability to detect hospitalizations). Regarding this metric, claimed results suggest that both simple and complex models achieve high sensitivity, although the complex model gets better results. The authors of this work suggest that the simple model could be a good substitute when certain hospitalization data is not available (e.g. to perform stratification outside the hospital).

A recent work by Yu et al. [5] presents an institution-specific readmission risk prediction framework. The idea beneath this approach is that most of the readmission prediction models have not sufficient accuracy due to differences between the patient characteristics of different hospitals. In this work an experimental study is performed, where a classification method (SVM) is applied as well as regression (Cox) analysis.

In [2] Besga et al. analyzed patients who attended Emergency Department of the Araba university Hospital (AUH) during June 2014. We exploit this dataset improving their results with further experiments.

3 Materials and Methods

The dataset, presented by Besga et al. in [2], is composed of 360 patients divided into four groups, namely: case management (CM), patients with chronic obstructive pulmonary disease (COPD), heart failure (HF) and Diabetes Mellitus (DM). For each patient a set of 97 variables were collected, divided into four main groups: (i) Sociodemographic data and baseline status, (ii) Personal history, (iii) Reasons for consultation/Diagnoses made at ED and (iv) Regular medications and other treatments. Dataset contains missing values.

In order to build our model following a binary classification approach, the target variable was set to *readmitted/not readmitted*. Those patients returning to ED within 30 days after being discharged are considered readmitted (value = 1), otherwise are seen as not readmitted (value = 0).

It is noteworthy that one patient returning the first day and another returning the 30th are both considered as *readmitted*. On the other hand, a patient returning the 31th day is considered as *not readmitted*, while in practice underwent a readmission. We believe that having the number of days passed before readmission would have been much more meaningful for identification and would have permitted even identifying a more accurate prediction, including the predicted time for readmission.

All the tests were conducted using 10-fold cross-validation. The evaluation metrics that we have used are: sensitivity, specificity and accuracy. In order to avoid any random number generation bias, we have conducted 10 independent executions with different random generating seeds and averaged the results obtained.

Table 1. Distribution of variables by category

Variable	No. (%) of variables n = 96
Sociodemographic and baseline status	4 (4.2)
Personal history	43 (44.8)
Reasons for consultation	16 (16.7)
Regular medications	33 (34.3)

According to the data shown in Table 1 our dataset has a high dimensional feature space. In this scenario we have carried out some feature selection techniques. The goal is to find a feature subset that would reduce the complexity of the model, so that it would be easier to interpret by physicians, while improving the prediction performance and reducing overfitting.

We are going to use the following approaches: filter methods and wrapper methods. Filter algorithms are general preprocessing algorithms that do not assume the use of a specific classification method. Wrapper algorithms, in the other hand, "wrap" the feature selection around a specific classifier and select a subset of features based on the classifier's accuracy using cross-validation [18]. Wrapper methods evaluate subsets of variables, that is, unlike filter methods, do not compute the worth of a single feature but the whole subset of features.

- **Filter method:** We have used Correlation-based Feature Selection (CBFS) method since it evaluates the usefulness of individual features for predicting the class along with the level of inter-correlation among them [19]. In this work we have used the implementation provided by Weka [8].

- **Wrapper method**: We have selected SVM as the specific classification algorithm and Area Under the Curve (AUC) as evaluation measure. Since an exhaustive search is impractical due to space dimensionality, we used heuristics, following a greedy stepwise approach. In this work we have used the implementation provided by Weka.

3.1 Support Vector Machine

Support vector machines (SVM) are supervised learning models which have been widely used in bioinformatics research and many other fields since their introduction in 1995 [7]. It is often defined as a non-probabilistic binary linear classifier, as it assigns new cases into one of two possible classes. In the readmission prediction problem, the model would predict whether a new case (the patient) will be readmitted within 30 days.

This algorithm is based on the idea that input vectors are non-linearly mapped into a very high dimensional space. In this new feature space it constructs a hyperplane which separates instances of both classes. Since there exist many decision hyperplanes that might classify the data, SVM tries to find the maximum-margin hyperplane, i.e. the one that represents the largest separation (margin) between the two classes.

In this work we have used the libSVM[1] implementation of the algorithm, which is the common implementation used for experimentation, and can be easily integrated to weka [8] using a wrapper. We have used a radial basis kernel function: $\exp(-\gamma*|u-v|^2)$ where $\gamma = 1/num\_features$ and $C = 1$.

3.2 Random Forest

Random Forest [6] is a classifier consisting of multiple decision trees trained using randomly selected feature subspaces. This method builds multiple decision trees at training phase. In order to predict the class of a new instance, it is put down to each of these trees. Each tree gives a prediction (votes) and the class having most votes over all the trees of the forest will be selected. The algorithm uses the bagging method, i.e. each tree is trained using a random subset (with replacement) of the original dataset. In addition, each split uses a random subset of features.

One of the advantages of random forests is that generally they generalize better than decision trees, which tend to overfitting and naturally perform some feature selection. They can also be run on large datasets and can handle thousands of attributes without attribute deletion. In this work we have used Weka's implementation of the algorithm.

---

[1] https://www.csie.ntu.edu.tw/~cjlin/libsvm/.

4 Results

In this section we analyze the prediction performance of different models on the emergency department short-time readmission dataset presented in [2]. As shown in Table 2 we have considered besides the original four subpopulations a fifth dataset that encompasses all of them.

Table 2. Comparative information about the subpopulations of the dataset

		Readmission within 30 days, no. (%) of patients	
	Overall no. of patients	No	Yes
	n = 360	n = 296 (82.2)	n = 64 (17.7)
Case management	94 (26.1)	73 (77.7)	21 (22.3)
Heart failure	70 (19.4)	62 (88.6)	8 (11.4)
Chronic obstructive pulmonary disease	80 (22.2)	64 (80)	16 (20)
Diabetes mellitus	116 (32.2)	97 (83.6)	19 (16.4)

4.1 Class Balancing

In readmission prediction analysis like in any other supervised classification problem, imbalanced class distribution leads to important performance evaluation issues and problems to achieve desired results. The underlying problem with imbalanced datasets is that classification algorithms are often biased towards the majority class and hence, there is a higher misclassification rate of the minority class instances (which are usually the most interesting ones from the practical point of view) [13].

As shown in Table 3, class imbalance is causing an accuracy paradox. If we just

Table 3. Confusion matrix of SVM on the diabetes mellitus dataset

		Predicted	
		Readmitted	Not readmitted
Actual	Readmitted	97	0
	Not readmitted	19	0

look at the accuracy of the model we get an 83.62 % although SVM just behaves as suing only the greater *a priori* probability to make the classification decision.

Resampling. There are several methods that can be used in order to tackle the class imbalance problem. Building a more balanced dataset is one of the most intuitive approaches. In our experiment we have used under-sampling as a preliminary approach and continued with an over-sampling using synthetic samples.

Under-sampling with random subsample. Given that there is a low number of samples for the minority-class, which is also the most relevant for classification, we can anticipate that reducing the amount of samples for the majority-class to be comparable to the minority-class and avoid the class imbalance will lead to a model with poor generalization capability.

Focusing on the diabetes mellitus subpopulation dataset, it is composed of 97 instances belonging to the *not-readmitted* class and only 19 of the *readmitted* class. An experiment consisting of subsampling the dataset to a distribution of 1:1.5 between the minority and majority classes, and then applying a Random Forest classifier shows the following results in Table 4.

Table 4. Comparison of performance evaluation metrics for RF over original and under-sampled versions of diabetes mellitus dataset

Dataset	Accuracy	Sensitivity	Specificity
Original	84.48	10.52	98.96
Under-sampled	61.7	31.57	82.14

As seen in Table 4, although the classification sensitivity has increased, it is still low (31.57 %) despite the sacrifice of both accuracy and specificity performance. Takin into account the low number of instances contained in our dataset, we don't consider under-sampling an effective approach.

Oversampling with SMOTE. We used Synthetic Minority Over-sampling Technique (SMOTE) [20] for oversampling the minority class. In order to avoid overfitting, we applied SMOTE (percentage of new instances equal to 200) at each fold of the 10-fold cross validation. If oversampling is done before 10-fold cross-validation, it is very likely that some of the newly created instances and the original ones are both in the training and testing sets, thus causing performance metrics being optimistic.

Our approach is to test the performance of two classifiers, namely SVM and Random Forests, using the over-sampled dataset, in order to compare it with the results obtained using the original imbalanced dataset. The choice of these two classifiers was based on the fact that both SVM and RF have been widely used in the literature, achieving good results [6]. On one hand, SVM has the advantage of been able to deal with data which is difficult to directly separate in the feature space, while on the other Random Forest has the advantage of the embedded feature selection process, which is helpful in high dimensional feature spaces. The experiment will be carried out by generating a model for each of the subpopulations on each of the specified scenarios. Table 5 shows the results of our experiment.

Results show that class-balanced dataset achieved better sensitivity than the original dataset. Nevertheless, both accuracy and specificity achieve worse results. It is worth noting that while performance is similar for both classifiers using the original dataset, SVM performs much better (in terms of sensitivity) when using the over-sampled version. At last, we observe that sensitivity improvement is rather small and it is obtained mainly at the expense of worsening both sensitivity and accuracy.

Table 5. Performance comparison using SVM and RF classifiers on original and over-sampled datasets

		Original			Over-sampled		
		Specificity	Sensitivity	Accuracy	Specificity	Sensitivity	Accuracy
Case management	SVM	1	0.42	0.87	0.98	0.42	0.86
	RF	1	0.42	0.87	1	0.42	0.87
Heart failure	SVM	1	0	0.88	0.90	0.12	0.81
	RF	1	0	0.88	1	0	0.88
COPD	SVM	1	0	0.80	0.81	0.37	0.72
	RF	1	0.37	0.87	1	0.43	0.88
Diabetes mellitus	SVM	1	0	0.83	0.88	0.15	0.76
	RF	1	0.10	0.85	0.96	0.10	0.82
All	SVM	1	0.21	0.86	0.78	0.40	0.71
	RF	1	0.28	0.87	0.99	0.28	0.86

4.2 Feature Selection

Our dataset has a high dimensional feature space. With the use of feature selection algorithms we want to find a feature subset that would reduce the complexity of the model (so that it would be easier to interpret by the physicians) while improving the prediction performance and reducing overfitting. For that purpose we are using a filter method, with Correlation-based Feature Selection [19] as metric and a wrapper method, with SVM as the specific classifier, both presented in Sect. 3.

The experiment consists in training a SVM and a RF classifier using the original feature set and the generated feature subsets. The performance of the classifiers will be compared in terms of sensitivity, specificity and accuracy for each of the subpopulations.

It's worth noting that the feature selection must be done using cross-validation. If full training set is utilized during attribute selection process, the generalization ability of the model can be compromised.

Table 6. Performance comparison of both feature selection methods

		Filter (CBFS)			Wrapper (SVM)		
		Specificity	Sensitivity	Accuracy	Specificity	Sensitivity	Accuracy
Case management	SVM	0.97	0.33	0.82	0.94	0.23	0.78
	RF	0.86	0.42	0.76	0.89	0.38	0.77
Heart failure	SVM	0.96	0.12	0.87	0.90	0.12	0.81
	RF	0.96	0.25	0.88	0.98	0	0.87
COPD	SVM	0.96	0.18	0.81	0.95	0.37	0.83
	RF	0.89	0.18	0.75	0.92	0.37	0.81
Diabetes mellitus	SVM	0.98	0	0.82	0.98	0.05	0.83
	RF	0.98	0	0.82	0.98	0.05	0.83
All	SVM	0.98	0.06	0.82	0.97	0.14	0.83
	RF	0.94	0.15	0.80	0.95	0.18	0.81

In Table 6 the results of the experiment are shown. According to these results, although in some cases the sensibility has been increased, overall the results are not as promising as expected. Actually, even though models are much simpler than the original model (i.e. the one using full feature set), the prediction performance has been reduced. Moreover, both feature selection methods have performed similarly, even if selected feature subsets differs considerably.

5 Conclusions and Future Work

This paper has presented a work on the prediction of 30-day readmission risk in Emergency Department. Several contributions have been presented regarding the enhancement of predictor's performance, with special focus on sensitivity, i.e. predictive power of the critical class of readmitting patients. First, we have conducted an experiment that shows the performance variations produced by class-balancing techniques. Second, we analyze different feature selection methods and metrics and evaluate their performance. Two classification algorithms have been used (SVM and Random Forest) in order to evaluate the different approaches.

According to the results of our analysis, we conclude that although class balancing improves sensitivity results, the dataset seems not to have enough minority-class instances. In addition, setting 30 days as the arbitrary threshold for assigning the binary class label may cause situations such as labelling a patient readmitted the 30th day as "readmitted" and another readmitted the 31st day as "not readmitted". This imposes a clear limitation to any generated model, since actually both patients should be treated as similar (in terms of readmission).

Future work will include addressing the problem with a regression approach, instead of supervised classification. Thus we want to avoid the mentioned arbitrary labelling problem. With a regression analysis approach we plan to predict not only the readmission risk but also the approximate readmission window (i.e. the time interval from hospital discharge and readmission).

We also plan to increase the size of the dataset, including more instances of the minority class. Extending the samples of the readmission class we expect to achieve better predictions and ultimately generate a better-generalizing model.

References

1. World Health Organization: Global health and ageing. World Health Organization, Geneva, Switzerland (2011)
2. Besga, A., Ayerdi, B., Alcalde, G., et al.: Risk factors for emergency department short time readmission in stratified population. BioMed Res. Int. 2015, 7 pages (2015). Article ID 685067, doi:10.1155/2015/685067
3. Van Walraven, C., et al.: Derivation and validation of an index to predict early death or unplanned readmission after discharge from hospital to the community. Can. Med. Assoc. J. **182**(6), 551–557 (2010)

4. Van Walraven, C., Wong, J., Forster, A.: LACE+ index: extension of a validated index to predict early death or urgent readmission after hospital discharge using administrative data. Open Med. **6**(3), 80–89 (2012)
5. Yu, S., Farooq, F., van Esbroeck, A., Fung, G., Anand, V., Krishnapuram, B.: Predicting readmission risk with institution-specific prediction models. Artif. Intell. Med. **65**(2), 89–96 (2015)
6. Ho, T.K.: Random decision forests. In: 1995 Proceedings of the Third International Conference on Document Analysis and Recognition, pp. 278–282. IEEE (1995)
7. Cortes, C., Vapnik, V.: Support-vector networks. Mach. Learn. **20**(3), 273–297 (1995)
8. Hall, M., Frank, E., Holmes, G., Pfahringer, B., Reutemann, P., Witten, I.H.: The WEKA data mining software: an update. SIGKDD Explor. **11**(1), 10–18 (2009)
9. Kansagara, D., Englander, H., Salanitro, A., Kagen, D., Theobald, C., Freeman, M., Kripalani, S.: Risk prediction models for hospital readmission: a systematic review. JAMA **306**(15), 1688–1698 (2011)
10. Health Quality Ontario - Early Identification of People At-Risk of Hospitalization. ISBN 978-1-4606-2908-6 (PDF) Queen's Printer for Ontario (2013). Accessed 09 Mar 2016. Enlace: https://secure.cihi.ca/free_products/HARP_reportv_En.pdf
11. Feachem, R.G., Dixon, J., Berwick, D.M., Enthoven, A.C., Sekhri, N.K., White, K.L.: Getting more for their dollar: a comparison of the NHS with California's Kaiser Permanente. BMJ **324**(7330), 135–143 (2002)
12. Chawla, N.V., Bowyer, K.W., Hall, L.O., Kegelmeyer, W.P.: SMOTE: synthetic minority over-sampling technique. J. Artif. Intell. Res. **16**, 321–357 (2002)
13. López, V., Fernández, A., García, S., Palade, V., Herrera, F.: An insight into classification with imbalanced data: Empirical results and current trends on using data intrinsic characteristics. Inf. Sci. **250**, 113–141 (2013)
14. Carpenter, C.R., Heard, K., Wilber, S., Ginde, A.A., Stiffler, K., Gerson, L.W., et al.: Research priorities for high-quality geriatric emergency care: medication management, screening, and prevention and functional assessment. Acad. Emerg. Med. **18**(6), 644–654 (2011)
15. Lopez-Aguila, S., Contel, J.C., Farre, J., Campuzano, J.L., Rajmil, L.: Predictive model for emergency hospital admission and 6-month readmission. Am. J. Manage. Care **17**(9), e348–e357 (2011)
16. Han, J.H., Zimmerman, E.E., Cutler, N., Schnelle, J., Morandi, A., Dittus, R.S., et al.: Delirium in older emergency department patients: recognition, risk factors, and psychomotor subtypes. Acad. Emerg. Med. **16**(3), 193–200 (2009)
17. New guidelines for geriatric EDs: guidance focused on boosting environment, care processes. ED Manage **26**(5), 49–53 (2014)
18. Phuong, T.M., Lin, Z., Altman, R.B.: Choosing SNPs using feature selection. In: 2005 IEEE Computational Systems Bioinformatics Conference (CSB 2005), pp. 301–309. IEEE (2005)
19. Hall, M.A.: Correlation-based feature selection for machine learning (Doctoral dissertation, The University of Waikato) (1999)
20. Chawla, N.V., Bowyer, K.W., Hall, L.O., Kegelmeyer, W.P.: SMOTE: synthetic minority over-sampling technique. J. Artif. Intell. Res. **16**, 321–357 (2002)

Use of Support Vector Machines and Neural Networks to Assess Boar Sperm Viability

Lidia Sánchez[1(✉)], Héctor Quintian[2], Javier Alfonso-Cendón[1], Hilde Pérez[1], and Emilio Corchado[2]

[1] Department of Mechanical, Computer and Aerospace Engineering, University of León, León, Spain
{lidia.sanchez,javier.alfonso,hilde.perez}@unileon.es
[2] Departamento de Informática y Automática, Universidad de Salamanca, Salamanca, Spain
{hector.quintian,es.corchado}@usal.es

Abstract. This paper employs well-known techniques as Support Vector Machines and Neural Networks in order to classify images of boar sperm cells. Acrosome integrity gives information about if a sperm cell is able to fertilize an oocyte. If the acrosome is intact, the fertilization is possible. Otherwise, if a sperm cell has already reacted and has lost its acrosome or even if it is going through the capacitation process, such sperm cell has lost its capability to fertilize. Using a set of descriptors already proposed to describe the acrosome state of a boar sperm cell image, two different classifiers are considered. Results show the classification accuracy improves previous results.

Keywords: Boar sperm cell · Acrosome state · Digital image processing · Classification

1 Introduction

In several works, digital image processing has been applied to boar sperm cell images in order to assess the acrosome state [1–6,11,13,17,20]. This property allows veterinarians to estimate the quality of a sample. Capacitation is a requirement for acrosome reaction and successful fertilization in mammals [16]. During such reaction, the anterior sperm head plasma membrane fuses with the outer membrane of the acrosome, exposing the contents of the acrosome. The released enzymes are required for the penetration of sperm through a layer of follicular (cumulus) cells that encase the oocyte. The acrosome reaction also renders the sperm capable of penetrating through the zone pellucida (an extracellular coat surrounding the oocyte) and fusing with the egg. For these reasons, veterinary experts believe that a semen sample with a high fraction of acrosome-damaged sperm (which present an acrosome reacting or detached) has low fertilizing capacity and cannot be used for artificial insemination [8]. For instance, if 30 % of spermatozoa of a semen sample present abnormalities, then such sample will have a

M. Graña et al. (eds.), *International Joint Conference SOCO'16-CISIS'16-ICEUTE'16*, Advances in Intelligent Systems and Computing 527, DOI 10.1007/978-3-319-47364-2_2

reduced fertility. The integrity of the plasma membrane of the spermatozoon is of crucial importance for sperm fertility [18].

Such information is useful in veterinary practice because artificial insemination centres require that provided samples present high probability to fertilize [10]. Although this analysis has been made by qualified staff by using staining techniques, specialized devices and even subjective visual estimation, such methods are expensive or subjective. Usually, stains and manual counting of stained spermatozoa are employed to estimate the fertility of a sample [15,19]. The goal is to avoid the use of conventional staining techniques that present some drawbacks like the slowness of the procedure and the requirement of expensive equipment, for instance, the fluorescence microscope [19]. They are also sensitive to temperature variations and manual manipulation of the samples, it is possible to make errors in pH adjustment and the stain preparation time is relatively long. Stains might be even toxics [9]. For that reason, Computer Aided Seminal Analysis (CASA) is an ongoing research field as it is showed by the increasing number of papers with this topic [21].

Neural networks have been used to classify sub cellular structures in fluorescence microscope images of cells [7] and also to assess semen quality [14]. Support Vector Machines are also employed to determine if a sample presents integrity in its acrosome or not [11]. In this paper, this techniques are considered to classify the set of descriptors that represents a sperm cell.

The paper is organized as follows. Section 2 explains the feature extraction. Experiments and results are discussed in Sect. 3. Finally, conclusions are gathered in Sect. 4.

2 Description of the Features

In [20] a set of features in order to describe the acrosome state of boar sperm cells is proposed. A subset of such features has been considered in [2] to assess the acrosome integrity. Here, the same dataset is used for the experiments. A bigger dataset that includes the considered samples is freely available[1] [12]. This dataset is described in the following paragraphs.

First, sperm cells are segmented from digital images of samples. After segmentation, the boundary of the sperm head image is obtained and 6 additional concentric contours ($c_i, i = 1, 2, ...6$) inside of the cell boundaries are computed. What defines the concentric contours is the Euclidean distance of each point of the boundary from the centroid (x_c, y_c), that is divided into 6 segments. So, each point at the boundary (x_p, y_p) produces 6 inner points (x_{inner}, y_{inner}) defined as the ones that are $d((x_p, y_p), (x_{inner}, y_{inner}))$ pixels far away from the point of the boundary (x_p, y_p). A logarithmic distance is used since the most of the information lays next to the border of the sperm cell. So, the closer the contours are, the closer the boundary is.

[1] http://pitia.unileon.es/varp/node/361.

$$d((x_p, y_p), (x_{inner}, y_{inner})) = d((x, y), (x_c, y_c)) * \frac{c_i}{c_N + 1} * (1 - 0.7 * \log_{10}(1 + c_N - c_i)) \quad (1)$$

being c_N the number of contours.

Some examples of the computed contours are shown in Fig. 1 for images of both types, with an intact acrosome and with an acrosome that has already reacted.

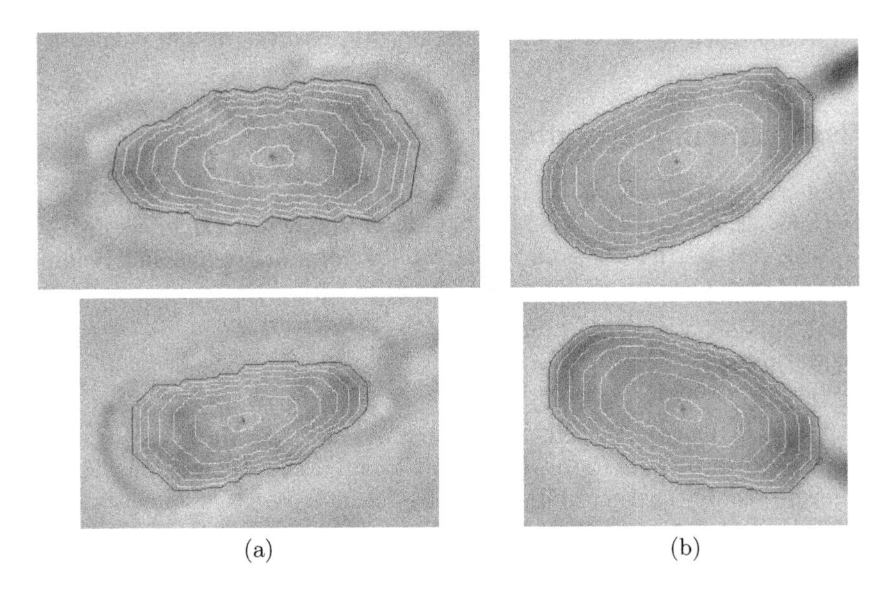

Fig. 1. Images of the boundaries and inner contours for both acrosome-damaged (a) sperm cells and acrosome-intact (b) sperm cells.

As it is explained in [2], different sets of texture features for a neighborhood of each point of the contours are computed:

- Local maximum gradient values: the maximum gradient value in a 5-by-5 neighborhood N after computing the scale-dependent gradient [17]:

$$v_i = \max_{(x,y)\in N} \left(\sqrt{\left(f * \frac{\partial g_\sigma}{\partial x}\right)^2 + \left(f * \frac{\partial g_\sigma}{\partial y}\right)^2} \right) \quad (2)$$

- Local mean of the grey level values: the mean of the grey levels of a pixel $I(x, y)$ in a 3-by-3 neighborhood N:

$$v_i = \frac{1}{9} \sum_{(x,y)\in N} I(x, y) \quad (3)$$

- Local standard deviation: the standard deviation of the grey levels in a n-sized neighborhood N (3-by-3), being μ_I the mean of the grey levels for that neighborhood: $\mu_I = \frac{1}{n}\sum_{(x,y)\in N} I(x,y)$:

$$v_i = \sqrt{\frac{1}{n-1}\sum_{(x,y)\in N}(I(x,y)-\mu_I)^2} \tag{4}$$

The features are computed from the point where tail starts along the contours clockwise. The obtained feature vector is interpolated to a constant size of 40 elements. So, each sperm cell produces a feature vector of 280 elements (40 features * 7 contours).

3 Experimental Results

Experiments have been carried out considering the same dataset as the one used in [2]. However, we have classified the dataset with different classifiers instead of using k-Nearest-Neighbor classification, class conditional means and Relevance Learning Vector Quantization which are used in [2]:

- Linear Support Vector Machine
- Quadratic Support Vector Machine
- A Neural Network.

Each method has classified the feature vectors formed by the considered descriptors: gradient, gray level mean and local standard deviation. The data set is formed by 360 sperm cell images, 210 of them have an intact acrosome and 150 have a reacted acrosome.

3.1 Linear and Quadratic Support Vector Machine

As the images have to be labelled into two classes, we have considered Support Vector Machines to the classification stage. We first train the SVM and then cross validate the classifier. In order to avoid overfitting, a k-fold cross-validation with k = 50 has been considered for the classification. We have considered both linear and quadratic Kernel functions.

3.2 Neural Network

A two-layer feed-forward network with 10 neurons in the hidden layer and sigmoid hidden and output neurons has been used for the classification. The training is carried out with scaled conjugate gradient backpropagation. Its performance has been evaluated using mean square error and confusion matrices.

We have considered two different experiments with different sizes of samples for training, validation and test, as it is shown in Table 1. So, for the experiment 1 we have split the set of feature vectors so a 60 % is used for training purposes, a 15 % of the samples are employed for validation and the remaining 25 % of the images are classified to test the network. The experiment 2 considers 70 % of the samples for training, 15 % for validation and the remaining 15 % for testing.

Table 1. Number of samples ($\sharp N$) used for training, validation and test in the Neural Network experiments.

	Experiment 1		Experiment 2	
	$\sharp N$	%	$\sharp N$	%
Training	216	60	252	70
Validation	54	15	54	15
Test	90	25	54	15

3.3 Discussion

Table 2 shows the percent error obtained for the different set of features using the three considered classifiers. Best results are obtained for the maximum value of the gradient and neural networks, that classifies the test set without error. This improves the results obtained in [2] where the hit rate was a 99 %. For the set of features formed by the local standard deviation, the results are also better than the ones in [2], obtaining a hit rate of 100 %. Finally, grey level features also provide a higher hit rate than in [2] since the hit rate achieved is a 98.15 % instead of the previous 94.9 %.

Table 2. Percent error $E\%$ obtained for the three considered classifiers.

	Linear SVM	Quadratic SVM	NN 25 % Test	NN 15 % Test
Grey levels	3.60 %	5.00 %	3.33 %	1.85 %
Max gradient	3.60 %	2.50 %	1.11 %	0 %
Std dev	1.70 %	1.70 %	1.11 %	0 %

4 Conclusions

In this paper a new classification scheme has been approached in order to determine if a feature vector that represents a boar sperm cell presents its acrosome intact or damage. Using a Neural Network and both Linear and Quadratic Support Machines, the classification stage is carried out. Obtained results show an improvement in the achieved hit rate comparing the results with other works that use the same dataset. For the three type of feature vectors that represent a boar sperm cell (grey level mean, local standard deviation and maximum gradient) the error decreases from a 1 % to a 1.5 %. Using Neural Networks the accuracy is a 100 %.

References

1. Alegre, E., Biehl, M., Petkov, N., Sanchez, L.: Automatic classification of the acrosome status of boar spermatozoa using digital image processing and LVQ. Comput. Biol. Med. **38**(4), 461–468 (2008)
2. Alegre, E., Biehl, M., Petkov, N., Sanchez, L.: Assessment of acrosome state in boar spermatozoa heads using n-contours descriptor and RLVQ. Comput. Methods Programs Biomed. **111**, 525–536 (2013)
3. Alegre, E., García-Olalla, O., González-Castro, V., Joshi, S.: Boar spermatozoa classification using longitudinal and transversal profiles (LTP) descriptor in digital images. In: Aggarwal, J.K., Barneva, R.P., Brimkov, V.E., Koroutchev, K.N., Korutcheva, E.R. (eds.) IWCIA 2011. LNCS, vol. 6636, pp. 410–419. Springer, Heidelberg (2011). doi:10.1007/978-3-642-21073-0_36
4. Alegre, E., Garcia-Ordas, M., Gonzalez-Castro, V., Karthikeyan, S.: Vitality assessment of boar sperm using NCSR texture descriptor in digital images. In: Vitrià, J., Sanches, J.M., Hernández, M. (eds.) IbPRIA 2011. LNCS, vol. 6669. Springer, Heidelberg (2011)
5. Alegre, E., Gonzalez-Castro, V., Alaiz-Rodriguez, R., Garcia-Ordas, M.: Texture and moments-based classification of the acrosome integrity of boar spermatozoa images. Comput. Methods Programs Biomed. **108**(2), 873–881 (2012)
6. Bijar, A., Pealver-Benavent, A., Mikaeili, M., Khayati, R.: Fully automatic identification and discrimination of sperm parts in microscopic images of stained human semen smear. J. Biomed. Sci. Eng. **5**, 384–395 (2012)
7. Boland, M., Murphy, R.: A neural network classifier capable of recognizing the patterns of all major subcellular structures in fluorescence microscope images of hela cells. Bioinformatics **17**(12), 1213–1223 (2001)
8. Chan, J., Krause, W., Bohring, C.: Computer-assisted analysis of sperm morphology with the aid of lectin staining. Andrologia **34**(6), 379–383 (2002)
9. Downing, T., Garner, D., Ericsson, S., Redelman, D.: Metabolic toxicity of fluorescent stains on thawed cryopreserved bovine sperm cells. J. Histochem. Cytochem. **39**(4), 485–489 (1991)
10. Fazeli, A., Hage, W., Cheng, F.P., Voorhout, W., Marks, A., Bevers, M., Colenbrander, B.: Acrosome-intact boar spermatozoa initiate binding to the homologous zona pellucida in vitro. Biol. Reprod. **56**, 430–438 (1997)
11. Garcia-Olalla, O., Alegre, E., Fernandez-Robles, L., Malm, P., Bengtsson, E.: Acrosome integrity assessment of boar spermatozoa images using an early fusion of texture and contour descriptors. Comput. Methods Programs Biomed. **120**(1), 49–64 (2015)
12. Gonzalez-Castro, V., Alaiz-Rodriguez, R., Alegre, E.: Class distribution estimation based on the hellinger distance. Inf. Sci. **218**, 146–164 (2013)
13. González-Castro, V., Alegre, E., García-Olalla, O., García-Ordás, D., García-Ordás, M.T., Fernández-Robles, L.: Curvelet-based texture description to classify intact and damaged boar spermatozoa. In: Campilho, A., Kamel, M. (eds.) ICIAR 2012. LNCS, vol. 7325, pp. 448–455. Springer, Heidelberg (2012). doi:10.1007/978-3-642-31298-4_53
14. Linneberg, C., Salamon, P., Svarer, C., Hansen, L.: Towards semen quality assessment using neural networks. In: Proceedings of IEEE Neural Networks for Signal Processing IV (1994)
15. Neuwinger, J., Behre, H., Nieschlag, E.: External quality control in the andrology laboratory: an experimental multicenter trial. Fertil. Steril. **54**(2), 308–314 (1990)

16. Oliva-Hernandez, J., Corcuera, B., Perez-Gutierrez, J.: Epidermal growth factor (EGF) effects on boar sperm capacitation. Reprod. Domest. Anim. **40**, 353 (2005)
17. Petkov, N., Alegre, E., Biehl, M., Sanchez, L.: LVQ acrosome integrity assessment of boar sperm cells. In: Proceedings of Computational Modelling of Objects Represented in Images: Fundamentals, Methods and Applications (CompIMAGE) (2006)
18. Petrunkina, A., Petzoldt, R., Stahlberg, S., Pfeilsticker, J., Beyerbach, M., Bader, H., Topfer-Petersen, E.: Sperm-cell volumetric measurements as parameters in bull semen function evaluation: correlation with nonreturn rate. Andrologia **33**, 360–367 (2001)
19. Pinart, E., Bussalleu, E., Yeste, M., Briz, M., Sancho, S., Garcia-Gil, N., Badia, E., Bassols, J., Pruneda, A., Casas, I., Bonet, S.: Assessment of the functional status of boar spermatozoa by multiple staining with fluorochromes. Reprod. Domest. Anim. **40**, 356 (2005)
20. Sanchez, L.: Boar sperm cell classification using digital image processing. Ph.D. thesis, University of Leon, Spain (2007)
21. Verstegen, J., Iguer-Ouada, M., Onclin, K.: Computer assisted semen analyzers in andrology research and veterinary practice. Theriogenology **57**, 149–179 (2002)

Learning Fuzzy Models with a SAX-based Partitioning for Simulated Seizure Recognition

Paula Vergara[1], José Ramón Villar[1], Enrique de la Cal[1(✉)], Manuel Menéndez[2], and Javier Sedano[3]

[1] Computer Science Department, University of Oviedo, Oviedo, Spain
{UO32599,villarjose,delacal}@uniovi.es
[2] Morphology and Cellular Biology Department, University of Oviedo, Oviedo, Spain
menendezgmanuel@uniovi.es
[3] Instituto Tecnológico de Castilla y León, León, Spain
javier.sedano@itcl.es

Abstract. Wearable devices are currently used in researches related with the detection of human activities and the anamnesis of illnesses. Recent studies focused on the detection of simulated epileptic seizures have found that Fuzzy Rule Base Classifiers (FRBC) can be learnt with Ant Colony Systems (ACS) to efficiently deal with this problem. However, the computational requirements for obtaining these models is relatively high, which suggests that an alternative for reducing the learning cost would be rather interesting. Therefore, this study focuses on reducing the complexity of the model by using a discretization technique, more specifically, the discretization proposed in the SAX Time Series (TS) representation.

Therefore, the very simple discretization method based on the probability distribution of the values in the domain is used together with the AntMiner+ and a Pittsburg FRBC learning algorithm using ACS. The proposal have been tested with a realistic data set gathered with participants following a very strict protocol for simulating epileptic seizures, each participant using a wearable device including tri-axial accelerometers placed on the dominant wrist.

The experimentation shows that the discretization method has clearly improved previous published results. In the case of Pittsburg learning, the generalization capabilities of the models have been greatly enhanced, while the models learned with this partitioning and the AntMiner+ have outperformed all the models in the comparison. These results represent a promising starting point for the detection of epileptic seizures and will be tested with patients in their own environment: it is expected to start gathering this data during the last quarter of this year.

Keywords: Time series · Symbolic aggregate approximation · Fuzzy rule based classifiers · Ant colony system · Discretize

M. Graña et al. (eds.), *International Joint Conference SOCO'16-CISIS'16-ICEUTE'16*, Advances in Intelligent Systems and Computing 527, DOI 10.1007/978-3-319-47364-2_3

1 Introduction

Wearable devices have gather the focus of the research community in the recent years due to their inherently ubiquity. Their application in a wide range of domains has been reported in the literature. For instance, tri-axial accelerometers (3DACC) have been used for human activity recognition and daily life movements classification [1,2]. The application of 3DACC in the detection and diagnosis of illnesses has been also acknowledged [3,4]. In some cases, the raw acceleration value needs processing in order to obtain the body acceleration -the acceleration of the part of the body where the sensor is placed-. Further preprocessing is also required to obtain the most remarkable acceleration transformation, like first to fourth statistical moments, the Signal Magnitude ration, among many others [3].

There are several concerns with the use of wearable sensors, though. For instance, the data stream would require an impressive amount of storage space, making virtually impossible to deploy solutions in the society. The huge amount of gathered data suggests not only the need of discretization to reduce storage space, but also to complete the optimization phase using symbolic algorithms [5,6].

The detection of epileptic seizures using 3DACC has been studied so far in the literature, e.g. [7,8]. However, the practical totality of these studies restrict the patients to in-room experiments, during the night sleeping or on the bed. There is a lack of studies of the detection of epileptic seizures in daily life. This study focuses on this problem, making use of a 3DACC and of a simulated epileptic seizure data set.

Previous research showed that Fuzzy Rule Based Classifiers (FRBC) can deal with the identification of these simulated epileptic seizures [4,9,10] when learned using the Pittsburg Ant Colony System proposed in [11]. Nevertheless, both the dispersion of the model performances and the high computational costs suggest the need of some simplifications and improvements in the modeling. In this research, we make use of SAX (Symbolic Aggregate Approximation) Time Series (TS) representation to obtain a fuzzy partitioning scheme. Furthermore, AntMiner+ and the Pittsburg FRBC learned with Ant Colony System are used for learning models to detect simulated epileptic seizures.

The organization of this manuscript is as follows. In the next section, the different techniques used in this study are outlined. Next, in Sect. 3, the main contribution of this paper is described. Section 4 deals with materials and methods, the experimentation and results. Finally, the main conclusions and the future work of this study are drawn.

2 Related Techniques

Over various years, ACO [12,13] has been applied to a variety of different problems such as scheduling [14], the traveling salesman problem [15,16], clustering [17,18] and classification tasks [19–21], which are the topic of interest in

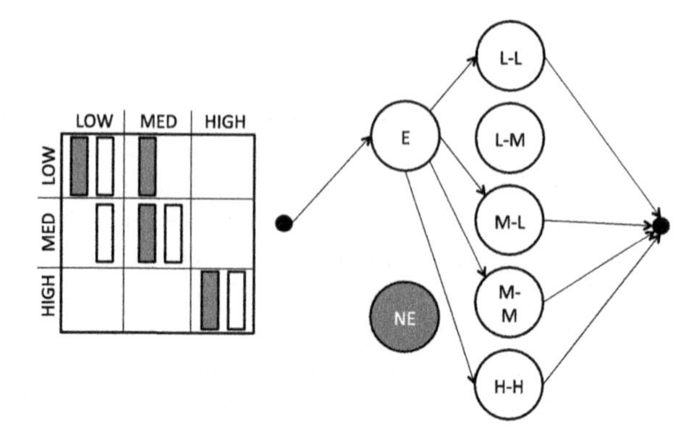

Fig. 1. The graph construction for a two-class problem. The table on the left represents the support of each possible rule according to the current data set. As the majority class is left as default, the E class is chosen. The graph on the right shows the valid paths -those with support-. Afterwards, two extra nodes are added with the candidates α and β pheromone updating parameters.

this study. The first application of ACO to classification rules was reported by Parpinelli [22,23], where the AntMiner algorithm is introduced, as well as other different strategies such as AntMiner2 [19], AntMiner3 [20] and AntMiner+ [21]. In this study we will use two different ACO-based classification techniques: a Michigan approach for learning FRBC using AntMiner+ and a Pittsburg approach for learning FRBC using ACS proposed in [11] and used in the context of this problem in [3,9]. These two techniques are briefly introduced in the next two subsections.

Furthermore, SAX is a Time Series (TS) representation technique that includes a statistical discretization method. As long as SAX is used for enhancing the learning process, the final subsection describes this technique in short.

2.1 AntMiner+

AntMiner+ [21,24] is an algorithm for sequentially discovering IF-THEN classification rules in discrete -either ordered or unordered- domains. Therefore, this method can be considered as a Michigan rule learning approach. AntMiner+ chooses the majority class as the default class, filtering all these examples from the data set. The algorithm represents a rule as an individual. To create an individual, the rule class is firstly randomly chosen and then the nodes for each variable are chosen according to the current pheromone values. Finally, the α and β parameters for the pheromone update are chosen from some tentative accepted values. For a more in-depth description of the algorithm, please refer to the original sources [21,24] (Fig. 1).

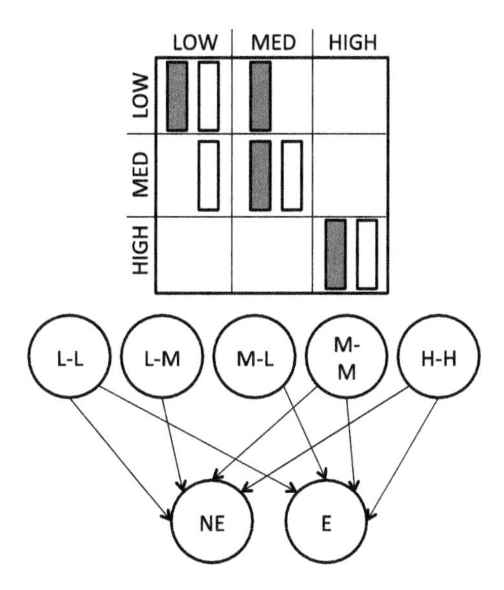

Fig. 2. A simplified graph following the Pittsburgh approach. This scheme uses two input variables in a two-class problem -classes EPILEPSY (E), NO_EPILEPSY(NE)- outlining the rule generation based on the support. This configuration leads to a set of 8 rules: Low-Low-NE, Low-Low-E, Low-Med-NE, Med-Low-NE, Med-Low-E, Med-Med-NE, Med-Med-E, and High-High-E. The graph is depicted at the bottom part of the figure.

2.2 Learning FRBC with ACS

Casillas et al. proposed the use of ACS to learn FRBC in a Pittsburg fashion [11,25,26]. In this approach, the fuzzy partitioning is firstly set up. Once given, the table of support is computed according to the data set. Each cell in this table represents a combination of the antecedents, which can be use as the antecedent part of an IF-THEN rule. The aim of this algorithm is to learn best association among the non-empty support antecedents and the available classes. Next step in the algorithm is the initial calculation of the pheromone for each path is computed, considering that each non-empty cell in the support table has a possible path -one for each of the non-empty support classes-. In this algorithm, each individual includes all the non-empty cells, assigning to each of them one of the possible classes, creating a complete FRBC -see Fig. 2. The evaluation of this model leads the pheromone update, changing the probabilities of each of the path as valid candidates. Again, we refer to interested readers to the sources for further details.

2.3 Simbolic Aggregation AproXimation

The SAX [6,27] method carries out the symbolic representation of the Time Series, meaning that dimensionality/numerosity may be reduced with no loss

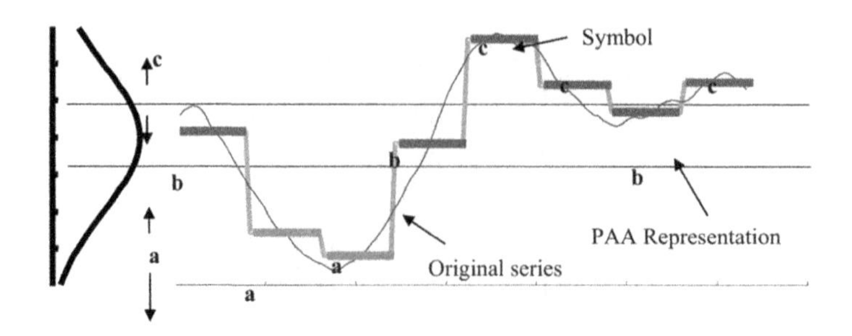

Fig. 3. SAX discretization: First we obtain a PAA of the time series. All coefficients greater than or equal to the smallest breakpoints and less than the second smallest breakpoints are mapped by the symbol "a". All coefficients greater than or equal to the second smallest breakpoints and less than the third smallest breakpoints are mapped by the symbol "b" and so on.

of structure. It also defines distance measurements on the symbolic domain, so that lower bound corresponding distance measures can be defined on the original series. To that end, the time series is divided into a finite predefined number of segments and the mean value is computed for each of them following PAA (Piecewise Aggregate Approximation). Each value is assigned the more statistically similar symbol. Each symbol represent an equiprobable interval in the range of the variable; the intersection of each pair of symbols is the empty set, while the union of all the symbols equals the variable range.

Figure 3 shows the main idea of SAX. Besides, the range is divided in equiprobable intervals, whose limits are specifically defined in the original paper based on remarkable probability thresholds; each of this equiprobable intervals are given with a symbol -see the left part of the figure-. The window to represent is normalized according to the mean and standard deviation of the values within the window. A TS is split in A subintervals and the mean value within each subinterval is computed. For each subinterval, the symbol assigned to the equiprobable interval that contains the mean value is chosen.

3 Introducing SAX in the Model Learning

Some slight modifications were performed on the SAX method in order to be used in the model learning. Firstly, the data normalization has been performed not in window basis but with the whole training data set. This way the data will not be affected by the variations from each subject and a generalized discretization is obtained. These mean and standard deviation are also used in the normalization of the testing and/or validation data sets. Secondly, instead of introducing a reduced number of symbols, we propose the use of a relatively high number of symbols, e.g., 7, 9 and 11 symbols. This high number of symbols would help in the discrimination of movements with very different intensity of movements.

Given a number of symbols and the above described data normalization, we can proceed with AntMiner+ as follows. Each input variable is initially

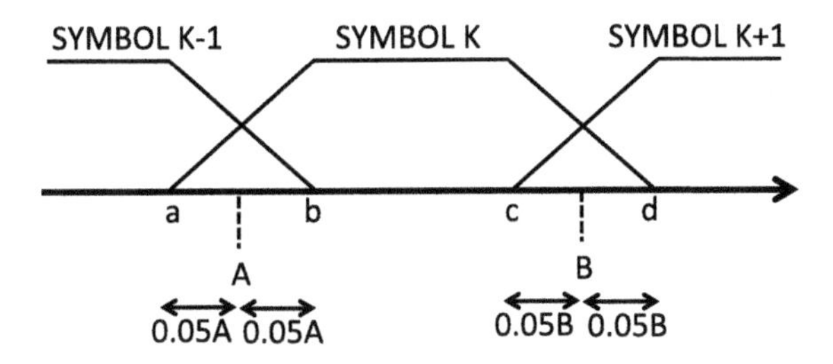

Fig. 4. Converting from an interval partition to a fuzzy partition. Further constrains should be revised; for instance, $a \leq b \leq c \leq d$.

partitioned in equiprobable intervals as proposed in SAX. These intervals are converted to fuzzy labels using trapezoidal membership functions as suggested in [10]; Fig. 4 depicts the method the fuzzy partitioning scheme is obtained. From now on, xAMP refers to AntMiner+ learning FRBC using the fuzzy partitioning scheme with x symbols.

Similarly, the fuzzy partitioning scheme described for AntMiner+ can also be used for learning Pittsburg FRBC models with ACS [3,9,11]. From now on, we refer as xACS this model learning using x SAX symbols.

4 Experiment and Results

4.1 Materials and Methods

The data used in this research form a realistic data set gathered following a previously defined and very strict protocol. In a previous study [9,28], a realistic experimentation was described with 6 participants, performing 10 trial runs of 4 activities (epilepsy convulsions simulation, running, sawing and walking). Each participant wore a 3DACC bracelet, sampling the acceleration at 16 Hz, the data set is also available online [4,29].

The data validation method based on participants (5×2 cross-validation) was taken from [3,4], where the participants were shuffled before they were placed in either the modeling or the validation data set. Once shuffled, the time-series set from each participant involved in modeling the data set was split in two: one for training -60 of the available time series for the participant- and the other for testing. The response of the models, for unseen time series from the participants involved in training, can be evaluated for the testing data set, while the validation data set allows evaluation of the generalized capabilities of the models. It is worth mentioning that 5×2 cross-validation perform better the Leave-One-Out in terms of measuring the generalization capabilities of the models [4].

The experiments include, on the one hand, the results of the group of well-know continuous classification algorithms. All of them have been extracted from a previous study on the same problem [4]: radial basis function support vector

Table 1. MAE results for each of the models included in this comparison.

Fold	RBFSVM	3-KNN	GFFSM	7AMP	9AMP	11AMP	FACS	7ACS	9ACS	11ACS
1	0.0555	0.0528	0.0370	0.0000	0.0003	0.0000	0.0281	0.0516	0.0516	0.0389
2	0.0453	0.0415	0.0286	0.0000	0.0000	0.0000	0.1144	0.0439	0.2036	0.0535
3	0.0486	0.0498	0.0261	0.0000	0.0000	0.0000	0.0292	0.0565	0.0827	0.0767
4	0.0471	0.0492	0.0418	0.0000	0.0000	0.0000	0.0172	0.0424	0.1167	0.0466
5	0.0451	0.0386	0.0374	0.0000	0.0000	0.0000	0.1000	0.0426	0.1146	0.0459
6	0.0395	0.0395	0.0310	0.0000	0.0001	0.0000	0.0253	0.0405	0.1494	0.4339
7	0.0440	0.0428	0.0233	0.0000	0.0000	0.0000	0.0274	0.0521	0.2556	0.0503
8	0.0597	0.0478	0.0358	0.0011	0.0005	0.0010	0.0185	0.0724	0.0402	0.0334
9	0.0396	0.0406	0.0242	0.0000	0.0000	0.0000	0.0305	0.0487	0.0573	0.0397
10	0.0490	0.0473	0.0331	0.0000	0.0000	0.0000	0.0978	0.0794	0.0477	0.5676
Mean	0.0473	0.0450	0.0318	0.0001	0.0001	0.0001	0.0488	0.0530	0.1119	0.1386
Median	0.0462	0.0450	0.0320	0.0000	0.0000	0.0000	0.0287	0.0502	0.0986	0.0485
Std	0.0060	0.0047	0.0059	0.0003	0.0002	0.0003	0.0366	0.0125	0.0687	0.1838

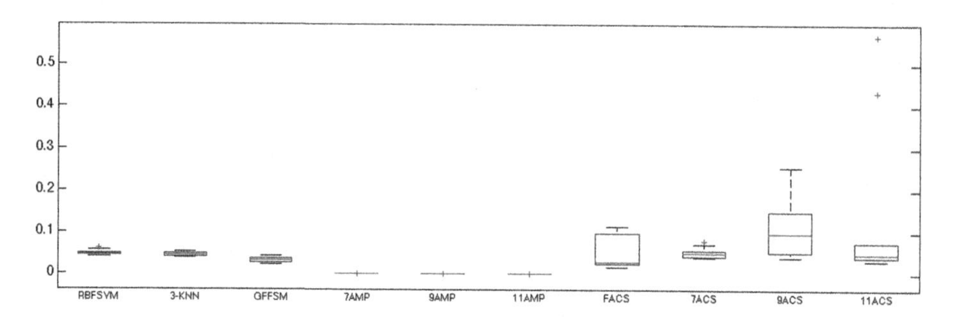

Fig. 5. MAE results boxplot. X-axis refers to the names of the algorithms under comparison.

machines (RBFSVM), K-nearest neighbor with K = 3 (3-KNN), genetic fuzzy finite state machines (GFFSM) and FRBC learned with ACS (FACS). On the other, the results of AntMiner+ and FRBC with ACS taking SAX-based partitioning (using 7, 9, 11 symbols). For the sake of comparison, we kept the same input feature set used in [4]: Signal Magnitude Area (SMA), Amount of Movement (AoM) and Time between Peaks (TbP) on the body acceleration.

The validation of the models was performed by using an error measurement (see MAE formula in Eq. 1), where N stands for the total number of samples in the corresponding data set, while o_i and $\hat{o}_i$ are the desired output and the real output of the model, and GM (see GM formula in Eq. 2) stands for the Geometric Mean where TP, TN, FP, and FN represent True Positive, True Negative, False Positive and False Negative, respectively. Moreover, SEN (Sensitivity) is the probability that a test will indicate 'True Positive' among those with the 'Positive', and SPEC (Specificity) is the fraction of the 'True negative' that will

Table 2. $GM = \sqrt{SensitivityXSpecificity}$ results for each of the symbolic models included in this comparison.

	SAX discretization					
Fold	7AMP	9AMP	11AMP	7ACS	9ACS	11ACS
1	1.0000	0.9985	1.0000	0.0000	0.0000	0.8165
2	1.0000	1.0000	1.0000	0.7852	0.7795	0.0000
3	1.0000	1.0000	1.0000	0.9306	0.3952	0.0000
4	1.0000	1.0000	1.0000	0.7303	0.8765	0.7626
5	1.0000	1.0000	1.0000	0.8750	0.7740	0.5234
6	1.0000	0.9995	1.0000	0.8693	0.7577	0.2012
7	1.0000	1.0000	1.0000	0.0000	0.7108	0.6283
8	0.9951	0.9976	0.9956	0.0000	0.7529	0.9422
9	1.0000	1.0000	1.0000	0.0000	0.6553	0.6635
10	1.0000	1.0000	1.0000	0.0000	0.8966	0.5514
Mean	0.9995	0.9996	0.9996	0.4190	0.6599	0.5089
Median	1.0000	1.0000	1.0000	0.3651	0.7553	0.5898
Std	0.0015	0.0008	0.0013	0.4220	0.2562	0.3162

have a 'negative' test result.

$$MAE = \frac{1}{N}\sum_{i=1}^{N} |o_i - \hat{o}_i| \tag{1}$$

$$GM = \sqrt{SENXSPEC} = \sqrt{\frac{TP}{TP+FN} X \frac{TN}{TN+FP}} \tag{2}$$

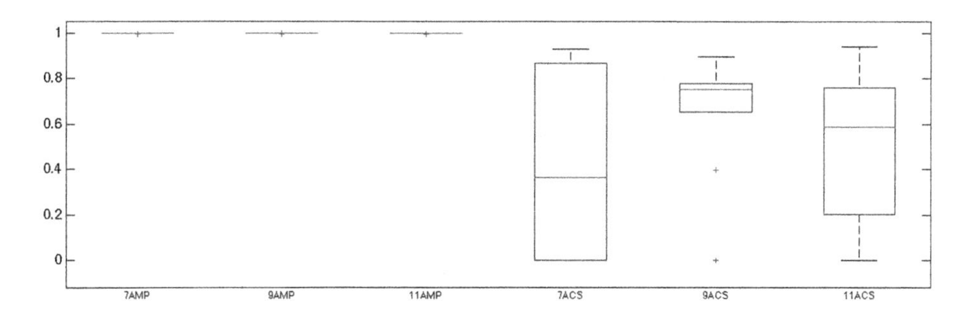

Fig. 6. GM results boxplot. The y-axis refers to the GM error, a plot for each compared method using the SAX-based partitioning.

4.2 Results and Discussion

Results from the experiments using MAE and GM are included in Tables 1 and 2, and in box plots in Figs. 5 and 6, respectively. Clearly, the performance of the FACS has been enhanced with the SAX-based partitioning, mainly when talking about robustness. This means that the generalization capabilities of the obtained models, though not better in terms of error, perform similarly independently of the studied population.

Regarding with the xAMP methods, they show an impressive performance that makes them the best methods from this comparison.

5 Conclusion

This study has focused on the detection of simulated epileptic convulsions using wearable devices and FRBC. A SAX-based partitioning scheme has been proposed, which has been used together with the AntMiner+ and the Pittsburg FRBC learned with ACS. Moreover, a comparison of the results with state of the art methods shows this partitioning scheme together with the AntMiner+ outperforming the remaining methods. This fact suggests this would be a very interesting method for the detection of epileptic seizures, which is part of the future work of this study. It is expected that in the final quarter of this year the experimentation with data from real patients in their everyday life will be carried out.

Acknowledgements. This research has been funded by the Spanish Ministry of Science and Innovation, under project MINECO-TIN2014-56967-R, and Junta de Castilla y León project BIO/BU01/15.

References

1. Fu, T.: A review on time series data mining. Eng. Appl. Artif. Intell. **24**, 164–181 (2011)
2. Mueen, A., Keogh, E., Zhu, Q. Cash, S., Westover, B.: Exact discovery of time series motifs. In: Proceedings of the 2009 SIAM International Conference on Data Mining, pp. 473–484 (2009)
3. Villar, J.R., González, S., Sedano, J., Chira, C., Trejo-Gabriel-Galan, J.M.: Improving human activity recognition and its application in early stroke diagnosis. Int. J. Neural Syst. **25**(4), 1–20 (2015). doi:10.1142/S0129065714500361
4. Villar, J.R., Vergara, P., Menéndez, M., de la Cal, E., González, V.M., Sedano, J.: Generalized models for the classification of abnormal movements in daily lige and its applicability to epilepsy convulsion recognition. Int. J. Neural Syst. **26**(6), 1650037 (2016)
5. Dimitrova, E.S., Licona, M.P.V., McGee, J., Laubenbacher, R.: Discretization of time series data. J. Comput. Biol. **17**(6), 853–868 (2010)
6. Lin, J., Keogh, E., Lonardi, S., Chiu, B.: A symbolic representation of time series, with implications for streaming algorithms. In: Proceedings of the 8th ACM SIGMOD Workshop on Research Issues in Data Mining and Knowledge Discovery, DMKD 2003, pp. 2–11. ACM, New York (2003)

7. Panayotopulos, C.P.: A clinical guide to epileptic syndromes and their treatment, 2nd edn. Springer, London (2007)
8. Schulc, E., Unterberger, I., Saboorc, S., Hilbe, J., Ertl, M., Ammenwerth, E., Trinka, E., Them, C.: Measurement and quantification of generalized tonic–clonic seizures in epilepsy patients by means of accelerometry—an explorative study. Epilepsy Res. **95**, 173–183 (2011)
9. Vergara, P., Villar, J.R., Cal, E., Menéndez, M., Sedano, J.: Fuzzy rule learning with ACO in epilepsy crisis identification. In: 11th International Conference on Innovations in Information Technology (IIT 2015), Dubai, UAE, November 2015
10. Vergara, P., Villar, J.R., Cal, E., Menéndez, M., Sedano, J.: Comparing ACO approaches in epilepsy seizures. In: Martínez-Álvarez, F., Troncoso, A., Quintián, H., Corchado, E. (eds.) HAIS 2016. LNCS (LNAI), vol. 9648, pp. 261–272. Springer, Heidelberg (2016). doi:10.1007/978-3-319-32034-2_22
11. Casillas, F.H.J., Cordón, O.: Learning fuzzy rules using ant colony optimization algorithms. University of Granada, pp. 13–21 (2000)
12. Dorigo, M., Maniezzo, V., Colorni, A.: The ant system: optimization by a colony of cooperating agents. IEEE Trans. Syst. **26**(1), 1–13 (1996)
13. Dorigo, M., Stützle, T.: Ant Colony Optimization, A Bradford Book, vol. 1 (2004)
14. Blum, C., Roli, A., Dorigo, M.: Hc-aco: the hyper-cube framework for antcolony optimization. In: MIC'2001 - 4th Metaheuristics International Conference, pp. 399–403 (2001)
15. Stützle, T., Hoos, H.: Max -min ant system. Future Gener. Comput. Syst. **16**, 889–914 (2000)
16. Dorigo, M., Gambardella, L.: Ant colony system: a cooperative learning approach to the traveling salesman problem. IEEE Trans. Evol. Comput. **1**(1), 53–66 (1997)
17. Abraham, A., Ramos, V.: Web usage mining using artificial ant colony clustering. In: The Congress on Evolutionary Computation, pp. 1384–1391 (2003)
18. Handl, J., Knowles, J., Dorigo, M.: Ant-based clustering and topographic mapping. Artif. Life **1**, 35–61 (2006)
19. Liu, B., Abbass, H.A., McKay, B.: Density-based heuristic for rule discovery with ant-miner. In: Proceedings of 6th Australasia-Japan Joint Work-Shop on Intelligent and Evolutionary Systems, pp. 180–184 (2002)
20. Liu, B., Abbass, H.A., McKay, B.: Classification rule discovery with ant colony optimization. In: Proceedings of IEEE/WIC International Conference Intelligent Agent Technology, pp. 83–88 (2003)
21. Martens, D., Backer, M.D., Haesen, R.: Classification with ant colony optimization. Trans. Evol. Comput. **11**, 651–665 (2007)
22. Parpinelli, R., Lopes, H., Freitas, A.: Data mining with an ant colony optimization algorithm. IEEE Trans. Evol. Comput. **6**(4), 321–332 (2002)
23. Parpinelli, R.S., Lopes, H.S., Freitas, A.A.: An ant colony algorithm for classification rule discovery. In: Data Mining: A Heuristic Approach, pp. 191–208 (2002)
24. Martens, D., Baesens, B., Fawcett, T.: Editorial survey: swarm intelligence for data mining. Mach. Learn. **82**, 1–42 (2011)
25. Cordón, O., Herrera, F., Hoffmann, F., Magdalena, L.: Genetic Fuzzy Systems. Evolutionary Tuning and Learning of Fuzzy Knowledge Bases. World Scientific Publishing, Singapore (2004)
26. Fernández, A., López, V.V., del Jesus, M.J., Herrera, F.: Revisiting evolutionary fuzzy systems: taxonomy, applications, new trends and challenges. In: Knowledge-Based Systems, February (2015) (In Press, Accepted Manuscript)

27. Shieh, J., Keogh, E.: iSAX: indexing and mining terabyte sized timeseries. In: KDD 2008 Proceedings of the 14th ACM SIGKDD International Conference on Knowledge Discovery and Data Mining, vol. 24, pp. 623–631 (2008)
28. Villar, J.R., Menéndez, M., de la Cal, E.A., González, V.M., Sedano, J.: General models for the recognition of epileptic episodes. In: Evaluation for BMC Bioinformatics (2015)
29. Villar, J.R.: Seizure simulation raw data (2016). http://di002.edv.uniovi.es/~villar/Jose_R._Villar/Public_Data_sets/Public_Data_sets.html. Accessed 01 May 2016

Real Prediction of Elder People Abnormal Situations at Home

Aitor Moreno-Fernandez-de-Leceta[1(✉)], Jose Manuel Lopez-Guede[2], Manuel Graña[3], and Juan Carlos Cantera[1]

[1] Sistemas Inteligentes de Control y Gestión Parque Tecnológico de Álava Leonardo Da Vinci, Instituto Ibermática de Innovación, 9 - 2°- Edificio E5, 01510 Miñano, Spain
ai.moreno@ibermatica.com

[2] Department of Systems Engineering and Automatic Control, University College of Engineering of Vitoria, Basque Country University (UPV/EHU), Nieves Cano 12, 01006 Vitoria, Spain
jm.lopez@ehu.es

[3] Faculty of Informatics, Department of Computer Science and Artificial Intelligence, Basque Country Univesrsity (UPV/EHU), Paseo Manuel de Lardizabal 1, 20018 San Sebastian, Spain
manuel.grana@ehu.es

Abstract. This paper presents a real solution for detecting abnormal situations at home environments, mainly oriented to living alone and elderly people. The aim of the work described in this paper is, first, to reduce the raw data about the situation of the elder at home, tracking only the relevant signals, and second, to predict the regular situation of the person at home, checking if its situation is normal or abnormal. The challenge in this work is to transform the real word complexity of the user patterns using only "lazy" sensor data (position sensors) in a real scenario over several homes. We impose two restrictions to the system (lack of "a priori" information about the behavior of the elderly and the absence of historic database) because the aim of this system is to build an automatic environment and study the minimal historical data to achieve an accurate predictive model, in order to generate a commercial produtc working fully few weeks after the installation.

1 Introduction

Society is facing a challenge related to growing life expectancy and the increasing number of people living in situations of dependency, leading to an increasing demand of support systems for personal autonomy. Ambient Assisted Living (AAL) is defined as using information and communication technologies in intelligent living environments reacting to the needs of the inhabitants by providing relevant assistance and helping them to live a full and independent life. Research in smart environments is often related to AAL and attempts to derive information about peoples wellbeing (sleeping, awake, daily rhythm, falls, level of general activity, etc.) from various sensors, often including cameras in a smart

M. Graña et al. (eds.), *International Joint Conference SOCO'16-CISIS'16-ICEUTE'16*, Advances in Intelligent Systems and Computing 527, DOI 10.1007/978-3-319-47364-2_4

environment [2]. The European Statistical Office predicts that in 2060, the ratio between working and retired people will have passed from four-to-one to two-to-one in the EU. In addition, nowadays EU Member States spend approximately a quarter of their Gross Domestic Product (GDP) on social protection [4]. Such a demographic and economic context raises the concern of whether these high standards can be maintained. Historically, advanced support and care applications are based on physiological sensors or cameras for the detection of home accidents, long-term behavior analysis, telecare and teleasistance. Hovewer, both kinds of sensors are invasive, hard to maintain, and sometimes, with important privacy restrictions and some commercial gaps to solve: distrust of users to invasive systems with cameras or complex hardware, and the resistance to very cumbersome or expensive systems.

In this paper we introduce a system which works using "lazy" sensors through very cheap hardware, but with a powerful "context Inference Engine", which minimizes false positives. The key is the automatic customization of the users behaviors automatically. Several systems were introduced in recent years to address some of the issues related to elder-care, principally, fall detection systems. However, most of the systems developed in this context are either too expensive for mass use or of low quality. Most commercial solutions are capable only of fall detection, meaning that they recognize only a small set of hazardous situations [9]. Complex systems using diferent kind sensors or other device inputs, inferring other context situations (sleeping, eating, watching TV, etc.) have been demonstrated with a great confidence [1], but it is complex to deploy them.

The main hypothesis of this paper is to demonstrate that simply by analyizing the different localizations of the inhabitants in the house, it is possible to teach to an automatic system the patterns of those users, and thus, to determine whether there is something strange in these behaviors. In addition, we need to determine what minimum historical data is required for models to be effective above a threshold of performance. So, only collecting information about the daily living of elder persons, we can to learn routines and detect abnormal situations, and sending the conclusions to a call center or to relatives rising alerts for early risk detection of health situations.

There are other approaches toward detecting and studying automatically behavioral patterns at home. One of them [3] uses, as our work, a home sensor network to track the user's motion and different stages at home. Using the home zone and the occupancy time, they create a set of codes that defined the zone, time of day, duration of the presence, in order to discover frequent sequences of codes in a 30 days dataset. Some authors have been studying data mining and machine learning approach to recognize activities of daily life, in order to finding the most common pattern in motion sensor data [11]. Other recent works, like [5,6] were focused on using Support Vector Machines (SVM) to classify anomalous behavior using a dataset based on door sensors at home, but manually annotated.

Previous works are done over well know datasets, for example, the obtained from WSU's CASAS project, with only one user annotated. We are working over real data getting in real homes with a number of real users living alone and getting sensors data daily. The target is to develop an universal algorithm to detect abnormal situations at home, but without the need to annotate what is the mean of the combination of sensor values and timing, like other works [7] those turning them into some different classes, codes or segments (breakfasting, sleeping, etc.).

Our aim is that simply knowing the location of the users at one moment and their habits, to determine whether the location at a given time instant is fine or it is an anomalous location, and if this anomaly occurs during a certain cadence, that event could be an evidence of some risk and the system could generate an alert. In this regard, several environmental factors that do not occur in laboratory environments must be taken into account, for example the ambient noise or error measures in sensors, activation of sensors at the same instant, and also, multiple wrong events when the inhabitants of the house are visited by friends, celebrations, attendees at home working at certain hours, long absences (vacation, travel), etc. That is why the ideal scenario seems to work with a temporal depth of the data of at least 25 months, in order to find seasonality and periodicity over the temporal series. But in a real product on the market, we cannot wait to have that level of data collection to begin analyze the processes. The system would be working fine from the first few weeks, with a frequency of events like one or two months from the start, although the system is prepared for the future, being able to analyze longer periods of time.

The remainder of the paper is organized as follows. Section 2 introduces the dataset description, extraction and reduce strategy. Sections 3 and 4 details about the processes about the developing the predictive algorithms and the Sect. 5 explains the final implantation of the intelligence system, the final method in order to detect abnormal scenarios and discuss the results. Finally, Sect. 6 presents our conclusions and future work.

2 Dataset Description, Extraction and Reduce Strategy

The proposed system works with Wireless Sensor Networks (WSN) that offer the possibility of home deployment without the need for professional installations. These wireless sensor networks, are constituted by a set of stand-alone devices called nodes, distributed in the area to be monitored, which are able to obtain environment information and forward it in a wireless communication to a central point or coordinator. Our dataset contains series of events recorded in 29 houses. The scenarios are the following ones: at each home, there will be living one person or several persons; they can have several visits, either caregivers or family members. Events are all detections of sensors measurements, and, this is a first project critical decision, only sensors status changes are saved in the database, in five minutes periods. If there are no changes, the same status is registered in the database each 5 min (unless the subject changes room, meaning that it is moving

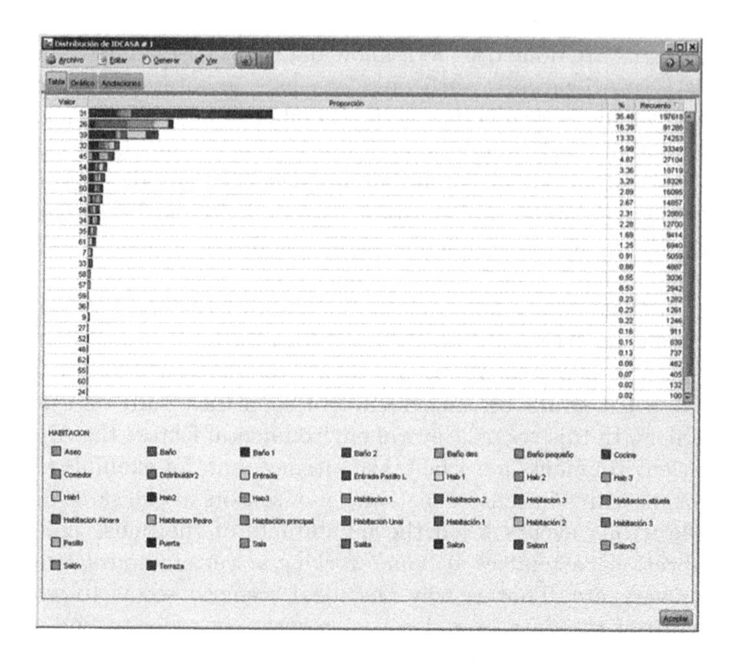

Fig. 1. Home events distribution

in that room, as there are different alerts, by stay long in a room, and not detect movement in the house, including the same room). There are 556,972 events, gathered between December 17th, 2011 and 3rd October, 2014 (nowadays the systems is working and collecting more data), with a very different distribution between diferentes homes, as can be seen in Fig. 1.

Various position sensors are installed in each home throughout the rooms and doors of the home, and each time somebody goes through the room, pass by the hall, or by the gateway, some sensors are activated. This means that there will be many signals at intermediate points, perhaps, not relevant, for example, the hallway, a lot of points in the sample, but with a very small frequency. From this simple information, we must be able to generate knowledge of user patterns. Human behavior and habits are characterized by three attributes of daily activities, namely time, duration and frequency [10]. The drifts in behavior can be identified looking at the changes in those attributes. For example, sleeping and napping behaviors of persons during a day. Any subtle change in sleeping or napping durations can be a sign of a serious disease, especially for the elderly, or an indicator in the progress of a health disease in the long term [12].

To do this, a process of modeling and transformation over the data input is carried out: first, the date is splitted into attributes of day, day of week, week, hour, and quarter hour in which the user is moving, and we are also extracting the frequency of this action taking into account the temporal difference between the last recorded action and the actual, so, the difference in minutes in changing from one state to another. Afterwards, we remove from the system memory events that have less frequency than 15 min. This data filtering decision means that we could only compare at future events if there are abnormalities lasting more than 15 min. But for our control purposes, it is a good measure, since only long events have information enough and evidence to avoiding false positives in the alarm system.

So, at final procress, we have a vector of attributes with the following structure:

$$event = (Room, Frecuency, TimeStamp, Previus\,Relevant\,Action) \quad (1)$$

In addition, frequent events in the same state at the same day and hour are aggregated per hour, so the more usual events. On the oher hand, the rooms are not the same at all homes, there are different rooms, with different names and uses, and each house has a diferenet numbers of rooms, too. So, each home event series differs in number of recorded events, rooms and patterns. This feature is important, because it forces us to approach each house separately as an individual task. This is due to fact, as our objective is to predict in which room will next event occurs, and to compare this estimation with the real room recorded by the system (event, obviously, with a frequency greater than 15 min). If there is any difference, it is possible that something is wrong.

The first objective is to predict a room when next signal will be detected, as an input data historical events is used. Intuitively, the length of the history of recorded events can have impact on the results (even if it is not clear what this impact can be). Therefore, we decided to tests our system for 5 different history length, i.e. 1,2,3,4,5 by a sliding window. From all attributes available in the original dataset we selected the following list of attributes to create both our datasets and the "historical" data: Room, Day of the month, Day of the week, Month, Hour, Frequency of Room.

3 Modeling the Elders Behaviour

The first approach is to collect those relevant events, in a pure state sequence, and try to correlate each result of events (room) with the time (day of week, hour) since we have no other relevant user information. In this case, we are using a hybrid system of classifiers, building by a set of classifiers: Naïve Bayes (NB), Support Vector Machines (with Sequential Minimal Optimization, SMO), Artificial Neural Networks (specifically Multi Layer Perceptron, MLP), and Random Trees (RandTree). The aim is to model the input data, i.e. the day, week, hour of earlier states, and with this information, predict, for the day, and timestamp, in which state or room should be the user, and check what the prediction really is. In a first static approach, using the above classifiers, (see Fig. 2), the results were very poor, even in the case of including previous states, when, in theory, it should improve the prediction. The best result is at the home 31th, with an 63 % of accuracy, a very poor result.

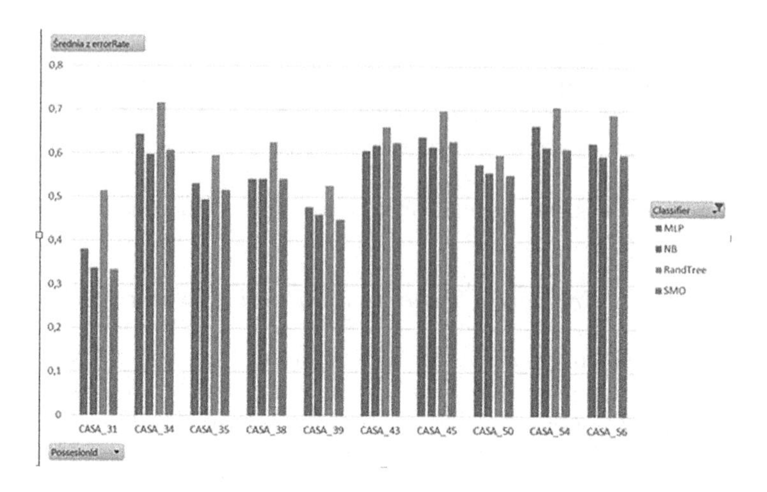

Fig. 2. Error Rate for some classifiers

However, several conclusions are extracted from this experiment:

- As expected, the day of the week prediction influences better than the day of the month.
- The month produces overfitting in classifiers. (It is clear, there not enough homes with more than two years for historical).
- The expected results do not depend largely on the volume of data. It is very important regarding to the final product.
- With the sliding window, the above target states, worsen predictions, unlike expected.

4 New Raw Data Vectors to Improve Behaviour Modeling

According to the state of art, some approaches to the problem are given by a segmentation or clustering over some vectors containing situation, time, frequency, and in the search anomalies in these segmentations. However, in this work, we are looking for anomalies in real time on patterns of previous modeling. So far we have only taken into account the time when the room change, or moments throughout the day when users are in different rooms is given, and based on a classifier, to determine where the user is going to be during the following fifteen minutes. But this approach is not valid with the tested dataset, so we had to devise a different approach. One of the interesting findings in the previous tests is that the level of predicted events success corresponds to each home, with the number of cases in which there is no room changes. That is, as we have the history of what is the last event in the last hour or time, when the event does not change in the in the next cycle, predictor succeeds, but in cases changing room, is when the system failure. As it can see in Fig. 3, if there was no room

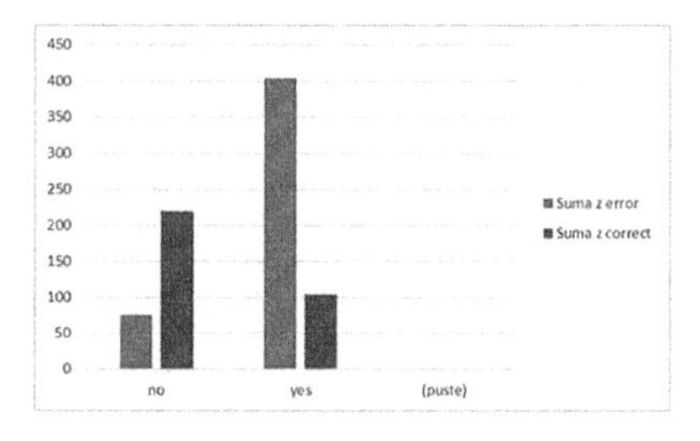

Fig. 3. Error in events changes

change, number of errors is very small and number of correct prediction is quite high. In other case (when the room was changed) system makes a lot of errors and classifies correctly only small number of events.

We understand that this situation occurs because in the data set there are many repetitive events at different times, but, in order to the product we are coming up, it is important to analyze every hour, or every quarter of an hour, if the situation in where the elderly is acting is common or not regarding his behavior, and what is the confidence of this conclusion, whether it is customary for being weekend, summer, morning or evening (i.e. without validating it with a query directly). That is why, as an approximation and under this axiom, we must first analyze whether there is a valid pattern, as is obvious, indicating in what situations there will be an event state change (e.g. in certain hours at night, the user stands up or some hours later, in summer, the user exits to take a walk). Keep in mind that, "a priori", we have no external data, which would help validate, for example, weather data, modeling patterns, as is clear from [8].

But we want to refute the hypothesis that only with the motion data we can build a good predictive model. Therefore, first, on the set of previous data, we generated a new field to indicate whether there is a change of scenery, and from this area, checking if we can develop a classifier to validate this proposal.

As detailed above, it is not only important to know what was the latest location of the user, (at where the user should be stay, or the user should not be), but also how long it has been in that location, and what was the previous location before, and how long the user has been in the previous location. It is clear that if we would build a predictive model including as input variable the time raw values, we can either overtrain the system, if we leave many learning cycles, because the input prediction data will being so specific, so we decided to discretize the usual values of permanence at each location. Keep in mind that this discretization is particular to each house, and the system generates the discretization table directly for each analysis automatically.

Once discretized these values, we predict, at first, if for the combination of the input vector (previous state, actual time frequency discretized, previous frequency discretized, day and hour), there is one pattern that can show us if the user should be in the same previous location or in at a new one, that is, if we can predict if in the actual state there will be an event change. So, we build a new target filed, that is true when was an event change, and false if the previous event is the same as the next event, and applying the same hybrid classifier, we have the following results.

The prediction model has a confidence between 75.82 % and 80.49 %. With these values, we can generate alerts with those events when the user are not in the same location as expected, or if there is not a changing where it is predicted. So, if it is usual to continue at the same stage (i.e. asleep), and there is a changed position and this event kept on more than 15 min, it is also likely to be any problems. At this point, simply with this classifier, we could develop a product able to determine abnormalities, at least on a state changes probability, based on user behavior patterns. However, to rebut the initial final hypothesis, we get to know if we are able to predict the next state of the user based on the results of this prior classifier.

5 The Hierarchical Supervised Classifier Learning

Once the classifier has determined whether change in the user status have been effective, we are able to predict the next state, only for those points where the system predicts that there is a change of state, because when there is no change it is obvious where the user is going to be. Thus, we have the first classifier, defined in the previous point, which indicates the probability of whether or not a change in the status of the user based on the frequency of the current state, the frequency of the previous state, and the stationary component day and time. If there is not prediction of change, we can infer the probability of this classifier that the user remains in the same previous state, and if there is a change probability, we will apply a second classifier to the process that only had learnt change patterns by the historical dataset, removing from the dataset the records where there is not an event change, that is, removing the continuous model sequences. Validating with the method of cross- validation this new hierarchical model, we see that confidence for different homes improvement significantly. If we take the two most representative sample, as in the begining, the 31th home, we see that we move from a 63 % chance of success to a 80 % accuracy in predicting the change of state. Obviously, there are some cases (homes), where the dataset is as chaotic, (a lot of people at home, or a very few historic), and this method do not improve results in these cases.

The last question is to know what is the minimal information gain, or the minimal number of records necesary to get a minimal accuracy. So, we are applying both classifier (event change classifier and event prediction classifier) to all homes with more than 1,000 events (usually, we could have 480 events per day, one each 15 min, so a minimal of two days), results are given in Fig. 4. We can extract some interesting conclusions:

- There is not a direct relation between rows numbers of events per home, and the accuracy.
- The first classifier (event change predictor) is a good classifier to detect if there will be a n event change, (without knowing what event will be).

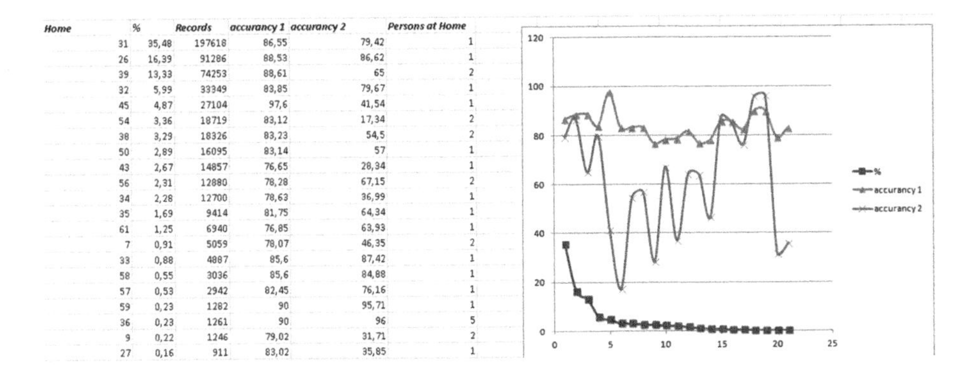

Home	%	Records	accurancy 1	accurancy 2	Persons at Home
31	35,48	197618	86,55	79,42	1
26	16,39	91286	88,53	86,62	1
39	13,33	74253	88,61	65	2
32	5,99	33349	83,85	79,67	1
45	4,87	27104	97,6	41,54	1
54	3,36	18719	83,12	17,34	2
38	3,29	18326	83,23	54,5	2
50	2,89	16095	83,14	57	1
43	2,67	14857	76,65	28,34	1
56	2,31	12880	78,28	67,15	2
34	2,28	12700	78,63	36,99	1
35	1,69	9414	81,75	64,34	1
61	1,25	6940	76,85	63,93	1
7	0,91	5059	78,07	46,35	2
33	0,88	4887	85,6	87,42	1
58	0,55	3036	85,6	84,88	1
57	0,53	2942	82,45	76,16	1
59	0,23	1282	90	95,71	1
36	0,23	1261	90	96	5
9	0,22	1246	79,02	31,71	2
27	0,16	911	83,02	35,85	1

Fig. 4. Home records numbers and accuracy

- The second classifier is very dependent on the nature of the data, so we must be very careful about the quality of them.
- In both cases, the system is independent of the number of people living at each home.

6 Use Cases and Conclusion

At beginning 2015, 60 homes are with our system running into, checking online the user's daily living, over three different customer segments: dependent elderly people, elderly whose habits are worsening due to the aging, and elderly people who are suffering the first symptoms of dementia. At said, the expert rules are running, but with the first month of raw data, the system is beginning to obtain behavior about the users. The challenge and the question were if is it possible detect abnormal situations only registering the users position at home along time, without incorporating external data on local weather or the use of home appliances, or the use of other more advanced sensors, like cameras or physiological sensors.

As conclusion, we can build a robust easily-deployable and cost-contained solution to ensure the safety of the elderly, only with the "lazy" position sensors, with a hierarchical system of classifiers, the first of which, as demonstrated, is very effective in predicting whether a change of state or not on the behavior of users, and can help us detect abnormal situations. On the other hand, a second classifier, trained with a subset of data from the first, can tell us what is the most likely state that should be the user, although this second classifier is much less effective. The ability to analyze these patterns to improve the system, and even as a starting point of widespread information to improve care systems for the governments and health agents, is a real fact, thanks to this research.

Our next steps, in the close future are, first, transforming this proposal in one reference at the telecare platforms at home, real and marketable, and on the other hand, that allows us greater control over the quality of the raw data extracted from the houses, and finally, improve this system accuracy incorporating a model of anomalies based on user segmentation and unsupervised associations between different events.

Acknowledgments. The research was supported by the REAAL project (CIP ICT PSP - 2012 - 325189).

References

1. Andre Chaaraoui, A., Ramon Padilla-Lopez, J., Javier Ferrandez-Pastor, F., Nieto-Hidalgo, M., Florez-Revuelta, F.: A vision-based system for intelligent monitoring: human behaviour analysis and privacy by context. Sensors **14**(5), 8895–8925 (2014)
2. Arif, M.J., El Emary, I.M., Koutsouris, D.D.: A review on thetechnologies and-services used in the self-management of health and independent living ofelderly. Technol. Health Care **22**(5), 677–687 (2014)
3. Bamis, A., Lymberopoulos, D., Teixeira, T., Savvides, A.: The behaviorscope framework for enabling ambient assisted living. Pers. Ubiquit. Comput. **14**(6), 473–487 (2010)
4. Eurobarometer, S.: Active ageing. dg comm research and speech writing unit, european comission, active ageing special eurobarometer 378, conducted by tns opinion & social at the request of directorate-general for employment, social affairs and inclusion. European Union (2012)
5. Gottfried, B.: Spatial health systems. In: Pervasive Health Conference and Workshops, pp. 1–7. IEEE (2006)
6. Jakkula, V.R., Cook, D.J.: Detecting anomalous sensor events insmart home datafor enhancing the living experience (2011). http://www.aaai.org/ocs/index.php/WS/AAAIW11/paper/view/3889
7. Men, L., Miao, C., Leung, C.: Towards online and personalized daily activity recognition, habit modeling, and anomaly detection for the solitary elderly through unobtrusive sensing. Multimedia Tools Appl. January 2016. Impact Factor: 1.35. doi:10.1007/s11042-016-3267-8
8. Lopez-Guede, J.M., Moreno-Fernandez-de Leceta, A., Martinez-Garcia,A., Grana, M.: Lynx: automatic elderly behavior prediction in home telecare (2015). http://dx.doi.org/10.1155/2015/201939
9. Kaluža, B., Mirchevska, V., Dovgan, E., Luštrek, M., Gams, M.: An agent-based approach to care in independent living. In: Ruyter, B., Wichert, R., Keyson, D.V., Markopoulos, P., Streitz, N., Divitini, M., Georgantas, N., Mana Gomez, A. (eds.) AmI 2010. LNCS, vol. 6439, pp. 177–186. Springer, Heidelberg (2010). doi:10.1007/978-3-642-16917-5_18
10. Noyes, J.: Human reliability analysis: context and control by Hollnagel. E. Ergonomics **38**(12), 2614–2615 (1995)
11. Spagnolo, P., Mazzeo, P., Distante, C.: Human Behavior Understandingin Networked Sensing: Theory and Applications of Networks of Sensors. Springer International Publishing (2014). https://books.google.es/books?id=gf85BQAAQBAJ
12. Suryadevara, N.K., Mukhopadhyay, S.C.: Determining wellness throughan ambient assisted living environment. IEEE Intell. Syst. **29**(3), 30–37 (2014)

SOCO 2016: Machine Learning

Assisting the Diagnosis of Neurodegenerative Disorders Using Principal Component Analysis and TensorFlow

Fermín Segovia(✉), Marcelo García-Pérez, Juan Manuel Górriz, Javier Ramírez, and Francisco Jesús Martínez-Murcia

Department of Signal Theory, Networking and Communications, University of Granada, Granada, Spain
fsegovia@ugr.es

Abstract. Neuroimaging data provides a valuable tool to assist the diagnosis of neurodegenerative disorders such as Alzheimer's disease (AD) and Parkinson's disease (PD). During last years many research efforts have focused on the development of computer systems that automatically analyze neuroimaging data and allow improving the diagnosis of those diseases. This field has benefited from modern machine learning techniques, which provide a higher generalization ability, however the high dimensionality of the data is still a challenge and there is room for improvement. In this work we demonstrate a computer system based on Principal Component Analysis and TensorFlow, the machine learning library recently released by Google. The proposed system is able to successfully separate AD or PD patients from healthy subjects, as well as distinguishing between PD and other parkinsonian syndromes. The obtained results suggest that TensorFlow is a suitable environment to classify neuroimaging data and can help to improve the diagnosis of AD and Parkinsonism.

Keywords: Multivariate analysis · Machine learning · TensorFlow · Principal component analysis · Alzheimer's disease · Parkinson's disease

1 Introduction

Alzheimer's disease (AD) and Parkinson's disease (PD) are the two most common neurodegenerative disorders in developed nations and have dramatic health consequences as well as socio-economic implications. Furthermore, the incidence of these diseases is increasing due to the growth of the older population and is expected to triple over the next 50 years [3,4].

The diagnosis of AD and PD is usually corroborated by means of structural and functional neuroimaging data, which are visually analyzed by experienced clinicians (see Fig. 1). In order to remove the human factor and take advantage of the huge amount of information provided by the neuroimages, several

M. Graña et al. (eds.), *International Joint Conference SOCO'16-CISIS'16-ICEUTE'16*, Advances in Intelligent Systems and Computing 527, DOI 10.1007/978-3-319-47364-2_5

computer-aided diagnosis (CAD) systems have been presented during the last years [12,19–22]. Two approaches have been proposed. On the one hand, the most familiar approach to the neuroimaging community concerns mass univariate statistical testing, which models data at the scale of individual voxel, i.e. each voxel is analyzed separately and relationships amongst distant voxels are not considered. The well-known Statistical Parametric Mapping (SPM) software [9] has become a standard to perform univariate analyses on neuroimaging data. However, SPM and other univariate approaches were not specifically developed to study a single image but to compare groups. Despite of this, univariate methods have been widely used to assist the diagnosis of AD and PD. For example, in [6] Foster and colleagues reported a diagnostic accuracy of 89.6 % and a very high specificity (97.6 %) in the differential diagnosis between AD and normal subjects and between AD and patients with frontotemporal degeneration.

On the other hand, multivariate approaches based on machine learning analyze a neuroimage as an single observation and explicitly consider the interrelationships across voxels. Here, effects due to brain structure or function as well as confounding and error effects are assessed statistically both at each voxel and as interactions among voxels [8]. Research in this field has benefited from recent advances in machine learning, which have allowed new algorithms with a higher generalization ability and able to deal with the small sample size problem

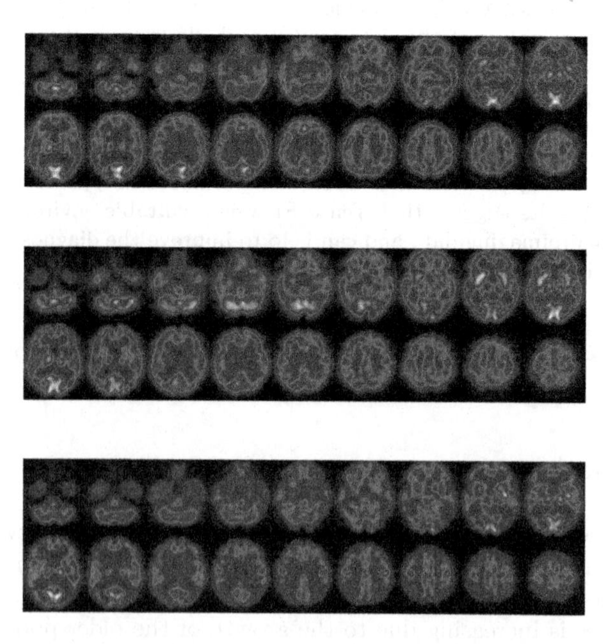

Fig. 1. Axial slices corresponding to three SPECT neuroimages from a neurologically healthy subject (top), from a AD patient at early stage (middle) and from a AD patient at advanced stage (bottom). Observe that differences between the first and third images are visually detectable, however distinguishing between the first and second neuroimage is very difficult. In these cases, it is recommended to use computer systems.

[5]. In this regard, the research branch of the multinational technology company, Google Inc., has recently released as open-source a machine learning environment known as TensorFlow [2]. Provided as an API for Python and C/C++, the purpose of this software is to train neural networks to detect and decipher patterns and correlations. TensorFlow is currently used by many Google services such as Google Search or Gmail but its effectiveness in other classification problems remains still poorly studied.

In this work we present a CAD system based on TensorFlow that provides high accuracy when analyzing neuroimaging data in order to assist de diagnosis of neurodegenerative disorders. We also propose to include a feature extraction procedure based on Principal Component Analysis (PCA) before the classification. This step allows reducing the dimensionality of the data, which helps to deal with the small sample size problem. The proposed approach based on PCA and TensorFlow was evaluated using three neuroimaging databases and we obtained high accuracy rates and good trade-off between sensitivity and specificity in all the experiments. To our knowledge this is the first time that a classification methodology based on TensorFlow is applied to assist the diagnosis of neurodegenerative disorders.

2 Materials and Methods

2.1 Data Description

Three different neuroimaging databases were used to evaluate the computer system proposed in this work:

- **^{99m}Tc-ECD SPECT database.** Ninety-seven (97) SPECT images from AD patients and healthy subjects were collected from a recent study carried out by the 'Virgen de las Nieves' hospital (Granada, Spain). The data were acquired by means of a Picker Prism 3000 gamma camera and reconstructed using a filtered back-projection (FBP) algorithm in combination with a Butterworth noise removal filter. The neuroimages were labeled by three experienced clinicians after visually analyzing the data. As a result, 41 images were labeled as 'normal', 30 as 'possible AD', 22 as 'probable AD' and 4 as 'certain AD'. In our experiments, patients belonging to any of the AD categories were considered as positive and the remaining subjects (labeled as 'normal') were considered negative.
- **DaTSCAN database.** Neuroimaging data from 189 subjects were used to differentiate between healthy subjects and patients with parkinsonian syndromes (PS). The images were acquired by the 'Virgen de la Victoria' hospital (Málaga, Spain) from January 2003 until December 2008. The data corresponds to patients that attended to the hospital redirected by primary care services and who had symptoms that could be considered as Parkinsonism. The images were obtained between 3 and 4 h after the intravenous injection of 185 MBq (5 mCi) of Ioflupane-I-123 with prior thyroid blocking with Lugols solution. A General Electric gamma camera (Millennium model) equipped

with a dual head was used to acquire the data. Then, the transaxial image slices were reconstructed using the FBP algorithm without attenuation correction and a Hanning filter (cutoff frequency equal to 0.7). Finally, the neuroimages were visually labeled by three nuclear medicine specialists from the hospital using only the information contained in the images, without any other medical information. As a result, 95 neuroimages were labeled as 'normal' and the remaining ones (94) as 'PS'. Although PD is the most representative pathology of the PS group, it contains other pathologies with similar symptoms such as multi-system atrophy (MSA), progressive supra-nuclear palsy (PSP) and corticobasal degeneration.
- **^{18}F-DMFP-PET database.** This database was collected in a longitudinal study carried out in the University of Munich [7]. Eighty-seven (87) patients with parkinsonism, previously confirmed by a ^{123}I-FP-CIT SPECT scan according to widely accepted criteria [13], were undergone $D_{2/3}$ receptor imaging with ^{18}F-DMFP. This radioligand allows us distinguishing between different parkinsonian syndromes. The neuroimaging data were acquired 60 min after the radiopharmaceutical injection using a ECAT EXACT HR$^+$ PET scanner (Siemens/CTI). As usual, the neuroimages were reconstructed by means of the FBP algorithm and a Hann filter with a cutoff frequency of 0.5 Nyquist and corrected for randoms, dead time, and scatter. The patients were clinically monitored during the following years. Two years after the data acquisition, the neuroimages were labeled by experienced clinicians on the basis of last observations. According to the United Kingdom Parkinson Disease Society Brain Bank Diagnostic Criteria for Parkinson Disease [11], the second consensus statement on the diagnosis of multiple-system atrophy [10] and the established criteria for the diagnosis of progressive supranuclear palsy [14], 3 groups were defined: idiopathic PD (39 subject), MSA (24 subjects) and PSP patients (24 subjects).

The neuroimaging data from all the three databases were spatially normalized to ensure that any given voxel in different images refers to the same anatomical position across the brains. To this end, the well-known template matching approach implemented in the SPM software (version 8) was used. Finally, the intensities of the neuroimages were also normalized (individually for each image) with respect to a value, I_{max}, computed by averaging the 3 % of the voxels with highest intensity [18].

2.2 Feature Extraction Based on Principal Component Analysis

PCA is a mathematical procedure that transforms the data to a new coordinate system such that the largest variance by any projection of the data comes to lie on the first dimension, the second largest variance on the second dimension, and so on. The feature extraction of neuroimaging data is carried out by projecting the voxel intensities over the eigenvectors (a.k.a. principal components), which are computed as follows [15]:

Let $\mathbf{X} = [\mathbf{x}_1, \mathbf{x}_2, ..., \mathbf{x}_n]$ be a set of n zero-mean neuroimages with unity norm and covariance matrix, $\mathbf{C}$, defined as:

$$\mathbf{C} = \frac{1}{n}\mathbf{X}\mathbf{X}^t \tag{1}$$

The eigenvector Γ and eigenvalue Λ matrices are computed so that $\mathbf{C}\Gamma = \Gamma\Lambda$. Since the dimensionality of the neuroimages is larger than n, diagonalizing $\mathbf{X}^t\mathbf{X}$ instead of $\mathbf{X}\mathbf{X}^t$ reduces the computational burden and the eigenvectors/eigenvalues decomposition is reformulated as:

$$(\mathbf{X}^t\mathbf{X})\Phi = \Phi\Lambda^* \tag{2}$$

$$\Gamma^* = \mathbf{X}\Phi \tag{3}$$

where $\Lambda^* = diag(\lambda_1, \lambda_2, ..., \lambda_n)$ and $\Gamma^* = [\Gamma_1, \Gamma_2, ...\Gamma_n]$ are the first n eigenvalues and eigenvectors respectively.

2.3 Classification Based on TensorFlow

TensorFlow is a machine learning library developed by Google and released as open source on November 2015 [2]. It provides an interface for expressing and executing machine learning algorithms. These algorithms are described by directed graphs composed of a set of nodes, which represent the instantiation of an operation and have zero or more inputs and zero or more outputs. The graph represents a dataflow computation, with extensions for allowing some kinds of nodes to maintain and update persistent state and for branching and looping control structures within the graph in a manner similar to Naiad [16].

One of the key feature of TensorFlow is its ability to run on multiple CPU and devices, reducing the computation times for complex problems. Additionally, it could use CUDA extensions for general-purpose computing on graphics processing units. The library is currently used in dozens of commercial Google products such as speech recognition, Gmail, Google Photos and Google Search [1]. It is available as a Python API as well as a C/C++ API.

In this work we used the Python API provided by the SkFlow project, which allows building deep neural networks using the interface of Scikit Learn [17]. Specifically, we used a 3 layers Deep Neural Network with 5 hidden units per layer. The Adagrad algorithm was used as optimizer. A block diagram of the overall computer system is shown in Fig. 2.

3 Experiments and Results

Several experiments were conducted in order to evaluate the proposed approach. Three diagnosis problems were addressed: (i) AD diagnosis, (ii) PS diagnosis and (iii) Distinguishing between PD and other parkinsonian syndromes. For each problem we used one of the three databases described in Sect. 2.1.

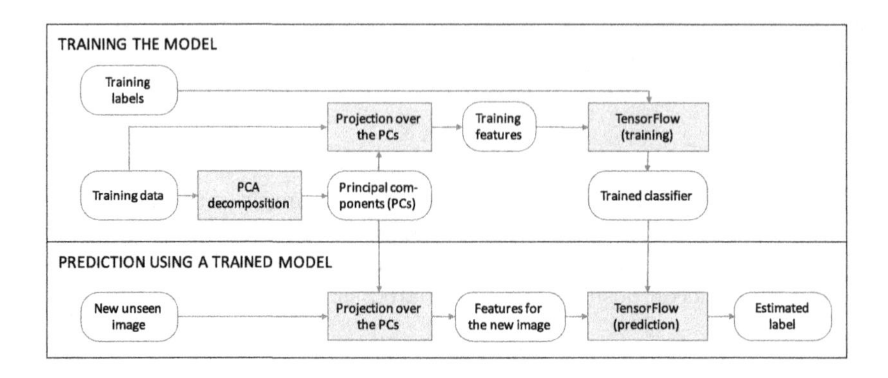

Fig. 2. Block diagram of the proposed CAD system. Rectangles with round corners represent data and rectangles with square corners represent the operations carried out with the data that *enter* to them.

The accuracy, sensitivity and specificity of the implemented systems was estimated using a k-fold cross-validation (CV) scheme (the parameter k was set to the commonly used value of $k = 10$). In order to avoid biased results, the feature extraction procedure based on PCA was carried out using only the training data, i.e. inside the CV loop, and therefore it was repeated for each fold. The obtained results, shown in Table 1, were compared with previous approaches based on Naive Bayes, Decision Trees and Support Vector Machine (SVM) classification.

As described in Sect. 2.2, the feature extraction algorithm used in this work performs an important dimensionality reduction of the data and summarizes several hundred thousand of voxel intensities into a few number of PCA scores. The number of PCA scores is equal to the number of neuroimages used in the decomposition. However, a stronger reduction of the dimensionality can be performed by selecting only the first PCA scores, which correspond with the dimensions of highest variance. Figure 3 shows the accuracy achieved by the proposed approach in function of the number of PCA scores selected. Peak accuracy rates of 90.72 %, 92.59 % and 71.26 % were obtained for the AD, PS and PD diagnosis problem respectively.

4 Discussion and Conclusions

In this work, we evaluated a computer system based on PCA and TensorFlow, and designed to analyze neuroimaging data. As shown in Table 1 and Fig. 3 the proposed system achieves successful results and can help to improve the diagnosis of neurodegenerative disorders such as AD and PD. The lower accuracy rates obtained when separating PD and other parkinsonian syndromes are explained by the difficulty of this problem. PD, MSA and PSP have very similar symptoms, especially at early stages. In addition, the criteria followed to label the neuroimages used in this problem (the ^{18}F-DMFP PET database) were different than those used to label the other databases. In this case the data were labeled

Table 1. Classification measures obtained with the proposed method when used to assist the diagnosis of AD, PS and PD. The results are compared with the ones obtained by other systems based on Naive Bayes, Decision Trees and Support Vector Machines classification. PL and NL stand for positive likelihood and negative likelihood respectively. Dataset 1, 2 and 3 contain ^{99m}Tc-ECD SPECT (AD diagnosis), DaTSCAN (PS diagnosis) and ^{18}F-DMFP-PET (PD diagnosis) data respectively.

		Accuracy	Sensitivity	Specificity	PL	NL
Dataset 1	Proposed system	83.51 %	95.12 %	75.00 %	3.80	0.07
	Naive Bayes	81.44 %	68.29 %	91.07 %	7.65	0.35
	Decision Trees	82.47 %	78.05 %	85.71 %	5.46	0.26
	SVM	82.47 %	82.92 %	82.14 %	4.64	0.20
Dataset 2	Proposed system	87.30 %	89.36 %	85.26 %	6.06	0.12
	Naive Bayes	75.66 %	86.17 %	65.26 %	2.48	0.21
	Decision Trees	77.25 %	74.46 %	80.00 %	3.73	0.32
	SVM	91.53 %	92.55 %	90.52 %	9.77	0.08
Dataset 3	Proposed system	70.11 %	77.08 %	61.54 %	2.00	0.37
	Naive Bayes	66.67 %	68.75 %	64.10 %	1.91	0.48
	Decision Trees	62.06 %	60.42 %	64.10 %	1.68	0.62
	SVM	67.82 %	70.83 %	64.10 %	1.97	0.45

using the observations gathered two years after the image acquisition. It is worth noting that most of the data used in this work correspond to early disease stages, thus the separation of the data is difficult. Furthermore, the results presented above were obtained in a full-automatic mode, i.e. there is no an initial region selection based on previous knowledge.

A direct comparison (using the same data) of the proposed method with previous approaches is given in Table 1. The system based on TensorFlow outperformed previous approaches for most of the datasets considered in this work. Only SVM classification with the DaTSCAN dataset yielded a higher accuracy rate. The difference is specially important for the third dataset where separating the groups is more difficult as mentioned above. Although the results of different works are not directly comparable due to differences in the disease stage and the neuroimaging modality, the results we obtained are also in line with previous studies based on machine learning. For example, in [22] the authors reported a peak accuracy rate of 79.9 % when diagnosing AD. For the same problem, Stoecket et al. achieved accuracy rates between 70 % and 90 % using different regions of SPECT neuroimages. In [21], PS patients and healthy subjects were separated using DaTSCAN data and a multivariate approach based on Singular Value Decomposition and Bayesian classification. In that work, an accuracy rate of 94.8 % was reported.

We showed that TensorFlow is a valuable tool to separate neuroimaging data with diagnosis purposes. However, TensorFlow is not a single classifier but also

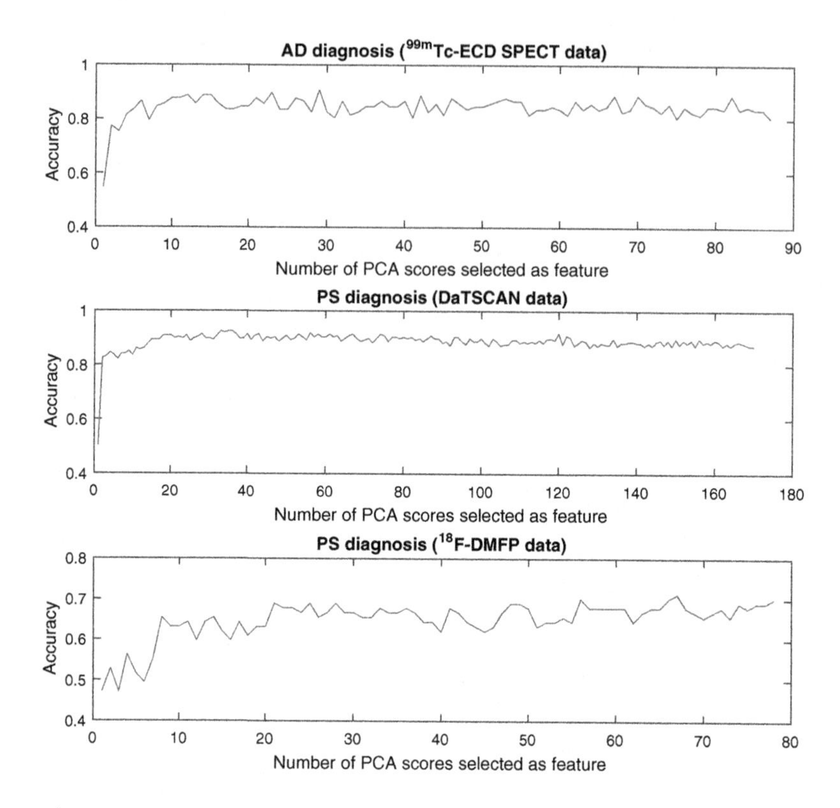

Fig. 3. Accuracy rates obtained by the proposed model when only the first PCA scores are selected.

a machine learning environment that provides a wide range of possibilities to analyze these data. Thus, this work try to encourage the neuroimaging community to go in depth with the TensorFlow library. As future work, we plan to evaluate custom models for the classification estimator that fit better our classification problem. In addition, we will evaluate other neuroimaging modalities as structural magnetic resonance imaging.

Acknowledgment. This work was supported by and the MINECO under the TEC2012-34306 and TEC2015-64718-R projects and the Ministry of Economy, Innovation, Science and Employment of the Junta de Andalucía under the Excellence Projects P09-TIC-4530 and P11-TIC-7103 and a Talent Hub project granted to FS (project approved by the Andalucía Talent Hub Program launched by the Andalusian Knowledge Agency, co-funded by the European Union's Seventh Framework Program, Marie Sklodowska-Curie actions (COFUND Grant Agreement no 291780) and the Ministry of Economy, Innovation, Science and Employment of the Junta de Andalucía).

References

1. TensorFlow - google's latest machine learning system, open sourced for everyone. http://googleresearch.blogspot.com.es/2015/11/tensor-googles-latestmachine_9.html
2. Abadi, M., Agarwal, A., Barham, P., Brevdo, E., Chen, Z., Citro, C., Corrado, G.S., Davis, A., Dean, J., Devin, M., Ghemawat, S., Goodfellow, I., Harp, A., Irving, G., Isard, M., Jia, Y., Jozefowicz, R., Kaiser, L., Kudlur, M., Levenberg, J., Mané, D., Monga, R., Moore, S., Murray, D., Olah, C., Schuster, M., Shlens, J., Steiner, B., Sutskever, I., Talwar, K., Tucker, P., Vanhoucke, V., Vasudevan, V., Viégas, F., Vinyals, O., Warden, P., Wattenberg, M., Wicke, M., Yu, Y., Zheng, X.: TensorFlow: large-scale machine learning on heterogeneous systems (2015). http://tensorflow.org/, software available from tensorow.org
3. Bach, J., Ziegler, U., Deuschl, G., Dodel, R., Doblhammer-Reiter, G.: Projected numbers of people with movement disorders in the years 2030 and 2050. Mov. Disord. **26**(12), 2286–2290 (2011)
4. Brookmeyer, R., Johnson, E., Ziegler-Graham, K., Arrighi, H.M.: Forecasting the global burden of Alzheimer's disease. Alzheimer's Dement. J. Alzheimer's Assoc. **3**(3), 186–191 (2007)
5. Duin, R.: Classifiers in almost empty spaces. In: Proceedings of 15th International Conference on Pattern Recognition, vol. 2, pp. 1–7 (2000)
6. Foster, N.L., Heidebrink, J.L., Clark, C.M., Jagust, W.J., Arnold, S.E., Barbas, N.R., DeCarli, C.S., Turner, R.S., Koeppe, R.A., Higdon, R., Minoshima, S.: FDG-PET improves accuracy in distinguishing frontotemporal dementia and Alzheimer's disease. Brain **130**(10), 2616–2635 (2007)
7. Fougère, C.I., Pöpperl, G., Levin, J., Wängler, B., Böning, G., Uebleis, C., Cumming, P., Bartenstein, P., Bötzel, K., Tatsch, K.: The value of the dopamine D2/3 receptor ligand 18F-Desmethoxyfallypride for the differentiation of idiopathic and nonidiopathic parkinsonian syndromes. J. Nucl. Med. **51**(4), 581–587 (2010)
8. Friston, K., Büchel, C.: Functional connectivity: eigenimages and multivariate analyses. In: Friston, K., Ashburner, J., Kiebel, S., Nichols, T., Penny, W. (eds.) Statistical Parametric Mapping, Chap. 37, pp. 492–507. Academic Press, London (2007)
9. Friston, K.J., Ashburner, J.T., Kiebel, S.J., Nichols, T.E., Penny, W.D.: Statistical Parametric Mapping: The Analysis of Functional Brain Images, 1st edn. Academic Press, Amsterdam, Boston (2006)
10. Gilman, S., Wenning, G.K., Low, P.A., Brooks, D.J., Mathias, C.J., Trojanowski, J.Q., Wood, N.W., Colosimo, C., Dürr, A., Fowler, C.J., Kaufmann, H., Klockgether, T., Lees, A., Poewe, W., Quinn, N., Revesz, T., Robertson, D., Sandroni, P., Seppi, K., Vidailhet, M.: Second consensus statement on the diagnosis of multiple system atrophy. Neurology **71**(9), 670–676 (2008)
11. Hughes, A.J., Daniel, S.E., Ben-Shlomo, Y., Lees, A.J.: The accuracy of diagnosis of parkinsonian syndromes in a specialist movement disorder service. Brain **125**(4), 861–870 (2002)
12. Illán, I.A., Górriz, J.M., Ramírez, J., Segovia, F., Jiménez-Hoyuela, J.M., Lozano, S.J.O.: Automatic assistance to parkinson's disease diagnosis in DaTSCAN SPECT imaging. Med. Phys. **39**(10), 5971–5980 (2012)
13. Koch, W., Radau, P.E., Hamann, C., Tatsch, K.: Clinical testing of an optimized software solution for an automated, observer-independent evaluation of dopamine transporter SPECT studies. J. Nucl. Med. **46**(7), 1109–1118 (2005)

14. Litvan, I., Agid, Y., Calne, D., Campbell, G., Dubois, B., Duvoisin, R.C., Goetz, C.G., Golbe, L.I., Grafman, J., Growdon, J.H., Hallett, M., Jankovic, J., Quinn, N.P., Tolosa, E., Zee, D.S.: Clinical research criteria for the diagnosis of progressive supranuclear palsy (Steele-Richardson-Olszewski syndrome): report of the NINDS-SPSP international workshop. Neurology **47**(1), 1–9 (1996)
15. Lopez, M., Ramirez, J., Gorriz, J., Salas-Gonzalez, D., Alvarez, I., Segovia, F., Puntonet, C.G.: Automatic tool for Alzheimer's disease diagnosis using PCA and bayesian classification rules. Electron. Lett. **45**(8), 389–391 (2009)
16. Murray, D.G., McSherry, F., Isaacs, R., Isard, M., Barham, P., Abadi, M.: Naiad: a timely dataflow system. In: Proceedings of the Twenty-Fourth ACM Symposium on Operating Systems Principles, SOSP 2013, pp. 439–455. ACM, New York (2013)
17. Pedregosa, F., Varoquaux, G., Gramfort, A., Michel, V., Thirion, B., Grisel, O., Blondel, M., Prettenhofer, P., Weiss, R., Dubourg, V., Vanderplas, J., Passos, A., Cournapeau, D., Brucher, M., Perrot, M., Duchesnay, E.: Scikit-learn: machine learning in python. J. Mach. Learn. Res. **12**, 2825–2830 (2011)
18. Saxena, P., Pavel, D.G., Quintana, J.C., Horwitz, B.: An automatic threshold-based scaling method for enhancing the usefulness of Tc-HMPAO SPECT in the diagnosis of Alzheimer's disease. In: Wells, W.M., Colchester, A., Delp, S. (eds.) MICCAI 1998. LNCS, vol. 1496, pp. 623–630. Springer, Heidelberg (1998). doi:10.1007/BFb0056248
19. Segovia, F., Górriz, J.M., Ramírez, J., Salas-Gonzalez, D., Álvarez, I., López, M., Chaves, R.: A comparative study of feature extraction methods for the diagnosis of Alzheimer's disease using the ADNI database. Neurocomputing **75**(1), 64–71 (2012)
20. Segovia, F., Bastin, C., Salmon, E., Górriz, J.M., Ramírez, J., Phillips, C.: Combining PET images and neuropsychological test data for automatic diagnosis of Alzheimer's disease. PLoS ONE **9**(2), e88687 (2014)
21. Towey, D.J., Bain, P.G., Nijran, K.S.: Automatic classification of 123I-FP-CIT (DaTSCAN) SPECT images. Nucl. Med. Commun. **32**(8), 699–707 (2011)
22. Trambaiolli, L.R., Lorena, A.C., Fraga, F.J., Kanda, P.A.M., Anghinah, R., Nitrini, R.: Improving Alzheimer's disease diagnosis with machine learning techniques. Clin. EEG Neurosci. **42**(3), 160–165 (2011)

Cyclone Performance Prediction Using Linear Regression Techniques

Marina Corral Bobadilla[1(✉)], Roberto Fernandez Martinez[2], Rubén Lostado Lorza[1], Fátima Somovilla Gomez[1], and Eliseo P. Vergara Gonzalez[1]

[1] Mechanical Engineering Department, University of La Rioja, Logroño, Spain
marina.corral@unirioja.es
[2] Electrical Engineering Department, University of Basque Country, Bilbao, Spain

Abstract. A wide range of industrial fields utilize cyclone separators and so, evaluating their performance according to different materials and varying operating conditions could contribute useful information and could also save these industries significant amounts of capital. This study models cyclone performance using linear regression techniques and low errors were obtained in comparison with the values obtained from real experiments. Linear regression and generalized linear regression techniques, simple and enhanced with Gradient Boosting techniques, were used to create linear models with low errors of approximately 0.83 % in cyclone performance.

Keywords: Cyclone performance · Linear regression · Generalized linear regression · Gradient boosting

1 Introduction

Cyclone separators are widely employed to control air contamination, separate solids and gases, sample aerosols and track emissions in industrial contexts. With the advantages of relative simplicity to fabricate, low cost to operate, and well adaptability to extremely harsh conditions, cyclone separators have become one of the most important particle removal devices which are preferably utilized in the field of both science and engineering [1]. Considering the numerous combinations of functions that cyclones may perform in different industrial settings, establishing their ideal dimensions for top performance depends on a wide range of factors and thus, is not an easy task. During the last years, several authors have developed models based on machine learning techniques for modeling and optimizing the cyclone performance. Thus, for example, Elsayed [2] modeled and optimized the pressure drop and the collection efficiency of a cyclone using a combination of Radial Basis Function Neural Networks (RBF INNS) and Genetic Algorithms (GA). Also, Elsayed [3] proposed two Support Vector Regression surrogates (SVR) combined with Computational Fluid Dynamics (CFD) for modeling and optimizing the collection efficiency of cyclones. Moreover, Lostado [4] proposed a linear regression model for the modeling of the cyclone performance. The linear regression model was improved by a regression algorithm called

M. Graña et al. (eds.), *International Joint Conference SOCO'16-CISIS'16-ICEUTE'16*,
Advances in Intelligent Systems and Computing 527, DOI 10.1007/978-3-319-47364-2_6

weighted majority vote [5], which allows the combination of several regressors supervised models to improve the accuracy of the regression model. To this end, the present study predicts cyclone performance based on different operating variables, while also including a wide range of operating options. Thus, various linear regression techniques are examined and utilized herein to predict performance and at the same time determine the influence of the different operating variables on performance in a straightforward and simple manner.

1.1 Operating Principle

Cyclone separators operate under the action of centrifugal forces. Fluid mixture enters the cyclone and makes a swirl motion and, due to the centrifugal forces, the dense phase of the mixture gains a relative motion in the radial direction and is separated from the main flow [6]. Dust-laden gas enters centrifugal devices through a tangential inlet which promotes a whirlpool effect. Particles stick to the device's interior walls and are transported by the flow of gas along the wall toward the lower end. Figure 1 depicts a standard cyclone, where once the gas has removed the solid particles; it follows a vertical trajectory to exit the device. The gas movement pattern in a cyclone is normally referred to as a "double vortex" since two clearly distinct vortexes are at work: the external, which moves downwards; and the internal, which moves upwards. Gas exits the upper part of the cyclone, whereas the particles transported by gas into the cyclone's inlet hit the device's walls and are dragged downwards by the airflow's centrifugal force.

1.2 Collection Efficiency

There is no simple theoretical method to accurately calculate efficiency (defined as the percentage of particle mass that enters the cyclone and is then removed by it). This problem is partly due to the fact that in practice, some small particles that should exit the cyclone along with the gas, are actually removed because of build-up, sweeping and contact with other bigger particles; meanwhile, large particles that ought to be trapped and removed by the cyclone may bounce off the walls or be captured by turbulence and escape through the cyclone inlet. Numerous theories propose how to calculate the theoretical efficiency of cyclone collection based on experimental results. These theories relate collection efficiency with particle size, geometric relationships, air flow, operating temperatures, and its influence on gas properties [7–9].

The theory proposed by Leith and Licht's is that which best incorporates experimental behavior. Their theory predicts the collection efficiency of particulate matter based on the physical properties of the particulate matter and the carrier gas, and the relationships between the cyclone's proportions. This theory is expressed as follows (1).

$$\eta_i = 1 - e \cdot \left[-2 \cdot \left(\frac{G \cdot T_i \cdot Q \cdot (n+1)}{D_c^3} \right)^{\frac{0.5}{(n+1)}} \right] \quad (1)$$

Where G is the cyclone configuration factor, T_i is the relaxation time for each particle (in seconds), Q is the gas flow (in m^3/s), n is the cyclone's vortex exponent and D_c is the cyclone diameter (in m). Another approach for estimating particle removal efficiency in cyclones is based on the number of rotations or vortexes that the gas flow undergoes inside the cyclone [10–12]. The number of vortexes is the same for each family of cyclones. Equation (2) relates cyclone efficiency with geometric parameters, flow properties, and operating conditions.

$$\eta_i = 1 - e \cdot \left[\frac{-\pi \cdot N \cdot \rho_p \cdot D_p^2 \cdot V_i}{9 \cdot \mu \cdot b} \right] \quad (2)$$

Where N is the number of rotations completed, D_p is particle diameter, ρ is particle density, V_i is gas velocity, μ is gas dynamic viscosity and b is the width of cyclone inlet.

2 Materials

This study utilized one cyclone to obtain the information that constitutes the database used to generate and validate the performance models. Therefore, the cyclone's measurements do not vary. Figure 1b lists the dimensions and illustrates the cyclone used herein.

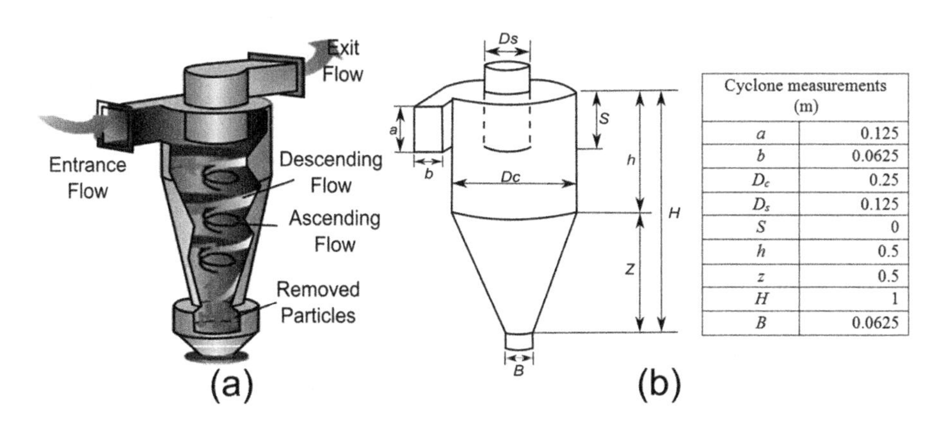

Cyclone measurements (m)	
a	0.125
b	0.0625
D_c	0.25
D_s	0.125
S	0
h	0.5
z	0.5
H	1
B	0.0625

Fig. 1. (a) Standard reverse-flow cyclone (b) Geometry of cyclone analyzed

Hence, in this case where the cyclone's dimensions do not change, the following variables were considered to predict performance: grading of the particles used in the study (μm), density of material (g/cm^3), velocity (m/s), all of which were used to predict the final performance of the cyclone. Table 1 shows some of the variables used in this study for modelling the performance of the cyclone.

Figure 2 depicts the relationships between the variables studied herein.

Table 1. Samples obtained experimentally for modelling the performance of the cyclone.

Inputs				Output
Sample	Grading µm	Density g/cm3	Velocity m/s	Cyclone performance
1	160–320	2.804	10	97.24
2	160–320	2.804	13	98.23
…	…	…	…	…
39	1250–1600	2.725	19	99.88
40	1250–1600	2.725	22	99.71

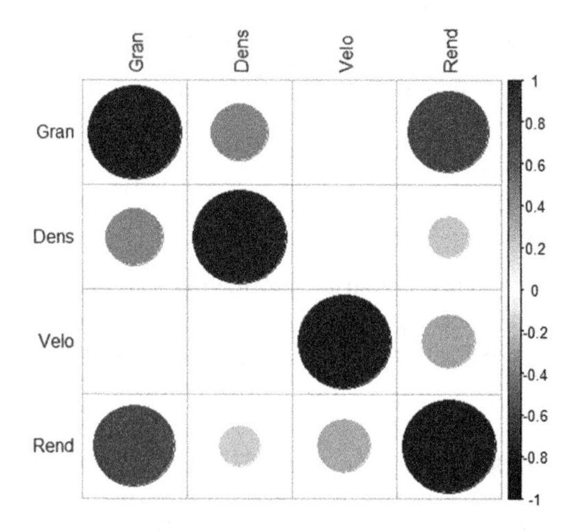

Fig. 2. Correlation of variables

3 Methodology

Considering the complexity and high number of variables that affect the calculation of cyclone performance, this study proposes a model that estimates performance with a greater success rate and incorporates a lesser number of variables. To this end, linear regression models were utilized because it is also interesting to determine the influence of the different variables on the model's final results.

3.1 Linear Regression

Multivariate linear regression (LR) [13, 14] is a technique that aims to provide a probabilistic model of the expected value of an output variable, based on the values of various input variables, according to (3).

$$\eta_i = \varepsilon_i + \beta_1 x_{i1} + \beta_2 x_{i2} + \ldots + \beta_n x_{in} \tag{3}$$

Where $i = 1, \ldots. n$ is the number of variables to be modeled; X_j, the known variables; η_i, the variables to be modeled; β_1 are the impact that each variable has on the modeled variable and indicates the influence of these variables; and ε_i is the error or noise.

3.2 Enhanced Linear Regression

Gradient Boosting [15, 16] was utilized herein to improve the precision of the linear regression models, which combines with the results of various different linear regressions to establish the cost function known as the boosted linear regression (LRB). The algorithm calculates the residues obtained, making them decrease according to a squared error loss function until convergence is achieved [17]. This study also considered the number of repetitions that maximized the algorithm's prediction (mstop), as well as the Shrinkage factor that indicates the impact of each additional regression (shrink) so as to avoid overtraining.

3.3 Generalized Linear Regression

Generalized linear regression (GLM) is a flexible generalization of ordinary linear regression. Herein, it relates the random distribution of the dependent variable with the systematic distribution (not random) through a linking function [18–20]. A generalized regression model consists of three components:

- A random component that specifies the conditional distribution of the response variable ηi, when the variables explaining the model are known. This is normally based on a distribution of the exponential family: Gaussian, binomial, gamma, etc.
- A linear predictor which is a function of linear regression (3).
- A reversible linking function that transforms the response variables according to (4). Functions such as inverse, logarithm, Poisson, etc. can be utilized.

$$\eta_i = g^{-1}(\mu_i) = g^{-1}(\varepsilon_i + \beta_1 x_{i1} + \beta_2 x_{i2} + \ldots + \beta_n x_{in}) \tag{4}$$

To use this technique, and considering the data incorporated in the present study, the distributions of variables were assumed to be Gaussian and the linking functions used herein are listed in Table 2.

3.4 Enhanced Generalized Linear Regression

The L2 Boosting technique was utilized [21, 22] to reduce the error from the descending gradient function. As with enhanced linear regression, in boosted generalized linear regression (GLRB) [23, 24], the algorithm iteratively calculates residues making them decrease until the determined loss function reaches a convergence. The

Table 2. Linking functions used for generalized linear regression

	$\eta_i = g(\mu_i)$	$\mu_i = g^{-1}(\eta_i)$
Logit	$\log_e \frac{\mu_i}{1-\mu_i}$	$\frac{1}{1+e^{-\eta_i}}$
Log	$\log_e \mu_i$	e^{η_i}
Inverse	μ_i^{-1}	η_i^{-1}

degree of error of the predictions is calculated based on error criteria between real and predicted data. In this case, the number of iterations that maximized the algorithm's prediction (mstop) was taken into consideration, and variable selection was conducted according to the Akaike criterion (AIC) (prune).

3.5 Method of Validation

In this study, the statistical software tool R v2.15.2, R development core team [25] was utilized to conduct the processing tasks and to model the data. Before generating the regression models, the original database was normalized between zero and one to improve the model's final prediction quality. Once the data had been normalized, 70 % of the instances were randomly selected to become part of the training database, while the remaining 30 % was used to test and validate the generated models during the training stage. During training, in order to determine which model has the best predictive performance, 10 fold cross-validation was applied. The models that performed the best during training were used in testing, and then efficiency criteria (Sect. 3.6) were employed to determine which is the most efficient.

3.6 Model Efficiency

When using models that make numeric predictions, the model's efficiency must be evaluated by certain measures of accuracy. However, there are various ways to measure accuracy, and each one presents its own nuances. In this study, the criteria employed are based on computational validation errors. These criteria indicate the differences between the values obtained by the model and actual values measured during experiments. In case of discrepancy, the lesser the error, the greater the accuracy of the model in predicting new values. Meanwhile, in the case of correlation, the degree of linear covariation is measured considering the real and modeled values, which indicates the degree of correlation between both sets of values. The following measurements were utilized:

- Mean absolute error (MAE) (5)

$$MAE = \frac{1}{n}\sum_{k=1}^{n}|m_k - p_k| \tag{5}$$

- Root mean squared error (RMSE) (6)

$$RMSE = \sqrt{\frac{1}{n}\sum_{k=1}^{n}(m_k - p_k)^2} \tag{6}$$

- Correlation coefficient (CORR) (7)

$$CORR = \frac{\sum_{k=1}^{n}\frac{(p_k-\bar{p})(m_k-\bar{m})}{n-1}}{\sqrt{\sum_{k=1}^{n}\frac{(p_k-\bar{p})^2}{n-1}\sum_{k=1}^{n}\frac{(m_k-\bar{m})^2}{n-1}}} \tag{7}$$

Where m and p are, respectively, the real and model-calculated values, n is the number of instances used from the database to validate the model,

$$\overline{m} = \frac{1}{n}\sum_{k=1}^{n} m_k \text{ and } \overline{p} = \frac{1}{n}\sum_{k=1}^{n} p_k$$

4 Results

The database on cyclone performance was obtained experimentally (Table 1) and normalized according to the methodology outlined in the above sections. This database of 45 instances was divided in two parts: 70 %, which equals 31 instances, to generate and train model; and 30 %, or 14 instances, was used for testing and validating. During training, the 10-fold cross validation method was used varying certain parameters (predefined in each case) to adjust and optimize the model's accuracy, in order to make the most accurate predictions (Fig. 3).

After analyzing these variations, the results obtained during training for those most accurate models were grouped according to each type of algorithm and are listed in Table 3.

Considering the efficiency criteria analyzed, the following models were selected according to the parameters included in Table 2. These models were then later tested with the data previously set aside for this task, thereby obtaining more significant efficiency rates of the entire range of possibilities examined according to the input variables, since the validation data was not included in the training stage. The results obtained in the testing and validation stages are included in Table 4.

All the errors mentioned thus far are percentages and in relation to the performance normalized during the methodology's initial steps. Several observations can be made based on the training and testing results. Firstly, the models that performed the best during training are not necessarily those that best performed during testing. This means that overtraining during the training stage causes the model to over-adjust to the data used to train it; but when new data (in this case data reserved for testing) was utilized,

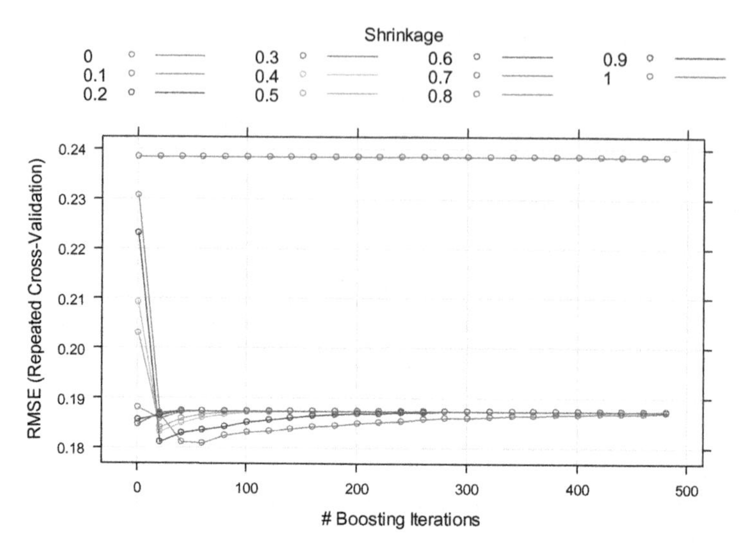

Fig. 3. RMSE obtained during the training stage for the boosted linear regression algorithm

Table 3. Results obtained during training stage

Method	Adjustments	RMSE (%)	RMSE SD (%)	CORR (%)	CORR SD (%)
LR		14.76	56	86.94	20.84
LRB	mstop = 61shrink = 0,1	18.10	9.18	78.30	29.90
GLM	Function = logit	5.89	2.88	95.60	10.19
GLMB	mstop = 81 prune = TRUE	14.50	5.73	87.70	23.10

Table 4. Results obtained from testing and validation

Method	MAE (%)	RMSE (%)	CORR (%)
LR	12.20	15.02	65.03
LRB	11.24	19.08	56.52
GLM	152.83	231.93	66.10
GLMB	10.96	14.49	67.28

the model made much less accurate predictions. One can observe this phenomenon in the GLM models. One can also observe that the methods that utilized the Boosting technique outperformed the algorithms that did not use this technique during the testing stage. This clearly demonstrates that this technique accounts for new possibilities in the input data to a greater extent, as opposed to the models that do not use boosting. And finally, one can also observe that the GLMB model presented the best performance according to the error measurements analyzed in the training and the testing stage as well. Therefore, this linear model can be deemed the most accurate predictor of cyclone performance with a RMSE of 14.49 % and a correlation of 67.28 % in the testing stage, according to the expression of performance of (8).

$$\begin{aligned}\eta_i &= 0,6117284 \cdot D_p + 0,1020288 \cdot \rho \\ &\quad + 0,2272984 \cdot V_i + 0,7177223\end{aligned} \tag{8}$$

Where D_p is particle diameter (cm), ρ material density (g/cm^3) and V_i is gas velocity.

5 Conclusions

This work use linear regression and generalized linear regression techniques, simple and enhanced with Gradient Boosting techniques (GLMB), for modeling the cyclone performance. The GLMB model presented the best performance according to the error measurements analyzed in the training and the testing stage as well (RMSE values of 14.5 % and 14.49 % respectively). The GLM model shows an equivalent error to approximately 0.83 % in the performance of the cyclone obtained experimentally.

References

1. Zhao, B., Shen, H., Kang, Y.: Development of a symmetrical spiral inlet to improve cyclone separation performance. Powder Technol. **145**(1), 47–50 (2004)
2. Elsayed, K., Lacor, C.: CFD modeling and multi-objective optimization of cyclone geometry using desirability function, artificial neural networks and genetic algorithms. Appl. Math. Model. **37**(8), 5680–5704 (2013)
3. Elsayed, K., Lacor, C.: Multi-objective surrogate based optimization of gas cyclones using support vector machines and CFD simulations. In: Application of Surrogate-Based Global Optimization to Aerodynamic Design, pp. 59–72. Springer, Heidelberg (2016)
4. Lostado-Lorza, R. Corral-Bobadilla, M., Escribano-Garcia, R., Fernandez Martinez, R., Alfonso Cendon, J.: XV Congreso Internacional de Ingeniería de Proyectos, Huesca, pp. 2219–2229 (2011)
5. Kuncheva, L.I.: Combining Pattern Classifiers: Methods and Algorithms. Wiley, Hoboken (2004)
6. Avci, A., Karagoz, I.: Effects of flow and geometrical parameters on the collection efficiency in cyclone separators. J. Aerosol Sci. **34**(7), 937–955 (2003)
7. Gimbun, J., Chuah, T.G., Choong, T.S.Y., Fakhru'l-Razi, A.: Prediction of the effects of cone tip diameter on the cyclone performance. Aer. Sci. **36**, 1056–1065 (2005)
8. Chuah, T.G., Gimbun, J., Choong, T.S.Y.: A CFD study of the effect of cone dimensions on sampling aerocyclones performance and hydrodynamics. Powder Technol. **162**, 126–132 (2006)
9. Raoufi, A., Shams, M., Farzaneh, M., Ebrahimi, R.: Numerical simulation and optimization of fluid flow in cyclone vortex finder. Chem. Eng. Process. **47**, 128–137 (2008)
10. Leith, D., Licht, W.: The collection efficiency of cyclone type particle collectors a new theoretical approach. AIChE Symp. **68**(126), 196–206 (1972)

11. Echeverri Londoño, C.A.: Diseño óptimo de ciclones. Rev. Ing. Univ. Medel. **5**(09), 123–139 (2006). App. Mat. Modelling, 35, 1952–1968
12. Cortes, C., Gil, A.: Modeling the gas and particle flow inside cyclone separators. Prog. Energy Comb. Sci. **33**, 409–452 (2007)
13. Friedman, J.H.: Greedy function approximation: a gradient boosting machine. Technical Report, Department of Statistics, Sequoia Hall, Stanford University, Stanford California 94305 (1999)
14. Friedman, J.H.: Stochastic gradient boosting. Technical Report, Department of Statistics, Sequoia Hall, Stanford University, Stanford California 94305 (1999)
15. Wang, Z.: HingeBoost: ROC-based boost for classification and variable selection. Int. J. Biostat. **7**(1), 1–30 (2011)
16. McCullagh, P., Nelder, J.A.: Generalized Linear Models. Chapman and Hall, London (1989)
17. Dobson, A.J.: An Introduction to Generalized Linear Models. Chapman and Hall, London (1990)
18. Hastie, T.J., Pregibon, D.: Generalized linear models. In: Chambers, J.M., Hastie, T.J. (eds.) Statistical Models in S. Wadsworth & Brooks/Cole, Pacific Grove (1992)
19. Venables, W.N., Ripley, B.D.: Modern Applied Statistics with S. Springer, New York (2002)
20. Fox, J.: Applied Regression Analysis and Generalized Linear Models, 3rd edn. McMaster University, SAGE Publications Inc, Thousand Oaks (2015)
21. Freund, Y., Schapire, R.E.: Experiments with a new boosting algorithm. In: Proceedings of the 13th International Conference on Machine Learning, San Francisco, CA, pp. 148–156 (1996)
22. Buehlmann, P.: Boosting for high-dimensional linear models. Ann. Stat. **34**, 559–583 (2006)
23. Buehlmann, P., Yu, B.: Boosting with the L2 loss: regression and classification. J. Am. Stat. Assoc. **98**, 324–339 (2003)
24. Buehlmann, P., Hothorn, T.: Boosting algorithms: regularization, prediction and model fitting. Stat. Sci. **22**(4), 477–505 (2007)
25. R development core team, R: A language and environment for statistical computing. R Foundation for Statistical Computing, Vienna, Austria (2012). http://www.R-project.org/

Time Analysis of Air Pollution in a Spanish Region Through *k*-means

Ángel Arroyo[1(✉)], Verónica Tricio[2], Álvaro Herrero[1], and Emilio Corchado[3]

[1] Department of Civil Engineering, University of Burgos, Burgos, Spain
{aarroyop, ahcosio}@ubu.es
[2] Department of Physics, University of Burgos, Burgos, Spain
vtricio@ubu.es
[3] Departamento de Informática y Automática, University of Salamanca, Salamanca, Spain
escorchado@usal.es

Abstract. This study presents the application of clustering techniques to a real-life problem of studying the air quality of the Castilla y León region in Spain. The goal of this work is to analyze the level of air pollution in eight points of this Spanish region between years 2008 and 2015. The analyzed data were provided by eight acquisition stations from the regional network of air quality. The main pollutants recorded at these stations are analyzed in order to study the characterization of such stations, according to a zoning process, and their time evolution. Four cluster evaluation and a clustering technique, with the main distance measures, have been applied to the dataset under analysis.

Keywords: Clustering · *K*-means · Air quality · Time evolution

1 Introduction

In recent years, our knowledge of atmospheric pollution and our understanding of its effects have advanced greatly. It has been accepted for some years now that air pollution not only represents a health risk. Systematic measurements in Spain, are fundamental due to the health risks caused by high levels of atmospheric pollution. The measurement stations acquire data continuously. Thanks to the open data policy promulgated by the public institutions [1] these data are available for further study and analysis.

Clustering can be defined as the unsupervised classification of patterns into groups [2]. Hence, clustering (or grouping) techniques divide a given dataset into groups of similar objects, according to several different "similarity" measures. These sets of techniques have been previously applied to air pollution data [3, 4]. A clustering method for the study of multidimensional non-stationary meteorological time series was presented in [3]. Principal Components Analysis (PCA) and Cluster Analysis (CA), were applied in [4] over a 3-year period to analyze the mass concentrations of Sulfur Dioxide (SO_2) and Particulate Matter (PM10) in Oporto.

M. Graña et al. (eds.), *International Joint Conference SOCO'16-CISIS'16-ICEUTE'16*,
Advances in Intelligent Systems and Computing 527, DOI 10.1007/978-3-319-47364-2_7

The main idea of present study is the analysis of the time evolution of the most important pollutant variables between the years 2008 and 2015. The data were recorded at eight data acquisition stations from four provinces of the region of Castilla y León, considering the zoning process stated by the European Union in [5]. Four clustering evaluation techniques [6] are applied in a first step to determine the optimal number of clusters existing in the data set. After this, *k*-means [7], combined with the most widely-used distance measures is applied to each one of the years in order to analyze the evolution of air pollution by taking into account the clustering results of the year-by-year analysis.

The rest of this paper is organized as follows. Section 2 presents the techniques and methods that are applied. Section 3 details the real-life case study that is addressed in present work, while Sect. 4 describes the experiments and results. Finally, Sect. 5 sets out the main conclusions and future work.

2 Clustering Techniques and Methods

Clustering is one of the most important unsupervised learning problems [8]. It can be defined as the process of organizing objects into groups whose members are similar in some way. A cluster is a collection of objects which are similar to those in the cluster and are dissimilar to those belonging to other clusters.

Those methods and measure distances are described in this section.

2.1 Cluster Evaluation Measures

Clustering validation evaluates the goodness of clustering results [6]. The two main categories of clustering validation are external and internal. The main difference is whether external information (for which *a priori* knowledge of the dataset is required) is used for clustering validation. Internal validation measures can be used to choose the best clustering algorithm, as can the optimal numbers of clusters, with no further information needed. The following four internal validation measures were all applied in the present work: Calinski-Harabasz Index [9], Silhouette Index [10], Davies-Bouldin Index [11] and Gap Index [12].

2.2 *k*-means Clustering Technique

The well-known *k*-means [13] is a partitional clustering technique for grouping data into a given number of clusters. Its application requires two input parameters: the number of clusters (*k*) and their initial centroids, which can be chosen by the user or obtained through some pre-processing. Each data element is assigned to the nearest group centroid, thereby obtaining the initial composition of the groups. Once these groups are obtained, the centroids are recalculated and a further reallocation is made. The process is repeated until there are no further changes in the centroids. Given the heavy reliance of this method on initial parameters, a good measure of the goodness of the grouping is simply the sum of the proximity Sums of Squared Error (SSE) that it attempts to minimize, Where *p()* is the proximity function, k is the number of the groups, c_j are the centroids, and n the number of rows:

$$SSE = \sum_{j=1}^{k} \sum_{x \in G_j} \frac{p(x_i, c_j)}{n} \tag{1}$$

In the case of Euclidean distance [14], the expression is equivalent to the global mean square error.

K-means technique takes distance into account to cluster the data. Different distance criteria were defined and the distance measures applied in the study are described in this subsection.

An mx-by-n data matrix X, which is treated as mx (1-by-n) row vectors x_1, x_2,..., x_{mx}, and my-by-n data matrix Y, which is treated as my (1-by-n) row vectors y_1, y_2,..., y_{my}.. are given. Various distances between the vector x_s and y_t are defined as follows:

Seuclidean Distance. In Standardized Euclidean metrics (Seuclidean), each coordinate difference between rows in X is scaled, by dividing it by the corresponding element of the standard deviation:

$$d_{st}^2 = (x_s - y_t) V^{-1} (x_s - y_t)' \tag{2}$$

Where V is the n-by-n diagonal matrix the jth diagonal element of which is $S(j)^2$, where S is the vector of standard deviations.

Cityblock Distance. In this case, each centroid is the component-wise median of the points in that cluster.

$$d_{st} = \sum_{j=1}^{n} \left| x_{sj} - y_{tj} \right| \tag{3}$$

Cosine Distance. This distance is defined as one minus the cosine of the included angle between points (treated as vectors). Each centroid is the mean of the points in that cluster, after normalizing those points to unitary Euclidean lengths:

$$d_{st} = 1 - \frac{x_s y_t'}{\sqrt{(x_s x_s')(y_t y_t')}} \tag{4}$$

Correlation Distance. In this case, each centroid is the component-wise mean of the points in that cluster, after centering and normalizing those points to a zero mean and a unit standard deviation.

$$d_{st} = 1 - \frac{(x_s - \bar{x}_s)(y_t - \bar{y}_t)'}{\sqrt{(x_s - \bar{x}_s)(x_s - \bar{x}_s)'} \sqrt{(y_t - \bar{y}_t)(y_t - \bar{y}_t)'}} \tag{5}$$

3 Real-Life Case Study

In present study, pollutant data recorded in eight different places in the region of Castilla y León are analyzed. This region is full of vegetation varieties and large natural areas to be protected; another chance is the compensation ratio among the number of urban stations and urban background traffic stations, of which virtually lacked Castilla y León. Some representative data acquisition stations for the air quality monitoring have been selected from four provinces of the region, being these four provinces which own more available data for the study. The main reason that determines the selection of the stations listed below is the characterization of the stations: four of them are assigned to the zone division oriented to the health protection, and the other four stations are assigned to the ozone protection, according to the zoning process in Castilla y León for the assessment of air quality [15].

A compendium of European legislation on air quality is the Directive 2008/50/EC of the European Parliament and of the Council of 21 May 2008 on ambient air quality and cleaner air for Europe [16]. This Directive established that air quality plans should be developed for zones and agglomerations within which concentrations of pollutants in ambient air exceed the relevant air quality target values or limit values, plus any temporary margins of tolerance. Two of these zones are: the ozone protection stations and the stations for the human health protection. The eight stations selected for this study, according to the information in [17] are:

1. Burgos 4. Fuentes Blancas, Burgos. Geographical coordinates: 03º38'10"W; 42º20'10"N; 929 meters above sea level (masl). Data acquisition station oriented to the health protection.
2. Salamanca 6. Aldehuela park, Salamanca. Geographical coordinates: 05º38'23"W; 40º57'39"N; 743 masl. Data acquisition station oriented to the health protection.
3. León 4. Escolar preserve, León. Geographical coordinates: 05º33'59"W; 42º34'31"N; 814 masl. Data acquisition station oriented to the health protection.
4. Medina del Campo. Bus station, Valladolid province. Geographical coordinates: 04º54'33"W; 41º18'59"N; 721 masl. Data acquisition station oriented to the health protection.
5. Burgos 5. Teresa de Cartagena Saravia St., Burgos. Geographical coordinates: 03º43'16"W; 42º20'44"N; 929 (masl). Data acquisition station oriented to the study of the ozone.
6. Salamanca 5. La Bañeza St., Salamanca. Geographical coordinates: 05º39'55"W, 40º58'45"N; 797 masl. Data acquisition station oriented to the study of the ozone.
7. León 1. The Pinilla neighborhood, León. Geographical coordinates: 05º35'14"W; 42º36'14"N; 838 masl. Data acquisition station oriented to the study of the ozone.
8. Valladolid 14. Regueral bridge, Valladolid. Geographical coordinates: 04º44'02"W; 41º39'22"N; 691 masl. Data acquisition station oriented to the study of the ozone.

From the timeline point of view, data are selected between years 2008 and 2015. There are a total of 715 samples containing monthly averages. These samples are distributed as described in Table 1 (corrupted or missing data are omitted):

Table 1. Number of samples by year and for each type of protection zone.

Zone	Year							
	2008	2009	2010	2011	2012	2013	2014	2015
Health protection	48	47	39	36	45	48	48	48
Ozone protection	56	36	48	48	48	48	46	46

For each one of the station and monthly sample, the following parameters (four air quality variables) were gathered and are considered in present study:

1. Nitric Oxide (NO) - μg/m^3, primary pollutant. NO is a colorless gas which reacts with ozone undergoing rapid oxidation to NO_2, which is the predominant in the atmosphere [18].
2. Nitrogen Dioxide (NO_2) - μg/m^3, primary pollutant. From the standpoint of health protection, nitrogen dioxide has set exposure limits for long and short duration [18].
3. Particulate Matter (PM10) - μg/m^3, primary pollutant. These particles remain stable in the air for long periods of time without falling to the ground and can be moved by the wind over long distances. Defined by the ISO as follows: *"particles which pass through a size-selective inlet with a 50 % efficiency cut-off at 10 μm aerodynamic diameter. PM10 corresponds to the 'thoracic convention' as defined in ISO 7708:1995, Clause 6"* [19].
4. Sulphur Dioxide (SO_2) - μg/m^3, primary pollutant. It is a gas. It smells like burnt matches. It also smells suffocating. Sulfur dioxide is produced by volcanoes and in various industrial processes. In the food industry, it is also used to protect wine from oxygen and bacteria [18].

4 Results and Discussion

The techniques described in Sect. 2 were applied to the case study presented in Sect. 3 and the results are discussed below. Table 2 shows the information on the cluster evaluation for the whole dataset (years from 2008 to 2015) performed by applying the different cluster evaluation measures. In this table, column '*k*' represents the optimum number of clusters estimated by each one of the measures from the 'InspectedK' parameter (taking values from 2 to 6), 'Time' is the execution time (in seconds) and 'Criterion Values' corresponds to each proposed number of clusters in 'InspectedK',

Table 2. Cluster evaluation for the whole dataset

Cluster evaluation measure	*K*	Time (s)	Parameters
Calinski-Harabasz	5	1.18	Criterion Values: [209.31 252.57 133.26 288.45 239.96]
Davies-Bouldin	2	1.31	Criterion Values: [0.63 0.84 1.12 1.17 1.05]
Gap	2	98.62	Criterion Values: [1.40 1.39 1.50 1.50 1.21]
Silhouette	2	1.51	Criterion Values: [0.77 0.61 0.52 0.48 0.42]

Table 3. Cluster evaluation distributed by years (years 2008 to 2010)

Year	Cluster evaluation measure	*K*	Time (s)	Parameters
2008	Calinski-Harabasz	4	0.65	Criterion Values: [42.52 35.93 52.81 41.83 29.29]
2008	Davies-Bouldin	2	0.67	Criterion Values: [0.64 0.91 2.08 1.80 1.43]
2008	Gap	2	48.95	Criterion Values: [0.84 1.05 0.52 1.02 0.87]
2008	Silhouette	2	0.69	Criterion Values: [0.79 0.78 0.33 0.17 0.47]
2009	Calinski-Harabasz	6	0.48	Criterion Values: [29.90 36.14 38.74 42.53 46.21]
2009	Davies-Bouldin	2	0.59	Criterion Values: [0.62 1.80 0.79 1.03 1.22]
2009	Gap	4	46.76	Criterion Values: [0.32 0.49 0.72 0.82 0.84]
2009	Silhouette	2	0.54	Criterion Values: [0.60 0.18 0.34 0.40 0.32]
2010	Calinski-Harabasz	4	0.45	Criterion Values: [39.25 33.68 42.28 34.04 38.86]
2010	Davies-Bouldin	2	0.46	Criterion Values: [0.48 0.82 0.97 0.87 1.06]
2010	Gap	3	49.35	Criterion Values: [0.62 0.88 0.70 0.62 1.06]
2010	Silhouette	2	0.58	Criterion Values: [0.77 0.39 0.37 0.32 0.28]

Table 4. Cluster evaluation distributed by years (years 2011 to 2015)

Year	Cluster evaluation measure	*K*	Time (s)	Parameters
2011	Calinski-Harabasz	2	0.42	Criterion Values: [31.07 25.34 24.40 28.83 29.83]
2011	Davies-Bouldin	2	0.41	Criterion Values: [0.48 0.82 0.97 0.87 1.06]
2011	Gap	2	48.62	Criterion Values: [0.87 0.52 0.49 0.62 0.65]
2011	Silhouette	2	0.51	Criterion Values: [0.54 0.37 0.35 0.32 0.27]
2012	Calinski-Harabasz	5	0.46	Criterion Values: [25.75 37.09 20.49 41.48 26.03]
2012	Davies-Bouldin	2	0.56	Criterion Values: [0.46 0.71 0.88 0.89 1.50]
2012	Gap	2	51.53	Criterion Values: [0.89 0.84 0.56 0.75 0.45]
2012	Silhouette	2	0.54	Criterion Values [0.81 0.59 0.43 0.23 0.49]
2013	Calinski-Harabasz	3	0.43	Criterion Values: [45.92 34.85 15.16 52.23 44.89]
2013	Davies-Bouldin	2	0.46	Criterion Values: [0.65 0.99 1.13 0.83 0.78]
2013	Gap	3	50.73	Criterion Values: [0.69 0.86 0.69 1.13 0.97]
2013	Silhouette	2	0.50	Criterion Values [0.76 0.36 0.46 0.24 0.37]
2014	Calinski-Harabasz	4	0.43	Criterion Values: [45.92 34.85 15.16 52.23 44.89]
2014	Davies-Bouldin	2	0.43	Criterion Values: [0.52 0.80 1.65 1.02 1.30]
2014	Gap	2	49.84	Criterion Values: [0.85 0.92 0.87 1.05 0.98]
2014	Silhouette	2	0.45	Criterion Values: [0.72 0.39 0.35 0.25 0.23]
2015	Calinski-Harabasz	2	0.56	Criterion Values: [65.89 63.80 28.56 42.26 21.86]
2015	Davies-Bouldin	4	0.63	Criterion Values: [2.80 0.98 0.76 1.40 1.27]
2015	Gap	3	50.46	Criterion Values: [0.84 1.09 0.63 0.89 1.21]
2015	Silhouette	2	0.62	Criterion Values: [0.78 0.64 0.54 0.19 0.37]

stored as a vector of numerical values. Each value of this vector is calculated according to the evaluation measure on cluster centroids, the number of points in each cluster, the sum of Squared Euclidean and the number of clusters.

The output of the four measures applied is $k = 2$ in all cases, except for the Calinski-Harabasz measure. This suggested value of $k = 2$ in three of four cases points to the usefulness of the k parameter, required as an input for the k-means subsequent

Table 5. k-means clustering results on yearly subsets of data (2008-2015).

Year	Distance	SumD	Cluster Samples Allocation (%)	
			Health Protection	Ozone Protection
2008	Seuclidean	[0.08 0.02]	[85 15]	[97 3]
2008	Cityblock	[3.91 0.53]	[85 15]	[100 0]
2008	Cosine	[0.98 0.62]	[77 23]	[39 61]
2008	Correlation	[2.40 3.55]	[33 67]	[64 36]
2009	Seuclidean	[0.05 0.04]	[49 51]	[14 86]
2009	Cityblock	[1.31 2.59]	[26 74]	[72 28]
2009	Cosine	[0.96 0.48]	[81 19]	[39 61]
2009	Correlation	[2.96 1.21]	[79 21]	[36 64]
2010	Seuclidean	[0.03 0.06]	[49 51]	[4 96]
2010	Cityblock	[2.70 1.49]	[46 54]	[90 10]
2010	Cosine	[0.81 0.57]	[74 26]	[29 71]
2010	Correlation	[2.03 2.98]	[33 67]	[77 23]
2011	Seuclidean	[0.05 0.04]	[44 56]	[88 13]
2011	Cityblock	[2.45 1.61]	[44 56]	[88 13]
2011	Cosine	[0.75 1.32]	[69 31]	[29 71]
2011	Correlation	[3.93 3.10]	[31 69]	[75 25]
2012	Seuclidean	[0.07 0.06]	[51 49]	[88 13]
2012	Cityblock	[2.87 1.99]	[44 56]	[85 15]
2012	Cosine	[1.47 1.30]	[71 29]	[25 75]
2012	Correlation	[4.54 5.15]	[33 67]	[73 27]
2013	Seuclidean	[0.04 0.10]	[35 65]	[2 98]
2013	Cityblock	[1.81 3.45]	[52 48]	[6 94]
2013	Cosine	[1.59 1.66]	[79 21]	[31 69]
2013	Correlation	[5.02 6.77]	[79 21]	[31 69]
2014	Seuclidean	[0.05 0.06]	[52 48]	[4 96]
2014	Cityblock	[2.13 2.30]	[31 69]	[87 13]
2014	Cosine	[1.39 0.82]	[25 75]	[80 20]
2014	Correlation	[4.02 2.66]	[29 71]	[85 15]
2015	Seuclidean	[0.07 0.04]	[69 31]	[94 6]
2015	Cityblock	[3.00 1.62]	[65 35]	[27 73]
2015	Cosine	[1.02 0.79]	[65 35]	[31 69]
2015	Correlation	[3.69 2.65]	[58 42]	[27 73]

clustering technique. This value of k provides information about the internal structure of the data. In this case study is equivalent to the two main subsets of data existing in the data set (health and ozone protection stations). The Gap evaluation measure was the slowest in terms of computing time.

Tables 3 and 4 shows the information on the cluster evaluation distributed by years, one data set for each year.

Applying the four cluster evaluation techniques to a subset of data for each year, the value of k equals 2 is selected in 65 % cases and in all the years of the case study. All the values in the range of k (2, 6) are selected at least one time.

Table 5 shows the results obtained for the k-means, distributed for each of the years between 2008 and 2015, with different distance criteria and a value of k equals 2 (value of k mostly selected in Table 2). In this table, 'Distance' is the distance criterion applied (see Sect. 2) and 'SumD' is the within-cluster sums of point-to-centroid distances in the k-by-1 vector. The Cluster Samples Allocation columns represent the percentage of samples from each one of the zones (Heath and Ozono) that are allocated to each one the clusters; e.g. [85 15] represents 2 clusters and 85 % of samples allocated to the first cluster and 15 % to the second one.

Some issues from the results in Table 5 are worth mentioning: for all the years under study, the best (minimum) value for parameter SumD is obtained when applying 'Seuclidean' distance, followed by 'Cosine'. Regarding with the sample process allocation, 'Seuclidean' distance allocates most of the samples in the same cluster in four of the eight years, despite the characterization (zoning) of its station. Clustering with 'Cosine' and 'Correlation' distances let us separate most of the samples in different clusters for all the years, according to the station characterization (zoning).

Figure 1 shows the evolution between the years 2008 to 2015 of the parameter SumD (Sums of point-to-centroid distance), when applying k-means (k = 2) and the different distance measures applied. It can be seen that the lowest value of SumD for all the years is obtained when applying 'Seuclidean' distance. This means a high level of compactness in the samples of data when applying this distance measure. Another important aspect to be highlighted is that the highest values for SumD are obtained in

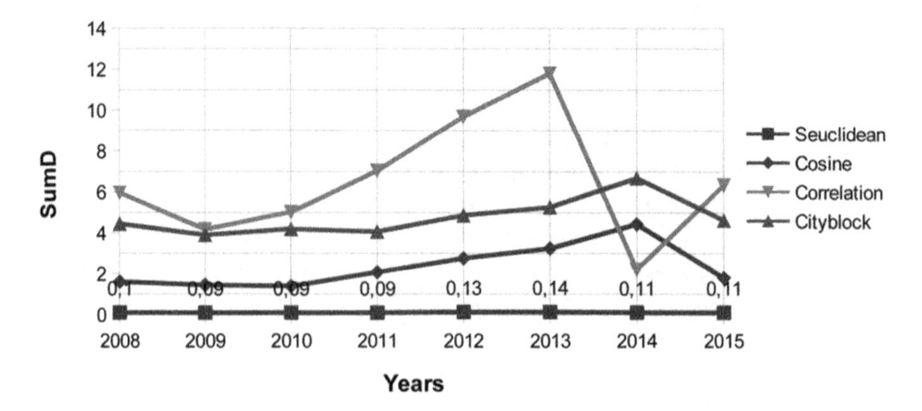

Fig. 1. Year evolution of the SumD parameter.

years 2012, 2013 and 2014. The 'Correlation' distance measure performs in a different way from the other three distance, presenting the highest value in year 2013, the lowest in 2014 and increasing in the last year when the other three distances decrease. This is because 'Correlation' depends of the typical deviation. Although the years from 2012 to 2015 present lower levels of air pollution in the pollutants analyzed, the typical deviation, especially in NO and NO_2, is bigger than in previous years, major pollution peaks exist in these years of low pollution in the region of Castilla y León.

5 Conclusions and Future Work

Main conclusions derived from obtained results (see Sect. 4) can be divided into two groups; at first, those regarding the analysis of air quality conditions in the case study considered. Secondly, those related to the behaviour of the two clustering techniques applied in the case study.

Talking about the air quality conditions in the eight selected places, grouped by the data acquisition station type, the average monthly levels of air pollution in the stations oriented to the ozone protection are lower than those recorded in the health oriented stations, especially in NO and NO_2. The evolution in the period of time analyzed (2008–2015) shows higher levels of air pollution between 2008 and 2011, when compared with the subsequent years. By working with monthly data average, the pollutant concentration levels are smoothed in both areas.

Regarding the applied clustering techniques, clustering measure techniques are a very useful set of techniques to determine the optimal value for parameter k (number of clusters). The four techniques applied obtained similar results, not being very appropriate the use of Gap Index with large datasets due to high elapsed time. When applying k-means with the different measure distance explained in Sect. 2, 'Seuclidean' distance is the best in terms of creating compact clusters of data, as parameter SumD takes the lowest values, but is not the best technique in the sample process allocation, where tends to keep samples from stations of different zones in the same cluster of data. 'Cosine' distance measure offers the best balance between a good sample allocation process and a not very high value for parameter 'SumD'.

Future work will consist of extending proposed analysis to a wider time period, data from different locations and some other clustering techniques.

References

1. Government of Spain - Aporta Project. http://administracionelectronica.gob.es
2. Jain, A.K., Murty, M.N., Flynn, P.J.: Data clustering: a review. ACM Comput. Surv. (CSUR) **31**(3), 264–323 (1999)
3. Kassomenos, P., Vardoulakis, S., Borge, R., Lumbreras, J., Papaloukas, C., Karakitsios, S.: Comparison of statistical clustering techniques for the classification of modelled atmospheric trajectories. Theoret. Appl. Climatol. **102**, 1–12 (2010)

4. Pires, J.C.M., Sousa, S.I.V., Pereira, M.C., Alvim-Ferraz, M.C.M., Martins, F.G.: Management of air quality monitoring using principal component and cluster analysis—Part I: SO2 and PM10. Atmos. Environ. **42**(6), 1249–1260 (2008)
5. European Commission - Air Quality Standards. http://ec.europa.eu/environment/air/quality/standards.htm
6. Liu, Y., Li, Z., Xiong, H., Gao, X., Wu, J.: Understanding of internal clustering validation measures. In: IEEE International Conference on Data Mining, pp. 911–916 (2010)
7. Jain, A.K.: Data clustering: 50 years beyond K-means. Pattern Recogn. Lett. **31**, 651–666 (2010)
8. Barlow, H.: Unsupervised learning. Neural Comput. **1**, 295–311 (1989)
9. Caliński, T., Harabasz, J.: A dendrite method for cluster analysis. Commun. Stat. Theory Methods **3**, 1–27 (1974)
10. Rousseeuw, P.J.: Silhouettes: A graphical aid to the interpretation and validation of cluster analysis. J. Comput. Appl. Math. **20**, 53–65 (1987)
11. Davies, D.L., Bouldin, D.W.: A cluster separation measure. IEEE Trans. Pattern Anal. Mach. Intell. **1**(2), 224–227 (1979)
12. Tibshirani, R., Walther, G., Hastie, T.: Estimating the number of clusters in a data set via the gap statistic. J. Roy. Stat. Soc.: Ser. B (Stat. Methodol.) **63**, 411–423 (2001)
13. Ding, C., He, X.: K-means clustering via principal component analysis. In: Proceedings of the Twenty-First International Conference on Machine Learning, p. 29 (2004)
14. Danielsson, P.E.: Euclidean distance mapping. Comput. Graph. Image Process. **14**, 227–248 (1980)
15. Government of Castilla y León - Zoning of the territory in Castilla y León. http://www.jcyl.es/
16. European Union Law - Directive 2008/50/EC of the European Parliament and of the Council of 21 May 2008 on ambient air quality and cleaner air for Europe. http://eur-lex.europa.eu/
17. Government of Castilla y León - Annual reports of the Air Quality. http://www.medioambiente.jcyl.es/
18. PubChem - PubChem Compounds. https://pubchem.ncbi.nlm.nih.gov/compound
19. ISO - International Organization for Standardization. PM10/PM2.5. https://www.iso.org/

Using Non-invasive Wearables for Detecting Emotions with Intelligent Agents

Jaime Andres Rincon[1(✉)], Ângelo Costa[2], Paulo Novais[2], Vicente Julian[1], and Carlos Carrascosa[1]

[1] D. Sistemas Informáticos y Computación, Universitat Politècnica de València, Valencia, Spain
{jrincon,vinglada,carrasco}@dsic.upv.es

[2] Centro ALGORITMI, Escola de Engenharia, Universidade do Minho, Guimarães, Portugal
{acosta,pjon}@di.uminho.pt

Abstract. This paper proposes the use of intelligent wristbands for the automatic detection of emotional states in order to develop an application which allows to extract, analyze, represent and manage the social emotion of a group of entities. Nowadays, the detection of the joined emotion of an heterogeneous group of people is still an open issue. Most of the existing approaches are centered in the emotion detection and management of a single entity. Concretely, the application tries to detect how music can influence in a positive or negative way over individuals' emotional states. The main goal of the proposed system is to play music that encourages the increase of happiness of the overall patrons.

1 Introduction

Over the last few years, research on computational intelligence is being conducted in order to emulate and/or detect emotional states [11]. The emulation of emotional states allow machines to represent some human emotions. This artificial representation of emotions is being used by machines to improve the interaction process with humans. In order to create a fluid emotional communication between human and machines, the machines need first to detect the emotion of the human with the final purpose of improving human-computer interactions [18]. To do this it is necessary to use different techniques such as: artificial vision [15], speech recognition [12], body gestures [21], written text [5] and biosignals [16].

Human beings perceive and analyse a wide range of stimuli in different environments. These stimuli interfere in our commodity levels modifying our emotional states. Before each one of these stimuli, humans generate several type of responses, like varying our face gestures, body movement or bio-electrical impulses. These variations in our emotional states could be used as a very useful information for machines. To do this, machines will require the capability of interpreting correctly such variations. This is the reason for the design of

M. Graña et al. (eds.), *International Joint Conference SOCO'16-CISIS'16-ICEUTE'16*, Advances in Intelligent Systems and Computing 527, DOI 10.1007/978-3-319-47364-2_8

emotional models that interpret and represent the different emotions in a computational way. In this case, emotional models such as *Ortony, Clore & Collins* model [6] and the *PAD (Pleasure-Arousal-Dominance)* model [20] are the most used ones to detect or simulate emotional states. Moreover, emotional states are a very valuable information, allowing to develop applications that help to improve the human being quality of life.

Nowadays, the detection of the joined emotion of an heterogeneous group of people is still an open issue. Most of the existing approaches are centered in the emotion detection and management of a single entity. In this work we propose to detect the social emotion of a group of people in an Ambient Intelligence (AmI) application with the help of wearables. Specifically, we show a system that controls automatically the music which is playing in a bar through the detection of the emotions of the patrons with the use of individual wristbands. Thus, the main goal of the proposed system is to play music that encourages the increase of happiness of the overall patrons. Each one of the individuals will have an emotional response according to his musical taste. This response will be detected and transmitted by the wristbands in order to calculate a social emotion of the set of individuals. This social emotion will be used to predict the most appropriated songs to be played in the bar.

2 State of the Art

The AmI area is rapidly gaining notoriety due to its usage on complex social environments like nursing homes and regular homes. By monitoring fragile users (like elderly or mentally challenged people) the available systems pose as an alternative to regular caregiving services while being cost-effective. Despite the several aims AmI projects have they can be clustered in five clusters of operational areas [1]:

- Daily living activities
- Fall and movement detection
- Location tracking
- Medication control
- Medical status monitoring

In terms of daily living activities there is the project Caregiver's Assistant, which uses RFID and a database with human activities events and a fast inference mechanism that allows the identifications of the actions within a given space [10]. It works by registering the RFID cards that the users carry, which in some cases they have to actively pass them through the readers due to they small communication range. Thus, it is very intrusive to the users of the system as they have to be actively aware of the procedures so that the system is able to correctly access the information.

In terms of fall detection or movement detection, most of the operation methods resort to use cameras to register the visual information and extract information from it, like the projects in [3,4,17]. Although they require no interaction with the users, these systems are very invasive, not to mention the possible loss of privacy, due to the permanent recording of the environment.

The location tracking systems like the one presented on [22] use mobile devices sensors to provide the current location of the user to an AmI system, more specifically to the caregivers. These systems require the constant monitoring of the localization, thus there is no guarantee of privacy, thus being very intrusive systems. This intrusion is not done directly but by allowing 3rd party users to constantly know the location of another person the system becomes very intrusive [13].

Medication control projects consist in systems that help the users to remind the medications that they have to take [26,28]. They play an important role on the users life, as most of the AmI and AAL projects users have some sort of cognitive disability and have trouble in remembering to do activities, such as taking medication. These systems are mostly recommenders and are able to only provide information without being disruptive or actively monitoring the users. Due to the simplicity of the projects premises it is available a large number of simple applications for mobile devices and desktop computers currently.

The medical status monitoring projects like the ones presented in [2,14,19,30] show platforms that are constituted by sensor systems that are directly in contact with the human body. These sensor systems create a body area network and provide information about the carrier vital signs and bodily functions. For instance, all three works presented capture electrocardiograms and reason that information to obtain knowledge about their specific domain. The [30] uses the electrocardiogram information and ballistocardiogram information to assert if the drivers are calm and concentrated or if they are stressed or having some kind of medical issue (as the project is directed to elderly people). The project ALADIN [19] presents a system that manages home lights (brightness and colour) according to the users physical state. The users carry a biosignal reading glove (that captures photoplethysmography) that sends the readouts to the server and according to their pulse/heart-rate the lights are dimmed or changed their colour. The aim of this project is to provide comfort and promote a peaceful living, adapting the environment to the user state or preferences. The glove has to be put by the users and limits its use to a confined range of actions (it cannot be wetted or be used to manage heavy objects) thus being quite invasive, possibly undermining the results as the users become actively aware of their status, thus allowing them to manipulate the system. The issue with these systems is that they require users to attach sensors on their own body (the case of [2,14]) or that the users are in a very controlled environment like the [30].

These projects are a small representation of the plethora of the existent projects and show the current lines of development. One common problem revealed of these projects is that they are interested in the implementation and the execution of their components but do not reveal any particular interest on the opinion of the users towards the devices that they are using. Only recently the theme of invasiveness has been considered due to the high reluctance of the users towards clumsy and complex apparatus [9].

In the line of ideas presented by [19], we are aiming to produce a system that increases the comfort level of the users (by managing the current music) through the use of a non-invasive wearable bracelet that performs medical status monitoring to attain the users' emotional status.

3 Problem Description

This application example is based on how music can influence in a positive or negative way over emotional states [25,31,33]. The application example is developed in a bar, where there is a DJ agent in charge of playing music and a specific number of individuals listening to the music. The main goal of the DJ is to play music making that all individuals within the bar are mostly as happy as possible. To get this, it is necessary to detect the human emotions, and there exist many techniques to do it. But in our case, we decided to use the bio-signals to detect the emotional change. This emotional change is used by the DJ agent to change the musical genre and try that the clients of the bar are mostly as happy as possible. Each one of the individuals will have an emotional response according to its musical taste. This response is reflected in a variation of the bio-signals [7,29]. This variation allows us to calculate the social emotion [23] of the people within the bar. Based on the metric of the social emotion, the DJ agent could change the music genre to move the social emotion to a target emotion. To capture the bio-signals we designed a prototype of an *Emotional Smart Wristband* (Fig. 1), that will be explained with more detail in Sect. 4.

Fig. 1. Prototype of an *Emotional Smart Wristband.*

4 System Proposal

This section explains the different components that constitute the multi-agent system which describes a way to detect emotions based on bio-signals through wearable devices. The main problem in the detection of human emotions is the information capture. This information is normally obtained using image processing, text analysis or voice analysis. These ways are invasive and in some of them is necessary to have the consent of the person. In currently years the use of wearable devices has been growing, devices such as *Samsung*[1] with the *Gear Fit, Gear S2* or *Apple*[2] with the *Apple Watch* are only some examples. These

[1] http://www.samsung.com.
[2] http://www.apple.com.

devices can measure heart rate beat or hand movement using the IMU (Inertial Measurement Unit). Based on these devices and using the currently technology in embedded systems, it is possible to create new smart bracelets which include other type of measures, such as the EEG, GSR, Photoplethysmogram or ECG allowing the acquisition of biosignals that can help for the detection of the human's emotions. Using signals of this kind along with the incorporation of complex algorithms based on machine learning techniques, it is possible to recognise how humans change their emotional states.

The proposed multi-agent system is formed by three types of agents. These agents are: the *Wristband agent*, the *Social Emotion Agent*, and the *DJ agent.* The Wristband agent is mainly in charge of: (i) capture some emotional information from the environment and specifically from a specific individual, this is done by interacting with the real world through the employed wristband. The agent captures the different bio-signals, that will be used to detect the emotion of a human being; and (ii) predict the emotional state of the individual from the processed biosignals. In order to analyze these changes and predict emotional states, the Wristband agent employs a classifier algorithm that will be later explained. Once the emotion has been obtained, it is sent to the agent which is in charge of calculating the social emotion of the agent group. This agent is called *Social Emotion Agent* or *SEtA*. The main goal of this agent is to receive the calculated emotions from all the *Wristband agents* and, using this information, generate a social emotional state for the agent's group (details of how this social emotion is calculated can be seen in [23]). Once this social emotion is obtained, the *SEtA* can calculate the distance between the social emotion and a possible target emotion (in this case the target emotion is happiness). This allows to know how far is the agent's group of the target emotion. This can be used by the system to try to reduce that distance modifying the environment. This modification of the environment is made by the *DJ agent.* This agent uses this social emotional value to calculate what is the next song to be played. After different executions, the DJ agent can evaluate the effect that the song has had over the audience. This will help the DJ to decide whether to continue with the same musical genre or not in order to improve the emotional state of the group of people.

Due to the limits of the paper, we only describe in detail the processes made by the *Wristband agent* which are the data acquisition process and the emotion recognition. Moreover, the physical components of the wristband prototype are also described.

4.1 Data Acquisition Process

This process made by the *Wristband agent* is responsible to capture the different needed bio-signals. To do this, the *Wristband agent* uses different sensors. The sensors used are: *GSR and Photoplethysmogram* (Fig. 2). The GSR measures the galvanic skin response. The measurement is performed by passing through the skin a very low current, and storing small variations in voltage. On the other hand, the Photoplethysmogram is a process of applying a light source

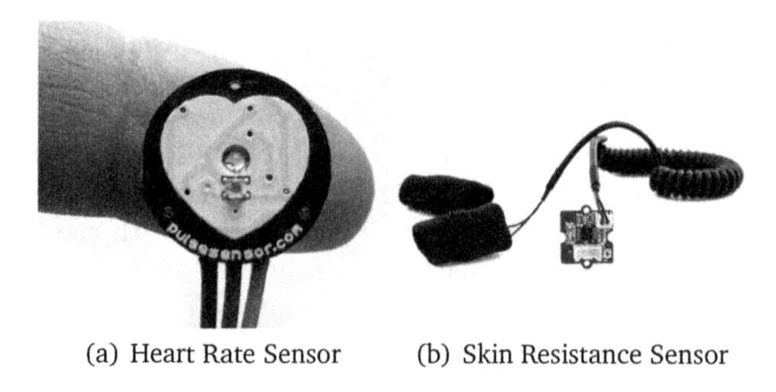

(a) Heart Rate Sensor (b) Skin Resistance Sensor

Fig. 2. View of the employed sensors.

measuring the light reflected by the skin. The received signal consists of pulses that reflect the change in vascular blood volume with each cardiac beat. The information captured by each one of these sensors is subsequently preprocessed. This last process allows to convert the measure captured for each sensor in the corresponding units. The GSR sensor converts the measurement of the skin conductance in Ohm and the Photoplethysmogram returns raw data that can be easily processed.

4.2 Emotion Recognition

Once the data has been obtained, it is necessary to implement a machine learning algorithm in order to identify the human emotions. To do this, the process has been divided into two subprocesses. The first one employs a *Fuzzy logic algorithm* in order to obtain the qualitative value (the name of the emotion) of the emotions stored in the employed dataset (which is below explained). The second process employs a *Neural Network* in order to classify new bio-signals inputs from the wristband into emotional values.

The dataset used to train and validate this model is the *DEAPdataset* [16]. The dataset contains physiological signals of 32 participants (and frontal face video of 22 participants), where each participant watched and rated their emotional response to 40 music videos along the scales of arousal, valence, and dominance, as well as their liking of and familiarity with the videos. This dataset integrates different bio-signals as: EEG, GSR, EOG, among other signals. All these signals are associated to the emotional changes using musical videos. Specifically, the authors identified 16 different emotions, which are the following:

1. Pride
2. Elation
3. Joy
4. Satisfaction
5. Relief
6. Hope
7. Interest
8. Surprise
9. Sadness
10. Fear
11. Shame
12. Guilt
13. Envy
14. Disgust
15. Contempt
16. Anger

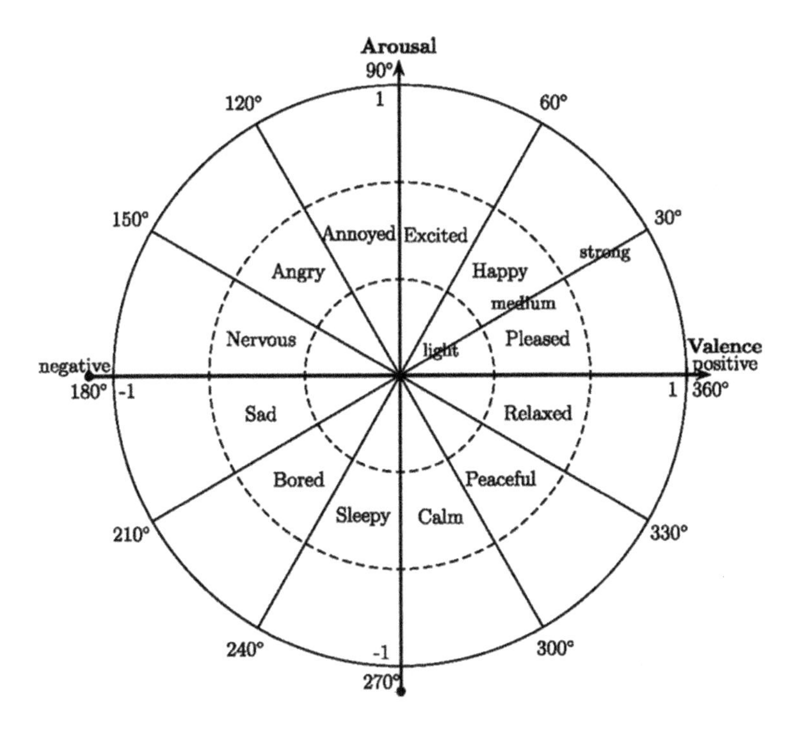

Fig. 3. Circle of emotions (arousal and valence).[27]

These emotions are represented following the circumflex emotional model [24,27,32]. This emotional model represents the emotions using three components *Valence, Arousal, Dominance.* In addition, the model evaluates every emotion in a trivalent scale: light, medium, and strong. As a result, the model has 36 possible emotional states (Fig. 3).

This model locates the emotions in twelve sub-quadrants, where each subquadrant is discretized in ranges of 30 degrees. The intensity of the emotion is the module of the vector composed by $\vec{E}(Ag) = [Arousal, Valence]$. The representation of emotions is done using a polar coordinate plane, where one takes into account the angle and the magnitude of the vector (see Eqs. 1 and 2).

$$r = \sqrt{Arousal^2 + +Valencd^2} \tag{1}$$

$$\theta = \begin{cases} arctan(\frac{Valence}{Arousal}) & if\ Arousal > 0 \\ \frac{\pi}{2} & if\ Arousal = 0 \\ arctan(\frac{Valence}{Arousal}) + \pi & if\ Arousal < 0 \end{cases} \tag{2}$$

Therefore, the emotion is represented as a tuple composed by the radius (r) and the angle (θ) $E(Ag) = \{r, \theta\}$ (all angles are in radians). Based on these data we employ a set of fuzzy logic rules in order to estimate the name of the emotion according to the input values stored in the dataset. These rules allow us to change a quantitative response to a qualitative response. This qualitative response is calculated and stored in the database for all the available registers. Once this has been calculated, the next step is the definition of a neural network, which allow us to identify the human emotions using only the data obtained from the GSR and Photoplethysmogram.

It is necessary to remark that, as each channel of the GSR and the Photoplethysmogram is formed by 8064 different values, it is impossible to build a neural network with these number of inputs. For this reason each channel was sub-sampled, converting each channel in an array of 252 values. Therefore, the neural network has 504 inputs (252 per each channel). The network has also has five hidden neurons and sixteen outputs (each output corresponds with a specific emotion). The architecture of our neural network is shown in Fig. 4.

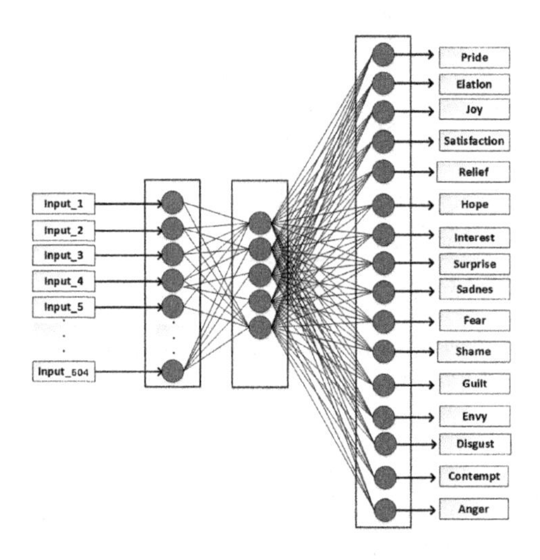

Fig. 4. Neural network architecture.

The ANN was trained using a supervised trained methodology, since the objective of the network is to classify the human emotion. Concretely, the training process employed a dataset composed by 1280 entries per channel. As before commented, this information was extracted from the *DEAPdataset*. From the selected dataset, the 20 % was used to test and the 80 % was used to train.

4.3 Wristband Prototype

This section describes the design of the physical wristband where the *Wristband agent* is executed. The wristband device was programmed in Python and it was embedded in the *Intel Edison*[3] computer-on-module. The Intel Edison (Fig. 5) is a new technology designed by Intel which contains a Dual Core IA-32 @ 500 MHz, a 32-bit CPU specially designed for Internet of Things (IoT) applications and wearable computing products.

[3] http://www.intel.la/content/www/xl/es/do-it-yourself/edison.html.

Fig. 5. Intel edison processor.

The Intel Edison supports *Yocto Linux*[4] and incorporates the *SPADE*[5] platform [8], which is a multi-agent system platform based on Python. As before commented, the prototype has been designed as a wristband in which is deployed the wristband agent. The bio-signals captured in the wristband are passed by an *Analog to Digital Conversion* or *ADC* allowing the discretization of the analogue signals. Figure 6 shows the different components of our wristband prototype which are the following.

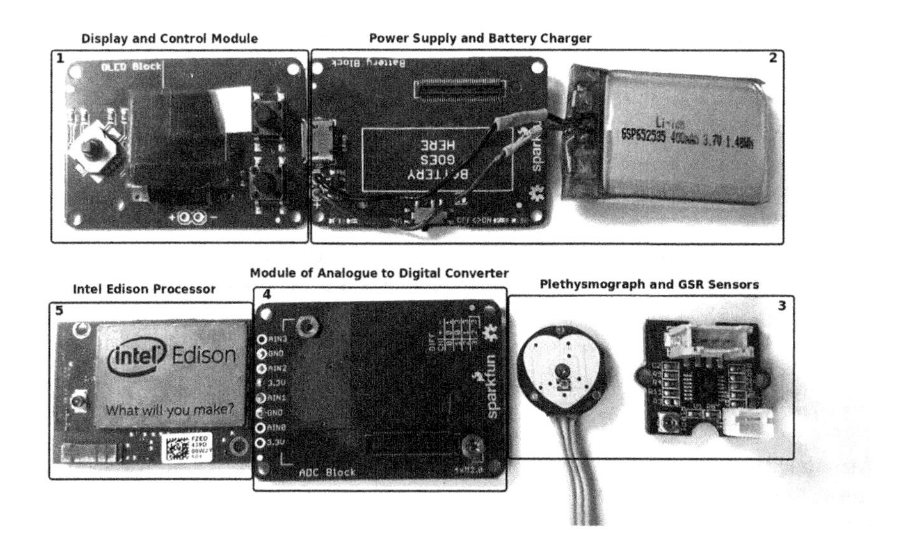

Fig. 6. Components of the smart wristband prototype.

1. Display and Control Module: This module is employed to show information of our emotional state through the LCD screen, and also controls the wristband.
2. Power Supply and Battery Charger: This is the power supply system. The wristband uses a 3.7 volt battery.

[4] https://www.yoctoproject.org/.
[5] https://github.com/javipalanca/spade.

3. Sensors: Sensors are responsible for carrying out the acquisition of the signals of GSR and Photoplethysmogram.
4. Analogue to Digital Convert: This module is responsible for digitizing the signals captured by the sensors.
5. Intel Edison Processor: This is the microprocessor where the agent is located. It is responsible for performing the processes of emotion recognition and communication with other agents (through the built-in Wi-Fi).

5 Conclusions and Future Work

This paper presents how to integrate non-invasive biosignals for the detection of human emotional states through an agent-based application. The identification and detection of human emotional states allow the enhancement of the decision-making process of intelligent agents. The proposed application allows extracting (in a non-invasive way) the social emotion of a group of persons by means of wearables facilitating the decision-making in order to change the emotional state of the individuals. As commented before, the application incorporates automatic emotion recognition using biosignals and machine learning techniques, which are easily included in the proposed system. The flexibility and dynamism of the proposed application allow the integration of new sensors or signals in future stages of the project. Moreover, as future work, we want to apply this system to other application domains, specifically the proposed framework fits with the industrial one, for instance representing production lines including the individuals and their emotional states as yet another elements to be considered in the production line.

Acknowledgements. This work is partially supported by the MINECO/FEDER TIN2015-65515-C4-1-R and the FPI grant AP2013-01276 awarded to Jaime-Andres Rincon. This work is supported by COMPETE: POCI-01-0145-FEDER-007043 and FCT – Fundação para a Ciência e Tecnologia within the projects UID/CEC/00319/2013 and Post-Doc scholarship SFRH/BPD/102696/2014 (A. Costa)

References

1. Alemdar, H., Ersoy, C.: Wireless sensor networks for healthcare: a survey. Comput. Netw. **54**(15), 2688–2710 (2010). http://dx.doi.org/10.1016/j.comnet.2010.05.003
2. Baig, M.M., GholamHosseini, H., Connolly, M.J., Kashfi, G.: Real-time vital signs monitoring and interpretation system for early detection of multiple physical signs in older adults. In: IEEE-EMBS International Conference on Biomedical and Health Informatics. IEEE (2014). http://dx.doi.org/10.1109/BHI.2014.6864376
3. Castillo, J.C., Fernández-Caballero, A., Castro-González, Á., Salichs, M.A., López, M.T.: A framework for recognizing and regulating emotions in the elderly. In: Pecchia, L., Chen, L.L., Nugent, C., Bravo, J. (eds.) IWAAL 2014. LNCS, vol. 8868, pp. 320–327. Springer, Heidelberg (2014). doi:10.1007/978-3-319-13105-4_46
4. Castillo, J.C., Serrano-Cuerda, J., Fernández-Caballero, A., Martínez-Rodrigo, A.: Hierarchical architecture for robust people detection by fusion of infrared and visible video. In: Novais, P., Camacho, D., Analide, C., El Fallah Seghrouchni, A., Badica, C. (eds.) Intelligent Distributed Computing IX. SCI, vol. 616, pp. 343–351. Springer, Heidelberg (2016). doi:10.1007/978-3-319-25017-5_32

5. Chuang, Z.J., Wu, C.H.: Multi-modal emotion recognition from speech and text. J. Comput. Linguist. Chin. **9**(2), 45–62 (2004). http://www.aclweb.org/anthology/O/O04/O04-3004.pdf
6. Colby, B.N., Ortony, A., Clore, G.L., Collins, A.: The Cognitive Structure of Emotions, vol. 18. Cambridge University Press, Cambridge (1989)
7. Coutinho, E., Cangelosi, A.: Musical emotions: Predicting second-by-second subjective feelings of emotion from low-level psychoacoustic features and physiological measurements. Emotion **11**(4), 921–937 (2011). (Washington, D.C.)
8. Escriva, M., Palanca, J., Aranda, G., García-Fornes, A., Julian, V., Botti, V.: A Jabber-based multi-agent system platform. In: Proceedings of the Fifth International Joint Conference on Autonomous Agents and Multiagent Systems (AAMAS 2006), pp. 1282–1284. Association for Computing Machinery, Inc. (ACM Press) (2006)
9. Fensli, R., Pedersen, P.E., Gundersen, T., Hejlesen, O.: Sensor acceptance model - measuring patient acceptance of wearable sensors. Method Inf. Med. **47**, 89–95 (2008). http://dx.doi.org/10.3414/ME9106
10. Fishkin, K.P., Jiang, B., Philipose, M., Roy, S.: I sense a disturbance in the force: unobtrusive detection of interactions with RFID-tagged objects. In: Davies, N., Mynatt, E.D., Siio, I. (eds.) UbiComp 2004. LNCS, vol. 3205, pp. 268–282. Springer, Heidelberg (2004). doi:10.1007/978-3-540-30119-6_16
11. Gratch, J., Marsella, S.: Tears and fears: modeling emotions and emotional behaviors in synthetic agents. In: Proceedings of the Fifth International Conference on Autonomous Agents, pp. 278–285. ACM (2001). http://dl.acm.org/citation.cfm?id=376309
12. Han, K., Yu, D., Tashev, I.: Speech emotion recognition using deep neural network and extreme learning machine. In: Fifteenth Annual Conference of Interspeech, pp. 223–227, September 2014. http://research.microsoft.com/pubs/230136/IS140441.PDF
13. Hert, P., Gutwirth, S., Moscibroda, A., Wright, D., Fuster, G.G.: Legal safeguards for privacy and data protection in ambient intelligence. Pers. Ubiquit. Comput. **13**(6), 435–444 (2008). http://www.springerlink.com/index/10.1007/s00779-008-211-6
14. Hristoskova, A., Sakkalis, V., Zacharioudakis, G., Tsiknakis, M., Turck, F.D.: Ontology-driven monitoring of patient's vital signs enabling personalized medical detection and alert. Sensors **14**(1), 1598–1628 (2014). http://dx.doi.org/10.3390/s140101598
15. Karthigayan, M., Rizon, M., Nagarajan, R., Yaacob, S.: Genetic algorithm and neural network for face emotion recognition. In: Affective Computing, pp. 57–68 (2008). http://cdn.intechopen.com/pdfs-wm/5178.pdf
16. Koelstra, S., Mühl, C., Soleymani, M., Lee, J.S., Yazdani, A., Ebrahimi, T., Pun, T., Nijholt, A., Patras, I.: DEAP: a database for emotion analysis; using physiological signals. IEEE Trans. Affect. Comput. **3**(1), 18–31 (2012)
17. Kuo, C.H., Chen, C.T., Chen, T.S., Kuo, Y.C.: A wireless sensor network approach for rehabilitation data collections. In: 2011 IEEE International Conference on Systems, Man, and Cybernetics. Institute of Electrical & Electronics Engineers (IEEE) (2011). http://dx.doi.org/10.1109/ICSMC.2011.6083773
18. Maaoui, C., Pruski, A.: Emotion recognition through physiological signals for human-machine communication. In: Cutting Edge Robotics, pp. 317–333 (2010). http://www.intechopen.com/source/pdfs/12200/InTech-Emotion_recognition_through_physiological_signals_for_human_machine_communication.pdf

19. Maier, E., Kempter, G.: ALADIN - a magic lamp for the elderly? In: Nakashima, H., Aghajan, H., Augusto, J.C. (eds.) Handbook of Ambient Intelligence and Smart Environments, pp. 1201–1227. Springer, Berlin, Heidelberg (2010). http://dx.doi.org/10.1007/978-0-387-93808-0_44
20. Mehrabian, A.: Analysis of affiliation-related traits in terms of the PAD temperament model. J. Psychol. **131**(1), 101–117 (1997). http://dx.doi.org/10.1080/00223989709603508
21. Piana, S., Odone, F., Verri, A., Camurri, A.: Real-time Automatic Emotion Recognition from Body Gestures. arXiv preprint arXiv:1402.5047, pp. 1–7 (2014). http://xxx.tau.ac.il/pdf/1402.5047.pdf
22. Ramos, J., Oliveira, T., Satoh, K., Neves, J., Novais, P.: Orientation system based on speculative computation and trajectory mining. In: Bajo, J., et al. (eds.) PAAMS 2016. CCIS, vol. 616, pp. 250–261. Springer, Heidelberg (2016). doi:10.1007/978-3-319-39387-2_21
23. Rincon, J.A., Julian, V., Carrascosa, C.: Social emotional model. In: Demazeau, Y., Decker, K.S., Bajo Pérez, J., de la Prieta, F. (eds.) PAAMS 2015. LNCS (LNAI), vol. 9086, pp. 199–210. Springer, Heidelberg (2015). doi:10.1007/978-3-319-18944-4_17
24. Salmeron, J.L.: Fuzzy cognitive maps for artificial emotions forecasting. Appl. Soft Comput. J. **12**(12), 3704–3710 (2012). http://dx.doi.org/10.1016/j.asoc.2012.01.015
25. Scherer, K.R., Zentner, M.R.: Emotional effects of music: production rules. In: Music and Emotion: Theory and Research, pp. 361–392 (2001). http://icquran.persiangig.com/weblog/schererzentner.pdf
26. Stawarz, K., Cox, A.L., Blandford, A.: Don't forget your pill! In: Proceedings of the 32nd Annual ACM Conference on Human Factors in Computing Systems, CHI 2014. Association for Computing Machinery (ACM) (2014). http://dx.doi.org/10.1145/2556288.2557079
27. Thayer, R.: The Biopsychology of Mood and Arousal. Oxford University Press, Oxford (1989)
28. Tran, N., Coffman, J.M., Sumino, K., Cabana, M.D.: Patient reminder systems and asthma medication adherence: a systematic review. J. Asthma **51**(5), 536–543 (2014). http://dx.doi.org/10.3109/02770903.2014.888572
29. Villarejo, M.V., Zapirain, B.G., Zorrilla, A.M.: A stress sensor based on galvanic skin response (GSR) controlled by ZigBee. Sensors **12**(5), 6075–6101 (2012). (Switzerland)
30. Walter, M., Eilebrecht, B., Wartzek, T., Leonhardt, S.: The smart car seat: personalized monitoring of vital signs in automotive applications. Pers. Ubiquit. Comput. **15**(7), 707–715 (2011). http://dx.doi.org/10.1007/s00779-010-0350-4
31. Whitman, B., Smaragdis, P.: Combining musical and cultural features for intelligent style detection. In: Ismir, pp. 5–10, Paris, France (2002). http://citeseerx.ist.psu.edu/viewdoc/download?doi=10.1.1.100.8383&rep=rep1&type=pdf
32. Yik, M., Russell, J.A., Steiger, J.H.: A 12-point circumplex structure of core affect. Emotion **11**(4), 705–731 (2011)
33. van der Zwaag, M.D., Westerink, J.H.D.M., van den Broek, E.L.: Emotional and psychophysiological responses to tempo, mode, and percussiveness. Musicae Scientiae **15**(2), 250–269 (2011). http://msx.sagepub.com/content/15/2/250.short

Impulse Noise Detection in OFDM Communication System Using Machine Learning Ensemble Algorithms

Ali N. Hasan(✉) and Thokozani Shongwe

Department of Electrical and Electronic Engineering Technology, University of Johannesburg, Doornfontein, P. O. Box 17011, Johannesburg 2028, South Africa
{alin, tshongwe}@uj.ac.za

Abstract. An impulse noise detection scheme employing machine learning (ML) algorithm in Orthogonal Frequency Division Multiplexing (OFDM) is investigated. Four powerful ML's multi-classifiers (ensemble) algorithms (Boosting (Bos), Bagging (Bag), Stacking (Stack) and Random Forest (RF)) were used at the receiver side of the OFDM system to detect if the received noisy signal contained impulse noise or not. The ML's ensembles were trained with the Middleton Class A noise model which was the noise model used in the OFDM system. In terms of prediction accuracy, the results obtained from the four ML's Ensembles techniques show that ML can be used to predict impulse noise in communication systems, in particular OFDM.

Keywords: Ensemble · Prediction · Bagging · Boosting · Stacking · Random forest · OFDM and impulse noise

1 Introduction

Orthogonal Frequency Division Multiplexing (OFDM) has become a popular modulation for both wireline and wireless communications. A block diagram of an OFDM system (including a ML block at the receiver) is shown in Fig. 1. An OFDM system has advantages of being robust against frequency selective fading and high data rate compared to single carrier systems, due to the transmission of data in multiple frequency carriers. However, OFDM can be adversely affected by impulse noise because the energy of an impulse is spread by the FFT such that it appears distributed across all the frequency carriers at the output of the FFT [1]. It is for this reason that most impulse noise mitigation schemes on OFDM focus on reducing the effect of the impulse noise before the FFT on the receiver side of the OFDM system (see [2, 10] for impulse noise mitigation schemes and impulse noise models). Such methods are termed clipping and/or nulling [4, 5], where thresholds are used to detect impulse noise in the time-domain and clip or null any time sample that is above the set threshold. Other impulse noise mitigation methods can be used together with the clipping/nulling scheme, for example: in [3, 8], the authors implemented the iterative impulse noise estimation technique with clipping/nulling. In [6, 7, 9], error correcting coding is used

M. Graña et al. (eds.), *International Joint Conference SOCO'16-CISIS'16-ICEUTE'16*,
Advances in Intelligent Systems and Computing 527, DOI 10.1007/978-3-319-47364-2_9

to combat impulse noise and it can still work together with the clipping or nulling scheme to obtain a powerful impulse noise combatting scheme.

In this paper, the focus is the clipping and/or nulling schemes using thresholds to detect impulse noise. In that regard machine learning multi-classifier or ensemble algorithms were used to estimate the amplitude (or power) of the impulse noise.

Machine learning (ML) is a subfield of artificial intelligence theory that was developed from the study of pattern recognition and computational learning theory [12]. Recently, Machine Learning algorithms have been utilised in prediction, classification, monitoring and optimisation tasks in many important applications such as medical science, engineering applications, intelligent control systems etc. [12, 13].

Ensembles or multi-classifier methods have recently become as a common learning method, not only because of their straightforward implementation, but also due to their outstanding predictive performance on practical and real-life problems [13]. An ensemble contains a set of individually trained classifiers (for example decision trees or neural networks) whose predictions are combined when classifying distinctive instances. Ensemble methods aim to improve the predictive performance of a given statistical learning or model fitting technique [13].

This work was conducted to examine the use of four popular and powerful multi-classifiers (ensembles) (Bag, Bos, Stack and RF) to predict, thus estimate impulse noise on OFDM. In this work we consider the conventional OFDM communication system employing PSK/QAM modulation as shown in Fig. 1, which we call PSK/QAM-OFDM in short. The OFDM system is discussed in detail in Sect. 2.

The ML classifiers are trained with the impulse noise statistics so that they should be able to predict the DFT samples (of OFDM) that contain impulse noise at the receiver. Once the samples with impulse noise are located using the ML classifiers, the impulse noise can be subtracted from the received signal, leaving an estimate of the transmitted signal plus additive white Gaussian noise (AWGN).

2 System Model

OFDM uses the power of the discrete Fourier transform (DFT) to transmit data in multiple frequencies as follows: Symbols from phase shift Keying (PSK) modulation or quadrature amplitude modulation (QAM) are taken as input to the OFDM transmitter. These symbols are processed by the inverse discrete Fourier transform (IDFT) at the transmitter. At the receiver side, a DFT is performed on the received symbols which would have been affected by channel noise. The PSK/QAM-OFDM system is shown in Fig. 1, where the transmitter side is shown together with the transmitted signal (Tx) which is affected by additive noise (AWGN and Impulse noise) as it passes through the channel.

The noise affected signal (Rx) is received at the receiver side for processing by the ML noise prediction tool which contains the ensemble algorithms before being fed to the DFT. The ML noise prediction tool task is to estimate the noise in the received signal and classify the signal as either containing only AWGN or impulse noise. The details of how the ML noise prediction ensemble algorithms are used to classify the noise are discussed in Sect. 4. For now the ML ensemble algorithms used in the prediction of noise are discussed, in the next section.

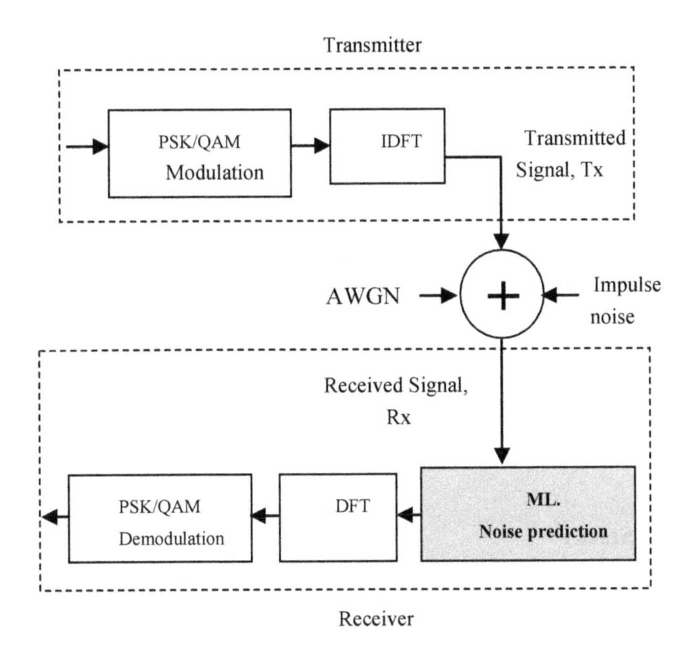

Fig. 1. OFDM communication system with Machine Learning for impulse noise estimation

3 Multi-classifiers (Ensembles) Algorithms

3.1 Bagging

Bagging or Bootstrap aggregating (Bag) is a popular way to obtain multiple classifiers. Bag. was proposed by Breiman in 1996 to improve the classification results by merging outputs of classifiers that are trained using randomly-generated training sets [14, 15].

Bag. is a "bootstrap" multi-classifier method that produces individuals for its ensemble by training each classifier on a random redistribution of the training set. Each classifier's training set is generated by randomly drawing, with replacement, X examples, where X is the size of the original training set; many of the original examples may be repeated in the resulting training set while others may be left out [14, 15]. Each single classifier in the ensemble is generated with a different random sampling of the training set [15].

3.2 Boosting (Bos)

Bos. algorithm was proposed by Schapire and Freund [16]. Boosting comprises a family of methods. The focus of those methods is to generate a series of classifiers. The training set used for each member of the series is chosen based on the performance of the earlier classifier(s) in the series [17]. In Bos, cases that are wrongly predicted by previous classifiers in the series are chosen more often than cases that were appropriately predicted. Thus Boosting tries to create new classifiers that are better able to predict cases for which the present ensemble's performance is poor. Note that in Bag. technique, the resampling of the training set is not dependent on the performance of the earlier classifiers [15–17].

3.3 Random Forest (RF)

The Random Forest method is based on bagging (bootstrap aggregation) models built using the Random Tree method, in which classification trees are grown on a random subset of descriptors [18]. The Random Tree method can be viewed as an implementation of the Random subspace method for the case of classification trees. Combining two ensemble learning approaches, bagging and random space method, makes the Random Forest method a very effective approach to build highly predictive classification models [19].

3.4 Stacking (Stack)

Stack. is historically one of the first ensemble learning methods. It combines several base classifiers, which can belong to absolutely different classes of machine learning methods, by means of a "meta-classifier" that takes as its inputs the output values of the base classifiers [19]. Although stacking is a heuristic method and does not guarantee improvement in all cases, in many practical studies it shows excellent performance.

4 Simulations

4.1 Simulation Set-up

The four machine learning ensemble techniques were used to classify thresholds of the received signals as either containing the transmitted signal, containing the transmitted plus AWGN or containing the transmitted signal plus AWGN plus impulse noise. To do this we create three classes which will be used by the four used ensemble techniques.

To set up the three thresholds (or classes) we use the following knowledge about signal transmission in an impulse noise channel. Impulse noise is usually of very high amplitude compared to the transmitted signal and AWGN. The transmitted signal is usually given a variance of one $\left(\sigma_s^2 = 1\right)$ and AWGN also has a variance of one $\left(\sigma_g^2 = 1\right)$. We can set the variance of impulse noise (σ_I^2) to any value greater than one. When employing the Middleton Class A noise model, it is customary to define the variance of impulse noise as function of $\sigma_g^2 = 1$, such that $\sigma_I^2 = K\sigma_g^2$, where $K > 1$. On average, we can note that the amplitude of the transmitted signal plus AWGN will be the value 2 $\left(\sigma_s^2 + \sigma_g^2 = 1 + 1\right)$. Therefore we set our first threshold to cover values from 0 to 1 ($T_0 = 0 - 1$). The second threshold is set to cover values from 1.1 to 2.1 ($T_1 = 1.1 - 2.1$). The third threshold is set to be values from 2.2 and above

Table 1. Signal, AWGN and Impulse noise level classes

Class	Threshold description	Numerical threshold level
1	Signal, T_0	0–1
2	Signal + AWGN, T_1	1.1–2.1
3	Signal + AWGN + IN, T_2	> 2.2

($T_2 \geq 2.2$). The threshold of 2.2 was used in [8] and was shown to be effective for detecting impulse noise. Table 1 shows a summary of the different classes of the received signal.

The split to train ratio for all used classifiers was 70 % to 30 % of the data.

Figure 2 illustrates the number of instances (count) and weight distribution for each class for noise classes.

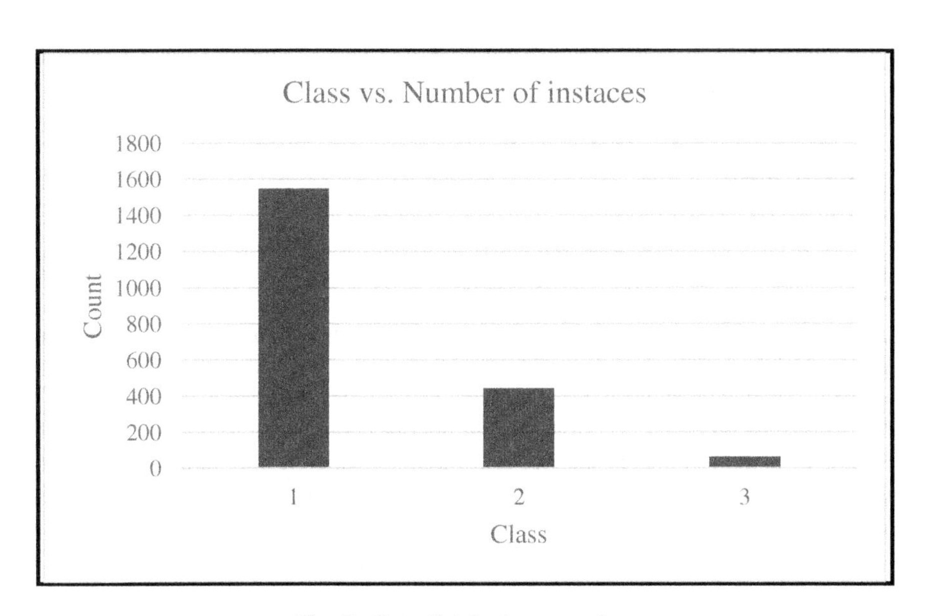

Fig. 2. Data distribution over class

It can be seen from Fig. 2 that 1546 instances were classified as class 1, 441 instances as class 2, and 61 instances as class 3. Almost 75 % of the data were classified as class 1 which cause the data to be imbalance, however the data in this experiment were dealt with collectively using cross validation by randomly choosing 70 % of the impulse noise generated data to train the classifiers and 30 % to test them.

4.2 Results Discussion

The main performance measure for this experiment is the prediction accuracy, however mean absolute error and root mean square error are included as a secondary

Table 2. Impulse noise prediction accuracies

Description	Bag	Bos	Stack	RF
Prediction accuracy	99.85 %	99.51 %	97.31 %	99.83 %
Mean absolute error	0.002	0.017	0.028	0.002
Root mean squared error	0.022	0.066	0.028	0.030

performance measures to provide more statistical information about each classifier performance. MATLAB simulator was used to classify the data, using default parameters for all classifiers. The simulation results are shown in Table 2.

In terms of prediction accuracy, it can be seen from Table 2 that Bag and RF barely outperformed the other two ensemble classifiers with a 99.85 % prediction accuracy. Bos and Stack also showed good performance with high prediction accuracy of 99.51 %, and 97.31 % respectively. In terms of the secondary performance measures, it can be noticed that Bag and RF also achieved the best performance and scored the lowest mean absolute error and root mean square error.

5 Conclusions

We have shown that ensemble or multi-classifiers techniques can be used for impulse noise prediction in OFDM systems affected by background noise (AWGN) and impulse noise. The results achieved from this investigation show that the Bag, RF, Stack and Bos algorithms can predict impulse noise with high level of confidence and accuracy. The four different ML's ensemble techniques were tested, and found to be effective at predicting impulse noise as Bag, Bos and RF realized more than 99.0 %, and Stack achieved 97.31 %. In terms of imbalanced data, this problem could be overcome by using techniques such as, re-sampling data, collecting more data. etc. The data imbalance tends to suit RF and Bagging, hence these methods could be used for predicting impulse noise. Statistically, Tukey multiple test shows that there no significant difference in performance between all classifiers.

References

1. Shongwe, T., Vinck, A.J.H., Ferreira, H.C.: On impulse noise and its models. In: Proceedings of the 2014 International Symposium on Power-Line Communications and its Applications, Glasgow, Scotland, March 30 - April 2, 2014, pp. 12–17 (2014)
2. Zhidkov, S.V.: Impulsive noise suppression in OFDM-based communication systems. IEEE Trans. Consum. Electron. **49**(4), 944–948 (2003)
3. Häring, J., Vinck, A.J.H.: OFDM transmission corrupted by impulsive noise. In: Proceedings of the 2000 International Symposium on Power-Line Communications and its Applications, Limerick, Ireland, April 5–7, 2000, pp. 5–7 (2000)

4. Zhidkov, S.V.: Performance analysis and, optimization of OFDM receiver with blanking nonlinearity in impulsive noise environment. IEEE Trans. Veh. Technol. **55**(1), 234–242 (2006)
5. Tseng, D.-F., Han, Y.S., Mow, W.H., Chang, L.-C., Vinck, A.J.H.: Robust clipping for OFDM transmissions over memoryless impulsive noise channels. IEEE Commun. Lett. **16** (7), 1110–1113 (2012)
6. Sargrad, D.H., Modestino, J.W.: Errors-and-erasures coding to combat impulse noise on digital subscriber loops. IEEE Trans. Commun. **38**(8), 1145–1155 (1990)
7. Li, T., Mow, W.H., Siu, M.: Joint erasure marking and viterbi decoding algorithm for unknown impulsive noise channels. IEEE Trans. Wireless Commun. **7**(9), 3407–3416 (2008)
8. Mengi, A., Vinck, A.J.H.: Successive impulsive noise suppression in OFDM. In: Proceedings of the 2009 IEEE International Symposium on Power Line Communications, Rio de Janeiro, Brazil, Mar. 5–7, 2009, pp. 33–37 (2009)
9. Faber, T., Scholand, T., Jung, P.: Turbo decoding in impulsive noise environments. Electron. Lett. **39**(14), 1069–1071 (2003)
10. Shongwe, T., Vinck, A.J.H., Ferreira, H.C.: A study on impulse noise and its models. SAIEE Afr. Res. J. **106**(3), 119–131 (2015)
11. Witten, I., Frank, E.: Data Mining, Practical Machine Learning Tools and Techniques, 2nd (2005). ISBN: 0-12-088407-0
12. Mitchell, T., McGraw, H.: Machine learning, 2nd, Chap. 1, January 2010
13. Hasan, A.N., Twala, B., Marwala, T.: Moving Towards Accurate Monitoring and Prediction of Gold Mine Underground Dam Levels. In: IEEE IJCNN WCCI, Beijing, China (2014)
14. Sun, Q., Pfahringer, B.: Bagging Ensemble Selection. The University of Waikato, Hamilton, New Zealand (2010)
15. Breiman, L.: Bagging predictors. Mach. Learn. **24**(2), 123–140 (1996)
16. Vemulapalli, S., Luo, X., Pitrelli, J., Zitouni, I.: Using bagging and boosting techniques for improving coreference resolution. Informatica **34**, 111–118 (2010)
17. Buhlmann, P.: Bagging, Boosting and Ensemble Methods. In: ETH Zurich, Seminar fur Statistik, HG G17, CH-8092 Zurich, Switzerland (2010)
18. Breiman, L.: Random Forests. Mach. Learn. **45**(1), 5–32 (2001)
19. Wolpert, D.H.: Stacked generalization. Neural Netw. **5**, 241–259 (1992)

SOCO 2016: Soft Computing Applications

A Hybrid Method for Optimizing Shopping Lists Oriented to Retail Store Costumers

Santiago Porras and Bruno Baruque(✉)

University of Burgos, Burgos, Spain
{sporras, bbaruque}@ubu.es

Abstract. In the present day, one of the most common activities of everyday life is going to a supermarket or similar retail spaces to buy groceries. Many consumers organizations like The European Consumer Organization [1], advise buyers to prepare a "grocery list" in order to be ready for this activity. The present work proposes a system that helps to develop this activity in several ways: Firstly, it enables the user to create lists with different levels of abstraction: from concrete products to generic ones (or families of products). Secondly, the lists are collaborative and can be shared with other users. Finally, it automatically determines the best store to buy a given product using the proposed optimization algorithm. Furthermore, the optimization algorithm assigns a part of the list to each user balancing the cost that every user has to pay and choosing the cheapest supermarket where they have to buy.

Keywords: Balanced shopping list · Purchase optimization · Collaborative list

1 Introduction

The purchase of grocery products in supermarkets is a common routine in our days. Numerous studies explain the purchase habits of the consumers, especially with healthy products [2–4]. Likewise, there are multitude of factors that can make a consumer choose to opt for one product over another. Between these, we can find the brand, the price, comparatives or the trust of the client in the product due to previous experiences. This can help to determine in advance which products the user really needs and an approximate price for the complete list in order to plan the household expenses.

Nowadays, numerous mobile applications have appeared that simplify the task of elaborate the shopping list to the user. More importantly, they let the user access all this information even at the moment of the purchase, avoiding the necessity of carefully planning the activity in advance. To the knowledge of the authors, there are not many publications regarding the optimization of low expense purchases, mainly dedicated to individual consumers rather than big retail operations.

In this study three main features are introduced to facilitate the process of completing a "grocery list" in an automated and flexible way which to the knowledge of authors have not yet been considered by commercial products: One objective is to introduce the concept of product category in the shopping lists, for example milk or beer, without specifying a particular brand or product, therefore making available the selection among the variety of products that will adapt better to the user's criteria. In

M. Graña et al. (eds.), *International Joint Conference SOCO'16-CISIS'16-ICEUTE'16*,
Advances in Intelligent Systems and Computing 527, DOI 10.1007/978-3-319-47364-2_10

family units or contexts as a shared apartment, several people can bring products to the shopping list. Hence arises the objective of collaborative shopping lists in which several people contribute to their elaboration. Finally, the proposed model can take all this information as input in order to complete a suggested best distribution among buyers and supermarkets to make the actual purchase of the desired items.

This paper presents a bi-objective approach to divide shopping lists. On one hand to minimize the total cost of purchase and on the other, that shopping lists are balanced in terms of price. This means that each participant person makes a purchase of a similar cost. The user can assign the weight he wants assign to each criterion. To achieve these objectives metaheuristic techniques are used, looking on the one hand the quality of the results and other fast calculations. These techniques are used because the consideration of categories can produce a combinatorial explosion.

2 Features of the System

Our purpose is the design and development of a collaborative shopping list application for mobile devices. The application will indicate the user where is more convenient to make the purchase of each of the products on the list, based on the criteria that the user has chosen.

Combination of Specific Products and Categories: The lists are composed of products that are on sale simultaneously in different supermarket chains or only in a particular supermarket chain. Each product has a certain price, depending on the chain in which it is offered.

Each product is listed in one category (or more). The system allows to include categories into a list, not just specific products. For example, it is possible that an element of the list is “whole milk” without specifying particular brand. The optimization process will select the specific product, which belongs to the category, which best fits the user preferences.

Collaborative Lists: Our aim is set a collaborative environment between users by including a network of contacts. The objective of this feature is to allow users to organize and coordinate a joint purchase. A consequence of the collaborative list is that will be more difficult to obtain an equilibrium in the purchases between all users.

Automated Optimization of Lists: A “shopping list” will serve as starting point for the user to make a “purchase list”. The purchase list consists on a list of the specific products that will be purchased in a given retailer and will be generated by the optimization process.

The fact that the purchases that are made at different supermarkets have a similar price is called “balancing” in this scenario. The philosophy is to collaborate between multiple users to buy all the products and categories listed. The aim is to develop an algorithm that, taking as input a list of products and categories with a maximum number of supermarkets, divides the list between supermarkets following user settings. Users can choose between obtaining the cheapest purchase in terms of price or getting a purchase in which the amount spent in each supermarket is similar, this is useful if the purchase is shared with other users.

Additionally, to provide satisfactory optimization results, the algorithm must deliver them in a reasonable amount of time.

3 Optimization Problem Description

The problem is how to divide a list of articles and categories among the selected supermarkets in order to get the articles that is more convenient to purchase in each supermarket satisfying the user requirements, minimizing the cost of the purchase and balancing the purchases among the supermarkets.

The main restrictions in our problem are two: The articles could be a specific article or a category. If it is a category, the algorithm has to choose the particular article that best fits into the solution from all articles available included in that category. If a product is only available in one supermarket, it is mandatory to make a purchase in that supermarket.

All the same items are treated as a batch, regardless of their quantity and are assigned to a single supermarket. To measure both objectives at the same scale the values of the prices of products previously normalized.

3.1 Mathematical Definition

Firstly, we denote our variables:

- Pi, i = (1…n) is the set of all products.
- Mi, i = (1…n) is the set of all supermarkets.
- Ci, i = (1…n) is the set of all categories.
- Ci, $\subseteq$ {Pi, … Pn}, i = (1…n) a category is a subset of products.
- L = {Pi … Pn } $\cap$ {Ci … Cn } $\subseteq$ {Pi, Ci, }, i = (1…n) is the subset (List) of products and categories selected by the user.
- Mi Pj, i, j = (1…n) is the cost of the product Pj in the supermarket Mi.

Our objective function is:

$$S = minimize\ \alpha * Balance + \beta * Cost \tag{1}$$

$$\text{Cost} = \sum\nolimits_{i,j=1}^{n} MiPj \forall Pi, Ci \in L \tag{2}$$

$$\text{Balance} = \sum\nolimits_{\substack{i,j=1 \\ i \neq j}}^{n} |cost(M_i) - cost(M_j)| \tag{3}$$

The cost function is the sum of the value of the products once assigned to a specific supermarket as well as of the categories once chose a particular product and assigned to a supermarket. The balance function measures the difference between the total costs of each supermarket in the solution.

Parameters α and β allow the user to choose the specific weight of each function. There are complementary and have a value between 0 and 1 so that $\beta = 1 - \alpha$. Also, we use another parameter Δ in the GRASP phase to determine the size of the Restricted Candidate List (RCL) defined in the next section.

4 Optimization Algorithms

GRASP [5] is a fast multi-start method in which local search is applied to the initial solutions constructed with a greedy randomized heuristic. In [6, 7] surveyed the application of GRASP to solve combinatorial optimization problems in various domains. GRASP can be hybridized in different ways, for instance by replacing the local search with another metaheuristic [8].

In our solution we propose a hybrid GRASP/VND replacing the local search of GRASP by a variable neighborhood descent (VND) method [9]. Variable Neighborhood Search (VNS) is a technique that tries to escape from a local optimum systematically changing the structure of the environment [10, 11].

Replacing the local search by a descent local search in VNS, we obtain a variant of VNS called Variable Neighborhood Descent (VND). The aim of this replace is change the neighborhood every time a local minimum is reached.

4.1 Proposed Algorithm

To solve the problem, we have designed a method that combines GRASP and VND. As input of the algorithm we need the List L of products and categories to optimize. As input parameters we use: α, β and Δ.

Initial phase: A cost matrix is constructed with supermarkets and the products that make the list, if the list contain a category, they include all products belonging to the category. A GRASP model is used for the construction of an initial solution. Given the constraints of the problem, initially the number of supermarkets which will form the solution will be limited to a minimum number of supermarkets where you have to buy a mandatory product of the list and the maximum number will be indicated by the user. Items that must be bought compulsorily at the supermarket are assigned directly as part of the solution.

Constructive phase: Initially all the possible combinations of supermarkets are calculated. For every combination the GRASP phase begin, constructing RCL. The RCL is built as follow: fist we calculate the average cost of a product in all supermarkets. If it is a category the average of all products belonging at the category is calculated for all supermarkets. After both are ordered in a descent way according to the average calculated.

The size of the RCL is delimited by the parameter Δ who take values between 0 and 1. The size is calculated as follow:

$$\text{Averagemin} + \Delta * (\text{Averagemax} - \text{Averagemin}) \tag{4}$$

This allows us to encourage diversification and intensification. Once a RCL is formed one element from it is randomly chosen and added to the solution. This process continues until all products from the list are assigned to a supermarket.

The whole process is repeated 10 times, in our case, for every supermarket combination. Finally, the best solution is selected to be improved in the next phase.

Improvement phase: Firstly, we are going to define two different movements: Movement 1–0 and Movement 1–1. These two movements define the neighborhood structures of the VND (Figs. 1 and 2).

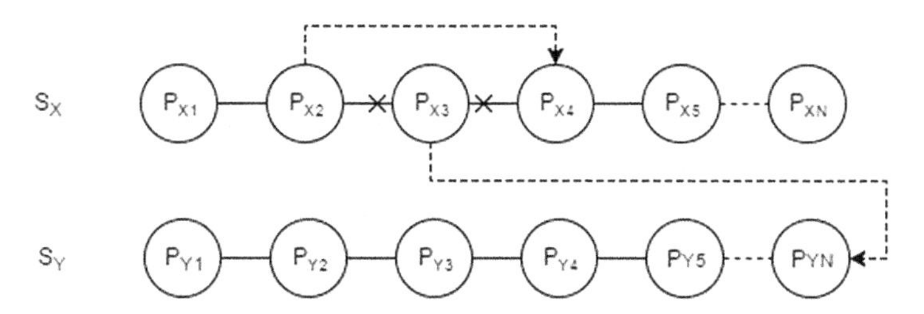

Fig. 1. Movement 0–1.

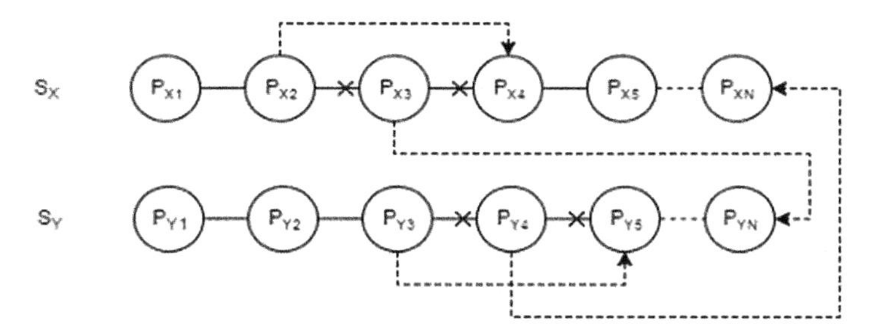

Fig. 2. Movement 1–1.

In the first one, the process evaluates if it is possible move a product or category from one supermarket to another one, if it is feasible and gives a better solution, the movement is executed eliminating the product or category from the host supermarket and adding it at the end of the destination supermarket. In the problem the order of the products in the supermarket is indifferent. In the second one, is an exchange of two products or categories, between two supermarkets, if the interchange is feasible and improve the solution the movement is done.

The usefulness of this hybrid method has been demonstrated in the literature, for instance in [12, 13].

Complete method:

```
GRASP/VND (L,M,n,α,Δ)
Var S //Solution
  For each P or C in L
    For each M
      CostMatrix(M,P)=MP (Build Cost Matrix)
  End
  For each P or C in L Mandatory
    Assign P to M
  End
  Calculate S
//GRASP
  For 1 to n
   GRASP Phase(S, L,M,n,α,Δ)
  End
  Return S //Best Solution.
//Improvement Phase VND
  VND(S)
  Return S // Best Solution
End
```

5 Experiments and Results

5.1 Test Dataset

Data is taken from a real price products of Spanish supermarkets. We have defined six different scenarios or Lists. Both categories and products are stored into a database, with his name, category, supermarket and other characteristics like the capacity depends of the type of product, calories… this database is ready to use for other studies like for example try to minimize the price of the purchase and the calories of the products. Also the number of products in each category may vary depending on the category, this is because of it is based in a real case. Summarizing we have 591 different products divided in 127 different subcategories, these subcategories are grouped in 29 main categories. The majority of subcategories (85 %) has 5 products or less, while fewer categories have between 5 and 10 products (9,5 %). Only 7 of them have more than 10 products, while 2 of them have more than 20 products. In Fig. 3 the volume of the data grouped by the main categories can be observed.

5.2 Parameters

Our procedure has three parameters, α, β and Δ. α and β are chosen by the user, and Δ allows us to encourage diversification and intensification.

To establish the best value of Δ inside of the application several test have been performed over a subset based in the maximum number of purchases of the lists

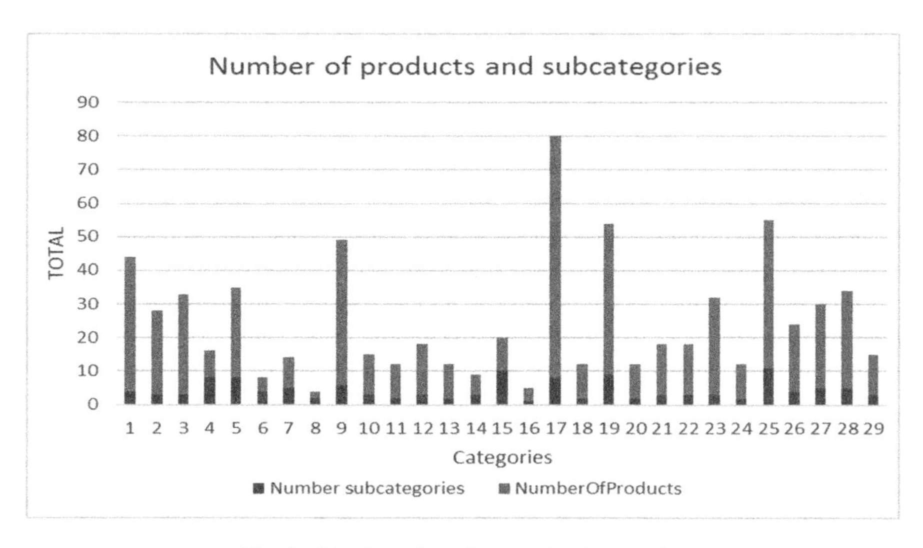

Fig. 3. Number of products and subcategories

detailed in Table 1. We can divide these scenarios in two groups, small (L1–L3) and big (L4–L6). This value is selected by the user. The values tested for all parameters are 0, 0.25, 0.5, 0.75 and 1. Results are calculated in base of the average of all α and β combinations. Solution average values (Sol) and time (T) in milliseconds are shown (Table 2).

Table 1. List description.

List	Number of products	Number of categories	Mandatory products	Total products	Max purchases
L1	2	0	1	3	1–3
L2	2	1	0	3	1–3
L3	1	1	1	3	1–3
L4	0	39	0	39	1–7
L5	3	3	3	9	2–7
L6	4	3	7	14	3–7

Table 2. Delta parameter study

Delta (Δ)									
0		0,25		0,5		0,75		1	
Sol	T	Sol	T	Sol	T	Sol	T	Sol	T
0,432	457	0,432	508	0,432	478	0,432	466	0,431	496

Table 2 represents results obtained in the solution. Solutions are very close to each other, and results are very close too in running times. In order to obtain an equilibrium between solution quality and execution time we have choose the value of delta 0.5.

5.3 Results

We show the average values obtained for every α and β combination and for all number of purchases considered (indicate beside list number) in Table 3.

Analyzing the results shown in Table 3, we can conclude: In small instances, if the whole weight of the objective function is given to the equilibrium (α = 0, β = 1), and there is only one purchase, the optimal result of the objective function would be 0.

Table 3. Algorithm results depending on the list and the parameters used

α		0		0,25		0,5		0,75		1	
β		1		0,75		0,5		0,25		0	
List		Sol	T	Sol	T	Sol	T	Sol	T	Sol	T
L1	1	0	15	0	13	0	15	0	14	0	17
	2	0,086	33	0,307	35	0,529	32	0,75	33	0,972	29
	3	0,32	52	0,487	59	0,653	52	0,807	54	0,962	60
L2	1	0	41	0,111	41	0,223	52	0,334	40	0,446	43
	2	0,119	117	0,195	102	0,27	122	0,34	123	0,409	106
	3	0,002	152	0,106	165	0,208	165	0,31	160	0,409	169
L3	1	0	58	0	39	0	42	0	54	0	59
	2	0,236	56	0,323	61	0,544	65	0,621	59	0,774	60
	3	0,107	90	0,295	97	0,476	85	0,63	103	0,774	91
L4	1	0	931	0,1	969	0,199	909	0,299	1117	0,398	1161
	2	0	2522	0,094	2235	0,187	2136	0,269	2121	0,345	2303
	3	0	4427	0,092	4562	0,177	3640	0,264	3409	0,343	4113
	4	0	4377	0,098	5217	0,185	4802	0,273	4527	0,343	4166
	5	0,001	4160	0,102	4156	0,191	4202	0,278	3741	0,343	3461
	6	0,001	2985	0,104	3655	0,201	2943	0,287	2297	0,343	1727
	7	0,009	2455	0,115	2430	0,21	3073	0,291	1923	0,343	1251
L5	2	0,004	126	0,2	115	0,382	120	0,563	122	0,737	137
	3	0,022	191	0,204	171	0,393	175	0,57	191	0,736	171
	4	0,018	255	0,213	238	0,399	243	0,581	242	0,735	230
	5	0,054	298	0,253	283	0,439	264	0,599	304	0,735	262
	6	0,143	220	0,315	222	0,465	272	0,614	207	0,735	207
	7	0,223	165	0,369	178	0,511	154	0,634	151	0,735	157
L6	3	0,034	139	0,247	158	0,45	146	0,662	195	0,855	161
	4	0,052	183	0,277	185	0,489	205	0,666	189	0,853	203
	5	0,121	276	0,344	253	0,526	245	0,694	243	0,853	221
	6	0,209	240	0,405	210	0,566	220	0,714	212	0,853	227
	7	0,302	172	0,456	176	0,603	175	0,735	171	0,853	166

This is because it is not necessary to divide the purchase. In these instances, as the number of different purchases increases, the result is worse, in almost all cases.

The next case is introducing weight to the price of the purchases looking for a saving in the final purchase price, but remaining the equilibrium the most important consideration ($\alpha = 0.25$; $\beta = 0.75$). In this case we observe that the best values are obtained dividing the list in two, three or four purchases. In big instances if the division is made in more purchases the value is worst.

The third case is when the price and the equilibrium have the same weight ($\alpha = 0.5$; $\beta = 0.5$). In this case the best solution is to divide the list in a few purchases: two or three.

The following case is ($\alpha = 0.75$; $\beta = 0.25$), where majority of the weight goes to the price saving. Initially we can think that more purchases, visiting more supermarkets will make a big save; but the more supermarkets you visit, the worse is the equilibrium. This is true for almost all instances.

Last case is evaluating only the price ($\alpha = 1$, $\beta = 0$), as you can think if you visit more supermarkets looking for the lowest price, the solution will be better. This is true in our instances but it is no necessary visit all supermarkets, with only three or four (depending on the list size), you achieve the lowest price.

Analyzing the running times, the algorithm converges quickly, the worst times are for big lists, especially with a big number of categories and an intermediate number of purchases. The values of α and β for the same number of purchases in the same list do not present a big difference among them.

A comparison between using only one of the two models included in the proposed system, and the system as a whole has not been included since some other previous studies prove that usually the use of both in combination yields better results than using them individually [12, 13].

Taking all in account, we can conclude that it is no necessary divide a huge list in a big number of purchases in order to achieve the best solution. Except in the cases in which we are forced to make more purchases caused by mandatory products. Including categories instead of products increases the computing time according to the high number of combination possibilities.

6 Conclusions and Future Work

The main objective was design an algorithm to automatically calculate the best distribution of a domestic grocery list's purchase. As it needs to be integrated in a real mobile application, computation time is crucial to the operation of the complete application so the use of techniques GRASP and VND was decided, as they are known to converge into a solution in a minimum time.

The times obtained in the computational experiences are all under a second, so we can say that we have achieved our goal of make a fast application and a fast algorithm.

As we can observe in results, parameters α and β serve their purpose in order to control the balance between equilibrium between purchases in different supermarkets and final aggregated price, and also allows the user to focus only in one of them or a combination of both.

Methodologically, analyzing various proposed combinations, splitting lists in a large number of purchases does not ensure a better solution. In the contrary, it adds more computation time and the same solution is reached.

The main lines of work in the future could be to elaborate a Pareto's Front in order to reach a better solution for every combination of α and β. Also working in other techniques like a genetic algorithm to compare the results achieved for all method, both in time and results.

Acknowledgments. This work was partially supported by FEDER, the Spanish Ministry of Economy and Competitiveness (Project ECO2013-47129-C4-3-R) and the Regional Government of Castilla y León (Project BU329U14), Spain.

References

1. T.E.C. Organization: The European consumer organization (2016). http://www.beuc.eu/
2. Ver Ploeg, M., Mancino, L., Todd, J.E., Clay, D.M., Benjamin, Scharadin: Where do Americans usually shop for food and how do they travel to get there? Initial Findings From the National Household Food Acquisition and Purchase Survey, U.S.D.o. Agriculture, Editor. Economic Research Service (2015)
3. Miller, M.A., et al.: Food purchasing habits of participants in the supplemental nutrition assistance program (SNAP) among shelby county, TN residents. J. Acad. Nutr. Diet. **115**(9, Supplement), A76 (2015)
4. Behrens, J.H., et al.: Consumer purchase habits and views on food safety: a Brazilian study. Food Control **21**(7), 963–969 (2010)
5. Feo, T.A., Resende, M.G.C.: Greedy Randomized Adaptive Search Procedures. J. Global Optim. **6**(2), 109–133 (1995)
6. Festa, P., Resende, M.G.C.: An annotated bibliography of GRASP - part I: algorithms. Int. Trans. Oper. Res. **16**(1), 1–24 (2009)
7. Festa, P., Resende, M.G.C.: An annotated bibliography of GRASP-part II: applications. Int. Trans. Oper. Res. **16**(2), 131–172 (2009)
8. Resende, M.G.C.: Metaheuristic hybridization with greedy randomized adaptive search procedures. Tutorials Oper. Res. 295–319 (2008)
9. Hansen, P., Mladenović, N.: Variable neighborhood search: principles and applications. Eur. J. Oper. Res. **130**(3), 449–467 (2001)
10. Mladenovic, N.: A variable neighborhood algorithm - a new metaheuristics for combinatorial optimization. In: Abstract of Papers Presented at Optimization Days, Montreal Canada, p. 112 (1995)
11. Mladenovic, N., Hansen, P.: Variable neighborhood search. Comput. Oper. Res. **24**(11), 1097–1100 (1997)
12. Nguyen, V.-P., Prins, C., Prodhon, C.: Solving the two-echelon location routing problem by a GRASP reinforced by a learning process and path relinking. Eur. J. Oper. Res. **216**(1), 113–126 (2012)
13. Villegas, J.G., et al.: GRASP/VND and multi-start evolutionary local search for the single truck and trailer routing problem with satellite depots. Eng. Appl. Artif. Intell. **23**(5), 780–794 (2010)

Estimation of Daily Global Horizontal Irradiation Using Extreme Gradient Boosting Machines

Ruben Urraca, Javier Antonanzas, Fernando Antonanzas-Torres, and Francisco Javier Martinez-de-Pison(✉)

EDMANS Group, University of La Rioja, Logroño, Spain
edmans@dim.unirioja.es, fjmartin@unirioja.es
http://www.mineriadatos.com

Abstract. Empirical models are widely used to estimate solar radiation at locations where other more readily available meteorological variables are recorded. Within this group, soft computing techniques are the ones that provide more accurate results as they are able to relate all recorded variables with solar radiation. In this work, a new implementation of Gradient Boosting Machines (GBMs) named XGBoost is used to predict daily global horizontal irradiation at locations where no pyranometer records are available. The study is conducted with data from 38 ground stations in Castilla-La Mancha from 2001 to 2013.

Results showed a good generalization capacity of the model, obtaining an average MAE of 1.63 MJ/m^2 in stations not used to calibrate the model, and thus outperforming other statistical models found in the literature for Spain. A detailed error analysis was performed to understand the distribution of errors according to the clearness index and level of radiation. Moreover, the contribution of each input was also analyzed.

Keywords: Global horizontal irradiation · Extreme gradient boosting · XGBoost · Solar radiation

1 Introduction

High-quality time series of solar radiation are demanded by the solar energy industry to spur the installation of new power plants. This data is required to reduce the uncertainty around power estimations, and hence, to reduce risks in the economic assessment. In particular, the variables required are the Direct Normal Irradiation (DNI) for concentrating solar systems, and the Global Horizontal Irradiation (GHI) or the Global Titled Irradiation (GTI) for most photovoltaic plants. In this study, we focus on the GHI, as it is a variable more commonly measured, and due to the fact that DNI can be subsequently derived with a global to direct decomposition model [10].

The straightforward way to obtain a GHI time series is via the installation of a pyranometer, which is the type of radiometer most commonly used

M. Graña et al. (eds.), *International Joint Conference SOCO'16-CISIS'16-ICEUTE'16*, Advances in Intelligent Systems and Computing 527, DOI 10.1007/978-3-319-47364-2_11

to record the global component of radiation. If well calibrated and carefully maintained, pyranometer records are the most accurate sources of solar radiation [13]. However, the availability of pyranometer records is limited both in time and mainly in space, due to the high installation costs associated. Even in most developed countries, the density of the pyranometer grid is quite coarse. As a result, several methods have been proposed to estimate GHI from other secondary variables more commonly measured. The most elementary ones are the empirical models. These type of models attempt to correlate different meteorological variables more readily available at meteorological stations with solar radiation. The correlations used can vary from simple parametric expressions [3,5], to more complex non-linear algorithms from machine learning and artificial intelligence fields. In relation to the second, different predictive techniques have been tested such as artificial neural networks [9], support vector machines [4], regression trees [16] or fuzzy logic [12]. However, most of these works focused on the development of local models, i.e., models are trained and tested with data from the same stations. These applications are limited to filling gaps or to the extrapolation of relatively short time series, always requiring the installation of a pyranometer in the site of study. In this work we focus on the ability of models to generate predictions in locations not used during the training process. This is considered the most promising application of these type of models, as it allows to extend the grid of GHI records to locations where no pyranometer records are available at all [6]. To this end, a new version of Gradient Boosting Machines (GBMs) named XGBoost is used due to its reported generalization ability [7]. The model is tested with on-ground records of daily GHI at 38 meteorological stations located in central Spain, for the period 2001-2013.

2 Data

Records of daily GHI from 38 meteorological stations of SIAR network [1] are used as reference data (Fig. 1). All stations are located in Castilla-La Mancha, an autonomous community of central Spain. The region, along with Andalucia and Extremadura, presents one of the highest potential for solar energy applications, with almost all stations recording on average more than 1800 kWh/m^2 per year. Indeed, it is the autonomous community with more installed photovoltaic power capacity with 923 MW. In all stations, GHI was recorded with SP1100 Skye Pyranometers (Campbell), a First Class pyranometer with an absolute accuracy of ±5 %.

Other secondary meteorological variables were available at SIAR stations such is the case of temperature (T), relative humidity (RH), wind speed (WS) and rainfall (R). Besides, the database was complemented with some computed variables: the extraterrestrial irradiance (I_{ex}), the logical variable of rainfall (M) and the daily temperature variation (ΔT).

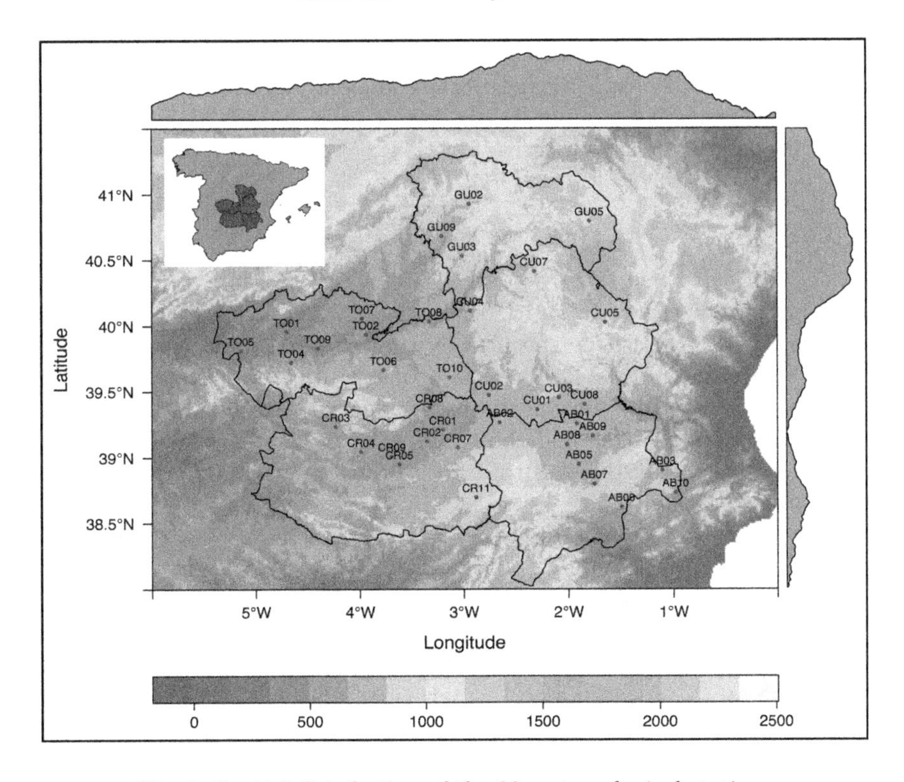

Fig. 1. Spatial distribution of the 38 meteorological stations

3 Methods

3.1 Estimation Model - XGBoost

The predictive technique selected is the so-called eXtreme Gradient Boosting (XGBoost) [8], one of the latest implementation of Gradient Boosting Machines. The model is based on the theory of boosting, so the predictions of several "weak" learners (models whose predictions are slightly better than random guessing), are combined to develop a "strong" learner. In GBMs [11], these "weak" learners are combined by following a gradient learning strategy. At the beginning of the calibration process, a "weak" learner is fit to the whole space of data, and then, a second learner is fit to the residuals of the first one. This process of fitting a model to the residuals of the previous one goes on until some stopping criterion is reached. Finally, the output of the GBM is a kind of weighted mean of the individual predictions of each weak learner. Traditionally, regression trees are selected as "weak" learners.

The main pitfall of boosting learning in general, and of GBMs in particular, is the risk of overfitting, as a consequence of the additive calibration process. Typically, this issue is faced by including different regularization criteria in order to control the complexity of the algorithm. However, the computational burden entailed by these type of mechanisms is considerably high. In this context, the main objective of XGBoost is to control the complexity of the algorithm without excessively increasing the computational cost. To this end, XGBoost is based on the following *Loss+Penalty* objective function:

$$Obj^{(t)} = \sum_{i=1}^{n} l(y_i, \hat{y}_i) + \sum_{i=1}^{t} \Omega(f_i) \tag{1}$$

where l is the predictive term and Ω the regularization term. The loss function for the predictive term can be specified by the user, and it is always truncated up to the second term of its Taylor series expansion for computational reasons. The regularization term is obtained with an analytic expression based on the number of leaves of the tree and the scores of each leaf. The key point of the calibration process of XGBoost is that both terms are ultimately rearranged in the following expression:

$$Obj^{(t)} = -\frac{1}{2}\sum_{j=1}^{T} \frac{G_j^2}{H_j + \lambda} + \gamma T \tag{2}$$

where G and H are obtained from the Taylor series expansion of the loss function, λ is the L2 regularization parameter and T, the number of leaves. This analytic expression of the objective function allows a rapid scan from left to right of the potential divisions of the tree, but always taking into account the complexity.

XGBoost has a wide range of tuning parameters as described in [8]. Moreover, the flexibility of the algorithm is enhanced by giving the chance to the user to include some self-defined parameters, such as the loss function or the metric used for validation and testing. XGBoost is freely available in GitHub in different programming languages (Python, R, Java, Scala and Julia).

3.2 Implementation and Evaluation

Data from the 38 stations was initially preprocessed by detecting and eliminating extreme samples (1 % of the initial number of samples). Subsequently, the database was split into calibration and testing sets following a leave-one-out (LOO) methodology. As the main application of these models is the estimation of solar radiation in places where no records from pyranometers are available, 38 different models were trained by using 37 stations for training and testing in the 38th in order to check the spatial generalization ability of the models. However, due to the high number of daily samples available to calibrate the model for training (37 stations and 13 years), only the 15 % of the available calibration set were used for training. These samples were obtained following a stratified random sampling process. The subgroups for the stratified sampling were generated by taking into account the three following variables: the station, the year and the month. Finally, the stratified calibration set was again split into training (70 %) and validation (30 %).

Table 1. XGBoost parameters.

General	Regularization	Tree learner	Subsampling
booster gbtree	*gamma* 0	*max_depth* [1, 12]	*colsample_bytree* 0.8
objective reg:linear	*lambda* 1	*min_child_weight* [1, 50]	*subsample* 1
eval_metric $RMSE_{val}$	*alpha* 1		
eta 0.01			
early.stop.round 15			

Initially, during the calibration of all 38 models, all available input variables were included: 8 meteorological variables - RH (avg, min, max), WS (avg, max), T (avg, min., max.), R -, 3 computed variables - I_{ex}, ΔT, M -, and 3 topographical variables - longitude, latitude and altitude. The ΔT of the days before and after the day studied was also included. Then, based on the relative gain added by each variable some of them were discarded following a backward selection procedure.

For the calibration of XGBoost, most parameters were manually set based on the dimensions of the existing database (see Table 1). Only the most sensitive parameters, *max_depth* and *min_child_weight*, were automatically set using a Random Search (RS) optimization procedure. The initial search space was [1, 12] for *max_depth* and [1, 50] for *min_child_weight*. At each iteration, 50 models were trained based on 50 different random combinations of the parameters being tuned. The 10 best performing models were selected based on the $RMSE_{val}$ to shrink the search space. The process was repeated 3 times until convergence was reached.

3.3 Software

The freely distributed statistical software `R` [15] was used to perform all calculations. In particular, the `zoo` [2] package was used to work with time series, the `solaR` [14] package was used to perform the different computations in relation to solar geometry, the `xgboost` [8] package was used to train the XGBoost, and the `ggplot2` [17] package to illustrate the results.

4 Results and Discussion

Table 2 shows the performance of the model, detailing the mean and the standard deviation for each metric. The bias of the model was very low, reaching a rMBE of $-0.21\,\%$. The standard deviation values shows that the small bias obtained was in part due to cancellation of errors of different sign. However, this still proves the absence of any systematic trend that could lead towards under- or overestimation. Results for MAE and RMSE showed a good degree of accuracy and low dispersion. The errors obtained with this technique (MAE of $1.63\,\mathrm{MJ/m^2}$) are lower than the errors found in the literature at Spain, proving the good generalization ability of XGBoost. For instance, MAEs of $1.86\ \mathrm{MJ/m^2}$

Table 2. Performance measurements (mean and standard deviation - sd) of the daily GHI obtained at all testing locations. Surface records from pyranometers are used as reference data.

	MAE $[MJ/m^2]$		rMAE [%]		RMSE $[MJ/m^2]$		rRMSE [%]		MBE $[MJ/m^2]$		rMBE [%]		$Frac_{tol}$ [%]	
	mean	sd	mean	sd	mean	sd	mean	sd	mean	sd	mean	sd	mean	sd
XGBoost	1.63	0.14	9.08	0.81	2.22	0.16	12.34	0.96	−0.04	0.34	−0.21	1.88	36.89	4.33

were reported with Support Vector Machines (SVM) [4] and MAEs of 1.74, 1.76 and 1.76 MJ/m^2 were obtained with SVMs, ANNs and ELMs, respectively [16]. The number of high quality predictions is measured with $Frac_{tol}$ metric, i.e., the number of samples withing the range of accuracy of the pyranometer used (5 %), which in this case is up to a 36.89 %.

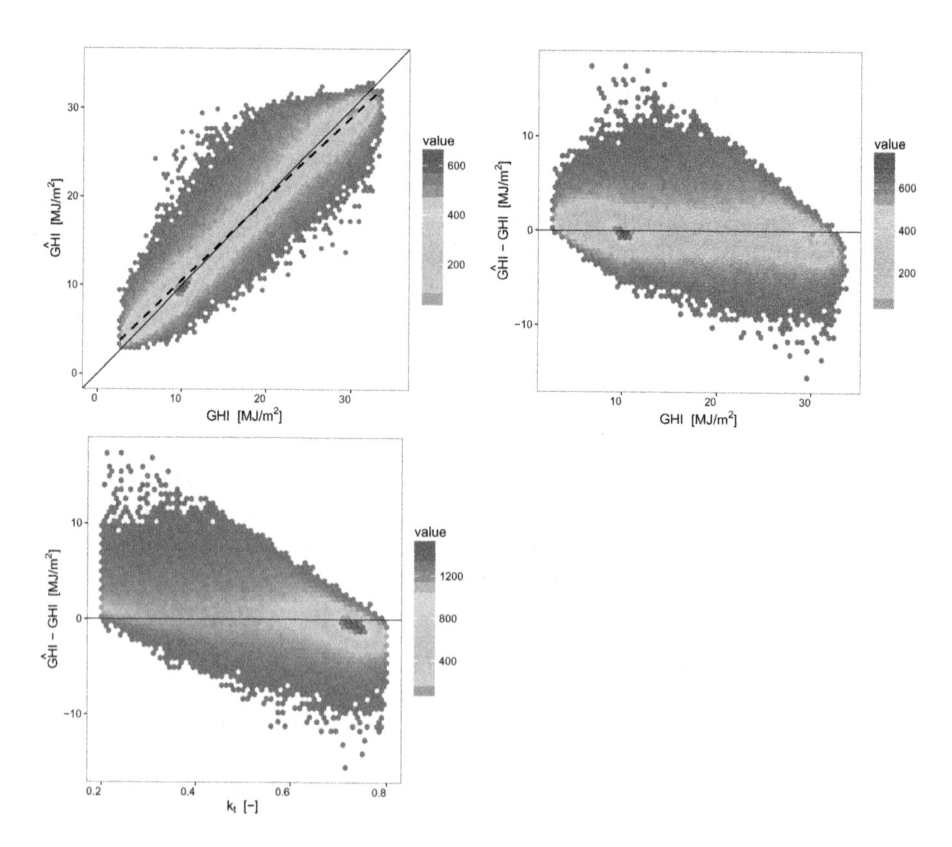

Fig. 2. Plots of predicted $(\hat{GHI})$ vs observed (GHI), residuals $(\hat{GHI} - GHI)$ vs observed (GHI), and residuals $(\hat{GHI} - GHI)$ vs clearness index (k_t) at the 38 testing stations

Figure 2 compiles different approaches to error characterization. The plot of predicted vs observed GHI shows a good agreement between predicted and observed data, with two areas of a higher density of points, around 10 and 30 $\mathrm{MJ/m^2}$. They correspond to clear sky days in winter and summer, respectively, when the model performed especially well. This is corroborated in the plot between residuals and the clearness index (k_t), a dimensionless metric of the state of cloudiness of the atmosphere. In this plot, the highest concentration of points and lower errors were located for k_t values higher than 0.7, which correspond days under cloudless conditions. Contrary, errors tend to increase under cloudy conditions (low k_t) values. This is a consequence of the limitations of empirical models, due to the limited correlation between meteorological variables and the state of cloudiness.

Moreover, a singular trend is observed in the error distribution, with high overestimations on cloudy situations and a slight underestimation of clear-sky days. That trend is clearly seen in the residuals vs observed GHI scatterplot (upper right plot). Therefore, the linear fit of observed vs estimated daily GHI showed a lower slope than the ideal. Also, the highest errors were found in cloudy skies (k_t<0.5), which again reveals the difficulty to model clouds with a reduced set of inputs.

The abovementioned correlation between meteorological variables and solar irradiation is further studied in Fig. 3, which depicts the relative importance of each variable in the final model. As revealed, most of the gain (%) was due to the inclusion of the extraterrestrial irradiance, accounting for 62 %. Hence, the remaining variables try to model the atmospheric transmissivity, i.e., the amount of radiation lost from outer atmosphere to earth, which is mainly influenced by the presence of clouds. From the rest of the 15 variables included, the most

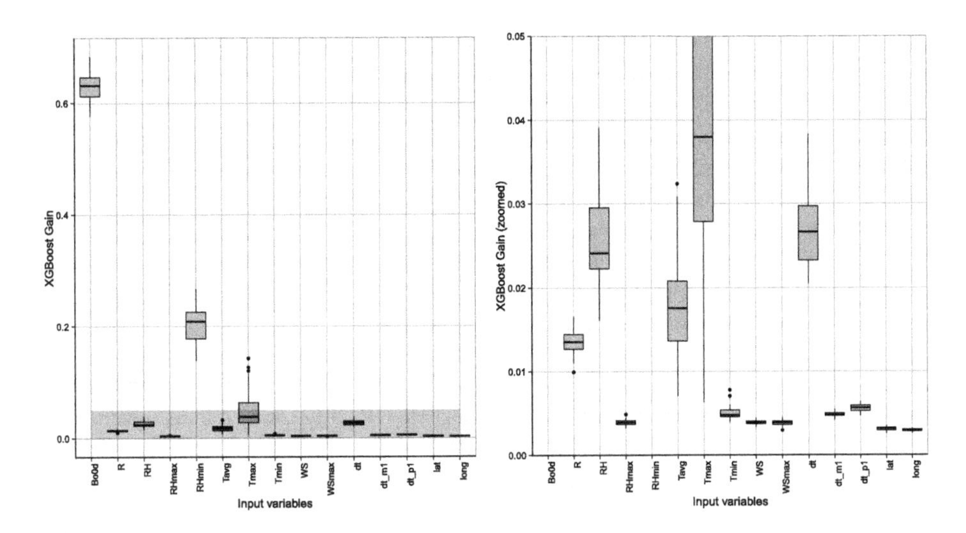

Fig. 3. Boxplots of the relative gain added by each variable at the 38 models calibrated

important variables were RH_{min} and T_{max}, explaining an additional 20 and 3.8 % respectively. The limited correlation of relative humidity and temperature to atmospheric transmissivity sets the foundations for the moderate accuracy of the model under cloudy days. Additional relevant information was provided by RH, T_{avg} and ΔT. The contribution of the rest of variables was discrete, being the relative gain of 8 of them lower than 0.6 %. The gain obtained by including latitude and longitude was close to null.

5 Conclusions and Future Work

Solar radiation measurements are necessary for a wide range of applications, from climate monitoring, to agriculture. In the field of renewable energies, they are specially useful for feasibility studies. Nevertheless, the relatively low density of meteorological stations in certain areas, combined with the fact that radiation measurements are not always recorded in them, creates the necessity of estimating solar irradiation by other means. With this aim, empirical models have been developed, which correlate solar radiation to other variables more readily available at meteorological stations. This paper presented an application of XGBoost to estimate solar radiation at 38 sites in central Spain.

The generalization capacity of the model was tested, since predictions of GHI were performed in stations not used during the training process. Results showed a satisfactory level of accuracy, outperforming other statistical models found in the literature. An average MAE of 1.63 MJ/m^2 was obtained, with relatively low dispersion of errors (0.81 %). Moreover, the number of days that fell within the accuracy of the pyranometer amounted to 36.9 %. A detailed study of the error distribution revealed that during clear sky days the model performed specially well. However, despite the overall satisfactory performance of XGBoost, the limited correlation between the input meteorological data available and the state of cloudiness of the atmosphere was evidenced. The most important variable in the model was the extraterrestrial irradiation, with a gain of 62 %, followed by the minimum relative humidity and the maximum temperature. This limited correlation led to a significant overestimation of cloudy situations and a slight underestimation of high levels of clearness index.

Future work will address a sensitivity analysis of the number of training stations and the length of the training time-series, and the inclusion of external variables with higher correlation with the atmospheric transmissivity.

Acknowledgments. J. Antonanzas and R. Urraca would like to acknowledge the fellowship FPI-UR-2014 granted by the University of La Rioja. F. Antonanzas-Torres would like to express his gratitude for the FPI-UR-2012 and ATUR grant No. 03061402 at the University of La Rioja. Finally, all the authors are greatly indebted to the Agencia de Desarrollo Economico de La Rioja for the ADER-2012-I-IDD-00126 (CONOBUILD) fellowship for funding parts of this research.

References

1. Servicio de Información Agroclimática para el Regadío (SIAR) (2015). http://eportal.magrama.gob.es/websiar/Inicio.aspx
2. Aeileis, A., Grothendieck, G.: zoo: S3 infrastructure for regular and irregular time series. J. Stat. Softw. **14**(6), 1–27 (2005). https://cran.r-project.org/package=zoo
3. Antonanzas-Torres, F., Martinez-de Pison, F.J., Antonanzas, J., Perpinan, O.: Downscaling of global solar irradiation in complex areas in R. J. Renew. Sustain. Energy **6**, 063105 (2014)
4. Antonanzas-Torres, F., Urraca, R., Fernandez-Ceniceros, J., Martinez-de Pison, F.J.: Generation of daily global solar irradiation with support vector machines for regression. Energy Convers. Manage. **96**, 277–286 (2015)
5. Besharat, F., Dehghan, A.A., Faghih, A.R.: Empirical models for estimating global solar radiation: a review and case study. Renew. Sustain. Energy Rev. **21**, 798–821 (2013)
6. Bojanowski, J.S., Vrieling, A., Skidmore, A.K.: A comparison of data sources for creating a long-term time series of daily gridded solar radiation for Europe. Solar Energy **99**, 152–171 (2014)
7. Chen, J.L., Li, G.S., Wu, S.J.: Assessing the potential of support vector machine for estimating daily solar radiation using sunshine duration. Energy Convers. Manage. **75**, 311–318 (2013)
8. Chen, T., He, T., Benesty, M.: XGBoost: eXtreme Gradient Boosting (2015). https://github.com/dmlc/xgboost, R package version 0.4-2
9. Dahmani, K., Notton, G., Voyant, C., Dizene, R., Nivet, M.L., Paoli, C., Tamas, W.: Multilayer perceptron approach for estimating 5-min and hourly horizontal global irradiation from exogenous meteorological data in locations without solar measurements. Renew. Energy **90**, 267–282 (2016)
10. Gueymard, C.A., Ruiz-Arias, J.A.: Extensive worldwide validation andclimate sensitivity analysis of direct irradiance predictions from1-min global irradiance. Solar Energy **128**, 1–30 (2016). http://www.sciencedirect.com/science/article/pii/S0038092X15005435, Special Issue: Progress in Solar Energy
11. Hastie, T., Tibshirani, R., Friedman, J.H.: The Elements of Statistical Learning: Data Mining, Inference, and Prediction. Springer, New York (2001)
12. Kisi, O.: Modeling solar radiation of Mediterranean region in Turkey by using fuzzy genetic approach. Energy **64**, 429–436 (2014)
13. Paulescu, M., Paulescu, E., Gravila, P., Badescu, V.: Solar radiation measurements. In: Paulescu, M., et al. (eds.) Weather Modeling and Forecasting of PV Systems Operation. Green Energy and Technology, pp. 17–42. Springer, London (2013)
14. Perpiñán, O.: Solar radiation and photovoltaic systems with R. J. Stat. Softw. **50**(9), 1–32 (2012). https://cran.r-project.org/web/packages/solaR/index.html
15. R Core Team: R: A Language and Environment for StatisticalComputing. R Foundation for Statistical Computing, Vienna, Austria (2014). http://www.R-project.org/
16. Urraca, R., Antonanzas, J., Martinez-de Pison, F.J., Antonanzas-Torres, F.: Estimation of solar global irradiation in remote areas. J. Renew. Sustain. Energy **7**(2), 1–14 (2015)
17. Wickham, H.: ggplot2: Elegant Graphics For Data Analysis. Springer, New York (2009). http://had.co.nz/ggplot2/book

The Control of the Output Power Gas Temperature at the Heat Exchanger

Martin Pieš[(✉)], Blanka Filipová, and Pavel Nevřiva

Faculty of Electrical Engineering and Computer Science,
Department of Cybernetics and Biomedical Engineering,
VSB-Technical University of Ostrava, 17. listopadu 15/2172,
70833 Ostrava, Czech Republic
{martin.pies,blanka.filipova,pavel.nevriva}@vsb.cz
http://www.fei.vsb.cz/en

Abstract. This paper deals with the control of the output power gas temperature at the Main Heat Exchanger (MHE). The MHE is the basic part of the Flexible Energy cogeneration System (FES) with the combined Brayton - Rankine cycle, which is designed and constructed at Vitkovice Power Engineering JSC. The FES burns solid fuel and generates electrical energy and thermal energy. The standard temperature control at MHE has two goals. The first control task is the protection of the heat transfer surfaces of the MHE against overheating. The second control task is the stabilization of the temperature of power gas at the output of the MHE which is performed, in principle, by the change of the flow rate of air generated by the compressor and by the change of the flow rate of fuel at the inlet of the FES combustion chamber. In this paper, the control of the temperature of the power gas without the overheating of the heat transfer surfaces of the MHE will be described and analyzed.

Keywords: Control systems · Heat exchanger · Protective control · Simulation · Systems with distributed parameters

1 Introduction

The Flexible Energy cogeneration System (FES) designed and developed by VITKOVICE POWER ENGINEERING joint-stock company operates with the combined Brayton - Rankine cycle. Its power is designed on 110 MW. In the FES, the heating medium is flue gas generated by the combustion of fuel. The heated medium is power gas, which is a gas mixture of air (75 %) and water steam (25 %). Power gas is superheated in the main heat exchanger (MHE) which is the basic part of the FES. The very high superheated power gas is led to two gas turbines. One gas turbine drives a 50 MW asynchronous generator, the second powers the compressor. The protective control protects the heat transfer surfaces of the MHE against damage by overheating. The algorithm of the protective control is integrated in the FES control system.

M. Graña et al. (eds.), *International Joint Conference SOCO'16-CISIS'16-ICEUTE'16*,
Advances in Intelligent Systems and Computing 527, DOI 10.1007/978-3-319-47364-2_12

The principle of the FES is shown in Fig. 1. The figure shows a combustion chamber **1** which burns the fuel. The walls of the combustion chamber have evaporative pressure water cooling. Water under pressure is supplied by a pump. The generated water steam is fed to the front mixer **2**, where it is added to atmospheric air supplied to the power circuit by the compressor **3**. The portion of the generated steam is available for the protective control of the MHE **4**, see below. The resulting power gas enters the MHE **4** where it is superheated by the flue gas from the combustion chamber. The superheated power gas expands in the gas turbines **5** and **6**. The compressor turbine **5** drives the compressor **3**. The generator turbine **6** drives the asynchronous generator that is linked to the 50 Hz electric network. The residual heat that power gas contains after the decompression in the gas turbines is used to generate electricity in a Rankine cycle. The heat exchanger **7** exploits the residual heat of the flue gas for the preheating of the combustion air. The discharged combustion products are cleaned in the purification plant. To support the development of the FES, VPE built an experimental FES in 2009–2013. A mathematical model and some basic experimental information concerning this unit were presented by [5,8]. Other parts of the FES control system are similar to conventional steam cogeneration energy sources.

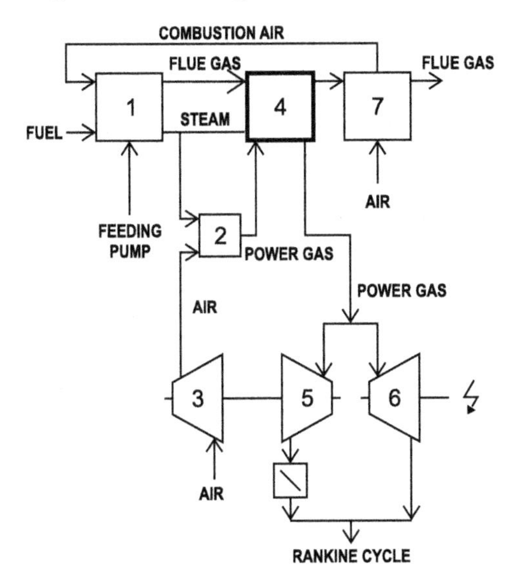

Fig. 1. The principle of FES.

2 The Standard Temperature Controls at MHE

The power gas at the input of the MHE is the mixture of compressed air and saturated steam, its temperature is about 200 °C. At the output of the MHE, the power gas is superheated to about 800 °C. The MHE is equipped with standard control functions and emergency control functions. Emergency temperature control is based on the injection of liquid water into the power gas. The discussion of emergency control lies beyond the scope of this paper. The standard temperature control at the MHE has two goals. The first is the stabilization of the temperature of the power gas at the output of the MHE. The second is the protection of the heat transfer surfaces of the MHE against overheating.

The control of the temperature of power gas at the output of the MHE is performed, in principle, by the change of the flow rate of fuel at the inlet of the FES combustion chamber and by the change of the flow rate of air generated by the compressor. The overheating of sections of the MHE could result from uneven temporary distributions of the flue gas flow rate along the MHE. The time constant of the local increase of the temperature of the MHE heat transfer surface may be less than 1,000 s. The protective control is the fast control, which keeps the temperature of the wall of the MHE heat transfer surfaces below the specified temperature limits. The limits depend on the materials used for the construction of MHE.

The heat exchanger is a system with distributed parameters. Its dynamics can be characterized by the dominant time constant. The dominant time constant of the FES MHE is about 5,000 s. The protective control is based on the technological decomposition of the MHE to partly controllable segments. In this paper the MHE divided into five segments is discussed; see Fig. 2. Each of the segments has a dominant time constant of about 1,000 s. Segments are connected in a series. Cool power gas enters the first segment. The superheated power gas is led from the fifth segment to the turbines.

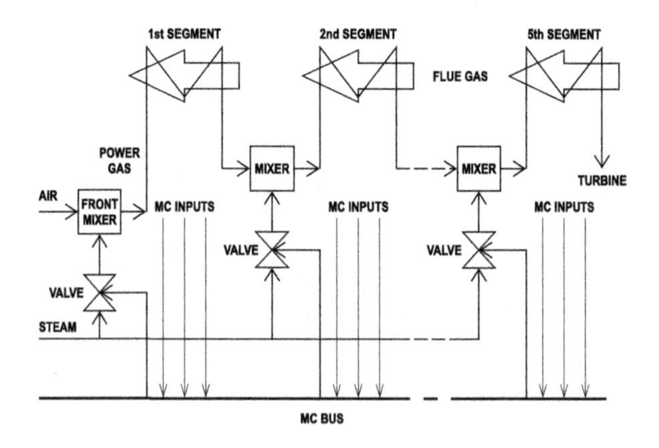

Fig. 2. The five segments of the MHE.

The MHE is designed as a counter flow. The walls of the heat transfer surfaces are realized from bundles of thin-walled tubes. Bundles of MHE segments are made from different appropriately chosen steels and alloys which have their own temperature limits. For each segment of the MHE, the limit is constant. Every segment may be heated below its specific temperature limit.

Protective control of the FES maintains temperature heat transfer surfaces of MHE segments under their technological limits. Controllers add the cooling steam to power gas at the inputs of all vulnerable segments. The lower temperature of the segment input power gas leads subsequently to the lower temperature of the segment heat transfer surface. To achieve minimal thermal losses, the algorithm prefers cooling of the first segments of MHE and minimizes the quantity of steam fed to the last segments of the MHE.

The protective control saves the heat transfer surfaces of the MHE but from the point of view of the control of the temperature of the power gas at the output of the MHE it represents a control error because it decreases the temperature of the output power gas. The MHE control system prefers the safety. In the situation described, it decreases the setpoint of the output temperature of the power gas. The decrease of the setpoint is the function of the flow rate of the cooling steam. The decrease of the temperature setpoint causes an increase of the flow rate of the compressed air and a decrease of the flow rate of fuel at the inlet of the FES combustion chamber. The temperature of all heat exchanging surfaces of the MHE slowly decreases and the cooling steam inserted by the protective control is minimized.

The overheating of sections of the MHE results from temporary uneven distributions of the flue gas flow rate along the MHE. These anomalies arise during changes of flue and/or generated power. See that the original setpoint is restored when the heating anomalies perish.

In this paper, the control of the temperature of power gas at the output of the MHE is analyzed in a situation where there is no overheating of the heat transfer surfaces of the MHE. The temperature control made by a change of the flow rate of air generated by the compressor is described. That means temperature control is made only by a change of the flow rate of the power gas and the temperature of the flue gas is constant. Long term overheating of the heat transfer surfaces leads to damages on the heat exchanger. Long term overheating of the heat transfer surfaces is handled by injecting steam before each heat exchanger, see Fig. 3. This combined temperature control will be described in a future paper.

3 The Control Loop

The simplified control loop discussed in this paper, see Fig. 3, consists of the MHE which is actuated by power gas from the front mixer and energized by flue gas from the combustion chamber, the front mixer which mixes the compressed air generated by the compressor with compressed steam, and the controller assembly which measures the actual temperature of power gas at the output of the MHE, compares it at the setpoint, and drives the valve which leads the steam to the compressor turbine which drives the compressor to minimize the control deviation.

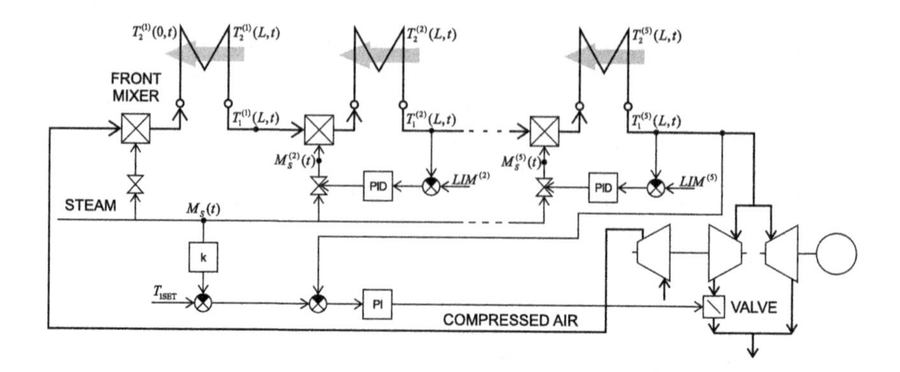

Fig. 3. The control loop.

3.1 Mathematic Model of the Heat Exchanger

The heat exchanger transfers heat energy from a heating media to a heated media. Heat from the heating media is transmitted to the heated media through the walls of the steel tubes.

The MHE is a series arrangement of five counter-flow segments. Overheating does not occur in the presented task, so the four mixers behind the first segment are neglected. Each counter-flow segment has the same structure. Segments differ from one another by the dimensional and material parameters.

The comprehensive mathematical model of a heat transfer process was discussed by [4]. The dominant time constants of the MHE are orders of magnitude larger than the time constant of the rest of the FES are. This enables simplifying the problem and describing the temperature dynamics of the heat exchanger by the set of three partial differential equations with admissible accuracy as follows

$$
\begin{aligned}
T_W(x,t) - T_1(x,t) &= \tau_1 \left[u_1 \frac{\partial T_1(x,t)}{\partial x} + \frac{\partial T_1(x,t)}{\partial t} \right] \\
T_W(x,t) - T_2(x,t) &= \tau_2 \left[u_2 \frac{\partial T_2(x,t)}{\partial x} + \frac{\partial T_2(x,t)}{\partial t} \right] \\
\frac{T_1(x,t) - T_W(x,t)}{\tau_{W1}} + \frac{T_2(x,t) - T_W(x,t)}{\tau_{W2}} &= \frac{\partial T_W(x,t)}{\partial t}
\end{aligned}
\tag{1}
$$

where

$T_1(x,t)$	temperature of power gas	°C
$T_2(x,t)$	temperature of flue gas	°C
$T_W(x,t)$	temperature of the wall of the heat transfer surface of the tube	°C
$u_1(x,t)$	velocity of power gas in the x direction	$\mathrm{m \cdot s^{-1}}$
$u_2(x,t)$	velocity of flue gas in the x direction	$\mathrm{m \cdot s^{-1}}$
τ_1	power gas parameter	s
τ_2	flue gas parameter	s
τ_{W1}	wall-power gas parameter	s
τ_{W2}	wall-flue gas parameter	s
x	the space variable along the active length of the wall of the heat transfer surface of the exchanges	m
t	time	s

where parameters τ are determined by the technical construction of the heat exchanger and the physical parameters of the heat transfer media and the heat transfer surface. Figure 4 shows the physical state variables at the counter-flow heat exchanger. L is the length of the heat exchanger.

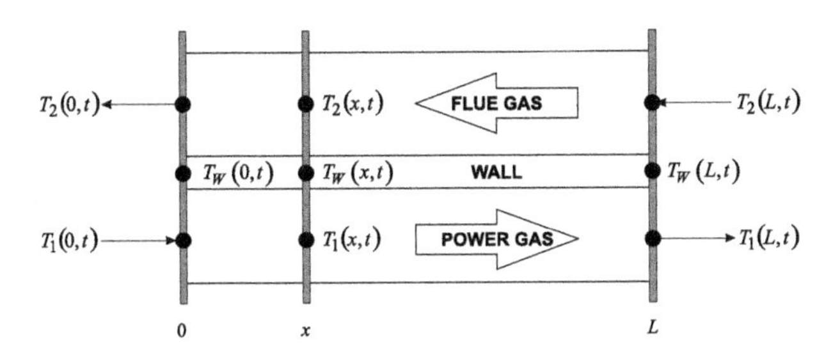

Fig. 4. Physical state variables at the counter-flow heat exchanger.

Every segment of the MHE was described in the mathematical model by the set of PDE (1). All PDE were then transformed onto sets of ODE. The method of finite differences was used, see [7], also [3]. Every PDE was approximated by a set of 20 ODE. The resulting set of ODE enabled evaluating and checking temperatures of the MHE tubes in $5 \times 20 = 100$. Numerical calculation was made in MATLAB by [1].

3.2 Front Mixer

The model of this mixer is embedded into the mathematical model of the power plant, serving as one of the actuating terms for the regulation of the power gas in the power plant. The mixer model is implemented as the S-function in Matlab Simulink. The mathematical model supposes that the input temperature of air/steam mixture coming to the injection front mixer is higher than the boiling point of the water at a given pressure. It means the mixture will contain no condensed water. The output of the front mixer is power gas whose composition is determined by the concentration of dry air w_{da} and by the concentration of the steam w_s. The following Table 1 describes the physical quantities involved in the mathematical model of the front mixer.

Table 1. Symbols occurring in the model

Symbol	Description	Unit
h	Enthalpy	$[\mathrm{kJ \cdot kg^{-1}}]$
M	Mass flow rate	$[\mathrm{kg \cdot s^{-1}}]$
p	Pressure	[Pa]
Q	Heat added/drained per second	$[\mathrm{kJ \cdot s^{-1}}]$
T	Temperature	[°C]
w	Concentration of mixture components	$[\mathrm{kg \cdot kg^{-1}}]$

The assumption to be fulfilled is that both incoming media have the same pressure p_{pg}, corresponding to the power gas pressure at the mixer output. The first step includes the determination of water steam concentration $w_{s,\mathrm{out}}$ and dry air concentration $w_{da,\mathrm{out}}$ in the power gas. The amount of water steam in the air/steam mixture is given by the ratio between the concentration of the steam in the input mixture $w_{s,\mathrm{in}}$ and the quantity of air/steam mixture $M_{pg,\mathrm{in}}$, and by the amount of injected steam M_{is}. This injected steam changes the ratio of concentration $w_{s,\mathrm{in}}$ to $w_{s,\mathrm{out}}$ and cools down the input air/steam mixture at the same time. The ratios of particular concentrations are illustrated in Fig. 5.

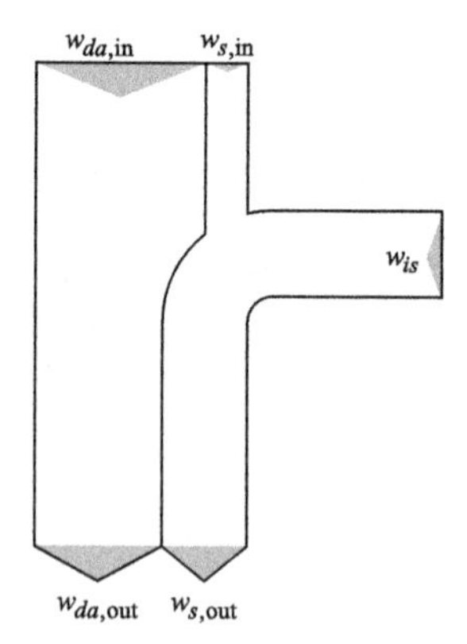

Fig. 5. Definition of power gas in the block representing the front mixer.

Indexes

da, in	correspond to dry air in incoming power gas
da, out	correspond to dry air in outgoing power gas
is	correspond to input steam injected to power gas
pg	correspond to power gas
pg, in	correspond to incoming power gas
pg, out	correspond to outgoing power gas
s	correspond to steam
s, in	correspond to steam in incoming power gas
s, out	correspond to steam in outgoing power gas

The total quantity of power gas is provided by the formula (2).

$$M_{pg,\text{out}} = M_{pg,\text{in}} + M_{is} \tag{2}$$

The concentration of steam coming to the front mixer, is provided by the formula (3).

$$w_{is} = \frac{M_{is}}{M_{pg,\text{in}} + M_{is}} \tag{3}$$

The overall concentration of water vapor in the mixture is provided by equation (4).

$$w_{s,\text{out}} = (1 - w_{is})w_{s,\text{in}} + w_{is} \tag{4}$$

The concentration of dry air in the power gas $w_{da,\text{out}}$ is then a supplement to one.

$$w_{da,\text{out}} = 1 - w_{s,\text{out}} \tag{5}$$

The partial pressure of the water steam and a dry air in power gas are determined by (6).

$$\begin{aligned} p_s &= p_{pg} \cdot \left[1 - \frac{w_{da,\text{out}} \cdot r_{da}}{w_{da,\text{out}} \cdot r_{da} + w_{s,\text{out}} \cdot r_s}\right] \\ p_{da} &= p_{pg} - p_s \end{aligned} \tag{6}$$

where r_{da} and r_s are specific gas constants of dry air and water vapor.

The enthalpy of power gas h_{pg}, created as a mixture of incoming power gas and the steam is composed of three enthalpy elements. The first one is the enthalpy of dry air h_{da}. This enthalpy can be determined by set of the tables stated in [6] using this command:

```
hda = humde(d,Tpgin)
```

d	relative humidity level [kg/kg] ($d = 0$)
$T_{pg,\text{in}}$	temperature of incoming power gas [°C]

The second element is the enthalpy of water steam h_s contained in humid air. This enthalpy can be determined by set of the tables stated in [2] using this command:

```
hs = xsteam('h_pT',psin,Tpgin)
```

$p_{s,\text{in}}$ partial pressure of the steam in a power gas [bar] (conversion Pa → bar necessary)

$T_{pg,\text{in}}$ temperature of incoming power gas [°C]

Partial pressure of the water steam in the incoming power gas $p_{s,\text{in}}$ expresses the partial pressure of the water steam in the power gas before mixing the water and power gas. This partial pressure is computed according to the formulas (7) and (8).

For $w_{is} = 0$ according (4) we get

$$w_{s,\text{out}} = (1 - 0) \cdot w_{s,\text{in}} + 0 \Rightarrow w_{s,\text{out}} = w_{s,\text{in}} \tag{7}$$

Then

$$\begin{aligned} p_{s,\text{in}} &= p_{pg} \cdot \left[1 - \frac{w_{da,\text{in}} \cdot r_{da}}{w_{da,\text{in}} \cdot r_{da} + w_{s,\text{in}} \cdot r_s} \right] \\ p_{da,\text{in}} &= p_{pg} - p_{s,\text{in}} \end{aligned} \tag{8}$$

The third component of the mixture enthalpy is the input steam enthalpy h_{is} being injected to the power gas. This enthalpy can be determined by set of the tables stated in [2] using this command:

```
his = xsteam('h_pT',pis,Tis)
```

p_{is} partial pressure of injected steam [bar]

T_{is} temperature of injected steam [°C]

Partial pressure of the input steam p_{is} means the difference of the partial pressures of the water steam before and after mixing. It can be expressed according (9).

$$p_{is} = p_s - p_{s,\text{in}} \tag{9}$$

Thus it is possible to say that particular enthalpies are functions of the following quantity:

$$\begin{aligned} h_{da} &= f\left(T_{pg,\text{in}}\right), \text{ where } d = 0 \\ h_s &= f\left(T_{pg,\text{in}}, p_{s,\text{in}}\right) \\ h_{is} &= f\left(T_{is}, p_{is}\right) \end{aligned} \tag{10}$$

3.3 Valve, Turbine, Compressor and Controller

The time constants of the valve at the input of the turbine which drives the compressor, of the compressor turbine, and of the compressor are due to their inertia moments smaller than 20 s and can be neglected when compared with the representative time constants of the MHE segments. In the MHE working point, the difference of the flow rate of the compressed air at the output of the compressor is approximately proportional to the difference of the signal at the input of the valve servomechanism. The resulting proportionality constant K adds to the amplification of the control loop.

The controller measures the actual temperature of the power gas at the output of the MHE, compares it at the setpoint, calculates the control deviation and drives the servomechanism of the valve by its PI output signal.

4 Simulation Results

Figures 6 and 7 show the temperature time responses of selected physical state variables of the MHE generated by step changes of input flue gas temperature $T_2^{(5)}(L,t)$, where the upper index in the round brackets stands for the number of the MHE segment. The time responses of the output power gas temperature $T_1^{(5)}(L,t)$, the output flue gas temperature $T_2^{(1)}(0,t)$, the wall temperature at the output of the MHE $T_W^{(5)}(L,t)$, and the power gas flow rate $M_1(t)$ are shown.

Figure 6 shows the temperature responses at the MHE that is not equipped with the control loop discussed. In the first task, illustrated in Fig. 6, the MHE has the constant power gas input which corresponds to the constant standard steady state output temperature of the power gas 790.5 °C.

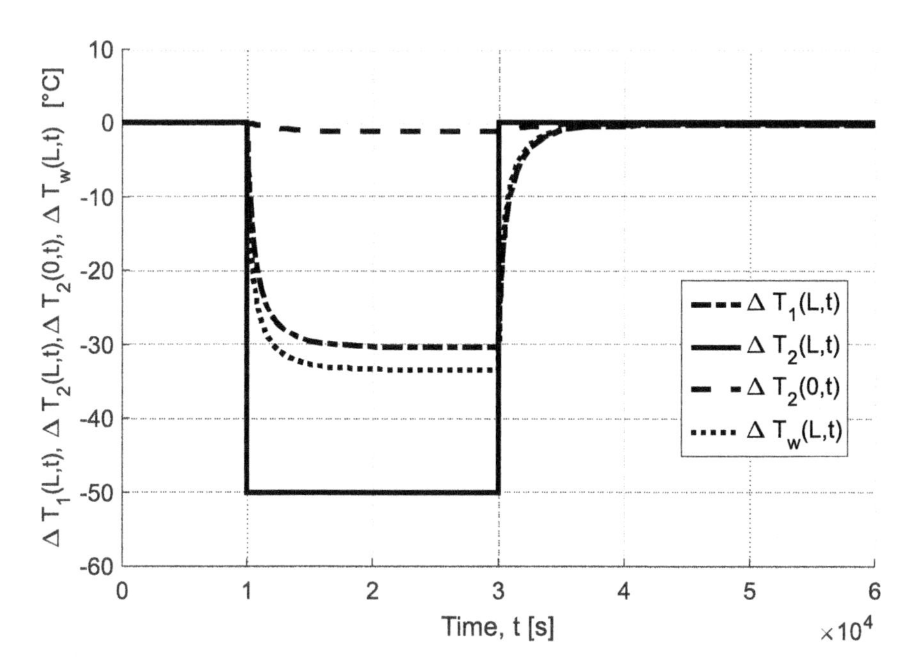

Fig. 6. The temperature responses at the MHE that is not equipped with the control loop.

Figure 7 shows the temperature responses at the MHE where the output temperature of the power gas is controlled by the control loop discussed. In the second task, illustrated in Fig. 7, the system has the constant output power gas temperature setpoint value 790.5 °C.

The initial state of the MHE is identical in both tasks. In time zero, the system is in a steady state. The temperature of the power gas at the input of the MHE is 201.5 °C and its output temperature is 790.5 °C. The initial flow

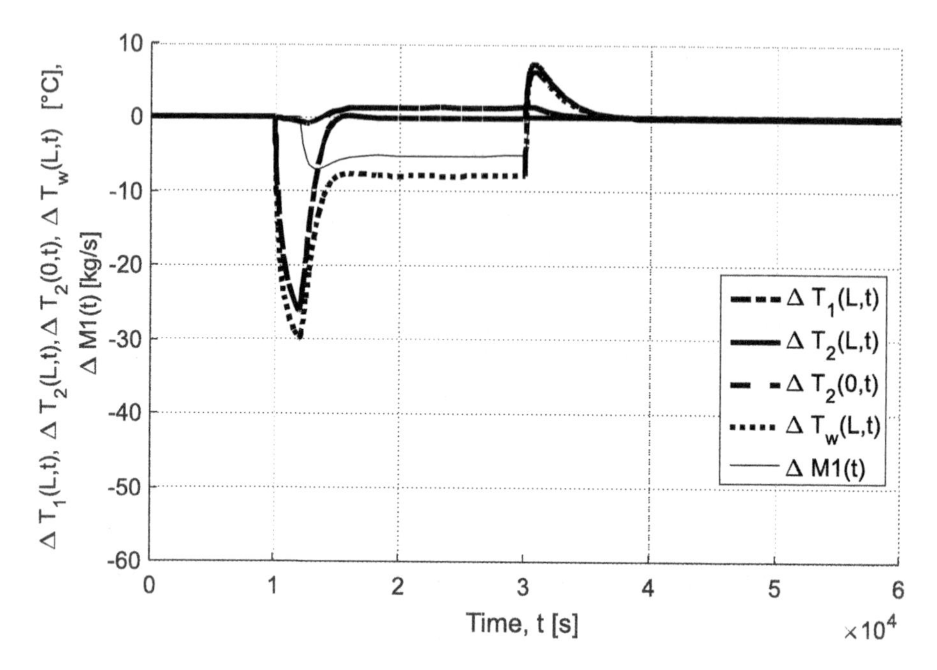

Fig. 7. The temperature responses at the MHE that with output power gas temperature control.

rate of power gas is $88.8\,\text{kg}\cdot\text{s}^{-1} = 66.6\,\text{kg}\cdot\text{s}^{-1}$ of air + $22.2\,\text{kg}\cdot\text{s}^{-1}$ of steam. The input flue gas temperature is 1176.6 °C.

In the time of 10,000 s, the temperature of the input flue gas drops from the standard 1176.6 °C to 1126.6 °C. The drop has a duration of 20,000 s, then the temperature returns to the standard value of 1176.6 °C. A comparison of Figs. 6 and 7 indicates the dynamic properties of PI control.

5 Conclusion

The paper describes the dynamics of output power gas temperature control at the flue gas to power gas heat exchanger. The presented paper was to show possibilities of temperature control of particular parts of the Flexible Energy System unit in order to design a controller. As an example, the temperature control of power gas at the outlet of the heat exchanger has been chosen. Mathematical models of the heat exchanger and air/steam mixer have been introduced in the paper. The control of the flue gas to the power gas heat exchanger is prepared for the flexible energy system FES Vitkovice Power Engineering JSC.

Acknowledgements. The work was supported by the grant project No. TA 04021687 of the Czech TACR agency and by the project SP2016/162, "Development of algorithms and systems for control, measurement and safety applications II" of the Student Grant System, VSB-TU Ostrava.

References

1. Bober, W., Tsai, C.T., Masory, O.: Numerical and Analytical Methods with MATLAB. CRC Press, Boca Raton (2009)
2. IF-97, I.: Thermodynamical properties of steam and water (1997). http://xsteam.sourceforge.net/. Accessed 2nd June 2016
3. Jaluria, Y., Torrance, K.E.: Computational Heat Transfer. Series in Computational and Physical Processes in Mechanics and Thermal Sciences, 2nd edn. CRC Press, Boca Raton (2002)
4. Nevriva, P., Ozana, S., Pies, M.: Simulation of power plant superheater using advanced simulink capabilities. Int. J. Circ. Syst. Signal Process. **5**(1), 86–93 (2011)
5. Pies, M., Ozana, S., Hajovsky, R., Vojcinak, P.: Modeling and simulation of partial blocks of flexible energy system in matlab & simulink for temperature control of steam/air mixture. In: Lecture Notes in Engineering and Computer Science, vol. 2, pp. 874–878 (2013)
6. Rice University, Department of Chemical and Biomolecular Engineering.: Chbe 301 - material & energy balances (2000). http://www.owlnet.rice.edu/ceng301/toc.html. Accessed on 2nd June 2016
7. Smith, G.D.: Numerical Solution of Partial Differential Equations. Oxford Applied Mathematics and Computing Science Series, 3rd edn. Clarendon Press, Oxford (1986)
8. Vilimec, L., Starek, K.: Power production process with gas turbine from solid fuel and waste heat and the equipment for the performing of this process, US Patent App. 12/607,800 (2010). http://www.google.com/patents/US20100199631

Industrial Cyber-Physical Systems in Textile Engineering

Juan Bullón Pérez[1(✉)], Angélica González Arrieta[2], Ascensión Hernández Encinas[3], and Araceli Queiruga-Dios[3]

[1] Department of Chemical and Textile Engineering, University of Salamanca, Salamanca, Spain
perbu@usal.es
[2] Department of Computer Science and Control, University of Salamanca, Salamanca, Spain
angelica@usal.es
[3] Department of Applied Mathematics, University of Salamanca, Salamanca, Spain
{ascen,queirugadios}@usal.es

Abstract. Cyber-Physical Systems (CPS) is an emergent approach of physical processes, computer and networking, that focuses on the interaction between cyber and physical elements. These systems monitor and control the physical infrastructures, that is why they have a high impact in industrial automation. The implementation and operation of CPS just like the management of the resulting automation infrastructure is of key importance to the industry. The evolution towards Industry 4.0 is mainly based on digital technologies. We present the integration of Industry 4.0 within the textile industry.

Keywords: Industry 4.0 · Cyber-Physical Systems · Textile industry

1 Introduction

Cyber–Physical Systems and their applicability in the industrial domain began as part of different programs like the Industrial Internet from General Electric [4], and the Spanish initiative about the Industrial Cyber–Physical Systems [13].

The industry is a basic sector to improve economy in the European Union (EU) and remains a driver of growth and employment. In particular, industry (which in this context means manufacturing and excludes mining, construction and energy) provides added value through the transformation of materials into products.

Some observers find this goal overly ambitious. But many others, believe that we are involved in a new industrial revolution, called Industry 4.0. This industry could boost the productivity and added value of European industries and could also stimulate the economy. In this new digital singe marke strategic, the European Commission wants to help all industrial sectors exploiting new technologies and managing a transition to a smart industrial system, that is industry 4.0.

M. Graña et al. (eds.), *International Joint Conference SOCO'16-CISIS'16-ICEUTE'16*, Advances in Intelligent Systems and Computing 527, DOI 10.1007/978-3-319-47364-2_13

This paper is organized as follows: Sect. 2 describes the basics concepts of the Industry 4.0. Section 3 introduces a case study of these concepts implemented in real textile industry scenario. Finally, some conclusions and perspectives are presented in Sect. 4.

2 Overview of Industry 4.0

Industry 4.0 is applied to a group of rapid transformations in the design, the manufacture, the operation and the services of manufacturing systems and products. The designation 4.0 is related to the world's fourth industrial revolution, see Fig. 1, the successor to the three previous industrial revolutions that caused quantum leaps in productivity and changed the lives of people throughout the world. In the words of the German Chancellor Angela Merkel to the OECD Conference, she said that Industry 4.0 is "the comprehensive transformation of the whole sphere of industrial production throughout the merging of digital technology and the Internet with conventional industry" [12], i.e., everything in and around a manufacturing operation (suppliers, plant, distributors, even the product itself) is digitally connected, providing a highly integrated value chain.

The term Industry 4.0 was originated in Germany, where it became a priority issue for research centers, companies, and universities [8]. In other European countries, it was labeled as Smart Factories, the Industrial Internet of Things (IIoT), Smart Industry, or Advanced Manufacturing (this is also used to refer to manufacturing techniques for new materials such as plastic, electronics and

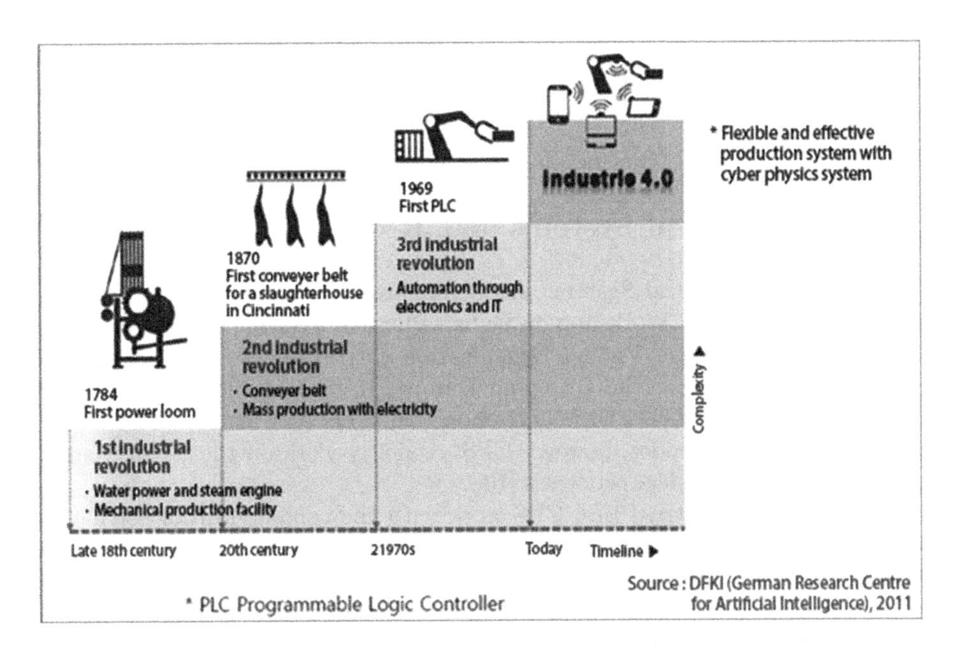

Fig. 1. Industry developments

composites) [7]. General Electric used the name Industrial Internet for the IIoT, and others called it the Internet of Everything and others Internet 4.0 or other variants [5]

The result of the fourth industrial revolution will be a so-called intelligent factory, or smart factory. Cyber-physical systems and the Internet of Things will be key technologies to reach this goal [9]. In fact, Industry 4.0 includes:

1. Smart products, procedures and processes. Smart factories constitute a key feature of Industry 4.0. They are capable of managing complexity, are less prone to disruption and are able to manufacture goods more efficiently. In the smart factory, human beings, machines and resources communicate with each other as naturally as in a social network.
2. CPS: physical and engineered systems whose operations are monitored, coordinated, controlled and integrated by a computing and communication core. This intimate coupling between the cyber and physical will be manifested from the nano-world to large-scale wide-area systems of systems, and at multiple time-scales [15].
3. The IoT which means that things which have identities and virtual personalities operating in smart spaces using intelligent interfaces to connect and communicate within social, environmental, and user contexts. Or what is the same: Interconnected objects having an active role in what might be called the Future Internet [1].

The IoT is an infrastructure, which collects information and controls by itself and other things in the physical space, while CPS creates synergy among the entities of the physical and cyber space, by integrating analogue and computational hardware, middleware, and cyber ware. This means, that the IoT will connect different products to each other, for example a smart watch with a smart phone. On the other hand, CPS uses the connection to the cloud and sensors to actively adjust a physical thing to a current state.

3 Cyber–Physical Systems and Textile Engineering

The term Cyber–Physical Systems was defined in 2006 in a high-level wor-king group, during the NSF Workshop on Cyber–Physical Systems organized by the Cyber–Physical Systems Virtual Organization (http://cps-vo.org/) composed by selected experts from the USA and Europe. It suggest the co-existence of cyber and physical elements with a common goal. This systems have been developed over the past decades, however CPS explicitly work on the integration of computation with physical processes [10].

In industrial infrastructures, CPS work with the combination of mechatronics, communication and information technologies. This is to control distributed physical processes and systems, designed as a network of interactive software and hardware devices and systems, many of them with higher level of decision making capabilities in both aspects: autonomic with self-decision processes [18] and collaborative with negotiation-based decision processes [11].

All this will have impact on the value creation, business models, downstream services and work organization. In order to achieve the goals of this CPS strategy the following features of Industry 4.0 should be implemented not only in textile industries bat in all industries adapted to 4.0 scheme:

- Horizontal integration through value networks: refers to topics that make business strategies and new value networks sustainable being supported by using CPS.
- End to end digital integration of CPS and software engineering across the entire value chain require appropriate Information Technology Systems (ITS) to provide end to end support to the entire value chain, from product development to manufacturing systems engineering, production, and service.
- Vertical integration and networked manufacturing systems require integration of CPS to create flexible and reconfigurable manufacturing systems which means that configuration rules will be defined to be used on a case-by-case basis to automatically build a structure of the machines of the shop floor for every situation, including all associated requirements in terms of models, data, communication, and algorithms.

Finally, in such industrial infrastructures, humans are seen as important stakeholders and CPS as human-centered technology that plays a special role [2]. The interaction with people can be achieved with mobile devices, such as smart phones, tablets or, more recently, the wearable devices. This behave uses the interfaces for a bidirectional communication. Technologies such as augmented reality can be combined with agents to support operators during the installation, operation and maintenance of automation systems in manufacturing plants. This improves the productivity and efficiency of operators by providing information related to historic and current status of the device/system and provides additional information, e.g., documentation, web pages or videos. Google Glass is an example of such augmented reality technology that can be used in industrial shop floor environments.

Textile and clothing activities present different subdivisions, each with its own traits. The length of the textile process and the variety of its technical processes lead to the coexistence (within the textile and clothing sector) of different sub-sectors in regards to their business structure and integration.

The textile industry is developing expert system applications slowly to increase production, improve quality and reduce costs. Such systems are surfacing in a variety of areas throughout the textile manufacturing process. This important decision scenario in the textile industry generates a sequence of production planning decisions necessary to produce a specific category of end product. This sequence begins with the decision to produce a particular type of end product, then the appropriate fiber type is chosen. Secondly, the appropriate yarn count group is chosen. Then the appropriate spinning system is chosen; and finally, the appropriate preparation method is selected. Each decision in the sequence depends on the combination of decisions made in the preceding stages.

Production planning is a complex area of any manufacturing operation. In textiles, planning is complicated as there exists different types of fibers, yam counts, spinning systems, preparation methods, and end products. All of these factors, combined with the customer's demands for correctly filled orders and short delivery times, make the production planning process more complicated. In addition, effective production planning in the textile industry has become more and more critical as intense foreign competition has impacted the market.

In the textile manufacturing era, the textile process chains in high-wage countries are mostly described along the production chain. To adopt these textile process chains on the Digital Manufacturing level (what we call Industry 4.0), information that flows through all levels of an enterprise needs to be connected to other entities of the textile process which enables a flexible and fast fabrication, feasible to deal with the order lot size. Thus, in the cyber-physical systems world, machine communicates to each other and the plant operation can be realized. They inform about their status and upcoming problems such as maintenance. In this case, the factory will reconfigure itself in order to fulfill the customer's production order. Textile machines with open interfaces will be highly flexible and able to independently adapt status based on an overall information platform. Can, core, and warp beam and fabric will become carriers of information which lead to autonomic textile process chains.

A main aspect of production in the cyber-physical systems world is the human–machine interaction. The use of smart personal devices, such as smart phones, tablets, or head-mounted displays, offers a huge potential for innovation. Smart personal devices can be used to make production more transparent by providing relevant production key parameters in a sophisticated way. In addition, guidance programs can lead to optimized production or faster act in case of machine breakdowns. Also aspects of tele-maintenance, such as repair of machine supported and produced by the machine itself.

Self-optimization of the warp tension of textile machines is one of the innovative digital technologies in Digital Manufacturing. The aim of this method is to enable the loom to set the warp tension automatically on a minimum level without reducing the process stability. Therefore, as part of the self-optimization, the model of the process has to be set up to enable the loom to set the warp tension automatically on a minimum level without reducing the process stability and implemented in the weaving process. So far the loom is able to create its process model for a given process domain independently, then the machine runs an experimental design, and automatically determines at respective test points the warp yarn tension. So the operating point can be determined with the aid of quality criterion where the warp tension becomes minimal [14].

In self-optimization method, further sensors are embedded in the weaving process. A system for automatic in-line flaw detection in industrial woven fabrics operates on low resolution ($\approx$ 200 ppi) image data, and the new cyber-physical system describes process flow to segment single yarns in high resolution ($\approx$ 1.000 ppi) textile images. This work is partitioned into two parts: First of all, mechanics, machine integration, vibration canceling, and illumination scenarios, based on the integration into a real loom. Secondly, the software framework for high-precision fabric defect detection. The system is evaluated on a database of 54 industrial fabric images, achieving a detection rate of 100 % with minimal false alarm rate and very high defect segmentation quality [17].

As the weaving machines are highly flexible production systems, it is usually possible to produce one article on many different types of machines, with different efficiencies and settings. But only few textile industries are able to use the complete potential of the production system (e.g. products for new markets). Despite the huge effort in weaving machine automation, the best machine settings can only be found by trial and error, best after a run-time of more than one week. But today the lot sizes are not big enough, so it is almost impossible to produce new articles with suitable efficiency.

An intelligent textil world will solve this shortcoming with a new method for determining the best machine settings for a given article with a given yarn material by using case based reasoning for storing and analyzing old situations and generating new, adapted machine settings [3].

Furthermore, textile products can act as CPS, in this case the products can be named as smart textiles.

Smart textiles can be described as textiles that are able to sense stimuli from the environment, to react to those stimuli and adapt to them by integrating functionalities in the textile structure. Advanced materials, such as breathing, fire-resistant or ultra strong fabrics, are according to this definition but are not considered intelligent, although they have high-technological characteristics. The extension of intelligence can be divided in three subgroups [19]:

- Passive smart textiles can only sense the environment, they are sensors;
- active smart textiles can sense the stimuli from the environment and also react to them, besides the sensor function, they also have an actuator function;
- finally, very smart textiles take a step further, having the gift to adapt their behaviour to the circumstances.

Therefore, two components need to be presented in the textile structure in order to bear the full mark of smart textiles: a sensor and an actuator, possibly completed with a processing unit which drives the actuator on the basis of the signals from the sensor.

The application possibilities offered by these materials are only limited by human imagination.

Processing these intelligent materials (in the form of fibres, threads, gels, liquids) into textiles or producing textiles from these intelligent materials results into an intelligent textile.

The CEN/TC 248/WG 31 published by a working group of the European Committee of Standardization (CEN) gives a definition of smart textiles. The document distinguishes between:

- Functional textile material:
    - Can be components of intelligent textile systems and hence functional textile materials which are relevant for intelligent textile systems.
    - It has a specific (passive) function.
    - Examples are: electrically, thermally or optically conductive materials, materials that release substances,fluorescent materials.
- Intelligent (smart) textile material:

    - Functional textile material, which interacts actively with its environment, i.e. it responds or adapts to changes in the environment.
    - Examples are: chromic, piezoelectric, phase change, electroluminescent, shape change, auxetic, electrolytic, dilating or shear thickening, capacitive materials.
- Smart textile systems:
    - A textile system which exhibits an intended and exploitable response as a reaction either to changes in its surroundings/environment or to an external signal/input
    - Comprises actuators, possibly completed by sensors
    - Contains an information management device that controls/manages the information within the textile system characterized by two functions: energy and external communication

Basically, five functions can be distinguished in a smart suit, namely: sensors, data processing, actuators, storage and communication.

In essence the smart textiles will interact with the body, the environment or themselves. All functions have to be integrated onto or into the textile for achieving the smart textile system without having any impact on comfort and ease of use. In addition, all processes to do so have to be compatible with current manufacturing technologies.

To allow the comparison of technologies, a set of five integration levels has been established, see Fig. 2. The levels relate to the integration of the electronic function starting at rigid electronics inserted by design, followed by rigid and flexible microelectronics attached to the textile substrate, and textronics permitting full integration. The latter achieves the highest level of integration from a textile engineering perspective and is referring to textronics [16].

The goal is to create clothing with integrated sensor function. To reach this goal three tracks can be followed:

- Incorporating: is the action of uniting (one thing) with something else that already existence for example attaching electronics onto textiles. Integration levels 1 to 3 (added-on) are related to incorporation.
- Embedding: is becoming an integral part of a surrounding whole, for example weaving an electronic tape into the textile structure. Integration Levels 3 (built-in) to 5 refer to the action of embedding.
- Integrating: is the action of bringing all parts together to unify them into a whole. The combination of the electronic function with the textile in generally, is called integration.

The first option is mainly driven by the electronics industry, and the last one by the textile industry.

As one can imagine, the integration level 5 is the most sophisticated level of integration and is challenging research in invisibly merging the two fields: textiles and electronics into textronics. The final product truly integrates textiles and electronics, making the separation impossible. Consequently, if electronics are the basis for generating the electronic function, the integration levels 1–3 have

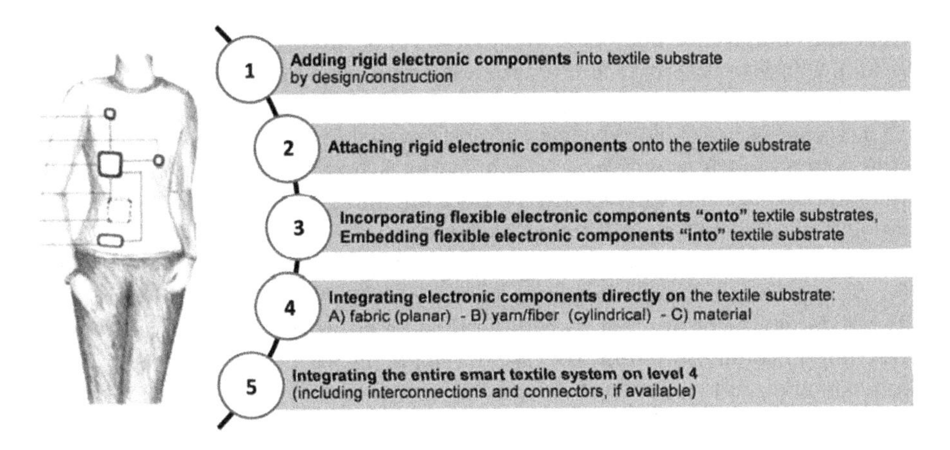

Fig. 2. Smart textile system component integration levels

to be taken into consideration. If the textile is the base for integration, level 4 is of concern. Level 5 can be reached in both ways, in textiles this would be full integrated in the entire system. In electronics, something like electronic skin could be an option [8]. Up to date, systems on integration level 5 are none existent for both substrate types. It is worth noting that smart textile products are systems. There is a good chance that the single e-textile components of the system are integrated on different levels. The highest level of integration, level 5 relates clearly to the entire smart textile system and not to the single etextile component. Therefore, it is evident that integration level 5 only can be reached if all components fulfill these preconditions. For single component the highest integration level that can be reached is level 3 or alternatively level 4.

4 Conclusions

The 4.0 era is accompanied by decentralized control paradigms that result from the system design of the distributed systems arising here, their targeted adaptivity to unexpected events, and the high changeability of the components. Decentralized control paradigms are not yet very well spread in the industrial context. On the one hand, this is because, while the underlying theoretical models are old, practical insights, evidence, and architectures have only been possible since the availability of high computing power. At the same time, however, it is precisely the decentralized control systems that open up room for new increases in efficiency through greater flexibility and adaptivity.

To make the dream of a smart factory come true, much more research has to be done. A standardized basis would lead to an efficient use of new technologies and should be established since Industry 4.0 will have an impact on almost any company. The provided funding could lead to a quick development of new standards and ensure a quick implementation of the next industrial revolution.

The use of digital technologies, and in a world of smart things, will change the way we produce textiles. Machines with additional sensors, actors and cognition acting in a network can lead to the fourth industrial revolution: Industry 4.0.

The scenarios are explored using currently available technology to build smart textile systems, such as wireless communications, textile antennas, chromic display materials, textile switches, textile pressure and gesture sensors, textile circuits, and micro-component welding technology.

Future industrial applications will need to be developed at a rapid pace in order to capture the agility required by modern businesses. Typical industrial software development approaches will need to be adjusted to the new paradigm of distributed complex system software development with main emphasis on collaboration and multi-layer interactions among systems of systems, which is challenging. To do so, some generic common functionality will need to be provided, potentially by a distributed service platform hosting common functionalities, following the service–oriented architecture approach.

References

1. Bassi, A., Horn, G.: Internet-of-Things in 2020, EC-EPoSS Workshop Report (2008)
2. Boy, G.A.: Orchestrating Human-Centered Design. Springer, London (2012)
3. Pérez, J.B., Arrieta, A.G., Encinas, A.H., Dios, A.Q.: Textile engineering and case based reasoning. In: Omatu, S., Semalat, A., Bocewicz, G., Sitek, P., Nielsen, I.E., García, J.A.G., Bajo, J. (eds.) DCAI 2016. AISC, vol. 474, pp. 423–431. Springer, Heidelberg (2016). doi:10.1007/978-3-319-40162-1_46
4. Evans, P.C., Annunziata, M.: Industrial internet: pushing the boundaries of minds and machines, general electric (2012). http://www.ge.com/docs/chapters/Industrial_Internet.pdf
5. Gilchrist, A.: Industry 4.0 Alasdair (2016)
6. Gloy, Y.S., Schwarz, A.: Cyber-physical systems in textile production, the next industrial revolution? http://www.textile-future.com/textile-manufacturing.phpreadarticle=1829
7. Griffiths, M.: Research briefings - advanced manufacturing. In: Parliamentary Office of Science and Technology (POST). http://researchbriefings.parliament.uk/ResearchBriefing/Summary/POST-PN-420
8. Hermann, M., Pentek, T., Otto, B.: Design principles for industrie 4.0 scenarios. In: IEEE 49th Hawaii Hawaii International Conference on System Sciences, pp. 3928–3937 (2016)
9. Kagermann, H., Wahlster, W., Helbig, J.: Recomendations for implementing the strategic initiative Industrie 4.0. ACATECH, National Academy of Science and Engineering (2013)
10. Lee, E.A., Seshia, S.A.: Introduction to Embedded Systems. A Cyber-Physical Systems Approach (2013). http://LeeSeshia.org
11. Marrón, P.J., Minder, D., Karnouskos, S.: The Emerging Domain of Cooperating Objects: Definition and Concepts. Springer, Berlin, Heidelberg (2012)
12. Merkel, A.: Federal government-speech by federal chancellor Angela Merkel to the OECD conference. www.bundesregierung.de/Content/EN/Reden/2014/2014-02-19-oecd-merkel-paris_en.htm

13. de Industria, M.: Energía y Turismo: Industria conectada 4.0. La Transformación Digital de la Industria Española. http://www.industriaconectada40.gob.es/
14. Möller, D.P.F.: Guide to Computing Fundamentals in Cyber-Physical Systems. Computer Communications and Networks. Springer, Heidelberg (2016)
15. Parvin, S., Hussain, F., Hussain, O., Thein, T., Park, J.: Multicyber framework for availability enhacement of cyber physical systems. Computing **95**(10–11), 927–948 (2013)
16. Rambausek, M.: Definition, development and characterization of fibrous organic field effect transistors. Ph.D. dissertation, Ghent University, Belgium, p. 375 (2014)
17. Schneider, D., Holtermann, T., Neumann, F., Hehl, A., Aach, T., Gries, T.: A vision based system for high precision online fabric defect detection. In: Proceedings of 7th IEEE Conference on Industrial Electronics and Applications, pp. 1494–1499 (2012)
18. Vasilakos, A.V., Manish, P., Stamatis, K., Witold, P.: Autonomic Communication. Springer, New York (2010)
19. Zhang, X., Tao, X.: Smart textiles: passive smart. Text. Asia **32**, 45–49 (2001)

Optimal Scheduling of Joint Wind-Thermal Systems

Rui Laia[1,2], Hugo M.I. Pousinho[1], Rui Melício[1,2(✉)], and Victor M.F. Mendes[2,3,4]

[1] IDMEC, Instituto Superior Técnico, Universidade de Lisboa, Lisbon, Portugal
ruimelicio@gmail.pt
[2] Departamento de Física, Escola de Ciências e Tecnologia, Universidade de Évora, Évora, Portugal
[3] Instituto Superior de Engenharia de Lisboa, Lisbon, Portugal
[4] C-MAST Center for Mechanical and Aerospace Sciences and Technology, Lisbon, Portugal

Abstract. This paper is about the joint operation of wind power with thermal power for bidding in day-ahead electricity market. Start-up and variable costs of operation, start-up/shut-down ramp rate limits, and ramp-up limit are modeled for the thermal units. Uncertainty not only due to the electricity market price, but also due to wind power is handled in the context of stochastic mix integer linear programming. The influence of the ratio between the wind power and the thermal power installed capacities on the expected profit is investigated. Comparison between joint and disjoint operations is discussed as a case study.

Keywords: Stochastic · Mixed integer linear programming · Wind-thermal

1 Introduction

Renewable energy sources play an important role in the need for clean energy in a sustainable society [1]. Renewable energy can partly replace fossil fuels, avowing anthropogenic gas emissions. Energy conversion from renewable energy has been supported by policies, providing incentive or subsidy for exploitation [2]. These polices have pushed the integration of renewable energy forward, but by an extra-market approach. The approach involves, for instances, legislative directives, feed-in tariffs, favorable penalty pricing and grid right of entry, and survives at reserved integration level. But as integration of renewable energy increases the approach is expected to be untenable [3]. Sooner or later, a wind power producer (WPP) has to face competition in a day-ahead electricity market. For instances, in Portugal, a WPP is paid by a feed-in tariff under the condition of a limited amount of time or of energy delivered. Otherwise, the route is the day-ahead market or by bilateral contracting [4].

Conversion of wind energy into electric energy to trade in the day-ahead electricity market has to face uncertainty, particularly, on the: availability of wind energy, energy price and imbalance penalty. These uncertainties have to be addressed to avoid dropping profit [5–7]. A stochastic programming addresses uncertainty by the use of modeling via scenarios, and is a suitable approach to aid a Wind-Thermal Power

M. Graña et al. (eds.), *International Joint Conference SOCO'16-CISIS'16-ICEUTE'16*,
Advances in Intelligent Systems and Computing 527, DOI 10.1007/978-3-319-47364-2_14

Producer (WTPP) in developing a joint bid strategy in a day-ahead market [8–12]. The problem formulation is approached in a way of approximating all expressions regarding the objective function and the constraints to describe the problem by a mixed integer linear program (MILP) one. The approximation is intended to use excellent commercial available for MILP. The uncertainties are treated by uncertain measures and multiple scenarios built by wind power forecast [13–15] and market-clearing electricity price forecast [16–18] applications.

A case study with data from the Iberian Electricity Market is used to illustrate the effectiveness of the proposed approach. The approach proves both to be accurate and computationally acceptable.

2 Problem Formulation

2.1 Market Balancing

System imbalance is defined as a non-null difference on the trading, i.e., between physical delivered of energy and the value of energy on contract at the closing of the market. If there is an excess of delivered energy in the power system, the system imbalance is positive; otherwise, the system imbalance is negative. The system operator seeks to minimize the absolute value of the system imbalance in a power system, using a mechanism based on prices penalization for producer imbalance, i.e., the difference of the physical delivery of energy from the one accepted due to the bid of the producer. If the system imbalance is negative, the system operator keeps the price for the physical delivery of bided energy for the producers with positive imbalance and pays a premium price for the energy produced above bid. The revenue R_t of the producer in hour t is given by [19]:

$$R_t = \lambda_t^D P_t^{offer} + I_t \tag{1}$$

In (1), P_t^{offer} is the power traded by the producer in the day-ahead market and I_t is the imbalance income resulting from the balancing process, $\lambda_t^D P_t^{offer}$ is the revenue that the producer collects from trading energy if there is no uncertainties. The deviation of the producer in hour t is given by:

$$\Delta_t = P_t^{act} - P_t^{offer} \tag{2}$$

In (2), P_t^{act} is the physical delivery of energy in hour t. Two ratio prices for positive and negative imbalances are, respectively, given by:

$$r_t^+ = \frac{\lambda_t^+}{\lambda_t^D}, \quad r_t^+ \leq 1; \; r_t^- = \frac{\lambda_t^-}{\lambda_t^D}, \quad r_t^- \geq 1 \tag{3}$$

In (3), λ_t^+ is the price paid by the market to the producer for a positive imbalance, λ_t^- is the price to be charged to the producer for a negative imbalance. The imbalance in (1) using (3) is given by:

$$I_t = \lambda_t^D r_t^+ \Delta_t, \ \Delta_t \geq 0; \ I_t = \lambda_t^D r_t^- \Delta_t, \ \Delta_t < 0 \tag{4}$$

A producer that needs to correct its energy imbalance in the balancing market incurs on an opportunity cost, because energy is traded at a more profitable price in the day-ahead market.

The imbalance in (2) will cause an opportunity cost given by:

$$C_t = \lambda_t^D (1 - r_t^+) \Delta_t, \ \Delta_t \geq 0; \ C_t = -\lambda_t^D (r_t^- - 1) \Delta_t, \ \Delta_t < 0 \tag{5}$$

The uncertainties are considered by a set of scenarios Ω for wind power, energy price and ratio prices for system imbalance. Each scenario ω will be weighted with a probability of occurrence π.

2.2 Thermal Production

The operating cost $F_{\omega it}$in scenario $\omega$for a thermal unit i in hour $t$is given by [20]:

$$F_{\omega it} = A_i u_{\omega it} + d_{\omega it} + b_{\omega it} + C_i z_{\omega it} \quad \forall \omega, \ \forall i, \ \forall t \tag{6}$$

In (6), the operating cost is composed by four terms, namely: A_i is a fixed operating cost; $d_{\omega it}$ is a variable cost, i.e., is a part of the cost incurred by the amount of fossil fuel consumed above the minimum power; $b_{\omega it}$ is the unit start-up cost; C_i is the unit shut-down cost. The typical non-differentiable and nonconvex functions used to quantify the variable costs of a thermal unit is replaced by a piecewise linear approximation in order to take the advantage of using MILP [5]. The piecewise linear approximations for the variable cost $d_{\omega it}$ is formulated by the statements given by:

$$d_{\omega it} = \sum_{l=1}^{L} F_i^l \delta_{\omega it}^l \quad \forall \omega, \ \forall i, \ \forall t \tag{7}$$

$$p_{\omega it} = p_i^{\min} u_{\omega it} + \sum_{l=1}^{L} \delta_{\omega it}^l \quad \forall \omega, \ \forall i, \ \forall t \tag{8}$$

$$(T_i^1 - p_i^{\min}) t_{\omega it}^1 \leq \delta_{\omega it}^1 \quad \forall \omega, \ \forall i, \ \forall t \tag{9}$$

$$\delta_{\omega it}^1 \leq (T_i^1 - p_i^{\min}) u_{\omega it} \quad \forall \omega, \ \forall i, \ \forall t \tag{10}$$

$$(T_i^l - T_i^{l-1}) t_{\omega it}^l \leq \delta_{\omega it}^l \quad \forall \omega, \ \forall i, \ \forall t, \ \forall l = 2, .., L-1 \tag{11}$$

$$\delta^{l}_{\omega it} \leq (T_i^l - T_i^{l-1}) t^{l-1}_{\omega it} \ \forall \omega, \ \forall i, \ \forall t, \ \forall l = 2, .., L-1 \tag{12}$$

$$0 \leq \delta^{L}_{\omega it} \leq (p_i^{\max} - T^{L-1}_{\omega it}) \, t^{L-1}_{\omega it} \quad \forall \omega, \ \forall i, \ \forall t \tag{13}$$

In (7), the variable cost is computed as the sum of the product of the slope of each segment F_i^l by the segment power $\delta^{l}_{\omega it}$. In (8), the power production of the unit i is given by the minimum power production plus the sum of the segment powers associated with each segment. The binary variable $u_{\omega it}$ ensures that the power production is equal to 0 if unit i is offline. In (9), if the binary variable $t^{l}_{\omega it}$ has a null value, then the segment power $\delta^{1}_{\omega it}$ can be less than the segment 1 maximum power; otherwise and in conjunction with (10), if the unit is on, then $\delta^{1}_{\omega it}$ is equal to the segment 1 maximum power. In (11), from the second segment to the second last one, if the binary variable $t^{l}_{\omega it}$ has a null value, then the segment power $\delta^{l}_{\omega it}$ can be less than the segment l maximum power; otherwise and in conjunction with (12), if the unit is on, then $\delta^{l}_{\omega it}$ is equal to the segment l maximum power. In (13), if the binary variable $t^{L-1}_{\omega it}$ has a null value, then the segment power L must be zero; otherwise is bounded by the last segment maximum power.

The exponential nature of a start-up cost of thermal units is modelled by an approximation of a non-decreasing stepwise function with a step [5] given by:

$$b_{\omega it} \geq K_i^{\beta} \left(u_{\omega it} - \sum_{r=1}^{\beta} u_{\omega it-r} \right) \quad \forall \omega, \quad \forall i, \quad \forall t \tag{14}$$

$$b_{\omega it} \geq 0 \quad \forall \omega, \quad \forall i, \quad \forall t \tag{15}$$

In (14), if in scenario ω unit i in hour t is online and has been offline in β preceding hours, the expression in parentheses is equal to one, implying that a start-up happen in hour t and the respective cost K_i^{β} is incurred.

The box constraints for the power production in scenario ω of unit i in hour t are given by:

$$p_i^{\min} u_{\omega it} \leq p_{\omega it} \leq p^{\max}_{\omega it} \ \forall \omega, \ \forall i, \ \forall t \tag{16}$$

$$p^{\max}_{\omega it} \leq p_i^{\max} (u_{\omega it} - z_{\omega it+1}) + SD \, z_{\omega it+1} \ \forall \omega, \ \forall i, \ \forall t \tag{17}$$

$$p^{\max}_{\omega it} \leq p^{\max}_{\omega it-1} + RU \, u_{\omega it-1} + SU \, y_{\omega it} \ \forall \omega, \ \forall i, \ \forall t \tag{18}$$

$$p_{\omega it-1} - p_{\omega it} \leq RD \, u_{\omega it} + SD \, z_{\omega it} \ \forall \omega, \ \forall i, \ \forall t \tag{19}$$

In (16), the power limits of the units are set. In (17) and (18), the upper bound of $p^{\max}_{\omega it}$ is set, which is the maximum available power in scenario ω for a thermal unit i in hour t. This variable considers the: actual power of a unit, start-up/shut-down ramp rate limits, and ramp-up limit. In (18)–(19), the relation between the start-up and shut-down variables of the unit are given, using binary variables and their weights. In (19), the ramp-down and shut-down ramp rate limits are considered.

The minimum down time constraint is imposed by a linear formulation given by:

$$\sum_{t=1}^{J_i} u_{\omega it} = 0 \ \forall \omega, \ \forall i \tag{20}$$

$$\sum_{t=k}^{k+DT_i-1} (1 - u_{\omega it}) \geq DT_i z_{\omega it} \ \forall \omega, \forall i, \forall k = J_i + 1 \ldots T - DT_i + 1 \tag{21}$$

$$\sum_{t=k}^{T} (1 - u_{\omega it} - z_{\omega it}) \geq 0 \ \forall \omega, \ \forall i, \ \forall k = T - DT_i + 2 \ldots T \tag{22}$$

$$J_i = \min\{T, (DT_i - s_{\omega i0})(1 - u_{\omega i0})\} \tag{23}$$

In (21), the minimum down time is satisfied for all the possible sets of consecutive hours of size DT_i, and in (22) is satisfied for the last $DT_i - 1$ hours.

The minimum up time constraint is imposed by linear formulation given by:

$$\sum_{t=1}^{N_i} (1 - u_{\omega it}) = 0 \ \forall \omega, \ \forall i \tag{24}$$

$$\sum_{t=k}^{k+UT_i-1} u_{\omega it} \geq UT_i y_{\omega it} \ \forall \omega, \ \forall i, \ \forall k = N_i + 1 \ldots T - UT_i + 1 \tag{25}$$

$$\sum_{t=k}^{T} (u_{\omega it} - z_{\omega it}) \geq 0 \ \forall \omega, \ \forall i, \ \forall k = T - UT_i + 2 \ldots T \tag{26}$$

$$N_i = \min\{T, (UT_i - U_{\omega i0}) u_{\omega i0}\} \tag{27}$$

In (24), the minimum up time is satisfied for all the possible sets of consecutive hours of size UT_i. In (25), the minimum up time will be satisfied for the last $UT_i - 1$. The relations between the binary variables to identify start-up and shutdown are given by:

$$y_{\omega it} - z_{\omega it} = u_{\omega it} - u_{\omega it-1} \ \forall \omega, \ \forall i, \ \forall t \tag{28}$$

$$y_{\omega it} + z_{\omega it} \leq 1 \ \forall \omega, \ \forall i, \ \forall t \tag{29}$$

The total power produced by the thermal units is given by:

$$p^g_{\omega t} = \sum_{i=1}^{I} p_{\omega it} \ \forall \omega, \ \forall t \tag{30}$$

In (30), I is the set of indexes for the thermal units, $p^{g}_{\omega t}$ is the total thermal power in scenario ω in hour t.

The total operating costs $F^{T}_{\omega t}$ of the thermal power system is given by:

$$F^{T}_{\omega t} = \sum_{i=1}^{I} F_{\omega i t} \ \forall\omega, \ \ \forall t \tag{31}$$

In (31), the operating cost $F_{\omega i t}$ in scenario ω for unit i in hour is given by (5).

2.3 Objective Function

The power in the bid submitted by the WTPP is the sum of the power from the thermal power system with the power from the wind power system and is given by:

$$p^{offer}_{\omega t} = p^{th}_{\omega t} + p^{D}_{\omega t}; \ p^{act}_{\omega t} = p^{g}_{\omega t} + p^{\omega d}_{\omega t} \ \forall\omega, \ \ \forall t \tag{32}$$

In (32), $p^{g}_{\omega t}$ is the actual power of the thermal power system and $p^{\omega d}_{\omega t}$ is the actual power of the wind power system produced for scenario ω. The expected revenue of the WTPP over the time horizon N_T is given by the solution of the following mathematical programing problem with the objective function given by:

$$\sum_{\omega=1}^{N_\Omega} \sum_{t=1}^{N_T} \pi_\omega \left[\left(\lambda^{D}_{\omega t} P^{offer}_{\omega t} + \lambda^{D}_{\omega t} r^{+}_{\omega t} \Delta^{+}_{\omega t} - \lambda^{D}_{\omega t} r^{-}_{\omega t} \Delta^{-}_{\omega t} \right) - F^{T}_{\omega t} \right] \tag{33}$$

Subject to:

$$0 \leq p^{offer}_{\omega t} \leq p^{M}_{\omega t} \ \forall\omega, \ \ \forall t \tag{34}$$

$$\Delta_{t\omega} = \left(p^{act}_{\omega t} - p^{offer}_{\omega t} \right) \ \forall\omega, \ \ \forall t \tag{35}$$

$$\Delta_{t\omega} = \Delta^{+}_{t\omega} - \Delta^{-}_{t\omega} \ \forall\omega, \ \ \forall t \tag{36}$$

$$0 \leq \Delta^{+}_{t\omega} \leq P_{t\omega} d_t \ \forall\omega, \ \ \forall t \tag{37}$$

$$p^{M}_{\omega t} = \sum_{i=1}^{I} p^{\max}_{\omega i t} + p^{E\max} \ \forall\omega, \ \ \forall t \tag{38}$$

In (34), $p^{M}_{\omega t}$ is the maximum available power (38), limited by the sum of the installed capacity in the wind power system, $p^{E\max}$, with the maximum thermal production. Some day-ahead markets require that the bidding to be submitted is given by:

$$(p^{offer}_{\omega t} - p^{offer}_{\omega' t})(\lambda^{D}_{\omega t} - \lambda^{D}_{\omega' t}) \geq 0 \ \ \forall\, \omega, \, \omega', \ \ \forall\, t \tag{39}$$

In (39), if the day-ahead market prices are equal for two scenarios ω and ω' then the power bid difference between the two scenarios is indifferent. Otherwise, the power bids have to be non-decreasing with the price. Non decreasing energy bids are assumed. Hence, when wind power and thermal power bids are disjoint submitted implies that each bid has to be a non-decreasing one. While, only one joint non-decreasing bid is submitted in the joint schedule.

3 Case Study

The simulations are carried out in Gams using the Cplex solver for MILP. The effectiveness of the stochastic MILP approach is illustrated by a case study using a set of data from the Iberian electricity market, comprising 10 days of June 2014 [21]. The scenarios for the energy prices and the energy availability for the wind power system are respectively in the left and right sides of Fig. 1.

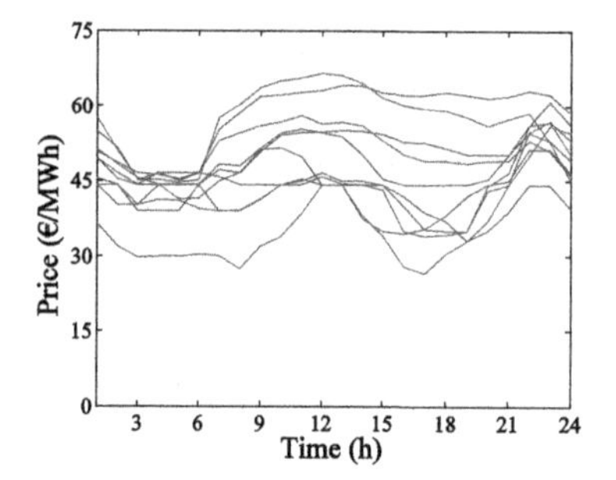

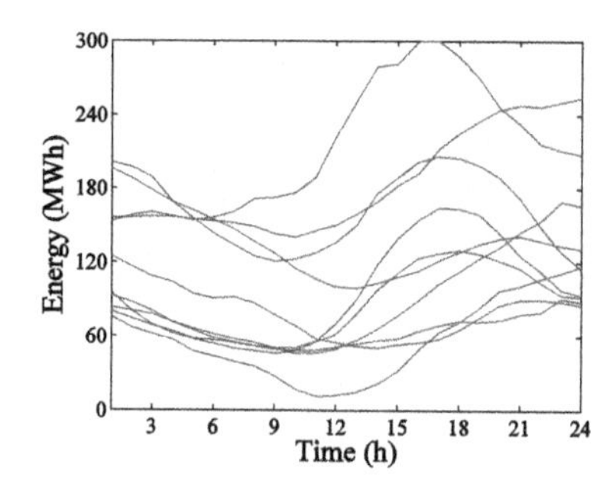

Fig 1. June 2014 (ten days); left: Iberian market price, right: wind energy.

The producer owns a wind power system with an installed capacity of 360 MW and a thermal power system with 8 units and a total installed capacity of 1440 MW. The variable costs of the thermal units are modelled by three segments in the piecewise linear approximation. Firstly, the simulations are carried out with the previous values of the installed capacities in order to find the expected profit and the expected imbalance cost without joint schedule, i.e., for the wind power and for the thermal power systems standing alone, and with joint schedule. The expected profit and the expected imbalance cost without and with joint schedule are shown in Table 1.

Table 1 Results without and with joint schedule

Case study	Profit (€)	Imbalance cost (€)
Wind system	119200	−17826
Thermal system	516848	229398
Disjoint wind and thermal systems	636047	…
Wind-thermal system	642326	3643
Gain (%)	0.99	…

In Table 1, the expected profit of the joint schedule is 0.99 % higher than the disjoint one and the processing is not a burden in computational resources in comparison with the disjoint one: the CPU time given by Gams is about the same for both schedules, since the wind power system schedule CPU time is irrelevant when compared with the thermal power system one in the disjoint schedule. Information for hour 15 regarding the sum of the wind power with the thermal power bid without and with joint schedule is shown in Fig. 2.

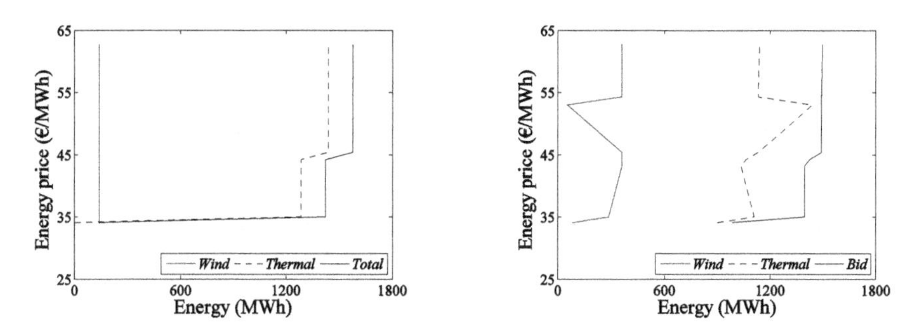

Fig 2. Bid of energy for hour 15; left: disjoint, right: joint.

In Fig. 2, note that wind and thermal power do not have to be non-decreasing per se.

The energy bids in scenario 3 for the disjoint and joint schedule are respectively in the left and right figures of Fig. 3.

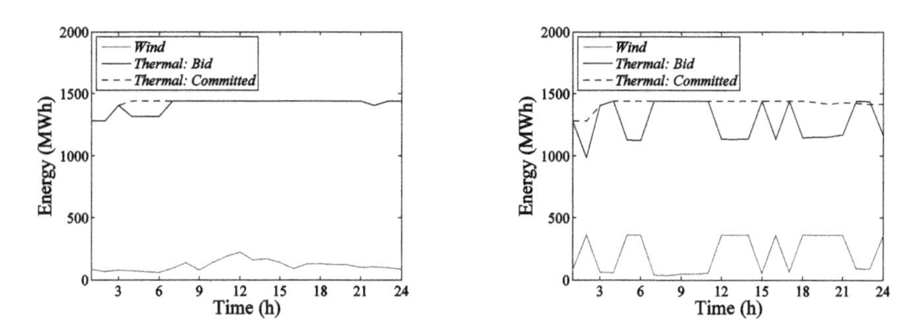

Fig 3. Bid of energy and committed in scenario 3; left: disjoint, right: joint.

The wind parcel of the energy bid is higher for the joint schedule and the thermal bid behavior tends to be the opposite of the wind behavior: when the wind parcel increases, the thermal one decreases. The higher values of the wind parcel of the energy bid is compensated by the decreasing of the thermal parcel of the energy bid, implying a lower imbalance. Secondly, the simulations are carried keeping constant the thermal power installed capacity, i.e., 1440 MW, with same thermal units and the same scenarios of Fig. 1. The expected profits and the gains are shown in Table 2.

Table 2. Gain in function of wind capacity

Wind power (MW)	Profit disjoint (€)	Profit joint (€)	Gain (%)
1440	993646	1012520	1.90
2160	1232045	1257004	2.03
2880	1470444	1499547	1.98
3600	1708843	1741753	1.93

Table 2 shows that the gain is dependent in a nonlinear manner of the ratio between the wind power system and the thermal power system installed capacity. The maximum gain, 2.03%, is achieved when the wind power system installed capacity is about 1.5 times the thermal power system installed capacity.

Finally, consider that each thermal unit power capacities are scaled down by the ratio given by the quotient of the thermal units installed capacity given in the first column of Table 3 by the initial installed capacity of 1440 MW. An equivalent conversion are performed on the ramp up/down, start-up and shutdown costs. The expected profit as a function of the thermal power installed capacity, keeping constant the wind power installed capacity, i.e., 360 MW, are shown in Table 3.

Table 3. Gain variation in function of thermal capacity

Thermal units (MW)	Profit disjoint (€)	Profit joint (€)	Gain (%)
940	351366	357749	1.82
890	318697	326988	2.60
840	287759	296958	3.20
780	252500	257529	1.99

Table 3 allows to conclude that the gain is dependent of the ratio between the wind power and the thermal power installed capacities. The maximum gain of 3.20% is achieved when the thermal power system installed capacity is about 2.3 times the wind power system installed capacity. Hence, there is not a fixed ratio between the wind power and the thermal power installed capacities that can be recommended independently of the power system total installed capacity.

4 Conclusion

Stochastic programming is a suitable approach to address parameter uncertainty in modelling via scenarios. Particularly, the stochastic MILP approach is well-known by being accurate and having greater computationally acceptance, since the CPU time scales up linearly with number of price scenarios, units and hours on the time horizon.

The joint bid of thermal and wind power by a stochastic MILP approach proved to provide better expected profits than the disjoint bids. The expected profit is dependent in a nonlinear relation of the ratio between the wind power and the thermal power systems installed capacities.

The joint schedule is not a burden in computational resources in comparison with the disjoint one: the CPU time is about the same for both schedules, since the wind power system schedule CPU time is irrelevant when compared with the thermal power system one.

Acknowledgments. This work is funded by Portuguese Funds through the Foundation for Science and Technology-FCT under the project LAETA 2015-2020, reference UID/EMS/50022/2013; FCT Research Unit nº 151 C-MAST Center for Mechanical and Aerospace Sciences and Technology.

References

1. Laia, R., Pousinho, H.M.I., Melício, R., Mendes, V.M.F.: Self-scheduling and bidding strategies of thermal units with stochastic emission constraints. Energy Convers. Manage. **89**, 975–984 (2015)
2. Kongnam, C., Nuchprayoon, S.: Feed-in tariff scheme for promoting wind energy generation. In: IEEE Bucharest Power Technical Conference, Bucharest, Rumania, pp. 1–6 (2009)
3. Bitar, E.Y., Poolla, K.: Selling wind power in electricity markets: the status today, the opportunities tomorrow. In: American Control Conference, Montreal, Canada, pp. 3144–3147 (2012)
4. Barros, J., Leite, H.: Feed-in tariffs for wind energy in Portugal: current status and prospective future. In: 11th International Conference on Electrical Power Quality and Utilization, Lisbon, Portugal,pp. 1–5 (2011)
5. Al-Awami, A.T., El-Sharkawi, M.A.: Coordinated trading of wind and thermal energy. IEEE Trans. Sustain. Energy **2**(3), 277–287 (2011)
6. Cena, A.: The impact of wind energy on the electricity price and on the balancing power costs: the Spanish case. In: European Wind Energy Conference, Marceille, France, pp. 1–6 (2009)
7. El-Fouly, T.H.M., Zeineldin, H.H., El-Saadany, E.F., Salama, M.M.A.: Impact of wind generation control strategies, penetration level and installation location on electricity market prices. IET Renew. Power Gener. **2**, 162–169 (2008)
8. Bathurst, G.N., Weatherill, J., Strbac, G.: Trading wind generation in short term energy markets. IEEE Trans. Power Syst. **17**, 782–789 (2002)
9. Matevosyan, J., Solder, L.: Minimization of imbalance cost trading wind power on the short-term power market. IEEE Trans. Power Syst. **21**, 1396–1404 (2006)
10. Pinson, P., Chevallier, C., Kariniotakis, G.N.: Trading wind generation from short-term probabilistic forecasts of wind power. IEEE Trans. Power Syst. **22**, 1148–1156 (2007)
11. Ruiz, P.A., Philbrick, C.R., Sauer, P.W.: Wind power day-ahead uncertainty management through stochastic unit commitment policies. In: IEEE/PES Power System Conference and Exposition, Seattle, USA, pp. 1–9 (2009)
12. Laia, R., Pousinho, H.M.I., Melício, R., Mendes, V.M.F.: Optimal bidding strategies of wind-thermal power producers. In: Camarinha-Matos, L.M., J. Falcão, A., Vafaei, N., Najdi, S. (eds.) DoCEIS 2016. IFIP AICT, vol. 470, pp. 494–503. Springer, Heidelberg (2016). doi:10.1007/978-3-319-31165-4_46
13. Fan, S., Liao, J.R., Yokoyama, R., Chen, L.N., Lee, W.J.: Trading wind generation from short-term probabilistic forecasts of wind power. IEEE Trans. Power Syst. **24**, 474–482 (2009)

14. Kusiak, A., Zheng, H., Song, Z.: Wind farm power prediction: a data-mining approach. Wind Energy **12**, 275–293 (2009)
15. Laia, R., Pousinho, H.M.I., Melício, R., Mendes, V.M.F., Reis, A.H.: Schedule of thermal units with emissions in a spot electricity market. In: Tomic, S., Graça, P., Camarinha-Matos, L.M. (eds.) DoCEIS 2013. IFIP AICT, vol. 394, pp. 361–370. Springer, Heidelberg (2013)
16. Catalão, J.P.S., Mariano, S.J.P.S., Mendes, V.M.F., Ferreira, L.A.F.M.: Short-term electricity prices forecasting in a competitive market: a neural network approach. Electr. Power Syst. Res. **77**, 1297–1304 (2007)
17. Coelho, L.D., Santos, A.A.P.: A RBF neural network model with GARCH errors: application to electricity price forecasting. Electr. Power Syst. Res. **81**, 74–83 (2011)
18. Amjady, N., Daraeepour, A.: Mixed price and load forecasting of electricity markets by a new iterative prediction method. Electr. Power Syst. Res. **79**, 1329–1336 (2009)
19. Morales, J.M., Conejo, A.J., Ruiz, J.P.: Short-term trading for a wind power producer. IEEE Trans. Power Syst. **25**(1), 554–564 (2010)
20. Laia, R., Pousinho, H.M.I., Melício, R., Mendes, V.M.F., Collares-Pereira, M.: Spinning reserve and emission unit commitment through stochastic optimization. In: IEEE SPEEDAM, Ischia, Italy, pp. 444–448 (2014)
21. http://www.esios.ree.es/web-publica/

ANN Based Model of PV Modules

Jose Manuel Lopez-Guede[1,5(✉)], Jose Antonio Ramos-Hernanz[2], Manuel Graña[3,5], and Valeriu Ionescu[4]

[1] Faculty of Engineering of Vitoria, Department of Systems Engineering and Automatic Control, Basque Country University (UPV/EHU), Nieves Cano 12, 01006 Vitoria, Spain
jm.lopez@ehu.es

[2] Faculty of Engineering of Vitoria, Department of Electrical Engineering, Basque Country University (UPV/EHU), Nieves Cano 12, 01006 Vitoria, Spain

[3] Faculty of Informatics, Department of Computer Science and Artificial Intelligence, Basque Country University (UPV/EHU), Paseo Manuel de Lardizabal 1, 20018 San Sebastian, Spain

[4] Faculty of Electronics, Communications and Computers, Department of Electronics, Computers and Electrical Engineering, University of Pitesti, Targu din Vale 1, 110040 Pitesti, Romania

[5] Computational Intelligence Group, Basque Country University (UPV/EHU), Vitoria, Spain

Abstract. In this paper authors address the practical problem of designing an empirical model for a commercial photovoltaic (PV) module (Mitsubishi PV-TD1185MF5) placed at the Faculty of Engineering of Vitoria (Basque Country University, Spain) based on artificial neural networks (ANN). This model obtains Ipv from Vpv, and the paper explains how the empirical data have been gathered and discusses the obtained results. The model reached an average accuracy of 0,15 A and a medium correlation value of R = 0,995.

1 Introduction

The paper addresses the building of two models of a real Mitsubishi PV-TD1185MF5 photovoltaic module placed at the Faculty of engineering of Vitoria (Spain) based on artificial neural networks (ANN). The derivation of a model of a real photovoltaic panel is very useful for making simulations based on it to tune some control algorithms, avoiding differences between ideal models and the real devices. The construction of different types of models has been addressed in the literature. In [4] a model based on mathematical equations that define the photovoltaic cell is specified, while in [3,5] the same type of model is tested, and several types of models have been compiled. The paradigm of modeling through ANN has been used previously used in the literature [1,2]. Each one of the models obtained in this paper is oriented to different seasons, i.e., winter and summer.

The remainder of the paper is organized as follows. Section 2 gives a brief background on key topics of the paper. A detailed description of all parts involved

M. Graña et al. (eds.), *International Joint Conference SOCO'16-CISIS'16-ICEUTE'16*, Advances in Intelligent Systems and Computing 527, DOI 10.1007/978-3-319-47364-2_15

in the experimental design is given in Sect. 3, while Sect. 4 discusses the obtained results. Finally, Sect. 5 presents the main conclusions of the paper and future work is addressed.

2 Background

2.1 Characteristic Curves

A theoretical model of photovoltaic modules are the I-V curves provided by manufacturers. These curves give the manufacturer's specification of the relation between the current (I_{PV}) and the voltage (V_{PV}) supplied by a particular photovoltaic module. In fact, this is the main instrument used by the manufacturers to explain to the customers the capabilities of a given photovoltaic module. In Fig. 1 the I-V curves of a commercial PV module are shown, at a specific temperature for a few irradiance values. Temperature and irradiance are relevant magnitudes in the relation between I_{PV} and V_{PV}, however manufacturers usually do not take them into account when drawing the curves, so there is a lack of relevant information in order to design efficient control of the photovoltaic regime.

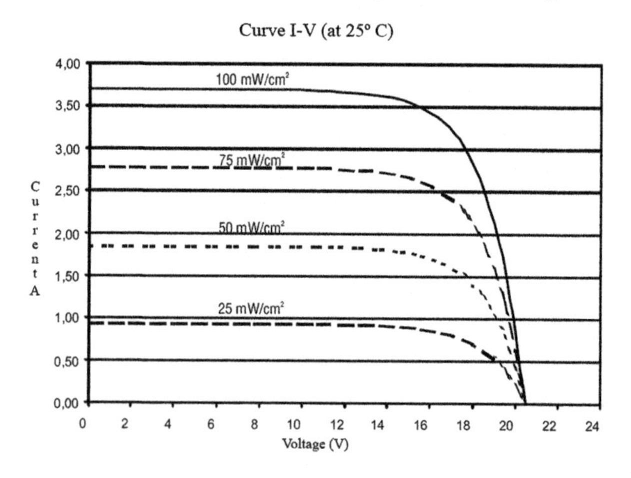

Fig. 1. I-V curve of a commercial PV module

2.2 Artificial Neural Networks

The use of ANNs is motivated by its ability to model systems [6]. These bio-inspired computational devices have several advantages, and among others, these are the most outstanding to our problem:

- Learning capabilities: If they are properly trained, they can learn complex mathematical models. There are several well known training algorithms and good and tested implementations of them. The main challenge concerning this issue is to choose appropriate inputs and outputs to the black box model and the internal structure.

- Generalization capabilities: Again, if they are properly trained and the training examples cover a variety of different situations, the response of a neural network in unseen situations (i.e., with unseen inputs) will probably be acceptable and quite similar to the correct response. So it is said that they have the *generalization property.*
- Real time capabilities: Once they are trained, and due to their parallel internal structure, their response is always very fast. Their internal structure could be more or less complex, but in any case, all the internal operations that must be done are several multiplications and additions if it is a linear neural network. This fast response is independent of the complexity of the learned models.

3 Experimental Design

3.1 Solar Module Characteristics

In this subsection we introduce the characteristics of the photovoltaic module of which we are going to obtain an empirical ANN model. The PV-TD1185MF5 PV module has 50 series connected polycrystalline cells. Table 1 shows their main characteristics. The performance of solar cell is normally evaluated under the standard test condition (STC), where an average solar spectrum at AM 1.5 is used, the irradiance is normalized to 1,000 W/m^2, and the cell temperature is defined as 25 °C.

Table 1. Physical properties of the photovoltaic module

Attribute	Value
Model	PV-TD185MF5
Cell type	Polycrystalline Silicon 156 mm × 156 mm
Maximum Power [W]	185
Open Circuit Voltage Voc [V]	30,60
Short circuit Current Isc [A]	8,13
Voltage, max power Vmpp [V]	24,40
Current, max power Impp [A]	7,58
Normal operating cell temperature (NOCT)	47,5 °C

3.2 Data Logging

In this subsection we describe the collecting task of the physical signals of the photovoltaic module to elaborate the datasets to elaborate the train and test the neural models.

On one hand, Fig. 2(a) shows the conceptual disposition of the measuring devices: the voltmeter is placed in parallel with the module and the amperemeter in series. Besides, there is a variable resistance to act as a variable load and obtain different pairs of voltage and current with the same irradiance and temperature. The variable resistance value is controlled according to our convenience, but the temperature and the irradiance depends on the climatological conditions. On the other hand, Fig. 2(b) shows the real devices that have been used to capture the data. The first device is the data logger Sineax CAM, and it was configured to generate records with the irradiance and temperature of the environment and the voltage and current supplied by the photovoltaic module under those environmental conditions. The second element is the multimeter TV809, which helps to isolate the data logging device from the photovoltaic module and converts voltage and current magnitudes to a predefined range. The third element is the irradiance sensor Si-420TC-T-K. It is placed outside close to the photovoltaic module and it is used to provide the irradiance and temperature conditions under which the module is working outside. Finally, the fourth element of the figure are current clamps Chauvin Arnous PAC12 used to measure direct currents provided by the module.

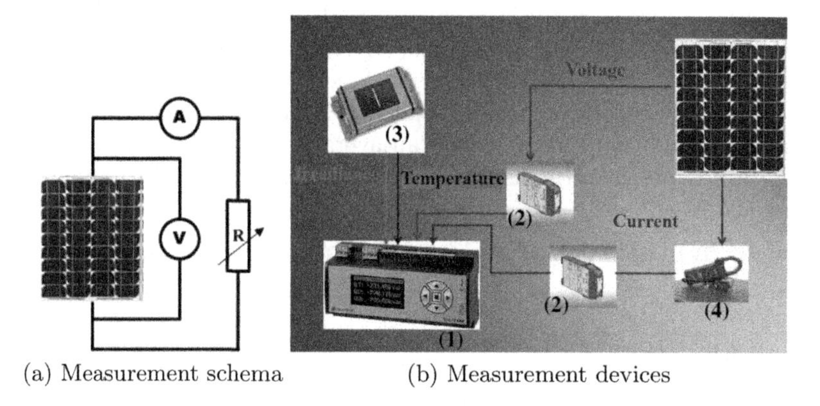

(a) Measurement schema (b) Measurement devices

Fig. 2. Real photovoltaic module data measurement

3.3 ANN Training

In this subsection we provide a detailed specification of the process followed for the training of the accurate approximation of the I-V curve by an ANN. The first step in the training process is to fix the structure of the neural network to train. We are going to use feed forward ANN, and as we stated before, we desire two neural models (one suited for summer and another for winter) consisting on one input (V_{PV}) and one output (I_{PV}).

To build both models as simple as possible, we have assumed that the temperature and the irradiance are quite constant values, and we have discarded them. The available data have been acquired with a temperature and irradiance described in Table 2 through their mean and standard deviation values. Winter suited data were collected in January 2014 while summer data in July 2014, in University College of Engineering of Vitoria-Gasteiz (Spain). The networks have one output neuron, because there is only one target value associated with each

Table 2. Mean and Standard deviation of the Temperature and Irradiance of the gathered data

		Winter	Summer
Temperature °C	Mean	11,86	54,93
	Standard deviation	0,45	0,28
Irradiance W/m^2	Mean	112,58	908,79
	Standard deviation	3,21	3,22

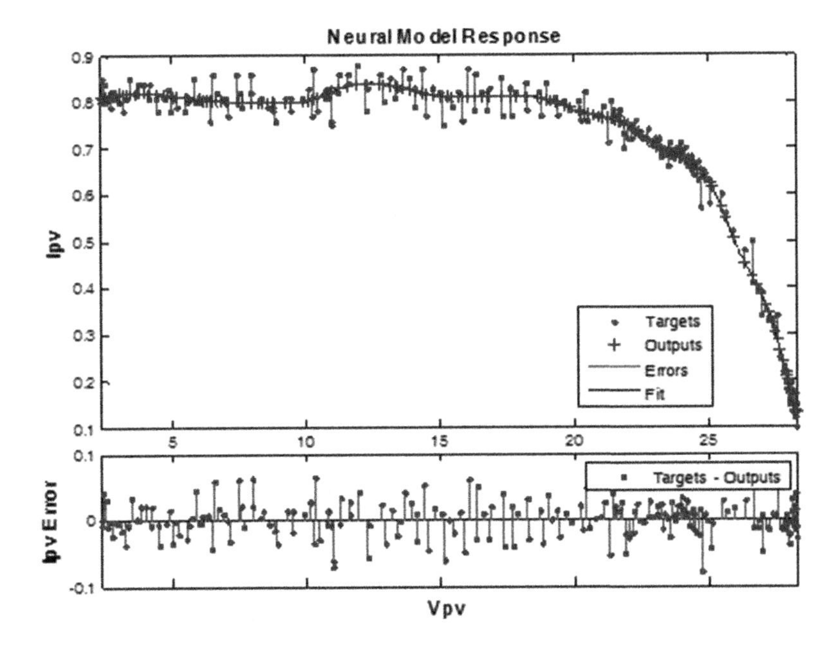

Fig. 3. Winter oriented neural model accuracy

input value. So, both the input and the output layers have only one neuron. Regarding to the hidden layers, we have considered to have only one layer for simplicity reasons, and this unique hidden layers will have 15 neurons in the model suited for summer, while only 10 neurons in the suited for winter, i.e., both are quite small networks. The networks will have the tan-sigmoid activation function in the hidden layer and linear activation function in the output layer. Once we have fixed the structure of the network, we have to choose the training algorithm. We have chosen the Levenberg-Marquardt algorithm due to speed reasons, despite of being very memory consuming because these are small ANNs. Five independent training/test processes have been performed for each network, to assess its generalization. Finally, we have to determine which data and how they are used to the training process. All the input vectors to the networks appear at once in a batch. We have used the raw data, i.e., without

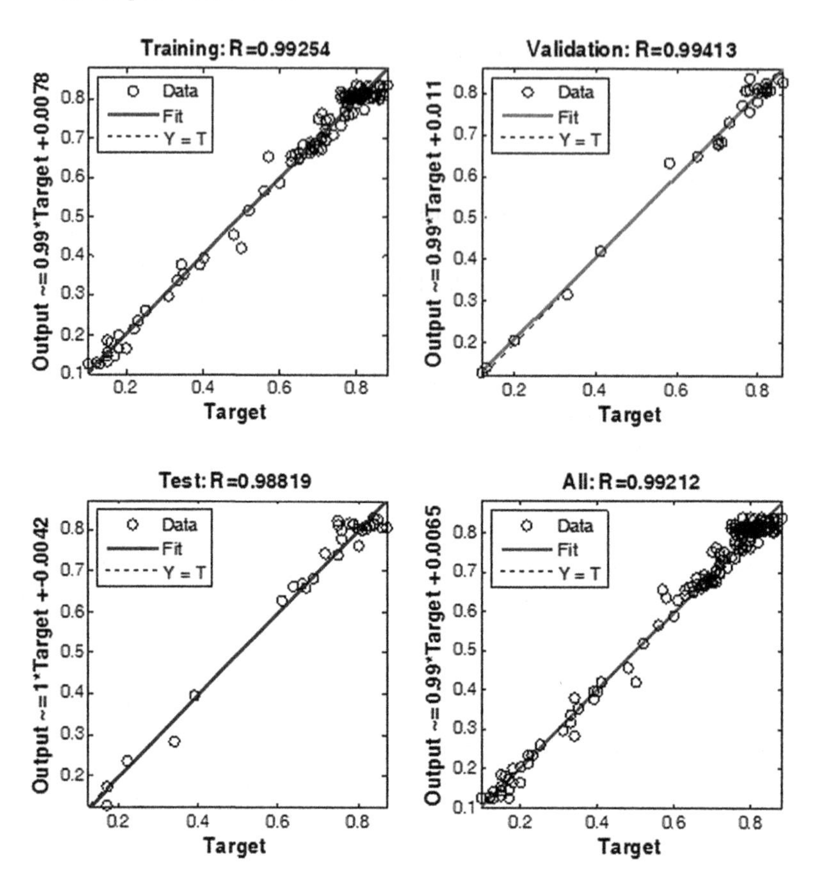

Fig. 4. Correlation between the winter oriented model response and the targets

normalization and in some cases, with more than one value of the output (I_{PV}) to one value of the input (V_{PV}). The input and target vectors of each dataset have been divided into three sets using interleaved indices as follows: 60 % are used for training, 20 % are used to validate that the network is generalizing and to stop training before overfitting, and finally, the last 20 % are used as a completely independent test of network generalization.

4 Experimental Results

4.1 Winter Oriented Neural Model

The training process stopped at iteration 12. As we can see in Fig. 3, the network response is reasonable because the outputs track the targets. Figure 3 shows several small errors, always less than 0,1 A, which can be explained by the following reasons:

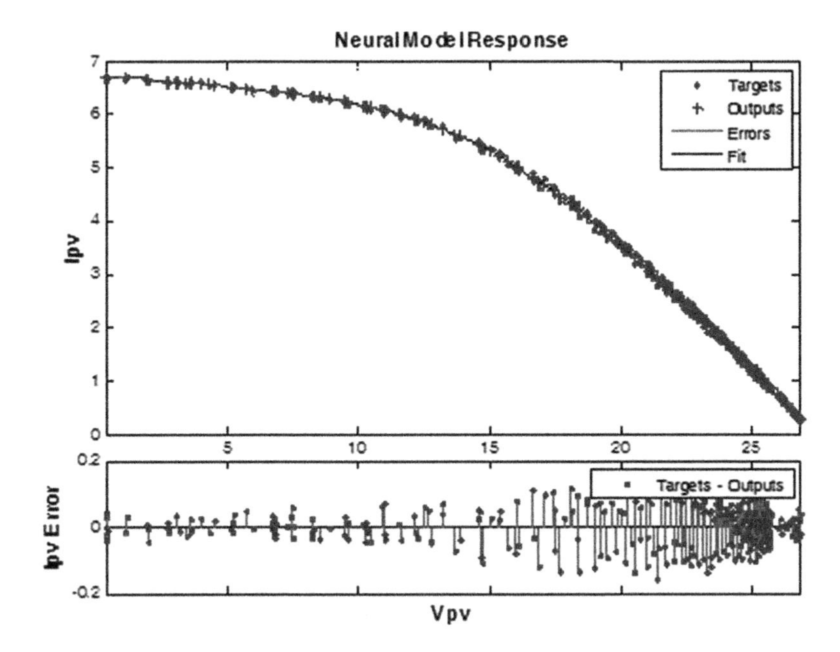

Fig. 5. Summer oriented neural model accuracy

- The dataset used to train the network is composed of raw and un-processed data.
- In several cases, for a unique value of the input (V_{PV}), we can see in Fig. 3 that the targets have several values for the output (I_{PV}). This means that we have tried to learn a multi-value function with a quite simple structure of neural network.
- We also have to take into account that the irradiance and the temperature magnitudes have been discarded to build the models as simple as possible, and it is well known that they exert influence over the performance of the photovoltaic modules.
- The distribution of the sample data of the dataset is not balanced. It is easy to see that there are few sampling data for input values V_{PV} around 26 V, while the ideal situation is that the samples were uniformly distributed over the input range.

On the other hand, Fig. 4 shows that the correlation coefficient (R-value) is over 0,99 for the total response, and taking into account that R = 1 means perfect correlation between the network response and the target, we can conclude that the model is quite accurate.

4.2 Summer Oriented Neural Model

In this case all the analysis carried out for the winter suited model is valid, only some small differences must the taken into account:

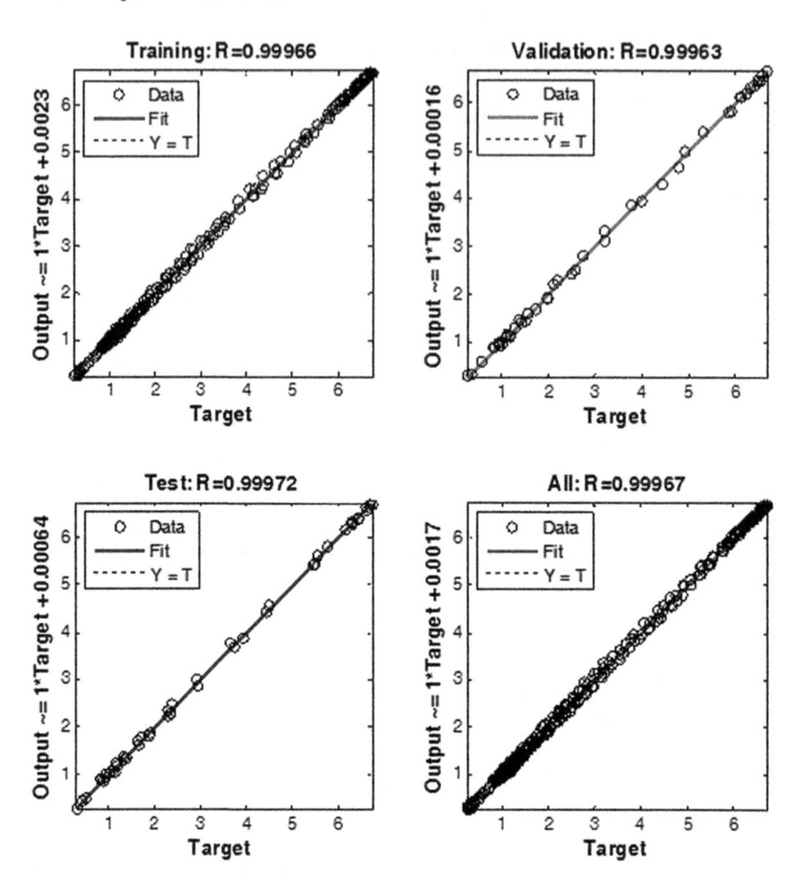

Fig. 6. Correlation between the summer oriented model response and the targets

- The training process stopped at iteration 15.
- As we can see in Fig. 5, the network response is good because the outputs track the targets with errors less than 0,2 A, which can be explained by the same reasons exposed when analyzing the previous model.
- The distribution of the sample data of the dataset is not balanced. It is easy to see that there are few sampling data for input values V_{PV} around 6 V, while there are a lot of them around 26 V. The ideal situation is that the samples were uniformly distributed over the input range.

Finally, Fig. 6 shows that the correlation coefficient (R-value) is over 0,999 for the total response, and taking into account that R = 1 means perfect correlation between the network response and the target, we can conclude that the model is very accurate.

5 Conclusions

The main objective of the paper was to build two models of a Mitsubishi PV-TD1185MF5 photovoltaic module installed at the University College of Engineering of Vitoria. To achieve that objective, first we have gathered some data through different devices during two seasons. Then we have trained two different ANNs which have obtained a quite good accuracy taking into account that the training data are not normalized and the irradiance and the temperature magnitudes are discarded.

Acknowledgments. The research was supported by the Computational Intelligence Group (Basque Country University, UPV/EHU), which is funded by the Basque Government with grant IT874-13.

References

1. Mellit, A., Benghanem, M., Kalogirou, S.: Modeling and simulation of a stand-alone photovoltaic system using an adaptive artificial neural network: proposition for a new sizing procedure. Renew. Energ. **32**(2), 285–313 (2007)
2. Mellit, A., Pavan, A., Kalogirou, S.: Application of artificial neural networks for the prediction of a 20-kwp grid-connected photovoltaic plant power output. In: Gopalakrishnan, K., Khaitan, S.K., Kalogirou, S. (eds.) Soft Computing in Green and Renewable Energy Systems. Studies in Fuzziness and Soft Computing, pp. 261–283. Springer, Heidelberg (2011)
3. Ramos, J., Zamora, I., Campayo, J., Larrañaga, J., Zulueta, E., Barambones, O.: Comparative analysis of different models for pv cell simulation. In: Proceedings of the 12th Portuguese-Spanish Conference on Electrical Engineering (2011)
4. Ramos, J., Zamora, I., Campayo, J.: Modelling of photovoltaic module. In: International Conference on Renewable Energies and Power Quality (ICREPQ10) (2010)
5. Ramos-Hernanz, J., Campayo, J., Larranaga, J., Zulueta, E., Barambones, O., Motrico, J., Gamiz, U.F., Zamora, I.: Two photovoltaic cell simulation models in matlab/simulink. Int. J. Tech. Phys. Probl. Eng. (IJTPE) **4**(1), 45–51 (2012)
6. Widrow, B., Lehr, M.: 30 years of adaptive neural networks: perceptron, madaline, and backpropagation. Proc. IEEE **78**(9), 1415–1442 (1990)

SCADA Network System for the Monitoring and Control of an Electrical Installation Supplied by a Hydro-Generator

Florentina-Magda Enescu, Cosmin Ştirbu, and Valeriu Ionescu(✉)

University of Pitesti, FECC, Pitesti, Arges, Romania
enescu_flor@yahoo.com, cosmin.stirbu@upit.ro,
manuelcore@yahoo.com

Abstract. The energy industry is one of the domains that need control and monitoring at several levels. This article presents a SCADA monitoring system targeting a large scale process which is in need of immediate and frequent interventions: the monitoring and control of an electrical installation supplied by a hydro-generator. Citect SCADA was used to design a flexible solution and can be used for small or large hydro-generators. The SCADA solution created will be used for the functioning optimization, transmission and supervising of functioning programs execution.

Keywords: SCADA · Human machine interface (HMI) · Hydro-generator · Reserve circuit (AR) · Citect SCADA · Mysql

1 Introduction

Today's energy industry is characterized by large structural conversions. Electricity generating companies aim to improve the efficiency and quality of service. Computerization is one of the premises for the basis of the increase in the efficiency and safety in the exploitation of the energy system.

Supervisory Control and Data Acquisition (SCADA) is computer software which monitors and controls a process that is used for industrial processes. The SCADA system has a multi-layered structure composed from basic functions and graphical user interfaces which are hardware and software supervised in real-time [1].

The functions of the SCADA system are the following [2–4]: it acquires data collected from the process; manages alarms; allows the needed actions for automation; stores and archives data; generates reports; allows the dispatcher to control the process via the HMI; allows the communication with user interface via the HMI using libraries with symbols, a connection between process and graphic elements, a collection of command's operators and multimedia features.

SCADA systems may be extended to a Large Scale System by architecture, maintenance, post-processing, decision support systems, and economic planning [5].

The paper presents the creation of an HMI interface for tracking and command of a hydro-generator. Chapter 2 presents the creation of interface accompanied by Citect SCADA connection between SCADA and database of MySQL. Hydro generator

M. Graña et al. (eds.), *International Joint Conference SOCO'16-CISIS'16-ICEUTE'16*,
Advances in Intelligent Systems and Computing 527, DOI 10.1007/978-3-319-47364-2_16

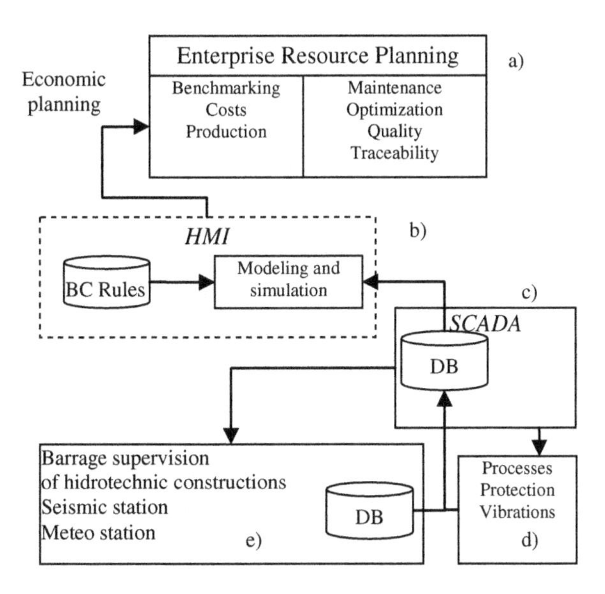

Fig. 1. Extended SCADA architecture with application in hydro energy has the components of (a) Enterprise planning, (b) HMI, (c) SCADA, (d) process and (e) access control

parameter values can be used both in this database and the values read from virtual devices presented interface. The following hydro-energy monitoring architecture is proposed, as shown in Fig. 1.

The components are the following: the physical components of process and access control; the interfacing made in SCADA with data captured from the process and stored in databases; the modeling a process simulation based on rules; the Human Machine Interface – HMI; managerial and operational decisions based on the gathered and interpreted data. Without the SCADA and the HMI interface, the economic planning would be based directly on data gathered from the lower (physical) components and would be slower to gather and analyze.

By using the Citect SCADA software we designed a wiring diagram for the services of a group hydro-generator (HG), capable of producing a power of 1100 kW. In the electrical diagram of the block generator transformer (TA) there are lines of 6 kV, which have a basic power supply and a reserve one and in between the two supplies there is a relation of Reserve Actuation (AR). The Reserve Actuation (AR) is necessary in the case of all groups used for the production of electricity, because when it is in operation the units block their dependents must be supplied with voltage at all times. The diagram includes a large number of motors (M1-M4 presented, but their number can be larger) that drive the fueling pumps, the circulation pumps, the air ventilation and other devices that are necessary in the technological process. In the case of hydro generators, their power consumption can actually reach 12 % of their nominal power. The HMI interface implemented is presented in Fig. 2. The data presented is real data (nominal power, current, power factor) and relates to the energy production of a hydro generator in Ramnicu Valcea, Valcea County, Romania.

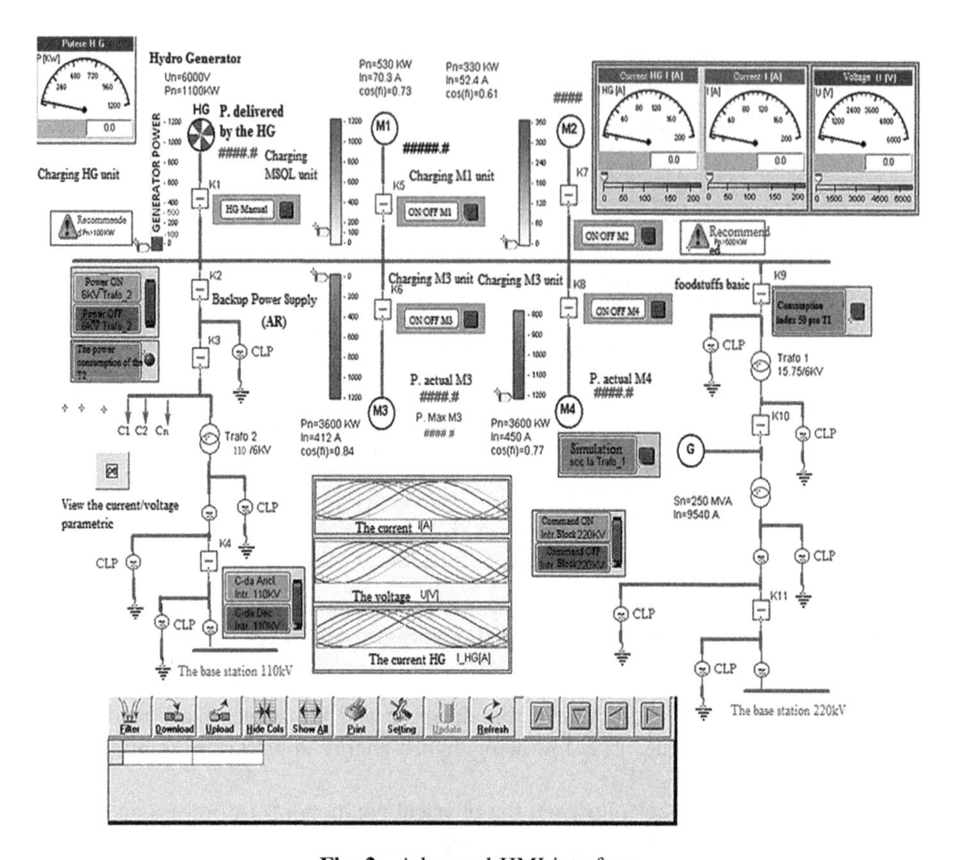

Fig. 2. Advanced HMI interface

2 Description of the Operating Mode of HMI Interface for Electrical Operation Wiring Diagram

A Citect SCADA project includes the following items: graphics, databases, Cicode programs. Graphics represents a graphic page which allows the monitor to display the graphical interface with control buttons. Databases allow the storage of process information collected for monitoring and controlling the system. These may be linked to the graphics page if desired. Cicode programs allow functionality and contain a number of useful functions stored. In order to design the application the following steps were made: identifying the elements used for data acquisition and those necessary for command and control; creating a project; defining the tags; creating the graphical data flow that will mimic the physical process flow that is connected to the data acquisition and control process; writing functions for the graphical elements; establishing the application users and their rights and finally testing and running the application.

The schematic of a Citect SCADA project looks like in the Fig. 3.

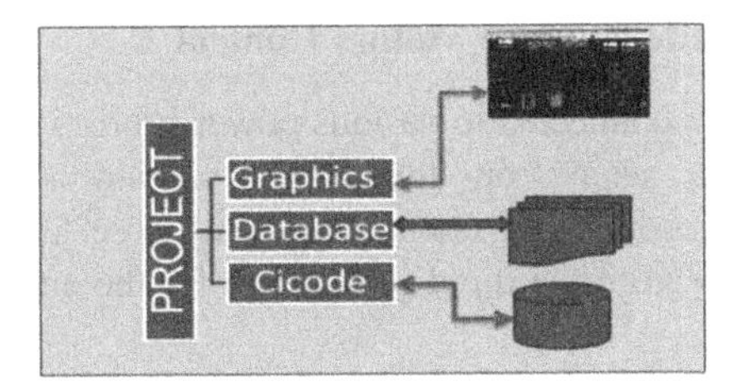

Fig. 3. Scheme of the *Citect SCADA* project

2.1 HMI Operation

For the HMI interface designed it is considered that the rated power of the block is of maximum of 120 MW. At the start of such a group, all units must be operated at nominal parameters and have the role of electricity transmission by the 4 engines of the installation.

The central element is the Hydro-Generator (HG) which was set up to provide electrical power only if reaches about 10 % of maximum power, i.e. 100 kW. If this power is not achieved (for example in the case when there is not enough water flow) a button, called "The power consumption of the T2", should be used to engage the AR. This will switch K1 in the closed position and K2 in the open position if Pn HG > 100 kW and if the power is below the threshold Pn HG < 100 kW then it will set K1 in the open position and K2 in the closed position. The powering of both the AR and the hydro generator is not allowed at the same time. (Figure 4) If the hydro generator needs to be disconnected there is a manual button "HG Manual" that allows this. The AR power is given via an 110 kW line.

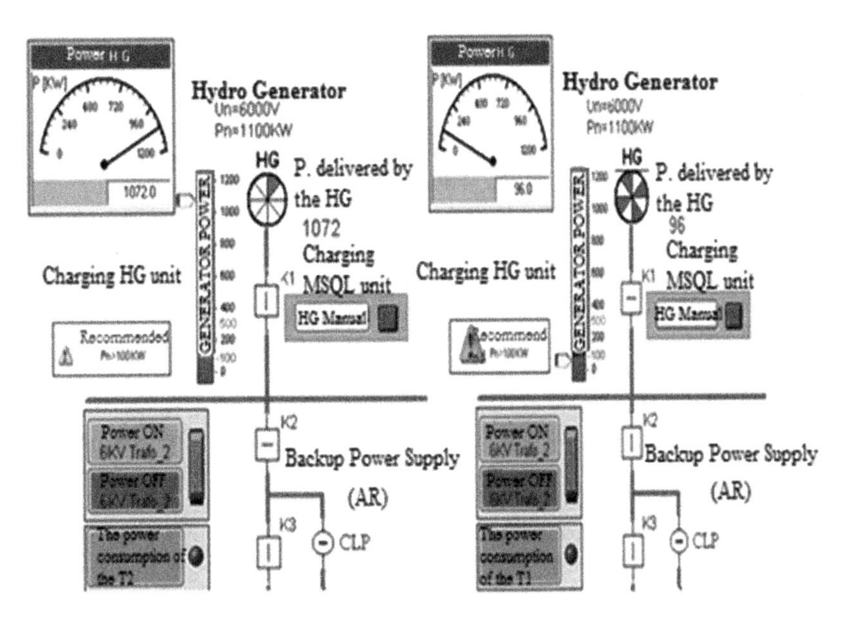

Fig. 4. Command HG/AR

2.2 Interface for Monitoring and Motors Control

The engines (M1-M4) are connected to various power sources according to their roles. The value of the power taken from the grid is set individually for each motor. As SCADA Citect applications can use databases to store the data, in this case the MySQL DB was used to store the input parameters for the engine (Fig. 5).

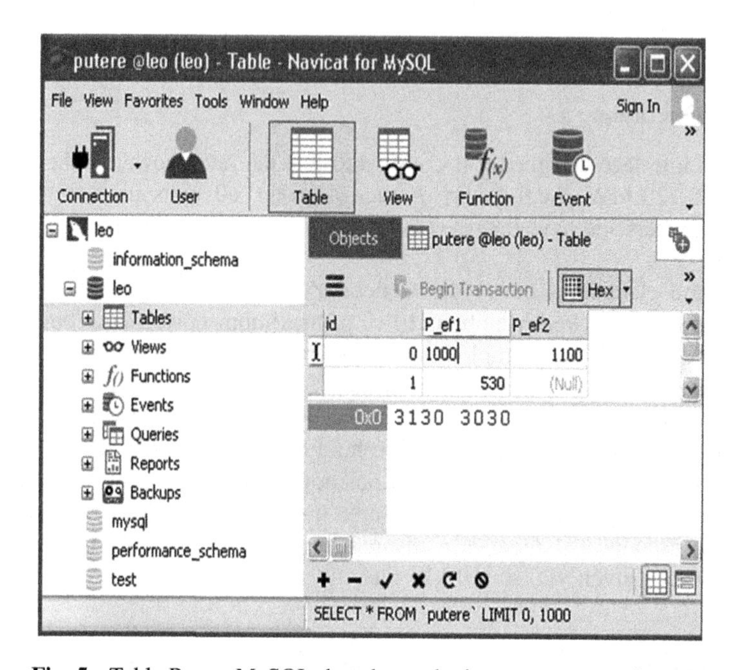

Fig. 5. Table Power MySQL that shows the input parameters for PM1

Citect SCADA does not natively use MySQL. The connection to a MySQL data base was made by using the Open Database Connectivity (ODBC) technology. ODBC provides a standard for the methods and procedures Application Programming Interface (API) software. The using of the ODBC ensures independence from the programming language, which is the basis for the data or the operating system. Most of the producers of databases offer drivers for ODBC connections.

The engine M1 can be powered from both the redundant network of the AR as well as from the supply line of hydro-generator. The amount of power absorbed (Fig. 6) is retrieved from the table of power, and can be set with different values in that table (Fig. 5). Power engine M1 does not depend on the power hydro-generator, being established at 530 kW [1, 6]. Switching off the engine M1 can be done manually with the button "n/Off M1" (Fig. 6).

The M2 can be powered only from the supply line of hydro-generator. The absorbed power is a function of the actual power value hydro-generator, maximum value (1200/3.33 = 360 (kW)) being established at (Fig. 7):

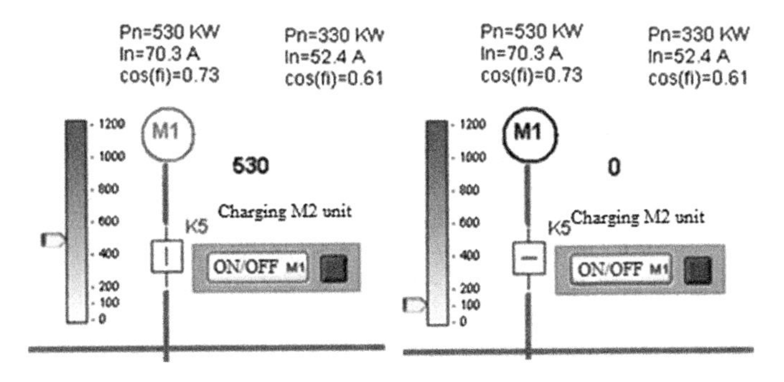

Fig. 6. The value PM1 /Disconnection M1

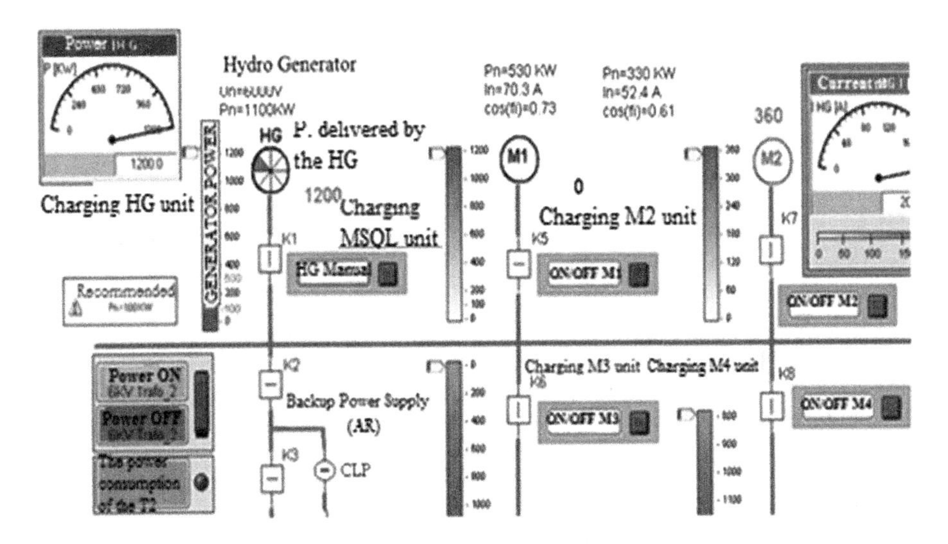

Fig. 7. Power M2 f(PnHG)

$$PM2 = \text{Pn } /3.33 = \text{P\_ef1}/3.33(\text{kW}) \quad (1)$$

As seen in Fig. 8, when switching on AR spare power, even if k7 is in the closed position, PM2 power value is zero.

The M3 (Fig. 9) can be powered only from the supply line of AR. So this is only possible if K2, K3, and K4 are closed. Control motor M3 is given by the on/off M3 that closes or opens the switch K6. The absorbed power is a function of the actual power value hydro-generator being established as the difference between maximum power of hydro-generator and its power to:

$$PM2 = Pmax - \text{P\_ef1}(\text{kW}) \quad (2)$$

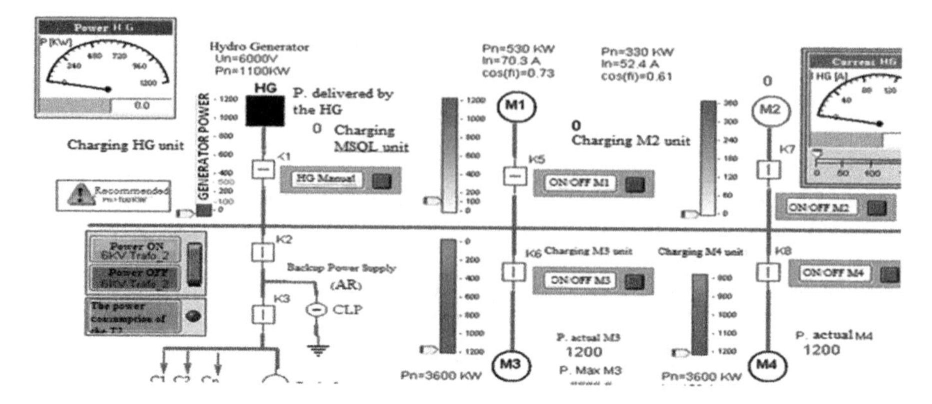

Fig. 8. M2 = 0 for AR

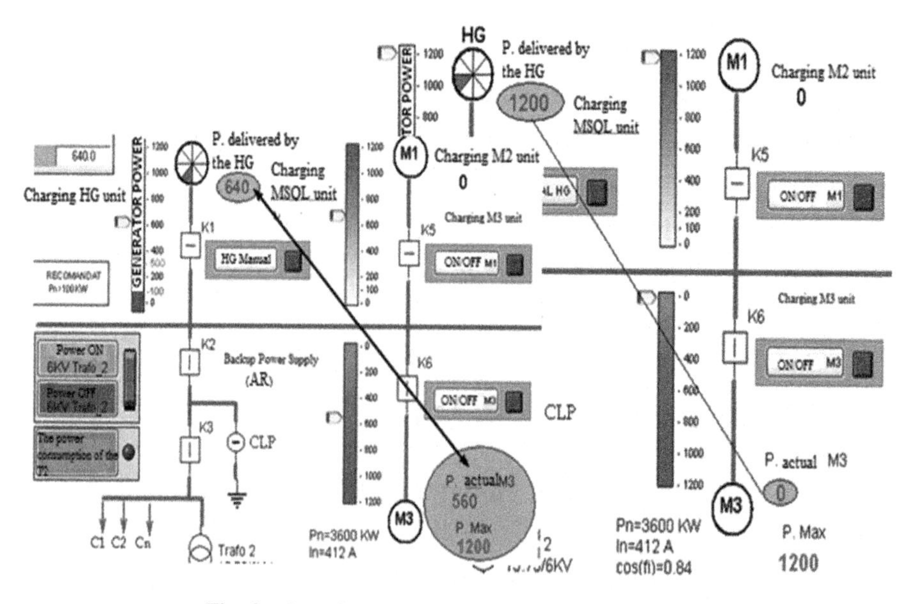

Fig. 9. Actual power M3 /PM3 at maximum PnHG

The power delivered by hydro-generator is at a maximum like is shown in the diagram below (Fig. 9 PM3 at maximum PnHG) (Fig. 11).

The M4 (Fig. 10), similar to M3 can be powered only from the supply line of AR. So this is only possible if K2, K3, and K4 are closed. The control of the engine M4 is given by the on/off M4 that closes or opens the switch K7. The absorbed power is established as the difference between maximum power of hydro-generator and a part of its effective power to:

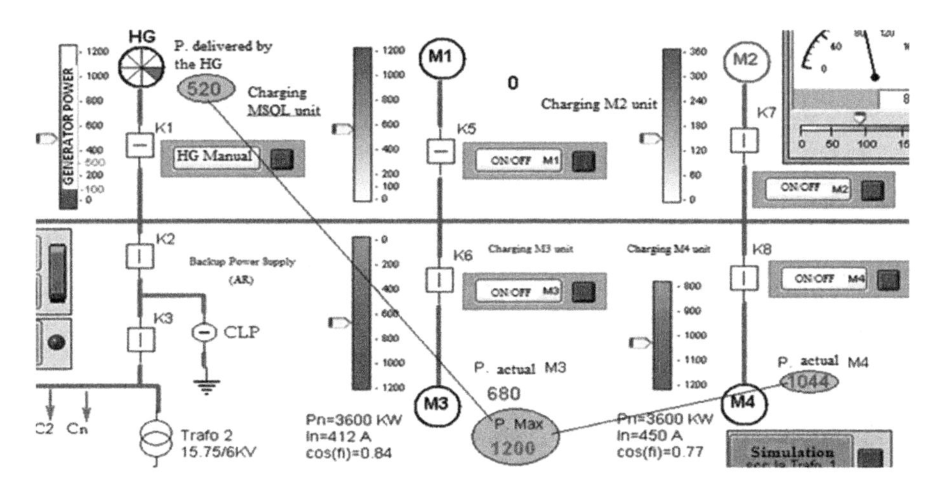

Fig. 10. Actual power M4

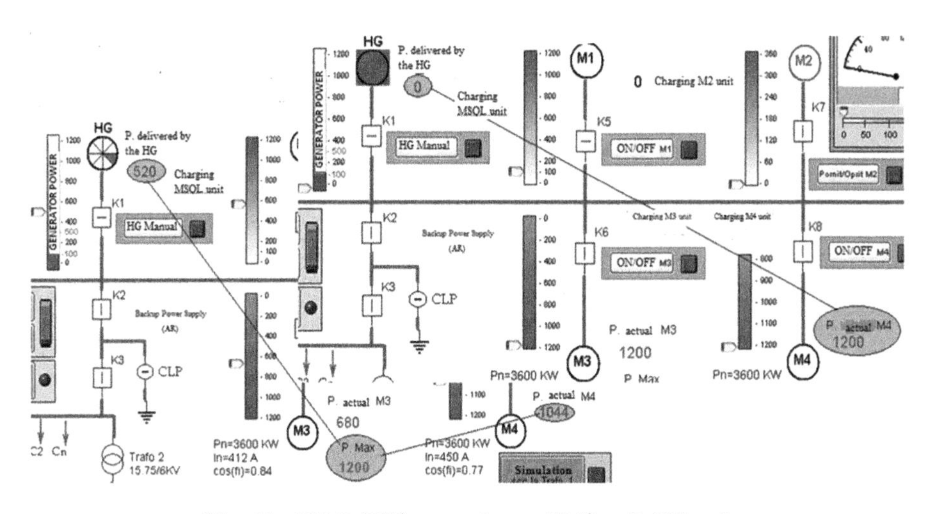

Fig. 11. PM PnHG$ to maximum /PM$ to PnHG = 0

$$PM2 = \text{Pmax} - \text{P\_ef1}/3.33(\text{kW}) \quad (3)$$

There were situations in which the conditions implemented by a simple IF are not sufficient; therefore it was necessary to implement our own functions. An example is presented below, necessary for the M4 engine operation. To add a new function the step presented in Fig. 12 were necessary.

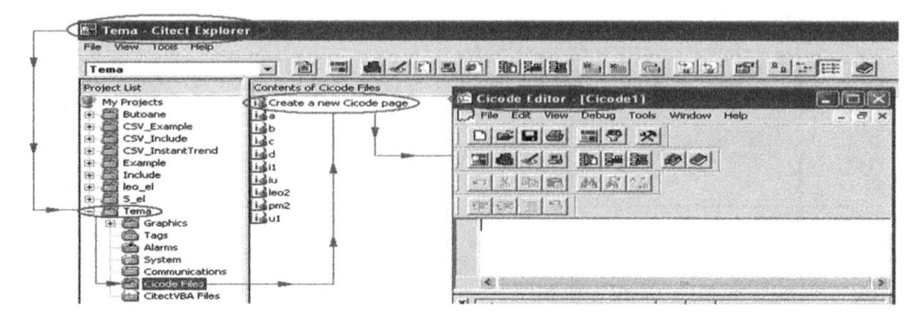

Fig. 12. Function implementation

The function implementation is as follows:

```
INT FUNCTION p_4()
IF I_T1=0 AND S_T1=1 AND C_daA=1 AND M_4=1 THEN
p4=1200-(P_ef1/3.33);
ELSE
p4=0;
END
RETURN p4;
END
REAL FUNCTION simul_i_1()
rad=rad+0.01;
i_alim=(55*(1+Sin(rad)))+90;
RETURN i_alim
END
```

3 Testing and Results

The testing scenarious will focus on the situation where the hydro-generator operates at Pn < 500 (kW) and supplies motors M1 and M2 are started when K1 is closed and k2 is open, as it is displayed in Fig. 13.

1. Supply motors M1 and M2 retrieved values from the table "power" that was made in MySQL. The acquired and stored values for these motors are 530 kW for M1 and 149 kW for M2. The power value for M2 motor will be computed using Eq. (4).

$$p2 = \text{P\_ef1}/3.33 \quad (4)$$

2. If Pn > 100 kW it is to be noted that the hydro-generator is disabled, but it remains on alert for tube current, for which Pn > 500 kW;

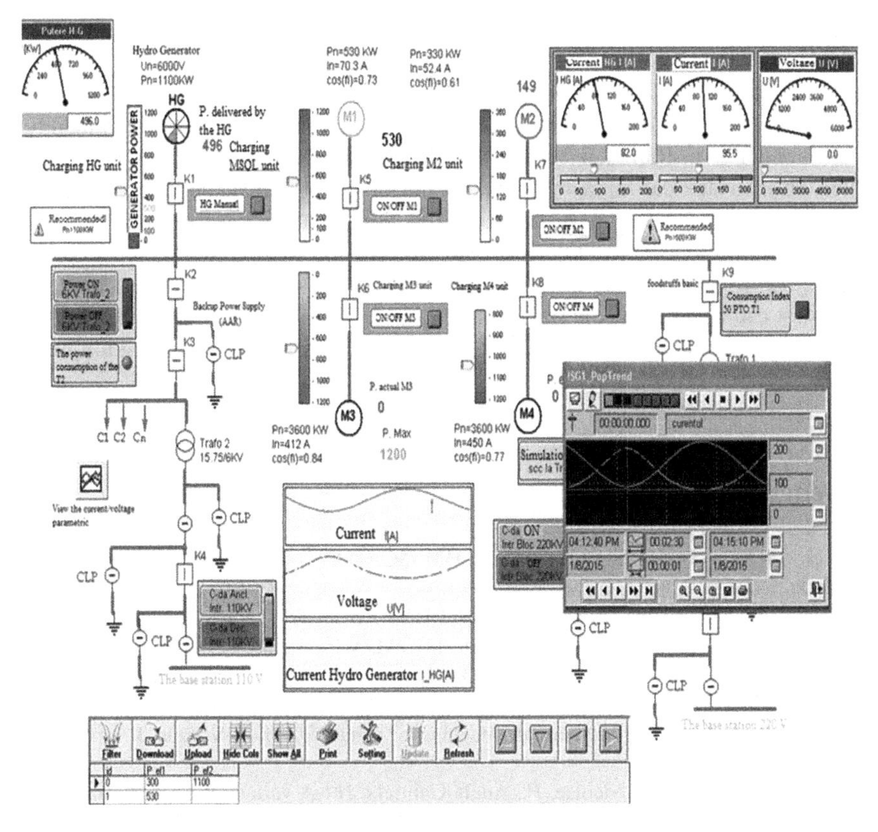

Fig. 13. Results HMI interface

3. The form and the value of the current i_alim debited from the reserve circuit AR, has the form sine, the amount of which shall be between 90 A and 200 A, which complies with the requirement design. In Eq. (5) it is presented the computed formula used for i_alim.

$$\text{i\_alim} = (55 * (1 + \sin(\text{rad}))) + 90 \tag{5}$$

4. The current i_alim_HG debited by hydro-generator HG has a linear form (so it is a direct current) and its amount it is determined by the ratio of power hydro-generator and voltage;
5. The supply voltage has a sinusoidal form (in the mirror with that of the intensity i_alim) and its value it is situated between 600 V for Pn < 500 kW and 6 kV for 1200 < Pn > 500.

The particulars entered in each of the motor wiring diagram, represents the characteristics of the various motors which are to be found in the context of such installations (nominal power, voltage, nominal current and power factor).

4 Conclusions

This paper presents the HMI implementation of the tracking and command functions of a hydro-generator. The data presented is real data from the energy production facility of a hydro generator in Ramnicu Valcea, Valcea County, Romania. The implementation of such an interface can limit the damage and help the human operator in the management of the process by implementing functions that will warn the user and act correctly in various system conditions. Using the HMI interface a full automation of the technological process can be achieved. The graphical user interface has a high flexibility and it can be implemented at a central heating located near a course of water. In the near future the user interface will be extended to the diagrams of operation with an increased number of operations.

References

1. Enescu, F.M.: Start designing with HMI/SCADA – Applications (2015). ISBN 978-606-560-425-4, Ed. Univ. Pitesti
2. Boyer, S.: SCADA: Supervisory Control and Data Acquisition, 3rd edn. ISA (2004)
3. Clarke, G., Reyders, D., Wright, E.: Practical SCADA Protocols. Elsevier (2004)
4. Figueiredo, J., Sá da Costa, J.: A SCADA system for energy management in intelligent buildings. Energy Buildings **49**, 85–98 (2012)
5. Choi, D.: Hybrid key management architecture for robust SCADA systems. J. Inf. Sci. Eng. **27**, 197–211 (2011)
6. Stan, M.-F., Cosmin, C., Nicolae, F., Adela-Gabriela, H.: A variant of a synchronous motor with two stators and high-energy permanent magnets disposed on the both rotor peripheries. In: ECAI 2015, Romania (2015). ISSN: 1843-2115; ISBN: 978-1-4673-6646-5

Co_2 and Idling Emission Estimation for Vehicle Routing Problem with Mid Way Halts

Ganesan Poonthalir[1](✉), R. Nadarajan[1], and S. Geetha[2]

[1] Department of Applied Mathematics and Computational Sciences, PSG College of Technology, Coimbatore, India
poonthalirk@gmail.com, nadarajan_psg@yahoo.co.in

[2] Department of Computer Science, Government Arts College, Udumalpet, India
geet_shan@yahoo.com

Abstract. Green Logistics are gaining importance due to green house gas emissions and its adverse impact on the environment. In this paper, we address the issues with vehicle routing and emissions. This paper reports the emissions that arise with Vehicle Routing Problem with Mid way Halts (VRPMH) and concentrates in finding low cost route for VRPMH using PSO with local exchange. Along with distance minimization, cruise and idling state emissions are reported. Computational experiments are carried out with green vehicle routing problem instances and the results are tabulated. The results project the impact of idling emissions and the need for its possible reduction.

Keywords: Green logistics · co_2 emission · Fuel consumption · Particle swarm optimization · Vehicle routing problem

1 Introduction

With the advent of emerging technologies and other reforms, there is a rapid growth in almost all walks of life. Since there is a high vehicle activity involved, green house gas emissions particularly carbon dioxide (co_2) are of great concern. Transport sector has a considerable share on these emissions. To the best of our knowledge the estimation of emission for VRPMH is not reported in the literature. VRPMH has a fleet of homogenous vehicles that are to be routed to serve customers. Apart from serving customers, they halt at facility center for some service. These facility centers can be a warehouse to load/unload vehicles, replenishment centers etc. The objective is to provide a low cost tour that in turn reduces co_2 emissions and is calculated for vehicles in cruise and in idle state.

Idling of vehicles is a severe problem and is prominent in the case of VRPMH as the vehicle halts for some service apart from servicing the customers. As specified in the works of Brodrick et al. [3] and Berg [2], engine tend to consume more fuel as the duration of idling increases than when it is in cruise. As reports suggest, the fuel consumed in 10 min of idling per day is more than 27 gallons of diesel per year. Also, idling emits harmful gases.

M. Graña et al. (eds.), *International Joint Conference SOCO'16-CISIS'16-ICEUTE'16*, Advances in Intelligent Systems and Computing 527, DOI 10.1007/978-3-319-47364-2_17

The contribution of this paper is to give an estimate of co_2 emission for VRPMH, to propose a unique method for calculating fuel consumption using air/fuel mixture ratio and to study the impact of fuel consumption by vehicle at idling stage.

There are a number of related studies available in literature. Sbihi and Eglese [13] discussed the relation between combinatorial optimization and green logistics. Kara et al. [9] reported an energy minimizing routing problem. Emission minimization routing was given by Figliozzi [7]. Suzuki [15] studied a truck routing problem that minimizes emissions. Xiao et al. [17] discussed a fuel consumption model for CVRP. A time dependent vehicle routing problem is demonstrated by Maden et al. [11]. Bektas and Laporte [1] introduced pollution routing problem that aim to reduce fuel consumption. A more detailed review regarding fuel consumption models and emission can be found in the works of Demir et al. [5] and Sbihi and Eglese [12]. Kuo [10] proposed a time dependent vehicle routing with minimal fuel consumption. A distance based approach for calculating emission is presented in Ubeda et al. [16]

2 Problem Description

VRPMH is defined on an undirected connected graph $G = (V, E)$ with vertices V and edges E connecting the vertices. The set V represents the set of vertices that includes customers, facility centers and depot. $V = C \cup F \cup D$, where $C = \{C_1, C_2, \ldots, C_N\}$ are the set of customers, $F = \{F_1, F_2, \ldots, F_L\}$ are the set of replenishment/facility centers and D is the depot. The total number of facility center is less than the set of customers to be served. D has a set of vehicles that aim to serve the customers with halts at facility centers and has constraints like, each customer is served by one vehicle, all vehicle has an upper time limit within which it should serve customers and reach the depot. The time includes the travel time in the arc between i and j where $i, j \in V$, service time at customer location for loading or unloading the goods and the service time at replenishment centers. The main objective is to reduce the total route cost with an aim to reduce the overall emission. The mathematical formulation of the problem is given as follows,

$$\min \sum_{i,j \in V, k \in M, i \neq j} d_{ij} x_{ijk} \tag{1}$$

$$\sum_{i \in V, i \neq j} x_{ijk} = 1, \forall j \in \mathrm{V},\ \mathrm{k} \in M \tag{2}$$

$$\sum_{j \in V \setminus \{F\}, i \neq j} x_{ijk} = 1, \forall i \in F,\ \mathrm{k} \in M \tag{3}$$

$$\sum_{i,q \in V, q \neq i} x_{iqk} - \sum_{j,q \in V, q \neq j} x_{qjk} = 0,\ \forall\ \mathrm{k} \in M \tag{4}$$

$$\sum_{j \in V \setminus \{D\}} x_{Djk} \leq \mathrm{M} \tag{5}$$

$$t_{VD} = p_{VD} = 0 \tag{6}$$

$$t_i + (t_{ij} + s_i)x_{ijk} - T(1 - x_{ijk}) \leq t_j, \forall\, i \in V,\; j \in V \backslash \{D\}, i \neq j \tag{7}$$

$$t_j \leq T - (t_{jD} + s_j), \forall j \in V \backslash \{D\} \tag{8}$$

$$x_{ijk} = \begin{cases} 1, & \textit{if} \text{ vehicle } k \text{ travels from i to } j,\; \text{i,j} \in \text{V} \\ 0, & \text{otherwise} \end{cases} \tag{9}$$

The primary objective of VRPMH is to minimize the overall route cost and is given in Eq. (1). A vehicle can visit any vertex from any other vertex as in Eq. (2). Equation (3) specifies that on visiting a facility center, the vehicle can visit either a customer vertex or a depot. Equation (4) specifies that the same vehicle enters and leaves a vertex. The vehicles that leaves the depot should not exceed the maximum available vehicles is given in Eq. (5). The service time is initially taken as 0 is given in Eq. (6). The time taken to reach any vertex j is specified in Eq. (7) and Eq. (8) is used to specify the time taken to reach depot from j. Equation (9) is about the decision variable x_{ijk}.

3 Emission Estimation

Fuel estimation models are used to calculate the total fuel consumed and assist to find emission. Detailed review is available in Demir et al. [4]. This paper discusses the Methodology for Calculating Transport Emission and Energy Consumption (MEET) proposed by Hickman et al. [8] and a proposed model for calculating fuel consumption using air/fuel mixture ratio.

3.1 Methodology for Calculating Transport Emission and Energy Consumption (MEET)

MEET is a successful model for calculating emission. Equation (10) is used to calculate emission for an unloaded vehicle with zero road gradient where e is the amount of co_2 emission in grams/kilometer.

$$e = K + aV + bV^2 + cV^3 + d/V + e/V^2 + f/V^3 \tag{10}$$

Table 1. Parameters used in MEET for different class of vehicle weight

Weight	K	a	b	c	d	e	f
$3.5 < wt \leq 7.5$	110	0	0	0.000375	8702	0	0
$7.5 < wt \leq 16$	871	−16.0	0.143	0	0	32031	0
$16 < wt \leq 32$	765	−7.04	0	0.000632	8334	0	0
$wt > 32$	1576	−17.6	0	0.00117	0	36067	0

Here V is the average vehicle speed and a, b, c, d, e, f and K are parameters used and is given in Table 1.

3.2 Proposed Method: Fuel Consumption Using Air/Fuel Mixture Ratio

Emission obtained using MEET is suited for determining emission in classical VRP. Demir et al. [4] studied various emission estimation models and has its own computational complexity. We propose a method for calculating emission based on air/fuel mixture ratio.

For an engine to run, it must have a proper mixture of air/fuel ratio. Air/fuel ratio has an important effect on engine power, efficiency and emissions. An air/fuel mixture, that has enough air to burn the fuel is said to be stoichiometric and it is given as 14.7:1 which is 14.7 parts of air to 1 part of fuel. Generally, complete combustion gives the best fuel economy as no fuel is wasted, but no internal combustion engine is 100 % efficient. If a vehicle fails to start, it gets less fuel and more air and are said to be lean air-fuel mixture. Rich mixtures are caused by too much of fuel and less air. Generally, vehicles that run with lean or rich mixtures are said to emit emissions higher. If the mixture is less with ratio 12:1 or 9:1, it is a rich mixture. If it is more like 17:1 or 19:1, then it is a lean mixture. When the vehicle speed is from 40 km/hr to 60 km/hr, a lean mixture is supplied as specified in Srinivasan [14].

Assuming that all fuel is used for combustion, the amount of emission is proportional to the amount of fuel combustion. First a low cost tour is got, and then co_2 emission estimation is made. The total fuel consumption by a vehicle is given in Eq. (11),

$$Total\_fuel = fuel\_cruise + fuel\_idle \tag{11}$$

Where $Total\_fuel$ is the total fuel spent, $fuel\_cruise$ is the fuel consumption by vehicle on cruise and $fuel\_idle$ is the fuel consumed by vehicle on idling. Let $a : b$ be the air/fuel mixture ratio at cruise and $a1 : b1$ be the air/fuel ratio at idle state of vehicle. Let k gallons of fuel be used per mile of travel and s be the speed of the vehicle in miles/hour. Then, the amount of fuel consumed per hour in cruise fc_c is given in Eq. (12),

$$fc\_c = k * s \tag{12}$$

fc_c is the amount of fuel spent in an hour of cruise. Equation (13) calculates the idle fuel rate fc_i using a linear relationship between fuel spent on cruise and idle.

$$fc\_i = (fc\_c/a) * a1 \tag{13}$$

3.3 Particle Swarm Optimization

VRPMH is solved using Particle Swarm Optimization (PSO). PSO is a population based Meta heuristic. It starts with a set of particles (solution) in a multi dimensional space. Each particle's best position is preserved as $personal_{best}$ and the best particle is identified as $global_{best}$. Each particle's position is updated using velocity. With basic PSO, the particle may trap in local minimum hence an inversion operator that act as a local exchange operator explores new areas in the search space. The algorithm is described below,

Algorithm 1. PSO with Inversion Operator (Local Exchange)

Initialize the particles $(y_{i1}, y_{i2}, \ldots, y_{iN})$ where y_{ij} is i^{th} particle with j^{th} dimension and each particle is a potential solution where the halts to fueling stations are included.
Calculate the objective function which is the fitness value.
Store each particle's best as $personal_{best}$ and each iteration best as $global_{best}$
Repeat till convergence
Update each particle's velocity using,
$v_{id}(t+1) = w \times v_{id}(t) + c1 \times r1 \times (personal_{best} - y_{id}) + c2 \times r2 \times (global_{best} - y_{id})$
where $c1$ and $c2$ are cognitive and social constants, w is inertia, $r1$ and $r2$ are random numbers.
Calculate inertia of particles using $w(t) = w2 + ((t-T)/(1-T)) \times (w1 - w2)$
Where t is the current iteration, T is the total number of iterations and $w1$ and $w2$ are the minimum and maximum inertia weights respectively.
Apply Inversion operator on the particle
Calculate fitness
Update $personal_{best}$ and $global_{best}$
End

Algorithm 2. Inversion operator
Generate a random number r
If $r > 0.5$
Choose two positions $p1$ and $p2$ in the particle
Invert the customer positions within $p1$ and $p1$
End

Algorithm 3 specifies the procedure to find the estimate using air/fuel mixture ratio where the cruise and idling emission are calculated.

Algorithm 3

Assumptions	*for* i=1 to size(route)-1
Let T total time taken in hours	{ { *if* route(i)=f_center
Let S speed of vehicle in miles/hr	$tot\_time = tot\_time + s\_facility$
Let f constant fuel rate in gallons/mile	else $tot\_time = tot\_time + s\_cust$ }
Let t_cruise is time taken in cruise	$tot\_time = tot\_time + t(i,i+1)$ }
Let t_idle be idle time taken	Initialize $idle\_time$ to 0
Let f_center is facility centre	*for i*=1 to *size*(*route*)
Let $route$ be route got using PSO local exchange	{ { *if* $route(i) = f\_center$
Let s_cust and $s\_facility$ be fraction of service time at customer and facility location respectively	$idle\_time = idle\_time + s\_facility$ *else* $idle\_time = idle\_time + s\_cust$ }}
Let $fuel\_cruise$ and $fuel\_idle$ are in litres	$time\_cruise = tot\_time - idle\_time$
Let tot_time be total time	$fuel\_cruise = time\_cruise * fc\_c$
Let fc_c and fc_i be fuel consumption at cruise and idle in gallons/hour	$fuel\_idle = idle\_time * fc\_i$
Let fcc e fuel conversion in gms/litre	$tot\_fuel = fuel\_cruise + fuel\_idle$
Let $t(i, j)$ e time taken from I to j	$emission = tot\_fuel * fcc$
	$emission\_cruise = fuel\_cruise * fcc$
	$emission\_idle = fuel\_idle * fcc$

4 Results and Discussion

The algorithm is tested on the instances of Green Vehicle Routing Problem (GVRP) proposed by Erdogan et al. [6]. GVRP has a fleet of vehicles stationed at depot and strive to serve a set of customers. These vehicles start with an initial capacity of fuel, if fuel level is minimal and not sufficient to serve the next customer, it is refueled in the nearby fueling station. Each vehicle has a maximum time limit within which it should serve customers and reach the depot.

The data set of GVRP has 4 sets, each with 20 customers. These data sets are classified as uniformly distributed customers and clustered customers with 3 refueling stations. Other two data sets are a combination of uniform and distributed customers and the refueling stations are 6 for the third data set and it increases from 2 to 10 for the fourth data set. Erdogan et al. solved GVRP using Modified Clarke and Wright Savings Algorithm (MCWS) and Density Based Clustering Algorithm (DBCA). Emissions are calculated with an average vehicle speed of 40 miles/hour and the fuel consumption

Table 2. Emission estimation of Co_2 using MEET

Data set	Total distance MCWS(kms)	Kg of co_2	Total distance DBCA(kms)	Kg of co_2	Total distance PSO-local exchange	Kg of co_2
1	27132.04	8903.38	26977.93	8852.80	**26103.18**	**8565.76**
2	31601.61	10370.07	31523.38	10344.4	**29127.57**	**9558.21**
3	30580.03	10034.84	30506.28	10010.64	**28732.62**	**9428.60**
4	25495.96	8366.49	25247.1	8284.83	**22750.31**	**7465.51**

rate is taken as 0.2 gallons per hour. The vehicle weight is taken in the range 3.5 tonnes to 7.5 tonnes. Routes obtained using PSO with local exchange is better than MCWS and DBCA and is reported in Table 2 along with an estimate of co_2 using MEET. The results show that the algorithm is able to minimize the total emission level. The estimation of vehicles in cruise and idling using air/fuel mixture ratio is tabulated from Tables 3, 4, 5 and 6.

Table 3. Co_2 emission for vehicle at cruise and idle for dataset 1

Data set	Fuel cruise (gallons)	Fuel idle (gallons)	Total fuel (gallons)	Emission(kg of co_2)		% co_2 at idling
				Total	Idling	
20c3su1	401.77	28.19	429.97	4285.81	280.99	6.56
20c3su2	376.27	25.63	401.91	4006.10	255.47	6.38
20c3su3	384.42	25.63	410.05	4087.25	255.47	6.25
20c3su4	361.86	28.38	390.24	3889.79	282.88	7.27
20c3su5	406.37	28.38	434.76	4333.54	282.88	6.53
20c3su6	385.21	25.63	410.84	4095.14	255.47	6.24
20c3su7	401.37	25.63	427.00	4256.25	255.47	6.00
20c3su8	406.92	28.38	435.29	4338.89	282.88	6.52
20c3su9	398.34	25.63	423.97	4226.01	255.47	6.05
20c3su10	323.36	20.13	343.49	3423.77	200.64	5.86

MEET use average speed as an estimate and calculates emission. But estimation of emission for idling of vehicles cannot be determined. All estimation models studied in literature tries for an exact estimation, but it is challenging and difficult. They use an extensive set of parameters, which requires proper tuning. The estimate is dependent on the kind and state of the vehicle, and cannot be uniformly applied to all vehicles of the same kind and with different manufacturing years. But, mostly an approximate estimation can be obtained. Hence, the estimation obtained using air/fuel ratio can be used for quick estimation by organization.

The speed of the vehicle is 40 miles/hr, consuming 0.2 gallons of fuel/mile. The air/fuel mixture is taken as 16:1 and 11:1 for cruise and idling state respectively.

All routes are obtained using PSO with local exchange. The results show the impact of vehicle at idling and its fuel consumption. In almost many cases, the fuel at idling

Table 4. Co_2 emission for vehicle at cruise and idle for clustered customers

Data set	Fuel cruise (gallons)	Fuel idle (gallons)	Total fuel (gallons)	Emission(kg of co_2)		% co_2 at idling
				Total	Idling	
20c3sC1	296.28	20.13	316.41	3153.93	200.64	6.36
20c3sC2	341.03	25.44	366.48	3653.00	253.66	6.94
20c3sC3	208.80	15.92	224.73	2240.06	158.76	7.09
20c3sC4	275.80	19.76	295.57	2946.20	197.03	6.69
20c3sC5	495.42	33.69	529.11	5274.09	335.89	6.37
20c3sC6	520.50	38.83	559.33	5575.32	387.10	6.94
20c3sC7	281.13	25.83	306.97	3059.81	257.55	8.42
20c3sC8	657.19	50.01	707.20	7049.25	498.55	7.07
20c3sC9	396.84	30.94	427.79	4264.13	308.48	7.23
20c3sC10	528.47	41.22	569.69	5678.57	410.89	7.24

Table 5. Co_2 emission for vehicle at cruise and idle for dataset 3

Data set	Fuel cruise (gallons)	Fuel idle (gallons)	Total fuel (gallons)	Emission(kg of co_2)		% co_2 at idling
				Total	Idling	
S1_2i6 s	588.21	22.88	611.09	6091.23	228.06	3.74
S1_4i6 s	346.59	25.63	372.22	3710.23	255.47	6.89
S1_6i6 s	374.96	28.38	403.34	4020.46	282.88	7.04
S1_8i6 s	649.87	28.38	678.25	6760.61	282.88	4.18
S1_10i6 s	322.52	20.13	342.65	3415.47	200.64	5.87
S2_2i6 s	382.62	28.38	411.00	4096.78	282.88	6.91
S2_4i6 s	355.17	25.44	380.62	3793.91	253.66	6.69
S2_6i6 s	617.20	50.38	667.58	6654.27	502.17	7.55
S2_8i6 s	483.96	35.90	519.87	5181.91	357.88	6.91
S2_10i6 s	777.66	20.13	797.79	7952.16	200.64	2.52

stage shares 6 % to 8 % on the total fuel consumption. This increase as the halts made increases. The figures are for a single day route and the amount is high when calculated for a year. Though the algorithm have produced routes that are already optimized for the number of halts, the fuel consumption at idling shares a considerable amount of emission. Hence it is important to ascertain the emission estimation based on idling of vehicles as substantial amount of reduction in emission can be realized. When the algorithm is not striving to minimize the total halts made the idling fuel rate and hence the emission can increase. Hence, idling of vehicles need to be minimized. At operational level itself proper reduction in fuel consumption has to be addressed.

Table 6. Co_2 emission for vehicle at cruise and idle for data set 4

Data set	Fuel cruise (gallons)	Fuel idle (gallons)	Total fuel (gallons)	Emission(kg of co_2)		% co_2 at idling
				Total	Idling	
S1_4i2 s	369.22	25.63	394.85	3935.75	255.47	6.49
S1_4i4 s	349.29	25.63	374.92	3737.10	255.47	6.84
S1_4i6 s	348.84	28.38	377.22	3760.04	282.88	7.52
S1_4i8 s	331.36	28.38	359.74	3585.81	282.88	7.89
S1_4i10 s	354.08	25.63	379.71	3784.93	255.47	6.75
S2_4i2 s	285.49	19.76	305.26	3042.77	197.03	6.48
S2_4i4 s	327.31	25.44	352.76	3516.22	253.66	7.21
S2_4i6 s	327.34	25.63	352.97	3518.35	255.47	7.26
S2_4i8 s	327.34	25.63	352.97	3518.35	255.47	7.26
S2_4i10 s	327.34	25.63	352.97	3518.35	255.47	7.26

5 Conclusion

An estimate of emission of co_2 for Vehicle Routing Problem with Midway Halts (VRPMH) is studied in this paper. This paper suggests a method of estimating emission for vehicle in cruise and idling using air/fuel mixture ratio to calculate co_2 emission for VRPMH and to find the impact of idling of vehicles. Idling of vehicles has a considerable share on the emission. To reduce this, vehicles may be directed to make minimum halts. Use of heterogeneous vehicles equipped with alternate fuel requirements and reducing idling of vehicle can have an impact on carbon emission reduction.

References

1. Bektaş, T., Laporte, G.: The pollution-routing problem. Transp. Res. Part B Methodol. **45**, 1232–1250 (2011)
2. Van den Berg, A.J.: Truckstop electrification: reducing CO2 emissions from mobile sources while they are stationary. Energy Convers. Manage. **37**, 879–884 (1996)
3. Brodrick, C.J., Dwyer, H.A., Farshchi, M., Harris, D.B., King Jr., F.G.: Effects of engine speed and accessory load on idling emissions from heavy-duty diesel truck engines. J. Air Waste Manage. Assoc. **52**, 1026–1031 (2002)
4. Demir, E., Bektas, T., Laporte, G.: A comparative analysis of several vehicle emission models for road freight transportation. Transp. Res. Part D Transp. Environ. **6**, 347–357 (2011)
5. Demir, E., Bektaş, T., Laporte, G.: A review of recent research on green road freight transportation. Eur. J. Oper. Res. **237**, 775–793 (2014)
6. Erdoğan, S., Miller-Hooks, E.: A green vehicle routing problem. Transp. Res. Part E Logistics Transp. Rev. **48**, 100–114 (2012)
7. Figliozzi, M.: Vehicle routing problem for emissions minimization. Transp. Res. Rec. J. Transp. Res. Board. (2010). doi:10.3141/2197-01

8. Hickman, J., Hassel, D., Joumard, R., Samaras, Z., Sorenson, S.: MEET-methodology for calculating transport emissions and energy consumption. European Commission DG VII Technical report (1999). http://www.transport-esearch.info/Upload/Documents/200310/meet.pdf
9. Kara, I., Kara, B.Y., Yetis, M.: Energy minimizing vehicle routing problem. In: Dress, A.W., Xu, Y., Zhu, B. (eds.) COCOA 2007. LNCS, vol. 4616, pp. 62–71. Springer, Heidelberg (2007)
10. Kuo, Y.: Using simulated annealing to minimise fuel consumption for the time-dependent vehicle routing problem. Comput. Ind. Eng. **59**(1), 157–165 (2010)
11. Maden, W., Eglese, R., Black, D.: Vehicle routing and scheduling with time varying data: A case study. J. Oper. Res. Soc. **61**, 515–522 (2010)
12. Sbihi, A., Eglese, R.W.: The relationship between vehicle routing and scheduling and green logistics-a literature survey (2007)
13. Sbihi, A., Eglese, R.W.: Combinatorial optimization and green logistics. Ann. Oper. Res. **175**(1), 159–175 (2010)
14. Srinivasan, S.: Automotive Engines. Tata McGraw Hill, New Delhi (2007)
15. Suzuki, Y.: A new truck-routing approach for reducing fuel consumption and pollutants emission. Transp. Res. Part D Transp. Environ. **16**(1), 73–77 (2011)
16. Ubeda, S., Arcelus, F.J., Faulin, J.: Green logistics at Eroski: A case study. Int. J. Prod. Econ. **131**, 44–51 (2011)
17. Xiao, Y., Zhao, Q., Kaku, I., Xu, Y.: Development of a fuel consumption optimization model for the capacitated vehicle routing problem. Comput. Oper. Res. **39**, 1419–1431 (2012)

Agent-Based Spatial Dynamic Modeling of Opinion Propagation Exploring Delaying Conditions to Achieve Homogeneity

Leire Ozaeta(✉) and Manuel Graña

Computational Intelligence Group, Department of CCIA,
University of the Basque Country, Leioa, Spain
lozaeta001@gmail.com

Abstract. Most computational models of influence spread nowadays are motivated by the need to identify the social actors with maximal influence, in order to achieve high penetration in the market with minimal effort. However, there are little literature on the mechanisms of influence propagation, i.e. computational models of how the social actors change their opinions. There are some works that relate the spatial distribution of the opinions with the mechanism by which an agent changes or maintains its opinions, but they assume a cell model, where agents have fixed spatial locations and neighbors. Here we explore the effect of spatial interaction of the agents, which are free to move in a given space, following attraction dynamics towards agents with similar opinions. The spatial distribution of opinions observed by the agent is used by the agent to decide about opinion changes. We report preliminary results of simulations carried out in Netlogo environment for the first three kinds of systems.

1 Introduction

The issue of influence of the minorities, how they can maintain their characteristics and even produce a change of the majority towards accepting their specificities has been an intriguing question in sociology and cognitive science [8], but there are little attempts to produce computational models which can be used to assess the value of the diverse mechanisms and hypothesis proposed.

Multi-agent and dynamic-network models have already established themselves as suitable methods for analyzing "complex social systems" and to formalize models of real-world systems [7], however hard to validate against real data. Therefore, they are one of the most popular techniques to study the social dynamics [1,4,6,9], even if other approaches as sociophysics [2], threshold models [5], and dynamic models [3] has been widely considered.

We focus on the approach of [6] which studies the spatial interaction of agents with their neighbors and the effect of several opinion change policies. Essentially, the model is a cellular automata model, which has the inconvenience of having fixed spatial relations. Nevertheless, the model achieved to reproduce

M. Graña et al. (eds.), *International Joint Conference SOCO'16-CISIS'16-ICEUTE'16*,
Advances in Intelligent Systems and Computing 527, DOI 10.1007/978-3-319-47364-2_18

interesting effects of the sociological models proposed of the interaction between majority and minority opinions. Following this lead, we narrow our problem to the particular issue of how to impede or delay the emergence of social system with a monolithic opinion situation, i.e. a homogeneous distribution of agent opinions, which is a problem similar to the prevention of infection spread covering the entire population, as presented in [5], if we match "infection" with the opinion being majority from the beginning, and that the "infected" agents are both curable and not vaccined.

Intended Contribution. The aim of the work in this article, which is in its initial stages, is to develop simulations where the early stages of the resistance against opinion homogeneity could be reproduced, in an attempt to better understand the underlying processes and the influence of the spatial behavior. Elaborating from the approach of [6], we added the agents' spatial mobility as we consider it essential to generate changes in the influence received and, therefore as relevant to the opinion spreading/resistance as it is the agents' opinion-changing process. We report preliminary results showing some differences in convergence to homogeneous opinion configuration of systems associated to agent's spatial behavior, in some cases showing significant resistance to majority opinion overtaking the entire system.

The paper contents are as follows: Section 2 comments the experimental designs of our simulations. Section 3 provides preliminary results of ongoing analysis. Section 4 gives some conclusions.

2 Experimental Design

In these experiments the idea was to observe the general movement of the agents and find the one that facilitates mostly a variety of opinions, conversely delaying or impeding homogeneity. To this end, we explore the reaction in the diversity of the systems to changes in the n value when selecting the n nearest neighbors for the local neighbors selection and changes in the θ distance value when searching for the agents in the radius for the influence neighbors selection.

The simulations were programmed in netlogo, carrying 10 runs of the model for each combination presented as follows. To observe the general movements of the agents groups we performed 7 simulation with each system, with a radius of 30, an initial population of 200 agents, 4 initial opinions, and changing the local neighbors size in each one of them using the following values: 10, 25, 50, 75, 100, 125, and 250. We did not consider more neighbors than 150, even with the initial population being bigger as the neighbors would be too far to be realistic to consider them attainable.

On the other hand, to consider the system reactions to changes in the influences radius and the number of nearest neighbors we consider a system with an initial population of 200 agents, once again and apply changes to all other variables. We used influence radius of 1, 5, 10, 15, 20, 25, 30 patches, and 1, 25, 50, 75, 100, 125, 150 nearest neighbors. For each pair we performed a simulation using 2, 3 and 4 initial different opinions, randomly spread in the arena and, therefore, the population.

From the 4th system and higher we considered a maximum path deviation degree of 30° to both sides. Also we added an opinion delay so the more complex spatial behaviour of the agents had a real impact in the opinion spread, this supposed that an agent moved 10 times for each opinion reconsideration.

The first four systems all were clearly prone to converge to a homogeneous opinion situation, therefore we run the models until all the agents have the same opinion and measured how long take them to achieve this homogeneity. However, in the last three systems the systems tend to oscilate without an opinion overrunning all the others which forced us to set an stoping parameter based in ticks, setting to 16,000 the maximinun of steps.

In an attempt to see whether the radius or the number of neighbors would affect the opinion spreading we consider the mean of the measures for all the radius variations and repeated the same process with the variations of the number of neighbors.

3 Results

The results of the simulations are presented in two ways. One is plot of the trajectories followed by the agents in a simulation, such as shown in Fig. 1. This visualization is helpful to achieve an understanding of the effect that the initial position of the agents may have on the final distribution of opinions and the changes in spatial behavior when considering changes in the number of neighbors considered. In this figures the path of the agents can be seen as the lines that part from them and changes in opinion are represented by color changes. The visualization is repeated for the Simple Peer Seeker System in Fig. 2 and the Ally/Enemy System in Fig. 3.

In those images it can be perceived the significant difference in behavior from the first two systems, the Simple Gregarious System and the Simple Peer Seeker System, to the third Ally/Enemy System. The first two have small difference between them, presenting more compact groups in the Simple Peer Seeker System. In the Ally/Enemy System, however, it can be appreciated a less clear grouping in the agent's spatial behavior and less if not none compact group exists in any final configuration. This mean that as it could be expected, less spatial "vision" in system with just attraction forces, considering "vision" the group of other agents a particular agent can "see", derives in smaller and more isolated groups whereas a wide "vision" supposes bigger compacted groups. However, in the system that considers reject forces too, this behavior is changed and wider "vision" supposes even less compacted groups than in the previous configurations, as the agent receives a bigger number of forces that force it to avoid more populated spots.

The other presentation of simulation results are the plots that show the mean disappearance of diversity taking account of changes both in the number of neighbours considered and in the influence radius. Regarding this plots we

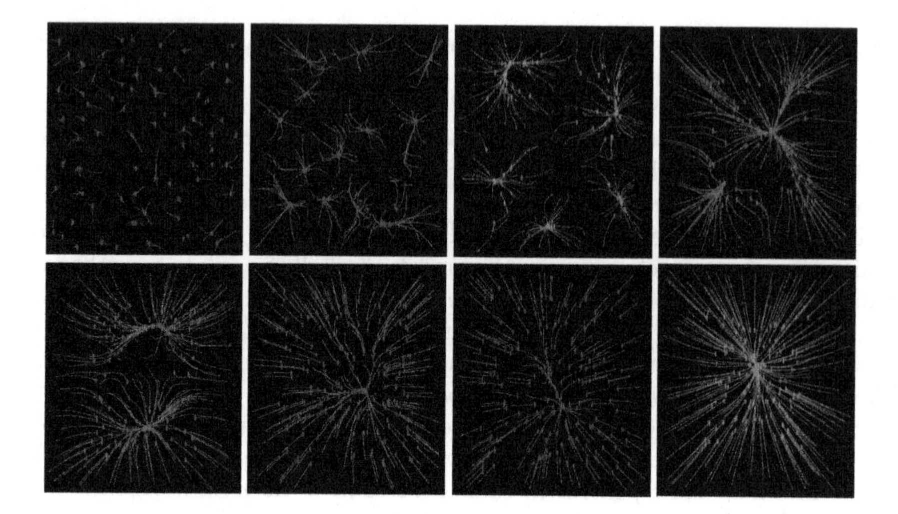

Fig. 1. Some dynamical evolutions of agents following the specification of System 1.

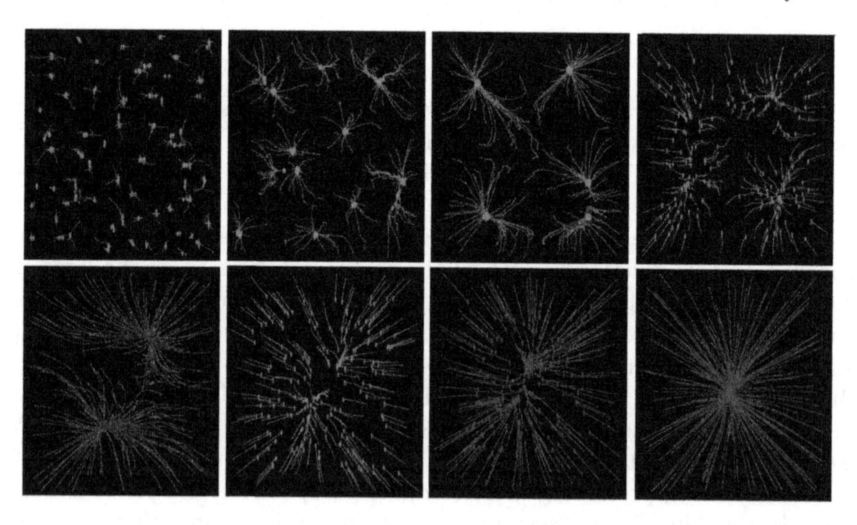

Fig. 2. Some dynamical evolutions of agents following the specification of System 2.

separate them in groups for the different start configuration: with two different opinions, three different opinions, and four different opinions. As it can be seen in the following figures the systems react in different ways to changes in both of the variables. However, in Fig. 4 it can be appreciated that there is an evident drop in the diversity endurance in some point. In the first system, this drop is when the number of neighbours considered is exactly half of the initial population or the considered influence radius is of five patches, what can be considered as little. For system two, there are two significant drops in both of the experiments: when considering the 50 or 125 nearest neighbours and when considering 10 or

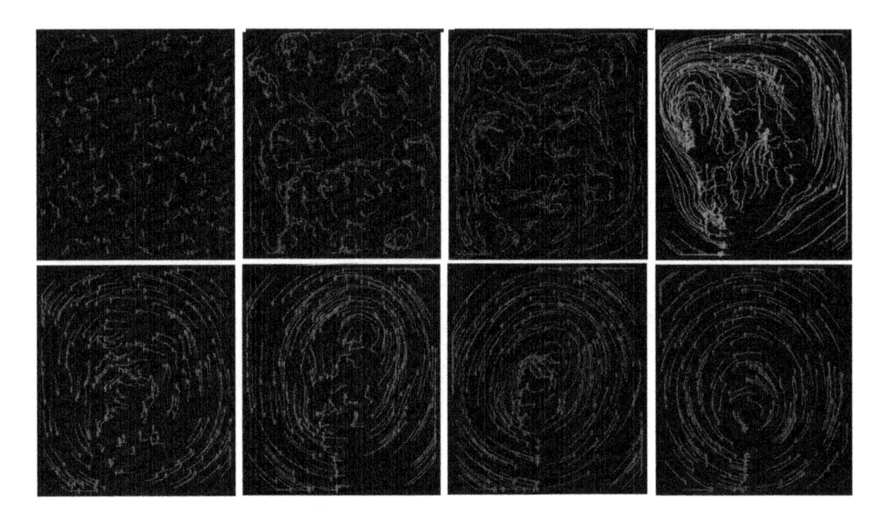

Fig. 3. Some dynamical evolutions of agents following the specification of System 3.

15 patches radius. This seems meaningful as in one of the cases a drop happens when considering a quarter of the initial population and in the other case a quarter of the arena. Also, there is a high peak resistance to homogeneity when considered the 25 nearest neighbours, just as in the previous system. The third system has a significantly different behaviour when changes in the number of nearest neighbours are applied. The plot approximately represents a stair, with less pronounced drops. When the changes are applied to the influence radio, however, it shows a drop similar to the one in the first system, when considered a 5 patch radius. In the forth system there are three drops when changing the radius, in 1, 10 and 30 patches considered. When observed the variations of nearest neighbours, however, it shows a downward trend with a slight peak at 50 nearest neighbours considered. For the fifth system, there is an evident change in behaviour with high peaks both when 15 patches influence radius considered and with 25 nearest neighbours taken into account, the last peak showing a higher endurance similar to the one present in the first two systems. The sixth system shows a drop in homogeneity resistance when the values in the middle are considered concerning the influence radius and the opposite behaviour when the nearest neighbours are the considered variable. The last system shows a peak in resistance when the 25 nearest neighbours are considered, as occurs in first, second, and fifth systems and a significant drop when 30 patches considered for the influence radius.

In the last group of simulations, starting with four different opinions, the figures in Fig. 5 show an slightly more stable behaviour. However, there are visible drops when considering the 100 nearest neighbours in the first and second systems, with a timid peak in the third system. Also, the second and third systems show a fall in the resistance when considering the 125 nearest neighbours, while the first system shows a peak in the same context. Both systems four and

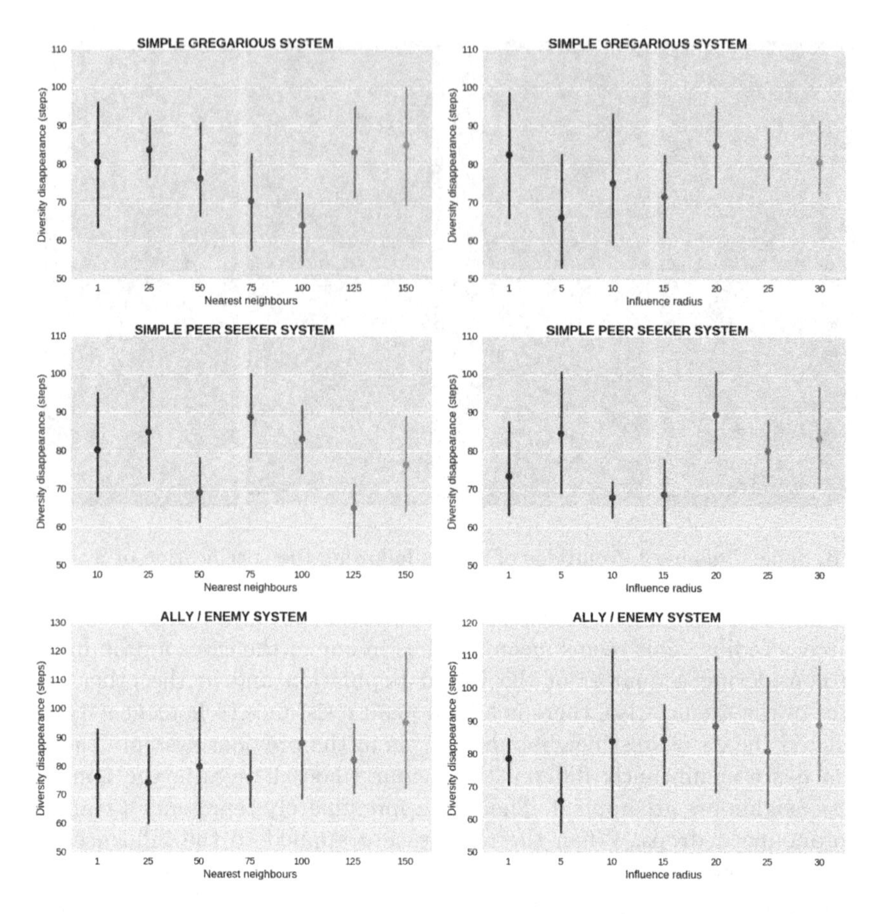

Fig. 4. The effect of the neighborhood size (left) and increasing radius (right) in the time to disappearance of the minority opinion, starting with two different opinions.

five show a upward trend in that abruptly drops in the last two cases, agains the peaks in the last two systems when considered the 150 nearest neighbours. Regarding the changes in influence radius there are peaks in the first and last systems when considering an 25 patch radius, whereas there is a drop in the same case for the second system. Similarly there is a resemblance in the behaviour if the first and second systems for the first few influence radius considered that is contrary to the behaviour in the third system. In the same fashion, fourth and seventh systems show a general behaviour opposite to the one of the fifth and sixth systems.

Regarding the value ranges, the simulations show higher endurance in every system with the increasing of the initial opinion number. From the forth system and beyond, however, there is a significant change in said values. While the three first systems needed around 50–140 steps to achieve homogeneity the forth system results vary from 400 steps to 1500. After this point, the systems were

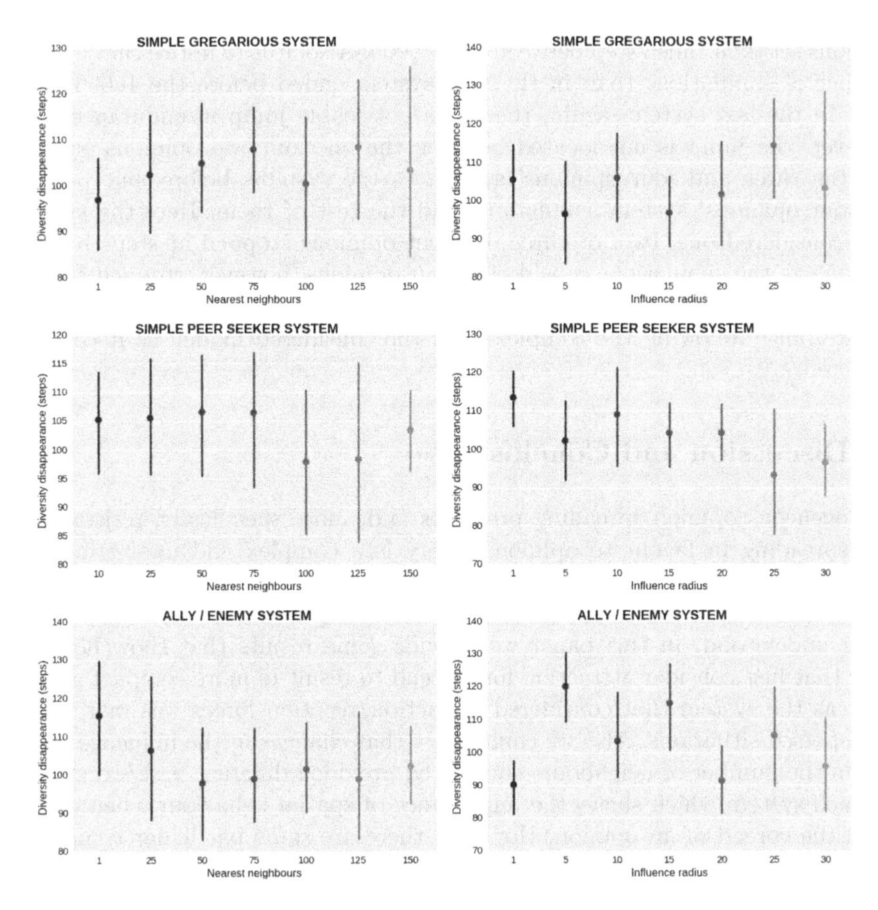

Fig. 5. The effect of the neighborhood size (left) and increasing radius (right) in the time to disappearance of the minority opinion, starting with four different opinions.

prone to oscilation and, thous, we set a stop at 16000 steps if stability was not achieved yet. We could observe that in the three remaining systems of agents, only three to six repetitions of each combination stopped before the 16000 steps and even in this situation there where significant differences in scale, even in the simulations of the same model, depending of the number of opinions considered. In the fifth system, with charismatic agents, the simulations starting with one or two opinions offered results that where in the 0–200 steps range, except for the situations where the 25 nearest neighbours where considered, where is a clear peak in the endurance, or a influence radius of 15 patches is considered, where another peak is visible. Regarding three and four opinions, however, the endurance varied from 100 to 1000 steps, which could be regarded as a wide range. For the sixth system, where the resistance to agents' opinion is added, there is a similar jump in endurance scale. While the simulations with two or three opinions have an stop at 500–2600 steps, the simulations of three and four opinions set their finish step between 1500–5000. Also, it is to notice that a bigger number of simulations than in the fifth system ended before the 16000 steps' stop. In the last system, again, there is a noticeable jump in endurance scale, however, the jump is not located between the one and two opinions' systems and the three and four opinions' system as the systems before, but between the four opinions' system simulations and the rest of them. Here the systems that considered one, two, or three different opinions stopped at steps between 1000–5500, the simulations considering four opinions, however, stopped at steps between 3500–7500. As it can be seen in all cases there seems to be a growing in endurance following the complexity of the considered model, as it could be expected.

4 Discussion and Conclusions

Homogeneous opinion spreading processes and, more specifically, resistance to this spreading in favour to opinion variety is a complex and interesting phenomena with a high number of details. The construction of sufficient complex agents and contexts it is yet to be done and the phenomena itself is far from being understood. In this paper we provide some results that show how systems that just consider attraction forces tend to result in more compact groups, whereas the system that considered attraction/reaction forces was more prone to dispersed situations. Also we could show that changes in the influence radius and in the number of neighbours considered provided different best/worst cases for each system which shows the importance of spatial behaviour when considering the spread of an opinion. However, there are some particular trends and behaviours, as it is the increasing proneness to oscilation of the systems and the growing gap between results in a very same system depending of opinion number. To extract an equality between the seven systems is highly complicated due to the basic differences between them, however, it is noticeable how the 20 patches influence radius with two initial opinions shows a higher endurance, and the 25 nearest neighbours with three initial opinions show a drop in resistance in the first three systems, whereas the third, sixth, and seventh systems show a similar general behaviour when considering changes in influence radius as well as the fourth and fifth systems show a similar general behaviour when considering changes in nearest neighbours number for four opinions. Following this line there is yet much more work to do, however we have provided a sound planning for our near future research, implementing the presented systems, and some interesting, even if preliminary, group of results.

Acknowledgments. Leire Ozaeta has been supported by a Predoctoral grant from the Basque Government.

References

1. Banisch, S., Lima, R., Araujo, T.: Agent based model and opinion dynamics as markov chains. Soc. Netw. **34**, 549–561 (2012)
2. Crokidakis, N.: Effects of mass media on opinion spreading in the sznajd sociophysics model. Phys. A Stat. Mech. Appl. **391**, 1729–1734 (2012)
3. Deffuant, G., Amblard, F., Weisbuch, G., Faure, T.: How can extremism prevail? A study based on the relative agreement interaction model. J. Artif. Soc. Soc. Simul. **5**(2) (2002)
4. Gil, S., Zanette, D.H.: Coevolution of agent and networks: opinion spreading and community disconnection. Phys. Lett. A **356**, 89–94 (2006)
5. Dreyer Jr., P.A., Roberts, F.S.: Irreversible k-threshold processes: graph-theorical threshold models of the spread of disease and of opinion. Discrete Appl. Math. **157**, 1615–1627 (2008)
6. Jung, J., Bramson, A.: An agent - based model of indirect minority influence on social change. In: ALIFE 14 (2014)
7. Louie, M.A., Carley, K.M.: The role of dynamic-network multi-agent models of socio-political systems in policy. Technical report, CASOS (2007)
8. Mucchi-Faina, A., Paclilli, M.G., Pagliaro, S.: Minority influence, social change and social stability. Soc. Pers. Psychol. Compass **4**, 1111–1123 (2010)
9. Rouly, O.C.: At the root of sociality: working towards emergent, permanent, social affines. In: Proceedings of The European Conference on Artificial Life, pp. 82–89 (2015)

SOCO 2016: Genetic Algorithms

Coevolutionary Workflow Scheduling in a Dynamic Cloud Environment

Denis Nasonov(✉), Mikhail Melnik, and Anton Radice

ITMO University, Saint Petersburg, Russia
denis.nasonov@gmail.com,
mihail.melnik.ifmo@gmail.com, antonradice@gmail.com

Abstract. In this paper, we present a new coevolutionary algorithm for workflow scheduling in a dynamically changing environment. Nowadays, there are many efficient algorithms for workflow execution planning, many of which are based on the combination of heuristic and metaheuristic approaches or other forms of hybridization. The coevolutionary genetic algorithm (CGA) offers an extended mechanism for scheduling based on two principal operations: task mapping and resource configuration. While task mapping is a basic function of resource allocation, resource configuration changes the computational environment with the help of the virtualization mechanism. In this paper, we present a strategy for improving the CGA for dynamically changing environments that has a significant impact on the final dynamic CGA execution process.

Keywords: Workflow scheduling · Coevolutionary algorithm · Genetic algorithm · Virtualization

1 Introduction

Today, multidisciplinary computational problems are complex in structure and often require a certain type of organization. For these purposes, workflow formalism is widely used to describe the structure of a composite application, especially in scientific computing, while the organization of its execution is normally handled by a workflow management systems (WMS) operating in a computing cloud. Environments like these have many differences compared to traditional clusters, grids or supercomputing structures (Korkhov et al. 2009; Krzhizhanovskaya and Korkhov 2007). One of the main differences is the ability to virtualize the computational resources in a cloud environment, creating new opportunities for optimization in terms of the energy efficiency and makespan. The scientific community offers many algorithms that solve this issue.

Guo et al. (2007) propose a coevolutionary algorithm called the Dynamical Coevolutionary Optimization Algorithm (DCOA). This algorithm mainly focuses on dynamically adapting the population size through a rule based approach to optimize the convergence of the algorithm. Before the coevolution step, a pretreatment step is executed in order to assign subspaces to the local optima. This allows for an ideal subspace division parallel initialization of sub-populations, which increases the performance in coevolution step. While this algorithm has been shown to find an optimal solution quicker than the macroevolutionary algorithm (MA) and it converges in about

M. Graña et al. (eds.), *International Joint Conference SOCO'16-CISIS'16-ICEUTE'16*,
Advances in Intelligent Systems and Computing 527, DOI 10.1007/978-3-319-47364-2_19

the same time as the simple genetic algorithm (SGA), it has not been applied truly dynamically to a workflow scheduling problem like we do here in this paper.

Rahman et al. (2013) propose a dynamic critical-path (CP) based adaptive workflow scheduling algorithm for the performance-driven grid applications. The featured solution is an extension of the dynamic critical path (DCP) algorithm that is optimized for adaptive resource environments. The extension maps critical tasks to resources, which provide minimal execution time and the CP length as well as the makespan for the workflow execution. The specific workflow scheduling problem that this paper attempts to solve is the most similar in nature to ours; however, we apply a coevolutionary approach to optimization that is not based on simply scheduling the longest execution path earliest.

Singh et al. (2013) describe a score-based deadline-constrained scheduling algorithm where the concept of score represents the capability of hardware resources in the cloud. This algorithm focuses on reducing the failure rate of workflow applications while meeting the user-defined deadline. However, this approach only focuses on the static optimization, whereas ours includes a coevolution step at every iteration to increase the efficiency where possible.

A metaheuristic optimization algorithm is applied by Liu et al. (2013) to the permutation flow shop scheduling problem (PFSSP). In this study, a modified version of particle swarm optimization (PSO) is presented. The key adjustment is a multi-population scheme used to increase the diversity, allowing division into subpopulations, where PSO can be applied separately and with the different local search structures. While this paper describes a meta-heuristic solution to the scheduling problem, we propose a coevolutionary approach to a different domain: workflow scheduling in the cloud computing environments.

Salimi et al. (2013) apply a non-dominated sorting genetic algorithm (NSGA) to the task scheduling in a grid computing environment. This is a standard heuristic multi-objective algorithm, but the authors expand it by introducing a variance based fuzzy crossover operator to increase the probability of an intelligent crossover operation, leading to the higher efficiency. In addition, the new algorithm prioritizes the load balancing as one of its objectives. It is shown through an experimental study that the proposed method has better performance and quality than similar algorithms, but our proposed workflow scheduling algorithm operates in a dynamic fashion with its main optimization goal being efficiency.

Using a game theory approach, Fard et al. (2013) developed an auction-based scheduling mechanism to schedule tasks in the cloud. The authors describe an extension of a biobjective scheduling strategy (BOSS) adapted for the dynamic workflow scheduling. The objectives of this algorithm are twofold: minimize the makespan and cost, in which data transfer time is implicitly covered. While this algorithm is dynamic, it does not consider a changing resource environment such as ours, where virtual machines can be reassigned to the different tasks to increase efficiency of the solution.

Pooranian et al. (2013) offer a hybrid meta-heuristic solution for the tasks scheduling in data grids. The base algorithm, which authors use, is the genetic algorithm (GA), but they adapt it to include gravitational emulation local search (GELS), since the GA is weak for local searches. However, this paper only describes the algorithm structure for the static environments where all the necessary data about tasks, resources, and the number of resources is specified before execution. Our workflow scheduling algorithm does not operate using this assumption.

In our work, we introduce a new extended scheme. The main idea of this work is that the proposed scheme can be applied to any metaheuristic algorithm for workflow scheduling and expand it to operate in a dynamically changing cloud environment based on virtualization mechanisms. As an example, in this paper extending scheme is applied for genetic algorithm and allows to optimize the computational environment in conjunction with the assignment of tasks and thus improve its performance relative to the base metaheuristic algorithm.

1.1 Problem Statement

Workflow scheduling is an NP-complete problem, which tries to find the best mapping of tasks to the computational resources according to some criteria and user-defined restrictions. First, for the dynamic scheduling problem, it is important to specify both the workload (workflow) and computational environment formally with the consideration of the time parameter. Typically, workflow is represented by two sets $WF = (T, E)$. The set T represents tasks in the form of their computational capacities $\{t_1, t_2, \ldots, t_n\}$, and the set E includes edges $\{e, e_2, \ldots, e_m\}$ that form an execution order in the WF and represent existing data dependences. The computational environment Env is also defined by two sets $Env = (R, C)$. The set R consists of the computational powers of the resources $\{r_1, r_2, \ldots, r_l\}$ and the set C contains the network bandwidth values between the resources $\{\mathrm{c}_1, \mathrm{c}_2, \ldots, \mathrm{c}_\mathrm{k}\}$. Resources represent virtual machines and can change their performance characteristics. Let $Sch_{WF,Env}$ be a list of ordered pairs

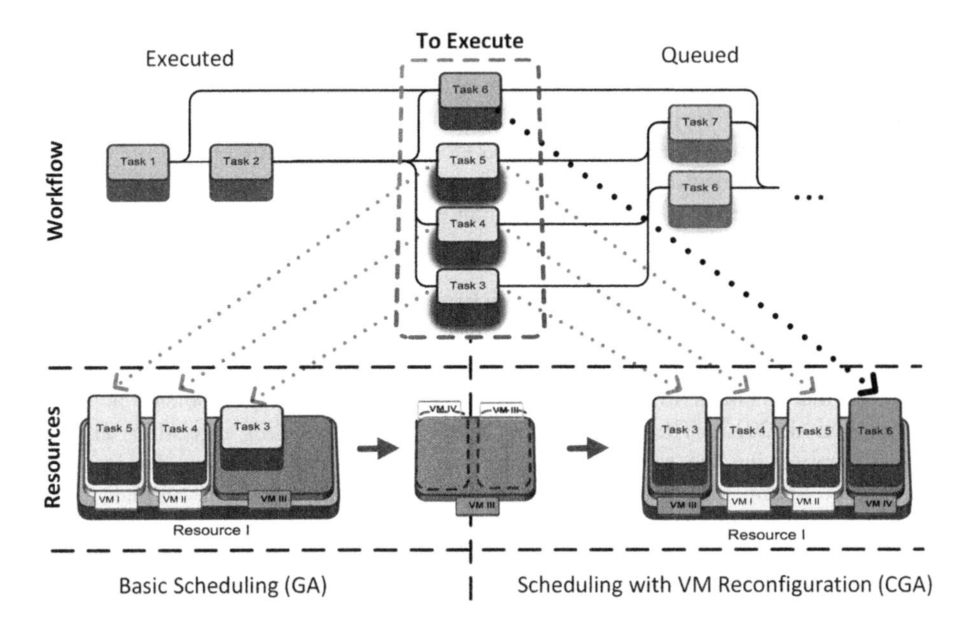

Fig. 1. Basic advantage of virtualization

$\{<t_1^*, r_1^*>, <t_2^*, r_2^*>, \ldots, <t_n^*, r_n^*>\}$, where each t_i^* is a task from a permutation of T with their assigned nodes r_i^*, which can be modified from the initial configuration by the search algorithm. In Fig. 1, the difference in capability between two algorithms is shown. On the left side, the basic genetic algorithm (GA) operating three resources simultaneously is shown. This algorithm chooses which tasks should be executed at the current moment. On the right side, the coevolutionary genetic algorithm (CGA) operating initially also three resources is shown. This algorithm transforms and utilizes 4 resources to map all currently available tasks at once. Let $M(Sch_{WF,Env})$ return the time (makespan) of overall WF task execution of the WF scheduled according to $Sch_{WF,Env}$ (Yu et al. 2005) Then, the optimization goal is to find an algorithm that produces $Sch_{WF,Env}$: $\forall Sch'_{WF,Env} \neq Sch_{WF,Env} \leftrightarrow M(Sch_{WF,Env}) \leq M(Sch'_{WF,Env})$.

As the environment dynamically changes during execution, the time parameter is taken into consideration. We consider dynamic changes of two types: environment modifications (resource and VM failures and new resource deployments) and workflow modifications (tasks fails/delays and newly submitted WFs). Let function $Cl(Al, WF, t_i)$ return the time that is needed to generate scheduling plan $Sch^{Al}_{WF,Env}$ by the algorithm Al at moment t_i. Therefore, the main optimization goal is to find an algorithm Al that $\forall Al' \neq Al \,\&\, \forall WF \,\&\, \forall Env$:

$$\sum_i \left[Cl(Al, WF_i) + M\left(Sch^{Al}_{WF_i,Env_i}, i\right)\right] \leq \sum_i \left[Cl(Al', WF_i) + M\left(Sch^{Al'}_{WF_i,Env_i}, i\right)\right] \quad (1)$$

Each part of the Eq. (1) contains both scheduling components and rescheduling sum of components that are enumerated through the iterator i, which is responsible for a time period t_i between two scheduling processes. Function for the makespan $M(Sch_{WF_i,Env_i}, i)$ with additional parameter i determines only a part of makespan inside current period of time t_i. Wf_i and Env_i represent the remaining set of tasks and current environment during this time period t_i. The Al can be given by a basic scheduling algorithm such as the HEFT (Topcuoglu et al. 2002), Harmony Search or GA algorithms (Melnik and Trofimenko 2015), or as a hybrid of the two (heuristic and metaheuristic scheme) based on the same GA or a more efficient CGA.

In this paper, a new dynamic coevolutionary genetic algorithm (DCGA) is presented and its efficiency is demonstrated in an experimental study.

2 Coevolutionary Genetic Algorithm (CGA)

The coevolutionary genetic algorithm (CGA) is an extension of the basic genetic algorithm that attempts to manage several workflow scheduling problems at once (Nasonov et al. 2015). The first problem, which many scheduling algorithms aim to solve, is optimizing the mapping of tasks to computational resources. The second problem is optimizing resources' configurations, usually based on virtualization mechanisms. This problem is of particular relevance today due to the increasing popularity of cloud computing, which uses virtualization as one of the core technologies in the computational environment organization. The CGA combines mapping tasks to resources and resources reconfiguration in one coevolutionary process that includes two different populations with one aggregate estimating function (fitness). In other words, an individual from one population can be estimated only when paired with an individual from the other population and not in any other way (see Fig. 2 below). The process of estimation is commensurate with population size by repeat count.

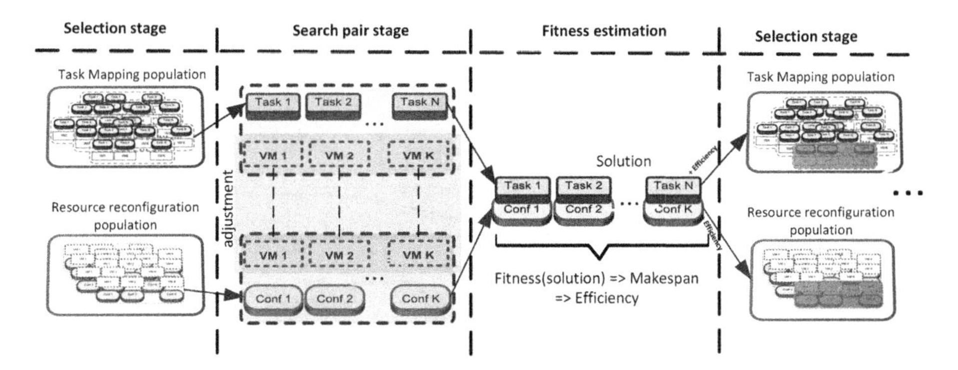

Fig. 2. Estimation of species in coevolution genetic algorithm

3 Dynamic Coevolutionary Genetic Algorithm (DCGA)

The dynamic coevolutionary genetic algorithm (DCGA) has certain design peculiarities that should be considered in the implementation (certain similarities can be found in our previous work (Nasonov et al. 2015). They stem from the fact that the computational environment is changing with time and the evolutionary process should be adapted to the resources and virtual machines failures, increasing and decreasing performance, task delays, unexpected operation system halts, and so on. This is partly demonstrated in Fig. 3. Before the time t_i occurs, the running background process of the DCGA tries to improve the previously generated and adapted schedule to the system through the cycled procedure, consisting of "Mutation", "Crossover", "Selection", and "Fitness" blocks for both evolutions. At the time t_i, a system event fires, informing that task 3 and 4 has completed the execution and task 6 is starting. This event sends a message to the DCGA and forces a population correction, changing the current state of all species in task 6 from "to Execute" to "Executing on VM 3". When this correction is complete, the DCGA continues its optimization search with the updated mapping and tries to find more efficient plan. A similar process of adaptation is executed for the evolution of resources if the occurred event changes the computational environment in any way.

3.1 Chromosome

The chromosome from the task mapping population is represented as a sequence of tasks and virtual machines, each indexed by the identification number (ID). This representation defines the ordering of tasks to be executed and the virtual machine assigned for each task.

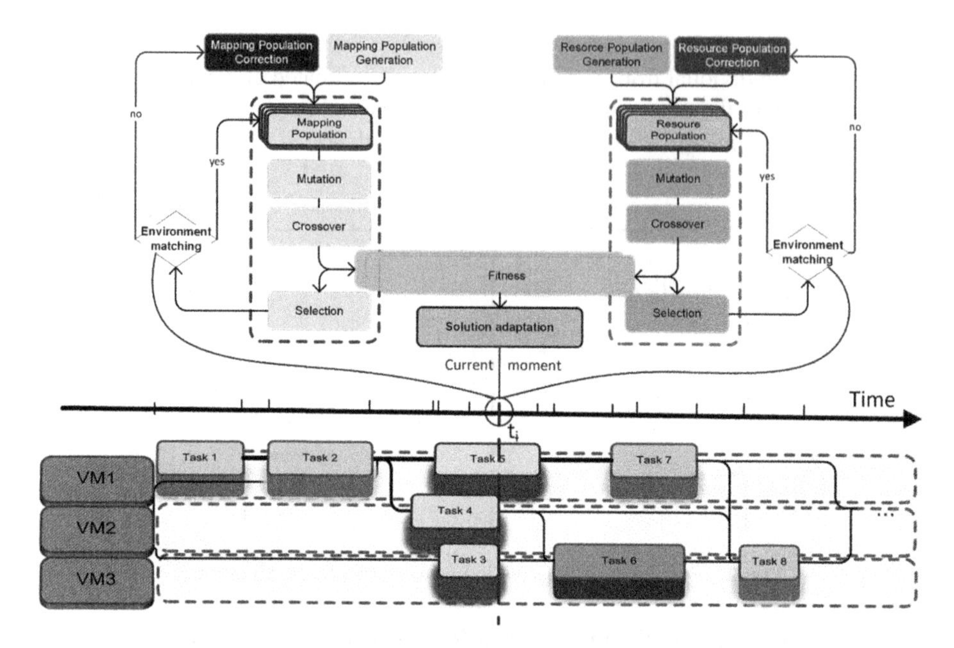

Fig. 3. The schema of DCGA execution.

The chromosome from the resource reconfiguration population is represented as a set of nodes and configurations, also indexed by ID. Configuration here may be defined as any set of parameters, such as the number of cores, size of ram etc. In our study, we use a single parameter to describe the configuration of a virtual machine: capacity. Additionally, it should be noted that we establish a static number of virtual machines for each resource. Thus, to shut down a virtual machine, its capacity must be set to zero.

Examples of these representations are shown in Fig. 4.

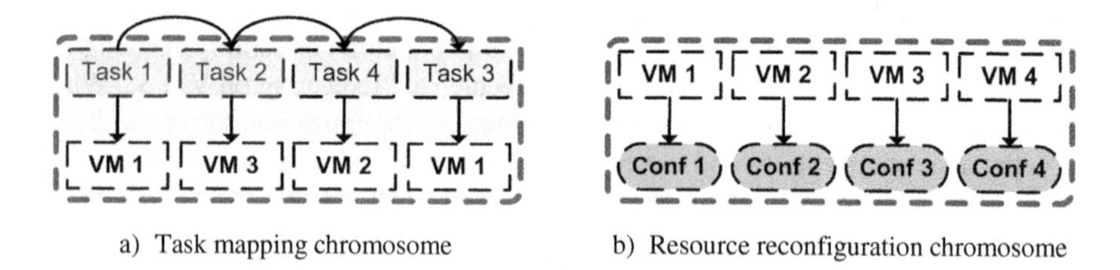

a) Task mapping chromosome

b) Resource reconfiguration chromosome

Fig. 4. Both types of chromosome representations

3.2 Crossover

For the crossover of chromosomes from the task mapping population, we use a 2-point crossover with 2 parents. A child inherits left and right parts from the first parent, while the middle part is inherited from the second parent. In Fig. 5a, the child received left

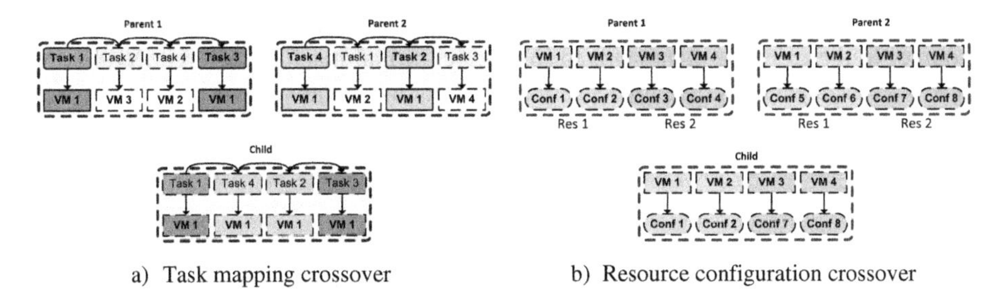

a) Task mapping crossover b) Resource configuration crossover

Fig. 5. Crossover operators for both populations

and right tasks (task 1, task 3) from the first parent. All other tasks (task 4, task 2) were received from the second parent and placed at the middle.

For the crossover of the resource reconfiguration population, the child solution receives the division for each computational resource into virtual machines randomly from one of two parents. In Fig. 5b, there are two physical resources, which are divided into two virtual machines (res 1: (vm 1, vm 2); res 2: (vm 3, vm 4)). The child received the division of the first resource from the first parent, and the configuration of the second resource from the second parent.

3.3 Mutation

Mutation operators are shown in Fig. 6.

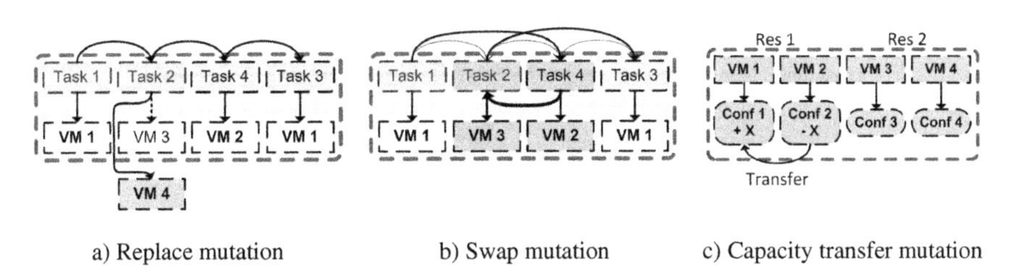

a) Replace mutation b) Swap mutation c) Capacity transfer mutation

Fig. 6. Mutation operators for both populations

There are two options when considering mutation for the chromosomes from the task mapping population. The first option randomly replaces one of the assigned nodes for a randomly chosen task. In Fig. 6a, vm4 was chosen as a new computational node for task 2.

The second type of mutation swaps two tasks in a chromosome together with their assigned nodes. In the example illustrated in Fig. 6b, gene (task 2: vm 3) is swapped with the gene (task 4: vm 2).

At the mutation step of the resource configuration mutation, computational resources are chosen randomly. Then, one of the virtual machines on this resource transfers part or full capacity to another virtual machine on the same computational resource. In the example demonstrated in Fig. 6c, virtual machine 2 is reduced by the value of X, while virtual machine 1 is extended by the same value X.

3.4 Fitness

As mentioned previously, the fitness function can be applied only for a complete paired solution (i.e. a chromosome from the task mapping population and a chromosome from the resource configuration population). The current execution state contains the current environment and schedule with fixed tasks which have already been executed, are running currently, or have failed. A new computational environment is created from the resource configuration chromosome. After that, each task from the task mapping chromosome must locate the time slot from the current schedule on the assigned node. When all tasks are assigned to the schedule, the last finish time represents the makespan of this specific schedule and the fitness value for the current pair of chromosomes.

However, since we can change the capacities of the virtual machines, tasks may not start immediately after the last task in the queue of the current virtual machine in the case that this virtual machine was extended. An example is shown in Fig. 7. After the failure of task 2, the algorithm fixed task 1 and task 3. Then, the algorithm found a solution where vm 2 was extended and vm 3 was reduced. In the following step, the next task (task 4) can't be placed after task 2 immediately.

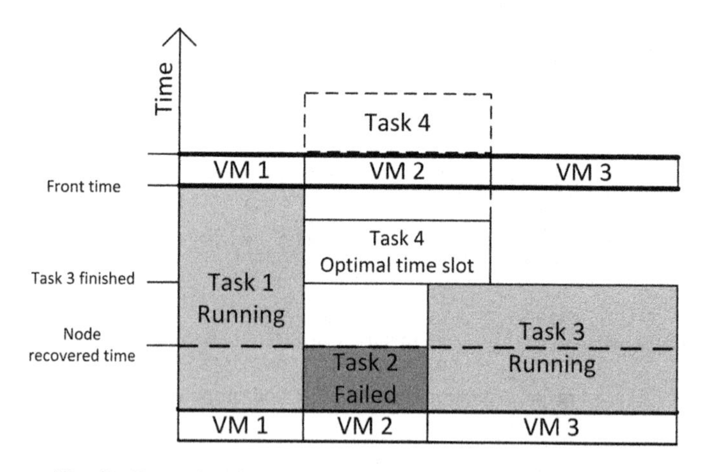

Fig. 7. Example of the boundary on the one physical resource

The easiest and most robust approach to avoid this problem is to find the last fixed task and use its end time (front time) as a minimal time slot for new tasks. However, our algorithm tries to lower tasks in the queue. To achieve this, we take end times of all fixed and already placed tasks and sort them in decreasing order. Iterating through this sorted list, free capacity at the current moment in time is calculated. If free capacity is enough for the placement of the task, then this task is lowered. Otherwise, the cycle is terminated.

3.5 Selection

Standard roulette wheel is chosen as a selection operator for each population.

4 Experimental Study

Experiments were conducted using our own workflow execution simulator. The computational environment is represented by a set of resources with their capacities and network bandwidth between resources (global bandwidth). Each computational resource is divided into several virtual machines. The bandwidth between virtual machines of one computational resource (local bandwidth) is much higher than the global bandwidth. For this experimental study, three well-known types of synthetic workflows – Montage (Mon), CyberShake (CbS) and Inspiral (Ins) – were generated using Pegasus workflow generator (Deelman et al. 2015) with a different number of tasks (25, 30 and 50). For example, Mon_25 means Montage workflow with 25 tasks in total.

The initial computational environment contains three physical computational resources. Each of the resources is divided into three virtual machines with capacities 10, 10, and 20. Capacity is measured in conventional units. Data transfer time between two virtual machines inside one physical resource (local bandwidth) is 10 Mbit/s, while the data transfer time between two physical resources (global bandwidth) is 100 Mbit/s. The workflow description file contains the run times for all of the tasks. A task's execution time is calculated by $execTime(task, node) = \frac{task.runTime}{node.capacity}$. Each computational node has 95 % probability of successful completion of a task. In case of the failure, the node is marked as failed for a specified period of time, and the scheduling process is launched again taking into account already finished tasks and failed resources.

First, the efficiency of our proposed strategy of boundary task placement in comparison with front time task placement is shown in Table 1. The values in the table below represent the profit of the optimal task placement for the full schedule's makespan. Thus, this small local optimization provides a profit of up to 5 %.

Table 1. Efficiency of the boundary optimization scheme.

	Mon_25	Mon_50	CbS_30	CbS_50	Ins_30	Ins_50
Profit, %	2.1	2.6	4.8	5.0	1.9	3.7

The next experiment compares three algorithms. The first is the popular heuristic algorithm HEFT, the second is our genetic algorithm with operators described in Sect. 3 (only with single task mapping population, however), and the third is our proposed DCGA algorithm. Since HEFT is a fast heuristic algorithm that can provide a suitable solution in a short period of time, the GA and DCGA are initiated with one initial chromosome, generated by HEFT. The results of these experiments are shown in Fig. 8.

It can be seen that the GA with the initial HEFT solution always outperforms simple HEFT. Nevertheless, with the ability to reconfigure computational resources, our DCGA outperforms the GA in all cases. For example, we can observe a huge profit

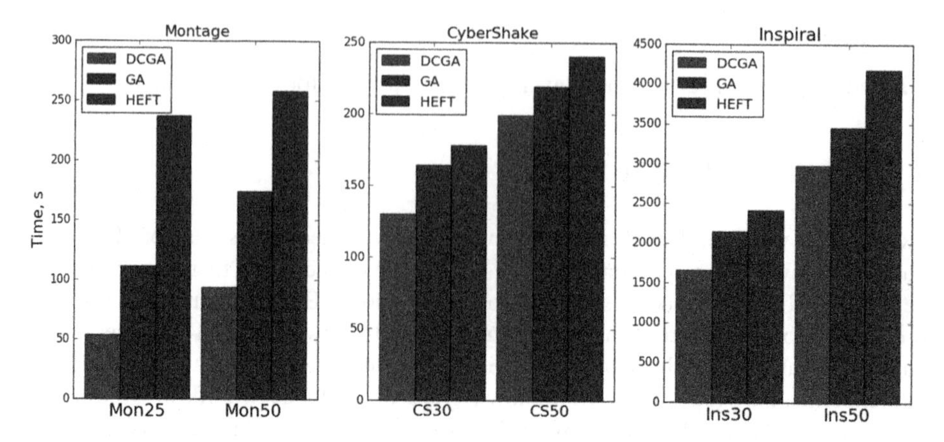

Fig. 8. Performance of the algorithms on different workflows

of almost 350 % in the case of Montage with 25 tasks, which is a data-intensive workflow. Thus, the result of applying our DCGA on Montage_25 is a solution where three initial virtual machines (10, 10, 20 capacities) were combined into one virtual machine with a capacity of 40, and all 25 tasks were assigned to this one virtual machine to avoid any data transfers.

In the next experiment, we investigated the dependence of the efficiency of each algorithm on the node recovery time. If a task failed, the virtual machine is marked as failed for a certain time, the node recovery time. The results of this experiment are presented in Fig. 9.

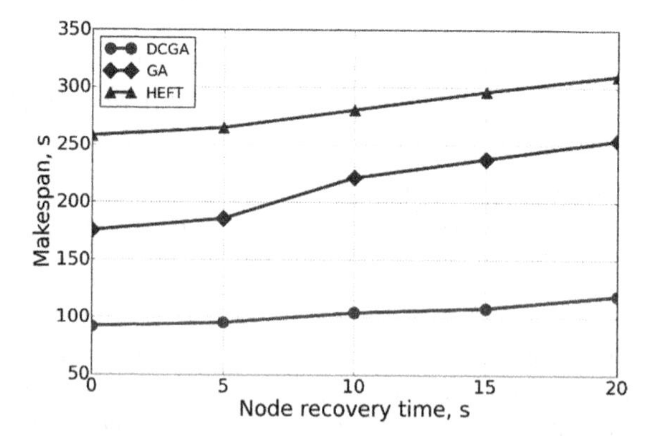

Fig. 9. Performance of the algorithms on Montage 25 dependent on the node recovery time

From this experiment, it is evident that the dependence of the DCGA on the node recovery time is much more stable than the GA and HEFT, due to the ability to reconfigure the computational environment and apply the change.

5 Conclusion

A novel dynamic coevolution genetic algorithm (DCGA) for workflow scheduling in a dynamic cloud environment was proposed in this paper. Based on the virtualization technology of computational resources, this approach allows physical computational resources to be divided into virtual machines. Thus, the classic workflow scheduling problem is extended by an additional optimization problem to consider resources reconfiguration. To resolve this, our coevolutionary approach was applied to divide two optimization problems into two different populations that evolve together over time. Since this extension only supplements the standard scheduling algorithm, the proposed scheme can be easily applied for any metaheuristic algorithm. In addition, our DCGA allows the optimization of task placement at the boundary, occurring when there are changes in the computational environment. Finally, experiments were conducted in the dynamically changing simulation environment, where failures and other changes might occur. In comparison to algorithms that do not include the ability to reconfigure the computational environment, our proposed scheme outperforms all of them, since it has no restrictions in the solution space.

Acknowledgements. This paper is financially supported by Ministry of Education and Science of the Russian Federation, Agreement #14.578.21.0077 (24.11.2014).

References

Butakov, N., Nasonov, D.: Co-evolutional genetic algorithm for workflow scheduling in heterogeneous distributed environment. In: 2014 IEEE 8th International Conference on Application of Information and Communication Technologies (AICT), pp. 1–5. IEEE, October 2014

Deelman, E., Vahi, K., Juve, G., Rynge, M., Callaghan, S., Maechling, P.J., Mayani, R., Chen, W., da Silva, R.F., Livny, M., Wenger, K.: Pegasus, a workflow management system for science automation. Future Gener. Comput. Syst. **46**, 17–35 (2015)

Fard, H.M., Prodan, R., Fahringer, T.: A truthful dynamic workflow scheduling mechanism for commercial multicloud environments. IEEE Trans. Parallel Distrib. Syst. **24**(6), 1203–1212 (2013)

Guo, Y., Cao, X., Yin, H., Tang, Z.: Coevolutionary optimization algorithm with dynamic sub-population size. Int. J. Innovative Comput. Inf. Control **3**(2), 435–448 (2007)

Liu, R., Ma, C., Ma, W., Li, Y.: A multipopulation PSO based memetic algorithm for permutation flow shop scheduling. Sci. World J. (2013)

Korkhov, V.V., Moscicki, J.T., Krzhizhanovskaya, V.V.: The user-level scheduling of divisible load parallel applications with resource selection and adaptive workload balancing on the grid. IEEE Syst. J. **3**(1), 121–130 (2009)

Krzhizhanovskaya, V.V., Korkhov, V.V.: Dynamic load balancing of black-box applications with a resource selection mechanism on heterogeneous resources of the grid. In: Malyshkin, V. (ed.) PaCT 2007. LNCS, vol. 4671, pp. 245–260. Springer, Heidelberg (2007). doi:10.1007/978-3-540-73940-1_26

Melnik, M., Trofimenko, T.: Polyrhythmic harmony search for workflow scheduling. Procedia Comput. Sci. **66**, 468–476 (2015)

Nasonov, D., Melnik, M., Shindyapina, N., Butakov, N.: Metaheuristic coevolution workflow scheduling in cloud environment. In: Proceedings of 7th International Conference on Evolutionary Computation Theory and Applications, pp. 252–260 (2015)
Pooranian, Z., Shojafar, M., Tavoli, R., Singhal, M., Abraham, A.: A hybrid metaheuristic algorithm for Job scheduling on computational grids. Informatica **37**(2), 157 (2013)
Rahman, M., Hassan, R., Ranjan, R., Buyya, R.: Adaptive workflow scheduling for dynamic grid and cloud computing environment. Concurrency Comput. Pract. Experience **25**(13), 1816–1842 (2013)
Salimi, R., Bazrkar, N., Nemati, M.: Task scheduling for computational grids using NSGA II with fuzzy variance based crossover. Advances in Computing **3**(2), 22–29 (2013)
Singh, R., Singh, S.: Score based deadline constrained workflow scheduling algorithm for Cloud systems. Int. J. Cloud Comput. Serv. Archit. (IJCCSA) **3**(6), 31–41 (2013)
Tang, M., Yusoh, Z.I.M.: A parallel cooperative co-evolutionary genetic algorithm for the composite SaaS placement problem in cloud computing. In: Coello, C.A., Cutello, V., Deb, K., Forrest, S., Nicosia, G., Pavone, M. (eds.) PPSN 2012, Part II. LNCS, vol. 7492, pp. 225–234. Springer, Heidelberg (2012). doi:10.1007/978-3-642-32964-7_23
Topcuoglu, H., Hariri, S., Wu, M.Y.: Performance-effective and low-complexity task scheduling for heterogeneous computing. IEEE Trans. Parallel Distrib. Syst. **13**(3), 260–274 (2002)

Searching Parsimonious Solutions with GA-PARSIMONY and XGBoost in High-Dimensional Databases

Francisco Javier Martinez-de-Pison(✉), Esteban Fraile-Garcia, Javier Ferreiro-Cabello, Rubén Gonzalez, and Alpha Pernia

EDMANS Group, University of La Rioja, Logroño, Spain
edmans@dim.unirioja.es, fjmartin@unirioja.es
http://www.mineriadatos.com

Abstract. EXtreme Gradient Boosting (XGBoost) has become one of the most successful techniques in machine learning competitions. It is computationally efficient and scalable, it supports a wide variety of objective functions and it includes different mechanisms to avoid over-fitting and improve accuracy. Having so many tuning parameters, soft computing (SC) is an alternative to search precise and robust models against classical hyper-tuning methods. In this context, we present a preliminary study in which a SC methodology, named GA-PARSIMONY, is used to find accurate and parsimonious XGBoost solutions. The methodology was designed to optimize the search of parsimonious models by feature selection, parameter tuning and model selection. In this work, different experiments are conducted with four complexity metrics in six high dimensional datasets. Although XGBoost performs well with high-dimensional databases, preliminary results indicated that GA-PARSIMONY with feature selection slightly improved the testing error. Therefore, the choice of solutions with fewer inputs, between those with similar cross-validation errors, can help to obtain more robust solutions with better generalization capabilities.

Keywords: XGBoost · Genetic algorithms · Parameter tuning · Parsimony criterion · GA-PARSIMONY

1 Introduction

EXtreme Gradient Boosting (XGBoost) [5] has been one of the most popular machine learning methods during the last years. In the last Kaggle [12] and KDD-Cup [13] challenges, many of the top-positioned competitors used XGBoost as their main tool to generate highly accurate models. However, with so many parameters a high human effort is necessary to solve the tuning stage with success. Therefore, Data Scientists are demanding methodologies to automatize the tuning process avoiding over-fitting and to seek models with good generalization capabilities.

M. Graña et al. (eds.), *International Joint Conference SOCO'16-CISIS'16-ICEUTE'16*,
Advances in Intelligent Systems and Computing 527, DOI 10.1007/978-3-319-47364-2_20

Between them, GA-PARSIMONY is a Soft Computing (SC) methodology proposed in [18], which tries to find good parsimonious models by using genetic algorithms (GA). The main objective of GA-PARSIMONY is to optimize a KDD-scheme by selecting the best features, transforming the skewed data, searching the best parameters for the machine learning algorithm, and selecting the best parsimonious models by means of a second selection of the best individuals, which is based on a complexity measurement.

The main objective of this study is to glimpse the possibilities of using GA-PARSIMONY with XGBoost to automatize the tuning process in order to obtain good parsimonious XGBoost solutions. In particular, this preliminary work tries to solve the following questions: (1) Is Feature Selection useful with XGBoost? (2) Which is the best metric to measure the complexity of XGBoost models?, and therefore, (3) Can we obtain better parsimonious solutions using GA-PARSIMONY with XGBoost in high dimensional datasets?

The article is organized as follows: in Sect. 2 a brief description of XGBoost and GA-PARSIMONY is presented. In the next point, Sect. 3, the experiments performed with six high dimensional datasets are described. Section 4 shows a discussion of the experimental results and finally, Sect. 5, presents the conclusions and suggestions for further research.

2 Materials and Methods

2.1 Extreme Gradient Boosting Machines

XGBoost [5] is based on the principle of gradient boosting machines (GBMs) proposed by Friedman [10], a technique that generates a boosting ensemble of weak prediction models by optimizing a differentiable loss function with gradient descent algorithm. However, XGBoost uses a more regularized model strategy to control over-fitting. Also, it is computationally more efficient, it is scalable and it consumes less memory.

XGBoost supports a lot of different objective functions that enable users to adapt it for any kind of problems. It also includes several mechanisms to avoid over-fitting and control the model complexity. For example, it incorporates Type L1 (Lasso) and L2 (Ridge) penalties into the loss function. It also contains pruning parameters to reduce the tree complexity by minimizing the depth of each tree, the sum of instances weight per leaf, the minimum loss reduction in each partition, etc. Moreover, it integrates "random subspaces" and "random subsampling" methods to reduce the variance and prevent over-fitting.

The tuning of these parameters is done during the training process, but the final model selection depends on the validation. Although the implementation of a validation process usually guarantees solutions with good generalization capabilities, it cannot be ensured that the best parsimonious model is obtained. Thus, there is an increasing tendency to include the model complexity into the selection process.

2.2 Seeking Good Parsimonious Models with GA-PARSIMONY

According with the "Principle of Parsimony" (Occam's Razor), between equally valid hypotheses, the simplest explanation is probably the best. The main idea is to select the least complex model among the solutions with good accuracy as it entails numerous advantages: it facilitates understanding the problem, models are more robust against perturbations or noise, and it simplifies future updating and exploiting stages with the consequence of reducing human and economic effort involved.

However, selecting the least complex model is a challenging task that depends on multiple factors. In particular, when trying to optimize these tasks, soft computing (SC) seems to be an interesting alternative to the classical searching methods [3,6,15,17,26]. Several authors have reported SC applications to different real fields where feature selection (FS), model parameters optimization (MPO) or data transformation (DT) are optimized with several bioinspired optimization methods [1,4,7,8,11,19,25].

In this context, we proposed a SC methodology named GA-PARSIMONY in [18] to automatically obtain good overall parsimonious models. This methodology uses Genetic Algorithms (GA) and generates parsimonious models, while performing the preprocessing of skewed data, parameter tuning and feature selection at a time. The flowchart is similar to a classical GA process but includes a model selection step arranged in two stages: first, best individuals are sorted by their fitness function based on an cross validation error metric like Root Mean Squared Error (RMSE), and second, models with similar accuracy are rearranged according to their complexities, promoting the simplest ones to the top positions (*ReRank* process). Testing error are only used to check the evolution of the models generalization capabilities.

GA-PARSIMONY has successfully been applied in the estimation of solar radiation [2], the mechanical characterization of T-sub components [9], the modeling of industrial steel processes [20], and hotel room demand forecasting [23]. Nevertheless, the methodology has not been yet evaluated with ensemble techniques like XGBoost.

3 Experiments

XGBoost was evaluated with seven GA-PARSIMONY configurations (Table 1) and six UCI high-dimensional datasets (Table 2). Two aspects were varied in the different implementations of the GA-PARSIMONY, (1) whether to use or not feature selection, and (2), the method used to compute the complexity of the models in the re-ranking process. The methods available to evaluate the complexity were:

- *None.* The re-rank process based on the complexity is not executed. Reference method.
- *Inner complexity of the model.* The complexity of the model equals the sum of the number of leaves (NLeaves) of all trees that conform the XGBoost.

- *Number of features (N_{FS}).* The complexity of the models defined as the number of inputs.
- *Generalized Degrees of Freedom (GDF).* The GDF is a metric that relates the complexity of a model to its ability to perform against random perturbations or noise. It was proposed by [27] and it is applicable to any modeling method. The intrinsic idea is that highly complex models easily fit perturbed data. Then, GDF is defined as the relation of sensitivities to perturbed data with respect to the original one:

$$GDF = \sum_{i=1}^{n} \frac{\Delta \tilde{y}_i}{\Delta y_i} = \frac{\tilde{y}_{i,pert} - \tilde{y}_i}{y_{i,pert} - y_i} \quad (1)$$

where n is the number of instances, $y_{i,pert}$ and y_i are the perturbed and original output for each i-th instance; and $\tilde{y}_{i,pert}$ and $\tilde{y}_i$ the predicted values. In practice, Seni and Elder [21] proposed the calculation of GDF in these steps: adding random noise to the output, training the model without and with perturbed data and measure the difference between fitted predictions of the model with original output variable and perturbed signal. The process is repeated m_{gdf} times and the sensitivity of each instance is evaluated by creating a linear regression model (LR) of $\Delta \tilde{y}_i$ vs. Δy_i. The average of the slopes of all LR models determines GDF. In this work, it was assumed $m_{gdf} = 10$.

Table 1. Configurations of the GA-PARSIMONY implemented.

		# Without FS	# With FS
Complexity	None	*None*	*None_FS*
	NLeaves	*NLeaves*	*NLeaves_FS*
	N_{FS}	-	N_{FS}*_FS*
	GDF	*GDF*	*GDF_FS*

3.1 GA-PARSIMONY Settings

Databases were split into a 70 % for the validation process and the other 30 % was used for checking the generalization capability of each model. In particular, the validation procedure was a 5 times 4-fold CV. The settings for the GA were as follows: The fitness function selected was the average CV error for the 5 runs ($J = RMSE_{val}$). The re-ranking process was conducted for individuals with their Js within a maximum difference of 0.1 %. An elitism percentage of 25 % was used. Selection and crossing methods were *random uniform* and *heuristic blending* respectively [14]. A mutation percentage of 10 % was used but the best two individuals were not mutated. The population size was set to $P = 64$ individuals and the maximum number of generations to $G = 40$. However, an early stopping strategy was implemented when the J of the best individual did not decrease more than a 0.1 % in $G_{early} = 10$ generations.

Table 2. Database description (Samples - *#Inst* -, and attributes - *#Attr* -) and results with GA-PARSIMONY without FS. *Gen* stands for the last generation and *Time* for the execution time.

Database			*None*			*NLeaves*			*GDF*		
Name	# Inst	# Attr	Gen	Time	$RMSE_{tst}$	Gen	Time	$RMSE_{tst}$	Gen	Time	$RMSE_{tst}$
Ailerons	13750	40	15	2397	0.0434	40	513	0.0451	18	4182	0.0447
Bank	8192	32	14	1369	0.1013	32	880	0.1038	19	3529	0.1036
Cpu	8192	21	18	2042	0.0231	21	422	0.0257	19	3576	0.0245
Elevators	16599	18	19	1933	0.0311	18	1163	0.0327	19	7433	0.0330
Pol	15000	26	24	7399	0.0394	26	1714	0.0410	18	12755	0.0402
Puma	8192	32	13	2424	0.0411	32	1136	0.0431	19	7747	0.0424

Table 3. Results with GA-PARSIMONY with FS. The number in parentheses stands for the number of inputs selected (Num_{FS}).

	None_FS			N_{FS}*_FS*			*NLeaves_FS*			*GDF_FS*		
Name	Gen	Time	$RMSE_{tst}$	Gen	Time	$RMSE_{tst}$	Gen	Time	$RMSE_{tst}$	Gen	Time	$RMSE_{tst}$
Ailerons	16	1503	(23) 0.0428	14	911	(9) 0.0438	20	759	(18) 0.0450	16	4871	(24) 0.0444
Bank	29	1235	(25) 0.0993	28	787	(7) 0.1014	40	1011	(20) 0.1013	24	9742	(16) 0.1015
Cpu	26	2191	(17) 0.0227	23	1282	(7) 0.0240	21	726	(17) 0.0247	33	4481	(13) 0.0239
Elevators	26	2904	(13) 0.0310	22	2010	(6) 0.0325	24	1719	(11) 0.0334	28	9742	(8) 0.0331
Pol	40	11837	(16) 0.0404	40	6352	(11) 0.0410	40	7984	(17) 0.0424	40	31361	(16) 0.0414
Puma	29	1801	(4) 0.0338	38	2990	(4) 0.0333	40	2501	(4) 0.0339	40	14231	(6) 0.0351

XGBoost parameters were defined between the following ranges: $nrounds = [010, 999]$, $eta = [0.01, 0.30]$, $max\_depth = [01, 20]$, $min\_child\_weight = [0.01, 0.99]$, $subsample = [0.80, 1.00]$, $colsample\_bytree = [0.40, 1.00]$. The random seed was fixed to 1234.

The representation of each individual (i) and generation (g) was a chromosome $\lambda_g^i = [nrounds, eta, max\_depth, min\_child\_weight, subsample, colsample\_bytree, Q]$, where the first six values were the XGBoost parameters and Q was a binary-coded array that includes the selected features.

All experiments were conducted with the statistical software `R` [16] and `XGBoost` package [5], with dual quad-core opteron servers (Intel ®Xeon ®CPU E5410 @ 2.33 GHz).

4 Results and Discussion

Tables 2 and 3 summarize the results for the six datasets with the seven GA-PARSIMONY configurations.

4.1 Is Feature Selection Useful with XGBoost?

The first important question is to analyze whether the use of feature selection into the optimization process can improve the model generalization capabilities. Although XGBoost performs well with many attributes, similar to other tree-based ensemble algorithms, the basic idea is that FS can remove some attributes

Table 4. Results of GA-PARSIMONY with vs without FS.

Database	Without FS		With FS		p-value
	$RMSE_{tst}^{mean}$	$RMSE_{tst}^{sd}$	$RMSE_{tst}^{mean}$	$RMSE_{tst}^{sd}$	
Ailerons	0.044394	0.001151	0.044010	0.000980	−(0.006714)
Bank	0.102912	0.000841	0.100874	0.001144	−(0.000122)
Cpu	0.024403	0.002224	0.023843	0.001027	=(0.107000)
Elevators	0.032269	0.000629	0.032488	0.000609	−(0.055360)
Pol	0.040186	0.001926	0.041322	0.001970	+(0.000610)
Puma	0.042214	0.001064	0.034012	0.000587	−(0.000061)

that are selected by the algorithm but could produce over-fitting or reduce accuracy.

Mean and standard deviation of $RMSE_{tst}$ for GA-PARSIMONY without and with FS are depicted in Table 4. Last column depicts the p-values for the Wilcoxon Signed-Rank [24] between both strategies. Minus or plus sign indicates that they are statistically significant different. Although there were small differences between them, the last column shows than GA-PARSIMONY with FS improved the $RMSE_{tst}$ in three out of six datasets tested. Also, p-value is close to the usual significance level of 0.05 in *elevators*.

4.2 Which Is the Most Useful Metric?

The Shaffer post-hoc test [22] was used to perform a multiple comparison between all 4 strategies which measure complexity. Adjusted p-values using the Shaffer post-hoc test are showed in Table 5. A "+" sign denotes that the method in the row is statistically better than the one in the column, a "−" indicates the contrary, and a "=" implies that there is no significant differences between both of them. Results show that GA-PARSIMONY with FS ($None\_FS$) performed better in *elevators* when compared against the other complexity metrics. This strategy also improved $NLeaves\_FS$ and GDF_FS in *ailerons* and *puma* respectively. Otherwise, in *bank* all p-values of $None\_FS$ were close to 0.05, similar to GDF_FS in *ailerons*. However, it has to be noted that the rearranging process implemented with N_{FS}_FS, $NLeaves\_FS$ and GDF_FS is constrained to a margin $RMSE_{val}$ of 0.1 %, and several differences of the $RMSE_{tst}$ with $None\_FS$ were below this margin.

In many real applications, it is more important to obtain robust and parsimonious solutions rather than solutions with lower generalization error in their third or fourth decimal numbers. A parsimonious solution with a reduced number of inputs, or with a less internal complexity, will be more robust against perturbations or noise. Also, they will be more simple to update or exploit, and they will require less effort in capturing and preprocessing the information. In this context, we can draw different conclusions between the proposed complexity metrics:

Table 5. Shaffer post-hoc test for the four strategies and six datasets.

	None_FS	*GDF_FS*	*NLeaves_FS*	N_{FS}*_FS*		*None_FS*	*GDF_FS*	*NLeaves_FS*	N_{FS}*_FS*	
None_FS	x	=(0.0533)	+(0.0184)	=(0.2571)	*ailerons*	x	+(0.0002)	+(0.0001)	+(0.0033)	*elevator*
GDF_FS	=(0.0533)	x	=(0.4120)	=(0.3373)		-(0.0002)	x	=(0.4593)	=(0.1108)	
NLeaves_FS	-(0.0184)	=(0.4120)	x	=(0.2571)		-(0.0001)	=(0.4593)	x	=(0.0791)	
N_{FS}*_FS*	=(0.2571)	=(0.3373)	=(0.2571)	x		-(0.0033)	=(0.1108)	=(0.0791)	x	
None_FS	x	=(0.0587)	=(0.0587)	=(0.0587)	*bank*	x	=(1.0000)	=(0.8210)	=(1.0000)	*pol*
GDF_FS	=(0.0587)	x	=(1.0000)	=(1.0000)		=(1.0000)	x	=(1.0000)	=(1.0000)	
NLeaves_FS	=(0.0587)	=(1.0000)	x	=(1.0000)		=(0.8210)	=(1.0000)	x	=(0.880)	
N_{FS}*_FS*	=(0.0587)	=(1.0000)	=(1.0000)	x		=(1.0000)	=(1.0000)	=(0.880)	x	
None_FS	x	=(0.4390)	=(0.1850)	=(0.4390)	*cpu*	x	+(0.0062)	=(0.8016)	=(0.4054)	*puma*
GDF_FS	=(0.4390)	x	=(1.0000)	=(1.0000)		-(0.0062)	x	-(0.0071)	-(0.0008)	
NLeaves_FS	=(0.1850)	=(1.0000)	x	=(1.0000)		=(0.8016)	+(0.0071)	x	=(0.4054)	
N_{FS}*_FS*	=(0.4390)	=(1.0000)	=(1.0000)	x		=(0.4054)	+(0.0008)	=(0.4054)	x	

- Number of leaves ($NLeaves\_FS$): Fig. 1 shows the evolution of optimization process for *ailerons* dataset, using FS and NLeaves as the complexity metric. The continuous line denotes $RMSE_{val}$ of the best individual of each generation while the dashed-dotted the $RMSE_{tst}$. White box-plots corresponds with the $RMSE_{val}$ evolution of the elitist individuals and $RMSE_{tst}$ is the gray-filled. The dashed line is the best individuals' Num_{FS} and the shaded area represents the Num_{FS} range of elitist models.
  In this case, the validation and testing errors started to increase within the error margin (0.001) after generation 9, obtaining worse solutions with a similar number of features. The same behavior was observed in the other datasets (data not shown). Therefore, we inferred that the number of leaves is not a practical metric because it did not lead to obtaining better parsimonious solutions.
- Generalized Degrees of Freedom (GDF_FS): the performance observed when using this metric was quite heterogeneous. In some databases, error decreased fast in the first generations but after that, $RMSE_{val}$ increased with the $RMSE_{tst}$ (similar to NLeaves). In other databases, the optimization was chaotic with several ups and downs appearing in the process. In brief, it seemed that the GA-PARSIMONY was not able to find parsimonious solutions with this metric. Moreover, the GDF calculation process multiplied by two, three or even four the elapsed time.
- Number of Features Selected (N_{FS}_FS): GA-PARSIMONY generated a $RMSE_{tst}$ slightly better with this metric compared to the other complexity proposals. Although it did not perform better than $None\_FS$, it found models with an error difference with $None\_FS$ lower or equal than the established error margin of 0.1 % in three data-bases. However, in five of the six datasets, an important reduction of Num_{FS} with a similar or less number of generations is obtained. For example, in *ailerons*, a high reduction from 23 to 9 Num_{FS} was obtained, in *bank* from 25 to 7, in *cpu* from 17 to 7, and in *elevators* from 13 to 6 (see Table 3). Therefore, we can conclude that use

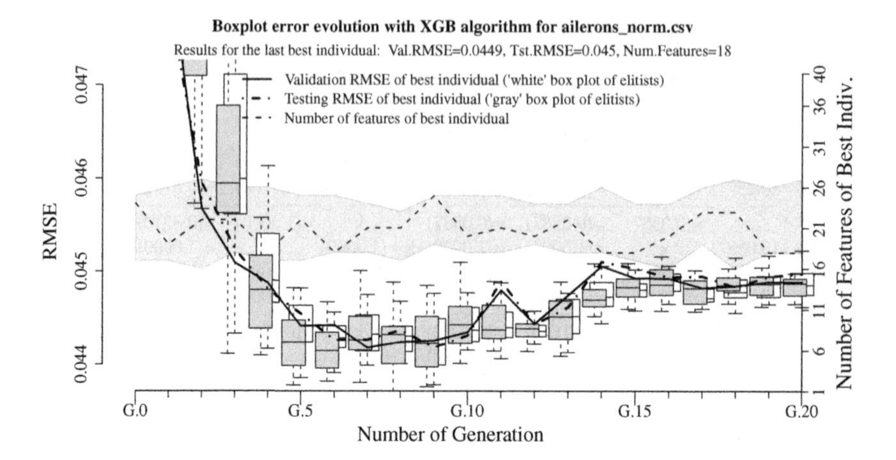

Fig. 1. Evolution of the optimization process using the GA-PARSIMONY with the total number of leaves (NLeaves) as complexity metric in the *ailerons* dataset.

GA-PARSIMONY with N_{FS}_FS could be an interesting configuration in the search of parsimonious models for real applications or Kaggle competitions if a correct error margin is established.

5 Conclusions and Future Work

In this paper, we present a preliminary study in which we implement the GA-PARSIMONY optimization methodology in high dimensional databases with XGBoost and different complexity metrics. The main objective was to determine whether feature selection and the use of rearranging models by complexity can improve the model generalization capabilities. For this purpose, three complexity metrics were used: the sum of model leaves (NLeaves), Generalized Degrees of Freedom (GDF) and the number of features selected (N_{FS}). They were benchmarked against the option of not using the rearrangement process based on complexity (reference method). All four strategies were implemented with and without feature selection.

Although acceptable results were obtained without feature selection (only tuning XGBoost parameters), $RMSE_{tst}$ improved in practically four out of six datasets using feature selection with the GA-PARSIMONY. Also, when GA-PARSIMONY was used with FS, the reference method (without complexity reranking) performed better in some datasets. However, an important reduction in the number of inputs was observed when N_{FS}_FS was implemented, while the $RMSE_{tst}$ remained close to the error margin of 0.1 % established. For example, a 61 % of features reduction was obtained in *ailerons* with similar accuracy, a 72 % in *bank*, a 59 % in *cpu*, or a 54 % in *elevators*. Therefore, the use of GA-PARSIMONY with N_{FS}_FS can be a good choice to obtain better parsimonious solutions in high dimensional datasets with GA-PARSIMONY and XGBoost. However, more experiments are needed with other complexity metrics, more high dimensional databases and other configurations to obtain more detailed conclusions.

Acknowledgments. The authors would like to acknowledge the fellowship APPI15/05 granted by the Banco Santander and the University of La Rioja.

References

1. Ahila, R., Sadasivam, V., Manimala, K.: An integrated PSO for parameter determination and feature selection of ELM and its application in classification of power system disturbances. Appl. Soft Comput. **32**, 23–37 (2015)
2. Antonanzas-Torres, F., Urraca, R., Antonanzas, J., Fernandez-Ceniceros, J., de Pison, F.M.: Generation of daily global solar irradiation with support vector machines for regression. Energy Convers. Manage. **96**, 277–286 (2015)
3. Caamaño, P., Bellas, F., Becerra, J.A., Duro, R.J.: Evolutionary algorithm characterization in real parameter optimization problems. Appl. Soft Comput. **13**(4), 1902–1921 (2013)
4. Chen, N., Ribeiro, B., Vieira, A., Duarte, J., Neves, J.C.: A genetic algorithm-based approach to cost-sensitive bankruptcy prediction. Expert Syst. Appl. **38**(10), 12939–12945 (2011)
5. Chen, T., He, T., Benesty, M.: xgboost: Extreme Gradient Boosting (2015). https://github.com/dmlc/xgboost, rpackageversion 0.4-3
6. Corchado, E., Wozniak, M., Abraham, A., de Carvalho, A.C.P.L.F., Snásel, V.: Recent trends in intelligent data analysis. Neurocomputing **126**, 1–2 (2014)
7. Dhiman, R., Saini, J., Priyanka: Genetic algorithms tuned expert model for detection of epileptic seizures from EEG signatures. Appl. Soft Comput. **19**, 8–17 (2014)
8. Ding, S.: Spectral and wavelet-based feature selection with particle swarm optimization for hyperspectral classification. J. Softw. **6**(7), 1248–1256 (2011)
9. Fernandez-Ceniceros, J., Sanz-Garcia, A., Antonanzas-Torres, F., de Pison, F.M.: A numerical-informational approach for characterising the ductile behaviour of the t-stub component. part 2: parsimonious soft-computing-based metamodel. Eng. Struct. **82**, 249–260 (2015)
10. Friedman, J.H.: Greedy function approximation: a gradient boosting machine. Ann. Stat. **29**(5), 1189–1232 (2001)
11. Huang, H.L., Chang, F.L.: ESVM: evolutionary support vector machine for automatic feature selection and classification of microarray data. Biosystems **90**(2), 516–528 (2007)
12. Kaggle: The home of data science. https://www.kaggle.com/
13. KDD-CUP: Annual data mining and knowledge discovery competition organized by ACM. http://www.kdd.org/kdd-cup
14. Michalewicz, Z., Janikow, C.Z.: Handling constraints in genetic algorithms. In: ICGA, pp. 151–157 (1991)
15. Oduguwa, V., Tiwari, A., Roy, R.: Evolutionary computing in manufacturing industry: an overview of recent applications. Appl. Soft Comput. **5**(3), 281–299 (2005)
16. Core Team, R.: R: A Language and Environment for Statistical Computing. R Foundation for Statistical Computing, Vienna, Austria (2013)

17. Reif, M., Shafait, F., Dengel, A.: Meta-learning for evolutionary parameter optimization of classifiers. Mach. Learn. **87**(3), 357–380 (2012)
18. Sanz-Garcia, A., Fernandez-Ceniceros, J., Antonanzas-Torres, F., Pernia-Espinoza, A., Martinez-de Pison, F.J.: GA-PARSIMONY: a GA-SVR approach with feature selection and parameter optimization to obtain parsimonious solutions for predicting temperature settings in a continuous annealing furnace. Appl. Soft Comput. **35**, 13–28 (2015)
19. Sanz-Garcia, A., Fernández-Ceniceros, J., Fernández-Martínez, R., Martínez-de-Pisón, F.J.: Methodology based on genetic optimisation to develop overall parsimony models for predicting temperature settings on annealing furnace. Ironmaking Steelmaking **41**(2), 87–98 (2014)
20. Sanz-García, A., Fernández-Ceniceros, J., Antoñanzas-Torres, F., Martínez-de Pisón, F.J.: Parsimonious support vector machines modelling for set points in industrial processes based on genetic algorithm optimization. In: Herrero, Á., et al. (eds.) International Joint Conference SOCO13-CISIS13-ICEUTE13. Advances in Intelligent Systems and Computing, vol. 239, pp. 1–10. Springer International Publishing, Heidelberg (2014)
21. Seni, G., Elder, J.: Ensemble Methods in Data Mining: Improving Accuracy Through Combining Predictions. Morgan and Claypool Publishers, Chicago (2010)
22. Shaffer, J.P.: Modified sequentially rejective multiple test procedures. J. Am. Stat. Assoc. **81**(395), 826–831 (1986)
23. Urraca, R., Sanz-Garcia, A., Fernandez-Ceniceros, J., Sodupe-Ortega, E., Martinez-de-Pison, F.J.: Improving hotel room demand forecasting with a hybrid GA-SVR methodology based on skewed data transformation, feature selection and parsimony tuning. In: Onieva, E., Santos, I., Osaba, E., Quintián, H., Corchado, E. (eds.) HAIS 2015. LNCS (LNAI), vol. 9121, pp. 632–643. Springer, Heidelberg (2015). doi:10.1007/978-3-319-19644-2_52
24. Wilcoxon, F.: Individual comparisons by ranking methods. Biometrics Bull. **1**(6), 80–83 (1945). http://dx.doi.org/10.2307/3001968
25. Winkler, S.M., Affenzeller, M., Kronberger, G., Kommenda, M., Wagner, S., Jacak, W., Stekel, H.: Analysis of selected evolutionary algorithms in feature selection and parameter optimization for data based tumor marker modeling. In: Moreno-Díaz, R., Pichler, F., Quesada-Arencibia, A. (eds.) EUROCAST 2011. LNCS, vol. 6927, pp. 335–342. Springer, Heidelberg (2012). doi:10.1007/978-3-642-27549-4_43
26. Xue, B., Zhang, M., Browne, W.N.: Particle swarm optimisation for feature selection in classification: novel initialisation and updating mechanisms. Appl. Soft Comput. **18**, 261–276 (2014)
27. Ye, J.: On measuring and correcting the effects of data mining and model selection. J. Am. Stat. Assoc. **93**(441), 120–131 (1998)

A K-means Based Genetic Algorithm for Data Clustering

Clara Pizzuti(✉) and Nicola Procopio

National Research Council of Italy (CNR),
Institute for High Performance Computing and Networking (ICAR),
Via P. Bucci 7/11, 87036 Rende (CS), Italy
{clara.pizzuti,nicola.procopio}@icar.cnr.it

Abstract. A genetic algorithm, that exploits the K-means principles for dividing objects in groups having high similarity, is proposed. The method evolves a population of chromosomes, each representing a division of objects in a different number of clusters. A group-based crossover, enriched with the one-step K-means operator, and a mutation strategy that reassigns objects to clusters on the base of their distance to the clusters computed so far, allow the approach to determine the best number of groups present in the dataset. The method has been experimented with four different fitness functions on both synthetic and real-world datasets, for which the ground-truth division is known, and compared with the K-means method. Results show that the approach obtains higher values of evaluation indexes than that obtained by the K-means method.

1 Introduction

Clustering is an unsupervised data analysis technique whose goal is to categorize data objects having similar characteristics in clusters. The applicability of this technique has been recognized in diverse fields such as biology, sociology, image processing, information retrieval, and many different methods have been proposed. Among the numerous existing clustering techniques, the *K-means* method is one of the most popular because of its efficiency and efficacy in grouping data. A plenty of heuristics have been presented to overcome some drawbacks inherent this method, such as the need to provide the number of clusters as input parameter, or the choice of initial centroids. In this context, evolutionary computation based approaches have been playing a central role because of their capability of exploring the search space and escaping from local minima during the optimization process. Several clustering algorithms based on Genetic Algorithms have been proposed in the last years. Among them, a number of approaches combine the ability of K-means in partitioning data with that of genetic algorithms of performing adaptive search process to find near optimal solutions for an optimization problem [7].

In this paper, a genetic algorithm that integrates the local search K-means principle into the genetic operators of crossover and mutation is proposed. The method, named *CluGA*, evolves a population of chromosomes, each representing

M. Graña et al. (eds.), *International Joint Conference SOCO'16-CISIS'16-ICEUTE'16*,
Advances in Intelligent Systems and Computing 527, DOI 10.1007/978-3-319-47364-2_21

a division of objects in a number of clusters not fixed a priori. Each individual is a string of length equal to the number of the dataset objects. The value of the i-th gene is the label of the cluster to which the i-th object belongs. Every individual is initialized with a random number in the interval $\{2, k_{max}\}$, where k_{max} is the maximum number of allowed clusters. The crossover operator first obtains the offspring by adopting a group-based strategy, then refines the offspring by performing the one-step K-means operator, introduced in [7]. Moreover, the mutation operator reassigns an object x belonging to a cluster C_i to another cluster C_j if x is closer to C_j. The method has been experimented with four different fitness functions on both synthetic and real-world datasets, and compared with the K-means method. Results show that the approach is very competitive with respect to K-means. It is able to find solutions having the exact number of clusters, without any prior knowledge, with high values of the evaluation indexes used to assess the performance of the method.

The paper is organized as follows. In the next section the clustering problem is introduced. In Sect. 3 an overview of the existing approaches that combine Genetic Algorithms and the K-means principles are reported. In Sect. 4 the algorithm *CluGA* is described in detail. Section 5 presents the results of the approach on real-world and synthetic datasets. Finally, Sect. 6 concludes the paper and outlines future developments.

2 The Clustering Problem

Let $\mathbf{X} = \{x_1, \ldots, x_n\}$ be a set of n data objects, also called points, where each $x_j, j = 1, \ldots, n$ is a d-dimensional feature vector representing a single data item. A partition, or clustering, of $\mathbf{X}$ is a collection $C = \{C_1, \ldots, C_k\}$ of k nonoverlapping subsets of $\mathbf{X}$ such that $\{C_1 \cup \ldots \cup C_k\} = \mathbf{X}$ and $\{C_i \cap C_j\} = \emptyset$ for $i \neq j$. The *centroid* of a cluster C_i is defined as the mean of the objects it contains. Independently of the clustering approach adopted, the main objective of the clustering problem is to partition data objects into a number of clusters such that both within cluster similarity and between cluster heterogeneity are high.

The K-means clustering method is a partitional clustering algorithm that groups a set of objects into k clusters by optimizing a criterion function. The technique performs three main steps: (1) selection of k objects as cluster centroids, (2) assignment of objects to the closest cluster, (3) updating of centroids on the base of the assigned data. Steps 2 and 3 are repeated until no object changes its membership to a cluster, or the criterion function does not improve for a number of iterations. The K-means method finds a partition that minimizes the *total within-cluster variance*, also known as *sum of squared error*, defined as

$$CV_W = \sum_{i=1}^{k} \sum_{x \in C_i} ||x - m_i||^2 \tag{1}$$

where m_i is the centroid of cluster C_i and $||.||$ denotes the Euclidean distance. In the following a brief overview of evolutionary approaches combined with the K-means method is reported.

3 K-means Based Genetic Algorithms

In the last years a lot of methods that integrate Genetic Algorithms and K-means in several different ways have been presented. One of the first proposals that hybridizes a genetic algorithm with the K-means technique is the GKA method of Krishna and Murty [7]. This method adopts the integer encoding scheme where a chromosome is a vector of n elements, i.e. the number of data objects, and each gene has value in the alphabet $\{1, \ldots, k\}$, with k the fixed number of clusters to find. At the beginning each gene receives a random number between 1 and k. The authors define a one-step K-means algorithm, called K-means operator KMO, that, given an individual, computes the centroids and reassigns each data object to the closest cluster. This operator is used as crossover in the algorithm to improve convergence. Moreover, a distance-based mutation that changes an allele value to a cluster number with probability proportional to the distance from the object to the cluster center is introduced. The fitness function minimizes the within-cluster variance. Because of the fixed number of clusters to find, both mutation and KMO can generate empty clusters, leading to what the authors call *illegal string (i.e. chromosome).* In these cases singleton clusters are introduced to re-obtain the fixed number of clusters.

Inspired by GKA, Lu et al. proposed *FGKA (Fast GKA)* [9] and *IGKA (Incremental)* [10]. The main difference between $FGKA$ and GKA consists in a different mutation operator, while $IGKA$ performs an incremental computation of centroids that makes the method faster. These methods have been experimented on gene expression data. It is worth to point out that GKA, $FGKA$ and $IGKA$ could better be classified as memetic evolutionary methods. In fact, they substitute the crossover operator with a local search which, in this case, is the K-means one-step operator.

Bandyopadhyay and Maulik [2] presented a method, named KGA, that uses the principles of K-means. Differently from GKA, a clustering solution is represented with a vector of k centroids, with k fixed a priori. The authors motivate this choice for efficiency requirements. The fitness function to minimize is defined as inverse of the sum of distances of each object to its centroid. Population is initialized by randomly selecting k points, then one-point crossover and a mutation operator based on the fitness value are applied. An extension of the approach that tries to avoid the problem of the parameter k is presented in [3]. Chromosomes are constituted by real numbers, representing the centroid coordinates, and a special symbol *don't care* meaning that a gene does not contain a centroid. The value of k is assumed between a minimum k_{min} and maximum k_{max} value, thus the chromosome length is $k_{max} - k_{min}$, and the presence of don't care symbols allow to have a variable number of clusters. The fitness function of this new algorithm is the Davis-Bouldin index [5]. Though the method works well on artificial data, for the well known *Iris* datasets the method merges two, out of the three clusters, having overlapping objects.

In the next section a method that exploits K-means principles inside a genetic algorithms is presented.

4 The *CluGA* Method

In this section a detailed description of *CluGA* is given, along with the genetic representation and operators adopted.

Encoding. The method adopts the label-based encoding of a clustering solution where a chromosome consists of a string of length equal to the number n of objects. Each object is located in a position i of the chromosome and is associated with an integer in the alphabet $\{2, \ldots, k_{max}\}$, where k_{max} is the maximum number of possible clusters and each integer is the label of a cluster. It is known that this coding is redundant because the same partitioning can be represented in $(k_{max}-1)!$ different ways, thus the search space can be very large. In order to improve the efficiency of the algorithm, we apply a renumbering procedure [6] after each application of genetic operators, avoiding, in this way, the presence of different strings representing the same solution. Figure 1 shows ten objects grouped in $k = 3$ clusters with the corresponding representation and that obtained after renumbering it.

Initialization. The simple strategy of generating k random cluster labels and assigning at random an object to a cluster can produce bad initial clusters that could overlap, thus increasing the convergence times. The initialization process employed by *CluGA* uses the approach of K-means++ [1] to select k centers, and then assigns each object to the closest center. The K-means++ seeding strategy has been shown by the authors to improve both accuracy and speed of K-means, thus it can be beneficial also for *CluGA*. For each element of the population, a random number k between 2 and k_{max} is generated. Then the following steps are performed. (1) Choose the first cluster center $c_1 = x$ uniformly at random from the dataset X. (2) Choose the next cluster center $c_i = x' \in X - \{x\}$ from the remaining data objects with probability proportional to its distance from

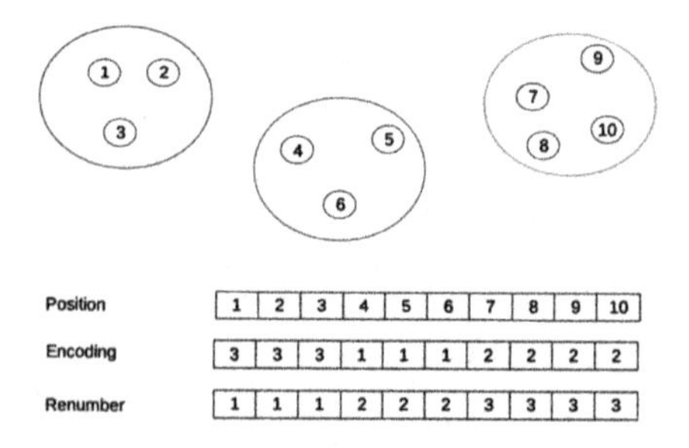

Fig. 1. Encoding of a clustering solution and its renumbering.

	p_1	p_2
Positions	1 2 3 4 5 6 7 8 9 10	1 2 3 4 5 6 7 8 9 10
Labels	1 **1** 2 3 2 4 4 2 3 1	1 **2** 2 1 1 1 3 3 3 2
Children	**2 2** 2 3 2 4 4 2 3 **2**	1 **1 1** 1 1 1 3 3 3 **1**
Renumber	1 1 1 2 1 3 3 1 2 1	1 1 1 1 1 1 2 2 2 1

Fig. 2. Group-based crossover where the random position $h = 2$ is selected, thus $l_1 = 1$ and $l_2 = 2$. The gene values of the parents, changed to generate the offspring, are highlighted in bold.

the closest center already chosen. (3) Repeat step (2) until k_{max} centers have been chosen.

Crossover. Standard one-point and two-point crossover have been shown to present drawbacks when applied to group problems [6]. In fact, it can happen that the offspring encode solutions too far from those of their parents. A cluster-oriented crossover operator can avoid these problems. *CluGA* performs a group-based crossover followed by the application of the K-means operator *KMO*. Given two individuals p_1 and p_2 of the population, the crossover operator performs the following steps: (1) Choose a random position h between 1 and n, (2) take the cluster labels $l_1 = p_1(h)$ and $l_2 = p_2(h)$, (3) generate the first offspring $child_1$ by substituting the value of all those genes of p_1 having cluster label equal to l_1 with l_2, (4) generate the second offspring $child_2$ by substituting the value of all those genes of p_2 having cluster label equal to l_2 with l_1. (5) Apply *KMO* to $child_1$ and $child_2$. (6) If any singleton cluster is present, assign its object to one of the existing clusters at random. An example of group-based crossover is shown in Fig. 2. If the random position $h = 2$ is selected, the members $\{1, 2, 10\}$ of the cluster in p_1 with label value 1 are assigned cluster label 2, while the members $\{2, 3, 10\}$ of the cluster in p_2 with label value 2 are assigned cluster label 1.

Mutation. The mutation operator performs a local search that reassigns objects to clusters at lower distance. For each object $x_i \in X$, it checks whether the distance from x_i to the cluster C_a it belongs to is higher than the distance from x_i to another cluster C_b. In such a case x_i is removed from C_a and added to C_b.

Fitness Function. The choice of an appropriate fitness function is a key point to obtain a good solution for the problem to solve. In the literature many indexes have been defined to evaluate clustering results. These indexes have been mainly used for cluster evaluation, in order to choose the best clustering method, and to determine the optimal number of clusters, when this information is not known, as it generally happens for real-world datasets. However, they have also been used by a number of clustering methods as functions to optimize when partitioning data objects. Some authors, like Bandyopadhyay and Maulik [2], proposed to minimize the cluster-within variance, analogously to the K-means method. However this criterion is not apt when using a genetic algorithm since it is biased towards a high number of clusters, i.e. the higher the number of clusters k found by a method, the lower the within-cluster variance value. Thus a genetic algorithm that optimizes this index has the tendency to prefer solutions with many clusters. The main motivation of this behavior is that CV_W takes into account only the closeness of points to centroids. However, to measure the goodness of a clustering structure, the concepts of *cluster cohesion and separation* must be both considered. The former suggests that clusters should be compact, that is objects in the same cluster must stay close. The latter defines how well clusters are separated. While compactness is generally measured through data variance, the lower the variance the better the cohesion, separation relies on the distances between cluster centroids or objects in different clusters.

Actually, minimizing CV_W corresponds to optimizing compactness. In order to take into account both compactness and separation, we consider three popular criteria as fitness functions: the *Calinski-Harabasz* [4], the *Silhouette* [12] and the *Davis-Bouldin* [5] indices. We show, in the experimental result section, that *CluGA*, endowed with one of these indices, is capable of obtaining clusterings very similar to the ground truth ones. The optimization of CV_W, instead, splits clusters in many small groups. In the following we give the definitions of these criteria.

The *Calinski-Harabasz* criterion, called also the *variance ratio criterion (VRC)* is defined as

$$VRC = \frac{CV_B}{CV_W} \times \frac{n-k}{k-1} \tag{2}$$

where CV_B is the total between-cluster variance, CV_W is the overall within-cluster variance, as defined by formula (1), k is the number of clusters, and n is the number of objects. CV_B is defined as:

$$CV_B = \sum_{i=1}^{k} n_i ||m_i - m||^2 \tag{3}$$

where m_i is the centroid of cluster i, n_i is the size of cluster i, m is the overall mean of the dataset, and $||.||$ denotes the Euclidean distance.

The *Silhouette* index, for each object, measures how similar that object is to elements in its own cluster, when compared to objects in other clusters. The silhouette criterion is defined as:

$$S = \frac{1}{k} \sum_{i=1}^{k} \frac{1}{n_i} \sum_{x \in C_i} \frac{b(x) - a(x)}{max(a(x), b(x))} \tag{4}$$

where $a(x)$ is the average distance from x to the other objects in the same cluster as x, and $b(x)$ is the minimum average distance from x to points in a different cluster, minimized over all clusters.

The silhouette value ranges from -1 to $+1$. A high silhouette value indicates that objects are well-matched in their own cluster, and poorly-matched to neighboring clusters.

The *Davies-Bouldin* criterion is based on a ratio of within-cluster and between-cluster distances. It is defined as:

$$DB = \frac{1}{k}\sum_{i=1}^{k} max_{i \neq j}\{\frac{\overline{d_i} + \overline{d_j}}{d_{i,j}}\} \tag{5}$$

where $\overline{d_i}$ and $\overline{d_j}$ are the average distances between each object in the i-th and j-th cluster, respectively, and the centroid of the own cluster, and $d_{i,j}$ is the Euclidean distance between the centroids of the i-th and j-th clusters.

The pseudo-code of the algorithm is shown in Fig. 3. In the next section the *CluGA* method is executed on synthetic generated and real-world datasets, and the clustering structures obtained by employing as fitness functions cluster-within variance, Calinski-Harabasz, Silhouette, and Davies-Bouldin criteria, are compared.

5 Experimental Results

This section provides a thorough experimentation for assessing the capability of *CluGA* in partitioning artificial and real-world datasets. The *CluGA* algorithm has been written in MATLAB 8.6 R2015b, by using the Genetic Algorithm Solver of the Global Optimization Toolbox. A trial and error procedure has been adopted for fixing the parameter values. Thus the crossover rate p_c has been fixed to 0.8, the mutation rate p_m to 0.2, population size 100, elite reproduction 10 % of the population size, number of generations is 50. In order to evaluate the method, since the ground-truth clusterings are known, we computed the well known measures *Adjusted Rand Index*, *Precision*, *Recall*, and *Fmeasure*, adopted in the literature to assess the capability of an algorithm in finding partitions similar to the true data division. We first present the results obtained by *CluGA* on randomly generated synthetic data sets. We consider three kinds of datasets of 500 objects each. The first one, named *Syn_3* contains three Gaussian clusters, distinct and well separated, with standard deviation of the centroids equal to (0.2, 0.2, 0.35) (Fig. 4(a)). In the second one, named *Syn_4*, there are four clusters close to each other with standard deviation of centroids equal to (0.2, 0.35, 0.45, 0.3) (Fig. 4(b)). The third one, named *Syn_6* is constituted by six mixed clusters with standard deviation of centroids equal to (0.2, 0.2, 0.35, 0.45, 0.1, 0.1) (Fig. 4(c)).

Table 1 shows the execution of the method on the synthetic data sets for the four fitness functions, along with the results of the K-means algorithm. Each method has been executed 10 times and average and best values of the ten runs, along with standard deviation, are reported. Notice that the K-means has been executed with input parameter k equal to the true number of clusters, while *CluGA* has been executed by fixing the maximum number of clusters to $\sqrt(n)$, which, in the literature, is considered a rule of thumb [11]. The table highlights the very good capabilities of *CluGA* in partitioning the datasets in a number of clusters very close to the ground truth, even if the algorithm does not know

Table 1. Comparison between *CluGA* and the K-means algorithm on synthetic datasets. *R*: Recall, *P*: Precision, *F*: Fmeasure. The best values, by excluding the CV_W index, are highlighted in bold.

		Syn_3					Syn_4					Syn_6				
		Fitness				K-Means	Fitness				K-Means	Fitness				K-Means
		CV_W	S	VRC	DB		CV_W	S	VRC	DB		CV_W	S	VRC	DB	
ARI	mean	0.2110	0.9994	0.9994	0.9988	0.9994	0.2612	0.8891	0.8654	0.8743	0.8972	0.2100	0.2907	0.3428	0.2871	0.3650
	best	0.2257	1	**1**	1		0.2791	0.9161	0.9161	**0.9174**		0.2423	0.3625	**0.4277**	0.3387	
	st. dev	(0.0088)	(0.0019)	(0.0019)	(0.0025)	(0.0019)	(0.0118)	(0.0299)	(0.0468)	(0.0435)	(0.0201)	(0.0142)	(0.0484)	(0.0519)	(0.0277)	(0.0457)
R	mean	0.9996	0.9989	0.9989	0.9978	0.9989	0.8937	0.8591	0.8519	0.8442	0.8630	0.5413	0.4645	0.4003	0.4907	0.4407
	best	1	1	**1**	1		0.9169	0.8832	0.8832	**0.8855**		0.5591	**0.5504**	0.5446	0.5308	
	st. dev	(0.0013)	(0.0035)	(0.0035)	(0.0047)	(0.0035)	(0.0141)	(0.0264)	(0.0247)	(0.0395)	(0.0233)	(0.0163)	(0.0926)	(0.0712)	(0.0724)	(0.0358)
P	mean	0.3706	0.9989	0.9989	0.9978	0.9989	0.4126	0.8519	0.8320	0.8431	0.8630	0.3289	0.4117	0.4778	0.3703	0.4480
	best	0.3786	1	**1**	1		0.4305	0.8833	0.8833	**0.8859**	0.3484	**0.6020**	0.5366	0.4423		
	st. dev	(0.0038)	(0.0035)	(0.0035)	(0.0047)	(0.0035)	(0.0086)	(0.0381)	(0.0497)	(0.0413)	(0.0230)	(0.0129)	(0.1004)	(0.0510)	(0.0303)	(0.0364)
F	mean	0.5407	0.9989	0.9989	0.9978	0.9989	0.5646	0.8554	0.8413	0.8436	0.8630	0.4092	0.4186	0.4284	0.4170	0.4443
	best	0.5492	1	**1**	1		0.5859	0.8832	0.8832	**0.8857**		0.4281	0.4583	**0.4710**	0.4509	
	st.dev	(0.0041)	(0.0035)	(0.0035)	(0.0047)	(0.0035)	(0.0106)	(0.0310)	(0.0341)	(0.0403)	(0.0232)	(0.0144)	(0.0254)	(0.0251)	(0.0280)	(0.0360)
CV_W	mean	13.2836	67.2093	67.2093	67.2291	67.2093	24.8088	**102.1898**	102.4346	106.6969	103.0274	12.6606	36.1794	58.3065	**24.4930**	44.9482
	best	11.5204	63.9562	63.9562	63.9562		23.4774	90.8306	91.4968	100.0645		11.2092	13.0521	14.6163	15.3501	
	st. dev	(0.9133)	(2.4072)	(2.4072)	(2.4305)	(2.4071)	(0.9599)	(4.5316)	(6.3204)	(7.2930)	(2.3689)	(0.9003)	(32.3026)	(21.0439)	(17.2006)	(3.5852)
k	mean	22.9000	3.0000	3.0000	3.0000	3	22.8000	4.1000	4.2000	4.1000	4	22.7000	14.1000	5.8000	17.3000	6
	best	22	3	**3**	3		22	4	4	**4**		21	3	**6**	5	
	st. dev	(0.3162)	(0)	(0)	(0)		(0.4216)	(0.3162)	(0.4216)	(0.3162)		(0.6749)	(7.8095)	(5.0947)	(4.7152)	

Table 2. Comparison between *CluGA* and the K-means algorithm on IRIS and CANCER datasets. *R*: Recall, *P*: Precision, *F*: Fmeasure. The best values, by excluding the CV_W index, are highlighted in bold.

		IRIS					*CANCER*				
		Fitness				K-Means	Fitness				K-Means
		CV_W	S	VRC	DB		CV_W	S	VRC	DB	
ARI	mean	0.3238	0.5418	**0.7223**	0.5681	0.7005	0.2583	0.7825	**0.8364**	0.8226	0.8337
	best	0.3417	0.5583	**0.7455**	0.5681		0.3053	0.8337	**0.8551**	0.8337	
	st. dev	(0.0128)	(0.0058)	(0.0148)	(0.0000)	(0.0941)	(0.043)	(0.0478)	(0.0102)	(0.0171)	(0)
R	mean	0.9237	0.5255	**0.7457**	0.5794	0.7298	0.8719	0.6665	**0.7289**	0.7115	0.7252
	best	0.9423	0.5537	**0.7869**	0.5793		0.8811	0.7252	**0.7543**	0.7252	
	st. dev	(0.0148)	(0.0099)	(0.0190)	(0.0000)	(0.0686)	(0.0066)	(0.0482)	(0.0135)	(0.0211)	(0)
P	mean	0.4212	0.8901	**0.7589**	1.0000	0.7542	0.2294	0.6914	**0.7373**	0.7227	0.7339
	best	0.4353	**0.9490**	0.8093	1	best	0.2390	0.7339	**0.7586**	0.7339	
	st. dev	(0.0094)	(0.0207)	(0.0229)	(0.0000)	(0.0341)	(0.0088)	(0.0335)	(0.011)	(0.0175)	(0)
F	mean	0.5785	0.6608	**0.7522**	0.7337	0.7413	0.3631	0.6786	**0.7331**	0.717	0.7295
	best	0.5955	0.6994	**0.7979**	0.7336		0.3746	0.7295	**0.7563**	0.7295	
CV_W	st. dev	(0.0115)	(0.0136)	(0.0209)	(0.0000)	(0.0533)	(0.011)	(0.0414)	(0.0122)	(0.0192)	(0)
	mean	21.6335	152.4457	**79.0029**	154.9470	85.2417	72.2422	200.1433	196.9676	197.4081	**196.9366**
	best	21.23349	152.348	78.85144	154.947		70.8324	196.9366	196.9366	196.9366	
	st. dev	(0.3481)	(0.3092)	(0.2705)	(0.0000)	(20.2076)	(1.0418)	(6.6177)	(0.0767)	(0.9443)	(0)
k	mean	13.0	2	3	2	3	26.6	2	2	2	2
	best	13	2	3	2		25	2	2	2	
	st. dev	(0.0000)	(0.0000)	(0.0000)	(0.0000)		(0.6992)	(0)	(0)	(0)	

the number of clusters to find, when using as fitness functions VRC (formula (2)), S (formula (4)), and DB (formula (5)). In fact, on *Syn_3* it reaches the same results of the K-means method. On *Syn_4* and *Syn_6*, though the average values of the measures are slightly lower, the best values, out of the ten runs, are always better. As regards the cluster-within variance, it is evident from the table the selection bias towards a high number of clusters, which confirms the experimentation of Liu et al. [8]. With this fitness function the method for the three datasets obtains an average of 22, 22, 21 clusters and average CV_W values 13.2, 24.8, 12.6, respectively.

Tables 2 and 3 compare *CluGA* and K-means on four well known real-world datasets of the UCI Machine Learning Repository. The *Iris* dataset consists of 150 samples of Iris flowers categorized into three species by four features. *Breast Cancer Wisconsin* dataset contains 683 objects with 9 features divided into two classes. The *Glass* dataset has 214 types of glass with 9 features, divided into 6 classes. The *Ecoli* dataset consists of 336 instances grouped with respect to 8 protein localization sites, described by seven attributes. The tables clearly point out that the optimization of the Calinski-Harabasz criterion obtains the best evaluation measure values on all the datasets, except for *Ecoli*. This fitness function allows the algorithm to outperform the K-means method on all the datasets with higher ARI value and lower within-cluster variance. For instance, on the *Iris* dataset *CluGA* finds the three clusters with the mean Adjusted Rand Index value equal to 0.7223, the best ARI equal to 0.7455, and cluster-within variance 79.0029, while the K-means obtains $ARI = 0.7005$ and $CV_W = 85.2417$. For the *Cancer* dataset the mean and best ARI values of *CluGA* are 0.8364 and 0.8551, respectively, while for the K-means $ARI = 0.8337$. Analogously for the

Table 3. Comparison between *CluGA* and the K-means algorithm on GLASS and ECOLI datasets. *R*: Recall, *P*: Precision, *F*: Fmeasure. The best values, by excluding the CV_W index, are highlighted in bold.

		GLASS					*ECOLI*				
		Fitness				K-Means	Fitness				K-Means
		CV_W	*S*	*VRC*	DB		CV_W	*S*	*VRC*	DB	
ARI	mean	0.2196	0.2062	**0.5803**	0.2087	0.5195	0.257	0.6189	0.573	**0.6805**	0.3876
	best	0.2446	**0.6762**	0.6064	0.6086		0.2785	0.6925	0.7065	**0.7277**	
	st. dev	(0.0309)	(0.2945)	(0.0316)	(0.2706)	(0.1411)	0.0128	0.1222	0.1397	0.0591	0.1599
R	mean	0.8405	0.1932	**0.4381**	0.1682	0.3575	0.7654	0.5005	0.4573	**0.6127**	0.6021
	best	0.8568	**0.7039**	0.5358	0.5336	0.7867	0.578	0.5667	**0.6877**		
	st. dev	(0.0146)	(0.2912)	(0.0726)	(0.2316)	(0.1009)	(0.0104)	(0.0995)	(0.1255)	(0.0717)	(0.2547)
P	mean	0.2321	0.2224	**0.4054**	0.2313	0.3900	0.4277	**0.7684**	0.7271	0.7215	0.5537
	best	0.2463	0.4094	**0.4310**	0.3933	0.4421	0.7859	**0.812**	0.7672		
	dt. dev	(0.0124)	(0.1653)	(0.0253)	(0.1382)	(0.0277)	(0.008)	(0.0183)	(0.0826)	(0.0631)	(0.0242)
F	mean	0.3636	0.1628	**0.4192**	0.1578	0.3660	0.5487	0.6007	0.5558	**0.6598**	0.6158
	best	0.3809	**0.5177**	0.4642	0.4528		0.5661	0.6544	0.6675	**0.6993**	
	st. dev	(0.0161)	(0.2218)	(0.0412)	(0.1909)	(0.0873)	(0.0085)	(0.0836)	(0.1191)	(0.0562)	(0.0235)
CV_W	mean	178.5231	932.6661	**752.9121**	929.1354	860.4198	8.7751	25.3293	26.9246	19.2687	**14.7483**
	best	173.3125	358.5322	589.0314	495.5036		8.5466	21.1775	23.261	13.5761	
	st. dev	(4.6974)	(361.2018)	(109.3977)	(324.6504)	(126.0378)	(0.2116)	(4.7125)	(5.6102)	(3.1077)	(0.6440)
k	mean	13.9000	3.0000	2.3000	2.8000	2	18.8	2.9	2.7	8.1	8
	best	13	2	2	2		18	4	3	8	
	st. dev	(0.3162)	(1.7638)	(0.4830)	(0.9189)		(0.4216)	(0.5676)	(0.483)	(2.7669)	(0)

Input: A set $\mathbf{X} = \{x_1, \ldots, x_n\}$ of n data objects, crossover probability p_c, mutation probability p_m
Output: A partition $C = \{C_1, \ldots, C_k\}$ of $\mathbf{X}$ in k groups

Method: **Perform** the following steps:
1 **Create** an initial population of individuals by applying the K-means++ approach
2 **Evaluate** all the individuals
3 **while** termination condition is not satisfied **do**
4 **for** each individual $I = \{g_1, \ldots, g_n\}$ in the population
5 **Perform** group-based crossover and *KMO* operator with probability p_c, **renumber** the individual
6 **Perform** local search based mutation with probability p_m, **renumber** the individual
7 **Evaluate** the fitness function
8 **end for**
9 **Select** solutions with the better fitness values
10 **end while**
11 **return** the individual having the best fitness value

Fig. 3. The pseudo-code of the *CluGA* algorithm.

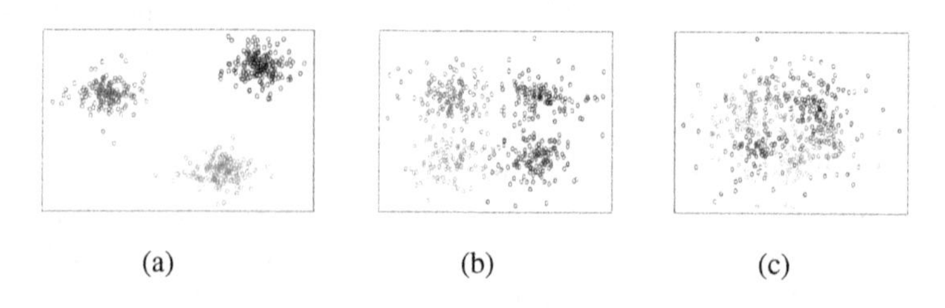

(a) (b) (c)

Fig. 4. Synthetic datasets with 500 objects: (a) k = 3, (b) k = 4, (c) k = 6.

Glass dataset, *CluGA* obtains $ARI = 0.5803$ and $CV_W = 752.9121$, while K-means finds $ARI = 0.5195$ and $CV_W = 860.4198$. As regards the *Ecoli* dataset, the Calinski-Harabasz and the Silhouette indexes merges some clusters, while the Davis-Bouldin criterion finds all the 8 clusters in almost all the executions. However it must be noted that this dataset contains two clusters with only two objects, and one cluster with 5 objects, which are not easy to find. In fact, the K-means finds 8 clusters with a precision value around 0.5, which is rather low, while that of the three indexes is above 0.7. Also for these real-world datasets it can be observed that the CV_W criterion prefers a number of clusters much higher than the ground-truth.

6 Conclusions

The paper proposed a genetic algorithm able to divide a dataset in a number of groups not known in advance. The method employs label-based representation, and exploits the K-means strategy to improve the offspring generated by the group-based crossover. Though the idea of combining genetic algorithms and K-means is not new, *CluGA* sensibly differs from the existing proposals. With respect to *GKA* [7], *FGKA* [9] and *IGKA* [10], it applies the one step operator of the K-means after having effectively performed the group-based crossover. The above methods, instead, substitute the crossover with the K-means operator. Moreover, *CluGA* does not need to fix the number of clusters. Each chromosome, in fact, is initialized with a random number in the interval $\{2, \ldots, k_{max}\}$, where k_{max} could be also n. As regards the *KGA* method [2], *CluGA* has different representation and completely different operators. Experiments on synthetic and real-world datasets by using four different fitness functions show that when the three popular evaluation criteria of *Calinski-Harabasz*, *Silhouette* and *Davis-Bouldin* are employed as fitness functions, the method obtains very good solutions and outperforms the K-means method. The cluster-within variance, instead, though used by many authors, is not apt for genetic algorithms since its optimization generates a bias towards a high number of clusters. The experiments also pointed out the *Calinski- Harabasz* criterion obtains the best results for all the datasets, except for *Ecoli*, where the 8 clusters are merged into three. Future work will evaluate the method on data sets coming from real-world applications.

Acknowledgment. This work has been partially supported by MIUR D.D. n 0001542, under the project *BA2KNOW – PON03PE_00001_1*.

References

1. Arthur, D., Vassilvitskii, S.: K-means++: the advantages of careful seeding. In: Proceedings of the Eighteenth Annual ACM-SIAM Symposium on Discrete Algorithms, SODA 2007, pp. 1027–1035 (2007)
2. Bandyopadhyay, S., Maulik, U.: An evolutionary technique based on k-means algorithm for optimal clustering in rn. Inf. Sci. Appl. **146**(1–4), 221–237 (2002)

3. Bandyopadhyay, S., Maulik, U.: Genetic clustering for automatic evolution of clusters and application to image classification. Pattern Rec. **35**, 1197–1208 (2004)
4. Calinski, T., Harabasz, J.: A dendrite method for cluster analysis. Commun. Stat. **3**(1), 1–27 (1974)
5. Davies, D., Bouldin, D.: A cluster separation measure. IEEE Trans. Pattern Anal. Mach. Intell. **1**(2), 224–227 (1979)
6. Falkenauer, E.: Genetic Algorithms and Grouping Problems. Wiley, New York (1998)
7. Krishna, K., Murty, M.N.: Genetic k-means algorithm. IEEE Trans. Syst. Man Cybern. Part B **29**(3), 433–439 (1999)
8. Liu, Y., Li, Z., Xiong, H., Gao, X., Wu, J.: Understanding of internal clustering validation measures. In: Proceedings of the 2010 IEEE International Conference on Data Mining, ICDM 2010, pp. 911–916 (2010)
9. Lu, Y., Lu, S., Fotouhi, F., Deng, Y., Brown, S.J.: Fgka: a fast genetic k-means clustering algorithm. In: Proceedings of the 2004 ACM Symposium on Applied Computing, SAC 2004, pp. 622–623 (2004)
10. Lu, Y., Lu, S., Fotouhi, F., Deng, Y., Brown, S.J.: Performance evaluation of some clustering algorithms and validity indices. BMC Bioinform. **5**(172), 1–10 (2004)
11. Pal, N.R., Bezdek, J.C.: On cluster validity for the fuzzy c-means model. IEEE Trans. Fuzzy Syst. **3**(3), 370–379 (1995)
12. Rousseeuw, P.: Silhouettes: a graphical aid to the interpretation and validation of cluster analysis. J. Comput. Appl. Math. **20**(1), 53–65 (1987)

Improvement in the Process of Designing a New Artificial Human Intervertebral Lumbar Disc Combining Soft Computing Techniques and the Finite Element Method

Rubén Lostado Lorza[1(✉)], Fátima Somovilla Gomez[1], Roberto Fernandez Martinez[2], Ruben Escribano Garcia[3], and Marina Corral Bobadilla[1]

[1] Mechanical Engineering Department, University of La Rioja, Logroño, Spain
ruben.lostado@unirioja.es
[2] Electrical Engineering Department, University of Basque Country, Bilbao, Spain
[3] Built Environment and Engineering, Leeds Beckett University, Leeds, UK

Abstract. Human intervertebral lumbar disc degeneration is painful and difficult to treat, and is often magnified when the patient is overweight. When the damage is excessive, the disc is replaced by a non-natural or artificial disc. Artificial discs sometimes have the disadvantage of totally different behavior from that of the natural disc. This affects substantially the quality of treated patient's life. The Finite Element Method (FEM) has been used for years to design an artificial disc, but it involves a high computational cost. This paper proposes a methodology to design a new Artificial Human Intervertebral Lumbar Disc by combining FEM and soft computing techniques. Firstly, a three-dimensional Finite Element (FE) model of a healthy disc was generated and validated experimentally from cadavers by standard tests. Then, an Artificial Human Intervertebral Lumbar Disc FE model with a core of Polycarbonate Polyurethane (PCU) was modeled and parameterized. The healthy and artificial disc FE models were both assembled between lumbar vertebrae L4-L5, giving place to the Functional Spinal Unit (FSU). A Box-Behnken Design of Experiment (DoE) was generated that considers the parameters that define the geometry of the proposed artificial disc FE model and the load derived from the patient's height and body weight. Artificial Neural Networks (ANNs) and regression trees that are based on heuristic methods and evolutionary algorithms were used for modeling the compression and lateral bending stiffness from the FE simulations of the artificial disc. In this case, ANNs proved to be the models that had the best generalization ability. Finally, the best geometry of the artificial disc proposed when the patient's height and body weight were considered was achieved by applying Genetic Algorithms (GA) to the ANNs. The difference between the compression and lateral bending stiffness obtained from the healthy and artificial discs did not differ significantly. This indicated that the proposed methodology provides a powerful tool for the design and optimization of an artificial prosthesis.

M. Graña et al. (eds.), *International Joint Conference SOCO'16-CISIS'16-ICEUTE'16*,
Advances in Intelligent Systems and Computing 527, DOI 10.1007/978-3-319-47364-2_22

Keywords: Finite elements method · Data mining techniques · Genetic algorithms · Biomechanics · Design of artificial intervertebral lumbar disc

1 Introduction

Human intervertebral lumbar disc degeneration is painful and difficult to treat, and often is magnified when the patient is overweight or has considerable stature. When the damage is excessive, the intervertebral lumbar disc is replaced by a non-natural or artificial disc. One of the main disadvantages of artificial lumbar discs is that their behavior differs entirely from that of a natural lumbar disc. This can greatly affect a patient's quality of life [1]. In recent years, the FEM has been used for prosthesis design as it provides the information (stresses, strains, displacements, etc.) that is necessary to ensure that the healthy disc's behavior and that of the artificial disc are as similar as possible [2]. A disadvantage of FEM is that it involves a high computational cost. This is especially true when the design process is based only on experience of the designer gained in simulations and trial and error tests. A combination of FEM and regression models has been used widely in recent years to model mechanical systems and components that have shown non-linear behavior [3]. In this regard, some recent studies have used a combination of FEM and data mining techniques to automate the process of adjusting the parameters that define a FE model of an intervertebral disc when submitting the disc to a combination of standard tests [4]. In this case, a regression technique that was based on support vector machines with different kernels was used to model the stiffness and bulges of the intervertebral lumbar disc when the parameters of the FE models were changed. The best combination of parameters that define an FE model of an intervertebral disc was achieved by applying evolutionary optimization techniques that are based on GA to the best, previously obtained regression models. Other authors have developed the optimal shape of an intervertebral body device by using a combination of FEM and GA [5]. The purpose of the work being discussed was to optimize the device's shape in order to reduce the subsidence resistance. The current paper proposes a method to design a new artificial human intervertebral lumbar disc by combining FEM and soft computing techniques when the patient's height and body weight are considered. Firstly, a three-dimensional healthy disc FE model was generated and validated experimentally from cadavers by standard tests. Then, a new artificial human intervertebral lumbar disc FE model with a core of PolyCarbonate Polyurethane (PCU) and a pair of plates made with a titanium alloy was parameterized. The healthy and artificial disc FE models were assembled on an FSU between the L4-L5 vertebrae. The healthy FE model that was assembled on the FSU was used to validate the compression and lateral bending stiffness obtained from the FSU with the artificial disc. A Box-Behnken DoE was generated that considered the parameters that define the geometry of the proposed artificial disc FE model and the load derived from the patient's height and body weight. Artificial Neural Networks (ANN) and Regression Trees (RT) that are based on heuristic methods and evolutionary algorithms were used to model the compression and lateral bending stiffness (displacement and angular distortion) obtained from the simulations of the FSU with the artificial disc FE model. Finally, the best parameters that define the geometry of the artificial disc that was proposed when the patient's height and body weight was considered was achieved by applying GA to the model that provided the greatest generalization capability. The difference between the compression and lateral bending stiffness that were obtained from the FSU with the healthy and artificial disc FE models were defined as objective functions to be used for the optimization process. This paper focuses on male patients of 30 years of age, who have heights of 160 to 190 cm; and weights of 70 to 120 kg.

2 FE Model for Modeling the Intervertebral Disc

2.1 Healthy Intervertebral Disc

The healthy human InterVertebral Disc (IVD) is a fibrocartilage structure that is located between the vertebrae of the spine. The spine is composed of complex structures, such as a nucleus pulposus, the annulus fibrosus and the cartilage endplates (Fig. 1a and 1b). The function of a healthy intervertebral disc is to provide mobility and spinal flexibility during movements of the human body. The cartilage endplate is a thin structure that surrounds all of the nucleus pulposus and has a composition that is similar to that of a particular cartilage, but with less water. The nucleus pulposus helps to distribute pressure evenly across the disc and avoid concentrations of stress that could damage the underlying vertebrae. The annulus fibrosus is composed of concentric layers of fibrosus tissue that surround the nucleus pulposus. It consists of a complex network of collagen fibers. Several models of behavior and their corresponding parameters have been used for decades to describe the behavior of the healthy FE model [6, 7]. In proposing the healthy FE model, this paper considered cartilage endplates with an isotropic formulation (Elastic modulus E and Poisson ratio μ). The nucleus pulposus was considered to be an incompressible and hyper-elastic Mooney-Rivlin, and was formulated according to the empirical constants C_{10} and C_0. The annulus fibrosus was assumed to be a matrix of four, homogenous, ground substances that were reinforced by collagen fibers (Fig. 1a). The fibers were organized in five different radial layers and oriented at an angle of $\pm$ 30° relative to the horizontal direction (Fig. 1b). The fibers were simulated by three-dimensional unidirectional line FE elements. The homogenous, ground substances were simulated by eight-node isoparametric solid FE elements. In this paper, the parameters that define an FE model of an intervertebral disc in a previous work were taken. It used a combination of FEM and data mining techniques to automate the process of adjusting the parameters of an intervertebral disc FE model [4]. Table 1 summarizes the ranges of eleven different parameters that have been considered in this paper to model the different tissues that compose the healthy intervertebral disc.

2.2 Artificial Intervertebral Disc

The artificial disc that is proposed in this work consists of a core of Polycarbonate Polyurethane DSM Biomedical Bionate ® 75D (PCU) and two plates from a titanium alloy, which bind it to the cartilage endplates. A gap between the core and the titanium plates has been defined in order to provide a variable stiffness to the artificial disc.

Table 1. Range of the material parameters proposed for defining the behavior of the healthy human intervertebral lumbar disc models based on FEM.

Nuc. Pulp.		Endplate		Annulus Fibrosus					
C_{10}	0.133	E	23.003	Fib. 1&2	515.352	Fib. 7&8	408.071	E	4.005
C_0	0.035	μ	0.373	Fib. 3&4	503.390	Fib. 9&10	360.151	μ	0.449
…	…	…	…	Fib. 5&6	455.511	…	…	…	…

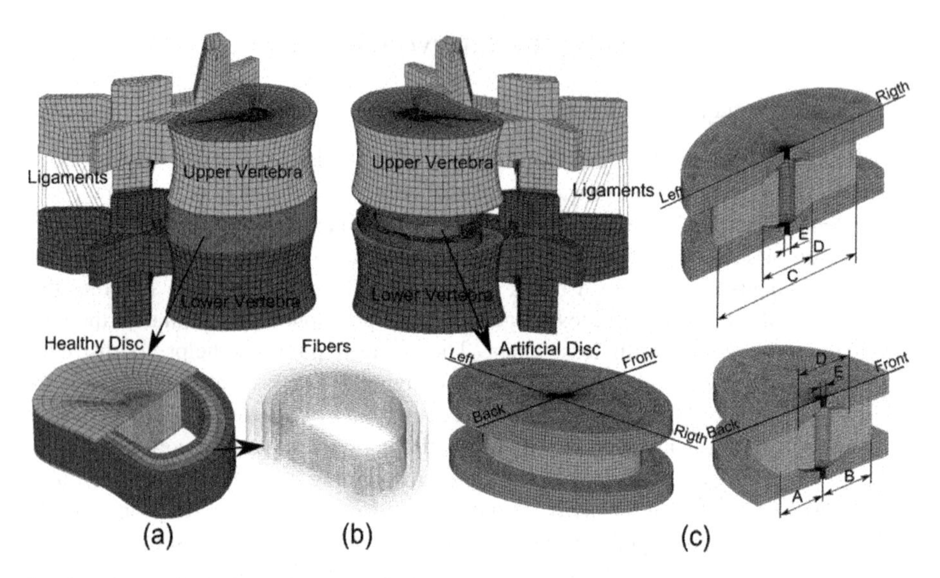

Fig. 1. **(a)** Detailed view of the composite structure formed by the nucleus pulpous, hyaline cartilage and four annulus fibrous. **(b)** Details of the orientation corresponding to the different five fiber layers in the healthy disc. **(c)** Details of the proposed artificial disc.

A mechanical contact with a small gap between the core and the titanium plates has been defined in order to make the stiffness of the artificial variable. This variable stiffness is achieved as the contact surface between the plates and the core increases when the load on the FSU also increases. In order to conduct modeling and optimization of the artificial disc seven variables or inputs were considered to take into account the patient's height and body weight. The five geometrical variables (Fig. 1c) were considered to reflect the geometry of the core and the gap thereof with the titanium alloy plates. However, there are also two body variables to take into account the patient's height and weight.

2.3 Vertebrae, Ligaments, Boundary and Loading Conditions

The FE models of the vertebras consider both its cortical and cancellous bone, and were modeled with eight-node isoparametric solid FE elements. Since the study focused on male patients of 30 years of age, the values of the elastic modulus and Poisson's Ratio for both cortical and cancellous bone were, respectively, E = 12,000 MPa; μ = 0.3 and E = 386 MPa; μ = 0.2 [8, 9]. Furthermore, three-dimensional unidirectional line FE elements [10] were considered in modeling the ligaments that attach the different parts that form the FSU. The body loads to apply to the healthy and artificial FSU were obtained by using the 3D Static Strength Prediction software™ (3DSSPP) and considering the patients' different heights and body weights. A Box-Behnken DoE was developed in this case to define the design matrix. Also, the average time to solve each of the simulated cases was about 12 h using computers with Intel Xeon Processor, CPU 2.5 GHz (8 processors) and 32.00 GB (RAM). Table 2 shows some of the fifty-six combination of geometrical (A, B, C, D, E) and body variables (weight and height) with the corresponding results obtained from the simulated FSU with the artificial disc FE models.

Table 2. Results of the simulation of the FE models when a combination of 56 geometrical and body variables are considered.

Run	A	B	C	D	E	Weight	Height	Displacement	Angle
1	14.45	12.75	42.5	15.3	5.1	70	175	−0.0525	2.159
2	14.45	12.75	42.5	20.4	5.1	70	175	−0.0629	2.194
…	…	…	…	…	…	…	…	…	…
56	14.45	12.75	51.00	15.30	6.8	95	160	−0.0749	2.301

3 Artificial Neural Networks and Regression Trees

Using R statistical software environment v2.15.2 [11] and based on Table 2, ANNs models and three types of RT, two built by heuristic methods and one built by soft computing methods, were used to predict the compression (displacement) and lateral bending (angle) stiffness. Both techniques have been used satisfactorily in many cases [12] and have provided successful results. ANNs are a powerful mathematical tool for modeling and finding patterns in data sets. They are based on the properties of biological neural systems. In this work, the applied ANNs were a multi-layer perceptron with one hidden layer based on Ripley′s work [13]. A general quasi-Newton optimization procedure based on the Broyden–Fletcher–Goldfarb–Shanno (BFGS) algorithm was proposed in this case for the weight adjustment of the ANNs instead of using a backpropagation procedure or a variant thereof. The second method that was proposed in this case was an RT that was based on a CART algorithm (Classification and Regression Tree). This method uses recursive partitioning methods to build the model in a forward stepwise search [14]. The third method (regression tree M5') improves the idea of a decision-tree induction algorithm by using linear regression as a way of making quantitative predictions where a real-valued dependent variable y is modeled as a linear function of several real-valued independent variables $x_1, x_2, \ldots, x_n$, plus another variable that reflects the noise, ε (Eq. 1).

$$y = \varepsilon + \beta_1 x_1 + \beta_2 x_2 + \ldots + \beta_n x_n \quad (1)$$

Each leaf of the regression tree M5'contains a linear regression model that is based on some of the initial attribute values. In that way, it combines a conventional decision tree with the possibility of linear regression functions at the nodes.

The last method uses an alternative way to search the parameter space of trees by using global optimization methods, like evolutionary algorithms [15]. These algorithms are inspired by natural Darwinian and are used to optimize a fitness function, such as error rate, by varying operators that modify the tree's structure. In this case, accuracy is measured by the Bayesian Information Criterion (BIC) (Eq. 2) [16].

$$BIC = -2\log\left(L\left(\hat{\vartheta}; y\right)\right) + K\log(n) \quad (2)$$

4 Model Selection Criteria

The proposed models were trained and tested and their results were compared to determine which of these machine learning techniques was most suitable for predicting the stiffness. Prior to generating the regression models, the 56 instances obtained from the FE simulations according to the Box-Behnken DoE were normalized to between 0 and 1. Subsequently, these 56 instances were used to train the models employing 56 times repeated ten-fold cross-validation. In order to test the models with new and data that had not been used during the training process, 10 new FE simulations were conducted. These were chosen randomly in an attempt to cover the entire range of possibilities of the problem from previous FE models that were obtained from the Box-Behnken DoE and thereby avoid overtraining on the models. Validation the models served to select the most accurate one. The criteria for the training and testing stage were the Mean Absolute Error (MAE), Root Mean Square Error (RMSE) and Correlation (CORR).

5 Results

During the training process, the most important parameters of each algorithm were fine-tuned to improve the prediction capability. For ANNs, the tuning was based on the number of hidden neurons, the weight decay (parameter to decrease the learning rate of the optimization function) and the number of iterations (maximum number of iterations in which the algorithm terminates if the quality of the best network does not improve further). To attempt to avoid a local minimum error, the models were built by varying the randomness of the parameters that the network uses as an initial weight during the training process. A total of 1000 ANNs per configuration were trained to predict both the compression and the lateral bending stiffness. In this case, ANNs with 18 and 20 neurons in the hidden layer were the models with the best generalization capacity for prediction of the compression and lateral bending stiffness. Using the 10 new FE simulations chosen randomly, the test process for the ANNs obtained an MAE of 5.63 %, RMSE = 6.38 %, and CORR = 0.98 for compression stiffness, and MAE = 3.17 %, RMSE = 3.56 %, and CORR = 0.99 for lateral bending stiffness. The trees that are proposed in this work were built using different a splitting index, but with a minimum number of four observations of a node in order to attempt a split. The complexity parameter was considered to be 0.01 with any split that does not decrease the overall lack of fit by this factor is not attempted. In addition, the maximum depth of any node of the final tree must be less than 5. Using CART methodology, where every class value is represented by the average number of instances that reach the leaf, the test process obtained an MAE = 11.8 %, RMSE = 14.4 %, CORR = 0.84 for compression stiffness (displacement) and an MAE = 14.6 %, RMSE = 17.79 % and CORR = 0.89 for the lateral bending stiffness (angle). The second kind of tree, where a linear regression model that predicts the class value of instances that reach the leaf, obtained in the test process an MAE = 19.03 %, RMSE = 21.32 % and CORR = 0.53 for compression stiffness, and an MAE = 14.18 %, RMSE = 16.87 % and CORR = 0.40 for lateral bending stiffness. In this case, the obtained model to predict the lateral bending stiffness contains the 16 linear models that belong to the 16 leaves that are, labeled LM1 through LM16. Finally, the third kind of trees that provides evolutionary methods for learning globally optimal regression trees obtained a test process of MAE = 13.59 %, RMSE = 16.36 % and CORR = 0.83 for compression stiffness and an MAE = 19.54 %, RMSE = 21.64 % and CORR = 0.75 for lateral bending stiffness. According to the criteria for the training and testing stage, the use of ANN with 18 and 20 neurons in the hidden layer are the models that have been shown to have the best generalization capacity for predicting compression and lateral bending stiffness.

6 Optimizing the Design of the Prosthesis

The regression models with the best generalization capacity obtained, ANN with 18 for the compression stiffness and ANN with 20 neurons for lateral bending stiffness, were used for search for the best geometry of the artificial disc (A, B, C, D and E) when a patient's height and weight were considered. The searching process was based in order to achieve the following objectives: The difference between the compression and lateral bending stiffness obtained from the FSU with the healthy and artificial disc were as low as possible. The objectives of the search was to minimize the difference between the compression and lateral bending stiffness from the FSU with the healthy and artificial discs. This search for the best combination of geometries was performed by applying GA. Optimization using GA usually involves the following six steps:

1. Coding/decoding–this involves changing the information from a binary chromosome of each individual to features' values that are converted to individual chromosomes. During this stage, if any value that defines an individual lies outside of the range that the DoE proposed, the individual is replaced by another.
2. Population initialization – Randomly selected individuals make up the initial population.

3. Evaluation-A fitness function evaluates the models' responses to decide which individuals should proceed to the next generation.
4. Selection-The individuals are sorted by fitness and the top 25 % are selected for the next generation.
5. Crossover-Parts of two parents of the current generation are combined to produce new offspring. These parts are selected according to two position and two longitudes. Offspring of crossovers from selected parents make up 60 % of the new population.
6. Mutation-From the 25 % that were selected and the 60 % that were obtained by crossovers, several individual are chosen on the basis of a uniform probability. A new individual is created by flipping a random element of the chromosome. Mutation accounts for 15 % of the new population.

In this case, the optimization process was conducted as follows: First, 1000 individuals, based on a combination of the geometry of the artificial disc (A, B, C, D and E) and the heights and body weights, were randomly generated to formed the initial generation or generation "0". Later, based on these individuals, the values of output features (compression and lateral flexion stiffness) were obtained by using the selected regression models (ANN with 18 and 20 neurons respectively). An objective function *F* (Eq. 3) was developed for selection of the best individuals of each generation. These individuals were those whose artificial disc geometry (A, B, C, D and E) was the most appropriate in order that the FSU behavior with the healthy and artificial discs were as similar as possible. *F* was defined as the minimum value of the difference between the values that were obtained from compression and lateral bending stiffness from the FSU with the healthy and artificial discs.

$$F = \min\left(\begin{array}{l} w_C |Comp.Stiff.Healthy - Comp.Stiff.Artificial|_{i,j} \\ + w_{FL} |Flex.LateralStiff.Healthy - Flex.LateralStiff.Artificial|_{i,j} \end{array} \right) \quad (3)$$

Also, each term of the equation was associated with its corresponding weights (w_c and w_{FL}) in order to consider different important values, depending on the requirements of design features. In this case, the weights were considered to be equally important. Thus, the weight that was assigned to each of them was for all generations that were studied equal to 1. Also, each of the subscripts *i, j* were defined as the corresponding values of the height and weight of the patient. The best individuals were those with the lowest values of *F*. Then, the next generations (first generation and subsequent generations) were created using selection, crossing and mutation. Table 3 shows the optimal artificial disc geometry that was obtained by using GA, and which also meets the objective function *F* when different weight and heights patient are considered. The first and second columns of the table show the patients' weights and heights. The next five columns shows the dimensions of the artificial disc (A, B, C, D and E) obtained with the methodology proposed. In these five columns, one can see that the dimensions C and E remains constant, whereas the other dimensions (A, B and D) vary in order to better adapt to the characteristics of the patients. Finally, the last four columns shows, respectively, the compression and lateral flexion stiffness obtained from the healthy FE FSU model (Stiffness healthy FE) and from the methodology proposed (Stiffness GA).

Table 3. Values obtained from the methodology proposed when different values of weight and height are considered.

Patient		Art. Disc Geometry					Stiffness Healthy		Stiffness GA	
Weight	Height	A	B	C	D	E	Disp.	Angle	Disp.	Angle
70	175	11.90	10.20	34.00	19.27	5.10	−0.313	4.006	−0.112	3.556
120	175	12.00	15.29	34.00	19.28	5.10	−0.519	5.905	−0.170	5.904
70	160	11.90	10.20	34.00	19.27	5.10	−0.313	3.760	−0.112	3.224
120	160	12.05	14.17	34.00	20.29	5.10	−0.519	5.550	−0.173	5.550
70	190	11.90	10.20	34.00	19.27	5.10	−0.313	4.242	−0.111	3.882
120	190	11.90	15.19	34.00	20.29	5.10	−0.519	6.228	−0.176	6.316
95	160	11.90	14.27	34.00	20.29	5.10	−0.418	4.680	−0.147	4.239
95	190	11.90	14.28	34.00	20.29	5.10	−0.418	5.280	−0.147	4.973
95	175	11.90	14.28	34.00	20.29	5.10	−0.418	4.975	−0.147	4.973

The figures in this table show that the stiffness values do not differ significantly. This indicates that the proposed methodology provides is a powerful tool for the design and optimization of artificial prosthesis.

7 Conclusions

This paper presents a methodology for the design of new artificial human intervertebral lumbar discs by combining FEM and soft computing techniques. First, a healthy disc FE model was generated and validated experimentally by standard tests. Then, a new artificial human intervertebral lumbar disc FE model composed of a core of PCU was proposed and parameterized in order to optimize it. ANN and RT that are based on heuristic methods and evolutionary algorithms were used to model the compression and lateral bending stiffness that were obtained from the simulations of the proposed artificial disc. In this case, the use of ANN provided the most accurate models for predicting the stiffness and lateral compression bending stiffness. Finally, the parameters that best define the geometry of the proposed artificial disc when the patient's height and body weight are considered were obtained by applying GA to the model based on ANN. With a desire to validate the proposed methodology, the compression and lateral flexion stiffness obtained from the healthy FE model and from the optimal artificial disc FE model were compared. The stiffness that was obtained did not differ significantly, demonstrating that the proposed methodology provides a powerful tool for the design and optimization of artificial prosthesis.

Acknowledgements. The authors wish to thank the University of the Basque Country UPV/EHU for its support through Project US15/18 OMETESA and the University of La Rioja for its support through Project ADER 2014-I-IDD-00162.

References

1. Lee, C.K., Goel, V.K.: Artificial disc prosthesis: design concepts and criteria. Spine J. **4**(6), S209–S218 (2004)
2. Van den Broek, P.R., Huyghe, J.M., Wilson, W., Ito, K.: Design of next generation total disk replacements. J. Biomech. **45**(1), 134–140 (2012)
3. Lostado, R., Martinez, R.F., Mac Donald, B.J., Villanueva, P.M.: Combining soft computing techniques and the finite element method to design and optimize complex welded products. Integr. Comput. Aided Eng. **22**(2), 153–170 (2015)
4. Gomez, F.S., Lorza, R.L., Martinez, R.F., Bobadilla, M.C., Garcia, R.E.: A proposed methodology for setting the finite element models based on healthy human intervertebral lumbar discs. In: Martínez-Álvarez, F., Troncoso, A., Quintián, H., Corchado, E. (eds.) HAIS 2016. LNCS, vol. 9648, pp. 621–633. Springer, Heidelberg (2016). doi:10.1007/978-3-319-32034-2_52
5. Hsu, C.C.: Shape optimization for the subsidence resistance of an interbody device using simulation-based genetic algorithms and experimental validation. J. Orthop. Res. **31**(7), 1158–1163 (2013)
6. Belytschko, T., Kulak, R.F., Schultz, A.B., Galante, J.O.: Finite element stress analysis of an intervertebral disc. J. Biomech. **7**, 276–285 (1974)
7. Panjabi, M.M., Brand, R.A., White, A.A.: Mechanical properties of the human thoracic spine. J. Bone Joint Surg. Am. **58**(5), 642–652 (1976)
8. Hoffler, C.E., Moore, K.E., Kozloff, K., Zysset, P.K., Goldstein, S.A.: Age, gender, and bone lamellae elastic moduli. J. Orthop. Res. **18**(3), 432–437 (2000)
9. Chen, H., Zhou, X., Fujita, H., Onozuka, M., Kubo, K.Y.: Age-related changes in trabecular and cortical bone microstructure. Int. J. Endocrinol. **3** (2013)
10. Tsouknidas, A., Michailidis, N., Savvakis, S., Anagnostidis, K., Bouzakis, K.D., Kapetanos, G.: A finite element model technique to determine the mechanical response of a lumbar spine segment under complex loads. J. Appl. Biomech. **28**(4), 448–456 (2012)
11. R Core Team: R: a language and environment for statistical computing. R Foundation for Statistical Computing, Vienna, Austria (2013). http://www.R-project.org/
12. Fernandez, R., Okariz, A., Ibarretxe, J., Iturrondobeitia, M., Guraya, T.: Use of decision tree models based on evolutionary algorithms for the morphological classification of reinforcing nano-particle aggregates. Comput. Mater. Sci. **92**, 102–113 (2014)
13. Ripley, B.D.: Pattern Recognition and Neural Networks. Cambridge University Press, New York (1996)
14. Breiman, L., Friedman, J.H., Olshen, R.A., Stone, C.J.: Classification and Regression Trees. Wadsworth International Group, Belmont (1984)
15. Grubinger, T., Zeileis A., Pfeiffer K.P.: evtree: Evolutionary Learning of Globally Optimal Classification and Regression Trees in R. Research Platform Empirical and Experimental Economics, Universitt Innsbruck (2011)
16. Schwarz, G.: Estimating the dimension of a model. Ann. Stat. **6**(2), 461–464 (1978)

SOCO 2016: Image and Video Analysis

Object Recognition by Machine Vision System of Inspection Line

Ondrej Petrtyl and Pavel Brandstetter(✉)

Department of Electronics, Faculty of Electrical Engineering
and Computer Science, VSB-Technical University of Ostrava,
17. listopadu 15, 70833 Ostrava, Czech Republic
{ondrej.petrtyl, pavel.brandstetter}@vsb.cz

Abstract. The paper deals with a machine vision for a recognition and identification of manufactured parts. The inspection line has been constructed. Stepper motors, direct current motors and electromagnetic coils have been used as actuators. Printed circuit boards have been designed and made for a power supplying and signal level conversion. A feedback is acquired by two–position switches. For the practical realization, various construction components have been used, for example buffers for objects prepared for the inspection and inspected objects, manipulation arms and manipulation platform. The main control unit is a compact programmable logical controller, which controls hardware parts during the inspection cycle. The image capturing has been done with a common web camera. Algorithms for the image processing has been programmed in MATLAB Simulink. Data between the control system and the image processing system are exchanged via a mutual communication protocol.

Keywords: Image processing · Machine vision · MATLAB · Programmable logical devices · Electric motor

1 Introduction

The paper is interested in an automatic quality inspection of produced parts by a machine vision system. The machine vision is usage of computer vision in industry applications. It processes information obtained from image and provides data to other devices. With an increasing power of graphic units, we can process a big amount of data in real-time. The machine vision is used for an automatic inspection, e.g. bar codes reading, final inspection of assembled products, inspection of quality and defects, labels and serial number reading, dimensions and shape recognition, pick & place applications etc. [1–7].

Each piece is inspected during a total quality control (TQC). It brings a high level of control of the production quality. In case of huge series, an effect of the TQC is enormous. Main advantages are a contactless measurement, flexibility and universality, fast processing following by communication and sending data to other peripherals or control systems. The inspection line or end of the line tester workplace contains parts for objects transfer from a previous operation or buffer. Then we have to transport parts

M. Graña et al. (eds.), *International Joint Conference SOCO'16-CISIS'16-ICEUTE'16*,
Advances in Intelligent Systems and Computing 527, DOI 10.1007/978-3-319-47364-2_23

by a conveyor, robot or pick-up arm to the position, where the inspection is executed. After that the final sorting to proper containers is done [7–10].

2 Hardware for Machine Vision

A choice of optimal components can minimize errors and disturbances during an image capturing and rapidly increase processing demands during an image pre-processing. Lighting elements have an important role for a proper contrast of inspected subject and minimized requirements for Software (SW) filters and image corrections.

Electronic sensors transform an original image made by photons to digital information. Optical elements polarize a luminous flux directly to the image sensor. Important parameters are a field of vision, resolution, image sharpness and distance of inspected object. Light filters are also used for an elimination of disturbances, reflections and filtering of different light wavelengths [8].

3 Mathematical Operations Used for Image Processing

A theoretical background of basic operations used for the image processing is shown in this part. Analysis of different approaches of the image processing for the object recognition has been done [11–15]. Finally Ref. [11] has been chosen as a reference source of equations for the image processing operations. An image is represented by matrix, where each element of this matrix shows the level of intensity for definite pixel. We work with gray scale images (8 bit). It brings one matrix with 256 intensity levels. All principles has been tested in MATLAB m-files first, based on discrete form of Eqs. (1)–(10), for a better understanding. Afterwards the more effective functions provided by MATLAB and Simulink have been used.

3.1 Normalized Histogram

The normalized histogram is a probability estimation of intensity occurrence level r_k in image. The histogram shows a number of pixels with one intensity level.

$$p_\mathrm{r}(r_\mathrm{k}) = \frac{n_\mathrm{k}}{MN} \text{ for } k = 1, 2, \ldots, L-1, \tag{1}$$

where n_k is a number of pixels with the kth intensity level r_k, L is a total number of intensity levels, M, N are dimensions of the matrix.

3.2 Histogram Equalization

The histogram equalization tries to balance a distribution of pixel intensity in the normalized histogram and maximize an image contrast (Fig. 1).

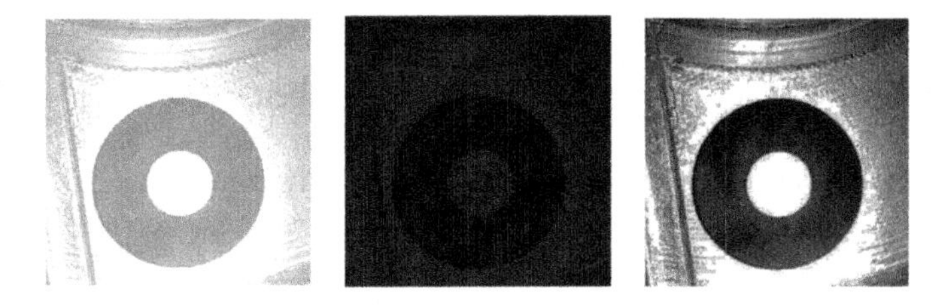

Fig. 1. Original - bright (left), dark (centre) and equalized image (right)

$$s_k = (L-1)\sum_{j=0}^{k} p_r(r_j) \quad \text{for } k = 1, 2, \ldots, L-1. \tag{2}$$

3.3 Two Dimensional Discrete Fourier Transform

The two dimensional (2D) Discrete Fourier Transform (DFT) represents a complex exponential function, this function transforms the spatial domain (x,y) to the frequency domain (u,v), where low frequencies are in centre and high on edges of transformed matrix.

$$F(u,v) = \sum_{x=0}^{M-1} \sum_{y=0}^{N-1} f(x,y) e^{-j2\pi\left(\frac{ux}{M}+\frac{vy}{N}\right)}, \tag{3}$$

where $f(x,y)$ is the original image.

3.4 Two Dimensional Inverse Discrete Fourier Transform

The 2D Inverse DFT (IDFT) function gives us the level of intensity for image. In fact, it is a transformation from the frequency domain to the spatial domain (Fig. 2).

$$f(x,y) = \frac{1}{MN} \sum_{u=0}^{M-1} \sum_{v=0}^{N-1} F(u,v) e^{j2\pi\left(\frac{ux}{M}+\frac{vy}{N}\right)}. \tag{4}$$

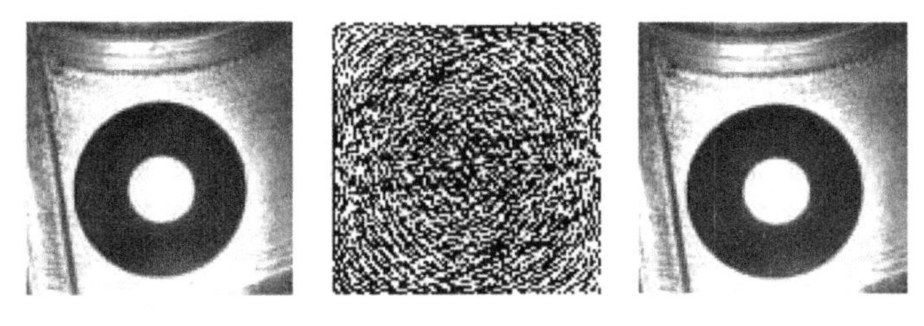

Fig. 2. Original image - spatial domain (left), result of 2D DFT - frequency domain (center), result of 2D IDFT - spatial domain (right)

3.5 Image Filtering Using Frequency Domain Filters

A filtering in the frequency domain is based on the modified Fourier transform. At first, an image is transformed by the 2D DFT to the frequency domain, there is correlated with a frequency filter and result of this operation is transform by the 2D IDFT to the spatial domain. The basic filtering equation in which we are interested has the following form.

$$g(x,y) = \Im^{-1}[\boldsymbol{H}(u,v)\boldsymbol{F}(u,v)], \tag{5}$$

where $\Im^{-1}$ is the IDFT, $\boldsymbol{F}$(u,v) is the DFT of an input $f(x,y)$, $\boldsymbol{H}(u,v)$ is filter function.

Smoothing. Low pass filters are used for this operation. A centre of filtered matrix (low frequencies) is near or equal to 1 and edges (high frequencies) are near or equal to 0. Result is cutting off high frequencies.

Sharpening. We use high pass filters for this operation, it cut off low frequencies. Edges are near or equal to 1 and the centre is near or equal to 0.

Ideal Filters. Low Pass (LP):

$$\boldsymbol{H}(u,v) = \begin{cases} 1 \text{ if } \boldsymbol{D}(u,v) \leq D_0 \\ 0 \text{ if } \boldsymbol{D}(u,v) > D_0 \end{cases}, \tag{6}$$

High Pass (HP):

$$\boldsymbol{H}(u,v) = \begin{cases} 0 \text{ if } \boldsymbol{D}(u,v) \leq D_0 \\ 1 \text{ if } \boldsymbol{D}(u,v) > D_0 \end{cases}. \tag{7}$$

The simplest LP and HP filters with two levels, 0 and 1. A centered circle with a radius D_0 is a boundary of these two values. $\boldsymbol{D}(u,v)$ is a distance between a point with coordinates (u,v) and a centre of the matrix, both in the frequency domain.

3.6 Gaussian Filters

Values between 0 and 1 are smoothly changed by the Gaussian curve. We can change steepness by transforming of parameter $\boldsymbol{D}_0$ [11] (Fig. 3).

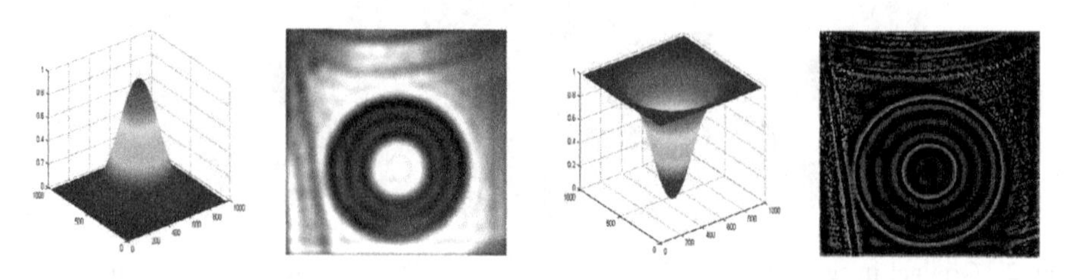

Fig. 3. From left: Gaussian LP, result of Smoothing, Gaussian HP, and result of Sharpening

LP:

$$H(u,v) = e^{-D^2(u,v)/2D_0}, \tag{8}$$

HP:

$$H(u,v) = 1 - e^{-D^2(u,v)/2D_0}. \tag{9}$$

3.7 Correlation

Correlation is a comparison of two images and the result is a level of concordance. A filter works with products summation of correlation filter coefficients and image pixels encompassed by the filter.

$$g(x,y) = \sum_{s=-a}^{a} \sum_{t=-b}^{b} w(s,t) f(x+s, y+t), \tag{10}$$

where $w(s,t)$ is the filter mask of size $m \times n$, $f(x,y)$ is original image, constants are $a = (m-1)/2$ and $b = (n-1)/2$.

3.8 Object Detection with Correlation

Object detection uses same principle like the spatial filtering, but in this case the mask $w(s,t)$ contains an example object. An inspected object shows different levels of concordance with compared examples after the correlation operation, the highest value shows which example is the most similar.

4 Hardware

4.1 Control System and Components

Control system - the compact programmable logic controller (PLC) Siemens SIMATIC CPU 314C-2DP (6ES7 314-6CH04-0AB0) has been chosen for the realization because of satisfactory parameters.

Supply voltage	24 V Direct current (VDC/DC) - 19.2–28.8 V, min. 4 A
Communication	Multi-point Interface (MPI), PROFIBUS
Inputs/Outputs	Digital 24/19(DI/DO), Analog 2/16 (AI/AO)
Work memory	192 kB
Memory card	Micro Memory Card (MMC)

4.2 Components

Construction - prefabricated profiles and construction elements (holders, gears and racks, gear-boxes, effectors, two position sensors etc.), buffers, manipulation platform have been produced by ourselves.

Drives - stepper motors (SM) MICROCON and SANYO with a step size 1.8°, supply voltage 24 VDC, DC motors (Fischertechnik) with supply voltage 9 VDC.

Optical system - web camera Genius (resolution 3200 × 2400, 8 mega pixels; 5 VDC, 0.05 mA) include lens wit manual focus.

Printed circuit boards (PCB) have been done for control system. Module 1 transforms the mutual supply voltage 24 VDC to the 9 VDC through use of voltage regulator for effectors and lights. Switching transistors for effectors and lights are placed here. Module 2 contains switching transistors controlled via signals from the PLC. A control of one SM is realized by Module 3 and current limiter is realized here. Signals from PLC switching particular SM through use of relays for 24 VDC. Module 4 transfers the mutual supply voltage to 9 VDC and controls another two SMs.

5 Software

5.1 Image Processing in MATLAB Simulink

MATLAB Simulink has been used as image processing software. Main advantage is high level of modularity as well as implemented functions. We used in particular Image Acquisition Toolbox and Computer Vision System Toolbox.

5.2 Principle of Image Processing

Firstly, image from the camera has been loaded and the equalization of the histogram has been done for a better contrast. After that examples and the inspected image has been resized. It brought a significant acceleration in the image processing. Then the edge detection has been done. It has been realized by a transformation to the frequency domain by the Fast Fourier Transform (FFT), through use of application of HP frequency filters and usage of the IDFT. The last step in the image processing has been the 2D correlation of inspected image and each example. Subsequently, a function has been applied for finding the maximum correlation coefficient designed by ourselves.

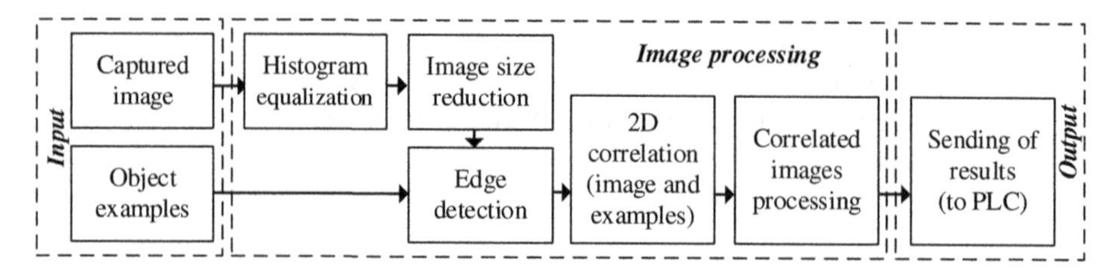

Fig. 4. Image processing in MATLAB Simulink

Finally, the example with the highest correlation coefficient has been set as the best fitting and has been sent to PLC (Fig. 4).

5.3 Control Algorithm

The control algorithm has been realized in a development environment TIA Portal V11, which is a complex tool for hardware (HW) configuration, programming, simulation and visualization. Step 7 (software tool provided by Siemens) for PLC programming is included as well. Different program languages have been used. Sequential Function Chart has been chosen for work cycle. Ladder Diagram has been chosen for particular components of program such as a control of motors, checking working and safety conditions, processing of input signals and communication. Structured Control Language has been used for an analyzing of results from the image processing.

5.4 Communication Between Systems

Object Linking and Embedding for Process Control (OPC) is a communication protocol, creating a united communication interface between HW and SW. Components produced by different vendors can be used in one project with a mutual communication interface. Here is only one condition, OPC interface has to be occurred on both sides, OPC Server for a used HW and OPC Client for SW. A communication between PC and PLC is executed via PLC programmer in this test rig. OPC communication tools in TIA are used for PLC and OPC Toolbox in MATLAB Simulink has been used in PC.

6 Experimental Rig

We can use also this method for a position detection. After correlation the searched object is the most likely in place with the highest value of concordance level. The inspection line for washers sorting has been constructed. The control system executes main control logic of the inspection, it controls all motors, effectors and lights. Input signals to this system are position sensor signals and communication with PC.

An image captured by the camera is send to the image processing system, which makes analysis of image, with recognition of two sizes and one type of deformation - washer without hole. Then send results to the control system. Finally, the control system evaluates information and finishes the inspection cycle. Figure 5 shows connections between all parts. Particular parts are described and all supply and signal connections are explained too.

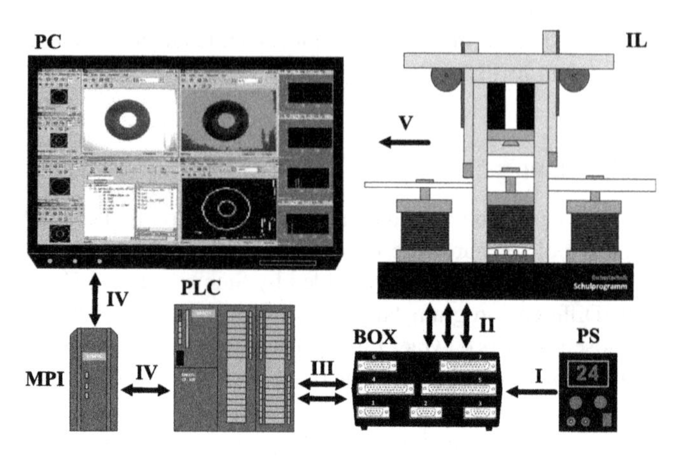

Fig. 5. Scheme of final experimental rig

Explanatory notes for Fig. 5:

BOX	Contains all modules	I	Supplying for all system (BOX, PLC, IL)
IL	Inspection line	II	Signals from PLC to IL and vice versa
MPI	PC adapter USB programming cable – MPI	III	DI and DO form PLC
PS	DC power supply	IV	Communication between PC and PLC
		V	Captured image from camera

The final realization of inspection line is shown on details in Fig. 6. This figure shows the complete rig. On right side we can see the power supply, BOX with Modules which are connected with other devices through Canon connectors and the control system PLC with programming PC adapter. Beside of this is the work place of the inspection line and on left side is PC for the image processing and PLC programming.

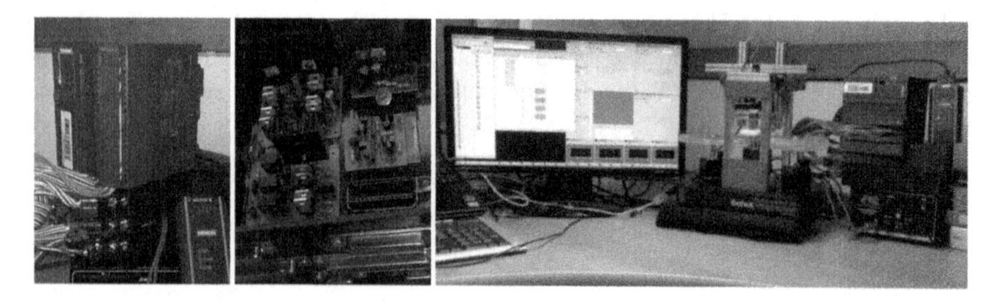

Fig. 6. Control system - PLC, BOX, power supply and programming PC adapter (left), interior of BOX (centre), complete system (right).

7 Experimental Results

A balance between the accuracy and the speed has to be found in the image processing. If full size images have been used (the captured image from the camera with the resolution 640 × 480 px, the samples of reference images for smaller washers 280 × 280 and for bigger washers 340 × 340) an operation time of one cycle has been over 10 min (laptop with Intel CORE i5). The cycle time has been 2 min for images with size 50 % of original, with 30 % approx. 20 s and with 10 % of the original size we have reached 2 s per cycle.

A deformation and a modification of the reference images with an increasing compression is demonstrated at the Fig. 7. Higher reduction than 10 % of the original image size has been shown as inapplicable.

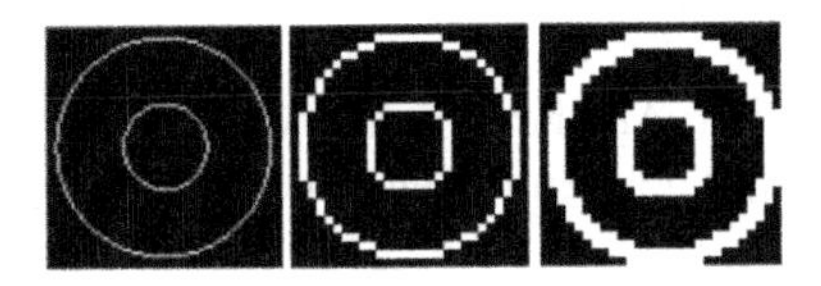

Fig. 7. Edge detection of reference image R2 (orig. size, resized to 10 % and 10 % with dilation)

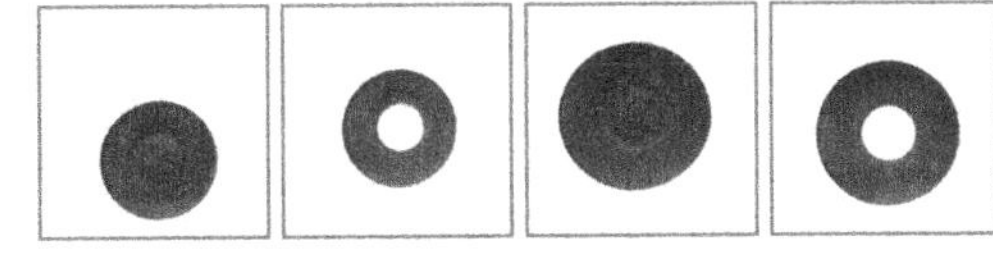

Fig. 8. Images captured from camera. From left: V1, V2, V3 and V4 (corresponds with R1, R2, R3 and R4 respectively).

The highest correlation coefficient has been calculated for each reference image. The coefficient indicates the number of white pixels in the image after the edge detection. Because we have used the compression, it is impossible to examine an exact size and small deformations, one pixel of the resized image corresponds with ten pixels at the original image. And also the edge detection brings an error, because it works with a different level of an intensity of bordering pixels. The real edge can be represented by different ways depending on the filter. For this case, the reference images after edge detection have been modified by a dilation and their edge curves are twice thicker. It can cover small deviations of edges after the detection, it is also the reason why some measured results are higher than the reference values in the Table 1.

Table 1. Result of experiments - correlation coefficient

	R1	R2	R3	R4
Max. corr. coef.	72	100	88	122
V1	**71,8**	**72,1**	31,1	31,3
V2	72,2	**99,7**	28,2	35,9
V3	25,9	26,4	**88,5**	**89,1**
V4	26,8	32,7	87,4	**119,3**

We multiply matrices which represent images by Boolean values (0, 1) during the correlation. If we have the object type V1 (see Fig. 8) and we compare it with the references R1 and R2, result is the same, because an outer perimeter is equal for both (total value 72). If we have the object type V2, outer perimeter is the same as V1 plus inner perimeter (total value 100). Identical principle works for V3, V4 and R3, R4.

The control program searches for the highest value of the correlation coefficient. It compares if the result is close to the size of washer with the same diameter (R1 with R2, R3 with R4). If the difference is not higher than 5 % washer without the hole (R1, R3) is chosen. If the value is lower than 20 it means there is no object for the inspection in the buffer. We had at least 10 pieces of each type and more than 100 tests have been done for each type. It means almost 500 cycles with this algorithm. Less than 10 mistakes appeared and all of these mistakes came from a wrong data transfer from the camera to the PC or parasitic light reflections. The average values of the experimental results are shown in Table 1, values related with the mistakes have been deleted from this statistic. Also other methods can be used, e.g. a threshold takes account of a whole area of the washes not only edges as the edge detection. Another possibility is a circle detection algorithm. If the higher accuracy is necessary, telecentric lens has to be used and also a camera with a high resolution.

8 Conclusions

The paper describes usage of the machine vision for object recognition. The positioning has been realized by the inspection line controlled by the PLC with the image processing in PC. The experimental rig has been developed and it works with an acceptable level of accuracy. High level of modularity is an advantage. The control units, modules for a signal level changing and drives can be replaced by elements of different brands with a various level of quality and professionalism. SW for the image processing can be replaced as well. Four modules have been done for a changing of signal levels, transformation of the mutual supply voltage 24 VDC to 9 VDC or transistor-transistor logic (TTL). They provide the supply voltage for DC and stepper motors, effectors and lights. Industrially produced control modules and SW for stepper motors have not been used, original and reliable PCBs and control algorithms have been produced. A unipolar control for SM without step loosing has been done, each phase of SM has been switched on by one DO from PLC, choice of SMs by relays also by PLC DO.

The control system is the compact PLC SIEMENS witch execute the control of mechanic parts. The image capturing is realized by the common web camera and the inspection has been done in MATLAB Simulink. All functions has been realized and tested as m-file in MATLAB based on the theory in the beginning and afterwards have been used components from MATLAB and toolboxes in Simulink. The image inspection contains the pre-processing, correlation of image and examples, finding of best match and results sending to PLC. The communication between PC and PLC has been realized via OPC. In the future research will be most important improving of control of SM by usage of professional control boards and SW. PLC can be replaced by faster digital signal processor (DSP) which is powerful enough for real-time image processing, than we execute control and image processing algorithms in one device.

The largest improvement would be achieved in the image processing. Also a number and quality of sensors would be increased.

Acknowledgements. The paper was supported by the projects: Center for Intelligent Drives and Advanced Machine Control (CIDAM) project, reg. no. TE02000103 funded by the Technology Agency of the Czech Republic, project reg. no. SP2016/83 funded by the Student Grant Competition of VSB-Technical University of Ostrava.

References

1. Hu, F., He, X., Niu, T.: Study on the key image processing technology in the inspection of packing quality for small-pack cigarettes. In: Second Workshop on Digital Media and Its Application in Museum & Heritages, Chongqing, pp. 67–71 (2007). doi:10.1109/DMAMH.2007.62
2. Islam, M.J., Ahmadi, M., Sid-Ahmed, M.A.: Image processing techniques for quality inspection of gelatin capsules in pharmaceutical applications. In: 10th International Conference on Control, Automation, Robotics and Vision, ICARCV 2008, Hanoi, pp. 862–867 (2008). doi:10.1109/ICARCV.2008.4795630
3. Arroyo, E., Lima, J., Leitão, P.: Adaptive image pre-processing for quality control in production lines. In: IEEE International Conference on Industrial Technology (ICIT), Cape Town, pp. 1044–1050 (2013). doi:10.1109/ICIT.2013.6505816
4. Mahale, B., Korde, S.: Rice quality analysis using image processing techniques. In: International Conference for Convergence of Technology (I2CT), Pune, pp. 1–5 (2014). doi:10.1109/I2CT.2014.7092300
5. Kuzu, A., Kuzu, A.T., Rahimzadeh, K., Bogasyan, S., Gokasan, M., Bakkal, M.: Autonomous hole quality determination using image processing techniques. In: IEEE 23rd International Symposium on Industrial Electronics (ISIE), Istanbul, pp. 966–971 (2014). doi:10.1109/ISIE.2014.6864743
6. Sahoo, S.K., Pine, S., Mohapatra, S.K., Choudhury, B.B.: An effective quality inspection system using image processing techniques. In: International Conference on Communications and Signal Processing (ICCSP), Melmaruvathur, pp. 1426–1430 (2015). doi:10.1109/ICCSP.2015.7322748
7. Fischer, R.B.: Dictionary of Computer Vision and Image Processing. Wiley, England (2005). ISBN 978-0-470-01526-1
8. Havle, O.: Automa. Machine Vision, 1st part (Czech). http://automa.cz/index.php?id_document=36550
9. Rusnak, J.: Design of a vision system with KUKA robot. M.S. thesis, Institute of Products Mach System and Robotics, Faculty of Mechanical Engineering, Brno University of Technology, Brno, Czech Republic (2011)
10. Graves, C.: Machine vision reaches top gear. IEE Rev. **44**(6), 265–267 (1998). doi:10.1049/ir:19980609
11. Gonzalez, R.C., Woods, R.E., Eddins, S.L.: The Digital Image Processing Using MATLAB, 827 p. Gatesmark Publishing, USA (2009). ISBN 978–0982085400
12. Belongie, S., Malik, J., Puzicha, J.: Shape matching and object recognition using shape contexts. IEEE Trans. Pattern Anal. Mach. Intell. **24**(4), 509–522 (2002). doi:10.1109/34.993558

13. Unser, M.: Splines: a perfect fit for signal and image processing. IEEE Signal Process. Mag. **16**(6), 22–38 (1999). doi:10.1109/79.799930
14. Banham, M.R., Katsaggelos, A.K.: Digital image restoration. IEEE Signal Process. Mag. **14** (2), 24–41 (1997). doi:10.1109/79.581363
15. Canny, J.: A computational approach to edge detection. IEEE Trans. Pattern Anal. Mach. Intell. **PAMI–8**(6), 679–698 (1986). doi:10.1109/TPAMI.1986.4767851

Pixel Features for Self-organizing Map Based Detection of Foreground Objects in Dynamic Environments

Miguel A. Molina-Cabello[1](✉), Ezequiel López-Rubio[1], Rafael Marcos Luque-Baena[2], Enrique Domínguez[1], and Esteban J. Palomo[1,3]

[1] Department of Computer Languages and Computer Science, University of Málaga, Bulevar Louis Pasteur, 35, 29071 Málaga, Spain
{miguelangel,ezeqlr,enriqued}@lcc.uma.es

[2] Department of Computer Systems and Telematics Engineering, University of Extremadura, University Centre of Mérida, 06800 Mérida, Spain
rmluque@unex.es

[3] School of Mathematical Science and Information Technology, University of Yachay Tech., Hacienda San José s/n., San Miguel de Urcuquí, Ecuador
epalomo@yachaytech.edu.ec

Abstract. Among current foreground detection algorithms for video sequences, methods based on self-organizing maps are obtaining a greater relevance. In this work we propose a probabilistic self-organising map based model, which uses a uniform distribution to represent the foreground. A suitable set of characteristic pixel features is chosen to train the probabilistic model. Our approach has been compared to some competing methods on a test set of benchmark videos, with favorable results.

Keywords: Foreground detection · Background modeling · Probabilistic self-organising maps · Background features

1 Introduction

Foreground object detection is a key problem in the design of computer vision systems. Algorithms to solve this problem must handle many difficulties which arise in real life videos. These inconveniences include illumination changes, shadow appearances in the foreground because of object lighting in the background or repetitive motions of background objects from the scene (waves of the sea, branches of the trees), among many others.

There are several approaches in the literature to model the background of a video sequence, employing different techniques like mixtures of Gaussians or probabilistic neural networks. In this paper we present a model based on probabilistic self-organising maps, with a suitable choice of characteristic pixel features.

The rest of the paper is structured as follows. The methodology from our proposal is described in Sect. 2. The experimental results are shown in Sect. 3. Finally we present our conclusions in Sect. 4.

M. Graña et al. (eds.), *International Joint Conference SOCO'16-CISIS'16-ICEUTE'16*, Advances in Intelligent Systems and Computing 527, DOI 10.1007/978-3-319-47364-2_24

2 Methodology

Our foreground detection system first computes the values of D features of each pixel of an incoming frame of size $NumRows \times NumCols$ pixels. The set of suitable features that we have considered is presented in [3]. After that, the feature vector $\mathbf{t} \in \mathbb{R}^D$ at pixel position $\mathbf{x} \in \{1, ..., NumRows\} \times \{1, ..., NumCols\}$ is provided as the input sample to a learning algorithm to adapt the parameters of a probabilistic mixture distribution with two mixture components ($Back$ for the background and $Fore$ for the foreground):

$$p_{\mathbf{x}}(\mathbf{t}) = \pi_{Back,\mathbf{x}} p_{\mathbf{x}}(\mathbf{t} \mid Back) + \pi_{Fore,\mathbf{x}} p_{\mathbf{x}}(\mathbf{t} \mid Fore) \tag{1}$$

The foreground values of the feature vector are modeled by the uniform distribution over the space of all possible feature vectors, so that any incoming foreground object can be represented equally well:

$$p_{\mathbf{x}}(\mathbf{t} \mid Fore) = U(\mathbf{t}) \tag{2}$$

$$U(\mathbf{t}) = \begin{cases} 1/Vol(\mathcal{S}) & \text{iff } \mathbf{t} \in \mathcal{S} \\ 0 & \text{iff } \mathbf{t} \notin \mathcal{S} \end{cases} \tag{3}$$

where $\mathcal{S}$ is the support of the uniform pdf and $Vol(\mathcal{S})$ is the D-dimensional volume of $\mathcal{S}$. The distribution of the background values of the feature vector is represented by means of a probabilistic self-organizing map:

$$p_{\mathbf{x}}(\mathbf{t} \mid Back) = \frac{1}{H} \sum_{i=1}^{H} p_{\mathbf{x}}(\mathbf{t} \mid i) \tag{4}$$

where H is the number of mixture components (units) of the self-organizing map, and the prior probabilities or mixing proportions are assumed to be equal. More details about the learning algorithm for the above defined mixture are given in [2].

The Bayesian probability that the observed sample (feature vector value) $\mathbf{t}$ is foreground is given by

$$R_{Fore,\mathbf{x}}(\mathbf{t}) = \frac{\pi_{Fore,\mathbf{x}} p_{\mathbf{x}}(\mathbf{t} \mid Fore)}{\pi_{Back,\mathbf{x}} p_{\mathbf{x}}(\mathbf{t} \mid Back) + \pi_{Fore,\mathbf{x}} p_{\mathbf{x}}(\mathbf{t} \mid Fore)} \tag{5}$$

However, $R_{Fore,\mathbf{x}}(\mathbf{t})$ is prone to noise due to isolated pixels that change their features randomly. The Pearson correlations $\rho_{\mathbf{x},\mathbf{y}}$ allow us to obtain a noise-reduced version of $R_{Fore,\mathbf{x}}(\mathbf{t})$ by combining it with the information from the 8-neighbours $\mathbf{y}$ of $\mathbf{x}$:

$$\widetilde{R}_{Fore,\mathbf{x}}(\mathbf{t}) = \operatorname{trunc}\left(\frac{1}{9} \sum_{\mathbf{y} \in Neigh(\mathbf{x})} \rho_{\mathbf{x},\mathbf{y}} R_{Fore,\mathbf{y}}(\mathbf{t}) \right) \tag{6}$$

where $Neigh(\mathbf{x})$ contains the pixel $\mathbf{x}$ and its 8-neighbours $\mathbf{y}$.

Table 1. Summary of the main model characteristics for each compared method.

Name	Model characteristics
WrenGA	One Gaussian distribution
GrimsonGMM	K Gaussian distributions
MaddalenaSOBS	Artificial neural networks
FSOM	Uniform distribution and probabilistic self-organizing map

3 Experimental Results

In this section we present the foreground detection performance of our method and a comparison with other algorithms of the state-of-art. Software and hardware used in our experiments are shown in Subsect. 3.1. We detail the tested sequences in Subsect. 3.2. The set of parameters by each method are specified in Subsect. 3.3. Finally the qualitative and quantitative results are reported in Subsect. 3.4.

3.1 Methods

Our method called FFSOM is based on the object detection method FSOM [2], which was previously developed by our research group and it is included in the comparisons. The code of this method can be downloaded for free[1].

The FFSOM and FSOM methods have been implemented in Matlab, using MEX files written in C++ for those quite time-demanding parts.

Additionally we have selected some reference methods of the literature. The first we used is the algorithm we note as WrenGA [7], which is the oldest and it features a single Gaussian. Other Gaussians approach method is the one we name GrimsonGMM [5], that uses two Mixture of Gaussians. Finally we have chosen an artificial neural networks approach method noted MaddalenaSOBS [4]. The main characteristics of all selected methods are shown in Table 1.

The implementation of these tested methods have been taken from the BGS libray version 1.3.0, which is accessible from its website[2].

Since our FFSOM and FSOM methods include a post-processing and the MaddalenaSOBS method has an implicit post-processing, we have added post-processing to all the other methods so as to make the comparisons as fair as possible.

The experiments reported in this paper have been carried out on a 64-bit Personal Computer with an eight-core Intel i7 3.60 GHz CPU, 32 GB RAM and standard hardware. The implementation of our method does not use any GPU resources, so it does not require any specific graphics hardware.

[1] http://www.lcc.uma.es/%7Eezeqlr/fsom/fsom.html.
[2] https://github.com/andrewssobral/bgslibrary.

Table 2. Considered parameter values for the competing methods, forming the set of experimental configurations.

Method	Parameters
FFSOM	Features, $F = \{[1\ 19\ 20]\}$
	Step size, $\alpha = \{0.01\}$
	Number of neurons, $N = \{12\}$
FSOM	Step size, $\alpha = \{0.01\}$
	Number of neurons, $N = \{12\}$
GrimsonGMM	Threshold, $T = \{12\}$
	Learning rate, $\alpha = \{0.0025\}$
	Number of Gaussians in the mixture model, $K = \{3\}$
MaddalenaSOBS	Sensitivity, $s_1 = \{75\}$
	Training sensitivity, $s_0 = \{245\}$
	Learning rate, $\alpha_1 = \{75\}$
	Training step, $N = \{100\}$
WrenGA	Threshold, $T = \{12\}$
	Learning rate, $\alpha = \{0.005\}$

3.2 Sequences

A set of videos have been selected from the 2014 dataset of the ChangeDetection.net website[3]. The sequences have been chosen are two videos from the Baseline category and other two from the Low Framerate category. The first one contains simple videos and the other one is composed by sequences with low frame rate. The video *Office* presents a room and a person who appears, he stays with low movements and then he goes out (360×240 pixels and 2050 frames), and *PETS2006* shows a train station with people moving on in (720×576 pixels and 1200 frames). This two videos are from the Baseline category. On the other hand, the two sequences selected from the Low Framerate category are *Tram-Crossroad*, a crossroad with cars driving for different ways (640×350 pixels and 900 frames); and *Turnpike* (320×240 pixels and 1500 frames), a highway with cars moving from left to right and vice versa.

3.3 Parameter Selection

We have defined a set of fixed values for the parameters of the methods to make the comparisons. The tuned values of each method are selected from the author's recommendations and they are shown in Table 2.

[3] http://changedetection.net/.

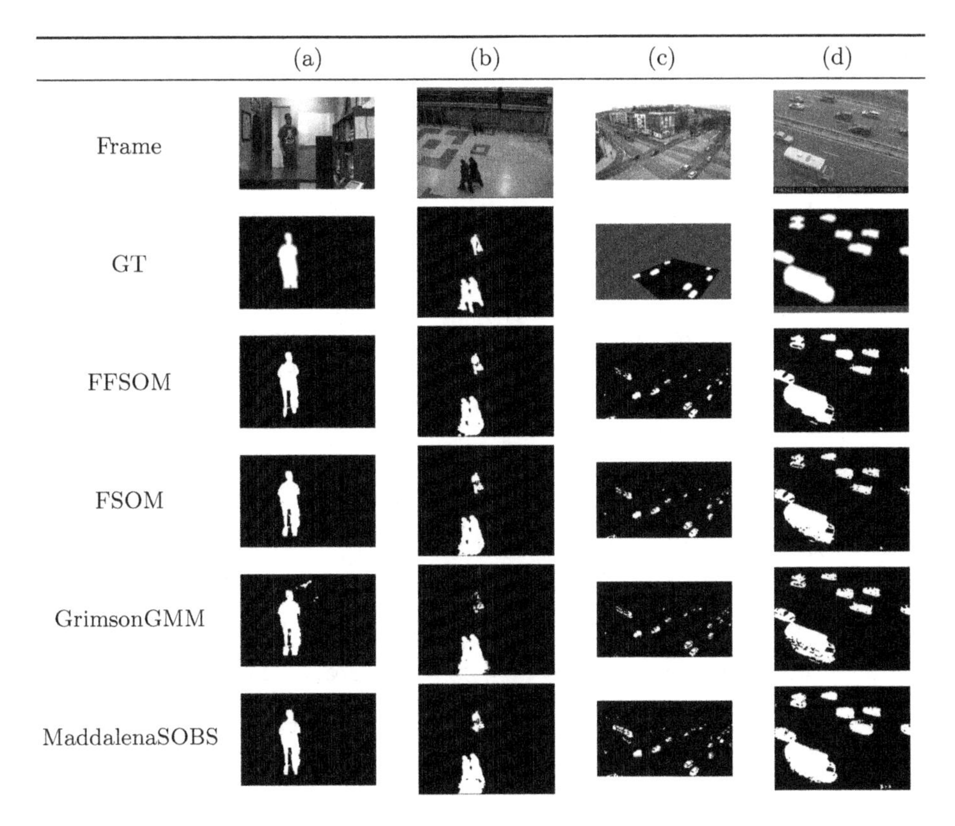

Fig. 1. Qualitative results for some benchmark scenes. From left to right: frame 1638 from Office (a), frame 956 from PETS2006 (b), frame 420 from TramCrossroad (c) and frame 958 from Turnpike (d) respectively. The first and second rows correspond to the original video frame and the ground truth. The remaining rows are the results given by the compared methods.

Table 3. Accuracy results (higher is better). Each column corresponds to a video and the rows indicate the methods. Each cell shows the mean and standard deviation of the accuracy over all tested configurations. Best results are highlighted in **bold**

Method	Office	PETS2006	TramCrossroad	Turnpike
FFSOM	0.569 ± 0.148	$\mathbf{0.679 \pm 0.077}$	$\mathbf{0.073 \pm 0.133}$	$\mathbf{0.317 \pm 0.321}$
FSOM	0.535 ± 0.147	0.658 ± 0.082	0.071 ± 0.129	0.298 ± 0.302
GrimsonGMM	0.285 ± 0.141	0.501 ± 0.157	0.069 ± 0.124	0.293 ± 0.297
MaddalenaSOBS	$\mathbf{0.701 \pm 0.118}$	0.638 ± 0.086	0.053 ± 0.101	0.299 ± 0.302
WrenGA	0.350 ± 0.154	0.448 ± 0.154	0.067 ± 0.122	0.262 ± 0.267

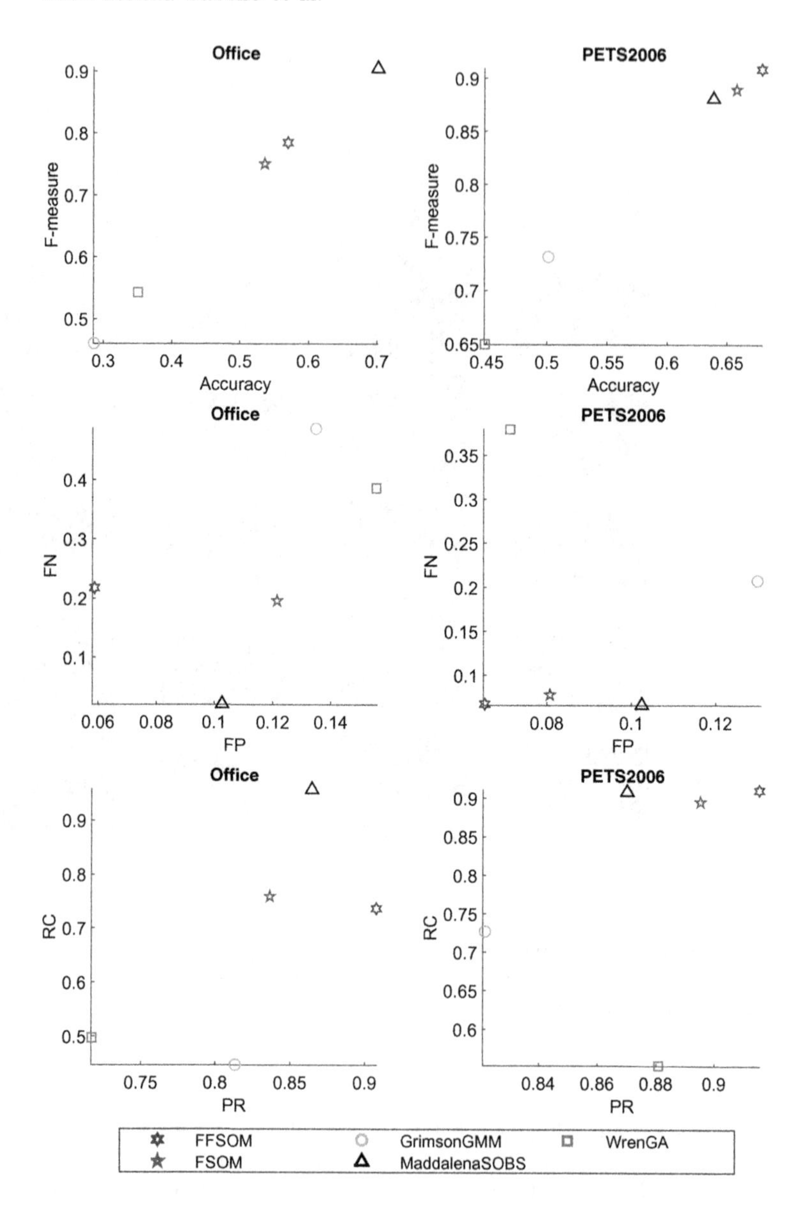

Fig. 2. Accuracy versus F-measure, false positives versus false negatives, and precision versus recall for each method. First column shows the office sequence and the second column corresponds to the PETS2006 video.

3.4 Results

On the one hand, from a qualitative point of view, in most cases the produced results by all compared methods are very similar, as it can be shown in Fig. 1.

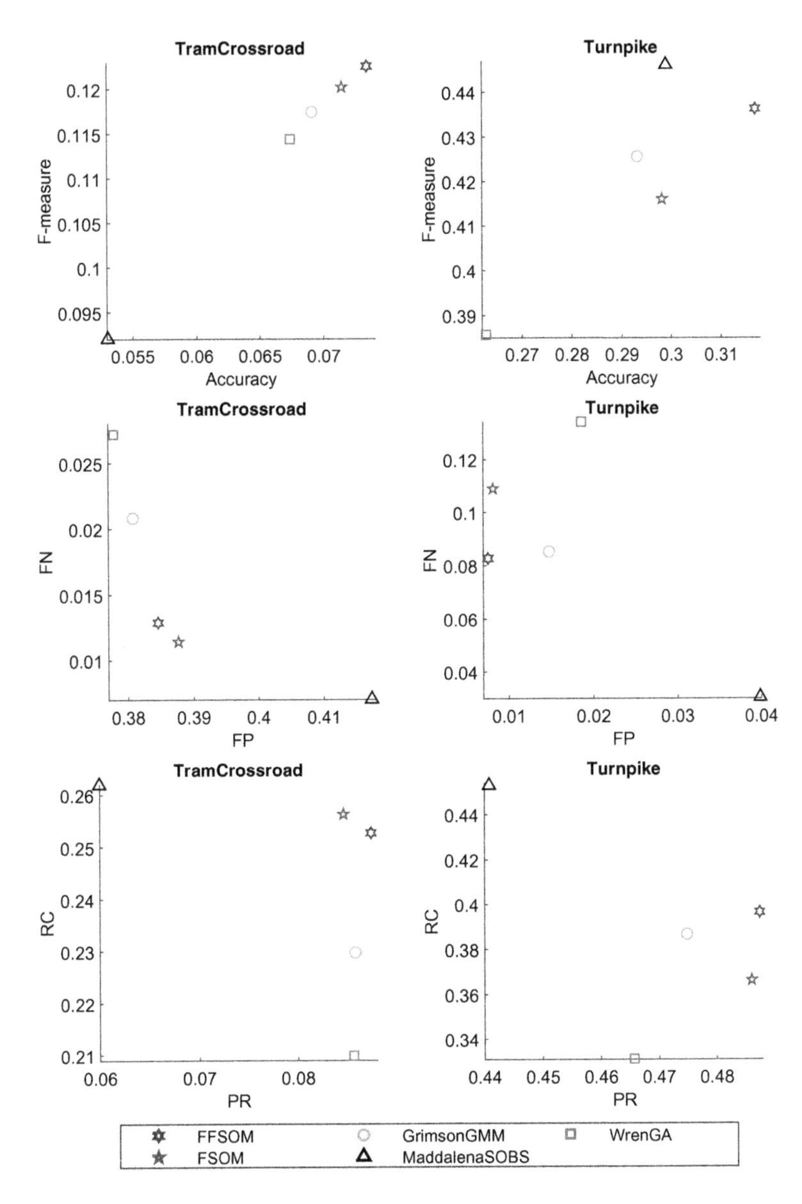

Fig. 3. Accuracy versus F-measure, false positives versus false negatives, and precision versus recall for each method. First column shows the TramCrossroad sequence and the second column corresponds to the Turnpike video.

On the other hand there are other scenes from the tested sequences for all the methods whose segmented images are obtained with noise and this promotes worst quantitative results. Furthermore there are other presented problems like camouflage (pixels from foreground and background are very similar) or sudden lighting changes in the scene.

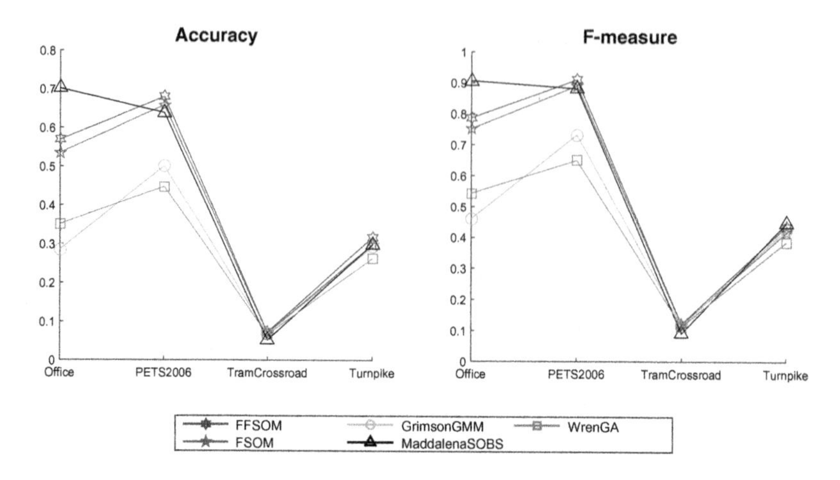

Fig. 4. Accuracy and F-measure for each method in each video. Please note that the values of each method are connected between them with lines to appreciate which method is better in each video, but this does not mean that the videos are related.

The goodness of a method and the comparison with others can be evaluated with different quantitative performance measures. One of them we have selected is the called *spatial accuracy*, which it has been used for the comparisons in other papers [1,6], and it is defined as follows:

$$AC = \frac{\mathrm{card}\,(A \cap B)}{\mathrm{card}\,(A \cup B)} \tag{7}$$

where 'card' stands for the number of elements of a set, A is the set of all pixels which belong to the foreground, and B is the set of all pixels which are classified as foreground by the analyzed method.

Furthermore, the F-measure is also employed, which is a proportion of the precision (PR) and recall (RC) metrics and it is defined as follows:

$$F - measure = 2 * \frac{PR * RC}{PR + RC} \tag{8}$$

The average accuracy with its standard deviation for the best configuration for each sequence are shown in Table 3. Furthermore some different results are presented in Figs. 2, 3 and 4.

FFSOM presents the best performance of all tested methods in three of the four analyzed videos. In addition, FFSOM obtained better results than FSOM.

Other significant aspect is the low accuracy presented in the TramCrossroad sequence by all methods. This is motivated because the ground truth of the different images from the video are not complete as we can see in Fig. 1, row GT, column (c).

robot's behavior when a risk of collision may happen) and the enabling technologies allowing acquisition data from sensor fusion and environmental data analyzing.

The second issue is addressed in this paper where we develop a monitoring system which guarantees human safety by achieving a reliable perception of the overall human-robot environment, precisely tracking the complete body of the human and activating safety strategies when the distance between them is too small. This paper presents the different techniques which have been implemented and integrated together under the Robot Operating System ROS [2] and shows its application in real human-robot collaborative tasks. In Sect. 2, the state of the art on workspace perception and human recognition and tracking is discussed. Later on, a reliable workspace perception is presented and implemented in Sect. 3 using point cloud fusion and filtering algorithms. A human detection and tracking is then performed in Sect. 4 and safety strategies are activated when the human-robot distance approaches predefined security thresholds. Finally, conclusions and future work are presented in Sect. 5.

2 State of the Art

Researchers and application developers have long been interested in performing automatic and semi-automatic recognition of human behavior from observations. Successful perception of human behavior is critical in a number of compelling applications, including automated visual surveillance and multimodal human–robot interaction (HCI) user interfaces that consider multiple streams of information about a user's behavior and the over-all context of a situation.

2.1 Workspace Perception

Most of the work on leveraging perceptual information to recognize human activities has centered on the identification of a specific type of activity in a particular scenario as the person's availability in an office [3]. Most work has focused on using 2D video [4,5] or RFID sensors placed on humans and objects [6]. The use of RFID tags is generally too intrusive because it requires a placement of RFID tags on the people. One common approach in activity recognition from a 2D video is to use space-time features to model points of interest in video [7,8].

Several authors have supplemented these techniques by adding more information to these features [6,9,10]. Other, less common approaches for activity recognition include filtering techniques [11], and sampling of video patches [12]. Also, GPS traces of a person were utilized through a model based on hierarchical conditional random fields (CRF) [13]. However, this model is only capable of off-line classification using several tens of thousands of data points. Wearable sensors are also used in [14] where authors combine the neural networks and the hidden Markov models. Multiple RGB-D sensors are simultaneously used in our case to ensure a 360° perception of the workplace and a reliable detection of humans working around the robot.

2.2 Human Recognition and Tracking

Workspace monitoring requires a detection and tracking of the human body. Therefore a human body model should be predefined to the system. This model gives to the algorithm the opportunity of exploiting the a priori information about the human body structure and, therefore, the search space related to possible body part configurations can be reduced through the definition of a set of constraints, such as human body proportions and limb.

Since the late 1970's more than 50 different human models have been developed. Early human models used only hands or arms to check clearances for tool manipulation. Today's models create whole-body representations using a basic "link" system resembling a human skeleton to enable posturing of the model within the work environment.

Representations have used either simple shape primitives (cylinders, cones, ellipsoids, and super-quadrics) or a surface (polygonal mesh, sub-division surface) articulated using the kinematic skeleton. A number of approaches have been proposed to refine the generic model shape to approximate a specific person. One example is the skeleton model defined in [15] which is a simplified representation of the human skeleton, named stick figure, where the joints are modeled as spheres or ellipsoids and the bones as cylinders or cones. Another approach is used in [3] to determine the parametric model of human from the images sequence. Moving object is separated from background by using background subtraction technique which small noises are removed by morphological opening and closing filters. The extracted foreground that supposed to be a human is then segmented into three regions representing the three important parts of human structure, for instance, head, body, and legs.

Furthermore, the OpenNI SDK module [16] provides, for the used RGBD Kinect sensors, a high-level skeleton tracking module, which can be used for detecting the captured human and tracking his body joints. More specifically, the OpenNI tracking module produces the positions of 15 joints, along with the corresponding tracking confidence. However, this module requires a-priori user calibration in order to infer information about the user's height and body characteristics. More specifically, skeleton calibration requires the captured user to stay still in a specific "calibration pose" for a few seconds to have accurate results. To avoid human calibration in front of all sensors, each time he enters in the specified workspace, which is not realistic for industrial collaborative applications, another technique is used in this paper. It consists on real time detection of standing/walking people on a ground plane using PCL Library [17,18].

Additionally, regarding human activity recognition, many techniques are targeted to recognize single, simple events as the person's availability in the workspace [3]. Others [19] used a simple human model and identified several basic actions. These actions were classified into two types: static and dynamic actions. The actions are considered static if only if there are at least one component which the velocity is null. By definition, the static actions are comprised of standing, bending and sitting. The actions are considered dynamic if only if all components of human model move.

3 Reliable Workspace Perception

To interpret a safe human-robot collaboration, a first activity is focused in achieving a reliable environment perception. As a foundation a system formed by several RGB-D cameras is used. They are installed in such positions to be able to capture robot and human activities consistently. After the calibration of the overall vision system, a consistent point cloud fusion algorithm is applied to develop a robust system that provides a unique point cloud with full human-robot environment perception avoiding occlusions.

3.1 Workspace Monitoring Setup

The setup is composed of multiple RGB-D sensors which are mounted in specific position/orientation to have a consistent perception of the overall human–robot environment (Fig. 1). Data redundancy caused by the simultaneous use of several 3D sensors with overlapping field of views allows a complete perception of the human activity, robot motions and environment's objects. Furthermore, it permits the avoidance of occlusions produced by the presence of the human's body near to the robot, and by the existence of industrial equipment or tools in the workplace.

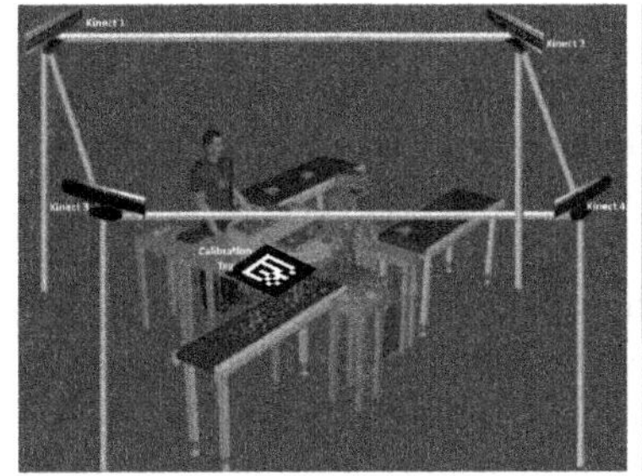

Fig. 1. Workspace monitoring setup with the calibration tag in simulation (left) and in the real setup (right).

Nevertheless, the efficiency of this system for human activity perception may highly increases if it was well adapted for the applied industrial tasks. The placement of these sensors should takes into consideration the executed task, the field of work of the human and its possible motions, in addition to the workspace of the robotic platform.

The presented techniques were implemented in the scope of an European project for Lean Intelligent Assembly Automation (LIAA) [20], the setup is composed of 4 RGB-D Kinect sensors consistently positioned in the 4 corners of a squared aluminum structure of dimension 4 m × 4 m with a height of 2.2 m (Fig. 1). Note that the position and orientation of these Kinects could be easily modified and adapted to the applied tasks and human-robot environment. In fact, due to the automatic extrinsic calibration, presented in the next section, sensors positioning can be freely and easily modified to be adapted for the applied tasks. This flexibility promotes the use of the presented technique in the frequently changing industrial applications.

3.2 Automatic System Calibration

In addition to the intrinsic calibration of the RGB-D sensors, an extrinsic calibration is necessary to determine the relative pose between these sensors. The results of these calibrations will be directly used to generate a complete point cloud of the environment.

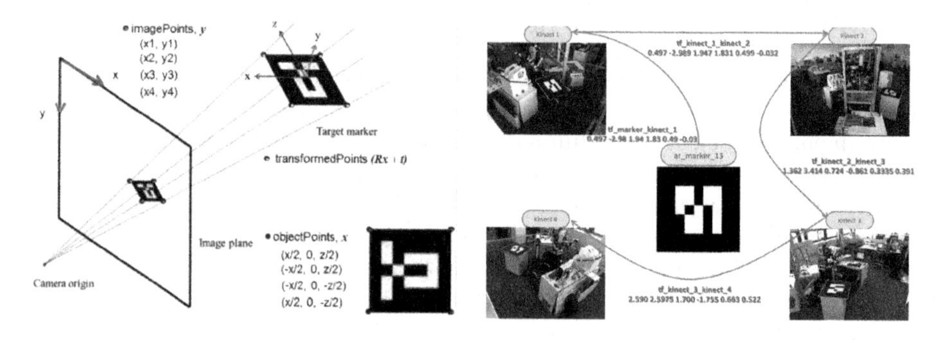

Fig. 2. A target marker is imaged allowing for the 4 corners of marker to be detected in the image. If the true distance between the corners is known, then there can be only one position along the back projected rays where marker of that size would produce the image [21] (left) and calibration results for the 4 Kinects system (right).

Extrinsic calibration of the system uses the ALVAR tracking library [22] which allow the detection and tracking of individual AR tags, and thus calculating the pose of these tags with respect to the different sensors using the formalism initially presented in [21] (Fig. 2). Several modifications were implemented to allow the online detection of the same tag by multiple sensors simultaneously; in addition to the integration of RGB-D depth data to have a better tag pose estimation.

Once the pose, of the used 587 mm squared tag, with respect to the different sensors is found with sufficient precision, the calculation of the relative transformations (tf) between the different RGB-D sensors is performed and the extrinsic calibration is done.

3.3 Point Cloud Fusion and Filtering

Later on, a point cloud fusion technique is needed to benefit from the presence of the different sensors to provide a complete perception of the human-robot environment. Therefore, an algorithm for merging the point cloud generated from multiple RGB-D sensors is implemented. It uses the results of the extrinsic calibration to reposition all the point cloud in a fixed frame. However, the dimension of the generated point cloud is relatively huge due to the superposition of the different sensors data. Therefore, down sampling and filtering actions are needed to optimize data dimension and then facilitate point cloud processing while performing human body detection and tracking.

PCL (Point Cloud Library) ROS interface stack [23] is a bridge for 3D applications involving n-D Point Clouds and 3D geometry processing in ROS. It is used to perform the filtering and optimization of the obtained point cloud. Down sampling uses a 3D voxel grid (3D boxes in space) over the input point cloud data. Then, in each voxel, all the points are approximated with their centroid. This approach is a bit slower than approximating them with the center of the voxel, but it represents the underlying surface more accurately.

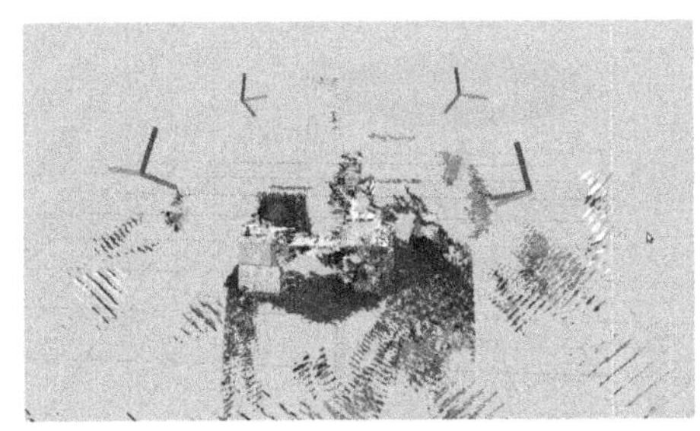

Fig. 3. Resulting merged point cloud before (left) and after (right) filtering and down sampling.

Specifications of the human working space and robot working area are used to decrease the dimension of the data to be processed later during human's detection and tracking. Therefore knowing the applied collaborative tasks, a predefinition of the human working area should be given to the system to limit the generated point cloud and limit the search area in the 3D scene for the worker.

As it could be seen in Fig. 3, the resulting point cloud from 4 Kinects is very big and contains a lot of irrelevant points outside of the bounding structure. The dimension of the resulted point cloud is very big and implies high time consumption for the computation process and data analysis. It also deleteriously affects the reactivity of the system to environment changes and human motions detection for example. Therefore a filtering and down sampling processes are implemented.

In our experiments, we consider that the human can move within all the mounted structure. Therefore we consider only the points which are inside the 3D structure (4 m $\times$ 4 m $\times$ 2.2 m). Despite the decrease of the dimension of the resulting point cloud, it contains of several superposed or very closed points. Therefore a down sampling action is performed with a predefined size of a leaf on X, Y and Z directions. A 10 mm voxel grid is used in our case. The resulting point cloud after filtering and down sampling is presented in Fig. 3. Compared to the original one, the final point cloud size was smaller by 90 %.

4 Safe Human Activity in the Workspace

4.1 Human Detection in System Workspace

Once the optimized point cloud is generated from the multi-Kinects system, a PCL library technique is used for real time detection of standing/walking people on a ground plane. This approach relies on selecting a set of clusters from the point cloud as people candidates which are then processed by a HOG-based people detector applied to the corresponding image patches (Fig. 4). The track initialization procedure allows to minimize the number of false positives and the online learning person classifier is used every time a person is lost, in order to recover the correct person ID even after a full occlusion.

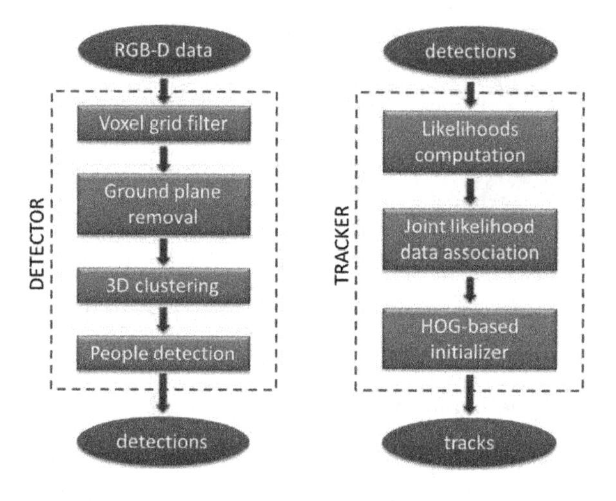

Fig. 4. Block diagram describing input/output data and the main operations performed by the detection and tracking modules [18].

In this scope, human model uses a preloaded database containing information about the geometry of the worker and specification about the workspace of the human. Several parameters are considered and adapted to have an efficient human detection as:

- **Minimum and maximum human height**: this information given by the human model is very useful for the algorithm and allows a decrease in the false positive detections.
- **Minimum and maximum number of point in the human point cloud.** These values should be adapted with respect to the workspace dimension and the applied tasks, because the number of points decreases proportionally to the distance of the human to the sensors. Furthermore, this number also depends on the dimension of the voxel grid used during the down sampling process.

Therefore these parameters should be carefully adapted and modified to have a more robust tracking and consequently a higher accuracy of the worker's position calculation, and finally improve the efficiency and reactivity of the workspace monitoring system. In our implementation, the human height is considered between 1.5 m and 2.1 m, and the human point cloud has between 280 and 650 points.

As it could be seen in Fig. 5 the human is detected while moving around the robot. It is identified by several Kinects simultaneously as shown in the images at the corners, the green box represents the detected human. The red arrow in the merged and filtered point could in the center of the figures, represents the final human pose in the environment. Knowing this position and the robot's one (from the predefined activity model of robot calibration), the human-robot distance is calculated and used in the safety strategies presented in the next section.

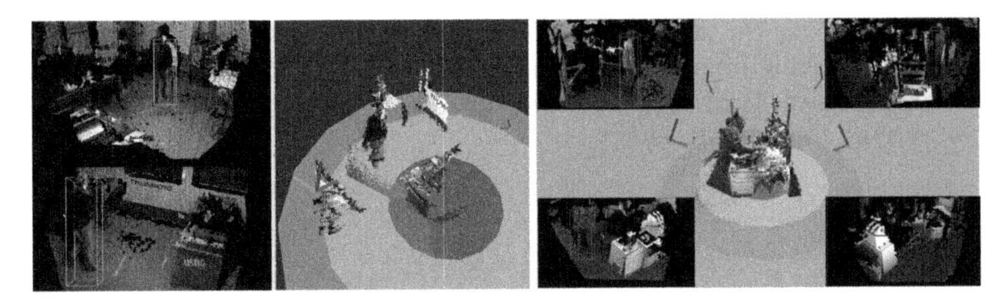

Fig. 5. Simultaneous human detection by several Kinects and representation of the human position in the final point cloud and within the defined safety zones.

Note that the detection of human from several sensors simultaneously, increases the precision of its position and makes the workspace monitoring system more reactive and robust to occlusions and sensor failures.

4.2 Implementation of Safety Strategies

Following the safety standards for collaborative robots about speed and separation monitoring (ISO 10218-1, 5.10.4, ISO/TS 15066) [1] which consist on reducing the risk by maintaining sufficient distance between the worker and robot in collaborative workspace. This goal is achieved by distance and speed supervision, by having a protective stop if minimum separation distance or speed limit is violated, and by taking account of the braking distance in minimum separation distance.

The developed system uses simple techniques to get a first robust coarse grain activity recognition. The first stage consists then to detect if the worker is inside or outside the workspace. In this part, the human activity recognition is simplified to recognize one single and simple event which the person's availability in the working place. In the second stage, the robot end-effector pose is used to evaluate the separation distance ($d_{robot/worker}$) between the detected worker and the robot to define 3 static activities:

- **Worker in safe zone**: The worker is the working area but far from the robot, no danger is present on the worker ($d_{robot/worker} > d_{warning\ zone}$).
- **Worker in red zone**: The worker is very close to the robot, a collision or interaction may be present between the robot and the worker ($d_{robot/worker} < d_{red\ zone}$).
- **Worker in warning zone**: The worker is between the safe zone and the red zone ($d_{red\ zone} < d_{robot/worker} < d_{warning\ zone}$).

At this stage, the activity model consists of two thresholds represented in Fig. 6 by $d_{red\ zone}$ and $d_{warning\ zone}$ that define the borders of these three levels. Furthermore, the robot's position in the environment is required to be able to calculation worker-robot distance, thus this information should be also predefined in the activity model or given online by other modules. The third stage consists of tracking the motion of the human; therefore four dynamic actions are defined: (1) Moving from safe zone to warning zone, (2) Moving from warning zone to red zone, (3) Moving from red zone to warning zone, and (4) Moving from warning zone to safe zone.

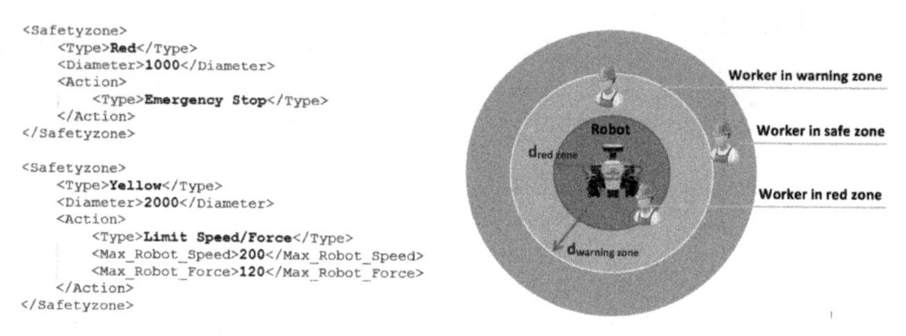

Fig. 6. Example of configuration file for safety zones (left) and representation of the working area with the safe, warning and red zones (right).

This method for human detection and tracking was integrated with other execution modules on different robotic systems, to activate several safety strategies when the human–robot distance approaches the predefined security thresholds in the configuration file (Fig. 6). These strategies consist on switching the robot motion mode between: normal speed when worker is in the safe zone, reduced speed/force when the worker approaches warning zone and finally robot enters in emergency stop when the worker arrives in the red zone[1].

5 Conclusions and Future Works

In this paper, we presented an implementation of a reliable vision system for full perception of the human-robot environment. A consistent and optimized point cloud of the system is generated and used for human detection in the defined

[1] An implementation of the presented techniques in the scope of LIAA project is available on https://youtu.be/AtZGeX2t51k.

safety zones. The presented work was implemented with several robotic systems in different configurations.

In addition to the ease of configuration and calibration, system's reactivity and robustness for occlusion promotes its use in flexible collaborative production industries with changing applications. This workspace monitoring system is an important step towards the implementation of a generic system for human-robot activity monitoring. It could be used to develop new methods for identifying static and dynamic actions of the human in the workspace. Furthermore, generated point cloud could be also used for dynamic collision anticipation and avoidance: for safe human robot co-working, the collision threat could be thus anticipated by either stopping the robot or generating of a new safe trajectory.

Acknowledgments. The research leading to these results has been funded in part by the European Union's seventh framework program (FP7/2007-2013) under grant agreements #608604 (LIAA: Lean Intelligent Assembly Automation).

References

1. ISO: ISO 10218–1: Robots and robotic devices-safety requirements for industrial robots-part 1: Robots. Geneva, Switzerland: International Organization for Standardization (2011)
2. Quigley, M., Conley, K., Gerkey, B., Faust, J., Foote, T., Leibs, J., Wheeler, R., Ng, A.Y.: Ros: an open-source robot operating system. In: ICRA Workshop on Open Source Software, vol. 3, p. 5 (2009)
3. Johnson, B., Greenberg, S.: Judging people's availability for interaction from video snapshots. In: Proceedings of the 32nd Annual Hawaii International Conference on Systems Sciences, HICSS-32, p. 9. IEEE (1999)
4. Ning, H., Han, T.X., Walther, D.B., Liu, M., Huang, T.S.: Hierarchical space-time model enabling efficient search for human actions. IEEE Trans. Circ. Syst. Video Technol. **19**(6), 808–820 (2009)
5. Gupta, A., Srinivasan, P., Shi, J., Davis, L.S.: Understanding videos, constructing plots learning a visually grounded storyline model from annotated videos. In: IEEE Conference on Computer Vision and Pattern Recognition, CVPR 2009, pp. 2012–2019. IEEE (2009)
6. Wu, J., Osuntogun, A., Choudhury, T., Philipose, M., Rehg, J.M.: A scalable approach to activity recognition based on object use. In: IEEE 11th International Conference on Computer Vision, ICCV 2007, pp. 1–8. IEEE (2007)
7. Laptev, I.: On space-time interest points. Int. J. Comput. Vis. **64**(2–3), 107–123 (2005)
8. Dollár, P., Rabaud, V., Cottrell, G., Belongie, S.: Behavior recognition via sparse spatio-temporal features. In: 2nd Joint IEEE International Workshop on Visual Surveillance and Performance Evaluation of Tracking and Surveillance, pp. 65–72. IEEE (2005)
9. Liu, J., Ali, S., Shah, M.: Recognizing human actions using multiple features. In: IEEE Conference on Computer Vision and Pattern Recognition, CVPR 2008, pp. 1–8. IEEE (2008)
10. Jhuang, H., Serre, T., Wolf, L., Poggio, T.: A biologically inspired system for action recognition. In: IEEE 11th International Conference on Computer Vision, ICCV 2007, pp. 1–8. IEEE (2007)

11. Rodriguez, M.D., Ahmed, J., Shah, M.: Action mach a spatio-temporal maximum average correlation height filter for action recognition. In: IEEE Conference on Computer Vision and Pattern Recognition, CVPR 2008, pp. 1–8. IEEE (2008)
12. Boiman, O., Irani, M.: Detecting irregularities in images and in video. Int. J. Comput. Vis. **74**(1), 17–31 (2007)
13. Liao, L., Fox, D., Kautz, H.: Extracting places and activities from GPS traces using hierarchical conditional random fields. Int. J. Robot. Res. **26**(1), 119–134 (2007)
14. Zhu, C., Sheng, W.: Human daily activity recognition in robot-assisted living using multi-sensor fusion. In: IEEE International Conference on Robotics and Automation, ICRA 2009, pp. 2154–2159. IEEE (2009)
15. Marcon, M., Pierobon, M., Sarti, A., Tubaro, S.: 3d markerless human limb localization through robust energy minimization. In: Workshop on Multi-camera and Multi-modal Sensor Fusion Algorithms and Applications, M2SFA2 2008 (2008)
16. The OpenNI Organization: Introducing openni, open natural interaction library. http://www.openni.org. Accessed: 30 Nov 2015
17. Munaro, M., Menegatti, E.: Fast RGB-D people tracking for service robots. Auton. Robots **37**(3), 227–242 (2014)
18. Munaro, M., Basso, F., Menegatti, E.: Tracking people within groups with RGB-D data. In: 2012 IEEE/RSJ International Conference on Intelligent Robots and Systems (IROS), pp. 2101–2107. IEEE (2012)
19. Noorit, N., Suvonvorn, N., Karnchanadecha, M.: Model-based human action recognition. In: Second International Conference on Digital Image Processing, p. 75460P. International Society for Optics and Photonics (2010)
20. LIAA: Lean intelligent assembly automation. http://www.project-leanautomation.eu. Accessed: 05 Jun 2016
21. Noonan, P.J., Anton-Rodriguez, J.M., Cootes, T.F., Hallett, W.A., Hinz, R.: Multiple target marker tracking for real-time, accurate, and robust rigid body motion tracking of the head for brain pet. In: 2013 IEEE Nuclear Science Symposium and Medical Imaging Conference (NSS/MIC), pp. 1–6. IEEE (2013)
22. Niekum, S.: ROS wrapper for alvar, an open source ar tag tracking library. http://wiki.ros.org/ar_track_alvar. Accessed: 30 Nov 2015
23. Kammerl, J., Woodall, W.: PCL (point cloud library) ros interface stack. http://wiki.ros.org/pcl_ros. Accessed: 30 Nov 2015

Forecasting Store Foot Traffic Using Facial Recognition, Time Series and Support Vector Machines

Paulo Cortez[1](✉), Luís Miguel Matos[1], Pedro José Pereira[1], Nuno Santos[2], and Duarte Duque[3]

[1] Department of Information Systems, ALGORITMI Centre, University of Minho, 4804-533 Guimarães, Portugal
pcortez@dsi.uminho.pt

[2] EXVA Technologies, Rua do Comércio, 6, 4710-820 Braga, Portugal
nuno@exva.pt

[3] ALGORITMI Centre, DIGARC - Polytechnic Institute of Cavado and Ave, 4750-810 Barcelos, Portugal
dduque@ipca.pt

http://www3.dsi.uminho.pt/pcortez

Abstract. In this paper, we explore data collected in a pilot project that used a digital camera and facial recognition to detect foot traffic to a sports store. Using a time series approach, we model daily incoming store traffic under three classes (all faces, female, male) and compare six forecasting approaches, including Holt-Winters (HW), a Support Vector Machine (SVM) and a HW-SVM hybrid that includes other data features (e.g., weather conditions). Several experiments were held, under a robust rolling windows scheme that considers up to one week ahead predictions and two metrics (predictive error and estimated store benefit). Overall, competitive results were achieved by the SVM (all faces), HW (female) and HW-SVM (male) methods, which can potentially lead to valuable gains (e.g., enhancing store marketing or human resource management).

Keywords: Data mining · Facial recognition · Time series forecasting · Support vector machine

1 Introduction

In this age of big data, the passive collection of human activities (without their explicit intervention), is becoming commonplace. In effect, there is a large number of Information and Communications Technologies (ICT) that facilitate such passive collection, including Radio-frequency identification (RFID) technology, digital cameras and other sensors. These technologies open room for what is known as modern retail management [16]. By data mining [20] the behavior of people, it is possible to detect interesting patterns and build data-driven models that allow the prediction of human activities. This data-driven knowledge can be used to enhance the management of shopping centers or individual stores.

M. Graña et al. (eds.), *International Joint Conference SOCO'16-CISIS'16-ICEUTE'16*, Advances in Intelligent Systems and Computing 527, DOI 10.1007/978-3-319-47364-2_26

This work focuses on a particular modern retail application, where a digital camera is used to detect foot traffic to a store and then time series and data mining methods are used to build store traffic forecasting models. Such models can have a potential benefit in diverse store management areas, including security, marketing campaigns and promotions to attack customers, store window and layout design, management of stocks and human resources.

There are several works that adopted digital cameras for tracking and predicting human trajectories. For instance, an intelligent system was proposed in [8] and that is able to detect in advance if a person is going to enter a region of interest (restricted area) by analyzing human movements in a stationary video camera that captures a wide outdoor scene. This system was devised for surveillance but it could also be used to track a shopping center or store incoming traffic by defining the physical entrance as the region of interest. A similar approach was adopted in [12], where digital cameras were combined with state-of-the-art semantic scene methods in order to forecasting of plausible human destination paths in outside car parks near buildings. More recently, [1] predicted crowd mobility in a transport hub that included a shop. The prediction was based on social affinity maps created from a large number of monitoring cameras placed through the hub main corridors.

Time series forecasting models a phenomenon based on its past temporal patterns and has been applied in distinct domains, such as grocery store sales [13] and Internet traffic [5]. In the 60s and 70s, several statistical methods were proposed for time series forecasting, including the popular Holt-Winters (HW) or AutoRegressive Integrated Moving Average (ARIMA) methodology [14]. More recently, Soft Computing methods, such as Support Vector Machines (SVMs), have also been proposed for time series prediction, often achieving competitive results when compared with statistical methods [7,17].

Recently, in 2013, we have executed a pilot project during a eight month period. In such pilot, a digital camera was set to capture the frontal view of the entrance of a Portuguese sports store. The camera was linked with a human facial recognition system that allowed the automatic detection of human faces. Each time a face is detected, the system also estimates its gender (female or male) and generates an event. In this paper, we explore these event data, in terms of daily counts. Thus, rather than modeling human trajectories, we predict store foot traffic in terms of three time series (all faces, female and male). Six forecasting methods are compared: a simple baseline, HW, ARIMA, a recent SVM approach [17] and two newly proposed hybrids HW-SVM and SVM-SVM that combine time series data with other features (e.g., weather and special events). The comparison assumes a robust and realistic rolling window evaluation scheme that considers from 1 to $h = 7$ daily ahead predictions and two metrics (predictive error and estimated store benefit).

The paper is organized as follows. Section 2 presents the collected data, forecasting methods and evaluation procedure. Next, Sect. 3 describes the experiments conducted and analyzes the obtained results. Finally, Sect. 4 draws the main conclusions and also mentions future work directions.

2 Materials and Methods

2.1 Collected Data

The data was collected from a pilot project conducted in a sports store, placed inside a shopping center in a Portuguese city. From April to December of 2013 (total of eight months), a digital camera captured the frontal view of the store unique entrance. Due to privacy issues, the camera did not record the video data. Yet, the video was fed to a facial recognition system, partially based in the technology proposed in [8] and that is property of the company EXVA Technologies (http://www.exva.pt/?lang=en). This system uses the Viola-Jones framework [19] to detect faces and a machine learning classifier to distinguish gender. The system has a high accuracy in facial and gender recognition and is is capable of detecting multiple faces within a single video frame. However, due to commercial issues, the full recognition system details cannot be disclosed. Since the goal of this paper is to measure the predictive accuracy of store foot traffic time series models, and not to validate the computer vision system, we assume that the collected facial event data is correct.

Each time a face is detected, the recognition system stores a face event that includes: a gender type {female, male, unknown}; and uptime, the total time (in seconds) that the face was detected. The unknown class is assigned when gender is more difficult to detect. This situation occurs very rarely. In effect, the percentages of collected gender types are: 46 % for female, 49 % for male and 5 % for unknown. In terms of data preprocessing, we discarded 235 events related with uptime values higher than 300 s and that might be related with anomalies (e.g., face in a marketing ad board). The result was a dataset with 888,376 event records. Then, we built the daily time series by aggregated the female, male and all faces events (including unknowns). After this aggregation, we detected a total of 16 days (12 consecutive days in September) with missing data and that were related with pilot system failures. The missing data only occurred during the training data period and were replaced by the previously known values for the same day of the week (e.g., a missing Monday value is replaced by the value from the previous Monday). The final three time series (all faces, female and male) include a total of 257 daily entries. For demonstration purposes, the all faces series is plotted in the right of Fig. 1.

To test the value of other features, we also collected six daily variables, including exterior weather conditions (maximum wind speed, average temperature, humidity and rain) in the city (from https://www.wunderground.com/) and special daily events (freeday – if weekend or holiday; special – if there is a major sports or entertainment event in the city), retrieved using Internet queries.

2.2 Forecasting Methods

A time series includes time ordered observations $(y_1, y_2, \ldots, y_t)$, where t is the time period. A time series model predicts a value for current time t based on past observations: $\hat{y}_t = f(y_{t-k_I}, ..., y_{t-k_1})$, where f is the forecasting function

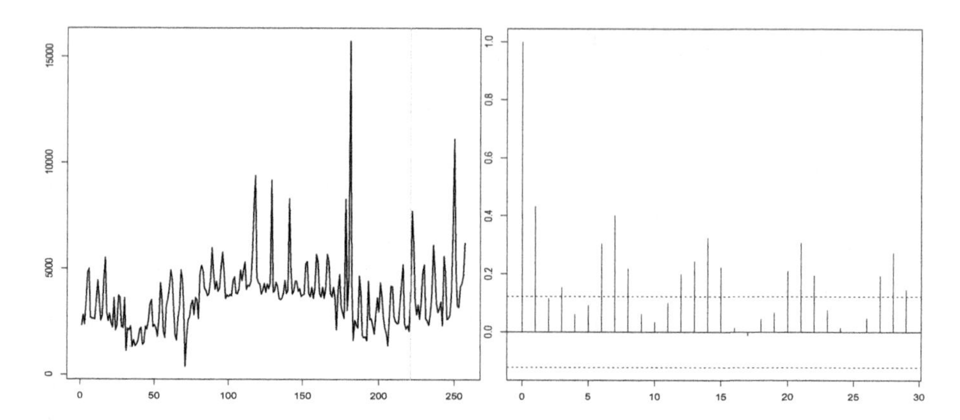

Fig. 1. Daily all faces time series (right, x-axis denotes the time period, y-axis the number of daily faces, vertical gray line splits the training and test data) and its autocorrelation values (left, x-axis denotes the time lags and y-axis the autocorrelations)

and the k_i values denotes the sliding window with I time lags. The k_i values can include any elements from the set $\{1, ..., t-1\}$. The horizon h is defined by the ahead time in which a prediction is executed. Multi-step ahead forecasts can be built by iteratively using 1-ahead predictions as inputs [5].

For example, let us consider the series $5_1, 10_2, 14_3, 17_4$ (y_t values). If the $\{1, 3\}$ window is adopted, then the prediction for time $t = 4$ is built using $\hat{y}_4 = f(5_1, 14_3)$. Multi-step ahead forecasts can be built by iteratively using 1-ahead predictions as inputs. If $h = 2$ and $t = 4$, then $\hat{y}_5 = f(10_2, \hat{y}_4)$.

In this work, we compare six time series approaches. Daily series often present a weekly seasonal period ($K = 7$), which is the case of the analyzed series. For example, the left of Fig. 1 shows higher autocorrelation values for the multiples of 7, which confirms a weekly cycle for the series. Thus, the baseline method is set to predict the future values (from $h = 1$ up to $h = 7$) using the known values from the previous week. It is assumed that this is the easiest method that could be used by the store manager. The HW is from the family of exponential smoothing methods and it is very popular for predicting trended and seasonal time series. We adopt the HW additive model with $K = 7$. For ARIMA, we adopt its multiplicative seasonal version, also known as SARIMA ($K = 7$).

Regarding the SVM model, we use the popular Gaussian kernel and ϵ-insensitive loss function for regression. The performance is affected by the choice of sliding window time lags (k_i, the model inputs) and the model hyperparameters (λ – the kernel parameter; C – a trade-off between fitting the errors and the flatness of the mapping; and ϵ – the width of the insensitive tube). In particular, we adopt a fast two stage input and hyperparameter selection methodology, similar to what was proposed in [3,17] and that has outperformed multilayer perceptrons in the forecasting of seasonal series [3]. Before fitting any SVM model, the data are first standardized to a zero mean and one standard deviation [10]. The first stage (input selection) is applied only once, using the first training dataset (of size W) with an internal ordered holdout split (with 70 % for fitting the model and 30 % for validation). Then, a fast backward selection feature selection procedure is used to discard irrelevant time lags. The feature selection starts with all k_i time lags ($i \in \{1, ..., I\}$) and computes its associated validation error. At this stage, the SVM hyperparameters are set to their default values ($\gamma = 2^{-4}, C = 1, \epsilon = 0.1$). Then, the sensitivity analysis proposed in [4] (e.g., DSA method, AAD measure of importance, 7 sensitivity levels) is used to measure input relevance and discard the least relevant time lag. This sensitivity analysis can measure the true global importance of an input, even when it includes interactions effects with other inputs. At the end of the first stage, the set of time lags with the best validation error is fixed and then the second stage (hyperparameter selection) is executed, in each iteration of the rolling windows evaluation scheme. Using the training data of each iteration, a similar ordered holdout split is used to execute a grid search of the best SVM hyperparameters (λ from 2^{-8} to 2^0, C from 2^{-1} to 2^6, ϵ from 2^{-8} to 2^{-1}), under a total of 13 searches as provided by a uniform design strategy [2].

Two additional hybrid models are proposed, with the goal of combining time lagged variables with other types of features (e.g., daily weather conditions). Since HW provided better predictions (when compared with ARIMA, see Sect. 3) we adopted it for the first hybrid, while the second hybrid is based on two SVM models. The HW-SVM model uses first HW to get the $\hat{y_t}$ predictions for the training data. The $\hat{y_t}$ input variable is merged with the six non lagged features (wind speed, temperature, humidity, rain, freeday, special) in order to create a dataset. Then, a SVM is trained (using the same hyperparameter selection procedure) to predict y_t based on the seven input features. When performing up to h ahead predictions, the HW-SVM model assumes the up to h ahead HW forecasts and previous knowledge for the other features (e.g., weather forecasts, special event schedules). The SVM-SVM hybrid uses the same input selection procedure of SVM to get the best set of time lags (k_i). Then, these time lags are merged with the six non lagged features and the combined dataset is used to train another SVM (set with the same hyperparameter selection). When executing predictions for test data, iterative 1-ahead predictions are used as inputs for the lagged inputs, as in SVM.

2.3 Evaluation

To measure the quality of the predictions, we applied a fixed-size rolling windows scheme [18] that allows a realistic training and testing of a large number of forecasting models. The rolling window assumes a training set of size W, from which from $h = 1$ to $h = H$ ahead predictions are executed at time $t - 1$ (first iteration, $i = 1$). Then training window is slided by discarding its oldest element and adding the value of t in order to retrain the model and predict from $h = 1$ to $h = H$ ahead predictions at time t (second iteration, $i = 2$), and so on. Thus, for a series of length L the rolling windows will have $U = L - (W + H - 1)$

model updates/iterations. For each forecasting method, we execute the rolling windows and store the $\hat{y}_{i,h}$ forecasts, where i denotes the rolling iteration.

To measure the predictive error, we selected the Normalized Mean Absolute Error (NMAE) metric [9]:

$$\begin{array}{rl} MAE_h &= \frac{1}{U}\sum_{i=1}^{U} |y_{T+i+h-1} - \hat{y}_{i,h}| \\ NMAE_h &= \frac{MAE_h}{y_{\text{High}} - y_{\text{Low}}} \end{array} \quad (1)$$

where T is the last known time period for $i = 1$, y_{High} and y_{Low} are the highest and lowest target values in the test set (of size U). The NMAE metric has the advantage of easy interpretation, since it expresses the error as a percentage of the full target scale. Also, it is a scale independent metric, which is a relevant issue since the all faces, female and male series have distinct scale ranges. Statistical significance of the forecasting differences will be evaluated by the Diebold-Mariano (DM) test [6], under pairwise comparison between the best method and the baseline for each h value. When comparing distinct forecasting methods, we use the Average $NMAE_h$ (ANMAE) values (h from 1 to H).

Also, we propose a new metric, the Estimated Store Benefit (ESB). The metric assumes that the store manager executes at time $t-1$ from $h = 1$ to $h = H$ ahead store traffic predictions, allowing her/him to set a plan with a better management of the store in the next seasonal cycle (with H days). For each individual daily absolute error, there is an average cost of c EUR, related with a missed opportunity (e.g., better marketing or human resource management). For a particular forecasting method M and plan i (related with iteration i of the rolling window), we define:

$$\begin{array}{rl} LOSS_i(M) &= c \times \sum_{h=1}^{h=H} |y_{T+i+h-1} - round(\hat{y}_{i,h})| \\ ESB_i(M) &= LOSS_i(\text{baseline}) - LOSS_i(M) \end{array} \quad (2)$$

where *round* is the rounding function, since we assume that face estimates are treated as integer numbers by the store manager (e.g., if $y_t - \hat{y}_t < 1$ then the loss is 0). Similarly to the NMAE metric, when comparing several time series methods, we use the Average $ESB_i(M)$ (AESB) values (i from 1 to U).

3 Results

All experiments were conducted using the R tool [15]. The statistical time series methods were executed using the **forecast** package [11], while the SVM based methods were implemented using the **rminer** package [2].

The rolling windows was set with $W = 220$ and a maximum ahead forecasts of $H = 7$ (one week). For each time series method, this allows the execution of $U = 31$ model trainings and predictions. For HW and ARIMA, the forecasting model is rebuilt using the W past observations. Regarding the ARIMA, we adopted the `auto.arima` function of the **forecast** package, which executes the automatic SARIMA model identification for each i-th iteration.

For the SVM time lagged models, we set the maximum number of time lags to $I = 28$ (four times the seasonal period). As explained in Sect. 2.2, the time lag model selection is only executed for the first rolling window iteration ($i = 1$), in order to reduce the computational effort. Once the best set of time lags is found, the hyperparameter selection is executed for each i-th iteration (similarly to the statistical methods). Under this setup, the SVM computation effort is relatively fast and affordable by current computers. For instance, the execution time for the all faces time series and full rolling window procedure was just around 16 s in a 2.5 GHz Intel Core i7 machine.

The obtained prediction errors are shown in Table 1, in terms of the ANMAE values. All methods present predictive errors lower than 13 %, which confirms that the three time series are predictable. In particular, the best forecasting methods obtain ANMAE values lower than 10 %: SVM for all faces (8.32 % error), HW for female (9.32 %) and HW-SVM for male (9.56 %). The predictive performance for these best models is detailed in Table 2, in terms of the NMAE values for distinct ahead predictions (h values). In this table, the selected time lags (k_i values) are shown for the SVM based methods. In both cases (all faces or male), feature selection procedure has kept time lags related with the seasonal period or its multiples (e.g., $k_i = 7$, $k_i = 14$). Table 2 also shows the p-values (<5 % or <10 %) when comparing the best method with the baseline using the DM test. For several horizon values (e.g., $h = 4$, $h = 5$), the differences are significant.

When estimating the impact of using these forecast models to improve store management, we opt two reasonable (and maybe conservative) scenarios, where the loss of one wrong visitor prediction is $c = 0.50$ and $c = 1.00$ EUR. Such

Table 1. Comparison of the forecasting errors ($ANMAE$ values, in %; best values in **bold**)

Series	Baseline	HW	ARIMA	SVM	HW-SVM	SVM-SVM
All faces	11.74	9.22	11.42	**8.32**	10.59	9.81
Female	11.37	**9.32**	11.25	9.36	10.86	11.53
Male	11.25	9.69	11.44	11.15	**9.56**	12.82

Table 2. Forecasting errors for the best methods and distinct horizons ($NMAE$ values, in %; * – p-value <5 %, ° – p-value <10 %)

Series	Method	$h = 1$	$h = 2$	$h = 3$	$h = 4$	$h = 5$	$h = 6$	$h = 7$
All faces	SVM ($k_i \in \{1,4,6,7,9,14,17,18\}$)	9.31°	8.85*	7.62*	7.50*	7.92*	8.54°	8.50°
Female	HW	9.71	10.32	9.29	8.80°	8.93*	8.88*	9.33*
Male	HW-SVM ($k_i \in \{1,14,16\}$)	11.13	10.91	9.61	8.46°	8.06°	9.18	9.58

Table 3. Comparison of the estimated store benefit gain ($AESB$ values, in EUR; best values in **bold**)

Series	c	Baseline	HW	ARIMA	SVM	HW-SVM	SVM-SVM
All faces	0.50	0	774	97	**1050**	353	593
All faces	1.00	0	1547	193	**2099**	707	1186
Female	0.50	0	**311**	17	305	78	-25
Female	1.00	0	**622**	34	609	155	-50
Male	0.50	0	241	-28	15	**261**	-241
Male	1.00	0	482	-56	31	**522**	-483

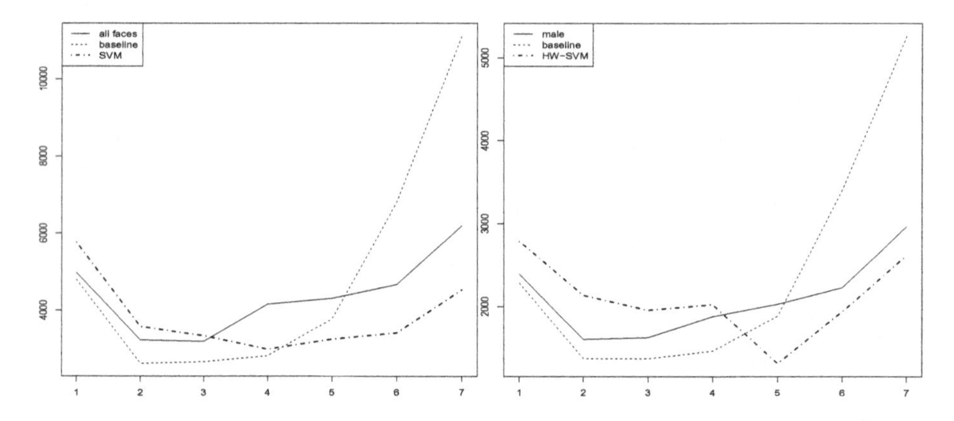

Fig. 2. Last rolling window weekly forecasts (i = 31) for all faces (right) and male (left) series (x-axis denotes the horizon, y-axis the number of daily faces)

impact is presented in Table 3, in terms of the AESB values. As expected, the best forecasting methods from Table 1 provide the best gains. When compared with the baseline method, the average weekly gain is: all faces (SVM), 1050 ($c = 0.50$) or 2099 ($c = 1.00$) EUR; female (HW), 311 ($c = 0.50$) or 622 ($c = 1.00$) EUR; and male (HW-SVM), 261 ($c = 0.50$) or 522 ($c = 1.00$) EUR.

For demonstration purposes, Fig. 2 shows the last rolling window iteration weekly ahead forecasts (i = 31, h from 1 to 7) for the all faces (left) and male (right) series. In both cases, the plots show an improved performance of the best forecasting methods when compared with the baseline.

4 Conclusions

In this paper, we explore six time series methods to predict foot traffic to a sports store, placed inside a shopping center of a Portuguese city. The data was collected from a pilot project, where a digital camera captured the store entrance during an eight month period, from April to December of 2013. The video data was fed into a facial recognition system that had a high accuracy when detecting faces and their genders, generating events. We preprocessed these events in order to get the daily incoming store traffic counts of all faces, female and male.

The obtained time series were modeled using: a simple baseline (use of last week values); statistical methods, namely Holt-Winters (HW) and AutoRegressive Integrated Moving Average (ARIMA); a Support Vector Machine (SVM) model; and two hybrid methods (HW-SVM and SVM-SVM), that combined time lagged variables with other data features (e.g., weather conditions, major sport events). The forecasting models were compared using a robust and realistic rolling windows evaluation that assumes up to $h = 7$ ahead predictions and two metrics: predictive error and estimated store benefit.

The best results were obtained by SVM for the all faces series (8.32 % error), HW for the female data (9.32 % error) and HW-SVM for male foot traffic (9.56 % error). We believe that these models can support a better store management (e.g., in terms of marketing or human resources). In effect, we have estimated a valuable impact with weekly gains of 1050 ($c = 0.50$) or 2099 ($c = 1.00$) EUR (SVM), 311 or 622 EUR (HW) and 261 or 552 EUR (HW-SVM).

This paper is the first study that analyzed the store pilot data. While competitive results were achieved, the forecasting models were built offline, after the store pilot execution. As such, in the future, we intend to apply the proposed forecasting models such that they could be embedded into a friendly decision support system for real usage in a store environment. This would allow us to get a valuable feedback from store managers. Also, we plan to adapt the forecasting methods to other time scales (e.g., hourly). Finally, further analysis will be devoted to measure the impact of the non lagged features (e.g., sport matches), in order to understand why HW-SVM gets better results for male store visits.

Acknowledgments. This work has been supported by COMPETE: POCI-01-0145-FEDER-007043 and FCT - Fundação para a Ciência e Tecnologia within the Project Scope: UID/CEC/00319/2013.

References

1. Alahi, A., Ramanathan, V., Fei-Fei, L.: Socially-aware large-scale crowd forecasting. In: 2014 IEEE Conference on Computer Vision and Pattern Recognition (CVPR), pp. 2211–2218. IEEE (2014)
2. Cortez, P.: Data mining with neural networks and support vector machines using the r/rminer tool. In: Perner, P. (ed.) ICDM 2010. LNCS (LNAI), vol. 6171, pp. 572–583. Springer, Heidelberg (2010). doi:10.1007/978-3-642-14400-4_44
3. Cortez, P.: Sensitivity analysis for time lag selection to forecast seasonal time series using neural networks and support vector machines. In: Proceedings of the International Joint Conference on Neural Networks (IJCNN 2010), Barcelona, Spain, pp. 3694–3701. IEEE, July 2010
4. Cortez, P., Embrechts, M.J.: Using sensitivity analysis and visualization techniques to open black box data mining models. Inf. Sci. **225**, 1–17 (2013)
5. Cortez, P., Rio, M., Rocha, M., Sousa, P.: Internet traffic forecasting using neural networks. In: Proceedings of the 2006 International Joint Conference on Neural Networks (IJCNN 2006), Vancouver, Canada, pp. 4942–4949. IEEE, July 2006

6. Diebold, F.X., Mariano, R.S.: Comparing predictive accuracy. J. Bus. Econ. Stat. **13**, 253–263 (2012)
7. Donate, J.P., Cortez, P., Sánchez, G.G., De Miguel, A.S.: Time series forecasting using a weighted cross-validation evolutionary artificial neural network ensemble. Neurocomputing **109**, 27–32 (2013)
8. Duque, D., Santos, H., Cortez, P.: Prediction of abnormal behaviors for intelligent video surveillance systems. In: CIDM, pp. 362–367. IEEE (2007)
9. Goldberg, K., Roeder, T., Gupta, D., Perkins, C.: Eigentaste: a constant time collaborative filtering algorithm. Inf. Retrieval **4**(2), 133–151 (2001)
10. Hastie, T., Tibshirani, R., Friedman, J.: The Elements of Statistical Learning: Data Mining, Inference, and Prediction. Springer, New York (2001)
11. Hyndman, R., Khandakar, Y.: Automatic time series forecasting: the forecast package for r 7, 2008 (2007). http://www.jstatsoft.org/v27/i03
12. Kitani, K.M., Ziebart, B.D., Bagnell, J.A., Hebert, M.: Activity forecasting. In: Fitzgibbon, A., Lazebnik, S., Perona, P., Sato, Y., Schmid, C. (eds.) ECCV 2012. LNCS, vol. 7575, pp. 201–214. Springer, Heidelberg (2012). doi:10.1007/978-3-642-33765-9_15
13. Ma, S., Fildes, R., Huang, T.: Demand forecasting with high dimensional data: the case of sku retail sales forecasting with intra- and inter-category promotional information. Eur. J. Oper. Res. **249**(1), 245–257 (2016)
14. Makridakis, S., Weelwright, S., Hyndman, R.: Forecasting: Methods and Applications, 3rd edn. Wiley, New York (1998)
15. R Core Team: R: A Language and Environment for Statistical Computing. R Foundation for Statistical Computing, Vienna, Austria (2015)
16. Segetlija, Z., et al.: New approaches to the modern retail management. Interdisc. Manage. Res. **5**, 177–184 (2009)
17. Stepnicka, M., Cortez, P., Donate, J.P., Stepnicková, L.: Forecasting seasonal time series with computational intelligence: on recent methods and the potential of their combinations. Expert Syst. Appl. **40**(6), 1981–1992 (2013)
18. Tashman, L.: Out-of-sample tests of forecasting accuracy: an analysis and review. Int. Forecast. J. **16**(4), 437–450 (2000)
19. Viola, P., Jones, M.: Rapid object detection using a boosted cascade of simple features. In: Proceedings of the 2001 IEEE Computer Society Conference on Computer Vision and Pattern Recognition, CVPR 2001, vol. 1, p. I-511. IEEE (2001)
20. Witten, I., Frank, E., Hall, M.: Data Mining: Practical Machine Learning Tools and Techniques, 3rd edn. Morgan Kaufmann, San Franscico (2011)

SOCO 2016: Special Session on Optimization, Modeling and Control Systems by Soft Computing

Using GPUs to Speed up a Tomographic Reconstructor Based on Machine Learning

Carlos González-Gutiérrez[1], Jesús Daniel Santos-Rodríguez[2(✉)], Ramón Ángel Fernández Díaz[3], Jose Luis Calvo Rolle[4], Nieves Roqueñí Gutiérrez[1], and Francisco Javier de Cos Juez[1]

[1] Department of Exploitation and Exploration of Mines, University of Oviedo, Oviedo, Spain
{gonzalezgcarlos,nievesr,fjcos}@uniovi.es

[2] Department of Physics, University of Oviedo, Oviedo, Spain
jdsantos@uniovi.es

[3] Department of Architecture and Technology of Computers, University of León, León, Spain
ramon.fernandez@unileon.es

[4] Department of Industrial Engineering, University of a Coruña, La Coruña, Spain
jlcalvo@udc.es

Abstract. The next generation of adaptive optics (AO) systems require tomographic techniques in order to correct for atmospheric turbulence along lines of sight separated from the guide stars. Multi-object adaptive optics (MOAO) is one such technique. Here we present an improved version of CARMEN, a tomographic reconstructor based on machine learning, using a dedicated neural network framework as Torch. We can observe a significant improvement on the training an execution times of the neural network, thanks to the use of the GPU.

Keywords: Neural networks · Torch · Adaptive optics

1 Introduction

Adaptive Optics is an essential tool when stellar observation is performed with grounded telescopes, since the atmosphere produces aberrations in the light that passes through it. This problem is first approached with the measurement of distortions produced in the wave-front of the incoming light and then, calculating the position that a deformable mirror has to adopt for compensating the aberrations in the wave-front as fast as possible due to the extremely changing nature of the atmosphere [1]. In order to measure more accurately the wave-front, guide stars are needed as a reference, where natural stars can be used when they are in the field of view of the object of interest, or even artificial stars that are created by laser scattering in the upper atmosphere. With the aim of correct the error from the atmospheric turbulences, computer tomography techniques are considered for compensating the astronomical image with deformable mirrors [2, 3].

M. Graña et al. (eds.), *International Joint Conference SOCO'16-CISIS'16-ICEUTE'16*, Advances in Intelligent Systems and Computing 527, DOI 10.1007/978-3-319-47364-2_27

Creation and development of algorithms that allow the obtainment of the deformations introduced by the atmosphere is one of the imperative issues of adaptive optics. Moreover, these algorithms have to control the real-control system, so the image can be accurately reconstructed. Some of the most common reconstructors are the Least Squared (LS) type matrix vector multiplication [4], the Learn and Apply (L&A) method [5], and the recently added Complex Atmospheric Reconstructor based on Machine lEarNing (CARMEN), which has shown some interesting results [6].

Nowadays, for large telescopes, modern AO systems are required, which relays on tomographic techniques in order to reconstruct the phase aberrations induced by the turbulent atmosphere [7]. CARMEN is a reconstructor, for Multi-object adaptive optics (MOAO), which is one of these techniques. It was initially developed using regression techniques as MARS [8–10], with promising results [2]. However, the use of machine learning techniques [11, 12] such as Supporting Vector Machines (SVM) [13] or Artificial Neural Networks (ANN) [14], have proved very successful in different fields, which led us to create a solution based on ANN. The neural network is trained with a large range of possible turbulent layer positions and therefore does not require any input of the optical turbulence profile. CARMEN has shown promising results in on-sky data, achieving a performance within 5 % in terms of Strehl ratio [15].

The development of large telescopes, in particular the future European Extremely Large Telescope (E-ELT), brings the inconvenience of the computational capability needed to process the enormous amounts of data [16]. Due to the larger number of subapertures and guide stars involved, tomography on ELT scales becomes computationally more difficult. ANN architecture allows its parallelism in neural processing. The use of Graphics Processor Units (GPUs) provides a solution to this problem, due to the parallelization of different calculations, and therefore, the speeding up of the processing times. There are some initial approximations to adapt some of the existing reconstructors, to the use of GPUs, like the Learn + Apply case [17, 18].

The main purpose of this paper, is to detail the implementation of CARMEN, in an adapted framework for neural networks based on GPU, and expose their training and execution times. Next section shows how AO systems work, and on the following one the architecture of CARMEN is shown as well as a small description of the framework used. The experiment is defined afterwards and different variables to compare different frameworks are proposed. Finally, results are analysed and conclusions are put forward.

2 Adaptive Optics Systems

The Shack-Hartmann Wave-front Sensor (SH WFS) is commonly used in astronomy to characterize an incoming wave-front. It consists of an array of lenses with the same focal length (called lenslets) each focused on a photon sensor. The incoming wave-front is divided into discrete areas and the local tilt of each lenslet can be measured as the deviation of the focal spot of the sensor from the positions due to a plane wave-front as shown in Fig. 1 creating a matrix of tilts characteristic of the wave-front aberration. With this data, the aberration induced in the wave-front by atmospheric turbulence can be approximated in terms of Zernike Polynomials.

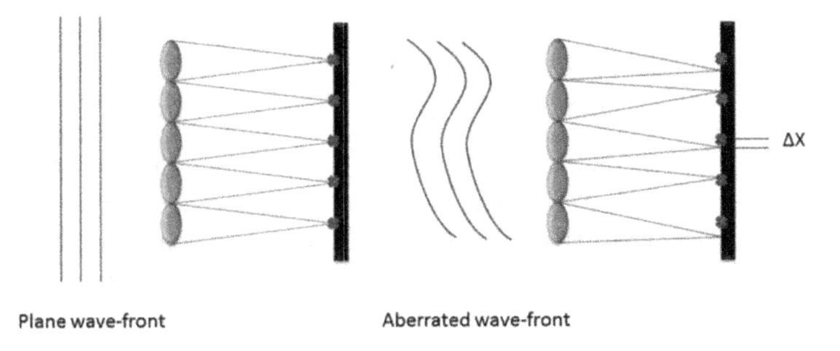

Fig. 1. Measurement of wave front tilts

CANARY is an adaptive optics (AO) on-sky demonstrator, principally intended for developing and testing AO concepts for the future 39 m European Extremely Large Telescope. It is operated on a Nasmyth platform of the 4.2 m William Herschel Telescope, one of the Isaac Newton Group of Telescopes (ING) of the Observatorio del Roque de los Muchachos (ORM), La Palma, Canary Islands, Spain.

There are severals configurations available:

- CANARY Phase B1 is designed to perform observations with 1 Laser Guide Star (LGS), and up to 4 Natural Guide Stars (NGS). It has a Shack Hartman Wavefront Sensor with 7 × 7 subapertures, although only 36 of them are functional.
- CANARY Phase C2 is designed for the study of Laser Tomography AO (LTAO) and Multi-Object AO (MOAO). There are 4 Rayleigh Laser Guide Stars, each with a 14 × 14 subaperture Shack Hartman Wavefront Sensor, where we only have 144 of them working.
- DRAGON aims to replicate CANARY concepts, to provide a single channel MOAO system with a woofer-tweeter DM configuration, 4 NGSs and 4 LGSs each with 30 × 30 subapertures. In this case, DRAGON is still a prototype, so we are going to use the worst case scenario where all the subapertures are functional, which gives as a total of 900 subapertures per star.

3 CARMEN Architecture

CARMEN is a tomographic reconstructor based on artificial neural networks, whose architecture is a multi-layer perceptron, with a single hidden layer. It is composed by two fully-connected layers, where each neuron is connected to all the neurons in the previous layer. The output of each neuron follows (1), where w is the weight of each connection, x is the value of the neurons in the previous layer, b is a constant value called bias, and f is an activation function.

$$\mathrm{Y} = f\left(\sum\nolimits_{i=0}^{n} (w_i \cdot x_i) + b\right) \tag{1}$$

The number of input, hidden and output neurons, is directly related to the optical instrumentation used, since it changes with the number of subapertures of the device. It is not the purpose of this paper to analyse the net size or quality, since it has been previously tested [19], so the number of neurons in the hidden layer will be equal to the number of neurons in the input layer. Also, the number of input and hidden neurons, depends on the number of stars (both natural and artificial), that we are observing. This means, that the size of the network is highly variable, and should be adapted to each observation. However, and to make it easier to compare, we are going to use the three configurations explained in the previous section, and assume that we can observe both natural and laser guide stars.

These rules make that the number of input neurons, matches the number of functional subapertures multiplied by 2-due to the two dimensional input in the lenslet array-, and also multiplied by the number of reference stars. The corrected deviations of the lenslet array will appear in the output, which is the number of functional subapertures multiplied by 2. The final size of the network can be summarized in Fig. 2.

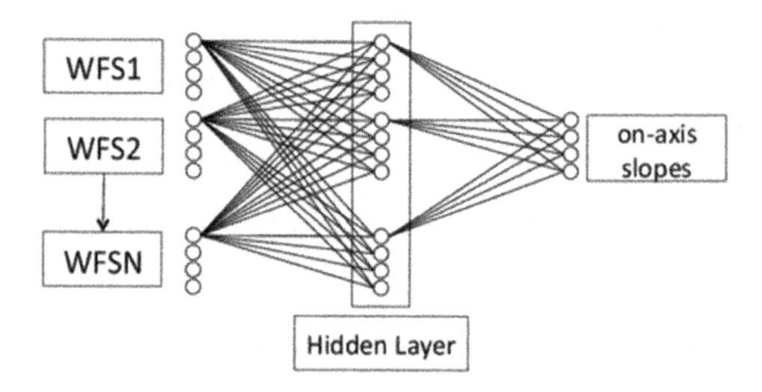

Fig. 2. CARMEN architecture

A network with 1 LGS and 2 NGS will be used in the case of CANARY Phase B1, that is, 216 input neurons. For CANARY Phase C2, with 4 Laser Guide Stars, there will be 1152 input neurons. In both cases the training data will be obtained from the CANARY simulator. Regarding DRAGON, the situation is different, due to the fact that it is still under development and there are still no simulated data, so the number of natural stars is unknown. In this case, laser guide stars will be used, and random data will be generated to train the network. This means there are 7200 input neurons for DRAGON. In Table 1 we can see a summary of the sizes of the different networks.

Table 1. Size of neural networks for adaptive optics systems

Name	Size	Number of training samples
CANARY-B1	216-216-72	350 k
CANARY-C2	1152-1152-288	1.5 M
DRAGON	7200-7200-1800	1 M

3.1 Neural Networks Frameworks

The growing popularity of Artificial Neural Networks in recent years, has caused the emergence of numerous frameworks, which help researchers in the use and development of more complex neural networks, without the need of hard programming efforts. There is a long list of existing neural networks frameworks[1], but for this paper, we are going to focus mainly in Torch.

Torch is a scientific computing framework with wide support for machine learning algorithms. It is written in Lua and C/CUDA to provide a fast execution and the possibility of importing modules to complete and accelerate the system. It is mainly used and maintained by some important companies, such as Google, Facebook, Twitter, etc.

4 Experiment Description

We have defined different networks for the optical instruments described above. Two different measurements have been considered to evaluate their performance. The different training times under specific conditions will be compared, and also the execution time of a network, which is crucial in the adaptive optics systems.

4.1 Training Benchmark

There is a large amount of parameters to fit when training a neural network, although only a few of them affect directly the training time. To simplify the comparison, we will only vary those parameters that are especially relevant for the different systems.

In every case the Stochastic Gradient Descent (SGD) will be used, with mini-batches and momentum, which is implemented in Torch. Two parameters are needed for this method: the learning rate and the momentum. Although these variables are crucial regarding the quality of the resulting network, they have no influence in the training times, so it will be assumed that both of them are optimized to achieve the best result as possible.

Another critical parameter is the size of the training data. However, is easy to notice that in a training method based on mini-batches, the time grows proportionally to the number of samples employed. Keeping this in mind, it is easy to calculate how much time will take a network to be trained when changing the training dataset. This idea makes the choice of the dataset size irrelevant for benchmarking purposes, although is crucial for obtaining a good network. In this case, the size of the training data is specified in Table 1.

Three different networks will be used, one per optical system, as shown in Table 1. The main parameter to be changed will be the network size and the size of the mini-batch, in order to compare the times. The number of samples in the mini-batch will be 16, 32, 64, 128 and 256.

---

[1] http://deeplearning.net/software_links/.

First the initialized weights and bias are copied to the GPU's VRAM, and all the training dataset is loaded in the main RAM, before starting the training. Then a timer starts and the program starts to copy the first mini-batch from RAM to VRAM, and to perform the loop to go over the entire dataset. This operation is repeated during 20 epochs, timing each of them individually. This procedure allows to obtain an average time for each epoch, and more reliable results, making it possible to see if there are significant time variations among different epochs.

4.2 Execution Benchmark

The same networks defined in Table 1 are used in this case. However, there are some important differences with the training benchmark that should be detailed.

First, the net is fed with a single input, instead of using mini-batches, simulating what happens in a real telescope. As the execution program is intended to be integrated in the telescope management system, is a fair assumption that all the variables are already initialized, and the weight matrices copied into the VRAM. The loading of input data from the SSD will be taken into account in the time measurement, as well as the copy from VRAM to the system RAM, and in the case of loading from SSD, the writing process to the disk. Each input is found in a separate file, in h5 format, and the output is written in a separate file. We will feed the system with 10000 inputs, one at a time, which allow us to promediate the execution time. We will compare the average time of the different frameworks, looking for which one is the fastest, and if there is any significant difference between them.

4.3 Experiment Equipment

The experiments are performed on a computer running on Ubuntu LTS 14.04.3, with Intel Xeon CPU E5-1650 v3 @ 3.50 GHz, 128 Gb DDR4 memory, Nvidia GeForce GTX TitanX, and SSD hard drive. We used CUDA 7.5 and cuDNN v3.

5 Results and Discussion

In this section obtained results are shown, from training and execution process. Times will be split through different adaptive optics systems, and possible explanations for the different results will be analysed.

5.1 CANARY-B1

As explained above, the network size is 216-216-72, and 350 k samples are used for training. As it is shown in

Figure 3, increasing the size of the batch has an obvious improvement on training times. We have managed to train the neural network in only a few minutes, which allow us to test different hyperparameters, improving the learning process of the

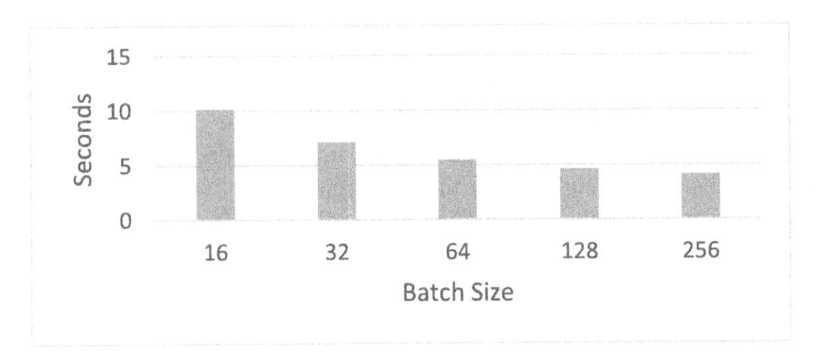

Fig. 3. CANARY-B1 s of training per epoch

network. Also, for smalls systems like this, the possibility of online training could be interesting, which allow us to adapt the reconstructor for the existing conditions during specific observations.

5.2 CANARY-C2

For CANARY-C2, the network size is 1152-1152-288, and use 1.5 M training samples. In this case Fig. 4, the reduction of batch size has more impact reducing the time, due to the bigger size of the network. It can be observed how, despite the increased number of operations for this network (about 28 $\times$ more weights and 4 $\times$ more data), training times does not increase proportionally. The reason for this behaviour could be that, in smaller networks, we don't provide enough workload to fill all the GPU cores, and an increasing amount of operations could be automatically parallelized.

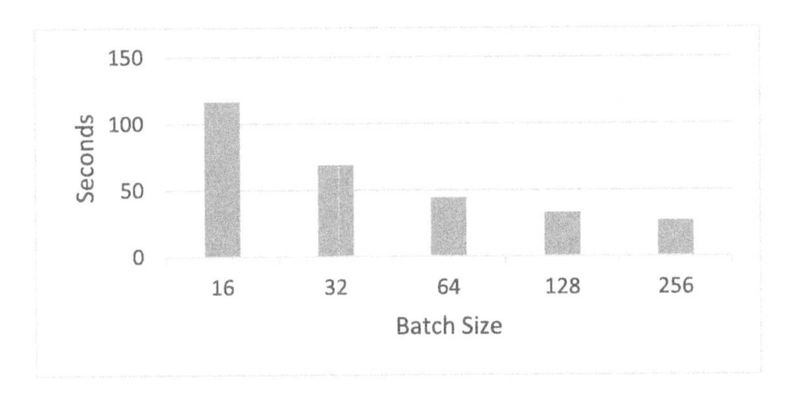

Fig. 4. CANARY-C2 s of training per epoch

5.3 DRAGON

For DRAGON, the largest network is employed, with 7200-7200-1800 neurons and 1 M samples for training. This case Fig. 5, is similar to the one shown in CANARY-C2, where the batch size has a great impact in the training process. However, for this net, training times start to grow significantly, which means that at this point, we are providing enough workload to fill all the GPU cores.

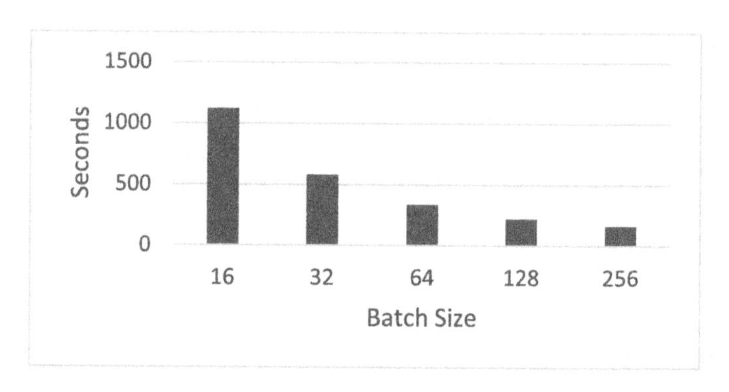

Fig. 5. DRAGON seconds of training per epoch

5.4 Execution Times

Regarding execution times shown in Table 2, different scenarios can be observed. On one hand, we can see that loading data from the main disk is much slower than loading the information from the main RAM. Although this seem obvious, it is interesting to measure the impact that this difference could have in real systems. For small networks, this difference could mean an increment between 3 × and 5 × times, that can be critical for this type of systems.

Table 2. Execution times

	CANARY	CANARY-C3	DRAGON
SSD	0.708 ms	0.935 ms	3.178 ms
RAM	0.154 ms	0.392 ms	2.632 ms

Also, we can see how increasing net size have a huge impact in execution times, despite we are using only one input per execution. In the case of DRAGON, we need more than 2.5 ms to obtain one single output. Although this is much faster than what we have seen in previous experiments [15], we still have to improve this time until the 2 ms mark, which is the traditional limit for adaptive optics systems.

6 Conclusions and Future Lines

Using a GPU based framework as Torch to train and execute the network, provide us a powerful tool to test and improve our reconstructor much faster than what we have been doing during previous experiments [15, 20]. Also, it is a necessary improvement due to the increasing size of the adaptive optics systems, especially the future E-ELT.

As this is still a work in progress, we still have a lot of different challenges to solve. One of them, is checking the performance of different frameworks, such as Theano or Caffe, or even develop a code directly in C/CUDA, and compare the training and execution times for all the different adaptive optics systems.

The use of a specific neural network framework, allow us to think about using convolutional neural networks or recurrent networks, which could provide some boost regarding the quality of the reconstruction. Although this could be done programming the code directly in any programming language, the use of Torch, or any of the existing frameworks could make much more easy to find a new architecture that could help CARMEN.

Also, we still have to address some possible challenges in the near future. It is expected that the number of inputs and outputs of the system, increase greatly during the next years. This will make that training the networks will cost much more time, and also could appear problems regarding the amount of GPU memory needed to deal with bigger weights matrices. A possible workaround for this, is to train an execute the network, using multi-GPU systems, where the computation and the memory could be splitted across different graphics cards.

References

1. Guzmán, D., De Cos Juez, F.J., Myers, R., Guesalaga, A., Lasheras, F.S.: Modeling a MEMS deformable mirror using non-parametric estimation techniques. Opt. Express **18**(20), 21356–21369 (2010)
2. Guzmán, D., de Cos Juez, F.J., Lasheras, F.S., Myers, R., Young, L.: Deformable mirror model for open-loop adaptive optics using multivariate adaptive regression splines. Opt. Express **18**(7), 6492–6505 (2010)
3. de Cos Juez, F.J., Lasheras, F.S., Roqueñí, N., Osborn, J.: An ANN-based smart tomographic reconstructor in a dynamic environment. Sensors (Basel) **12**(7), 8895–8911 (2012)
4. Ellerbroek, B.L.: First-order performance evaluation of adaptive-optics systems for atmospheric-turbulence compensation in extended-field-of-view astronomical telescopes. J. Opt. Soc. Am. A: **11**(2), 783 (1994)
5. Vidal, F., Gendron, E., Rousset, G.: Tomography approach for multi-object adaptive optics. J. Opt. Soc. Am. A Opt. Image Sci. Vis. **27**(11), A253–A264 (2010)
6. Osborn, J., de Cos Juez, F.J., Guzman, D., Butterley, T., Myers, R., Guesalaga, A., Laine, J.: Using artificial neural networks for open-loop tomography. Opt. Express **20**(3), 2420–2434 (2012)

7. Basden, A.G., Atkinson, D., Bharmal, N.A., Bitenc, U., Brangier, M., Buey, T., Butterley, T., Cano, D., Chemla, F., Clark, P., Cohen, M., Conan, J.-M., De Cos, F.J., Dickson, C., Dipper, N.A., Dunlop, C.N., Feautrier, P., Fusco, T., Gach, J.L., Gendron, E., Geng, D., Goodsell, S.J., Gratadour, D., Greenaway, A.H., Guesalaga, A., Guzman, C.D., Henry, D., Holck, D., Hubert, Z., Huet, J.M., Kellerer, A., Kulcsar, C., Laporte, P., Le Roux, B., Looker, N., Longmore, A.J., Marteaud, M., Martin, O., Meimon, S., Morel, C., Morris, T.J., Myers, R.M., Osborn, J., Perret, D., Petit, C., Raynaud, H., Reeves, A.P., Rousset, G., Lasheras, F.S., Rodriguez, M.S., Santos, J.D., Sevin, A., Sivo, G., Stadler, E., Stobie, B., Talbot, G., Todd, S., Vidal, F., Younger, E.J.: Experience with wavefront sensor and deformable mirror interfaces for wide-field adaptive optics systems. Mon. Not. R Astron. Soc. **459**(2), 1350–1359 (2016)
8. Antón, J.C.A., Nieto, P.J.G., de Cos Juez, F.J., Lasheras, F.S., Viejo, C.B., Gutiérrez, N.R.: Battery state-of-charge estimator using the MARS technique. IEEE Trans. Power Electron. **28**(8), 3798–3805 (2013)
9. De Cos Juez, F.J., Lasheras, F.S., Nieto, P.J.G., Suárez, M.A.S.: A new data mining methodology applied to the modelling of the influence of diet and lifestyle on the value of bone mineral density in post-menopausal women. Int. J. Comput. Math. **86**(10–11), 1878–1887 (2009)
10. Nieto, P.J.G., Fernández, J.R.A., Lasheras, F.S., de Cos Juez, F.J., Muñiz, C.D.: A new improved study of cyanotoxins presence from experimental cyanobacteria concentrations in the Trasona reservoir (Northern Spain) using the MARS technique. Sci. Total Environ. **430**, 88–92 (2012)
11. Casteleiro-Roca, J.L., Quintián, H., Calvo-Rolle, J.L., Corchado, E., del Carmen Meizoso-López, M., Piñón-Pazos, A.: An intelligent fault detection system for a heat pump installation based on a geothermal heat exchanger. J. Appl. Logic **17**, 36–47 (2015)
12. Casteleiro-Roca, J.L., Calvo-Rolle, J.L., Meizoso-López, M.C., Piñón-Pazos, A.J., Rodríguez-Gómez, B.A.: Bio-inspired model of ground temperature behavior on the horizontal geothermal exchanger of an installation based on a heat pump. Neurocomputing **150**, 90–98 (2015)
13. Nieto, P.J.G., García-Gonzalo, E., Lasheras, F.S., De Cos Juez, F.J.: Hybrid PSO–SVM-based method for forecasting of the remaining useful life for aircraft engines and evaluation of its reliability. Reliab. Eng. Syst. Saf. **138**, 219–231 (2015)
14. Vilán, J.A.V., Fernández, J.R.A., Nieto, P.J.G., Lasheras, F.S., de CosJuez, F.J., Muñiz, C.D.: Support vector machines and multilayer perceptron networks used to evaluate the cyanotoxins presence from experimental cyanobacteria concentrations in the trasona reservoir (Northern Spain). Water Resour. Manage. **27**(9), 3457–3476 (2013)
15. Osborn, J., Guzman, D., De CosJuez, F.J., Basden, A.G., Morris, T.J., Gendron, E., Butterley, T., Myers, R.M., Guesalaga, A., Lasheras, F.S., Victoria, M.G., Rodríguez, M.L. S., Gratadour, D., Rousset, G.: Open-loop tomography with artificial neural networks on CANARY: On-sky results. Mon. Not. R. Astron. Soc. **441**(3), 2508–2514 (2014)
16. Ramsay, S.K., Casali, M.M., González, J.C., Hubin, N.: The E-ELT instrument roadmap: a status report. p. 91471Z (2014)
17. Ltaief, H., Gratadour, D.: Shooting for the Stars with GPUs. In: GPU Technology Conference (2015). http://on-demand.gputechconf.com/gtc/2015/video/S5122.html. Accessed 14 Mar 2016
18. Marichal-Hernández, J.G., Rodríguez-Ramos, L.F., Rosa, F., Rodríguez-Ramos, J.M.: Atmospheric wavefront phase recovery by use of specialized hardware: graphical processing units and field-programmable gate arrays. Appl. Opt. **44**(35), 7587–7594 (2005)

19. Osborn, J., De Cos Juez, F.J., Guzman, D., Butterley, T., Myers, R., Guesalaga, A., Laine, J.: Open-loop tomography using artificial nueral networks. Adapt. Opt. Extrem. Large Telesc. II (2011)
20. Casteleiro-Roca, J.L., Calvo-Rolle, J.L., Meizoso-Lopez, M.C., Piñón-Pazos, A., Rodríguez-Gómez, B.A.: New approach for the QCM sensors characterization. Sens. Actuators A Phys. **207**, 1–9 (2014)

An Intelligent Model for Bispectral Index (BIS) in Patients Undergoing General Anesthesia

José Luis Casteleiro-Roca[1(✉)], Juan Albino Méndez Pérez[2], José Antonio Reboso-Morales[2], Francisco Javier de Cos Juez[3], Francisco Javier Pérez-Castelo[1], and José Luis Calvo-Rolle[1]

[1] Department of Industrial Engineering, University of A Coruña, Avda. 19 de febrero s/n, 15495 Ferrol, A Coruña, Spain
jose.luis.casteleiro@udc.es

[2] Dpto. de Ingeniería de Sistemas y Automática y Arquitectura y Tecnología de Computadores, University of La Laguna, Avda. Astrof. Francisco Sánchez s/n, 38200 S/C de Tenerife, Spain

[3] Department of Mining Exploitation, University of Oviedo, Calle San Francisco, 1, 33004 Oviedo, Spain

Abstract. Nowadays, the engineering tools play an important role in medicine, regardless of the area. The present research is focused in anesthesiology, specifically on the behavior of sedated patients. The work shows the Bispectral Index Signal (BIS) modeling of patients undergoing general anesthesia during surgery. With the aim of predicting the patient BIS signal, a model that allows to know its performance from the Electromyogram (EMG) and the propofol infusion rate has been created. The proposal has been achieved by using clustering combined with regression techniques and using a real dataset obtained from patients undergoing general anesthesia. Finally, the created model has been tested also with data from real patients, and the results obtained attested the accuracy of the model.

Keywords: EMG · BIS · Clustering · SOM · MLP · SVM

1 Introduction

Automatic control of anesthesia has been attracting the interest of many researchers in the past years. Although signal-based controllers like PID can provide satisfactory performance in many cases, more advanced techniques are based on the use of reliable models to predict patient response. According to the prediction, drug dosing can be adjusted to the specific patient needs.

This work is based on a previous one [8], where a hybrid model was created for predicting the EMG signal from the BIS signal and the propofol infusion rate. Besides that the EMG signal prediction is very important for many applications, under a clinical point of view, and taking into account the special case of the anesthesia problem, it is even more important to know the BIS behaviour forecast during surgeries.

M. Graña et al. (eds.), *International Joint Conference SOCO'16-CISIS'16-ICEUTE'16*, Advances in Intelligent Systems and Computing 527, DOI 10.1007/978-3-319-47364-2_28

The present research shows a new advance in this sense, specifically with the BIS when a patient is undergoing surgery with anesthesia. With this index it is possible to measure the hypnotic state. Its value varies between 0 (no electrical activity) and 100 (awake state). The target for general anesthesia is normally established in 50.

The hypothesis in this work is that the BIS is correlated to EMG and infusion rate. Then, the objective is to predict the BIS value in terms of EMG and propofol rate. The BIS signal measures the level of consciousness form the electroencephalogram of patients during general anesthesia [28].

For the BIS prediction, many different methods can be considered. The accepted regression methods are typically based on Multiple Regression Analysis (MRA) techniques, that are very usual in applications in different fields [1,4,5,7,10,14,21,22,29]. However, these methods have limitations and do not provide a good performance [5,9,26]. In order to increase this feature, many new proposals have been developed. These proposals are based on Soft Computing techniques, both simple or hybrid. As it is shown in [2,3,6,11–13,19,20,25] these techniques improve the first ones mentioned above.

This study implements a hybrid model to predict the BIS signal from the EMG signal and the propofol infusion rate. To develop the model, K-means clustering algorithm is used to create groups of data with similar behavior. Then, several regression methods were verified for each group to select the best one based on the lowest Mean Squared Error (MSE) reached.

This paper is structured in the following way. After the present section, the case of study is described, the Bispectral index. Then, the model approach and the tested algorithms taken into account in the research are shown. The results section shows the best configuration achieved by the hybrid model. After the results, the conclusions and future works are presented.

2 Case of Study

When a patient is undergoing surgery with general anesthesia, a proper dose of propofol should be administrated to achieve an adequate hypnosis level [18]. To monitor the anesthesia level, the BIS is measured [23,27]. This index will vary depending on the concentration of drug in the patient, that is related to the infusion rate. It will be also affected by the value of the EMG signal. The studied problem could be represented as shown in Fig. 1.

After induction, the automatic controller is activated. It will decide the correct dose of drug ($mg/Kg/h$) to achieve the desired BIS level. In this situation, small variations of EMG could result in variations in the BIS signal. When surgery is completed propofol infusion is stopped and the patients wakes up.

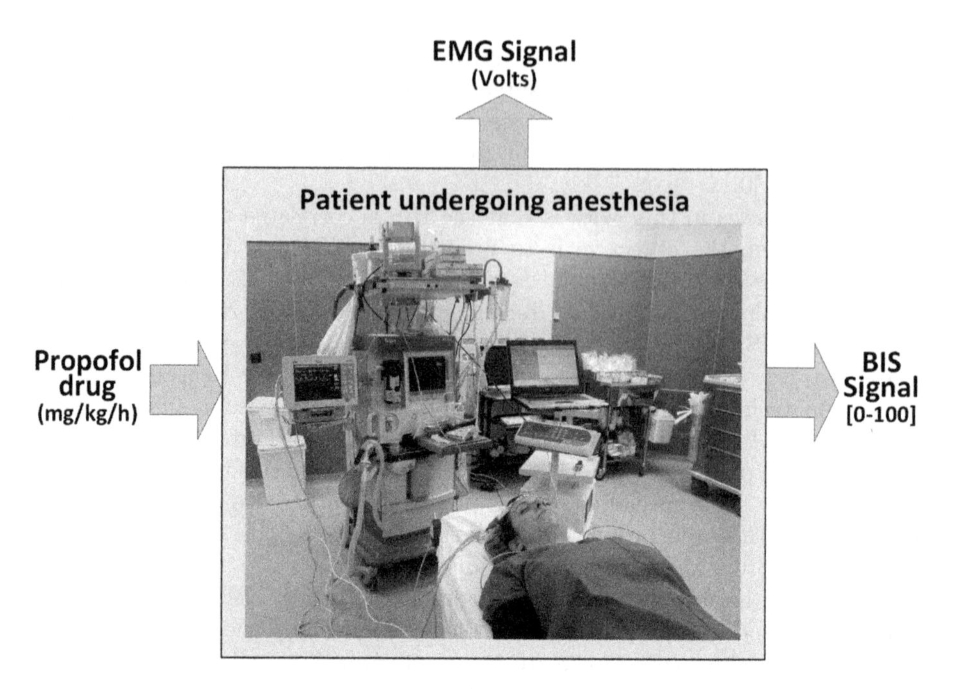

Fig. 1. Case of study. Input/Output representation

3 Model Approach

The scheme defined for the model approach is shown in Fig. 2. It is possible to divide the dataset in several operation ranges, taking into account the system behavior and the test accomplished to the dataset. Consequently, some clusters are created and, for each one, a regression model is implemented to calculate the output. As shown in Fig. 2, the global model has two inputs (the $-propofol-$ drug and the Electromyogram signal $-EMG-$) and one output (Bispectral index $-BIS-$). The cluster selector is a block designed to selected the right model according with the operating point. This block connects the chosen models with the output, based on the Euclidean distance between the input and the centroids on each cluster. On each cluster block, after testing different algorithms configuration, only the best model is implemented.

The modeling process is shown in Fig. 3. The dataset has been processed by using cross-validation (hold-out) to ensure the best results for the achieved model; the data was divided for training and testing as shown in the figure.

3.1 The Dataset Obtaining and Description

The dataset has been obtained from several patients undergoing general anesthesia with propofol drug during surgery. The three variables used on this research (BIS, EMG and propofol infusion rate) have been monitored during surgeries. A

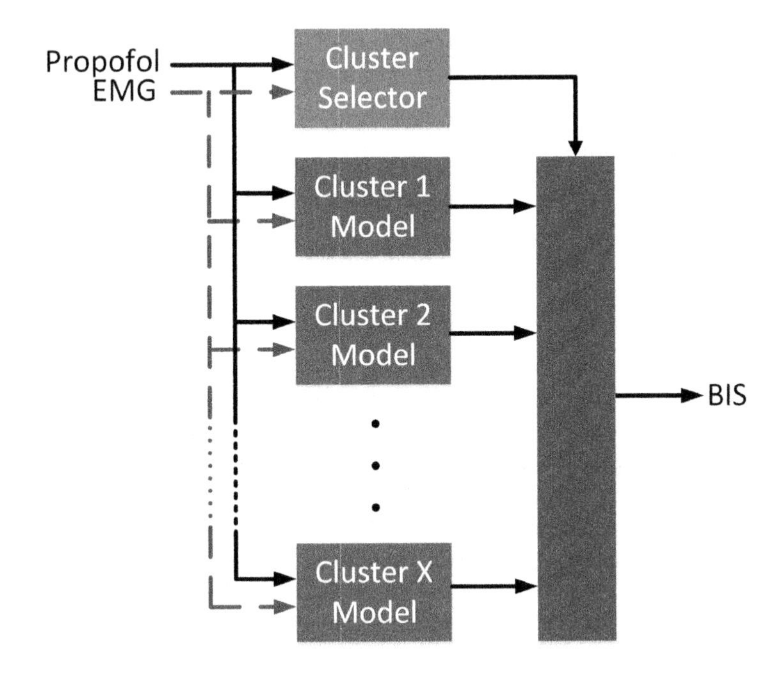

Fig. 2. Model approach

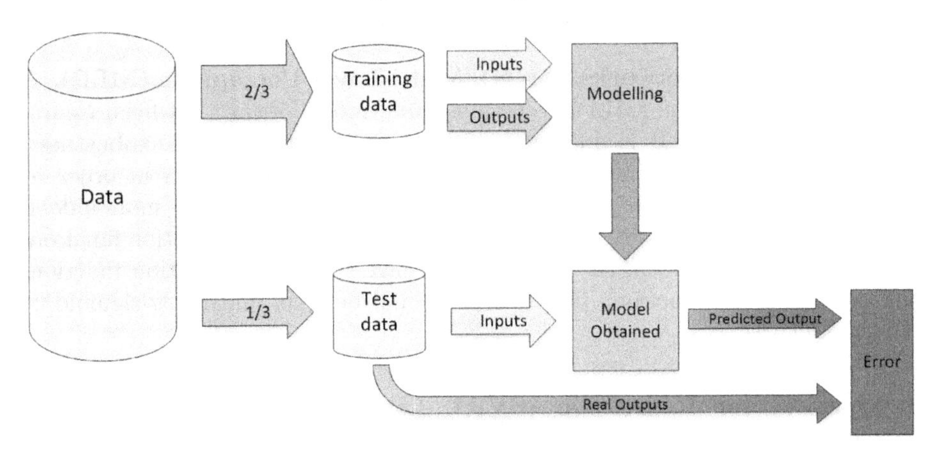

Fig. 3. Modeling process

preconditioning stage was considered for BIS and EMG. The dataset is composed with the data of a total of 50 patients, recording new set of values with a sample time of 5 s. Due to the slow variation of signals at the acquisition phase, a low pass filter was implemented to avoid the undesired noise. The induction phase and the recovery phase were not considered in this study; only the maintenance phase of surgery has been used. Thus, the results obtained are only valid to predict EMG in the phase when the BIS has been controlled. With the conditions exposed above, the employed dataset contains 42788 samples.

3.2 Used Techniques

The techniques used in this study, with the aim of achieving the best model, are described below.

The procedure to obtain the hybrid model was using K-means algorithm to clustering the dataset. Below, different intelligent regression techniques are tested in the clusters. Only the best one is chosen based on MSE criteria.

Data Clustering - The K-Means Algorithm. Clustering techniques make data grouping measuring the similarity between samples [16,24]. These algorithms organize unlabeled data in groups; the samples within a cluster are similar to each other [16]. K-means is a frequently used clustering algorithm with square-error criterion, which minimizes a specific error function.

The obtained clusters depend on the initial centroids and on the K value (number of groups). The most critical election is the choice of K value because it needs certain knowledge of total clusters present in the data and, sometimes, it is extremely uncertain. The K-means clustering algorithm is computationally effective, it works well when the data are close to its cluster, the cluster is hyperspherical in shape and they are well-separated in the hyperspace.

Artificial Neural Networks (ANN), Multi-Layer Perceptron (MLP). A Multi-Layer Perceptron (MLP) is the most known feedforward Artificial Neural Network (ANN) [32]. It is due to its simple configuration and its robustness. Despite of that, the ANN architecture must be carefully chosen in order to achieve satisfactory results. MLP is made of one input layer, one or more hidden layers and one output layer. The layers have neurons with an activation function. In a typical configuration, all layer neurons have the same activation function, but this is not a restriction. This function could be step, linear, log-sigmoid or tan-sigmoid.

Support Vector Regression (SVR), Least Square Support Vector Regression (LS-SVR). Support Vector Regression (SVR) is based on the algorithm of the Support Vector Machines (SVM) for classification. In SVR the data is mapped into a high-dimensional feature space F through a nonlinear plotting and linear regression is done in this space [30].

The Least Square algorithm of SVM is called LS-SVM. The solution estimation is obtained by solving a system of linear equations, and it is similar to SVM in terms of performance generalization [16]. The use of LS-SVM algorithm to regression is well-known as LS-SVR (Least Square Support Vector Regression) [17]. In LS-SVR, the insensitive loss function is replace by a classical squared loss function, which makes the Lagrangian by solving a linear Karush-Kuhn-Tucker (KKT).

Polynomial Regression. Usually, a polynomial regression model [15,31,33] could also be defined as a linear summation of basis functions. The quantity of basis functions depends on the number of inputs of the system, and the degree of the employed polynomial. The model becomes more complex when the degree rises.

4 Results

The model was obtained using not only the current value of EMG signal, and the propofol infusion rate quantity of drug (propofol); to include the dynamic of the modeled system, the last two previous values of the inputs were included to train the models. Also, the previous value of the desired output (the BIS signal) were included.

We used 2 previous states at the inputs and one at the output, because it is considered more than sufficient to model the dynamic response of the system. Tests have been conducted to confirm this, and have found that the inclusion of more previous values is not representative to improve the model.

To perform the clustering section, it is taking into account that the K-means algorithm performance depends on the initial state. The process was performed 20 times with a random initialization, and finally the best clusters was stored. As the number of clusters is not previously known, the model was trained with different configuration of clusters, created by using the K-means algorithm. The configuration was ranged from 2 to 10 clusters with the aim to create 9 different topologies. The global model was taking into account too.

It is remarkable that not always it is possible to divide the dataset in the number of groups that it is looking for. This restriction is based on the minimum number of samples that should have every cluster. The procedure to create the clusters detects when a cluster is smaller than 15 samples, and then the biggest group is divide in two different groups. This technique is repeated as times as necessary to achieve the desire 10 clusters.

In each cluster, it is used hold-out validation, 2/3 of the samples were used to train the models, and the other samples were used to calculate the MSE to select the best algorithm.

The MLP-ANN regression algorithm was trained for different configurations; always with one hidden layer, but the number of neurons in the hidden layer varies from 2 to 15. The activation function of this neurons was tan-sigmoid for all tests, and the output layer neuron had a linear activation function. The training algorithm used was Levenberg-Marquardt; gradient descent was used as learning algorithm, and the performance function was set to mean squared error.

The LS-SVR was trained with the self auto-tuning implemented in the toolbox for MatLab developed by KULeuven-ESAT-SCD. The kernel of the model was set to Radial Basis Function (RBF), and the type was 'Function Estimation' to perform regression. The optimization function is 'simplex' and the cost-criterion is 'leaveoneoutlssvm' with 'mse' as a performance function.

Table 1. Best MSE for each cluster

N° of Clusters	Cluster 1	Cluster 2	Cluster 3	Cluster 4	Cluster 5	Clusters MSE
Global model	0.1834	-	-	-	-	0.1834
2 Clusters	0.0916	0.1325	-	-	-	0.0663
3 Clusters	35.4018	0.1622	0.1397	-	-	0.0466
4 Clusters	6.3063	0.1350	0.1316	0.1254	-	0.0314
5 Clusters	56.2806	0.1563	0.1360	0.1195	0.1715	0.0343
6 Clusters	0.3434	0.0774	0.1077	0.1376	0.1328	0.0288
7 Clusters	1.0476	0.0384	0.3217	0.0938	0.1279	0.0178
8 Clusters	3.1567	0.0080	0.3774	0.1667	0.1031	0.0140
9 Clusters	**0.3269**	**0.0509**	**0.0321**	**0.2638**	**0.1746**	**0.0090**
10 Clusters	0.8473	0.0132	0.0522	0.1006	0.2847	0.0108
N° of Clusters	Cluster 6	Cluster 7	Cluster 8	Cluster 9	Cluster 10	Clusters MSE
Global model	-	-	-	-	-	0.1834
2 Clusters	-	-	-	-	-	0.0663
3 Clusters	-	-	-	-	-	0.0466
4 Clusters	-	-	-	-	-	0.0314
5 Clusters	-	-	-	-	-	0.0343
6 Clusters	0.1726	-	-	-	-	0.0288
7 Clusters	0.1129	0.1244	-	-	-	0.0178
8 Clusters	0.0889	0.1014	0.1121	-	-	0.0140
9 Clusters	**0.0918**	**0.0918**	**0.1264**	**0.0809**	-	**0.0090**
10 Clusters	0.2258	0.1352	0.1197	0.0773	0.1084	0.0108

For Polynomial regression, the order of the polynomial trained varies from 1st to 3rd order.

Table 1 shows the best MSE achieved for each cluster with the corresponding test data (with all the different configurations tested). Moreover, in the last column is shown the mean MSE achieved for each configuration. This mean MSE is calculated taking into account the number of samples in each cluster to ensure a real measure of the MSE for each configuration.

Table 2 shows the best regression technique used and its configuration for each cluster.

The best configuration achieved for the model was the one that divides the data in 9 different clusters, as is shown in Tables 1 and 2.

Table 2. Best regression technique for each cluster

N° of Clusters	Cluster 1	Cluster 2	Cluster 3	Cluster 4	Cluster 5
Global model	ANN-06	-	-	-	-
2 Clusters	ANN-07	ANN-14	-	-	-
3 Clusters	ANN-13	ANN-11	ANN-03	-	-
4 Clusters	ANN-13	ANN-03	ANN-09	ANN-07	-
5 Clusters	ANN-13	LS-SVR	ANN-02	LS-SVR	ANN-07
6 Clusters	ANN-12	ANN-03	ANN-05	ANN-02	ANN-09
7 Clusters	ANN-12	ANN-03	ANN-04	ANN-07	ANN-03
8 Clusters	ANN-02	ANN-08	ANN-07	LS-SVR	ANN-11
9 Clusters	**ANN-09**	**ANN-12**	**LS-SVR**	**ANN-05**	**LS-SVR**
10 Clusters	ANN-09	ANN-03	LS-SVR	ANN-01	ANN-03
N° of Clusters	Cluster 6	Cluster 7	Cluster 8	Cluster 9	Cluster 10
Global model	-	-	-	-	-
2 Clusters	-	-	-	-	-
3 Clusters	-	-	-	-	-
4 Clusters	-	-	-	-	-
5 Clusters	-	-	-	-	-
6 Clusters	ANN-05	-	-	-	-
7 Clusters	ANN-05	Poly-01	-	-	-
8 Clusters	ANN-02	ANN-16	Poly-01	-	-
9 Clusters	**ANN-08**	**ANN-03**	**ANN-03**	**ANN-05**	-
10 Clusters	ANN-03	ANN-06	ANN-08	ANN-03	ANN-06

5 Conclusions

This study provides a precise way of modeling the BIS. The accomplished model predicts the BIS from the EMG signal and the propofol drug quantity provided to the patient.

This model was obtained from a real dataset. The approach is based on a hybrid intelligent system, by combining different regression techniques on local models. After some tests, the analysis of the results shows that the best model configuration has 9 clusters. The regression techniques employed on the clusters were ANN with different configurations (between 3 and 12 neurons in the hidden layer), and the LS-SVR algorithm. The best mean MSE obtained with this configuration was 0.0090.

This analysis could be applied to several different systems with the aim of improving other specifications like: efficiency, performance, features of the obtained material. It is important to emphasize that quite satisfactory results have been obtained with the approach proposed in this research.

Acknowledgments. This study was conducted under the auspices of Research Project *DPI*2010 – 18278, supported by the Spanish Ministry of Innovation and Science.

References

1. Alaiz Moretón, H., Calvo Rolle, J., García, I., Alonso Alvarez, A.: Formalization and practical implementation of a conceptual model for pid controller tuning. Asian J. Control **13**(6), 773–784 (2011)
2. Antón, J.C.Á., Nieto, P.J.G., de Cos Juez, F.J., Lasheras, F.S., Viejo, C.B., Gutiérrez, N.R.: Battery state-of-charge estimator using the MARS technique. IEEE Trans. Power Electron. **28**(8), 3798–3805 (2013)
3. Calvo-Rolle, J.L., Casteleiro-Roca, J.L., Quintián, H., del Carmen Meizoso-Lopez, M.: A hybrid intelligent system for PID controller using in a steel rolling process. Expert Syst. Appl. **40**(13), 5188–5196 (2013)
4. Calvo-Rolle, J.L., Fontenla-Romero, O., Pérez-Sánchez, B., Guijarro-Berdinas, B.: Adaptive inverse control using an online learning algorithm for neural networks. Informatica **25**(3), 401–414 (2014)
5. Calvo-Rolle, J.L., Quintian-Pardo, H., Corchado, E., del Carmen Meizoso-López, M., García, R.F.: Simplified method based on an intelligent model to obtain the extinction angle of the current for a single-phase half wave controlled rectifier with resistive and inductive load. J. Appl. Logic **13**(1), 37–47 (2015)
6. Casteleiro-Roca, J., Calvo-Rolle, J., Meizoso-Lopez, M., Piñón-Pazos, A., Rodríguez-Gómez, B.: New approach for the QCM sensors characterization. Sens. Actuators A Phys. **207**, 1–9 (2014)
7. Casteleiro-Roca, J.L., Calvo-Rolle, J.L., Meizoso-López, M.C., Piñón-Pazos, A.J., Rodríguez-Gómez, B.A.: Bio-inspired model of ground temperature behavior on the horizontal geothermal exchanger of an installation based on a heat pump. Neurocomputing **150**, 90–98 (2015)
8. Casteleiro-Roca, J.L., Pérez, J.A.M., Piñón-Pazos, A.J., Calvo-Rolle, J.L., Corchado, E.: Modeling the electromyogram (EMG) of patients undergoing anesthesia during surgery. In: Herrero, Á., Sedano, J., Baruque, B., Quintián, H., Corchado, E. (eds.) 10th International Conference on Soft Computing Models in Industrial and Environmental Applications. Advances in Intelligent Systems and Computing, vol. 368, pp. 273–283. Springer, Heidelberg (2015)
9. Casteleiro-Roca, J.L., Quintián, H., Calvo-Rolle, J.L., Corchado, E., del Carmen Meizoso-López, M., Piñón-Pazos, A.: An intelligent fault detection system for a heat pump installation based on a geothermal heat exchanger. J. Appl. Logic **17**, 36–47 (2015)
10. Crespo-Ramos, M.J., MachóN-GonzáLez, I., LóPez-GarcíA, H., Calvo-Rolle, J.L.: Detection of locally relevant variables using SOM-NG algorithm. Eng. Appl. Artif. Intell. **26**(8), 1992–2000 (2013)
11. De Cos Juez, F.J., Lasheras, F.S., García Nieto, P., Suárez, M.S.: A new data mining methodology applied to the modelling of the influence of diet and lifestyle on the value of bone mineral density in post-menopausal women. Int. J. Comput. Math. **86**(10–11), 1878–1887 (2009)
12. García, R.F., Rolle, J.L.C., Castelo, J.P., Gomez, M.R.: On the monitoring task of solar thermal fluid transfer systems using NN based models and rule based techniques. Eng. Appl. Artif. Intell. **27**, 129–136 (2014)

13. García, R.F., Rolle, J.L.C., Gomez, M.R., Catoira, A.D.: Expert condition monitoring on hydrostatic self-levitating bearings. Expert Syst. Appl. **40**(8), 2975–2984 (2013)
14. Ghanghermeh, A., Roshan, G., Orosa, J.A., Calvo-Rolle, J.L., Costa, A.M.: New climatic indicators for improving urban sprawl: a case study of tehran city. Entropy **15**(3), 999–1013 (2013)
15. Heiberger, R., Neuwirth, E.: Polynomial regression. In: R Through Excel. Use R, pp. 269–284. Springer, New York (2009)
16. Kaski, S., Sinkkonen, J., Klami, A.: Discriminative clustering. Neurocomputing **69**(13), 18–41 (2005)
17. Li, Y., Shao, X., Cai, W.: A consensus least squares support vector regression (LS-SVR) for analysis of near-infrared spectra of plant samples. Talanta **72**(1), 217–222 (2007)
18. Litvan, H., Jensen, E.W., Galan, J., Lund, J., Rodriguez, B.E., Henneberg, S.W., Caminal, P., Villar Landeira, J.M.: Comparison of conventional averaged and rapid averaged, autoregressive-based extracted auditory evoked potentials for monitoring the hypnotic level during propofol induction. J. Am. Soc. Anesthesiologists **97**(2), 351–358 (2002)
19. Machón-González, I., López-García, H., Calvo-Rolle, J.L.: A hybrid batch SOM-NG algorithm. In: The 2010 International Joint Conference on Neural Networks (IJCNN), pp. 1–5 (2010)
20. Manuel Vilar-Martinez, X., Aurelio Montero-Sousa, J., Luis Calvo-Rolle, J., Casteleiro-Roca, J.L.: Expert system development to assist on the verification of "tacan" system performance. Dyna **89**(1), 112–121 (2014)
21. Nieto, P.G., Garcia-Gonzalo, E., Lasheras, F.S., de Cos Juez, F.J.: Hybrid PSO-SVM-based method for forecasting of the remaining useful life for aircraft engines and evaluation of its reliability. Reliab. Eng. Syst. Saf. **138**, 219–231 (2015)
22. Osborn, J., Juez, F.J.D.C., Guzman, D., Butterley, T., Myers, R., Guesalaga, A., Laine, J.: Using artificial neural networks for open-loop tomography. Opt. Express **20**(3), 2420–2434 (2012)
23. Pérez, J.A.M., Torres, S., Reboso, J.A., Reboso, H.: Estrategias de control en la práctica de anestesia. Revista Iberoamericana de Automática e Informática Industrial RIAI **8**(3), 241–249 (2011)
24. Qin, A., Suganthan, P.: Enhanced neural gas network for prototype-based clustering. Pattern Recogn. **38**(8), 1275–1288 (2005)
25. Quintián, H., Calvo-Rolle, J.L., Corchado, E.: A hybrid regression system based on local models for solar energy prediction. Informatica **25**(2), 265–282 (2014)
26. Rolle, J., Gonzalez, I., Garcia, H.: Neuro-robust controller for non-linear systems. Dyna **86**(3), 308–317 (2011)
27. Sánchez, S.S., Vivas, A.M., Obregón, J.S., Ortega, M.R., Jambrina, C.C., Marco, I.L.T., Jorge, E.C.: Monitorización de la sedación profunda. el monitor BIS. Enfermería Intensiva **20**(4), 159–166 (2009)
28. Sigl, J.C., Chamoun, N.G.: An introduction to bispectral analysis for the electroencephalogram. J. Clin. Monit. **10**(6), 392–404 (1994)
29. Turrado, C.C., López, M.C.M., Lasheras, F.S., Gómez, B.A.R., Rollé, J.L.C., Juez, F.J.C.: Missing data imputation of solar radiation data under different atmospheric conditions. Sensors **14**(11), 20382–20399 (2014)
30. Vapnik, V.: The Nature of Statistical Learning Theory. Springer, New York (1995)
31. Wu, X.: Optimal designs for segmented polynomial regression models and web-based implementation of optimal design software. State University of New York at Stony Brook, Stony Brook, NY, USA (2007)

32. Zeng, Z., Wang, J.: Advances in Neural Network Research and Applications. Springer Publishing Company, Incorporated, Heidelberg (2010)
33. Zhang, Z., Chan, S.C.: On kernel selection of multivariate local polynomial modelling and its application to image smoothing and reconstruction. J. Signal Process. Syst. **64**(3), 361–374 (2011)

Detection of Stress Level and Phases by Advanced Physiological Signal Processing Based on Fuzzy Logic

Unai Zalabarria(✉), Eloy Irigoyen, Raquel Martínez, and Asier Salazar-Ramirez(✉)

University of the Basque Country (UPV/EHU), Bilbao, Spain
uzalabarria001@ikasle.ehu.eus, {eloy.irigoyen, Raquel.martinez, asalazar030}@ehu.eus

Abstract. Stress has a big impact in the current society, being the cause or the incentive of several diseases. Therefore, its detection and monitorization has been the focus of a big number of investigations in the last decades. This work proposes the use of physiological variables such as the electrocardiogram (ECG), the galvanic skin response (GSR) and the respiration (RSP) in order to estimate the level and classify the type of stress. On that purpose, an algorithm based on fuzzy logic has been implemented. This computer-intelligent technique has been combined with a structured processing shaped in state machine. This processing classifies stress in 3 different phases or states: alarm, continued stress and relax. An improved estimation of stress level is obtained at the end, considering the last progresses made by different authors. All this is accompanied by stress classification, which is the novelty compared to other works.

Keywords: Fuzzy logic · State machine · Stress · Physiological signal

1 Introduction

The number of studies on emotional intelligence has increased considerably aiming to improve people's life quality and to increase the scientific knowledge of human emotions [1]. The close relationship existing between the different emotions and the physiological reactions has been derived from those studies [2–6].

There are physiological phenomena that can negatively affect the organism, such as those coming from continued stress [7]. Stress is a physiological response that occurs in new situations. It prepares the organism to fight or escape [8, 9]. Despite being a necessary reaction, its influence has awaken the interest of several research groups, in which the study of the physiological patterns related with stress has resulted in different publications, including: [10–12].

Due to last decades' development of software processing [13], innovative results have been achieved by treating digitally physiological signals based on intelligent computing techniques. In this way, highly accurate estimations of stress level have been obtained [14–17].

M. Graña et al. (eds.), *International Joint Conference SOCO'16-CISIS'16-ICEUTE'16*, Advances in Intelligent Systems and Computing 527, DOI 10.1007/978-3-319-47364-2_29

This article proposes a stress level detection-algorithm considering the information obtained from 3 physiological signals. Those signals are characterized by their non-invasive acquisition: the electrocardiogram, the galvanic skin response and the respiration. Those signals have shown to be representative of stress level, as indicated in [18–21]. From these 3 signals 7 physiological parameters have been derived: heart rate (HR), variation of the heart rate (VHR), GSR level, variation of the GSR (VGSR), integral of the GSR negative variation (IGSR), area resulting from the difference between the GSR and the first-order function derived from the gradient (AGSR) and the product between the heart rate variability (HRV) and the standard deviation of the frequency correlation of the RSP (HSCR). Those parameters will be the fuzzy algorithm inputs [22] configured for the detection of the stress level with values between 0 and 1 [14, 16]. The algorithm will also count on a state machine. This machine is able to identify and differentiate the alarm states of continued-stress periods and, at the same time, do the same with relaxation phases.

2 Experimental Phase

First, an experimental phase has been designed in order to obtain physiological signals that are stress and relax phase-representative. This experimental phase is based on already-made experiments by authors such as Gross [23]. Those experiments use videos, as well as 3D puzzles, to produce the different stress and relax phases.

To collect the signals of the experiment the BIOPAC MP36 (Biopac Systems Inc, USA) professional data acquisition system has been used, sampling at a speed of 1000 Hz. The Biopac Acqknowledge 3.7.1 (Biopac Systems Inc, USA) has been used as software of this device. The ECG and RSP signals have been measured in milliVolts (mV) and the GSR in microSiemens (mS). In Fig. 1, these 3 acquired physiological signals are shown.

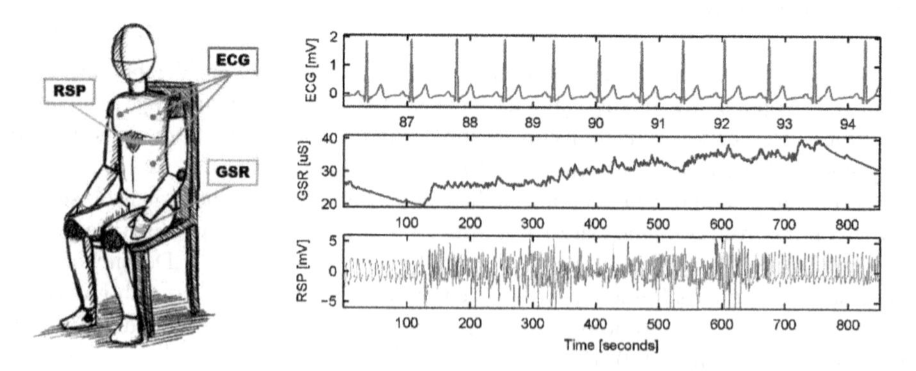

Fig. 1. Placement of the physiological sensors and the acquired signals.

A total of 68 volunteers (51 males and 17 females) aged between 19 and 45 (average age: 22, 88; standard deviation: 3, 1) participated in the experiment. The experiment was made using non-invasive sensorization. The ECG was obtained by putting 3 electrodes in certain points of the chest. Two electrodes were put for the GSR, on the index and the middle fingers, in order to measure the conductivity between them. Finally, the RSP was obtained with a strain gage located in a band placed around the person's chest. Figure 1 depicts the chosen sensor positioning.

As it has been mentioned, this experiment is characterized by the use of videos and 3D puzzles. The experiment is divided in 3 consecutive phases (see Fig. 2) and only runs once per person: a first relax phase, approximately lasting 2 min; a second phase, lasting 10 min, in which the volunteer tries to solve a 3D puzzle; and finally, a third relax phase, lasting also 2 min, in which the volunteer returns to the initial relax situation. The first and the third phases consist in the visualization of a relaxing video.

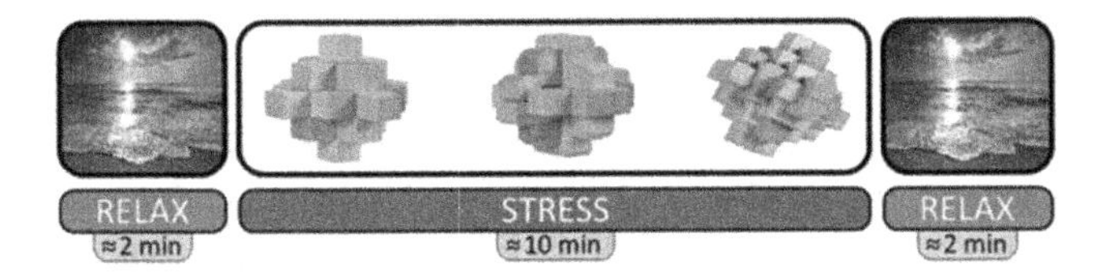

Fig. 2. Stress experiment phases and times that were made.

At the end of the experiment the volunteers were asked to fill a questionnaire (Self-Assessment Manikin, SAM [24]) where they explained how they had felt during the experiment.

3 Employed Physiological Signals

In order to obtain the previously mentioned physiological parameters the ECG, GSR and RSP signals need to be processed. The processing of the signals has been done with 20 s sliding windows on a 5 s intervals, as proposed in [18].

3.1 ECG Signal

One of the reasons to process the ECG is to obtain the HR signal. This processing can be relatively complex as the ECG does not provide information about the HR by itself. The first step is to calculate the position of the R peaks, which normally correspond, but not necessarily, to the highest values of each ECG period. The algorithm proposed in [25] was used in order to do that. This algorithm processes the ECG signal with a series of robust iterative methods implying different levels of analysis which comprise different robustness and computational charges depending on the signal quality in each moment. The algorithm includes a simultaneous evaluation of the intensity level of the noise and of the signal in each moment. It also performs wavelet-based techniques, as in [14]. In this filtering, the algorithm decides, considering the two intensity parameters, whether the signal segment has to be processed with wavelets or not. This enables to overcome several difficulties, such as the noise, offset variations and signal intensity variations produced by the loss of contact of sensor with the skin or by electromagnetic interferences (see Fig. 3).

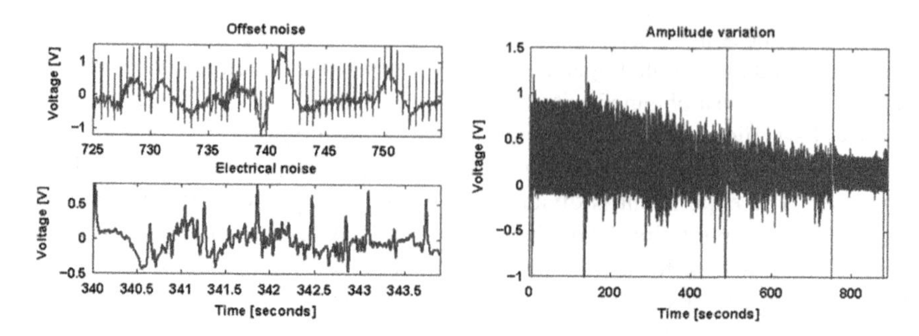

Fig. 3. Different artifacts found in the ECG signal.

Knowing the position of R peaks it is possible to calculate the interval between subsequent peaks and from there, the HR in beats per minute (bpm). The HR is normally affected by the respiration [26, 27] and other disturbances that must be eliminated. On this purpose, the values of the HR were obtained calculating the average values within a 20 s sliding window. This window size has been found to be long enough to omit disturbances and to allow to detect significant changes in stress while keeping a manageable computational cost. This way, a filtered HR was obtained, which is representative of the HR level (see Fig. 4).

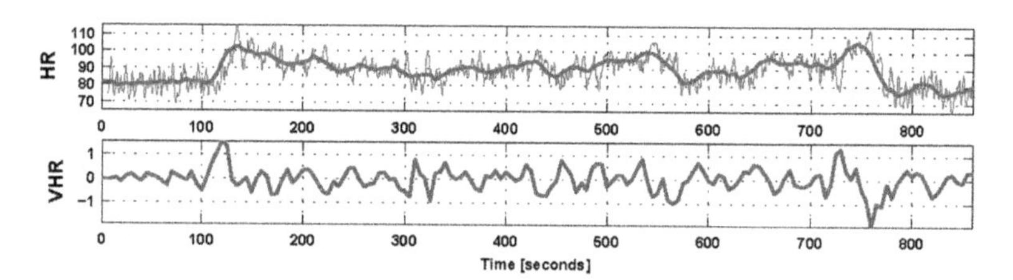

Fig. 4. 1st graph: HR (pink) and filtered HR (red). 2nd graph: VHR. (Color figure online)

The second physiological parameter to be extracted from the ECG is the VHR. To do so, the gradient corresponding to a first order function closed to each HR section will be calculated in each 20 s sliding window. This way, the result shown in Fig. 4 is obtained.

3.2 GSR Signal

The GSR is directly related to the sympathetic system [8, 10]. This relation makes the GSR to be a signal of great interest. A total of 4 physiological parameters have derived from the GSR, and all of them are calculated in each 20 s-data window.

The first one is the level of the GSR. In order to obtain it, the average value for each section, obtaining an artifact-free signal (see Fig. 5). One of the problems of using this signal is its slow drying, which results in different magnitudes for a same stress level.

This makes necessary to obtain a relative value like the VGSR. The VGSR is obtained in the same way as the VHR: approaching each section of the GSR to a leading role and extracting the pendent for each window. This way, the results of Fig. 5 are obtained.

In order to calculate the IGSR, only the negative values of the VGSR will be taken into account. Those values will be added as long as they are successive. The signal will be reseted to 0 in the moment that the VGSR changes to a positive value. This signal is very useful in order to identify the sections where relax has been continued (see Fig. 5).

As the last physiological parameter obtained from the GSR, the AGSR has been calculated. The importance of this parameter lies in its similarity with the stress level, which has been empirically demonstrated. This parameter is defined by the value corresponding to the distance between the line defined by the first order function calculated to obtain the VGSR and the own GSR. This calculation is derived from the conduct of the GSR. The absence of variations in the sections where the person is relaxed and a large presence of variations in the sections where the people is stressed can be appreciated. In order to normalize the signal, it has been divided with the value of the GSR, as it has been shown that in situations of identical stress, the magnitude of the AGSR is higher when the absolute level of perspiration is higher. As a result, the signal of Fig. 5 is obtained:

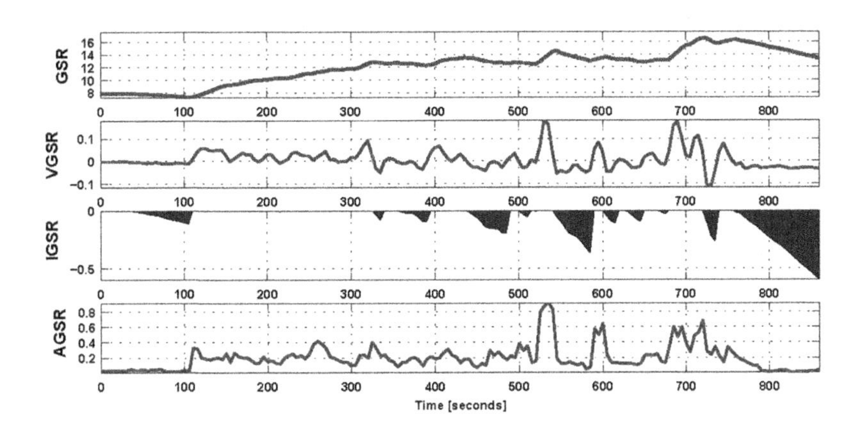

Fig. 5. Filtered GSR in the 1st graph and the 3 physiological parameters derived from GSR.

3.3 RSP Signal

Considering the way the respiration is performed in both relax and stress phases, it has been shown that it presents a different harmonicity depending on which phase the person is. In the periods of relax, the respiration tends to be more harmonic and concentrated in lower frequencies. In stress situations, though, the respiration is usually more chaotic and all kinds of frequencies are predominant.

On this basis, the decision was to do a frequency analysis of the signal around a frequency range comprised between 0.01 Hz and 0.5 Hz. To do so, a correlation between each 20 s sliding window of the respiratory signal and sinusoidal signals corresponding to the specified frequency range has been done. This range has been selected considering a frequency study about the RSP signals of every volunteer.

Figure 6 shows the frequency correlation over the RSP signal of a volunteer. The dispersion is higher in the stress phase (middle section) than in the relax section (first and last section).

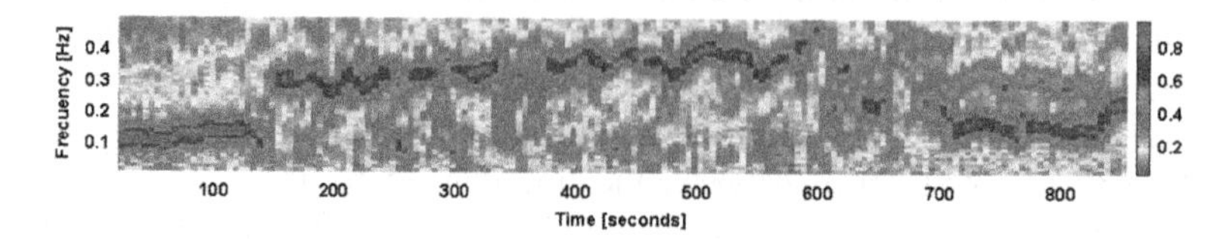

Fig. 6. Temporary development of the frequency correlation of a volunteer.

The calculation of the standard deviation of the frequency correlation (SCR) of each window would be a useful parameter in the estimation of the stress state of the person considering the RSP (see Fig. 7, green line).

Empirical evidences have demonstrated that, as the SCR shows elevated values in the relax sections and lower values in the stress sections, the HRV presents also those same symptoms. In order to strengthen the SCR, a product of both signals has been performed. This way, emphasizing the difference between the stress and relax phases has been an achievement. The improvement made by incrementing the difference between both states can be seen in Fig. 7:

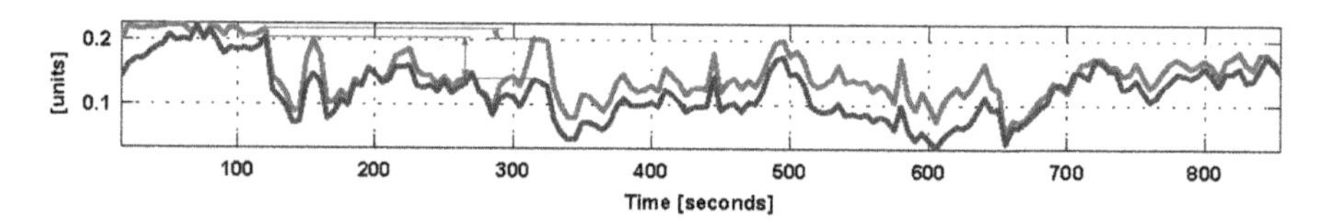

Fig. 7. SD Comparison between SCR (green) and the HSCR (purple). (Color figure online)

In the following section, the proposed algorithm to obtain the level of stress considering the physiological parameters obtained in this section will be developed.

4 Proposed Algorithm

In some cases the available information can be too imprecise to relate physiological parameters to stress. In addition, the differences between subjects and their physiological signal ranges gives the system certain ambiguity when classifying the state of a signal. This makes fuzzy logic the ideal methodology as it brings manageability, robustness, low cost of the solution and a better relation with the reality [22].

The fuzzy algorithm has been done from the study presented by Santos Sierra [16] and Salazar-Ramirez [14]. New parameters have been added, reaching a total amount of 7, which are: the HR, the VHR, the VGSR, the IGSR, the AGSR and the HSCR. The result obtained with the fuzzy processing is complemented by the identification of the stress type that occurs in every moment. This identification has been performed with a state machine, which classifies the stress level considering 3 different states: alarm, continued stress or relax. The first being explained will be the defined configuration for each of the membership functions of the fuzzy algorithm, followed by the configured rules for the obtaining of the stress variable at the output. Finally, the last to be shown will be the results obtained in the detection of the stress level combined with the classification given by the state machine.

4.1 Membership Functions in the Input

This work proposes the division of each parameter by 3 different membership functions: low (L), middle (M) and high (H). The level of the GSR has only been applied the low and high functions, due to the fact that its absolute value presents big tolerances for a same state. The parameters IVGRS and ADGRS have a unique membership function, simplifying them to presence or absence of stress. The parameters of the HR and the one of the GSR have been normalized to values between 0 and 1. This way, the adaptation of the fuzzy algorithm to the physiological features of any person has been achieved. On this purpose, every person is defined with 4 main parameters: HR(min), HR(max), GSR(min) and GSR(max). The stress output has been normalized the same way to values between 0 and 1. Due to the quantity of rules, it has been configured with 5 membership values: low, low-middle (LM), middle, middle-high (MH) and high. The table below presents the details of the performed configuration (Table 1):

Table 1. Specifications of the designed membership functions.

Variable	Definition	States	Form	Shape edges	Variable	Definition	States	Form	Shape edges
(Input) FC	[0,1]	L	Sen.	[0.1 0.6]	(In.) IGSR	[−1,0]	L	Sen.	[−.4 0]
		M	Trap.	[0.1 0.4 0.6 0.9]	(In.) AGSR	[0,0.5]	L	Sen.	[.02 .2]
		H	Sen.	[0.4 0.9]	(Input) HSCR	[0,0.2]	L	Trap.	[−1 −1 .07 .11]
(Input) VFC	[−2,2]	L	Sen.	[−1 0]			M	Trap	[0.07 0.09 0.11 0.13]

(*Continued*)

Table 1. (*Continued*)

Variable	Definition	States	Form	Shape edges	Variable	Definition	States	Form	Shape edges
		M	Gaus.	[0.4 0]			H	Trap	[0.09 0.13 1 1]
		H	Sen.	[0 1]	(Ouput) Stress	[0,1]	L	Trap	[−0.4 −0.2 0.2 0.4]
(Input) GSR	[0,1]	L	Trap.	[−1 −1 0.1 0.5]			LM	Trap	[0 0.2 0.4 0.6]
		H	Trap.	[0 0.6 50 50]			M	Trap	[0.2 0.4 0.6 0.8]
(Input) VGSR	[−0.3,0.3]	L	Sen.	[−0.05 0]			MH	Trap	[0.4 0.6 0.8 1]
		M	2xGaus.	[0.02 −0.02 0.03 0.05]			H	Trap	[0.6 0.8 1.2 1.4]
		H	Sen.	[0 0.1]					

4.2 Rule Inference System

Once the membership functions have been configured, the rules that will shape the stress output signal have been generated. The table shows the selected configuration (Table 2).

4.3 State Machine

This part of the algorithm uses the signals VHR, VGRS and AGSR to define the state the person is: alarm, continued stress or relax. The algorithm is fed back; it uses the states of the 3 named instants in order to differentiate the alarm phases from the stress ones. It is considered that 3 followed alarm states (15 s) mean the entry in a continued stress period. The alarm is activated when the VHR and the VGSR exceed simultaneously a certain value. Relax is activated when the VGSR and the AGSR reach another certain value. In order to abandon periods such as the continued stress phase, the stress level derived from the fuzzy processing is also taken into account. Figure 8 is the representation of this state machine's operate diagram.

Table 2. Configured rules of the diffuse algorithm.

FC	L	M	L							L	L	L	M	H	L	M	H	~L	L	L	M	M	H	H	H	M	L	H	M	L
VFC																									H	H	H	H	H	H
GSR				L						L	L									H	L	H	L	H						
VGSR																		M	M						H	H	H	H	H	H
IGSR								L				L	L	L	L	L	L													
AGSR									L			L	L	L																
HSCR					L	M	H				L																	H	H	H
Stress	L	M	L	L	L	M	H	L	L	L	L	L	LM	M	L	LM	M	MH	M	LM	LM	MH	M	H	H	MH	M	H	MH	M
Power	0.3	0.3	0.3	0.5	0.2	0.2	0.2	0.3	0.3	0.8	1	1	1	1	0.7	0.7	0.7	1	0.3	0.3	0.5	0.3	0.5	0.5	0.8	0.8	0.8	1	1	1

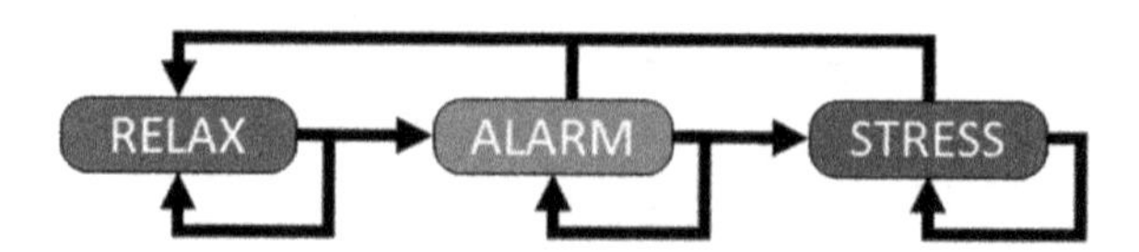

Fig. 8. Representative scheme of the state machine operation.

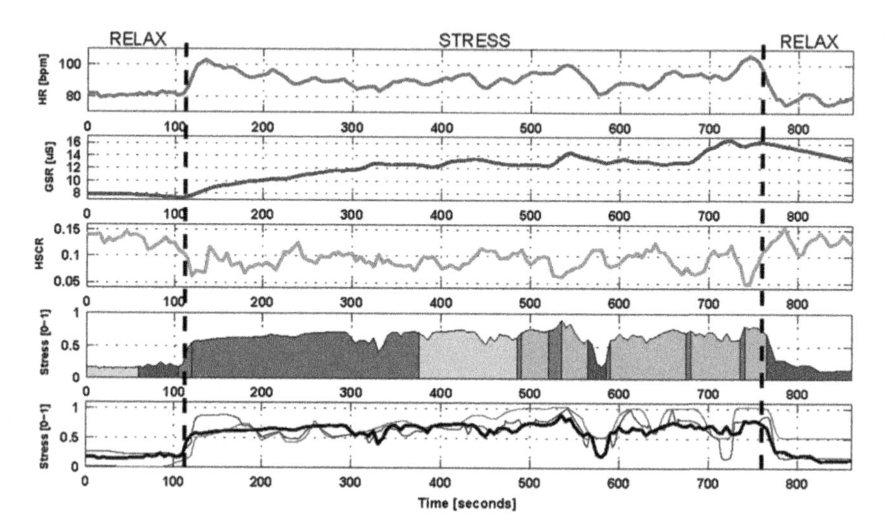

Fig. 9. Results of the level and stress type estimation and comparison between the proposed algorithm (black line) with the one proposed by Santos Sierra (red line) and Salazar-Ramírez (blue line). (Color figure online)

4.4 Comparative Results

The last step is the system validation. Figure 9 shows the results obtained in one of the signals. One of the improvements regarding to the algorithms proposed by Santos Sierra [16] and Salazar-Ramirez [14] is the addition of the stress-type interpretation by using the state machine.

About the stress level estimation, the results offered by the fuzzy algorithm being the only base, a more complex algorithm has been achieved. It offers improved results in comparison with the already mentioned authors (see Fig. 9). The correlation between the physiological variables and the stress estimation is better, mainly due to the higher quantity of physiological parameters used and to the higher complexity of the proposed algorithm.

5 Conclusions and Future Works

Several Works propose the use of intelligent computer techniques in order to obtain the stress level. In this work, besides achieving improvements in the stress estimation in regard to authors that have worked with the same techniques [14, 16], this estimation has been classified in 3 different phases.

The improvements in the stress level estimation are largely due to the addition of new physiological parameters. They have proved to be of great interest in the stress level estimation. Among them, the RSP and the variables derived from the GSR, such as the IGSR and the AGSR must be highlighted. With a higher quantity of input parameters, the used fuzzy algorithm has proved to have a higher grade of complexity than in other cases. This, between with a good attunement of the rules, has resulted in a better stress level estimation.

The stress-type classification is another great incentive for the contributions of this work: the alarm phases can be differentiated from the continued stress's ones. Those, at the same time, are differentiated from the relax phases.

Finally, the use of variables such as HR_min, HR_max, GSR_min and GSR_max has to be underlined. Those are important because they normalize the main physiological parameters (HR and GSR). Although their synchronization has been manually made, in future jobs there is the possibility of creating an algorithm that can synchronize those parameters considering the data collected during those long time periods. This would be made by adapting the program to each person and situation.

Acknowledgements. This work has been performed partially thanks to the support of the Foundation Jesús de Gangoiti Barrera, to which we are deeply grateful. It would not have been possible to perform it without the involvement of the biomedical investigation group of GICI, to which we also thank its effort and dedication.

References

1. Morris, C.G., Maisto, A.A.: Introducción a la Psicología. Pearson Educación, Mexico (2005)
2. Kreibig, S.D.: Autonomic nervous system activity in emotion: A review. Biol. Psychol. **3**(3), 394–421 (2010)
3. Lee, C.K. et al.: Using neural network to recognize human emotions from heart rate variability and skin resistance. In: 27th Annual International Conference of the IEEE Engineering in Medicine and Biology Society, pp. 5523–5525 (2006)
4. Porges, S.W.: The polyvagal theory: phylogenetic substrates of a social nervous system. Int. J. Psychophysiol. **42**(2), 123–146 (2001)
5. Bloch, S., et al.: Specific respiratory patterns distinguish among human basic emotions. Int. J. Psychophysiol. **11**(2), 141–154 (1991)
6. Ekman, P., et al.: Autonomic nervous system activity distinguishes among emotions. Science **221**(4616), 1208–1210 (1983)
7. De Rivera, J.G., et al.: La valoración de sucesos vitales: adaptación española de la escala de Holmes y rahe. Psiquis **4**(1), 7–11 (1983)
8. De Camargo, B.: Estrés, síndrome general de adaptación o reacción general de alarma. Revista Médico Científica **17**(2), 78–86 (2010)

9. Nelson, R.J.: An introduction to behavioral endocrinology. Sinauer Associates, Sunderland (2005)
10. Cacioppo, J.T., et al.: Handbook of Psychophysiology. Cambridge University Press, Cambridge (2007)
11. Healey, J.A., Picard, R.W.: Detecting stress during real-world driving tasks using physiological sensors. IEEE Trans. Intell. Transp. Syst. **6**(2), 156–166 (2005). IEEE
12. Cannon, W.B.: Stresses and strains of homeostasis. Am. J. Med. Sci. **189**(1), 13–14 (1935). LWW
13. Wozniak, M., et al.: A survey of multiple classifier systems as hybrid systems. Inf. Fusion **16**, 3–17 (2014). Elsevier
14. Salazar-Ramirez, A., Irigoyen, E., Martinez, R.: Enhancements for a robust fuzzy detection of stress. In: de la Puerta, J.G., Ferreira, I.G., Bringas, P.G., Klett, F., Abraham, A., de Carvalho, A.C., Herrero, A., Baruque, B., Quintián, H., Corchado, E. (eds.) International Joint Conference SOCO 2014-CISIS 2014-ICEUTE 2014. AISC, vol. 299, pp. 229–238. Springer, Heidelberg (2014)
15. Chang, C.-Y., et al.: Physiological emotion analysis using support vector regression. Neurocomputing **122**, 79–87 (2013)
16. De Santos Sierra, A., et al.: A stress-detection system based on physiological signals and fuzzy logic. IEEE Trans. Ind. Electron. **58**(10), 4857–4865 (2011). IEEE
17. Sakr, G.E., et al.: Support vector machines to define and detect agitation transition. IEEE Trans. Affect. Comput. **1**(2), 98–108 (2010). IEEE
18. Martinez, R.: Diseño de un sistema de detección y clasificación de cambios emocionales basados en el análisis de señales fisiológicas no intrusivas. University of the Basque Country (2016)
19. Pauws, S.C., et al.: Insightful stress detection from physiology modalities using learning vector quantization. Neurocomputing **151**, 873–882 (2015). Elsevier
20. Subramanya, K., et al.: A wearable device for monitoring galvanic skin response to accurately predict changes in blood pressure indexes and cardiovascular dynamics. In: 2013 Annual IEEE India Conference (INDICON), pp. 1–4 (2013)
21. Martinez, R., et al.: First results in modelling stress situations by analysing physiological human signals. In: Proceedings of IADIS International Conference on e-Health, pp. 171–175 (2012)
22. Zadeh, L.A.: Fuzzy logic = computing with words. IEEE Trans. Fuzzy Syst. **4**, 103–111 (1996). IEEE
23. Gross, J.J., Levenson, R.W.: Emotion elicitation using films. Cogn. Emot. **9**(1), 87–108 (1995). Taylor & Francis
24. Bradley, M.M., Lang, P.J.: Measuring emotion: the self-assessment manikin and the semantic differential. J. Behav. Ther. Exp. Psychiatry **25**(1), 49–59 (1994). Elsevier
25. Zalabarria, U., et al.: Procesamiento robusto para el análisis avanzado de señales electrocardiográficas afectadas por perturbaciones. Actas de las XXXVI Jornadas de Automática, pp. 807–814 (2015)
26. Bari, V., et al.: Nonlinear effects of respiration on the crosstalk between cardiovascular and cerebrovascular control systems. Phil. Trans. R. Soc. A **374**(2067), 20150179 (2016). The Royal Society
27. Porta, A., et al.: Conditional symbolic analysis detects nonlinear influences of respiration on cardiovascular control in humans. Phil. Trans. R. Soc. A **373**(2034), 1–21 (2015). The Royal Society

Reinforcement Learning for Hand Grasp with Surface Multi-field Neuroprostheses

Eukene Imatz-Ojanguren[1,2(✉)], Eloy Irigoyen[1], and Thierry Keller[2]

[1] Intelligent Control Research Group,
UPV/EHU - University of the Basque Country,
Alameda Urquijo, 48013 Bilbao, Spain

[2] TECNALIA Research and Innovation, Neurorehabilitation Area,
Mikeletegi Pasealekua 1-3, 20009 Donostia-San Sebastián, Spain

eukene.imatz@tecnalia.com

Abstract. Hand grasp is a complex system that plays an important role in the activities of daily living. Upper-limb neuroprostheses aim at restoring lost reaching and grasping functions on people suffering from neural disorders. However, the dimensionality and complexity of the upper-limb makes the neuroprostheses modeling and control challenging. In this work we present preliminary results for checking the feasibility of using a reinforcement learning (RL) approach for achieving grasp functions with a surface multi-field neuroprosthesis for grasping. Grasps from 20 healthy subjects were recorded to build a reference for the RL system and then two different award strategies were tested on simulations based on neuro-fuzzy models of hemiplegic patients. These first results suggest that RL might be a possible solution for obtaining grasp function by means of multi-field neuroprostheses in the near future.

Keywords: Neuroprostheses · Functional electrical stimulation · Grasp · Reinforcement learning · Modeling and control

1 Introduction

The human hand is a complex instrument that allows us to complete a wide variety of Activities of Daily Living (ADL). The neural and biomechanical complexity of the hand involves approximately 20 DOF controlled by more than 30 muscles located on the hand (intrinsic) and forearm (extrinsic) [1]. However, biomechanical and neural factors constrain the independent control of these DOF, resulting in a reduced dimensionality of the human hand [2], which suggests that there is a fixed synergistic control of the hand [3]. In healthy subjects the grasp is composed of two phases, the grip aperture that occurs during reaching an object, and the gradual closure of the fingers until matching the object size and shape [4]. However, people suffering from neural disorders, such as stroke, present modified movement patterns of reaching and grasping [5]. Grasp movements of stroke subjects in general imply slower grip aperture velocities, deficient timing of grasp formation and inaccurate scaling of grip aperture, with difficulties in

M. Graña et al. (eds.), *International Joint Conference SOCO'16-CISIS'16-ICEUTE'16*,
Advances in Intelligent Systems and Computing 527, DOI 10.1007/978-3-319-47364-2_30

extending the fingers. These impairments prevent stroke subjects from performing successful grasps, which can reduce considerably their ability to perform ADL and affect their quality of life [6].

In order to assist people suffering from these type of neurological impairments, limb neuroprostheses are used to bridge interrupted neural pathways and perform functional movements. These neuroprostheses are based on Functional Electrical Stimulation (FES), a technique that consists in delivering electrical pulses to the peripheral nerves in order to elicit muscle contractions that lead to functional movements [7]. The main components of the neuroprostheses are the electrical stimulator, which generates the electrical pulses, and the electrodes, which delivers the electrical pulses to the nerves. Depending on the location of these components within the body, the neuroprostheses can be classified as implanted or superficial. In the former case, both the electrodes and the stimulator are placed under the skin, and thus, require surgery to attach the electrodes to the targeted nerves. On the other hand, surface systems are external to the body and the electrodes are attached superficially over the skin. The latter type of neuroprostheses are preferred for therapeutic applications for being non-invasive and easy to don/doff [8]. In fact, multi-field surface electrodes, consisting of an array of electrodes [9], allow increased target nerve selectivity compared to conventional surface electrodes [10], which is essential in applications such as grasping, where precise activation of multiple motor nerves at different depths and positions over the forearm is required.

Main FES applications on upper-limbs involve neuroprostheses for reaching and grasping functions. However, upper-limb FES control is still in an early stage because, due to its complex nature, it should face with challenges such as the variability of inter-subject response to FES (differences on physiology/pathology), variability of task-dependent movements, high amount of DOF, and difficulty of measuring kinematic or dynamic data, among others [11]. Indeed, the complex anatomy of the forearm and hand, with small muscles located and overlapped at different depths, makes selective surface stimulation challenging.

Most closed-loop control applications of the upper-limb focus on reaching tasks, by applying FES to motor nerves that innervate muscles acting on shoulder and elbow joints [12]. For hand grasp, most of the proposed FES control systems are based on implanted [13] and/or open-loop systems [14]. With respect to surface multi-field electrodes, most research focused in a first stage on calibration or motor point search, for selecting specific fields, followed by fixed open-loop stimulation patterns applied to the previously selected fields [15]. Recently, an approach based on an iterative learning control system showed good results for estimating fields and stimulation parameters to achieve hand postures [16].

However, the proposed approaches for FES-induced hand grasp are either aimed at implanted systems; open-loop; or use fixed or predefined electrode application sites during hand grasp. One of the main challenges of surface FES grasping lies in reduced selectivity, due to the spreading of the current into neighboring tissues instead of reaching the target motor nerves. Thus, the electrode configuration and FES application site is key for achieving more selective stimulation. Indeed, the relative position of the nerves and the skin change during arm and hand movements [17]. Thus, we believe that the control of both the stimulation parameters and the spatial application sites of FES is needed during hand movements to achieve a good grasp precision. Due to the complexity of the grasping application; the dynamic characteristics of the surface FES-induced grasping; repetitive goal-oriented therapies aimed at stroke patients [18]; and the capacity of Reinforcement Learning (RL) systems to learn from exploration and adapt to non-stationary problems [19,20]; we decided to give the first steps towards a RL based FES-induced grasping system. However, the dimensionality of this application is an important drawback that can affect its performance.

Therefore, the aim of the present work is to check the feasibility of using RL for obtaining FES-induced grasps, which are tested on simulations with previously developed neuro-fuzzy models of surface multi-field neuroprostheses [21]. In the next section a brief description of the data acquisition session is given, followed by the information of the proposed RL system. Finally, obtained preliminary results are presented and discussed.

2 Case of Study

Different grasps from 20 healthy subjects were first recorded in order to construct a reference for the RL system described in the next section. Kinematic data of the hand from healthy subjects were recorded while performing the different grasps present in the Action Research Arm Test (ARAT), which is an assessment tool for assessing upper-limb function and consists of grasping objects of different shape, size and weight and releasing them at different distances and heights from the subject [22]. All subjects signed an informed consent for participating in the session. The data acquisition session lasted about 20 min and consisted in carrying out the different grasp tasks comprised in the ARAT.

The sensor system for measuring hand and finger flexion/extension consisted of a combination of two separated systems based on inertial sensors and optic fiber based sensors. For the measurement of finger flexion/extension the 5Data instrumented glove from Fifth Dimension Technologies was used, which contained 5 optic fiber sensors, one for each finger, and provided the percentage of curvature of both metacarpophalangeal and proximal interphalangeal joints with respect to a previously defined maximum value. In this experiment, the maximum and minimum values were defined by the passive range of motion (PROM) measured at the beginning of the sessions. Wrist flexion/extension was measured with two 3-Space wireless inertial sensors from YEI Technology, as shown in Fig. 1. Data from all sensors were collected at 25 Hz.

First, the donning stage was carried out, where subjects were seated on a chair and the glove was donned on their left hand. Then, the registration of PROM ranges was carried out. Finally, the subjects were asked to keep their right arm supported on the table during all the session, and they were instructed to perform all the exercises of the ARAT with their left hand, starting each exercise with the palm facing down on the table. Grasp time was defined as the time when

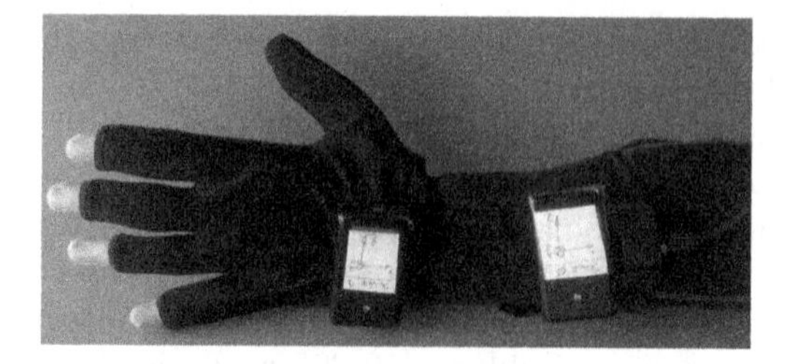

Fig. 1. Sensor system: glove with 5 optic fiber sensors and 2 inertial sensors.

hand-object contact occurred, and release time as the time when hand-object contact was not present anymore.

In order to temporally align grasps of different subjects, time normalization was carried out by the picewise linear length normalization method [23], taking grasp and release times as our points of interest (POI). Once normalized, average flexion/extension values were calculated for the wrist and each of the fingers. Examples of the time-normalized average values are shown for the cup grasp trials in Fig. 2, where wrist is represented on angles, with positive and negative angles representing flexion and extension respectively. Finger flexion is a normalized value respect to the PROM, where 0 is full extension and 1 full flexion.

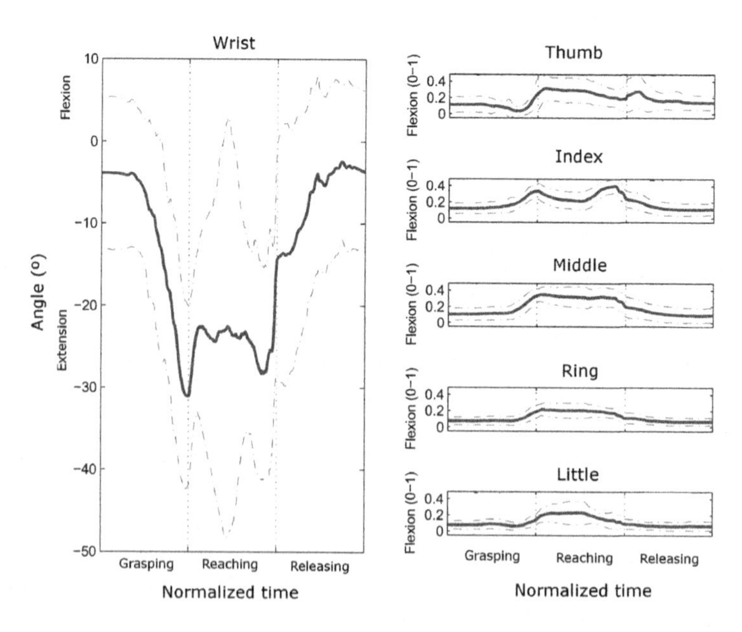

Fig. 2. Grasp of the cup - mean flexion/extension values of wrist and fingers. Blue line represents the mean values and red lines represent the +/- standard deviation ranges. (Color figure online)

Stages of the trials are marked, which represent the grasp stage, followed by the reaching or object displacement stage, and finally the release stage.

3 Reinforcement Learning System

Once we had the reference grasps from healthy subjects, we could test an RL system for achieving FES-induced grasps testing them on a previously developed neuro-fuzzy model that emulates the behavior of a surface multi-field grasping neuroprostheses applied on a post-stroke hemiplegic forearm [21]. With this purpose, we trained a forward model that was able to predict finger and wrist flexion/extension from stimulation amplitudes and activation sites on the forearm, therefore, models were provided with 3 inputs and 6 outputs. One of the inputs represented the amplitude, and the other two inputs represented the coordinates of the activated field on the arm in proximal-distal and medial-lateral dimensions. Regarding the outputs, they represented the flexion/extension of wrist and fingers (Fig. 3).

The objective of the RL system was to learn the electrode activation (distal-medial coordinates) and amplitude patterns to perform a specific grasp. For this preliminary tests, the grasp of the cup (cylindrical) was selected for being one of the most common grasps in ADL [24]. In these preliminary tests, the two reward strategies described below were tested on the same subject model.

Regarding the RL system, the SARSA method was selected because it is a model-free on-line policy method [19]. Replacing type eligibility traces were also included, with the trace-decay parameter $\lambda = 0.9$. The discount factor was set to $\gamma = 1$ (no discount) because we were facing episodic tasks.

The learning agent could choose among 96 actions in each state, consisting on possible combinations of three sub-actions: distal coordinate of active

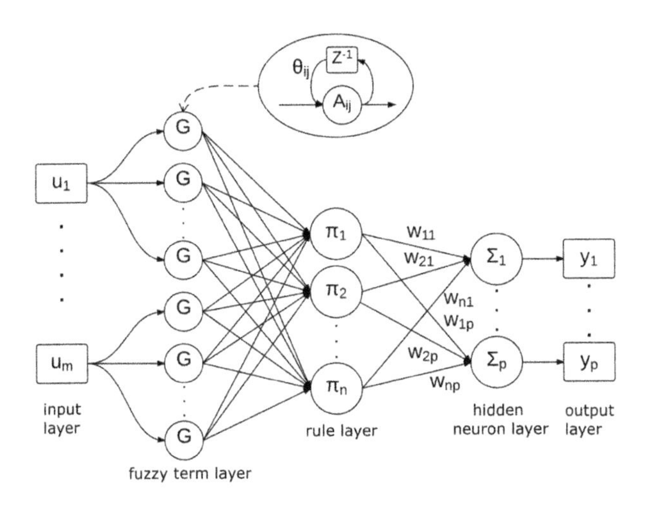

Fig. 3. RFNN model.

electrode (4 possibilities), medial coordinate of active electrode (8 possibilities), and amplitude (decrease 1 mA, no change, or increase 1 mA).

Due to the large continuous state space, value function approximation was carried out. For these preliminary tests tile coding was used due to its ease of implementation. The six-dimensional state space was divided by 32 tilings of 9×9 tiles, each of them being offset with a random fraction of a tile. Besides, hashing was applied to reduce memory costs. The learning parameter of the gradient descent for function approximation was set to $\alpha = 0.003$ as it is suggested to be approximately $\alpha = 1/10.ntiles$ to ensure a proper generalization [19].

Regarding policy, the ϵ-greedy method was selected for its capacity of ensuring a continuous exploration of the possible actions. Different ϵ values were selected to see their effect on the RL performance, these were $\epsilon_1 = 0.01$, $\epsilon_2 = 0.05$, $\epsilon_3 = 0.1$, and $\epsilon_4 = 0.25$.

The main difficulty of applying RL to the simulations relied on the definition of rewards, as they should provide information regarding successful or not successful grasps. In its practical application, this could be defined by the therapist, or even with sensorized objects. However, our current model was built upon kinematic data of the hand, so two different reward strategies were defined to describe successful grasps. These were based on POI, which were the open-hand state corresponding to the beginning of the reference curves and the closed-hand state corresponding to the end of the reference curves.

POI approach 1. This strategy evaluated the grasp success based on the errors that were calculated with respect to two POI. These were the open-hand POI ($POI^{open} = ref(1)$), and the closed-hand POI ($POI^{close} = ref(end)$). The aim of the error values was to detect if a hand opening occurred during the first half of the grasp, and a hand closing occurred during the second half of the grasp. These errors were calculated with respect to the POI as follows:

$$\begin{aligned} err_{open}(i) &= \frac{1}{4}\sum_{out} |y_{out}(i) - POI_{out}^{open}| \qquad \text{for } i = 1, 2, ..., n/2 \\ err_{close}(i) &= \frac{1}{4}\sum_{out} |y_{out}(i) - POI_{out}^{close}| \qquad \text{for } i = n/2+1, n/2+2, ..., n \end{aligned} \tag{1}$$

where *out* referred to the system output (wrist, thumb, index or ring finger), i to the sample number, n to the total number of samples, POI^{open} to the reference open-hand POI, POI^{close} to the reference closed-hand POI and y to the RL output. Reward was set to 0 for all the time-steps except for the last step of each episode. At this point, the total reward was $r_{tot} = r_{open} + r_{close}$, and the independent rewards were defined as

$$r_{open} = \begin{cases} -\,min(err_{open}) & \text{if } min(err_{open}) \geq 0.02 \\ 1 & \text{else} \end{cases} \tag{2}$$

where min is a function that takes the minimum value from err_{open}. The reward r_{close} was analogous but based on err_{close} instead.

POI approach 2. This strategy was also based on the same open-hand and closed-hand POIs and errors from the previous strategy, but rewards were defined differently. In this case, rewards were given at every time-step. The value of the reward was negative and proportional to err_{open} at every time step until the first time a hand opening was achieved, at this point a reward of 5 was given. After this, reward was negative and proportional to err_{close} until a hand closing was achieved, where the reward value was again 5. Formally it would be denoted as

$$r(t) = \begin{cases} -\,err_{open}(t) & \text{if } err_{open}(t) \geq 0.02 \text{ and } err_{open}(t_{prev}) \geq 0.02 \\ 5 & \text{if } err_{open}(t) < 0.02 \text{ and } err_{open}(t_{prev}) \geq 0.02 \\ -\,err_{close}(t) & \text{if } err_{close}(t) \geq 0.02 \text{ and } err_{open}(t_{prev}) < 0.02 \\ 5 & \text{if } err_{close}(t) < 0.02 \text{ and } err_{open}(t_{prev}) < 0.02 \end{cases} \tag{3}$$

where t represents the actual time-step and t_{prev} any time-step previous to t.

Due to the random character of the exploratory actions, each of the strategies and each of the ϵ values were repeated 5 times, and results show the mean value of these 5 iterations. Finally, it should be noted that instead of leaving the RL system learn trough a large number of episodes, learning was truncated at 1000 episodes. This was done because the future aim of this RL system is to be applied on stroke rehabilitation, so the RL should learn in a limited number of episodes. Although 1000 grasps were still too many to be performed in a rehabilitation session, this number could show us if it was able to learn in a reasonable amount of episodes, which might be improved in later tests.

For analysis of results obtained in this study, the error values were calculated as follows:

$$err_{POI} = \frac{2}{n}\sum_{i=1}^{n/2} err_{open}(i) + \frac{2}{n}\sum_{i=n/2}^{n} err_{close}(i) \tag{4}$$

where err_{open} and err_{close} are the errors previously described for each approach, i is the sample number and n is the total number of samples.

4 Results

In this section we summarize the preliminary results obtained when applying the different ϵ values and both reward strategies to the described RL application. Both approaches and the different ϵ values presented a similar behavior over the episodes, as it can be seen in Fig. 4 for the POI approach 1.

Moreover, in this figure we can see a descending trend over episodes for all ϵ values, where both approaches showed similar error values. Still, an example of the outputs and actions from the last episode of one of the iterations with POI approach 1 are shown in Fig. 5, where it shows that the system behavior is still far from desired. On the top figure the dotted lines represent the open-hand (first

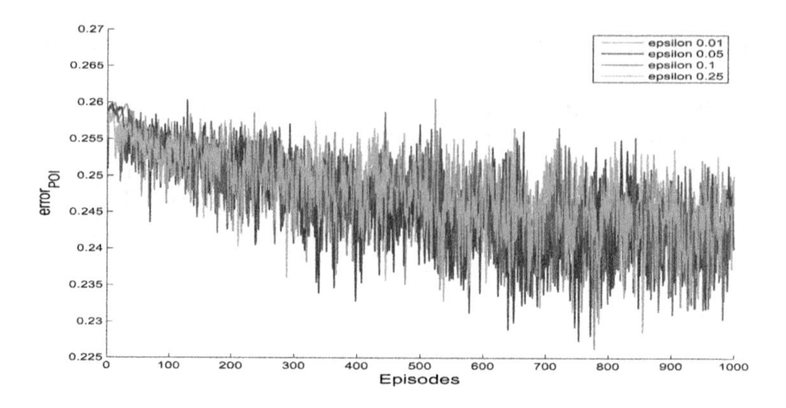

Fig. 4. POI approach 1 - error over episodes for different ϵ values.

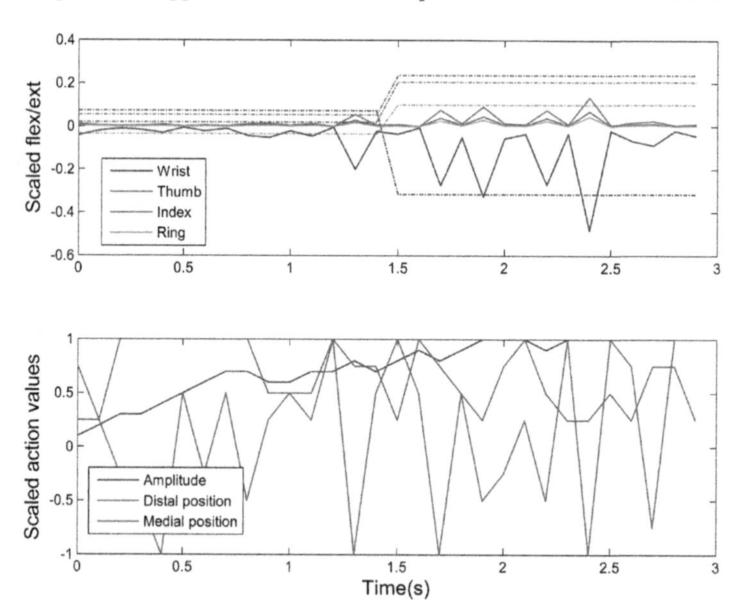

Fig. 5. POI approach 1 - example of resulted outputs and taken actions with $\epsilon = 0.01$, where dotted lines represent the open-hand and closed-hand POI references.

half) and closed-hand (second half) POI references, whereas the straight lines represent outputs of the RL system. On the bottom figure, the actions taken by the RL system, which led to the RL outputs above, are shown. It could be an over-exploratory effect caused by the negative reward (punishment) approach and the large action-state space, or because there was no electrode field and amplitude combination for achieving the target POIs on this specific subject (model). In any case, a further analysis would be needed to find its cause.

5 Conclusions

In the present work we checked the feasibility of using a RL system for its use in hand neuroprosthesis applications, specifically for performing a cylindrical grasp. The SARSA method with eligibility traces and tile coding function approximation was used in these preliminary tests, where ϵ-greedy policy with different ϵ values and different reward strategies were tested. Qualitative analysis showed that RL systems might be able to reduce error values along episodes. However, the system was not able to achieve the desired targets properly with any of the strategies. It showed random-like action selection behavior even in the last episodes, which suggested that the system was still exploring the action-state space. This hypothesis was reinforced by the fact that the ϵ values were not affecting the performance of the system, because due to the negative rewards approach, it was already in an exploration mode, selecting unexplored actions randomly. This effect could be caused by different reasons and combinations of them. One of the reasons could be that the dimensionality of the presented problem was too high for the selected SARSA RL system, which was shown successful on simpler examples [19]. Another reason could be that the negative reward (punishment) approach was forcing the system to explore until all the action-state space was visited, which in this case was extremely large due to the continuous character of the states. Finally, it could also be that there was no electrode field or amplitude or combinations of them on the presented surface FES system that could mimic the reference curves on this specific subject (model). A further analysis should be carried out to draw clear conclusions and find new solutions that could reach the optimal behavior in a reduced amount of episodes. In any case, the tested approaches showed a reducing error trend over episodes, which led us to think that it may be feasible to apply RL systems for FES-induced grasps in the near future. For this reason, other RL systems and methods should be tested to confirm the feasibility of successfully using RL systems for FES-induced grasp applications. As an example, future work could imply simplifying the state-action space, and the problem statement in general, or, on the contrary, designing more sophisticated RL systems in combination with other CI techniques.

Acknowledgements. Authors would like to thank to Intelligent Control Research Group of UPV/EHU for giving the means of carrying out this work and to the Health Division of Tecnalia for its continuous support.

References

1. Jones, L.A., Lederman, S.J.: Human Hand Function. Oxford University Press, Oxford (2006)
2. Schieber, M.H., Santello, M.: Hand function: peripheral and central constraints on performance. J. Appl. Physiol. **96**, 2293–2300 (2004)
3. Ingram, J.N., Kording, K.P., Howard, I.S., Wolpert, D.M.: The statistics of natural hand movements. Exp. Brain Res. **188**, 223–236 (2008)

4. Castiello, U.: The neuroscience of grasping. Nat. Rev. Neurosci. **6**, 726–736 (2005)
5. Roh, J., Rymer, W.Z., Perreault, E.J., Yoo, S.B., Beer, R.F.: Alterations in upper limb muscle synergy structure in chronic stroke survivors. J. Neurophysiol. **109**, 768–781 (2013)
6. Nichols-Larsen, D.S., Clark, P.C., Zeringue, A., Greenspan, A., Blanton, S.: Factors influencing stroke survivors? quality of life during subacute recovery. Stroke **36**, 1480–1484 (2005)
7. Peckham, P.H., Knutson, J.S.: Functional electrical stimulation for neuromuscular applications. Annu. Rev. Biomed. Eng. **7**, 327–360 (2005)
8. Popović, D.B.: Advances in functional electrical stimulation (FES). J. Electromyogr. Kinesiol. **24**, 795–802 (2014)
9. Keller, T., Kuhn, A.: Electrodes for transcutaneous (surface) electrical stimulation. J. Autom. Control. **18**, 35–45 (2008)
10. Westerveld, A.J., Schouten, A.C., Veltink, P.H., van der Kooij, H.: Selectivity and resolution of surface electrical stimulation for grasp and release. IEEE Trans. Neural Syst. Rehabil. Eng. **20**, 94–101 (2012)
11. Rau, G., Disselhorst-Klug, C., Schmidt, R.: Movement biomechanics goes upwards: from the leg to the arm. J. Biomech. **33**, 1207–1216 (2000)
12. Jagodnik, K.M., Blana, D., van den Bogert, A.J., Kirsch, R.F.: An optimized proportional-derivative controller for the human upper extremity with gravity. J. Biomech. **48**, 3692–3700 (2015)
13. Ethier, C., Oby, E.R., Bauman, M.J., Miller, L.E.: Restoration of grasp following paralysis through brain-controlled stimulation of muscles. Nature **485**, 368–371 (2012)
14. Popovic, M.B.: Control of neural prostheses for grasping and reaching. Med. Eng. Phys. **25**, 41–50 (2003)
15. Popovic, D.B., Popovic, M.B.: Automatic determination of the optimal shape of a surface electrode: selective stimulation. J. Neurosci. Meth. **178**, 174–181 (2009)
16. Freeman, C., Rogers, E., Burridge, J.H., Hughes, A., Meadmore, K.: Iterative Learning Control for Electrical Stimulation and Stroke Rehabilitation. Springer, Heidelberg (2015)
17. Wright, T.W., Glowczewskie, F., Cowin, D., Wheeler, D.L.: Radial nerve excursion and strain at the elbow and wrist associated with upper-extremity motion. J. Hand Surg. **30**, 990–996 (2005)
18. Gillen, G.: Stroke Rehabilitation: A Function-Based Approach. Elsevier Health Sciences, St. Louis (2015)
19. Sutton, R.S., Barto, A.G.: Reinforcement Learning: An Introduction. MIT press, Cambridge (1998)
20. Izawa, J., Kondo, T., Ito, K.: Biological arm motion through reinforcement learning. Biol. Cybern. **91**, 10–22 (2004)
21. Imatz-Ojanguren, E., Irigoyen, E., Valencia-Blanco, D., Keller, T.: Neuro-fuzzy models for hand movements induced by functional electrical stimulation in able-bodied and hemiplegic subjects. Med. Eng. Phys. (2016). (In Press)
22. Yozbatiran, N., Der-Yeghiaian, L., Cramer, S.C.: A standardized approach to performing the action research arm test. Neurorehab. Neural Repair **22**, 78–90 (2008)
23. Helwig, N.E., Hong, S., Hsiao-Wecksler, E.T., Polk, J.D.: Methods to temporally align gait cycle data. J. Biomech. **44**, 561–566 (2011)
24. Vergara, M., Sancho-Bru, J.L., Gracia-Ibanez, V., Perez-Gonzalez, A.: An introductory study of common grasps used by adults during performance of activities of daily living. J. Hand Ther. **27**, 225–234 (2014)

Fuzzy Candlesticks Forecasting Using Pattern Recognition for Stock Markets

Rodrigo Naranjo(✉) and Matilde Santos

Computer Science Faculty, University Complutense of Madrid,
28040 Madrid, Spain
rnaranjo.ina@gmail.com, msantos@ucm.es

Abstract. This paper presents a prediction system based on fuzzy modeling of Japanese candlesticks. The prediction is performed using the pattern recognition methodology and applying a lazy and nonparametric classification technique, k-Nearest Neighbours (k-NN). The Japanese candlestick chart summarizes the trading period of a commodity with only 4 parameters (open, high, low and close). The main idea of the decision system implemented in this article is to predict with accuracy, based on this vague information from previous sessions, the performance of future sessions. Therefore, investors could have valuable information about the next session and set their investment strategies.

Keywords: Trading · Fuzzy logic · K-NN · Forecasting · Candlesticks · Stock market

1 Introduction

In stock markets, different artificial intelligence techniques have been used to make a prediction on future stock prices. There are some articles that extract the information provided by the candlesticks, representing them by fuzzy logic [1] or by traditional crisp sets [2]. Other articles describe the stock market evolution using technical indicators commonly used by investors, such as MACD, RSI, … Moreover, artificial intelligence techniques, such as fuzzy logic, have been applied to define and work with these indexes, obtaining valuable information such as the possibilities of buying, selling or holding shares and the time and amount to enter into the market [3, 4].

Machine learning has been also applied to predict how the future market will evolve. Some works use neural networks [5, 6], perform different types of analysis for the prediction [7], or even make stock price forecasting applying SVM on web financial information sentiment analysis [8], among others. Other works apply different pattern recognition techniques to the classic candlesticks, such in [9], where k-NN is applied, or in [10], where authors use K-means clustering.

The aim of this article is to accurately predict future market behaviour from basic and vague information represented by fuzzy logic, which has been obtained from historical candles, and apply k-Nearest Neighbours (k-NN) to make the prediction for future stock market sessions.

In stock markets, candlesticks are used to summarize the price evolution of commodities or indexes in each session. For that, candlesticks are defined by 4 parameters:

M. Graña et al. (eds.), *International Joint Conference SOCO'16-CISIS'16-ICEUTE'16*,
Advances in Intelligent Systems and Computing 527, DOI 10.1007/978-3-319-47364-2_31

Open (value at the beginning of the trading session or period considered), High (maximum value reached by the commodity in the session), Low (minimum value reached during the session), and Close (value at the end of the trading session). The open value can be higher than the close one (or vice versa). Candlesticks can represent this fact in two ways. The first one is to fill the candlestick body with colour when the open value is higher than the close one, and the opposite when it is empty (Fig. 1a). The second method indicates the open values by a mark on the left and on the right side of the candlestick for close values (Fig. 1b).

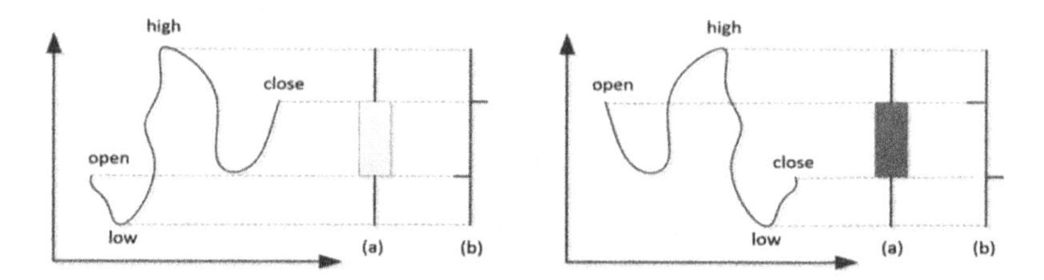

Fig. 1. The two representation of a session by Japanese candlestick theory

Using this representation of the candles, three parts can be distinguished (Fig. 2): the body of the candlestick, which represents the variation between the open and the close values of the trading session considered; the upper shadow, which is the price difference between the maximum value (high) and the open or close value (depending on whichever is higher); and the lower shadow, which is given by the price variation between the minimum price (low) and the close or open value (depending on whichever is lower).

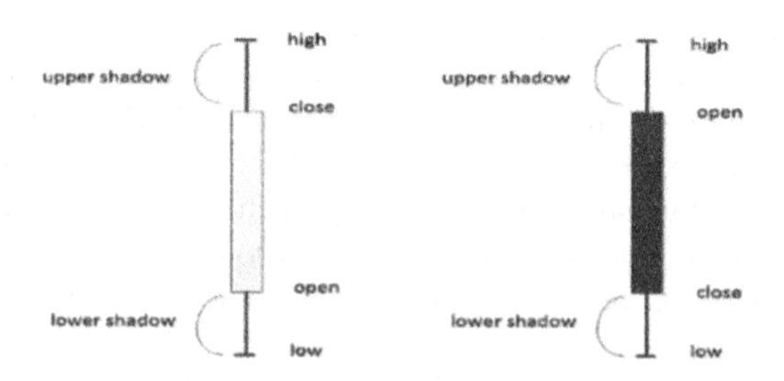

Fig. 2. Parameters and parts of the candle

In this work, a two-step methodology has been developed. First, the candlesticks of the last session and of all the previous available sessions go through a fuzzyfication process. Secondly, using the pattern recognition approach, the more similar fuzzy candlestick is retrieved from the historical database. In our case we have applied the nearest neighbour technique. Figure 3 shows the complete process developed in this work.

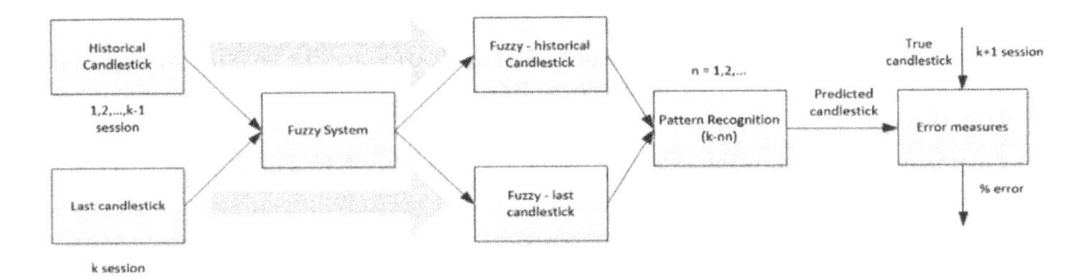

Fig. 3. Block diagram of the proposed system

2 Candlesticks Fuzzy Modeling

In this paper, a fuzzy modeling of the candlestick is proposed using the parameters that describe the candlestick itself. In this section, the procedure used to obtain and represent the inputs and outputs of the fuzzy system and the fuzzy rules are detailed.

2.1 Fuzzy Inputs

As it is said in the introduction, in the candlestick representation there are three parts that can be distinguished (upper shadow, body and lower shadow). As a result, the candlestick can be described by 4 parameters (open, high, low and close). As in [1], three variables (Lupper, Llower and Lbody) are considered as inputs of the fuzzy system, representing the shadows (upper and lower) and the body lengths. They allow to identify a candlestick pattern and to determine the efficiency of its identification. The formulas used to define these fuzzy inputs are:

$$L_{upper} = \frac{high - \max(open, close)}{open}, L_{lower} = \frac{\min(open, close) - low}{open} \tag{1}$$

$$\mathrm{L_{body}} = \frac{\max(\mathrm{open, close}) - \min(\mathrm{open, close})}{\mathrm{open}} \tag{2}$$

Unlike in [1], where there was a multiplicative factor of 100 to normalize the values of body and shadows between 0 and 14, because the price fluctuations of the Taiwanese stock market are limited to 14 percent, in this paper there is not any restriction, and therefore this approach can be applied to any stock markets. Therefore, the maximum and minimum values that can be obtained with the formulas (1) and (2) are not known. To solve this problem, a correction has been applied to the variables Lupper, Llower and Lbody to ensure the obtained values are in the range (0–100). This correction is also used in the field of image processing for stretching the histogram. It is expressed as:

$$g(x) = \frac{f(x) - f(x)_{min}}{f(x)_{max} - f(x)_{min}} \cdot (MAX - MIN) + MIN \tag{3}$$

Where $f(x)_{max}$ and $f(x)_{min}$ are the maximum and minimum values of the candlesticks used as historical data of the system, and *MAX* and *MIN* are the maximum and minimum of the range (in our case *MIN* = 0 and *MAX* = 100).

Thus, from now on, variables Lupper, Llower and Lbody include the correction (3). Four membership functions have been defined for each one: NULL, SHORT, MIDDLE and LONG (Fig. 4).

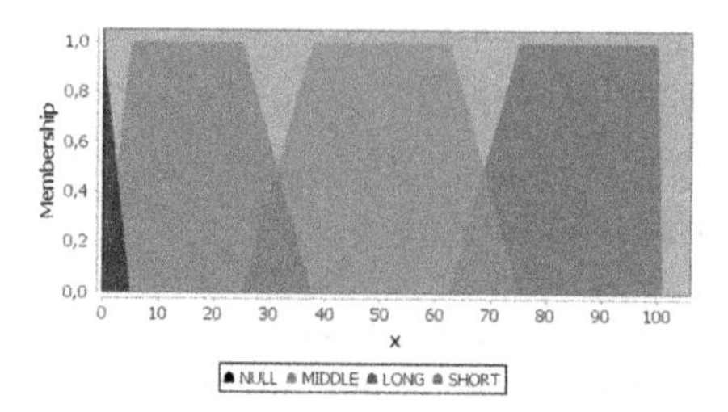

Fig. 4. Membership functions for Lupper, Llower and Lbody

The definitions of the membership functions NULL, SHORT, MIDDLE and LONG for Lupper, Llower and Lbody variables are as follows:

$$NULL(x : a, b) = \begin{cases} 1 & x<0 \\ \frac{b-x}{b-a} & 0 \leq x \leq 5 \\ 0 & x > 5 \end{cases} \tag{4}$$

$$SHORT(x : a, b, c, d) = \begin{cases} 0 & x<0 \\ (x-a)/(b-a) & 0 \leq x<5 \\ 1 & 5 \leq x<25 \\ (d-x)/(d-c) & 25 \leq x<37.5 \\ 0 & x \geq 37.5 \end{cases} \tag{5}$$

$$MIDDLE(x : a, b, c, d) = \begin{cases} 0 & x<25 \\ \frac{x-a}{b-a} & 25 \leq x<37.5 \\ 1 & 37.5 \leq x<62.5 \\ \frac{d-x}{d-c} & 62.5 \leq x<75 \\ 0 & x \geq 75 \end{cases} \tag{6}$$

$$LONG(x : a, b) = \begin{cases} 0 & x<62.5 \\ (x-a)/(b-a) & 62.5 \leq x<75 \\ 1 & x \geq 75 \end{cases} \tag{7}$$

2.2 Fuzzy Outputs

Two variables have been chosen as outputs of the fuzzy system, Rsize and Rpos. These variables refer to the existing relationship between the body size and the total size of

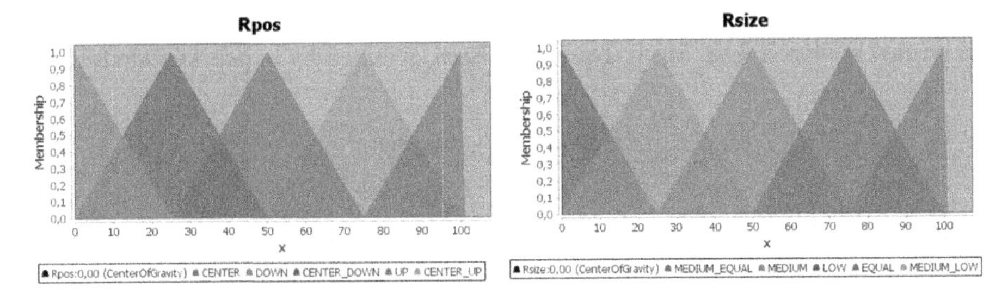

Fig. 5. Membership functions of Rpos (left) and Rsize (right)

the candle (Rsize), and to the position of the body within the candlestick (Rpos). Five membership functions have been assigned to Rsize (LOW, MEDIUM_LOW, MEDIUM, MEDIUM_EQUAL, EQUAL) and five to Rpos (DOWN, CENTER_DOWN, CENTER, CENTER_UP, UP) (Fig. 5).

The definitions of the membership functions of Rsize and Rpos are as follows:

$$DOWN, LOW(x:a,b) = \begin{cases} 1 & x<0 \\ (b-x)/(b-a) & 0 \le x \le 25 \\ 0 & x>25 \end{cases} \tag{8}$$

$$CENTER\_DOWN, MEDIUM\_LOW(x:a,b,c) = \begin{cases} 0 & x<0 \\ (x-a)/(b-a) & 0 \le x < 25 \\ (c-x)/(c-b) & 25 \le x < 50 \\ 0 & x \ge 50 \end{cases} \tag{9}$$

$$CENTER, MEDIUM(x:a,b,c) = \begin{cases} 0 & x<25 \\ (x-a)/(b-a) & 25 \le x < 50 \\ (c-x)/(c-b) & 50 \le x < 75 \\ 0 & x \ge 75 \end{cases} \tag{10}$$

$$CENTER\_UP, MEDIUM\_EQUAL(x:a,b,c) = \begin{cases} 0 & x<50 \\ (x-a)/(b-a) & 50 \le x < 75 \\ (c-x)/(c-b) & 75 \le x < 100 \\ 0 & x \ge 100 \end{cases} \tag{11}$$

$$UP, EQUAL(x:a,b) = \begin{cases} 0 & x<75 \\ (x-a)/(b-a) & 75 \le x \le 100 \\ 1 & x>100 \end{cases} \tag{12}$$

2.3 Fuzzy Rules

We have used a set of IF-THEN rules with a defuzzyfication method based on the centroid or centre of gravity. Because there are three fuzzy input variables and each one has four membership functions, there are 64 rules for each fuzzy output variable. The fuzzy rules are listed in Table 1. They have been obtained by expert knowledge.

Table 1. Fuzzy rules set

	Lupper	Llower	Lbody	Rsize	Rpos		Lupper	Llower	Lbody	Rsize	Rpos
1	LONG	LONG	LONG	MEDIUM_EQUAL	CENTER	33	SHORT	LONG	LONG	MEDIUM_EQUAL	UP
2	LONG	LONG	MIDDLE	MEDIUM	CENTER	34	SHORT	LONG	MIDDLE	MEDIUM	UP
3	LONG	LONG	SHORT	MEDIUM_LOW	CENTER	35	SHORT	LONG	SHORT	MEDIUM_LOW	UP
4	LONG	LONG	NULL	LOW	CENTER	36	SHORT	LONG	NULL	LOW	UP
5	LONG	MIDDLE	LONG	MEDIUM	CENTER_DOWN	37	SHORT	MIDDLE	LONG	MEDIUM_EQUAL	CENTER_UP
6	LONG	MIDDLE	MIDDLE	MEDIUM_LOW	CENTER_DOWN	38	SHORT	MIDDLE	MIDDLE	MEDIUM	CENTER_UP
7	LONG	MIDDLE	SHORT	LOW	CENTER_DOWN	39	SHORT	MIDDLE	SHORT	MEDIUM_LOW	CENTER_UP
8	LONG	MIDDLE	NULL	LOW	CENTER_DOWN	40	SHORT	MIDDLE	NULL	LOW	CENTER_UP
9	LONG	SHORT	LONG	MEDIUM_EQUAL	DOWN	41	SHORT	SHORT	LONG	EQUAL	CENTER
10	LONG	SHORT	MIDDLE	MEDIUM	DOWN	42	SHORT	SHORT	MIDDLE	MEDIUM_EQUAL	CENTER
11	LONG	SHORT	SHORT	MEDIUM_LOW	DOWN	43	SHORT	SHORT	SHORT	MEDIUM	CENTER
12	LONG	SHORT	NULL	LOW	DOWN	44	SHORT	SHORT	NULL	LOW	CENTER
13	LONG	NULL	LONG	MEDIUM_EQUAL	DOWN	45	SHORT	NULL	LONG	EQUAL	CENTER_DOWN
14	LONG	NULL	MIDDLE	MEDIUM	DOWN	46	SHORT	NULL	MIDDLE	EQUAL	CENTER_DOWN
15	LONG	NULL	SHORT	MEDIUM_LOW	DOWN	47	SHORT	NULL	SHORT	MEDIUM_EQUAL	CENTER_DOWN
16	LONG	NULL	NULL	LOW	DOWN	48	SHORT	NULL	NULL	LOW	CENTER_DOWN
17	MIDDLE	LONG	LONG	MEDIUM	CENTER_UP	49	NULL	LONG	LONG	MEDIUM_EQUAL	UP
18	MIDDLE	LONG	MIDDLE	MEDIUM_LOW	CENTER_UP	50	NULL	LONG	MIDDLE	MEDIUM	UP
19	MIDDLE	LONG	SHORT	LOW	CENTER_UP	51	NULL	LONG	SHORT	MEDIUM_LOW	UP
20	MIDDLE	LONG	NULL	LOW	CENTER_UP	52	NULL	LONG	NULL	LOW	UP
21	MIDDLE	MIDDLE	LONG	MEDIUM_EQUAL	CENTER	53	NULL	MIDDLE	LONG	EQUAL	UP
22	MIDDLE	MIDDLE	MIDDLE	MEDIUM	CENTER	54	NULL	MIDDLE	MIDDLE	MEDIUM_EQUAL	UP
23	MIDDLE	MIDDLE	SHORT	MEDIUM_LOW	CENTER	55	NULL	MIDDLE	SHORT	MEDIUM_LOW	UP
24	MIDDLE	MIDDLE	NULL	LOW	CENTER	56	NULL	MIDDLE	NULL	LOW	UP
25	MIDDLE	SHORT	LONG	MEDIUM_EQUAL	CENTER_DOWN	57	NULL	SHORT	LONG	EQUAL	CENTER_UP
26	MIDDLE	SHORT	MIDDLE	MEDIUM	CENTER_DOWN	58	NULL	SHORT	MIDDLE	EQUAL	CENTER_UP
27	MIDDLE	SHORT	SHORT	MEDIUM_LOW	CENTER_DOWN	59	NULL	SHORT	SHORT	MEDIUM_EQUAL	CENTER_UP
28	MIDDLE	SHORT	NULL	LOW	CENTER_DOWN	60	NULL	SHORT	NULL	MEDIUM	CENTER_UP
29	MIDDLE	NULL	LONG	EQUAL	DOWN	61	NULL	NULL	LONG	EQUAL	CENTER
30	MIDDLE	NULL	MIDDLE	MEDIUM_EQUAL	DOWN	62	NULL	NULL	MIDDLE	EQUAL	CENTER
31	MIDDLE	NULL	SHORT	MEDIUM_LOW	DOWN	63	NULL	NULL	SHORT	MEDIUM_EQUAL	CENTER
32	MIDDLE	NULL	NULL	LOW	DOWN	64	NULL	NULL	NULL	MEDIUM	CENTER

3 Forecasting System

Once the defuzzyfication results of the two fuzzy output variables (Rsize and Rpos), are obtained, they are used as inputs of the prediction system. To implement the forecasting system a nonparametric and lazy classification method, the k-Nearest Neighbours (k-NN), has been chosen. The forecasting system basically consists of comparing the candle of the last available session, characterized by the output variables Rsize and Rpos, with a group of n candles, made up with the last known candlestick and the n-1 previous candlesticks. The most similar candlestick to the current one is retrieved. Indeed, the classification system implemented finds the nearest neighbour, i.e., we have implemented a 1-NN system ($k = 1$).

To determine the similarity between neighbours, the Euclidean distance is applied:

$$D_k(\{C_{1\ldots n}\},\{C_{1k\ldots nk}\} \\ = \sqrt{(R_{pos1}-R_{pos1k})^2+(R_{size1}-R_{size1k})^2+\ldots+(R_{posn}-R_{posnk})^2+(R_{sizen}-R_{sizenk})^2} \quad (13)$$

Where the candlestick group used as pattern is made up of candlesticks $\{C_1, C_2, \ldots, C_n\}$ and the neighbours to compare with are the candlesticks $\{C_{1k}, C_{2k}, \ldots, C_{nk}\}$. This operation is repeated m times, where m is determined by:

$$m = n^{\circ} lastcandles - n + 1 \tag{14}$$

The previous sessions candlesticks are in the historical database of the system. Therefore, the neighbour with smallest distance (lowest error) will be chosen as the closest one. Therefore the next candlestick, that is, the one of the next session, is the one found as the most similar.

4 Experimental Set up and Result Discussion

To test the implemented forecasting system we have chosen the Nasdaq-100 stock market. Specifically, we have used the first 15 commodities. The selected time periods are shown in Table 2:

Table 2. Input and test time periods

	Input data		Test data	
	Start	End	Start	End
Period	1-Jan-11	31-Dec-11	1-Jan-12	31-Dec-12

The system has been run with different number of neighbours candlesticks, specifically n has varied from 1 to 10. To measure the effectiveness of the forecasting system, the error percentage of the predicted parameters of candle and the real parameters of the present candlestick are calculated. That is, the error percentage for the values of open, high, low and close parameters is calculated using:

$$\%error = \frac{|P_{TC} - P_{PC}|}{P_{TC}} \cdot 100 \tag{15}$$

Where P_{TC} represents the value of the parameter (open, high, low, close) of the true candlestick and P_{PC} is the same parameter of the predicted one. These error percentages have been calculated using two types of candlesticks. The first kind of candlesticks (indicated by subscript 1) refers to those obtained directly as a result of the application of the forecasting system. The second type of candlesticks (indicated by subscript 2) are those which have been obtained matching the open value of the estimated and the present candlestick and, consequently, the candlestick (result of the prediction) has been shifted that difference. That is, prediction candlestick values of the second type are:

$$open_{PC2} = open_{PT} \tag{16}$$

$$high_{PC2} = high_{PC1} + (open_{PT} - open_{PC2}) \tag{17}$$

$$low_{PC2} = low_{PC1} + (open_{PT} - open_{PC2}) \tag{18}$$

$$close_{PC2} = close_{PC1} + (open_{PT} - open_{PC2}) \tag{19}$$

Table 3 shows an example of the results obtained for the AAPL Nasdaq-100 commodity for n from 0 to 10. Average error percentages during the period studied, x, and standard deviation with the Bessel correction, s, have been calculated, being nc the total number of candlesticks of the commodity for the period considered.

Table 3. Results of the commodity AAPL for different values of n

Commodity	n	x(o1)	s(o1)	x(h1)	s(h1)	x(l1)	s(l1)	x(c1)	s(%c1)	x(h2)	s(h2)	x(l2)	s(l2)	x(c2)	s(c2)
AAPL	1	31,93	14,23	31,98	14,19	31,98	14,14	32,05	14,07	0,83	0,93	0,88	0,88	1,43	1,39
	2	28,92	14,30	28,89	14,42	28,85	14,24	28,86	14,36	0,78	0,87	0,99	0,86	1,48	1,23
	3	29,56	13,64	29,40	13,80	29,42	13,64	29,35	13,74	0,74	0,78	1,00	0,89	1,39	1,26
	4	28,53	13,17	28,41	13,32	28,46	13,15	28,34	13,27	0,77	0,77	0,97	0,90	1,38	1,16
	5	30,01	12,64	29,95	12,67	29,87	12,55	29,87	12,67	0,75	0,79	0,95	0,88	1,35	1,16
	6	30,16	12,58	30,08	12,63	30,02	12,57	29,98	12,65	0,79	0,79	0,97	0,91	1,39	1,15
	7	31,13	12,70	30,99	12,86	30,93	12,75	30,78	12,97	0,75	0,76	0,98	0,92	1,35	1,15
	8	30,81	12,67	30,61	12,91	30,61	12,66	30,46	12,92	0,76	0,78	1,00	0,93	1,41	1,10
	9	30,56	12,49	30,46	12,59	30,36	12,43	30,34	12,51	0,78	0,81	0,95	0,96	1,36	1,14
	10	29,77	13,58	29,72	13,61	29,66	13,42	29,69	13,44	0,74	0,79	0,96	0,93	1,34	1,11

$$\bar{x} = \frac{1}{nc} \cdot \sum\nolimits_{1}^{nc} \%\,error_i \qquad s = \sqrt{\frac{1}{nc-1} \cdot \sum\nolimits_{1}^{nc} (\%\,error_i - \bar{x})^2} \tag{20}$$

Where o1, h1, l1, c1 refer to the open, high, low and close percentage (respectively) of candlestick type 1, and o2, h2, l2, c2 to the same values of candlestick type 2.

Table 4 shows the maximum and minimum values of the average and standard deviation obtained for the commodities for the best and worst value of n.

Some interesting conclusions can be drawn from the analysis carried out in this work. First, as it is well known, the bullish stochastic trend in the market produces, among others effects, a significant depreciation of the currency. In our system, despite it uses as input and output variables the size ratio and positions of the candlestick, this effect can be observed in the error percentage obtained for open values (o1) and, consequently, in the obtained errors for the rest of parameters (h1, l1 and c1). This will allow us to predict this undesirable effect on the market.

Another conclusion is that, even if the estimated error is big, it is not related to the shape of the predicted candle in comparison to the true candle. Therefore, the obtained error can be considered as an offset. To show this effect, the second group of parameters (h2, l2 and c2) were shifted an offset given by the open value (o1), that is, the open values have been matched (o1 = o2). As can be observed in Table 4, the errors are now considerably lower than in the first case.

Currently there are many forecasting techniques for stock markets, such as those based on the segmentation of time series. One of the disadvantages of these techniques is the high demanding pre-processing they required, due to the high volume of input

Table 4. Minimum and maximum errors

Commodity		x(o1)	s(o1)	x(h1)	s(h1)	x(l1)	s(l1)	x(c1)	s(%c1)	x(h2)	s(h2)	x(l2)	s(l2)	x(c2)	s(c2)
AAPL	MIN	28,53	12,49	28,41	12,59	28,46	12,43	28,34	12,51	0,74	0,76	0,88	0,86	1,34	1,10
	MAX	31,93	14,30	31,98	14,42	31,98	14,24	32,05	14,36	0,83	0,93	1,00	0,96	1,48	1,39
ADBE	MIN	6,42	5,07	6,37	5,19	6,49	5,20	6,46	5,22	0,81	0,67	0,84	0,68	1,35	1,00
	MAX	8,19	7,69	8,27	7,67	8,35	7,91	8,34	7,81	1,08	0,81	1,01	0,86	1,77	1,31
ADI	MIN	4,55	3,41	4,51	3,41	4,68	3,57	4,63	3,57	0,80	0,69	0,75	0,66	1,21	0,94
	MAX	5,64	4,77	5,46	4,65	5,72	4,94	5,56	4,82	0,89	0,74	0,87	0,73	1,39	1,06
ADP	MIN	6,21	3,99	6,08	3,91	6,31	4,02	6,17	3,92	0,49	0,39	0,46	0,43	0,82	0,64
	MAX	7,98	5,87	7,74	5,65	7,99	5,80	7,77	5,57	0,57	0,51	0,54	0,54	0,88	0,73
ADSK	MIN	14,96	11,01	14,73	10,77	15,09	11,06	14,85	10,90	1,21	1,14	1,31	1,20	1,94	1,62
	MAX	16,72	11,79	16,52	11,56	16,64	12,10	16,54	11,83	1,39	1,25	1,49	1,70	2,25	1,92
AKAM	MIN	14,75	11,80	14,60	11,55	14,83	11,92	14,65	11,69	1,07	0,92	1,19	1,06	1,80	1,43
	MAX	23,18	16,83	24,03	17,74	23,59	17,09	24,36	17,91	2,08	2,06	1,40	1,43	2,57	2,21
ALTR	MIN	13,77	11,63	13,86	11,68	13,77	11,58	13,96	11,74	1,11	1,04	1,17	1,02	1,96	1,50
	MAX	20,29	14,43	20,13	14,32	20,18	14,27	20,10	14,17	1,34	1,30	1,36	1,17	2,09	1,74
ALXN	MIN	16,88	13,57	16,64	13,45	16,91	13,61	16,68	13,44	1,00	0,89	1,09	1,09	1,65	1,40
	MAX	21,83	15,95	21,61	15,87	21,85	16,02	21,54	16,02	1,20	1,15	1,43	1,27	1,86	1,62
AMAT	MIN	14,91	12,48	14,91	12,50	14,82	12,40	14,95	12,51	0,98	0,86	0,94	0,85	1,60	1,27
	MAX	18,00	13,98	17,90	14,10	18,16	14,05	17,97	14,10	1,10	1,06	1,17	1,25	1,82	1,55
AMGN	MIN	20,04	9,94	20,03	9,86	20,10	9,87	20,11	9,85	0,65	0,56	0,57	0,57	0,99	0,82
	MAX	21,30	11,29	21,24	11,30	21,34	11,34	21,35	11,38	0,69	0,62	0,63	0,65	1,11	0,88
AMZN	MIN	12,65	8,58	12,55	8,56	12,73	8,53	12,55	8,57	0,90	0,84	0,85	0,73	1,54	1,23
	MAX	14,25	9,62	14,20	9,43	14,37	9,64	14,24	9,48	1,06	0,99	1,04	0,89	1,72	1,45
ATVI	MIN	5,27	3,95	5,34	3,85	5,29	3,96	5,38	3,78	0,82	0,69	0,95	0,81	1,39	1,08
	MAX	5,69	4,37	5,63	4,37	5,75	4,47	5,73	4,48	0,93	0,88	1,08	0,95	1,53	1,25
AVGO	MIN	7,92	6,00	7,76	5,97	8,15	6,21	7,89	6,09	1,08	0,96	1,17	1,04	1,90	1,49
	MAX	11,54	7,38	11,56	7,57	11,94	7,59	11,53	7,38	1,31	1,21	1,44	1,35	2,09	1,71
BBBY	MIN	12,58	8,18	12,57	8,19	12,66	8,20	12,65	8,13	0,82	0,77	0,85	0,78	1,31	1,10
	MAX	16,08	9,45	16,11	9,49	16,11	9,45	16,14	9,46	0,92	0,87	0,99	0,96	1,50	1,24
BIDU	MIN	14,36	11,72	14,36	11,68	14,19	11,64	14,34	11,60	1,30	1,12	1,65	1,57	2,21	1,79
	MAX	15,98	14,16	16,04	14,20	15,89	14,18	16,01	14,21	1,49	1,37	1,87	1,94	2,50	2,04

data and the difficulty to rightly choose the segments, in order to extract the most representative and important points of the time series. In this paper, the pre-processing is not necessary because we have started with 4 values (parameters of the candle) per session. Another difference with other works is that the segmentation techniques they use are focused on the detection input/output market signals [11], or the detection of specific patterns [12–14], therefore the number of predictions is limited to the number of found patterns. For example in [12], the number of found patterns in the Hang Seng Index market from 1st January 2003 to 31st December 2012 (2506 sessions) varies between 36 and 81, depending on the window size. In this article, each session calculates a prediction of the next session and the mean errors for each parameter of the candle, which can be very useful for establishing input/output market rules by the investors.

5 Conclusions and Future Works

In this article, fuzzy modeling of candlestick is used to design a forecasting system. A pattern recognition technique, in our case k-NN, has been applied to retrieval the estimated future value (given by the candlestick parameter values) of the next session. The predictions obtained for the stock market can be very useful for investors.

Indeed, the system proposed in this paper can be a very valuable tool to increase the efficiency of the decision making process in stock markets even when the investors have already sent their orders once the session has started. In fact they can use it to obtain a framework in which to set boundaries and limits to the session in which they are investing and, therefore, to improve their strategies. Furthermore, it can be used as a starting point for other stock market modeling using fuzzy pattern recognition.

Future works include the application of different fuzzy measurement of distances, and the comparison with the results obtained by the Euclidean one. In this way we will be able to notice the advantages and disadvantages of applying fuzzy logic to the different steps of the forecasting system [15].

References

1. Lee, C.H.L., Liu, A., Chen, W.S.: Pattern discovery of fuzzy time series for financial prediction. IEEE Trans. Knowl. Data Eng. **18**(5), 613–625 (2006)
2. Arroyo, J.: Forecasting candlesticks time series with locally weighted learning methods. In: Locarek-Junge, H., Weihs, C. (eds.) Classification as a Tool for Research, pp. 603–611. Springer, Heidelberg (2010)
3. Naranjo, R., Meco, A., Arroyo, J., Santos, M.: An intelligent trading system with fuzzy rules and fuzzy capital management. Int. J. Intell. Syst. **30**, 963–983 (2015)
4. Ijegwa, A.D., Rebecca, V.O., Olusegun, F., Isaac, O.O.: A predictive stock market technical analysis using fuzzy logic. Comput. Inf. Sci. **7**(3), 1 (2014)
5. Ravichandra, T., Thingom, C.: Stock price forecasting using ANN method. In: Satapathy, S.C., Mandal, J.K., Udgata, S.K., Bhateja, V. (eds.) Information Systems Design and Intelligent Applications. AISC, vol. 435, pp. 599–605. Springer, Heidelberg (2016). doi:10.1007/978-81-322-2757-1_59
6. Wang, J., Wang, J.: Forecasting stock market indexes using principle component analysis and stochastic time effective neural networks. Neurocomputing **156**, 68–78 (2015)
7. Chen, Y.J., Chen, Y.M., Lu, C.L.: Enhancement of stock market forecasting using an improved fundamental analysis-based approach. Soft Comput. 1–23 (2016)
8. Cao, R., Liang, X., Ni, Z.: Stock price forecasting with support vector machines based on web financial information sentiment analysis. In: Zhou, S., Zhang, S., Karypis, G. (eds.) ADMA 2012. LNCS, vol. 7713, pp. 527–538. Springer, Heidelberg (2012). doi:10.1007/978-3-642-35527-1_44
9. Chmielewski, L., Janowicz, M., Kaleta, J., Orłowski, A.: Pattern recognition in the Japanese candlesticks. In: Wiliński, A., El Fray, I., Pejaś, J. (eds.) Soft Computing in Computer and Information Science. AISC, vol. 7713, pp. 227–234. Springer, Heidelberg (2015). doi:10.1007/978-3-319-15147-2_19
10. Chmielewski, L.J., Janowicz, M., Orłowski, A.: Prediction of trend reversals in stock market by classification of Japanese candlesticks. In: Burduk, R., Jackowski, K., Kurzyński, M., Woźniak, M., Żołnierek, A. (eds.) Proceedings of the 9th International Conference on Computer Recognition Systems CORES 2015. AISC, vol. 403, pp. 641–647. Springer, Heidelberg (2015). doi:10.1007/978-3-319-26227-7_60
11. Yin, J., Si, Y. W., Gong, Z.: Financial time series segmentation based on Turning Points. In: Proceedings of the 2011 International Conference on System Science and Engineering, pp. 394–399. IEEE (2011)

12. Wan, Y., Gong, X., Si, Y.W.: Effect of segmentation on financial time series pattern matching. Appl. Soft Comput. **38**, 346–359 (2016)
13. Banavas, G.N., Denham, S., Denham, M.J.: Fast nonlinear deterministic forecasting of segmented stock indices using pattern matching and embedding techniques. In: Computing in Economics and Finance, p. 64 (2000)
14. Si, Y.W., Yin, J.: OBST-based segmentation approach to financial time series. Eng. Appl. Artif. Intell. **26**(10), 2581–2596 (2013)
15. López, V., Santos, M., Montero, J.: Fuzzy specification in real estate market decision making. Int. J. Comput. Intell. Syst. **3**(1), 8–20 (2010)

Analysing Concentrating Photovoltaics Technology Through the Use of Emerging Pattern Mining

A.M. García-Vico[1], J. Montes[2], J. Aguilera[2], C.J. Carmona[3(✉)], and M.J. del Jesus[1]

[1] Department of Computer Science, University of Jaén, 23071 Jaén, Spain
{agvico,mjjesus}@ujaen.es
[2] Department of Electronics and Automatization Engineering, University of Jaén, 23071 Jaén, Spain
aguilera@ujaen.es
[3] Department of Civil Engineering, University of Burgos, 09006 Burgos, Spain
cjcarmona@ubu.es

Abstract. The search of emerging patterns pursues the description of a problem through the obtaining of trends in the time, or characterisation of differences between classes or group of variables. This contribution presents an application to a real-world problem related to the photovoltaic technology through the algorithm EvAEP. Specifically, the algorithm is an evolutionary fuzzy system for emerging pattern mining applied to a problem of concentrating photovoltaic technology which is focused on the generation of electricity reducing the associated costs. Emerging patterns have discovered relevant information for the experts when the maximum power is reached for the cells of concentrating photovoltaic.

Keywords: Emerging pattern mining · Concentrating photovoltaics · Evolutionary fuzzy system · Supervised descriptive rule discovery

1 Introduction

In data mining process there are two inductions clearly differentiated, predictive and descriptive induction. However, the last years there has been a great interest in the community about supervised descriptive rule discovery [18]. Latter includes a group of techniques for describing a problem through supervised learning such as subgroup discovery [4,14] or emerging pattern mining (EPM) [9], amongst others.

This contribution presents the application of the EPM technique to a real-world problem. The main objective of this data mining technique is to search for patterns with the ability to find large differences between datasets or classes. This property has led to the use of EPM in predictive induction with good results but not in descriptive induction although they were defined for this purpose. Specifically, the EPM algorithm employed in this contribution is the EvAEP algorithm

M. Graña et al. (eds.), *International Joint Conference SOCO'16-CISIS'16-ICEUTE'16*, Advances in Intelligent Systems and Computing 527, DOI 10.1007/978-3-319-47364-2_32

which is an evolutionary fuzzy system (EFS) [13]. These systems are based on evolutionary algorithms [11] which offer advantages in knowledge extraction and in rule induction process. In addition, they use fuzzy logic [23] with the use of fuzzy sets with linguistic labels in order to represent the knowledge allowing to obtain representation of the information very close to the human reasoning [16].

The algorithm is applied to a Concentrating Photovoltaic (CPV) problem which is an alternative to the conventional Photovoltaic for the electric generation. CPV technology is based on using concentrated sunlight to produce electricity in a cheaper way by means of high efficiency multi-junction solar cells, specifically designed for this type of technology. The efficiency of this type of solar cells has experienced a fast evolution in the last decade and it has a very strong potential of increasing along next years. Despite of these expectations, several obstacles to develop CPV technology currently still remain, as the lack of CPV normalisation and standardisation, the lack of knowledge of the influence of the meteorological parameters on the performance of high efficiency multijunction solar cells. Therefore it is necessary to deepen in the study and knowledge of CPV technology. Results obtained in this contribution are very promising when maximum power is obtained by cells analysed.

The paper is organised as follows: Sect. 2 describes the background of the contribution with the presentation of EPM and CPV. Next, Sect. 3 shows the main properties and features of the algorithm EvAEP that is the first EFS for extracting EPs throughout the literature. Section 4 outlines the experimental framework, shows the results obtained, and an analysis about these results. Finally, some concluding remarks are outlined.

2 Background

2.1 Emerging Pattern Mining

The EPM was defined in 1999 [8] as itemsets whose support increase significantly from one dataset to another in order to discover trends in data. In this way, an itemset is considered as emerging when the growth rate (GR) is upper than one, i.e.:

$$GR(x) = \begin{cases} 0, & IF\ Supp_{D_1}(x) = Supp_{D_2}(x) = 0, \\ \infty, & IF\ Supp_{D_2}(x) = 0\ \wedge\ Supp_{D_1}(x) \neq 0, \\ \frac{Supp_{D_1}(x)}{Supp_{D_2}(x)}, & another\ case \end{cases} \tag{1}$$

where $Supp_{D_1}(x)$ is the support for the pattern x in the first dataset and $Supp_{D_2}(x)$ is the support with respect to the second dataset, i.e. $Supp_{D_1}(x) = \frac{count_{D_1}(x)}{|D_1|}$ and $Supp_{D_2}(x) = \frac{count_{D_2}(x)}{|D_2|}$. This concept could be generalised for one dataset with different classes, for example, $D_1 \equiv Class$ and $D_2 \equiv \overline{Class}$.

The most representative algorithm is DeEPs [19] which is based on the borders concept. A border is a pair of minimal and maximal patterns $< L, R >$ in order to represent all patterns within this border. As can be intuited the search space could become huge considering a complex problem. Therefore, authors have used different heuristics and techniques in order to reduce the search space and obtaining better results.

To facilitate to the experts and the community the analysis of the knowledge extracted, this is represented through the use of rules (R) with the following representation:

$$R : Cond \rightarrow Class$$

where *Cond* is commonly a conjunction of attribute-value pairs as definitions mention, and *Class* is the analysed value for the class, i.e. the class with a high support in front of the remaining classes for the dataset.

2.2 CPV Technology

Photovoltaic technology has experienced a major boost because it is a method of generating electrical power by converting solar radiation into direct current electricity using solar panels composed of a number of solar cells containing a semiconductor material. A variant of this technology is the CPV which is based on using concentrated sunlight to produce electricity in a cheaper way by means of high efficiency multi-junction solar cells, specifically designed for this type of technology. The efficiency of this type of solar cells has experienced a fast evolution[1]. In addition, CPV technology needs to use solar trackers, allowing an important increment of the energy generated by the system with a lower cost. Despite of these expectations, several significant obstacles to the development of CPV technology currently still remain:

- The lack of CPV normalisation and standardisation.
- The complexity and variety of solar cells.
- The lack of knowledge of the influence of meteorological parameters on the performance of high efficiency multi-junction solar cells. In fact, in real projects the productivity of this type of technology has been below expectations.
- The lack of detailed experimental and operational data about real outdoor performance.
- The development of complex regression models for performance.

The most interesting parameter to analyse is the Maximum Module Power (P_m). For each kind of solar cell, the manufacturer measures the CPV module under certain atmospheric conditions (called Standard Test Conditions, STC) and provides the $P_{m,STC}$. Nevertheless, in a real operation these conditions are not satisfied and the performance of the CPV module can be very different from that indicated by the manufacturer. It is known that P_m is highly influenced by atmospheric conditions, but it is necessary to know what happens with the combination of real atmospheric conditions. This knowledge can be very useful for predicting energy production in a certain period of time.

[1] http://www.nrel.gov/ncpv/images/efficiency_chart.jpg.

The DNI is considered as the main atmospheric parameter which influences the outdoor electric performance of a CPV module. Using the DNI as the integration value along the whole wavelength range for a specific photovoltaic device is a common practice. Nevertheless, we can consider the DNI value for each wavelength value, obtaining the *solar spectrum distribution*. As has been widely demonstrated, the DNI, as well as its spectral distribution, have an important influence on the electric performance of multi-junction solar cells. It is well known that the multi-junction solar cells temperature affect to their electric performance. In this sense, the temperature has an almost negligible positive effect on the short circuit current delivered by the multi-junction solar cell, and a negative predominant effect on both the open circuit voltage and P_M [12,17]. The same behaviour is observed when analysing the impact of the temperature (T_A) on the electric performance of CPV modules equipped with multi-junction solar cells [21]. However, the own disposition of the multi-junction solar cells inside the CPV module makes it very difficult to measure their temperature. In this work, T_A is considered as influential factor, given a direct relation between cell temperature and T_A [1,2]. The consideration of the wind (W_S) as one of the influential factors whose contribution must be added to the study because it can perform a positive refrigerating effect on the electric performance of a CPV system, cooling the multi-junction solar cells which compose the module down, and obtaining therefore a better behaviour [7]. However, high W_S values can also exert a negative effect of misalignment on the tracker [20], displacing the multi-junction solar cells from their optimum arrangement in the solar beam direct trajectory. Finally, for this contribution have been considered the incident global irradiance (G) and the spectral irradiance distribution of G described through the average photon energy (APE) and the spectral machine radio (SMR). Both parameters intend to define the shape of the solar spectrum in an easy way. The use of one or the other depends of the monitorised parameter during the experimental campaign.

3 EvAEP: Evolutionary Algorithm for Extracting Emerging Patterns

This section describes the algorithm Evolutionary Algorithm for extracting Emerging Patterns (EvAEP) presented in [6]. This algorithm is able to extract emerging fuzzy patterns in order to describe a problem from supervised learning. The main objective of the EvAEP is the extraction of an undetermined number of rules to describe information with respect to an interest property for the experts. It is important to note that a target variable is able to have different values or classes, in this way the algorithm obtain patterns for all values of the target variable because it is executed once for each value.

The algorithm is an EFS [13] that is a well-known hybridisation between a fuzzy system [23] and a learning process based on evolutionary computation [10]. EvAEP employs an evolutionary algorithm with a codification "Chromosome = Rule" where only the antecedent part of the rule is represented, and the antecedent is composed by a conjunction of pairs variable-value. Figure 1 represents the phenotype and genotype for a chromosome = rule in the EvAEP algorithm. As can be observed, the value 0 represents the absence of a variable in the representation of a rule.

Genotype					*Phenotype*
x_1	x_2	x_3	x_4		
3	∅	1	∅	⇒	IF $(x_1 = 3)$ AND $(x_3 = 1)$ THEN $(x_{Obj} = ValorObjetivo)$

Fig. 1. Representation of a chromosome = rule for the EvAEP algorithm

On the other hand, if the variable has a continuous range, the algorithm employs a fuzzy representation with fuzzy sets composed by linguistic labels defined with uniform triangles forms.

The algorithm uses a mono-objective approach with an iterative rule learning (IRL) [22] which is executed once for each value of the target variable, i.e., for each class the best individuals are obtained in an iterative process. In this way, the algorithm iterates in order to obtain emerging patterns until a non-emerging pattern is obtained. Moreover, the algorithm stops if all instances for the class are covered for the patterns obtained previously or a pattern with null support is obtained.

The main operation scheme for the EvAEP algorithm is shown in Fig. 2.

The main elements of the algorithm are described in the following subsections.

BEGIN
Set of Emerging Patterns = ∅
repeat
  repeat
    Generate P(0)
    Evaluate P(0)
    repeat
      Include the best individual in P(nGen+1)
      Complete P(nGen+1): Cross and Mutation for individuals of P(nGen)
      Evaluate P(nGen+1)
      nGen ← nGen + 1
    until Number of evaluations is reached
    Obtain the best rule (R)
    Emerging Patterns ∪ R
    Marks examples covered by R
  until (GrowthRate(R) ≤ 1) OR (R has null support) OR (R not cover new examples)
until Class = ∅
return Emerging Patterns
END

Fig. 2. Operation scheme for EvAEP

3.1 Biased Initialisation

EvAEP generates an initial population (P_0) with a size determined through external parameter. The objective of this function is to create a part of the individuals with a maximum percentage of variables being part of the rule. Specifically, the algorithm creates a population with 50 % of the individuals generated in a random way completely, and the remaining individuals of the population must have at least one variable with a value and a maximum number of variables (80 %) with values.

This operator allows the obtaining of a first population with a wide generalisation in order to explore the major area in the search space. In the evolutionary process the main idea is to maximise precision of the rules.

3.2 Genetic Operators

The population of the next generation is generated through some genetic operators widely used throughout the literature. The algorithm employs an elitism size equal to one, in this way the best individual is saved directly in the population of the next generation. The best individual is measured through an aggregation function such as:

$$NSup(R) * 0.5 + Fitness(R) * 0.5 \tag{2}$$

where $NSup(R)$ is the number of examples covered for the rule R non-covered for the previous patterns obtained, divided by the number of remaining examples to cover for the class. The $Fitness$ is detailed in the following section. In case of tie, the individual with less number of variables is considered as the winner.

On the other hand, the algorithm employs the operator multi-cross point operator [15] and a biased mutation introduced in one algorithm of subgroup discovery [3].

3.3 Fitness Function

It is the key concept of the algorithm because the main objective is to search for emerging patterns with high values in confidence and precision, with the maximum generalisation possible, and finally, an interesting gain of accuracy for the community. The fitness employed by EvAEP is defined below:

$$Fitness(R) = \sqrt{TPr * TNr} \tag{3}$$

where a geometric average for an individual is used in order to maximise the precision in the class and non-class in a balance way. The first component (TPr) is known as *True Positive rate* or sensitivity and it measures the percentage of examples correctly classified for the class, and the second (TNr) is known as *True Negative rate* where the measurement of examples non-covered correctly is considered.

4 Experimental Study

The data were obtained from one model of CPV modules, whose main characteristics are solar cells type $Multijunction - GaInP/Ga(In)As/Ge$ with 25 solar cells and a concentration factor of 550. The measures were acquired at the rooftop of the Higher Polytechnical School of Jaén during the period between March 2013 and November 2013, forming a whole dataset composed of 8780 samples. The characteristics of data collected, recorded every 5 min, are shown in Table 1. In summary, this system is able to simultaneously measure P_m of the CPV modules and the outdoor atmospheric conditions that influence the performance of the module.

Table 1. Characteristics of data collected by the Automatic Test & Measurement System

Variable	Name	Range	Unit
DNI	Direct normal irradiance	[600, 1040]	W/m^2
T_A	Ambient temperature	[6, 44]	°C
W_S	Wind speed	[0, 25]	m/s
G	Incident global irradiance	[650, 1370]	W/m^2
APE	Spectral irradiance distribution of G, described through average photon energy	[1.72, 2.28]	
SMR	Spectral irradiance distribution of G, described through spectral machine radio	[0.65, 1.25]	

It is important to remark that the experimentation process is performed through a separation between training and test dataset. In this case, we use 80 % (training) of the whole dataset to calculate the EPs. Otherwise, 20 % (test) of the whole dataset was used to validate the descriptive capacity of the proposed model.

For the type of solar module under study (Fig. 3), P_m values under 64.5 W are not significant, and the samples of the dataset with these values have been removed. P_m values have been discretised in three different intervals according to the $P_{m,STC}$ provided by the manufacturer for this kind of module (150 W) and the expert criteria. These intervals are depicted in Table 2, where P_m *range* are the values of the intervals defined on P_m (in W) and % P_m *range* is the percentage with respect to the maximum power established by the manufacturer (150W). In addition, the percentage of instances for each values is shown.

The results obtained by the algorithm EvAEP have been summarised in Table 3 where the rule (R) and its quality measures GR, TPr and FPr are shown. As we have mentioned previously, the GR is the growth rate of the rule, TPr is the true positive rate and FPr is the false positive rate that measures the ratio between the examples covered incorrectly and the number of examples for the non-class.

Fig. 3. Solar tracker at High Technical School of the University of Jaen

Table 2. Intervals defined by experts for P_m variable

Interval	P_m range	% P_m range	% of instances
2	[64.5, 93]	(43 %, 62 %]	20 %
3	(93, 121.5]	(62 %, 81 %]	68 %
4	(121.5, 150]	(81 %, 100 %]	12 %

Table 3. Results obtained in *Concentrating Photovoltaic Module* dataset

Rule	GR	TPr	FPr
R_1: IF DNI = Very Low THEN $P_m = 2$	5.30	0.524	0.098
R_2: IF DNI = Low THEN $P_m = 2$	75.89	0.362	0.004
R_3: IF G = Medium THEN $P_m = 3$	1.35	0.889	0.656
R_4: IF APE = Low THEN $P_m = 3$	1.05	0.993	0.942
R_5: IF DNI = Medium THEN $P_m = 3$	4.33	0.397	0.091
R_6: IF DNI = High AND APE = Low AND T_A = Medium AND Ws = Medium THEN $P_m = 4$	3.87	0.037	0.009
R_7: IF DNI = High AND APE = Low THEN $P_m = 4$	9.92	0.134	0.013
R_8: IF APE = Low AND T_A = Low AND W_S = Medium AND SMR = High AND G = High THEN $P_m = 4$	31.00	0.033	0.001
R_9: IF APE = Low AND T_A = Low AND W_S = Low AND SMR = High AND G = High THEN $P_m = 4$	5.81	0.025	0.004
R_{10}: IF APE = Low AND T_A = Very Low AND SMR = High THEN $P_m = 4$	21.31	0.046	0.002

As can be observed in the results obtained in this contribution could be performed different analysis conditioned to the power analysed:

- Low power: For this class are obtained the rules with the highest GR. In addition, the rules extracted have a good balance between TPr and FPr where the number of examples covered are in almost all cases of the class analysed.
- Medium power: In this class there is an emerging pattern that should be discarded, the rule 4 because is very general and the ratio between TPr and FPr is very similar and close to the 100 %, i.e. this rule covers all examples for the dataset both positive and negatives examples in the same ratio. On the other hand, rule 3 and 5 continue the tendency of the analysis where a major value for DNI is synonyms of major power.
- High power: For this class are obtained very specific rules but with good values in GR and relationships between TPr and FPr. As can be observed, all rules have a low value for APE which is a very interesting value, however it is important to note that this value is combined with a high value for DNI or SMR, rules 6, 7 or 8, 9, 10, respectively.

In general, a direct relationship between DNI and the P_m can be observed in the study. This assumption confirms the results obtained in a previous analysis performed through subgroup discovery [14] in the paper [5].

5 Conclusions

The CPV technology has been analysed from a new point of view in this contribution, the EPM data mining technique which is a descriptive induction based on supervised learning. The analysis has been performed through the EFS called EvAEP.

Results obtained in this study confirm the possible relationships between atmospheric variables and P_m suspected by CPV experts, as well as some new knowledge. The knowledge extracted on interval of $P_m = [43\,\%, 62\,\%]$ and $P_m = (62\,\%, 81\,\%]$ confirm the existing relations between DNI and P_m. However, there is a new interest obtained when the P_m is maximum because there is a relation in all rules, the values for APE which belongs to the linguistic label *Low*. In this way, there is an interesting further analysis with respect to the influence of this variable in the performance of the CPV module with maximum P_m and low values for APE.

Acknowledgment. This work was supported by the Spanish Science and Innovation Department under project ENE2009-08302, by the Department of Science and Innovation of the Regional Government of Andalucia under project P09-TEP-5045, and by Spanish Ministry of Economy and Competitiveness under project TIN2015-68454-R (FEDER Founds).

References

1. Almonacid, F., Pérez-Higueras, P., Fernández, E., Rodrigo, P.: Relation between the cell temperature of a hcpv module and atmospheric parameters. Sol. Energy Mater. Sol. Cells **105**, 322–327 (2012)
2. Antón, I., Martínez, M., Rubio, F., Núñez, R., Herrero, R., Domínguez, C., Victoria, M., Askins, S., Sala, G.: Power rating of CPV systems based on spectrally corrected DNI, vol. 1477, pp. 331–335 (2012)
3. Carmona, C.J., González, P., del Jesus, M.J., Herrera, F.: NMEEF-SD: non-dominated multi-objective evolutionary algorithm for extracting fuzzy rules in subgroup discovery. IEEE Trans. Fuzzy Syst. **18**(5), 958–970 (2010)
4. Carmona, C.J., González, P., del Jesus, M.J., Herrera, F.: Overview on evolutionary subgroup discovery: analysis of the suitability and potential of the search performed by evolutionary algorithms. WIREs Data Min. Knowl. Disc. **4**(2), 87–103 (2014)
5. Carmona, C.J., González, P., García-Domingo, B., del Jesus, M.J., Aguilera, J.: MEFES: an evolutionary proposal for the detection of exceptions in subgroup discovery. An application to Concentrating Photovoltaic Technology. Knowl. Based Syst. **54**, 73–85 (2013)
6. Carmona, C.J., Pulgar-Rubio, F.J., García-Vico, A.M., González, P., del Jesus, M.J.: Análisis descriptivo mediante aprendizaje supervisado basado en patrones emergentes. In: Proceedings of the VII Simposio Teoría y Aplicaciones de Minería de Datos, pp. 685–694 (2015)
7. Castro, M., Domínguez, C., Núez, R., Antón, I., Sala, G., A. K.: Detailed effects of wind on the field performance of a 50 kw CPV demonstration plant. In: AIP Conference Proceedings, vol. 1556, pp. 256–260 (2013)
8. Dong, G.Z., Li, J.Y.: Efficient mining of emerging patterns: discovering trends and differences. In: Proceedings of the 5th ACM SIGKDD International Conference on Knowledge Discovery and Data Mining, pp. 43–52. ACM Press (1999)
9. Dong, G.Z., Li, J.Y.: Mining border descriptions of emerging patterns from dataset pairs. Knowl. Inf. Syst. **8**(2), 178–202 (2005)
10. Eiben, A.E., Smith, J.E.: Introduction to Evolutionary Computation. Springer, Heidelberg (2003)
11. Eshelman, L.J., Schaffer, J.D.: Real-coded genetic algorithms and interval-schemata. In: Foundations of Genetic Algorithms 2, pp. 187–202. Kaufmann Publishers (1993)
12. Helmers, H., Schachtner, M., Bett, A.: Influence of temperature and irradiance on triple-junction solar subcells. Sol. Energy Mater. Sol. Cells **116**, 144–152 (2013)
13. Herrera, F.: Genetic fuzzy systems: taxomony, current research trends and prospects. Evol. Intel. **1**, 27–46 (2008)
14. Herrera, F., Carmona, C.J., González, P., del Jesus, M.J.: An overview on Subgroup Discovery: Foundations and Applications. Knowl. Inf. Syst. **29**(3), 495–525 (2011)
15. Holland, J.H.: Adaptation in Natural and Artificial Systems. University of Michigan Press, Ann Arbor (1975)
16. Hüllermeier, E.: Fuzzy methods in machine learning and data mining: status and prospects. Fuzzy Sets Syst. **156**(3), 387–406 (2005)
17. Kinsey, G., Hebert, P., Barbour, K., Krut, D., Cotal, H., Sherif, R.: Concentrator multijunction solar cell characteristics under variable intensity and temperature. Prog. Photovoltaics Res. Appl. **16**(6), 503–508 (2008)
18. Kralj-Novak, P., Lavrac, N., Webb, G.I.: Supervised descriptive rule discovery: a unifying survey of constrast set, emerging pateern and subgroup mining. J. Mach. Learn. Res. **10**, 377–403 (2009)

19. Li, J.Y., Dong, G.Z., Ramamohanarao, K., Wong, L.: DeEPs: a new instance-based lazy discovery and classification system. Mach. Learn. **54**(2), 99–124 (2004)
20. Lin, C.-K., Fang, J.-Y.: Analysis of structural deformation and concentrator misalignment in a roll-tilt solar tracker. In: AIP Conference Proceedings, vol. 1556, pp. 210–213 (2013)
21. Peharz, G., Ferrer Rodríguez, J., Siefer, G., Bett, A.: Investigations on the temperature dependence of CPV modules equipped with triple-junction solar cells. Prog. Photovoltaics Res. Appl. **19**(1), 54–60 (2011)
22. Venturini, G.: SIA: a supervised inductive algorithm with genetic search for learning attributes based concepts. In: Brazdil, P.B. (ed.) ECML 1993. LNCS, vol. 667, pp. 280–296. Springer, Heidelberg (1993). doi:10.1007/3-540-56602-3_142
23. Zadeh, L.A.: The concept of a linguistic variable and its applications toapproximate reasoning. Parts I, II, III. Inform. Sci. **8-9**, 199–249, 301–357, 43–80 (1975)

Mobile Wireless System for Outdoor Air Quality Monitoring

Anton Koval[1(✉)] and Eloy Irigoyen[2]

[1] Department of Automation Control of Technology Processes and Computer Technologies, Zhytomyr State Technological University, 103, Chernyakhovskogo str., Zhytomyr 10005, Ukraine
koval.anton@gmail.com

[2] Computational Intelligence Group, Department of Systems Engineering and Automation, University of the Basque Country, Alda. Urquijo s/n, 48013 Bilbao, Bizkaia, Spain
eloy.irigoyen@ehu.eus

Abstract. Outdoor air quality monitoring plays crucial role on preventing environment pollution. The idea of use of unmanned aerial vehicles (UAV) in this area is of great interest cause they provide more flexibility than ground systems. The main focus of this work is to propose alternative, competitive outdoor wireless monitoring system that will allow to collect pollution data, detect and locate leakage places within petrol, gas and refinery stations or in hard to reach places. This system should be lightweight, compact, could be mounted on any UAV, operate in GPS denied environments and should be easily deployed and piloted by operator with minimal risk to his health. This paper presents the system, configured on a commercial UAV AR.Drone, embedding gas sensor to it, where as a ground station stands Robot Operation System. Conducted first stage experiments proved capabilities of our system to operate in real-world conditions and serve as a basis to carry out further research.

Keywords: AR.Drone · Pollution · Gas · ROS

1 Introduction and Problem Formulation

Air quality has direct influence on human health and quality of life. This causes a big number of research lines in this field. Mainly they could be divided into monitoring indoor and outdoor air quality. In our study we will focus on the second line cause outdoor air quality directly influences on the indoor.

Main outdoor pollutants in urban areas that influence on a human health are gases that emit from petrol and gas vehicles, stations and refineries [1]. For the last years population growth, especially in developed and industrial countries, caused rise of a number of vehicles that in its turn increased number of petrol and gas stations. Research made by Terrés et al. [2] showed that the impact of petrol stations on its surroundings is around 75 m but despite on that and even on law regulations, urban growth brings petrol stations that are located in urban areas.

M. Graña et al. (eds.), *International Joint Conference SOCO'16-CISIS'16-ICEUTE'16*, Advances in Intelligent Systems and Computing 527, DOI 10.1007/978-3-319-47364-2_33

Taking into account air pollution risks investigated in [3–5], especially impact of gases and their concentrations on human health, for instance one gases can cause cancer another respiratory diseases, besides that some of them create stink in neighbourhood areas. So, it is important to implement reliable gas monitoring system to prevent these problems.

The basic approach is to minimize gas emissions from petrol stations. To do this we need system that could control its concentration in immediate vicinity to the leakage place. But location and structure of petrol stations and its refineries varies and all these leads to different control strategies and monitoring scenarios. This means that monitoring system must fit following requirements: flexibility, reliability, accuracy, cost efficiency and easy deployment. Besides that is desirable to have unmanned measurement system with remote human control. At present most of existing solutions are ground based and can not fulfil formulated requirements. So it is necessary to review other possible platforms.

Nowadays unmanned aerial vehicles (UAV) or simply drones are used not only in military but also in a wide variety of civilian applications like surveillance, remote sensing, mapping, search and rescue, etc. Drones configuration and size mainly depend on their missions and sensing strategies. Typically they are piloted by operator via remote controller or a ground station [6].

Analyzing possible mission objectives and requirements we consider that implementation of outdoor drone based air pollution monitoring system or wireless monitoring system (WMS) remains to be a challenging task.

In this study we propose to use drone based system to monitor pollution level in places with possible gas leakages.

1.1 Related Work

There are several research lines in the field of air quality monitoring where appears idea of WMS but most of them are related to indoor air quality monitoring. For example Lozano et al. [7] proposed wireless sensor network that consists of a base station with internet connection and autonomous nodes equipped with different sensors to measure temperature, humidity, light and air quality. Later similar studies were made by Tsang-Chu Yu et al. [8], and Jun Li et al. [9].

A huge research in outdoor drone hazardous gases search was made by Neumann et al. [10,11]. This team presented a drone based system capable to execute missions in a variety of scenarios of gas emissions. Later they improved their system considering unpredictable nature of gas dispersion [12,13]. A similar systems were implemented by Rossi et al. [14] and by Croiz et al. [15] where a gas detection element was mounted on a commercial drone.

In our study we propose ground station - drone system that is performing in GPS denied environment unlike upper mentioned drone systems. For autonomous flight is used monocular SLAM framework PTAM presented by Klein and Murray [16].

1.2 Wireless Monitoring System

Precisely proposed WMS in general consists of:

- Low weight, low cost drone with inertial measurement unit (IMU), altimeter, frontal camera and is capable to carry additional sensors;
- Ground station;
- Human operator.

Nowadays most of the drones contain all upper mentioned sensors. IMU and altimeter allow us to estimate UAV attitude and navigation parameters [6]. However, measurements from low cost sensors are affected with bias errors. To compensate these errors in outdoor missions drones are usually equipped with GPS module. But in case of GPS-denied environments or when straightforward proximity is required (up to 1 m or less) exist other approaches for precise navigation and operation, as presented by Engel et al. [17] which uses off-board processing that minimizes on-board calculations and has minimum effect on flight endurance.

Thus, in our study vision-based system for autonomous navigation proposed in [17] is used.

Depending on different objectives and limitations (time, money) one can create own drone or to use existing solutions. Embedding extra sensors on UAV gives possibility to transform drone into unit capable to make different missions: environmental monitoring (in our study - gas), surveillance, etc. [6].

While extending drones functionality it is important to take into account UAV payload, since it has a direct influence on its performance. Preferably each additional item should have minimum influence on drones stabilization, control and flight endurance. Hence, additional sensors must be low power, light weight and compact.

Considering that this system has to transfer data from various sensors and to make off-board calculations, we also had to choose capable ground station [6].

General scheme of proposed WMS is depicted in Fig. 1

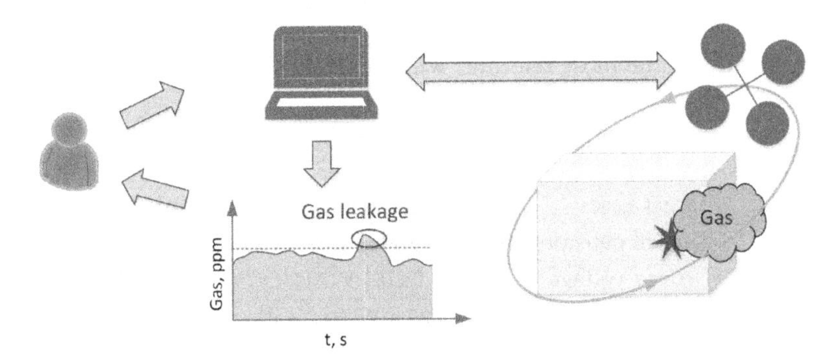

Fig. 1. General operation scheme of the proposed outdoor wireless monitoring system

Rest of this paper is structured as follows: In Sect. 2 we propose required components, namely UAV, gas sensor, ground station and embedded solution. Section 3 presents experimental design, plan, results and their analysis. Last section contains our conclusions and further research plans.

2 Components and System

2.1 Unmanned Aerial Vehicle/Flying Platform

At present exist huge number of UAVs of different sizes and configurations. In order to fulfil our mission requirements: it should have hovering capabilities, should be cheap as one drone corresponds one sensing module, it should have around 18 min or more of flight endurance with on-board payload capacity at least 50 g [18].

Comparison showed that one of the most affordable drones that fits our requirements is AR.Drone by Parrot, illustrated in Fig. 2. It is commercial autonomous quadrotor UAV with vertical take-off and landing capabilities. This drone is based on linux autopilot and is equipped with 3-axis accelerometer, 2-axis gyroscope (for pitch and roll), 1-axis high precision vertical gyroscope (for yaw), altimeter and two cameras (front and vertical) [19,20]. Worth mention that AR.Drone has also good crash robustness and could be easily repaired.

Besides that it is also compatible with Robot Operation System (ROS) [21] via Autonomy Labs driver [22].

Thus, in this work, we will use AR.Drone as a flying platform.

2.2 Pollution Sensor

Evidently that using UAV as a measuring platform we have to take into consideration its payload limits. Hence one of our objectives was to choose gas sensors with best relation weight-measuring performance. Market analysis and study [23] showed that catalytic sensors have large detection range with the highest detection concentration, good stability, reliability, simple equipment, low cost and weight, and could be used in industrial measurements.

Regarding these factors we selected catalytic sensor TGS6812-D00 shown in Fig. 2 by Figaro as the most relevant, which specifications are shown in Table 1.

Table 1. Specifications of Figaro Gas Sensor TGS6812-D00

Target gases	Hydrogen, Methane, LP gas
Typical detection range	0–100 % LEL of each gas
Driving voltage	3.0±0.1 V AC/DC
Power consumption	525 mW (typical)
Weight	Approx. 1.5 g

2.3 Ground Station

Robot operation system [21] is a powerful software framework to control robots. Nowadays it is actively used in numerous research projects [24–26] as it is modular, free, open source system that has support of different programming languages and provides powerful visualization tools.

Our ground station implementation is based on the ROS Indigo and Ubuntu 14.04 because of its functionality and compatibility with AR.Drone. As a connection layer between drone and a ground station we use ardrone_autonomy package, developed by Autonomy Lab of Simon Fraser University [22] and other contributors from github.com community. This package provides access to all data from AR.Drone's on-board sensors and allows to send control signals.

Another package that is used is tum_ardrone developed by Department of Computer Science of Technical University of Munich [27]. This system allows camera-based autonomous navigation for our UAV.

2.4 Embedded System

AR.Drone has limited number of on-board sensors so in our study we had to enable support of additional sensors to be able to embed a gas sensor.

Huge study of this drone was made by Daugaard and Thyregod [28,29], by Ardudrone project [30] and by contributors from blog.perquin.com, embedded-software.blogspot.com.es, ardrone-flyers.com and rcgroups.com.

Based on these prior studies we were able to enable USB port on AR.Drone with firmware 1.11.5, which is compatible with latest driver for ROS Indigo [28] and after we implemented connection layer between drone and sensor using Arduino Pro Mini 5V, 16 MHz with FTDI adapter [30]. All these makes possible to connect to AR.Drone lots of different devices and sensors.

The final hardware setup is shown on Fig. 2

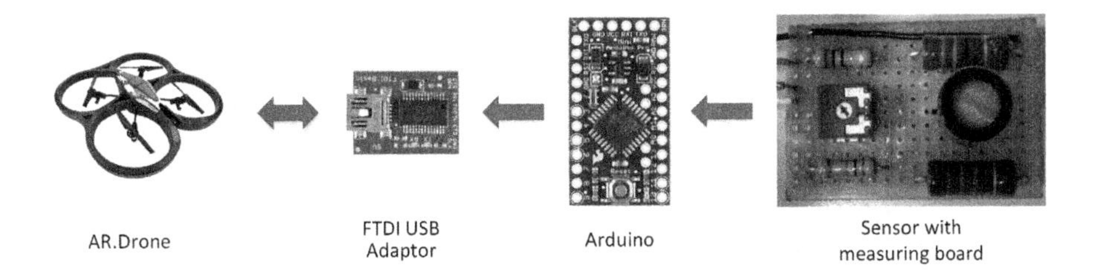

Fig. 2. Integration of Figaro Gas Sensor TGS6812-D00 in AR.Drone 1.0

Activation of hardware layer requires implementation of communication link similar to [30] or [31] that will work in parallel with AR.Drone data link. To do this we compiled drone kernel modules to enable USB port and support of FTDI adapter, implemented a simple program which will read sensor data and send it to AR.Drone through USB port. Based on [30,31] we implemented on-board proxy client that reads sensor data and sends them to the ground station proxy server.

General working scheme is presented in Fig. 3.

Fig. 3. Scheme of communication link between drone and ground station

3 Experiments

In this part we present experimental setup, control strategies and results.

3.1 Planning Experiment/Experimental Setup

In Fig. 4 hardware setup of our system with integrated gas sensor and linking modules is shown.

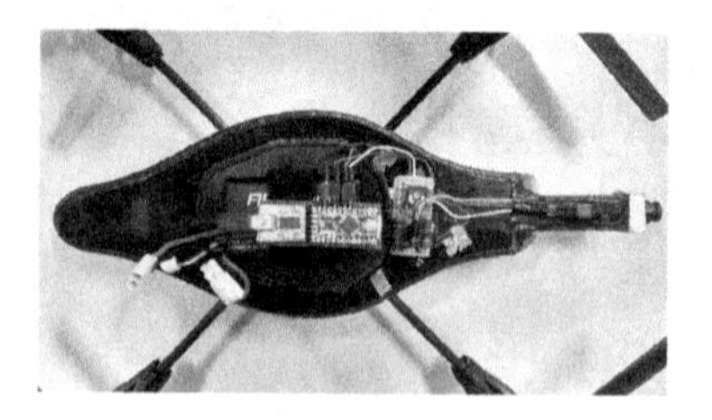

Fig. 4. Experimental setup of AR.Drone. Left: Top view. Right: Bottom view.

During flight AR.Drone continuously receives data with sampling rate set to 5 Hz from gas sensor and sends them to the ground station. One of our challenges was to take into account sensor response time. Considering that we conducted preflight experiments to get sensor properties.

3.2 Laboratory Tests

According to requirements of our missions we have to determine sensor's sensitivity to concentration that is caused by flow from leakage place.

To do this we conducted our experiments as follows: sensor was placed close to the gas source, each 60 s step of Methane flow was increased by 25 ml/min (up to max step 250 ml/min). Results are depicted in Fig. 5.

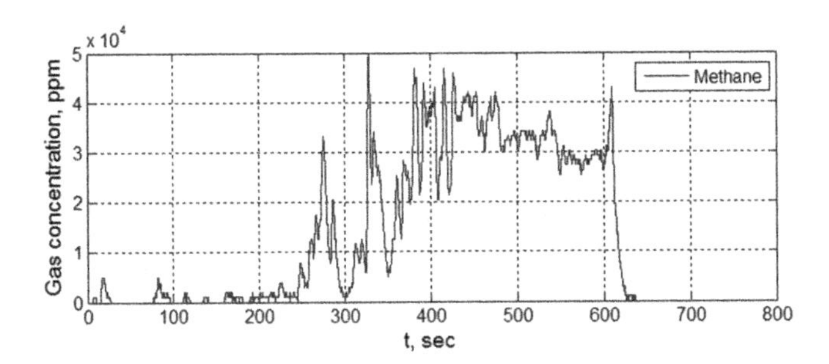

Fig. 5. Laboratory experiment. Measured concentration caused by gas flow in immediate vicinity of sensor to gas source

Experiments in laboratory conditions showed that sensor has sufficient sensitivity threshold and detection range for Methane. But added several restrictions in carrying out field tests, namely sensor has a time lag which means that we need longer hovering time in order to detect gas. Another restriction is that sensor does not react to combustible gases not balanced with oxygen, so we have to create mixed gas flow at least for the first field tests.

This lets us plan drone flight strategies and select appropriate gas source with reasonable flow speed for our experiments.

3.3 Outdoor Tests

To validate our system performance in conditions close to real-world, all outdoor experiments were carried out in the university patio, taking into account that we deal with hazardous gases.

In order to simulate gas leakage, experimental setup consisted of: Butane-2 balloon as a gas source, which flow direction and speed were regulated with an exhaust propeller.

Prior to flight tests were conducted static outdoor measurements shown in Fig. 6, where drone was placed at the distance of 1 m from the gas source. These tests showed good continuous concentration detection rate when sensor appears in the gas flow. The next step was to carry out field tests and to obtain measurements during autonomous flight.

In general flight strategy could be divided into two stages: drone initialization and trajectory flying. During first stage drone takes-off, goes up and down in order to have good scale estimation [17]. In the second stage drone flies generated trajectory towards the gas source, it moves to initial point and after follows path, that consists of 3 intervals of 0.5 m with hovering time 3 s in each. After reaching last trajectory point that has distance up to 1 m from gas source, drone hovers 20 s and lands.

Experiments were performed successfully and their results are depicted in Fig. 6.

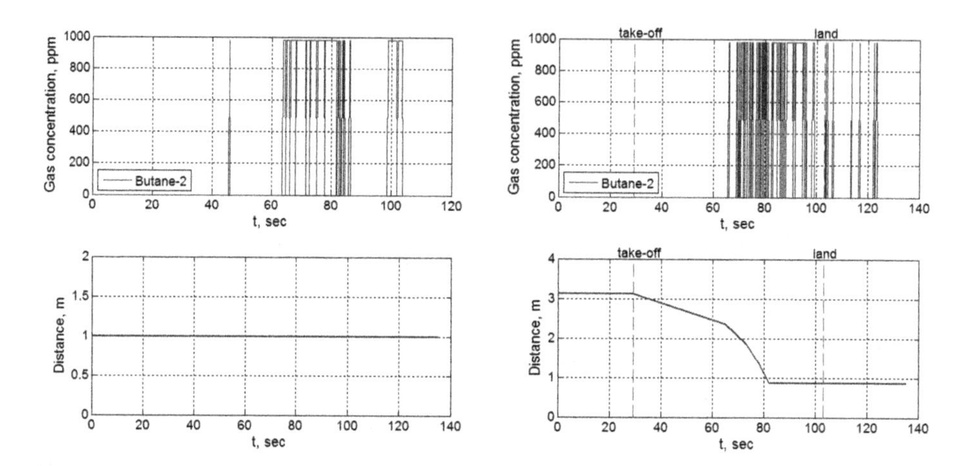

Fig. 6. Outdoor tests. Left Top and Bottom: AR.Drone in static conditions, fixed distance from gas source. Right Top and Bottom: Autonomous flight.

Obtained results showed that proposed system is capable to perform assigned tasks. Analyzing received dependencies one can say that as far as drone approaches to the gas source, measurements become more stable and with longer detection intervals (what reflects gas concentration) and this correlates with static experiments.

4 Conclusions and Future Plans

In this study we presented a platform for autonomous gas leakage detection that is based on AR.Drone with embedded gas sensor and controlled via Robot Operation System. Conducted preliminary experiments in laboratory conditions showed that sensor has good detection range but it also has a time lag which means that we need more deep investigation of sensor's response speed and how response depend on the gas concentration. Outdoor tests first static and after dynamic proved possibility of gas detection. In further steps we plan to continue development of this platform in order to optimize its performance and improve robustness. As this is ongoing work, in this paper we propose WMS in general. As a next steps we plan to conduct more experiments in gas measurements, source location, flow speed, etc. and battery performance.

Acknowledgements. This work comes under the framework of the project IT874-13 granted by the Basque Regional Government. The authors would like to thank the Erasmus Mundus Action 2 ACTIVE fellowship program, and the participating colleagues from the SUPREN research group, Environment and Chemical Engineering Department of the University of the Basque Country.

References

1. Edokpolo, B., Yu, Q.J., Connell, D.: Health risk characterization for exposure to benzene in service stations and petroleum refineries environments using human adverse response data. Toxicol. Rep. **2**, 917–927 (2015)
2. Terrés, I.M.M., Miñarro, M.D., Ferradas, E.G., Caracena, A.B., Rico, J.B.: Assessing the impact of petrol stations on their immediate surroundings. J. Environ. Manage. **91**(12), 2754–2762 (2010)
3. Correa, S.M., Arbilla, G., Marques, M.R.C., Oliveira, K.M.P.G.: The impact of BTEX emissions from gas stations into the atmosphere. Atmos. Pollut. Res. **3**(2), 163–169 (2012)
4. Kountouriotis, A., Aleiferis, P.G., Charalambides, A.G.: Numerical investigation of voc levels in the area of petrol stations. Sci. Total Environ. **470**, 1205–1224 (2014)
5. Sairat, T., Homwuttiwong, S., Homwutthiwong, K., Ongwandee, M.: Investigation of gasoline distributions within petrol stations: spatial and seasonal concentrations, sources, mitigation measures, and occupationally exposed symptoms. Environ. Sci. Pollut. Res. **22**(18), 13870–13880 (2015)
6. Valavanis, K.P., Vachtsevanos, G.J.: Handbook of Unmanned Aerial Vehicles. Springer Publishing Company, Incorporated, Dordrecht (2014)
7. Lozano, J., Suárez, J.I., Arroyo, P., Manuel, J.: Wireless sensor network for indoor air quality monitoring. Chem. Eng. **30**, 319–324 (2012)
8. Yu, T.-C., Lin, C.-C., Chen, C.-C., Lee, W.-L., Lee, R.-G., Tseng, C.-H., Liu, S.-P.: Wireless sensor networks for indoor air quality monitoring. Med. Eng. Phys. **35**(2), 231–235 (2013)
9. Li, J., Xin, J., Li, M., Lai, B., Ma, Q.: Wireless sensor network for indoor air quality monitoring. Sens. Transducers **172**(6), 86–90 (2014)
10. Bartholmai, M., Neumann, P.: Micro-drone for gas measurement in hazardous scenarios via remote sensing. In: Proceedings of 6th WSEAS International Conference on Remote Sensing (REMOTE 2010) (2010)
11. Neumann, P., Bartholmai, M., Schiller, J.H., Wiggerich, B., Manolov, M.: Micro-drone for the characterization and self-optimizing search of hazardous gaseous substance sources: a new approach to determine wind speed and direction. In: 2010 IEEE International Workshop on Robotic and Sensors Environments (ROSE), pp. 1–6. IEEE (2010)
12. Neumann, P.P., Bennetts, V.H., Lilienthal, A.J., Bartholmai, M., Schiller, J.H.: Gas source localization with a micro-drone using bio-inspired and particle filter-based algorithms. Adv. Robot. **27**(9), 725–738 (2013)
13. Neumann, P., Asadi, S., Schiller, J.H., Lilienthal, A.J., Bartholmai, M.: An artificial potential field based sampling strategy for a gas-sensitive micro-drone. In: IROS Workshop on Robotics for Environmental Monitoring (WREM), pp. 34–38 (2011)
14. Rossi, M., Brunelli, D., Adami, A., Lorenzelli, L., Menna, F., Remondino, F.: Gas-drone: portable gas sensing system on uavs for gas leakage localization. In: 2014 IEEE SENSORS, pp. 1431–1434. IEEE (2014)

15. Croizé, P., Archez, M., Boisson, J., Roger, T., Monsegu, V.: Autonomous measurement drone for remote dangerous source location mapping. Int. J. Environ. Sci. Dev. **6**(5), 391 (2015)
16. Klein, G., Murray, D.: Parallel tracking and mapping for small ar workspaces. In: 6th IEEE and ACM International Symposium on Mixed and Augmented Reality, 2007, ISMAR 2007, pp. 225–234. IEEE (2007)
17. Engel, J., Sturm, J., Cremers, D.: Accurate figure flying with a quadrocopter using onboard visual and inertial sensing. In: IMU, vol. 320, p. 240 (2012)
18. Liu, Z., Li, Z., Liu, B., Xinwen, F., Ioannis, R., Ren, K.: Rise of mini-drones: applications and issues. In: Proceedings of the 2015 Workshop on Privacy-Aware Mobile Computing, pp. 7–12. ACM (2015)
19. Parrot.: Ar.drone 1.0 (2010). http://ardrone2.parrot.com/support-ardrone-1/
20. Krajník, T., Vonásek, V., Fišer, D., Faigl, J.: AR-Drone as a platform for robotic research and education. In: Obdržálek, D., Gottscheber, A. (eds.) EUROBOT 2011. CCIS, vol. 161, pp. 172–186. Springer, Heidelberg (2011). doi:10.1007/978-3-642-21975-7_16
21. Quigley, M., Conley, K., Gerkey, B., Faust, J., Foote, T., Leibs, J., Wheeler, R., Ng, A.Y.: Ros: an open-source robot operating system. In: ICRA Workshop on Open Source Software, vol. 3, p. 5 (2009)
22. Monajjemi, M., et al.: Ardrone autonomy: a ROS driver for AR.Drone 1.0 & 2.0 (2015). http://github.com/AutonomyLab/ardrone_autonomy
23. Liu, X., Cheng, S., Liu, H., Hu, S., Zhang, D., Ning, H.: A survey on gas sensing technology. Sensors **12**(7), 9635–9665 (2012)
24. Mercado, D.A., Castillo, P., Lozano, R.: Quadrotor's trajectory tracking control using monocular vision navigation. In: 2015 International Conference on Unmanned Aircraft Systems (ICUAS), pp. 844–850. IEEE (2015)
25. Nguyen, T., Mann, G.K.I., Gosine, R.G., Vardy, A.: Appearance-based visual-teach-and-repeat navigation technique for micro aerial vehicle. J. Intell. Robot. Syst. pp. 1–24 (2016). doi:10.1007/s10846-015-0320-1
26. Li, P., Garratt, M., Lambert, A., Lin, S.: Metric sensing and control of a quadrotor using a homography-based visual inertial fusion method. Robot. Auton. Syst. **76**, 1–14 (2016)
27. Engel, J., Sturm, J., Cremers, D.: Scale-aware navigation of a low-cost quadrocopter with a monocular camera. Robot. Auton. Syst. **62**(11), 1646–1656 (2014)
28. Daugaard, M.: Semi-autonom indendørs navigation for luftbåren robot. Ph.D. thesis, Aarhus Universitet, Datalogisk Institut (2012)
29. Thyregod, T., Daugaard, M.: Navigation for robots with wifi and cv (2012)
30. Nosaari. Ardudrone (2011). https://code.google.com/archive/p/ardudrone/
31. Gunnarsson, G.: Udp client/server system (2012). https://www.abc.se/m6695/udp.html

SOCO 2016: Special Session on Soft Computing Methods in Manufacturing and Management Systems

ANN-Based Hybrid Algorithm Supporting Composition Control of Casting Slip in Manufacture of Ceramic Insulators

Arkadiusz Kowalski(✉) and Maria Rosienkiewicz

Wrocław University of Technology, Wrocław, Poland
{Arkadiusz.Kowalski,Maria.Rosienkiewicz}@pwr.edu.pl

Abstract. Published research on manufacturing processes of ceramic insulators concerns mostly material examinations. Little has been done in the field of assuring proper quality of insulators based on analysis of production data. This is why the paper discusses a new approach to supporting quality control in manufacture of ceramic insulators, based on regression analysis and ANN modeling. The proposed algorithm enables the user to control addition of raw aluminum oxide (and its graining) in order to obtain its desired grain-size composition in the mass and thus to reduce the number of defects to acceptable levels.

Keywords: ANN · Quality management · Ceramic insulators · Production control

1 Introduction

Ceramic insulators, being materials preventing or regulating current flow in electrical circuits, play an important role in electrically isolating a conductor from power and provide mechanical support for line conductors [12]. Insulators require several properties including high resistivity, high dielectric strength, low loss factor, good mechanical properties, dissipation of heat and protection of conductors from severe environment, like humidity and corrosiveness [8]. High-strength ceramic insulators require not only higher amount of alumina but, among others, fine and uniform distribution of particles. Analysis of recent demands on porcelain insulators indicates that improved mechanical strength comparable to that of firebricks and lowered thermal conductivity, similar to that of fibrous insulators, are highly desirable [6]. It can be observed that standard electrical porcelain has been replaced by advanced porcelains to prevent insulators from too frequent breakages [10]. With the purpose of enhancing properties of insulators, "the mainly used modern ceramics are: high strength electrical porcelain (50 % alumina porcelain) and high alumina porcelain (> 85 % alumina porcelain)" [15]. Porcelain is one of the most complex ceramics, which is "formulated from a mix of clay, feldspar and quartz and sintered to conform a glass-ceramic composite" [3].

For manufacturers of ceramic insulators, ensuring high-quality products is a difficult challenge. There is no possibility to carry out continuous laboratory inspection of manufactured ceramic mass, which results in unpredictable defect levels. Therefore, solutions utilizing the "data mining" approach should be searched-for in order to

M. Graña et al. (eds.), *International Joint Conference SOCO'16-CISIS'16-ICEUTE'16*,
Advances in Intelligent Systems and Computing 527, DOI 10.1007/978-3-319-47364-2_34

support quality management in manufacture of ceramic insulators, based on forecasting with no necessity to perform complicated and expensive material examinations. One of possible research directions in this field is utilizing artificial neural networks (ANN). For some time now, artificial neural networks, being mathematical models representing biological processes of human brain [14], have been recommended as an alternative to traditional statistical forecasting methods. AAN applications may be grouped in three main categories: speech processing, image processing and decision making [16]. ANN, being one of the most popular tools of artificial intelligence, can be applied in many areas, e.g. in forecasting of sale, planning of overhauls, diagnostics of machines or analysis of production problems. ANN have been studied as a control support in production systems [1]. In the field of quality management, two common configurations of utilizing ANN can be distinguished: ANN "*learns* the process dynamics using input/output data pairs" or ANN "is incorporated into the process control loop". In the latter case, ANN "provides control actions that adapt to changes in the system output during the process run" [9]. When it comes to application of ANN as a tool of forecasting the defect level of ceramic insulators to support quality control, it has to be stated that, in fact, little work has been done in this area.

Irrespective of analysis of the defects very significantly affecting mechanical strength, three basic theories are distinguished to describe influence of phase structure of a body on mechanical properties of porcelain [2–4]: mullite theory, matrix reinforcement theory and dispersive reinforcement theory. In order to obtain high quality of ceramic insulators, they are produced of C130 porcelain. Composition of the material is similar to that of traditional hard porcelain, but aluminum oxide (Al_2O_3) is used in place of quartz. This results in higher mullite content in the ceramic, as well as in higher mechanical parameters after baking. Some publications are available in that influence of grain-size distribution measured by BET method (m^2/g) on sinterability, piroplastic deformation, microstructure and mechanical bending strength is determined [5, 11]. Analysis of the literature on manufacturing processes of ceramic insulators, made by the Authors, concerns mostly material examinations. However, few papers consider the solutions aimed at assuring proper quality of insulators based on analysis of production data. This is why the Authors suggest a new approach to supporting quality control in manufacture of ceramic insulators, based on regression analysis and ANN modeling.

2 Problem Description

During manufacture of ceramic insulators, as early as at the drying stage, alarming increase of defective products number was found. At that time no change of composition of the ceramic mass was made, parameters of baking and other manufacturing stages also remained unchanged, so searching the reasons of increased reject number was focused on examinations of grain composition (grain size fractions) of aluminum oxide in the ceramic mass. Production specificity of ceramic mass, considering 30 % of fresh mass and 70 % of recycled mass prepared from rejects, hampered determination of aluminum oxide fraction in the mass designed currently for manufacture of insulators. Slow rate of changing components (once a week) in the mass caused also that

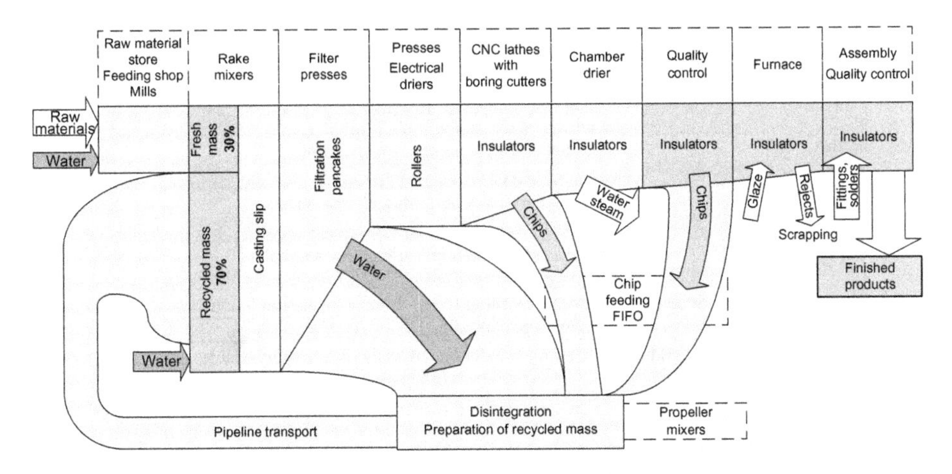

Fig. 1. Sankey diagram showing flow of ceramic mass in production of insulators, with use of recycled mass

the effect of possible corrections of aluminum oxide grain composition was far distant in time and thus difficult to be grasped (Fig. 1).

The main goal of this paper is presenting a new hybrid ANN-based approach supporting process control in manufacture of ceramic insulators in order to ensure the desired product quality. The merit of this paper is not suggesting a new ceramic reinforcement theory, but finding a correlation between grain-size distribution of aluminum oxide and the number of quality defects in manufacture of ceramic insulators. Therefore, it is possible to control addition of raw aluminum oxide (and its graining) in order to obtain its desired grain-size composition in the mass and thus to reduce the number of defects including cracks (on bodies, on sheds, on face surfaces and in holes), twists and disturbed structure to acceptable levels.

3 Solution

The Authors decided to consider the above-mentioned problem not on the grounds of experimental material examinations, but by using the hybrid approach based on data analysis considering regression modeling and ANN. The performed examinations made it possible to develop the hybrid algorithm supporting control of material fractions in a casting slip oriented to minimization of reject level. The authors propose the combined application of two forecasting techniques, to obtain a robust method for forecasting level of defects. The suggested solution is presented in the chart below (Fig. 2).

The proposed algorithm assumes that hybrid approach (defined by the Authors as combination of two different methods), which consists of two models is usually gaining better results than a single one. The developed hybrid algorithm, joining regression analysis and ANN, consists of five stages. The first three of them are: selection of raw material to be analyzed in the preparation phase, determination of a charge matrix,

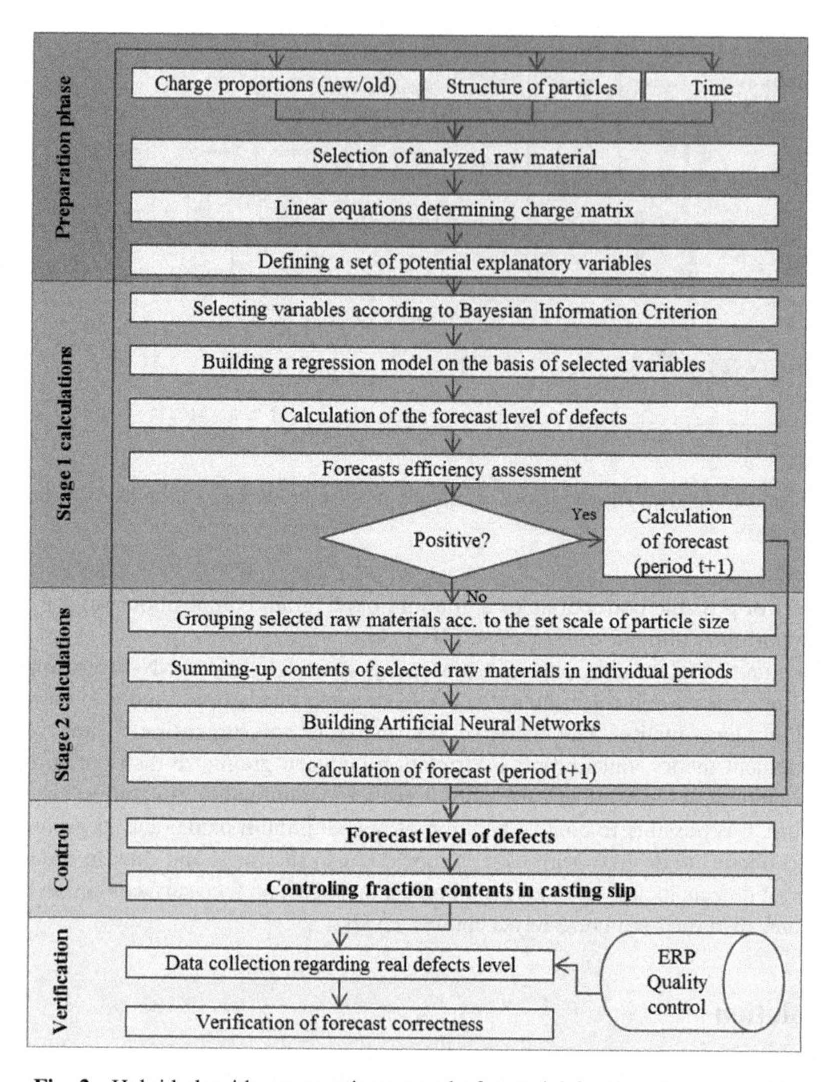

Fig. 2. Hybrid algorithm supporting control of material fractions in casting slip

definition of a set of potential explanatory variables (x_1, x_2, …, x_m). In order to determine the charge matrix, composition of the ceramic mass (total of fresh and recycled masses) may be written as $x + y = 100\%$, where x means fresh mass and y means recycled mass. If percentages of components in the mix (water-based suspension) are designated as m, quantity of the substance A that remains after subsequent additions of fresh mass can be described by the formula for i-th element of a geometrical progression, where i means a subsequent time period in that the next portion x of fresh mass is added:

$$a_i = xmy^{i-1} = xm(1-x)^{i-1}. \tag{1}$$

Total quantity of the substance A with percentage m after j additions can be expressed by the formula for sum of a geometrical progression:

$$\sum\nolimits_{i=1}^{j} a_i = a_1 \frac{1-y^j}{1-y} = xm\frac{1-(1-x)^j}{1-(1-x)} = m(1-(1-x)^j). \tag{2}$$

Replacement rate of fresh mass can be described by the function f:

$$f(x) = 1-(1-x)^i. \tag{3}$$

Practical application of this function makes it possible to determine the time after that the assumed level of mass replacement will be reached, e.g. for mass replacement in the relation 30 % new to 70 % old, it is obtained after 10 replacements:

$$f(x) = 1-(1-0,3)^{10} = 0,972. \tag{4}$$

This means that the 97-% replacement level of ceramic mass composition for fresh components is reached. At the next stage of the algorithm, an effective method of selecting explanatory variables for the model was applied: Bayesian information criterion (BIC), also called the Schwarz criterion. Gideon Schwarz suggested the criterion according to that the model should be selected, for that the following value is the highest [13]:

$$M_j(X_1,\ldots,X_n) - k_j \log n. \tag{5}$$

Detailed calculations are presented in the article *Estimating the dimension of a model*. Based on the BIC criterion, an optimum set of explanatory variables is selected and next a regression model is built on that ground forecasts of rejects level are constructed. In the next step, effectiveness of the forecasts is evaluated on the basis of the determination coefficient R^2. If this coefficient value is satisfactory (e.g. 90 %), the so verified model can be used for calculating a forecast. If the R^2 value is too low, according to the algorithm (stage 2 calculations) the selected raw materials should be grouped according to the presumed scale of particle size, e.g. in the ranges 0.76–0.81; 0.82–0.87; 0.88–0.93; 0.94–1.00; 1.01–1.05 m^2/g (BET), then fractions of selected raw materials in individual periods should be summed-up and finally, after proper data processing, ANN should be applied. There are many types of ANNs, but basic ANN classification groups divide them into those dedicated to classification problems or to regression problems. In regression problems, which apply to the issue discussed in this paper, the objective is to estimate the value of an output variable, given the known input variables. In the conducted research, a multi-layered perceptron (MLP) neural network was used to forecast the level of defects. On the grounds of the built network, a forecast reject level in the period t + 1 can be determined. The obtained result should be considered at controlling contents of fractions in casting slip. The last stage is acquiring real data in the period t + 2 from the period t + 1 and continuously completing databases in order to monitor and verify effectiveness of the model or ANN.

4 Case Study

4.1 Description of Manufacturing Processes

The process of manufacturing ceramic insulators (in this case of high-clay ceramic C130) begins from preparation of ceramic mass. The previously prepared components must be weighed: aluminum oxide, feldspar and kaolin. To obtain ceramic mass, the prepared raw materials – in proper proportions – are milled as water-based suspension in grinding mills. This way, fresh mass is obtained. Similarly obtained is also ceramic glaze necessary at further process stages. In manufacture of ceramic mass, chips created during turning insulators and rejects appeared after drying are used as well. Chips and disintegrated rejects are mixed with water in mixers and recycled mass is created this way. Fresh mass and recycled mass are mixed in proportion 30 % to 70 % in subsequent mixers and transferred to pug mills. After cleaning the mass in order to remove undesirable impurities (e.g. iron compounds), excessive water is filtered-out on filter presses. The mass obtained at this stage is shaped by the filter in form of pancakes with humidity of ca. 20 %. This way prepared mass is desecrated on presses in order to remove air bubbles that deteriorate its properties. The mass formed as cylinders is then cut for determined lengths and the cylinders are subjected to drying. After drying, their humidity is reduced to a dozen percent and their hardness is high enough to enable machining. Turning the insulators is performed on various horizontal and vertical lathes, with use of boring cutters. Construction of a cutter depends on the shape of insulator sheds. At this stage, most of chips are created, utilized for preparing recycled mass, as presented in the layout of manufacturing process of ceramic insulators, see Fig. 3.

At high temperature (1300°C), many physico-chemical transformations occur in ceramic mass and its crystallization stabilizes the shape given to the insulators, as well as ensures proper mechanical and electrical properties (transfer of bending and tensile loads and resistance to flashovers at nominal voltage). At subsequent stages, the

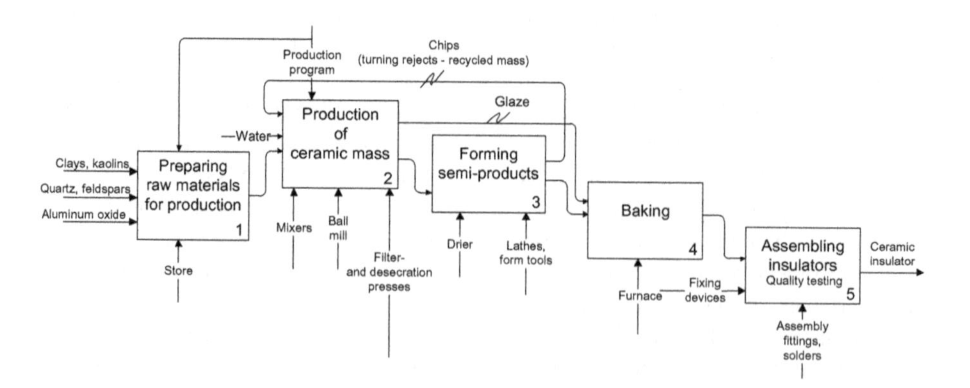

Fig. 3. Manufacturing process of ceramic insulators written acc. to IDEF0 notation

insulators are subjected – as required – to cutting and grinding, as well as fixing devices are installed. The last stage before packing and shipment are inspection and acceptance examinations in order to verify the insulator properties.

4.2 Application of the Suggested Method

The presented algorithm was applied for the data from a company manufacturing ceramic insulators. In this paper, exemplary results for aluminum oxide type A (13.6 % of body composition) and aluminum oxide type B (9.0 % of body composition) are presented. Application of the formulae 1 and 2 made it possible to determine changes of aluminum oxide fractions in ceramic mass, see Fig. 4. To that end, raw material with random fraction was added to current production of fresh mass, while the fraction values were recorded. According to the BIC criterion, the variables x_2, x_3, x_7, x_8, x_9, x_{11}, x_{13}, x_{15}, x_{16} were chosen from the set of potential explanatory variables $x_1 \div x_{16}$ to the model type A and the variables x_2, x_3, x_3, x_4, x_5, x_6, x_7, x_8, x_9, x_{11}, x_{12}, x_{13}, x_{15}, x_{16}, x_{17}, x_{19}, x_{20}, x_{21}, x_{22} were chosen from the set x_1–x_{22} to the model type B.

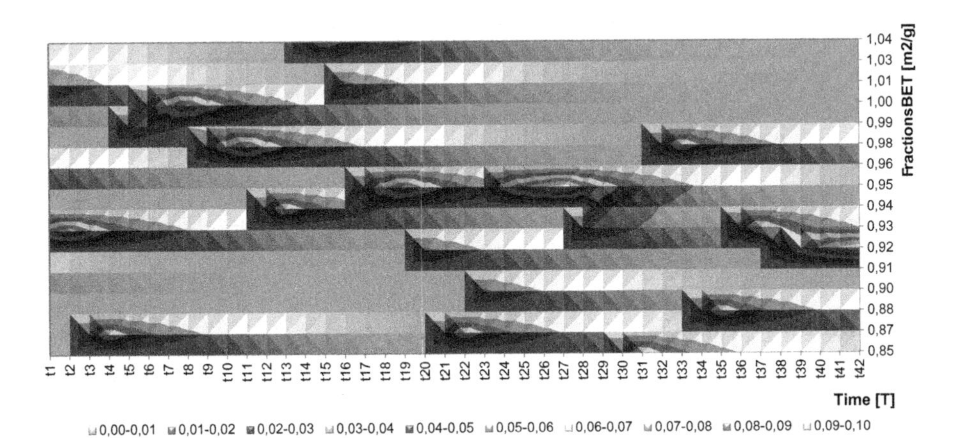

Fig. 4. Grain-size distribution of aluminum oxide in ceramic mass during the analyzed time

Then, parameters of linear models were estimated with the least squares method. The regression model of forecasting a reject level for raw material Al_2O_3 type A is:

$$\begin{aligned}\hat{y} = {} & 0,11 - 0,80x_2 - 0,72x_3 - 0,10x_4 - 0,87x_5 - 0,84x_6 - 0,53x_7 - 0,56x_8 \\ & - 0,56x_9 - 0,50x_{11} - 0,71x_{12} - 0,74x_{13} - 0,53x_{15} - 1,06x_{16} - 0,89x_{17} \\ & - 0,34x_{19} - 0,67x_{20} + 491,82x_{21} - 660,04x_{22}. \end{aligned} \quad (6)$$

In turn, the regression model of forecasting a reject level for Al_2O_3 type B is:

$$\hat{y} = 0,07 - 0,68x_2 - 0,43x_3 - 0,47x_7 - 0,62x_8 - 0,67x_9 - 0,51x_{11} - 0,81x_{13} - 5,02x_{15} - 0,34x_{16}. \quad (7)$$

In each model, $\hat{y}$ is the forecast level of defects. All the parameters are statistically significant. In the next step, forecasts and then values of determination coefficients were calculated. For Al_2O_3 type A, R^2 value is 0.9121, which means that the model is very well fitted to empirical data and explains variability of rejects level in over 91 %. So, according to the algorithm, if the determination coefficient $R^2 = 90$ % is accepted as the limit value, the obtained value may be considered as satisfactory and effectiveness of the model as sufficient. For Al_2O_3 type B, R^2 value is 0.761, which means that the model explains variability of reject level in 76 % only and thus can not serve as an effective tool of forecasting rejects levels. According to the algorithm (stage 3) an ANN should be construed. Results obtained for the ANN (MLP 5-7-1, learning algorithm - BFGS 71, error function – SOS, activation hidden/output - logistic) are given in Table 1 below. Quality values of learning, testing and validation are high, which indicates good effectiveness of the network. For the ANN, R^2 value was 0.995, which shows that, in this case, the ANN is much more effective than the regression model.

Tab. 1. Artificial neural network (MLP 5-7-1) – results

Quality			Error		
Training	Testing	Validation	Training	Testing	Validation
0.974	0.759	0.921	0.000007	0.000043	0.000009

Therefore, it can be an effective tool of forecasting reject levels and thus can support the process of controlling a fraction content in casting slip.

5 Summary

Planning of the manufacturing process, being one of the most important actions in the domain of the technical production preparation [7], is crucial for efficient production organization and quality assurance. This paper is aimed at presenting a new hybrid ANN-based approach supporting control of manufacturing process of ceramic insulators in order to guarantee desired quality of products. The developed algorithm permits the added fractions of aluminum oxide Al_2O_3 to be controlled in order to maintain their desired content in ceramic mass, which makes it possible to control the number of rejects. The examinations carried-out within the case study demonstrate that ANNs for selected kinds of aluminum oxide show higher effectiveness in comparison to regression models. Moreover, the results indicate that, in some cases, application of regression models is sufficient and building an ANN is not necessary. Within the research, a set of aluminum oxide fractions resulting in a low reject level was identified. Thanks to that, it is possible to control manufacturing process of insulators effectively, in spite of long replacement time of ceramic mass, by identification of changing content of fresh mass in the whole casting slip. The presented algorithm can make an effective tool supporting manufacturing process in the companies producing ceramic insulators.

References

1. Burduk, A.: The role of artificial neural network models in ensuring the stability of systems. In.: 10th International Conference on Soft Computing Models in Industrial and Environmental Applications. Advances in Intelligent Systems and Computing, vol. 368, pp. 427–437. Springer (2015)
2. Carbajal, L., Rubio-Marcos, A., Bengochea, M.A., Fernandez, J.F.: Properties related phase evolution in porcelain ceramics. J. Eur. Ceram. Soc. **27**(13–15), 4065–4069 (2007). Elsevier
3. Carbajal, L., Rubio-Marcos, F., Bengochea, M., Fernandez, J.: Properties related phase evolution in porcelain ceramics. J. Eur. Ceram. Soc. **27**, 4065–4069 (2007). Elsevier
4. Carty, W., Senapati, U.: Porcelain—raw materials, processing, phase evolution, and mechanical behavior. J. Am. Ceram. Soc. **81**(1), 3–20 (1998). Wiley
5. Cavalcante, T., Dondi, P.M., Ercolani, M., Guarini, G., Melandri, G., Raimondo, M.: The infuence of microstructure on the performance of white porcelain stoneware. Ceram. Int. **30** (6), 953–963 (2004). Elsevier
6. Fukushima, M., Yoshizawa, Y.: Fabrication and morphology control of highly porous mullite thermal insulators prepared by gelation freezing route. J. Eur. Ceram. Soc. **36**(12), 2947–2953 (2015). Elsevier
7. Grabowik, C., Krenczyk, D., Kalinowski, K.: The hybrid method of knowledge representation in a CAPP knowledge based system. In: Corchado, E., Snášel, V., Abraham, A., Woźniak, M., Graña, M., Cho, S.-B. (eds.) HAIS 2012, Part II. LNCS, vol. 7209, pp. 284–295. Springer, Heidelberg (2012)
8. Islam, R., Chan, Y., Islam, F.: Structure–property relationship in high-tension ceramic insulator fired at high temperature. Mater. Sci. Eng., B **106**, 132–140 (2004). Elsevier
9. Kahraman, C., Yanik, S.: Intelligent Decision Making in Quality Management. Theory and Applications. Springer, Switzerland (2015)
10. Menéndez, F., Gómez, A., Voces, F., García, V.: Porcelain insulators in electrostatic precipitator. J. Electrostat. **76**, 188–193 (2015). Elsevier
11. Partyka, J., Gajek, M., Gasek, K., Lis, J.: Effect of alumina grain size distribution on mechanical properties and microstructure of the alumina porcelain type 130. Ceram. Mater. **65**(3), 283–289 (2013). Poland
12. Reddy, S., Verma, R.: Novel technique for electric stress reduction across ceramic disc insulators used in UHV AC and DC transmission systems. Appl. Energy. Elsevier (2016)
13. Schwarz, G.: Estimating the dimension of a model. Ann. Stat. **6**(2), 461–464 (1978)
14. Segall, R.: Some mathematical and computer modelling of neural networks. Appl. Math. Model. **19**(7), 386–399 (1995)
15. Tao, L., Baoshan, Z.: Technical and economical comparisons between different kinds of materials and different geometries for electrostatic precipitator insulators. In.: International Society for Electrostatic Precipitation ICESP, Australia (2006)
16. Yegnanarayana, B.: Artificial Neural Networks. PHI Learning Pvt. Ltd., India (2004). p. 280

Genetic Algorithm Adoption to Transport Task Optimization

Anna Burduk(✉) and Kamil Musiał

Mechanical Department, Wrocław University of Technology, Wrocław, Poland
anna.burduk@pwr.wroc.pl,
kamil.musial@student.pwr.wroc.pl

Abstract. The paper presents an optimization task of transportation - production solved with genetic algorithms. For the network of processing plants (factories) and collection centers the cost-optimal transportation plan will be established. Plan is regarding to raw materials to the relevant factories. Task of transportation - production regard to the milk transport and processing will be investigated. It is assumed that the functions defining the costs of processing are polynomials of the second degree. Genetic algorithms, their properties and capabilities in solving computational problems will be described and conclusions will be presented. The program that uses genetic algorithms written in MATLAB will be used to solve an investigated issue.

Keywords: Optimization of production systems · Generic algorithms

1 Introduction

Nowadays challenges to transport are more and more advanced and more and more advanced methods need to be adapted. To proceed transport processes fluently and affective each company has to take under consideration the principles of logistics management and focus on system organization and synchronization of physical flow of materials from manufacturers or wholesalers to consumers through all phases of the process [1, 4, 6, 10].

Expense spent on transport is a large part of total costs and one of factors causing the difference between the cost of goods producing and price paid by the consumer. There are many types of products that need various procedures. Fresh products- due to limited period of validity- are a part of the most sensible group. In this case, transport should take place as quickly as possible and temperature and humidity conditions have to be meet. These and many more reasons tend to pay attention to the problem of transport optimizing [9]. To reach these requirements and calculate so complex and multi-threaded tasks the new methods that can be successfully used in solving transport problem should be deployed. It is possible because of continuous technological and information advances. Transport issues are most often used to [5, 8]:

- optimal products transport planning, taking into account the minimization of costs, or time of execution

M. Graña et al. (eds.), *International Joint Conference SOCO'16-CISIS'16-ICEUTE'16*,
Advances in Intelligent Systems and Computing 527, DOI 10.1007/978-3-319-47364-2_35

- optimization of production factors distribution in order to maximize production value, profit or income.

The main rule of genetic algorithms functioning is imitation of biological processes, namely the processes of natural selection and heredity. Because of their properties they receive more and more followers and are in use in many areas of applications in the scientific and engineering as an effective tool for effective and efficient searching [1, 4]. Genetic algorithms belong to group of stochastic algorithms.

They may be used where it is not well defined or understood how to solve the problem, but it is known way to evaluate the quality solution. Investigating in this paper transport problem, where to find the chipper processing way is such a problem. Evaluation of the quality of the proposed solution is rapid, while finding the optimal route qualifies for the class NP-hard problems. Genetic algorithms are also used in finding approximations extremes of functions that cannot be calculated analytically.

They are a powerful tool to search for solutions and at the same time very simple. GA simplicity stems from the fact they are free from the essential constraints imposed by the strong assumptions about the search space. Is is not necessary to know i.e. existence of derivatives, continuity, modality of the objective function etc. A standard problem solved through genetic algorithm consists of [2, 3, 5, 7]:

- optimization problem – searching for the best solution of all allowable,
- a set of allowable solutions - the set of all possible solutions to the task (not only the optimal),
- the evaluation function (adjustment) - the function that determines each possible solution quality and establishes an ordinal relationship on the set of feasible solutions. Because of this function, all the solutions of the problem may be sorted from best to worst,
- coding method - a function that represents each acceptable solution in the form of string code, which is in the form of a chromosome. Basically, more interesting is its inverse function, that, on the basis of chromosome, creates a new feasible solution. In fact, to solve the task the only basic function is required, analogous to nature. Nature has the specific 'knowledge' on how to create an adult from an embryo. I would be enough that each of the individuals will be stored in the memory of genotype, on the basis of which he was created.

2 The Mathematical Model of Transport – Production Task

In following example optimization of transport - production tasks for the processing of milk will be considered. For network of collection points and processing plants, according to its algorithm, the cost-optimal distribution of transport tasks has been determined. It is assumed that the functions defining the costs of conversion are polynomials of the second degree. To solve the problem a program written in MATLAB based on the genetic algorithm, equalizing marginal costs is used.

Enterprise processing a uniform material has m collection points and n plants processing this material. Additional information should be known:

- unit cost of transportation from any collection point to individual processing plants,
- amount of material collected at each point of supply,
- functions defining the cost of the material processing at each plant, depending on the size of the processing.

Features defining the costs of conversion are convex and square functions. They take into account only the variable costs, which depend on the size of production. The entire acquired material must be transported to the plant and converted there. It is assumed that plants are able to process the supplied amount of material (the possibility of processing by plants are known). This increases the production capacity of plants, but also results in an increase in the unit cost of production. Rising costs of conversion are a natural limitation of the size of production of each establishment.

It is needed to establish a plan of material supplies to individual plants and processing of raw materials in these plants, so the total costs of transport and processing were minimal. The following designations have been adopted:

i - the number of the collection (supplier number),

j - number of the processing plant (recipient number),

x_{ij} - the amount of raw material transferred from the i-th supplier to the j-th recipient,

x_j - the amount of raw material processed by the j-th recipient,

a_i - the amount of raw material, which must be send by i-th supplier,

c_{ij} - the unit cost of transport from the i-th supplier to the j-th recipient,

$f_j(x_j)$ - the cost of processing x_j units of raw material in the j-th plant (at j-th recipient).

Furthermore assumed that the convex cost function fi is a second degree polynomial of the form:

$$f_j(x_j) = e_j x_j^2 + c_j x_j,\ c_j, e_j > 0 \tag{1}$$

where:

c_j - describes the minimum unit cost of processing,

e_j - determines the growth rate of unit cost.

The first derivative of this function is determined by the marginal cost of processing:

$$F_J'(\mathrm{X_J}) = 2\ \mathrm{E_J X_J} + \mathrm{C_J} \tag{2}$$

while the second derivative - the rate of increase in the marginal cost:

$$F_J''(\mathrm{X_J}) = 2\ \mathrm{E_J} \tag{3}$$

The average cost of processing the j-th plant is determined by the formula:

$$K_J^P(X_J) = E_J X_J + C_J \tag{4}$$

The problem of determining the optimal supply plan of raw material and its processing can be presented in the form of a non-linear decision task.

Variables x_{ij} and x_j are sought that:

$$\sum_{i=1}^{m}\sum_{j=1}^{n} c_{ij}x_{ij} + \sum_{j=1}^{n} f_j(x_j) \rightarrow \min \quad (5)$$

By the conditions:

$$\sum_{j=1}^{n} x_{ij} = a_i; (i = 1, \ldots, m), \quad (6)$$

$$\sum_{i=1}^{m} x_{ij} = x_j; (j = 1, \ldots, n) \quad (7)$$

$$x_{ij}, x_j \geq 0; (i = 1, \ldots, n) \quad (8)$$

The objective function (5) minimizes the total cost of transport and processing. Condition (6) provides that each supplier will send all owned raw material. Condition (7) forces the processing in the j-th plant of all the raw material to which it is delivered. Task (5–7) is the task of quadratic programming with a special - transport structure. It can be solved by using an algorithm equalizing the marginal cost, which is based on genetic algorithm.

Marginal cost, the cost of which the manufacturer incurs due to the increased size of production of the good by one unit. It is the increase in total costs associated with producing an additional unit of a good. If the plant increases its production by one unit, then the total cost of production will increase. The difference in the size of the costs manufacturer incur earlier and costs incurred after the increase in production is a marginal cost. It is, therefore, the cost of producing an additional unit of a good.

The concept of marginal cost can also be formulated in relation to the consumer and is then taken as the cost of acquiring an additional unit of a good. The marginal cost is an important micro-economic category. It was observed that for typical business processes marginal costs initially decrease with the increase in production until the technological minimum is reached. Further increase of production over a minimum of technology, however, increases the unit cost of further increases in production and thus rising marginal costs. This observation is important in microeconomic analysis of the behavior of the manufacturer and determining the optimal level of production. According to economic theory, marginal cost cannot be negative. This means that the increase in production may entail reducing the total cost.

Method of equalizing the marginal cost based on genetic algorithm consists of:

- determination of the best possible, an acceptable solution output,
- improvement of new solutions X^1, X^2, …, by offset equalizing marginal costs.

A string of new obtained solutions X^1, X^2, …, X^r, …., does not need to be finished. It is therefore interrupted at some point of calculations. It is important, however, that the final solution does not deviate too far away (in terms of objective function value) from the optimum solution. JCC algorithm comes down to the following steps:

1. Determine the initial solution:
   (a) for the *i*-th supplier (i = 1,…, m) the route with minimal marginal cost is set,
   (b) on the selected route an entire supply of *i*-th supplier is located,
   (c) update the marginal costs in the column of the selected route.

   Then move on to the next vendor, and repeating steps (a)–(c) until the supply is disposed for all suppliers.
2 Make sure the current solution Xr meets the criterion of optimality. If so, the final solution is optimal. If not- return to step 3.
3 Make sure the solution Xr is ε - accurate. If so, finish the calculations. If not, go to step 4.
4 Improving the solution by shifting equalizing marginal costs and return to step 2.

Having designated solution Xr and the matrix of marginal costs Kr for each supplier, lets settle the differences between the maximum realized cost and the minimal cost.

3 Solving the Problem of Transport and Production Using a Genetic Algorithm

The study involved the delivery of milk (about 5620 m^3/month). The task was formulated as follows:

Six suppliers (in 8 cities): D1, D2, D3, D4, D5, D6, D7, D8 supplies milk to two factories - S1, S2, with restrictions:

S1: can accept and process 2100 or 3000 m^3 of milk,

S2: can accept and process 4500 m^3 of milk.

Data are summarized in Table 1 below and include unit transportation costs (in PLN per km), offered monthly deliveries A_i (m^3) and monthly demand of factories B_j(m^3).

The task is solved with the help of developed in MATLAB genetic algorithm equipped with a graphic interface GUI. The aim of the task is to determine the optimal marginal cost of “material processed” transport from any supplies, to one of two factories, taking into account its processing capacity.

Table 1. Unit transportation costs, supply and demand

Suppliers	Supply A_i [m^3]	Factories			
		Variant v1		Variant v2	
		S1	S2	S1	S2
D1	600	15	180	15	180
D2	1200	120	180	120	180
D3	1000	210	45	210	45
D4	700	210	15	210	15
D5	420	300	150	300	150
D6	500	185	420	185	420
D7	700	230	100	230	100
D8	500	300	240	300	240
Demand B_j [m^3]		2100	4500	3000	4500

3.1 Solving the Transport- Production Problem of in the General Scheme of Genetic Algorithm

In each issue investigated with genetic algorithms there are a few specific parameters that may affect the result and have to be determined: coding method (population may be coded as a vector of genes, tree structure data, etc.), the evaluation function (calculated for each solution on the basis of a model of the solving problem), selection methods (roulette method, ranking methods etc.), crossover method (determined by the encoding of chromosomes and problem specification and percentage of generation that mutate.

Population. The simple "vector of genes" coding method has been applied. The first step is to number all suppliers. Created chromosome has a length such as the number of suppliers. In the following genes another supplier is saved. For example, if the gene number 1 "represents" city X and gen number 2 city Y it means that the provider moves from town X to Y. In this way, genes in the chromosome are arranged exactly as the city cycle. For a sample of eight cities connected with road, a sample chromosome mapping cycle may look like this:
[1 4 8 6 2 3 7 5]

The evaluation function (cost). The function of evaluation is the total minimum cost of transport and processing- solutions with the lowest cost are favored.

Selection. Simulation using the roulette wheel assigns the probability of choosing each individual directly on the basis of a single evaluation function. The sum of the probabilities assigned to each chromosome is equal to 1, which means that if the area of a circle is the sum of the values of the objective function of population individuals, then each of them is associated with a circle section. In the current example, it eight fields exist.

Crossover. One-point crossover. Individuals are combined sequentially in pairs:

5.39	5.8
5.8	6.51

Parent 1: [3 4 7 6 2 8 1 5]
Parent 2: [4 1 5 3 7 2 8 6]
Among the offspring duplicates of existing individuals may appear. Duplicates do not bring anything to the database all the genes are subjected to forced mutation.

Mutation. After giving birth to offspring, approx. 50 % of generation mutates.

3.2 Solution to the Problem in MATLAB

After entering all necessary data, information about the correct solution of the problem is obtained (Fig. 1). On the y-axis the accuracy of the solution (in %), and on the x-axis the number of iterations (max = 5) is given.

If the solution is optimal- accurate results in a table showing the following information:

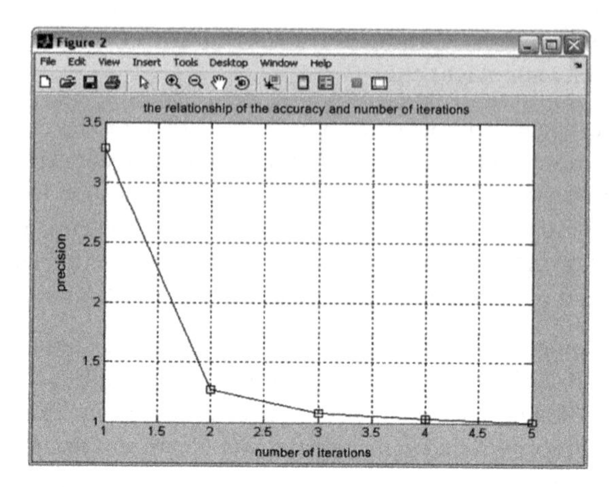

Fig. 1. The dialog box of the program - the relationship of the accuracy and the number of iterations

- amount of processing at the individual plants,
- the total cost of the transportation and processing of milk,
- the cost of transport,
- the cost of processing,
- average costs,
- marginal costs,
- way of the deployment of milk.

The calculations examined several variants: supply, demand (processing capacity), and processing costs (description of function) were variable. For example, variant v1 adopted by Table 1 the following demand: for factory S1-2100 m^3/month, and for the

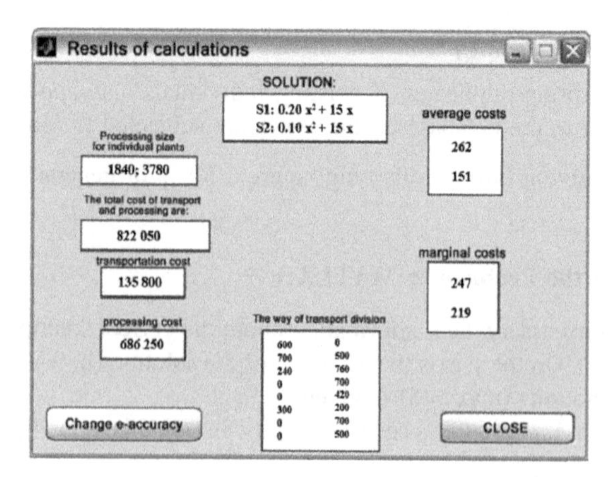

Fig. 2. Terms of the calculation in the program for variant v1

factory S2-4500 m^3/month. Also, based on studies carried out it was assumed that the processing functions have the following form:

$$f_1(x_1) = 0,2\ x_1^2 + 15x_1 \text{ and } f_2(x_2) = 0,1x_1^2 + 15x_2$$

The results of calculations are shown in Fig. 2 (screen of end results). In the case of variant V2 (data according to Table 1) adopted the demand: for factory S1-3000 m^3/month and for the factory S2-4500 m^3/month. Processing functions have the form:

$$f_1(x_1) = 0,2\ x_1^2 + 10x_1 \text{ and } f_2(x_2) = 0,1x_1^2 + 10x_2$$

The results are shown in Fig. 3.

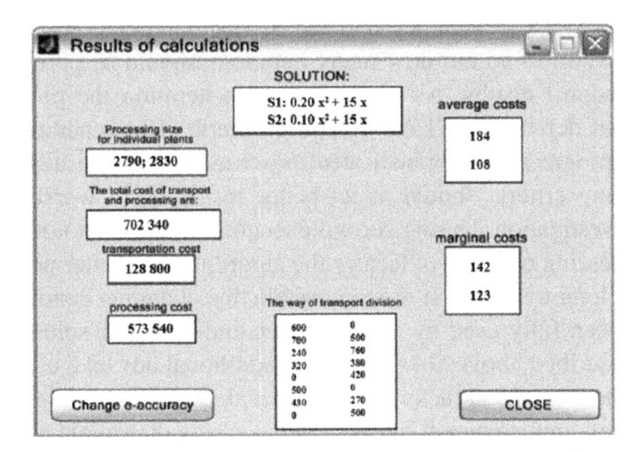

Fig. 3. The results of the calculations in the JCC for the variant v2

3.3 Summary of Results

Collective summary of results is shown in Table 2. There is placed sample of simulation results conducted for several variants, which are varied in parameters of processing function, transportation costs, the possibility of processing by individual factory. Statement contains the best results.

The chart shows that the total cost of task for variant v2 is 720 340 PLN and it is more than 100 000 PLN lover than the costs of the variant v2. This difference is due to the fact that for small factories processing costs are higher than for larger ones and extending the factory processing capacity an algorithm had chance for better suit.

As mentioned above: basic genetic algorithms do not guarantee finding the best solution. Process stops when stopping criteria are met e.g. when time is out or when a determined number of iterations does not give progress. Result also highly depends on coding method, evaluation function, selecting methods and number of gene that mutate. Limiting the mutation it is higher probability to achieve extremum (local maximum or minimum) rather then global optimum. From the other hand, to high mutation gives an effect opposite than assumed and next generations are more various rather than

Table 2. Results for the task

Suppliers	Factories			
	Variant v1		Variant v2	
Processing [m3]	1840	3780	2790	2830
Stock of processing [m3]	260	720	210	1670
Transport costs [PLN]	135 800		128 800	
Processing costs [PLN]	686 250		573 540	
Total costs of transport and processing [PLN]	822 050		702 340	
Average costs [PLN/m3]	262	151	184	108
Marginal costs [PLN/m^3]	247	219	142	123

convergent to optimum. Choosing coding and selecting method affects the accuracy of next generation.

The relationship of the accuracy and the number of iterations (Fig. 1) has been obtained and showed to find out how many iterations should be performed to achieve adequate (not random) results. As seen, above 5th iteration the progress disappears what let claim that default MATLAB stopping criteria (100 generations and 50 generations with no progress) are sophisticated to get meaningful results and it has been assumed. Extending criteria should affect better result or, in worst case, should be transparent for investigation. During second execution it has been noticed that because extending of processing capacity of factory the algorithm had better possibility to avoid losses because of long transport. It was noticeable that changing conditions (bottleneck liquidation) has been fully used by algorithm for finding better solution.

Because of described above GAs feature an additional advantage is clearly seen. In tasks in which time is a critical factor, the classical algorithm will finish or not and must be earlier adequately implemented. In case of time excess the simple classical algorithm is not able to improve its results. GA performs iterations- repeats the same procedure achieving incrementally more and more desired result. It may be easy programmed to work till next higher-priority operation has to be executed and interrupting GA calculation process. Following the relationship of the accuracy and the number of iterations (Fig. 1) it is noticeable that in case of time regime and lower-class result required an interrupting after 5th iteration may be taken under consideration. It is noticed that vector coding method and roulette selection method are appropriate for these type of transport tasks and make GA very efficient. Comparing with default stopping criteria this process would take less than 10 % based computing power.

4 Summary

The proposed work approach using solving the problem of production and transport costs with convex function should facilitate decision-making processes of transport and production management. Using genetic algorithms with specific settings the required conditions have been met: optimum (global, not local extremum) has been found with both cases. GAs fully confirmed its effectiveness for the problem, minimizing total costs of transport and processing.

References

1. Ayough, A., Zandieh, M., Farsijani, H.: GA and ICA approaches to job rotation scheduling problem: considering employee's boredom. Int. J. Adv. Manuf. Technol. **60**, 651–666 (2012)
2. Chodak G., Kwaśnicki W.: Genetic algorithms in seasonal demand forecasting. In: Information Systems Architecture and Technology 2000, Wrocław University of Technology, pp. 91–98 (2000)
3. Govindan, K., Jha, P.C., Garg, K.: Product recovery optimization in closed-loop supply chain to improve sustainability in manufacturing. Int. J. Prod. Res. **54**(5), 1463–1486 (2016)
4. Guvenir, H.A., Erel, E.: Multicriteria inventory classification using a genetic algorithm. Eur. J. Oper. Res. **105**(1), 29–37 (1998)
5. Jachimowski, R., Kłodawski, M.: Simulated annealing algorithm for the multi-level vehicle routing problem, Logistyka 4 (2013)
6. Krenczyk, D., Skolud, B.: Transient states of cyclic production planning and control. Appl. Mech. Mater. **657**, 961–965 (2014)
7. Nissen, V.: Evolutionary algorithms in management science. An overview and list of references. Papers on Economics & Evolution, Report No. 9303, European Study Group for Evolutionary Economics (1993)
8. Sahu, A., Tapadar, R.: Solving the assignment problem using genetic algorithm and simulated annealing. Int. J. Appl. Math. **36**, 1 (2007)
9. Yusoff, M., Ariffin, J., Mohamed, A.: Solving vehicle assignment problem using evolutionary computation. In: Tan, Y., Shi, Y., Tan, K.C. (eds.) ICSI 2010, Part I. LNCS, vol. 6145, pp. 523–532. Springer, Heidelberg (2010)
10. Zegordi, S.H., Beheshti Nia, M.A.: A multi-population genetic algorithm for transportation scheduling. Transp. Res. Part E: Logist. Transp. Rev. **45**(6), 946–959 (2009)

Detecting Existence of Cycles in Petri Nets

An Algorithm that Computes Non-redundant (Nonzero) Parts of Sparse Adjacency Matrix

Reggie Davidrajuh(✉)

Department of Electrical and Computer Engineering,
University of Stavanger, Stavanger, Norway
Reggie.Davidrajuh@uis.no

Abstract. Literature study reveals the existence of many algorithms for detecting cycles (also known as circuits) in directed graphs. Majority of these algorithms can be classified into two groups: (1) traversal algorithms (e.g. variants of depth-first-search algorithms), and (2) matrix-based algorithms (manipulation of the adjacency matrix and its power series). Adjacency matrix based algorithms are computationally simple and more compact than the traversal algorithms. However, a Petri net, due to its bipartite nature, possesses sparse matrix as adjacency matrix. Hence, the matrix based algorithms become inefficient as the algorithm work through redundant data most of its running time. In this paper, we take a closer look into the structure of the adjacency matrix of a Petri net and then propose an algorithm for detecting existence of cycles; the proposed algorithm is computationally efficient as its works through non-redundant data only, as well as simple, easy to understand and easy to implement.

Keywords: Cycles detection · Petri nets · Directed graph

1 Introduction

During analysis of Petri nets, detection of cycles becomes important. This is because, some of the techniques for analysis of Petri nets are based on the cycles in the Petri net. For example, in marked graphs (a subclass of Petri nets), performance analysis and finding bottlenecks are based on the cycles in the graph [1]; topological ordering of activities in Petri nets are also depend on the cycles in the Petri net.

This paper proposes an efficient adjacency matrix-based algorithm for cycle detection in Petri nets. Firstly, this paper shows that this is a difficult problem as the adjacency matrices of Petri nets are sparse, due to the bipartite nature; a sparse matrix is one in which most of the elements are zero. Secondly, this paper proposes an efficient algorithm after taking a closer look into the adjacency matrix.

In this paper: Sect. 2 presents a short literature study on detecting cycles in directed graphs. Section 3 sheds more details into matrix-based algorithms for cycle detection. Section 4 presents a closer look into the adjacency matrix of a Petri net, followed by a proposal for an efficient algorithm in Sect. 5. Section 6 presents a simulation study, to prove the efficiency of the new algorithm.

M. Graña et al. (eds.), *International Joint Conference SOCO'16-CISIS'16-ICEUTE'16*,
Advances in Intelligent Systems and Computing 527, DOI 10.1007/978-3-319-47364-2_36

2 Cycle Detection in Directed Graphs

Detecting the simple cycles in a directed graph is of polynomial computational complexity. There are several algorithms and methods that provide polynomial running time, with worst-case timing of $O(n^3)$, where n is the number of vertices. These algorithms can be classified into the following groups:

1. Traversal algorithms,
2. Special algorithms, and
3. Adjacency matrix-based algorithms

Under the traversal algorithms group, the simplest method for detecting existence of cycles in a directed graph is to use depth-first-search (DFS). A single DFS traversal takes $O(n^2)$ time; performing DFS from a single source vertex may identify some cycles [2]; if not, repeating DFS from all other vertices is necessary until any cycles are found; this repetition takes $O(n^3)$ time. There are several traversal algorithms (e.g. [3, 4]) that provide better running times, and the running time can be reduced to quadratic or even to linear timing (if the graph is sparsely connected) [5].

Under the special algorithms group, there are algorithms based on linear programming and hybrid approaches. Hruz and Zhou [6] suggests the use of linear programming for finding cycles. However, it is not clear how easy it is to formulate a directed graph as linear programming problem, as discussed in [7].

Adjacency matrix based algorithms are computationally simple and more compact than the traversal algorithms. In the next section, we take a closer look into the matrix-based algorithm for cycle detection.

3 Adjacency Matrix-Based Algorithm

The Fig. 1 shows a simple directed graph $\boldsymbol{G} = (\boldsymbol{V}, \boldsymbol{E})$, where $\boldsymbol{V}$ is the set of seven vertices (***a*** to ***g***), and ***E*** is the set of twelve edges (such as ***ab***, ***ad***, ***be***, ***bg***, ***cf***, ***da***, ***dc***, ***ef***, ***gc***, ***gd***, and ***ge***). The adjacency matrix (*A*) of the graph ***G***, also shown in the Fig. 1, indicates the destination sources, if we were to take one-step from a source. E.g. the first row of the matrix ***A*** indicates that from the source ***d***, we can reach the destinations ***a*** or ***c*** in one-step.

The transpose of the adjacency matrix ($\boldsymbol{A}^T$) indicates that from which sources we can reach a particular destination; e.g. the first row of $\boldsymbol{A}^T$ indicates that the vertex ***a*** can be reached from the vertex ***d*** in one-step. Finally, the cycle matrix (***C***) is obtained by conjuncting (element by element *AND* operation) ***A*** and $\boldsymbol{A}^T$ ($\boldsymbol{C} = \boldsymbol{A} \wedge \boldsymbol{A}^T$).

Matrix ***C*** indicates the 2-step cycles in the graph ***G***. E.g. the first row of the matrix ***C*** indicates that the vertices ***a*** and ***d*** are involved in 2-step cycle ***a-d-d-a***; the third row of the matrix ***C*** indicates that vertices ***c*** and ***f*** are also involved in 2-step cycle ***c-f-f-c***. Similarly, $\boldsymbol{C}^2$ (obtained by $\boldsymbol{A}^2 \wedge \boldsymbol{A}^{2T}$) indicates the 4-step cycles in ***G***, $\boldsymbol{C}^3$ (obtained by $\boldsymbol{A}^3 \wedge \boldsymbol{A}^{3T}$) indicates the 6-step cycles in ***G*** and so on. This means, for detecting existence of cycles, we need to compute cycle matrices $\boldsymbol{C}^r$ until we find a $\boldsymbol{C}^r$ that becomes zero-matrix or r becomes n.

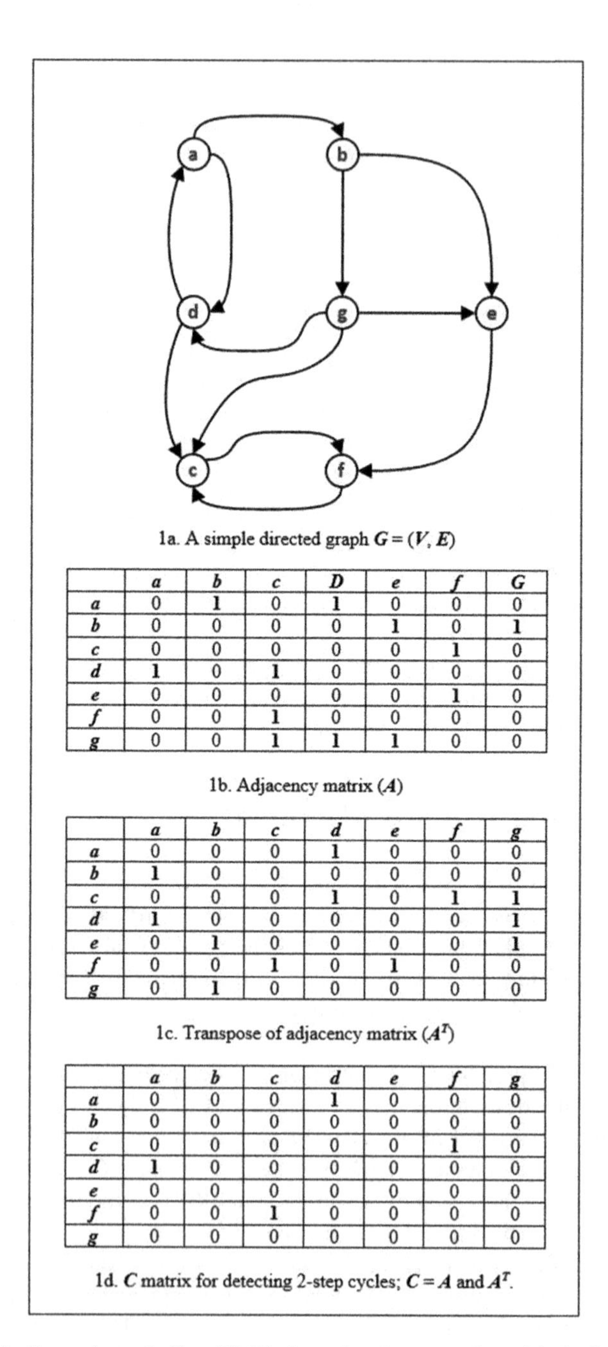

1a. A simple directed graph $G = (V, E)$

	a	*b*	*c*	*D*	*e*	*f*	*G*
a	0	1	0	1	0	0	0
b	0	0	0	0	1	0	1
c	0	0	0	0	0	1	0
d	1	0	1	0	0	0	0
e	0	0	0	0	0	1	0
f	0	0	1	0	0	0	0
g	0	0	1	1	1	0	0

1b. Adjacency matrix (A)

	a	*b*	*c*	*d*	*e*	*f*	*g*
a	0	0	0	1	0	0	0
b	1	0	0	0	0	0	0
c	0	0	0	1	0	1	1
d	1	0	0	0	0	0	1
e	0	1	0	0	0	0	1
f	0	0	1	0	1	0	0
g	0	1	0	0	0	0	0

1c. Transpose of adjacency matrix (A^T)

	a	*b*	*c*	*d*	*e*	*f*	*g*
a	0	0	0	1	0	0	0
b	0	0	0	0	0	0	0
c	0	0	0	0	0	1	0
d	1	0	0	0	0	0	0
e	0	0	0	0	0	0	0
f	0	0	1	0	0	0	0
g	0	0	0	0	0	0	0

1d. C matrix for detecting 2-step cycles; $C = A$ and A^T.

Fig. 1. A simple directed graph $G = (V, E)$; detecting 2-step cycles with the help of C matrix.

It will be clear in the next section that cycles in a Petri net always involve even numbered steps, due to its bipartite nature. Hence, non-zero elements in a $\boldsymbol{C}^r$ matrix indicate existence of $2r$-steps cycles. Finding any cycle matrix ($\boldsymbol{C}^r$, $r = 1 \ldots n$), takes $O(n^3)$ time. Thus the series ($\boldsymbol{C}, \boldsymbol{C}^2, \ldots, \boldsymbol{C}^r$) takes $r \times O(n^3)$, depending on the value of r.

4 Adjacency Matrix of a Petri Net

The Fig. 2 shows a simple Petri net with six places and four transitions.

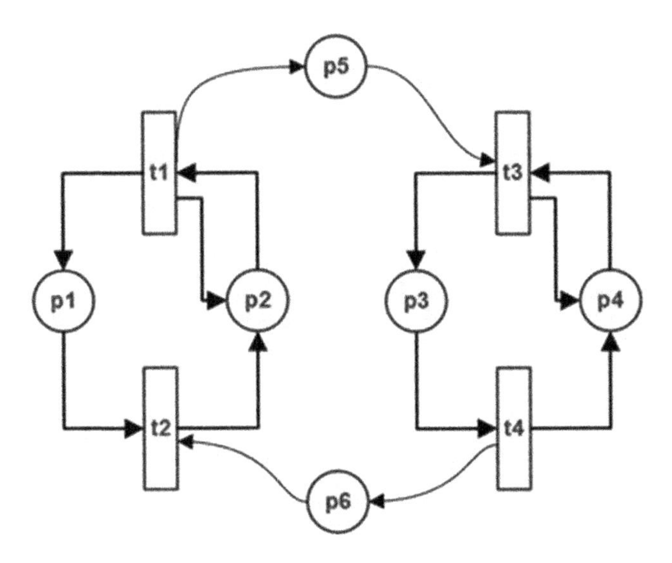

Fig. 2. A simple Petri net.

The Fig. 3 shows the input adjacency matrix (***Ai***) and the output adjacency matrix (***Ao***) of the Petri net. The Fig. 3c shows the adjacency matrix ***A*** as a composition of the matrices ***Ai***, ***Ao***, and two zero matrices. Being a bipartite graph, the first quadrant is a zero matrix as it represents the lack of connections between the transitions; the fourth quadrant is also a zero matrix as it represents the lack of connections between the places. Due to these two zero matrices, ***A*** is considered as a sparse matrix. Hence, computing cycle matrix $\boldsymbol{C}^r$ by conjuncting $\boldsymbol{A}^r$ and $\boldsymbol{A}^{rT}$ is inefficient due to the zero matrices involved in computations.

The Fig. 4 shows the matrices $\boldsymbol{A}$, $\boldsymbol{A}^T$, C, $\boldsymbol{A}^2$, $\boldsymbol{A}^{2T}$, $\boldsymbol{C}^2$, and $\boldsymbol{A}^3$, $\boldsymbol{A}^{3T}$, and $\boldsymbol{C}^3$. Comparing $\boldsymbol{C}$, $\boldsymbol{C}^2$, and $\boldsymbol{C}^3$, it is evident that there is a pattern in these matrices such that there are always two zero matrices and two non-zero matrices (non-redundant matrices). In addition, in the even-powered matrices ($\boldsymbol{C}^2$, $\boldsymbol{C}^4$, …), the first and the fourth quadrants are occupied by non-zero matrices, whereas in the odd-powered matrices ($\boldsymbol{C}$, $\boldsymbol{C}^3$, …), the second and the third quadrants are occupied by non-zero matrices.

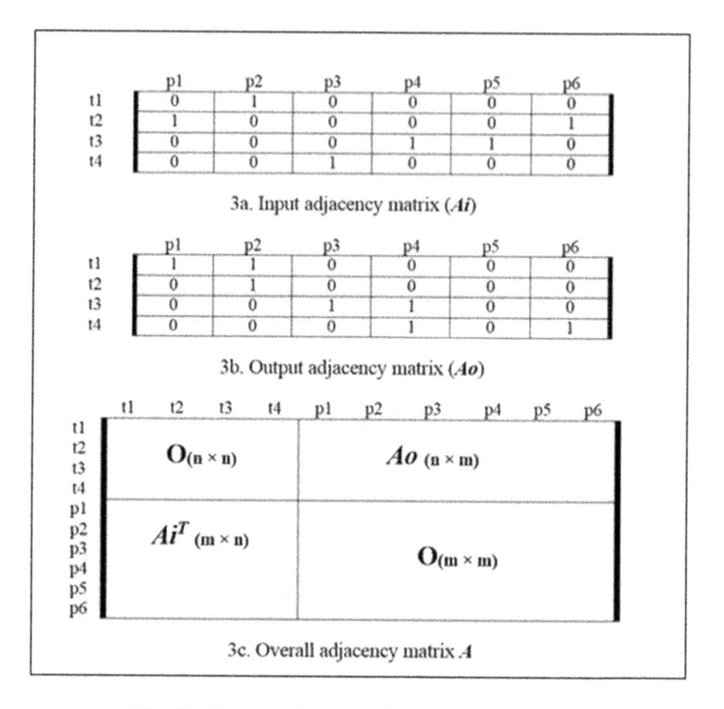

	p1	p2	p3	p4	p5	p6
t1	0	1	0	0	0	0
t2	1	0	0	0	0	1
t3	0	0	0	1	1	0
t4	0	0	1	0	0	0

3a. Input adjacency matrix (*Ai*)

	p1	p2	p3	p4	p5	p6
t1	1	1	0	0	0	0
t2	0	1	0	0	0	0
t3	0	0	1	1	0	0
t4	0	0	0	1	0	1

3b. Output adjacency matrix (*Ao*)

3c. Overall adjacency matrix *A*

Fig. 3. Input and output dependency matrices.

5 An Efficient Algorithm for Cycle Detection

From these observations (summarized in the Fig. 4), a new algorithm for computing the successive C^r matrices is shown in the Fig. 5. The algorithm excludes the redundant zero matrices.

The algorithm has a main loop that creates a higher order of C in each iteration. The loop terminates when a cycle is found or the maximum number of iterations (n – the total number of places and transitions) is reached. Within the main loop, there are three blocks:

- The first block creates C^r when r is an odd number; in this case, only the 2nd and the 3rd quadrants are computed (1st and 4th quadrants are zero matrices).
- The second block creates C^r when r is an even number; in this case, only the 1st and the 4th quadrants are computed (2nd and 3rd quadrants are zero matrices).
- Finally, in the third block, the old quadrant values are updated with newer values for computing C^{r+1} in the next iteration.

Due to brevity, the Fig. 5 only shows the core of the algorithm. The complete algorithm (coded in MATLAB M-files) is given in the web page [8]; interested readers are encouraged to visit the web page for downloading the codes and experimenting with them.

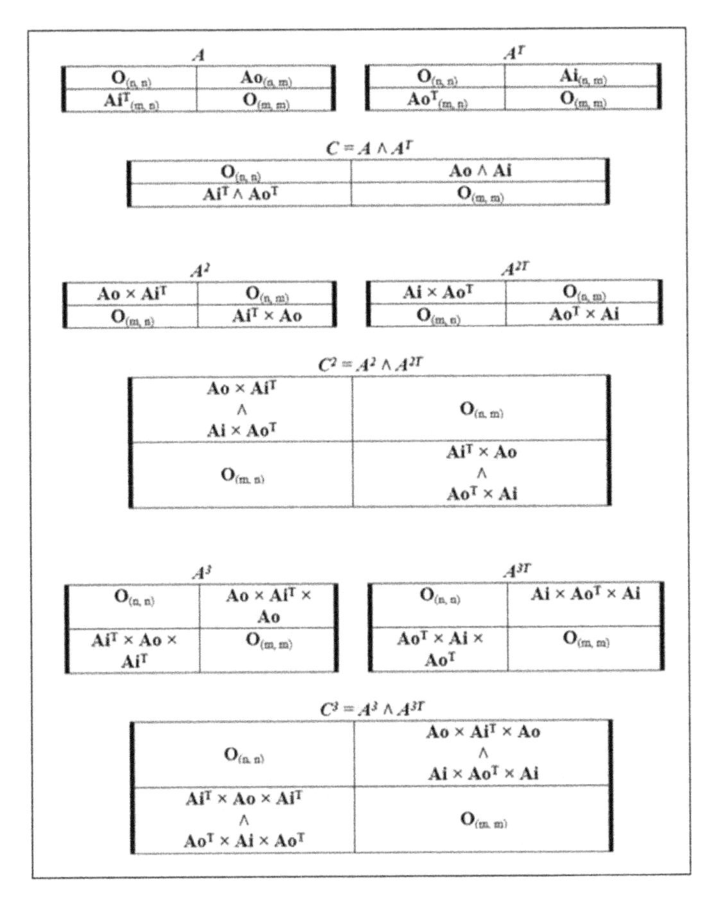

Fig. 4. The power matrices of $\boldsymbol{A}$ and $\boldsymbol{C}$.

6 Simulation

To show the efficiency of the algorithm, let us consider an extraordinary Petri Net, in which all the vertices (the places and the transitions) are connected in a series to form a single circle, as shown in the Fig. 6. The extraordinary structure (one complete cycle) is purposely is chosen so that the cycle will be detected only in $\boldsymbol{C}^n$, the last iteration of $\boldsymbol{C}^r$. Alternative, we could have also selected a Petri net without any cycles on it, so that all the iterations of $\boldsymbol{C}^r$ ($\boldsymbol{C}^1$, …, $\boldsymbol{C}^n$) has to be done to conclude non-existence of cycles. We believe that by performing all the iterations exhaustively, the execution times for the traditional algorithm (that uses the overall sparse adjacency matrix) and the new algorithm proposed in this paper (that uses only the non-redundant ***Ai*** and ***Ao*** parts of the adjacency matrix) will clearly show the performance improvement by the new algorithm.

Let's experiment executing the traditional algorithm (that process the sparse matrices $\boldsymbol{A}, \boldsymbol{A}^2$, and so on), and then compare the timing with the execution of the newer algorithm, for a variety of n (the number of places and the number of transitions).

```
While ( (r ≤ n)  and  not(cycles_found) )
{
   r = r + 1

   If (r is odd),
   {  % computing A^(r+1)
      new_quadrant_2 = Ao  × old_quadrant_4   % A^(r+1)
      new_quadrant_3 = Ai^T  × old_quadrant_1  % A^(r+1)
      new_quadrant_2T = transpose(new_quadrant_2)
      new_quadrant_3T = transpose(new_quadrant_3)

      % computing C^(r+1)  (1st and 4th quadrants are zero)
      C_quadrant_2 = and(new_quadrant_2, new_quadrant_2T)
      C_quadrant_3 = and(new_quadrant_3, new_quadrant_3T)

      Cycle_found = (C_quadrant_2 is not zero) or
                    (C_quadrant_3 is not zero)
   }

   If (r is even),
   {  % computing A^(r+1)
      new_quadrant_1 = Ao  × old_quadrant_3   % A^(r+1)
      new_quadrant_4 = Ai^T × old_quadrant_2   % A^(r+1)
      new_quadrant_1T = transpose(new_quadrant_1)
      new_quadrant_4T = transpose(new_quadrant_4)

      % computing C^(r+1)  (2nd and 3rd quadrants are zero)
      C_quadrant_1 = and(new_quadrant_1, new_quadrant_1T)
      C_quadrant_4 = and(new_quadrant_4, new_quadrant_4T)

      Cycle_found = (C_quadrant_1 is not zero) or
                    (C_quadrant_4 is not zero)
   }

   % update old quadrant with new quadrants values
   old_quadrant_x = new_quadrant_x;   x = 1,2,3, and 4
}
```

Fig. 5. The new algorithm.

Petri nets of very smaller size ($n < 10$) are excluded in the experiment, as the results will be inconclusive due to their very small size.

6.1 Experimenting with Small to Medium Sized Petri Nets

Table 1 given below shows the timings for small to medium sized Petri nets, where the size n varies from 10 to 100.

The Table 1 clearly shows that starting with the value of $n = 20$, the new algorithm outperforms the traditional algorithm. Specifically, for the value of hundred ($n = 100$), there is 69.5 % improvement.

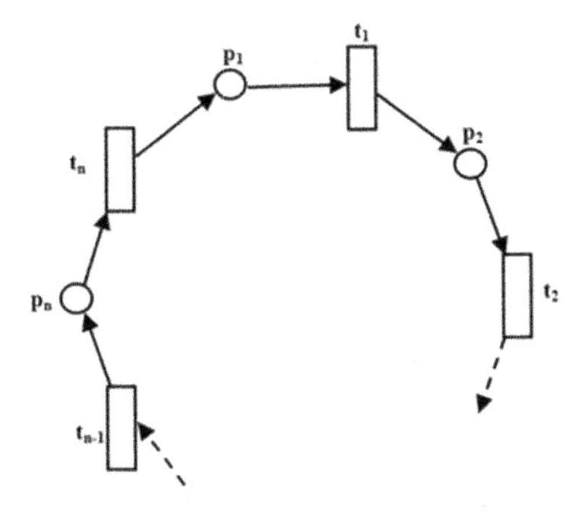

Fig. 6. The Petri net for simulation

Table 1. Execution time for small to medium sized Petri nets.

Size (n)	Timing in seconds		Size (n)	Timing in seconds	
	TA	NA		TA	NA
10	0.0033	0.0055	60	0.0312	0.0101
20	0.0008	0.0006	70	0.0487	0.0176
30	0.0030	0.0021	80	0.0795	0.0239
40	0.0103	0.0024	90	0.1160	0.0389
50	0.0180	0.0060	100	0.1726	0.0527

6.2 Experimenting with Large and Very Large Petri Nets

Table 2 shows the execution time for large to very large Petri nets.

The execution times presented in Table 2, also plotted in the Fig. 7, clearly shows the efficiency of the new algorithm. For the large Petri net of size 250 to 1000, the performance improvement seems to stabilize around 73 %.

Table 2. Execution time for large to very large Petri nets.

Size (n)	Timing in seconds		Performance improvement
	TA	NA	
250	6.577	1.765	73.2 %
500	96.5435	26.334	72.7 %
750	469.7175	126.7972	73.0 %
1000	1452.1988	378.0681	73.9 %

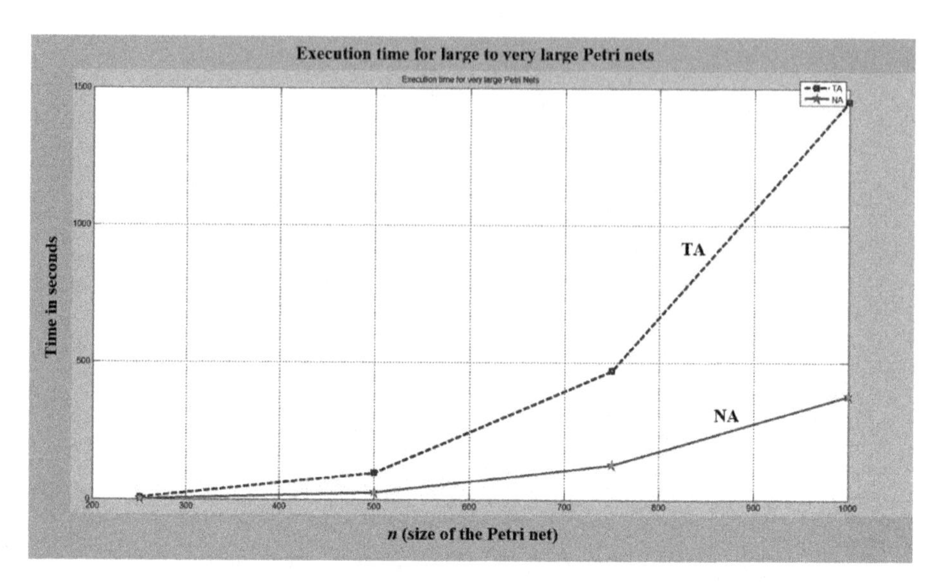

Fig. 7. Execution time of traditional algorithm (TA) and the new algorithm (NA) for very large Petri nets.

7 Conclusion

This paper proposes a new algorithm for detecting cycles in a Petri net. The proposed algorithm works faster as it only process non-redundant parts of the adjacency matrix, whereas the traditional algorithm processes the whole adjacency matrix which possess redundant zero matrices. The redundant zero matrices are always part of the adjacency matrix, due to the bipartite nature of the Petri nets. The simulation study presented in this paper shows that up to 73 % performance improvement can be achieved by the new algorithm.

The running time of the proposed algorithm is still $O(n^3)$, thus is not suitable for some specific situations (e.g. big data) that demand quadratic or linear running time. However, the constant factors hidden in the $O(n^3)$ notation are quite small. In addition, the proposed algorithm is simple as it is based on simple iterations. Hence, this algorithm is a suitable algorithm to be integrated into a Petri net simulator such as General Purpose Petri net Simulator (GPenSIM) [9].

The issue of sparse matrices has become a very important topic for research as we have entered the era of big data. Literature study reveals works on applying sparse matrices for graph algorithms [10], the matrix-to-matrix multiplication involving sparse matrices [11], sparse matrix applications in statistical analysis [12], etc. However, the treatment of sparse matrix in Petri nets due to its bipartite nature is completely new. There isn't a single work that is dedicated to this issue. Hence, we believe that this paper will be an interesting addition to the literature.

Further work: the simulation study given the paper reveals that for large Petri nets (consisting of a few hundred places and transitions) the performance improvement of the proposed algorithm over the traditional algorithm asymptotically converges to 73 %.

We don't know why the performance improvement converges to this value. Perhaps this issue should be solved in a more analytical paper.

References

1. GPenSIM V9. User Manual (2016). http://davidrajuh.net/gpensim/
2. Cormen, T.H.: Introduction to algorithms. MIT Press, Cambridge (2009)
3. Tarjan, R.: Enumeration of the elementary circuits of a directed graph. SIAM J. Comput. **2** (1973), 211–216 (1973)
4. Teirnan, J.C.: An efficient search algorithm to find the elementary circuits of a graph. Commun. ACM **13**(1970), 722–726 (1970)
5. Johnson, D.B.: Finding all the elementary circuits of a directed graph. SIAM J. Comput. **4** (1), 77–84 (1975)
6. Hruz, B., Zhou, M.C.: Modeling and Control of Discrete-event Dynamic Systems. Springer, Heidelberg (2007)
7. Yamada, T., Kataoka, S.: On some LP problems for performance evaluation of timed marked graphs. IEEE Trans. Autom. Control **39**, 696–698 (1994)
8. GPenSIM code for Job Scheduling in Grid Computing: http://www.davidrajuh.net/gpensim/2016-SOCO-Cycle-Detection
9. Davidrajuh, R.: Developing a new petri net tool for simulation of discrete event systems. In: Proceedings of the Second Asia International Conference on Modeling & Simulation (AMS), pp. 861–866. IEEE (2008)
10. Buluç, A., Gilbert, J.R., Shah, V.B.: Implementing sparse matrices for graph algorithms. In: Graph Algorithms in the Language of Linear Algebra, pp. 22, 287 (2011)
11. Buluç, A., and Gilbert, J. R.: New ideas in sparse matrix-matrix multiplication. In: Graph Algorithms in the Language of Linear Algebra, pp. 22, 315 (2011)
12. Slavakis, K., Giannakis, G.B., Mateos, G.: Modeling and optimization for big data analytics: (statistical) learning tools for our era of data deluge. IEEE Signal Process. Mag. **31**(5), 18–31 (2014)

An Instance Generator for the Multi-Objective 3D Packing Problem

Yanira González(✉), Gara Miranda, and Coromoto León

Dpto. de Ingeniería Informática y de Sistemas, Universidad de La Laguna,
Avda. Astrofísico Fco. Sánchez s/n, 38271 La Laguna, Santa Cruz de Tenerife, Spain
{ygonzale,gmiranda,cleon}@ull.es

Abstract. Cutting and packing problems have important applications to the transportation of cargo. Many algorithms have been proposed for solving the 2D/3D cutting stock problems but most of them consider single objective optimization. The goal of the problem here proposed is to load the boxes that would provide the highest total volume and weight to the container, without exceeding the container limits. These two objectives are conflicting because the volume of a box is usually not proportional to its weight. This work deals with a multi-objective formulation of the 3D Packing Problem (3DPP). We propose to apply multi-objective evolutionary algorithms in order to obtain a set of non-dominated solutions, from which the final users would choose the one to be definitely carried out. For doing an extensive study, it would be necessary to use more problem instances. Instances to deal with the multi-objective 3DPP are non-existent. For this purpose, we have implemented an instance generator.

Keywords: 3D Packing Problem · Multi-objective optimization · Instance generator

1 Introduction

The 3D Packing Problem (3DPP) belongs to an area with numerous applications in the real world. The most commonly cited applications are the container loading or truck loading in the transportation sector and distribution industries. When solving the 3DPP, the objective is to locate a set of 3D rectangular pieces (boxes) inside one large rectangular object (container) such as to maximize the total volume of packed boxes. However, a rather common aspect in the scope of this problem is the weight limit of the containers, since they normally can't exceed a certain weight for their transportation. The rented trucks to transport the shipment are paid according to the total weight they can transport regardless of the total volume. Give a solution for these problems is interesting. To determine the most efficient way of packing items inside the container, several heuristics have been developed.

According to Wäscher et al. [5], cutting and packing problems can be grouped by dimensionality, assortment of large items, assortment of small

M. Graña et al. (eds.), *International Joint Conference SOCO'16-CISIS'16-ICEUTE'16*,
Advances in Intelligent Systems and Computing 527, DOI 10.1007/978-3-319-47364-2_37

items and objective. In Zhao et al. [30] a classification of cutting and packing problems is proposed: if the objective is one of output maximization, then the aim is to pack a subset of boxes giving the highest value to the fixed set of containers. There may be a single container or multiple containers. In our case of study, we are particularly concerned with the Single Large Object Placement Problem (SLOPP), where there is a single container and weakly heterogeneous boxes. Nevertheless, our problem definition is a variant of the SLOPP, because we consider both homogeneous boxes and heterogeneous boxes.

The 3D packing problem has been widely studied. In the literature, isolated exact algorithms [16,20] have been proposed to solve it. Most results on this scope are developed using heuristics and metaheuristics, because, computationally, the 3DPP is a NP-hard problem [29]. The heuristics can be developed taking into account different specific distributions: wall-building [1, 27], cuboid arrangement [3,10], stack-building [12,17], or guillotine cuts [9]. Although there are many heuristic algorithms [1,15,28], in recent years the attention to the metaheuristics such as genetic algorithms [13,21], simulated annealing [11,19], tabu searches [6,26], and hybrid algorithms [4,23,25], has been increased. Most approaches deal with single-objective formulations of the 3DPP, in which, the most commonly objective is to maximize volume utilization or to minimize the unused volume inside the container. Unlike the large number of approaches proposed for the single-objective formulation, the multi-objective formulation has been less studied. In our case, we consider the 3DPP which simultaneously tries to maximize two objectives: the weight and volume utilization. So, the 3DPP can be stated as a multi-objective optimization problem, trying to optimize the pieces layout inside the container so that the volume is maximized at the same time that the weight, without exceeding the container weight limit.

For the computational studies, some instances have been proposed in the literature to deal with the single-objective formulation of the problem. However, there is a real lack of compatible instances for the multi-objective formulation. In [24], a set of 15 problem instances - denoted as *LN* instances - has been proposed. Each LN instance contains weakly heterogeneous boxes to be packed into a single container. The number of different types of boxes varies between 6 and 10; while the total number of boxes varies between 100 and 250. Boxes may be left out of the packing solution, if the single container is full. Other approaches has been tested on benchmark instances proposed in [2,7]. Such a benchmark consists of a total set of 1500 instances. These tests are classified into 15 classes including 100 instances each one, denoted as *BR* instances. The boxes can be classified from weakly heterogeneous to strongly heterogeneous. A standard ISO container has been considered in order to generate the dimensions of the items for all the instances. The length, width and height of the boxes are integer numbers in the range of $[30, 120]$, $[25, 100]$ and $[20, 80]$, respectively. For all these instances, the weight hasn't been considered. The objective is only focused on the maximization of the container volume utilization. In [18] the objective is minimized the number of containers used to locate all the boxes. This benchmark has 47 instances, denoted as *IMM*. For the multi-objective formulation, the only known instances that addresses this problem, is the one proposed by Dereli et al. [8]. The problem instance was collected from the company that distributed Procter & Gamble's products as well as their own products (paper, towels, toilet tissues, paper napkins, etc.). This benchmark included 12 different products. Each product has its quantities, dimensions ($length \times width \times height$), and weights of the boxes. So, in a previous work [14], we have used the real test problem instance proposed by Dereli. However, we consider that a single instance isn't enough for the development of a computational study. It's strongly necessary to generate a wide set of instances which would allow to compare different approaches for the problem here analyzed. For this reason, we have developed a software to generate 3DPP instances. It's essential that the weight of the generated items is not necessarily proportional to their volume. In opposition that you might think a priori, the total volume maximization doesn't imply a maximization of weight. That is, a box can have large dimensions and a lighter content than a smaller box. For this reason, we can state that both goals have at least some degree of conflict.

The remaining content of this paper is organised as follows. The description of the instance generator for the multi-objective 3DPP is given in Sect. 2. In the Sect. 3, a description of the generated instances is presented. The experimental results are provided in Sect. 4. Finally, the conclusions and some lines of future work are given in Sect. 5.

2 Instance Generator

To develop the instance generator (3D-MGenerator), we have defined the *Simple 3D Packing Problem* as the problem of identifying the boxes layout into a container. In our case of study, we want to solve the 3DPP as a multi-objective problem, in which trying to maximize the total volume and total weight at the same time. We deal with a variant of the SLOPP, in which we allow different box types: identical, weakly heterogeneous and strongly heterogeneous. So, a set of boxes is identical when all boxes are equal or very similar. If a set of boxes is weakly heterogeneous then the problem instance has many boxes of each type but only few types of different boxes. Finally, a set of boxes is strongly heterogeneous when there is only few boxes of each type but many type of different boxes.

In most optimization problems, the different objectives are conflicting among them. For the 3DPP, in most real cases, the dimensions of the boxes to be packed is not proportional to its weight. So, if we want to generate problem instances, we may take into account this feature, in order to keep at least some degree of conflict among objectives.

In order to define problem instances, the instance generator requires a set of input parameters. In the multi-objective 3DPP, we deal with a container with known dimensions ($L = length$, $W = width$, $H = height$) and a set of N 3D rectangular boxes. Each box belongs to one of the different boxes

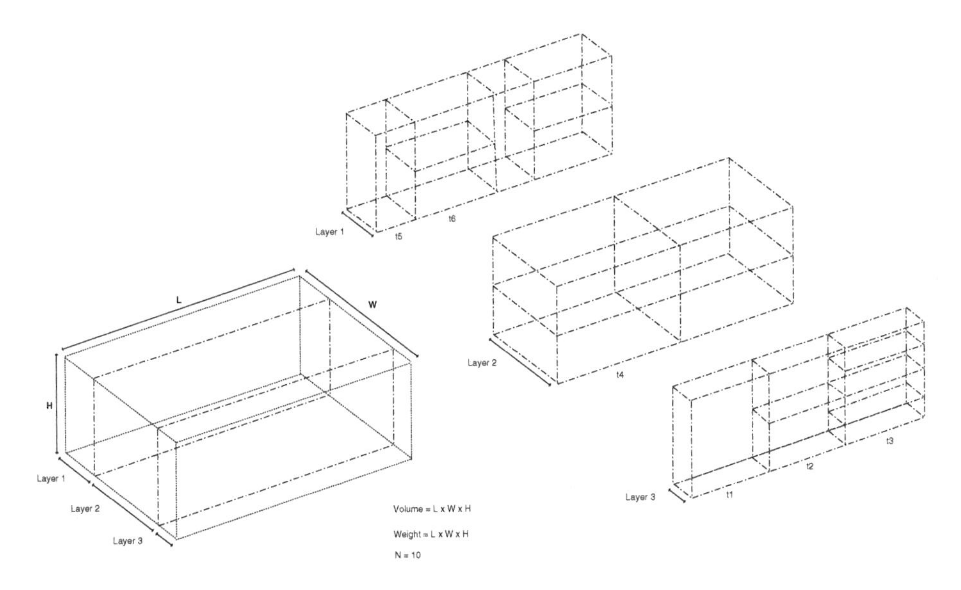

Fig. 1. Generation of the solution with maximum volume (Sol_1)

types N_i, in which each one has its own dimensions (l_i, w_i, h_i), weight or profit p_i and rotation o_i. Considering such a problem definition, the dimensions of the container (L, W, H) will form the first input parameters for the instance generator. The number of box types, N, will constitute another important input parameter for the generator. The box set may vary from small to large sized boxes. To determine the dimensions of each type of box, the generator requires two input parameters (D_l and D_h). These parameters determine the lowest and highest dimensions that a box type can have with respect to the dimensions of the container.

In order to obtain valuable instances for the multi-objective formulation of the problem, the designed problem generator is able to provide the optimal solution for each considered objective (volume and weight). Thus, the generator will give to the users the definition of a problem instance and two solutions for such a problem instance: a solution with maximum volume (Sol_1) and a solution with maximum weight (Sol_2). Since a problem instance will have N different box types, a subset of M ($M < N$) types will be used to create Sol_1 and a subset of $N - M$ type of boxes will be used to create Sol_2.

This way, two solutions are found (best in volume and best in weight) while the boxes and thus, the problem instance, is being generated. For the generation of Sol_1, the generator completely fills the container, so the solution with maximum volume will have a total volume of $L \times W \times H$. The weight of Sol_1 will be equal to its volume, since the boxes used for the creation of this solution will be assigned a weight proportional to the box volume.

For the generation of Sol_2, the total volume of the container is slightly reduced using for that the input parameter SL. SL represents the length, width

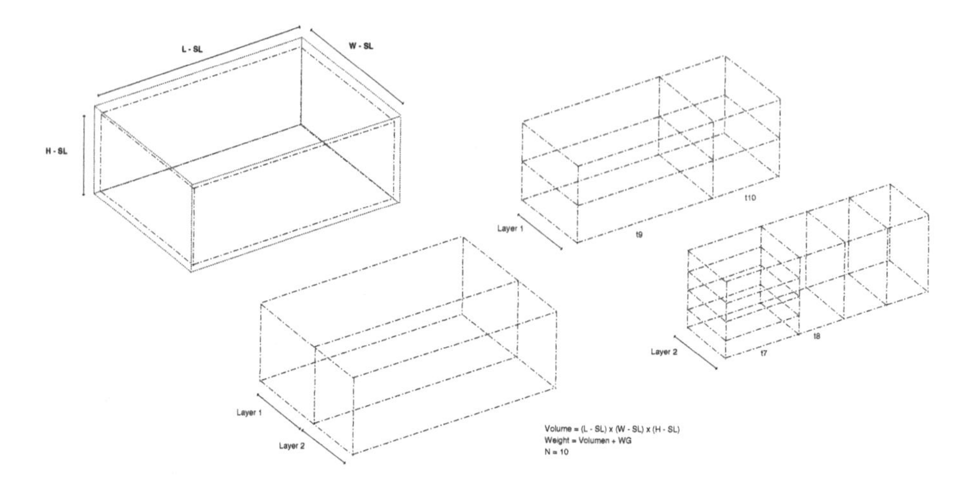

Fig. 2. Generation of the solution with maximum weight (Sol_2)

and height of the container that will remain unused for the creation of Sol_2. The input parameter WG will represent the weight gain of Sol_2 with respect to the weight of Sol_1, i.e. WG represents an increment on the container overall volume. Since Sol_2 will have a lower volume than Sol_1 and a higher weight than Sol_1, the instances generated will maintain at least some degree of conflict among their objectives. The instance generator distributes the total weight gain WG of Sol_2 among the $N - M$ box types used for Sol_2.

In order to decide the particular (dimensions and availability) of the boxes used in Sol_1 and the boxes used in Sol_2, the instance generator applies a filling heuristic whose design is based on the creation of layers. The layers create empty spaces which identify areas where is possible to allocate different types of boxes. The layers are parallel guillotine cuts on the different axis $(x, y, and\ z)$.

With regards to the allocation of boxes within the container, the problem instances will be generated using the following assumptions:

- Each box is placed in the container floor or on top of another box.
- The loaded boxes can not overlap.
- The stability of the distribution of the boxes is not considered.

For the generation of Sol_1 the filling heuristic works as shown in Fig. 1. First, the width of the boxes are determined by the width of the generated layers. The number of layers depends on M, which is the number of pieces available for building Sol_1, and also depends on the input parameters D_l and D_h, which determines a lower bound for the width of the layers and an upper bound for the width of the generated layers. Then, each layer is divided through its length, thus defining the length of the boxes in the problem instances. Finally, each obtained sublayer is also divided through the y axis, thus determining the height of the boxes in the problem instances. Again, in these two stages, the number of partitions - through length and height - within a layer will depend on M but

Table 1. Output file format

L W H	Dimensiones (length, width, and height) of the container
P_{max}	Maximum weight of the container
N_i	Number of different type of boxes
l_0 or_1 w_0 or_2 h_0 or_3 p_0 b_0	Length orientation1 width orientation2 height orientation3. Weight and number of boxes of type N_i
.	
.	
l_i or_1 w_i or_2 h_i or_3 p_i b_i	

also on the size of the new generated sublayers which dimensions will range in the interval $[D_l, D_h]$.

For the generation of Sol_2 the filling heuristic works following the same procedure but considering $N - M$ available pieces instead of N. The other difference for Sol_2 is that the filling process doesn't start from the whole container. It begins the partition process from a smaller container $(L - SL \times W - SL \times H - SL)$ in order to obtain a solution with lower volume than Sol_1 as explained before. The generation process for Sol_2 is shown in Fig. 2.

As a result of executing the instance generator, an output file is obtained. Table 1 shows the content and format of the output file. Each output file defines the dimensions of the container (L, W, H). A maximum weight P_{max} (specified by the user as an input parameter), and a set of N 3D rectangular boxes. These boxes belong to one of the N_i box types. Each type of box is characterized by its dimensions (l_i, w_i, h_i), weight p_i, a demand b_i and a number of orientations allowed $o_i \in [0, 5]$. According to the values taken in or_i, the boxes may rotate on any or all sides.

3 Problem Instances

In order to generate a set of problem instances for the multi-objective 3DPP, we have selected possible values for the input parameters. We have used two types of container according to the ISO specifications. Specifically, a 1C-20' container whose dimensions are 600 cm × 250 cm × 260 cm for length (L), width (W) and height (H), and a IA-40' container with dimensions 1200 cm × 250 cm × 260 cm. When we consider a container of type *1C-20'*, its stowage losses is of 12 % ($SL = 12$). If the container is of type *IA-40'*, then it has a stowage losses of 8.9 % ($SL = 8.9$). The set of lowest and highest dimensions used was the following: $[D_l - D_h] = [5 - 10]$ %, $[15 - 20]$ %, $[25 - 30]$ %. So, we consider that a set of boxes is small sized when its dimensions are between, i.e. $[5 - 10]$ % with respect to the length, height and width of the container. The WG can take three

Table 2. Problem instances for the 1C-20' container

Instance	Different type boxes	$[D_l - D_h]$	WG	$V_{max}(cm^3)$	$P_{max}(kg)$	*Avg. volume* (%)	*Avg. weight* (%)
$1C - 20'\_03$	5	$[5 - 10]$	30	$3.9e^{07}$	$5.07e^{07}$	98.76 %	94.03 %
$1C - 20'\_06$	5	$[15 - 20]$	30	$3.9e^{07}$	$5.07e^{07}$	97.15 %	77.96 %
$1C - 20'\_09$	5	$[25 - 30]$	30	$3.9e^{07}$	$5.07e^{07}$	96.70 %	77.65 %
$1C - 20'\_12$	8	$[5 - 10]$	30	$3.9e^{07}$	$5.07e^{07}$	96.64 %	87.66 %
$1C - 20'\_15$	8	$[15 - 20]$	30	$3.9e^{07}$	$5.07e^{07}$	97.83 %	79.25 %
$1C - 20'\_18$	8	$[25 - 30]$	30	$3.9e^{07}$	$5.07e^{07}$	95.69 %	81.43 %
$1C - 20'\_21$	10	$[5 - 10]$	30	$3.9e^{07}$	$5.07e^{07}$	96.44 %	94.75 %
$1C - 20'\_24$	10	$[15 - 20]$	30	$3.9e^{07}$	$5.07e^{07}$	94.54 %	93.05 %
$1C - 20'\_27$	10	$[25 - 30]$	30	$3.9e^{07}$	$5.07e^{07}$	93.95 %	86.61 %

Table 3. Problem instances for the IA-40'container

Instance	Different type boxes	$[D_l - D_h]$	WG	$V_{max}(cm^3)$	$P_{max}(kg)$	*Avg. Volume* (%)	*Avg. Weight* (%)
$IA - 40'\_02$	5	$[5 - 10]$	20	$7.93e^{07}$	$1.03e^{08}$	94.86 %	98.27 %
$IA - 40'\_05$	5	$[15 - 20]$	20	$7.93e^{07}$	$1.03e^{08}$	96.93 %	83.90 %
$IA - 40'\_08$	5	$[25 - 30]$	20	$7.93e^{07}$	$1.03e^{08}$	96.53 %	81.83 %
$IA - 40'\_11$	8	$[5 - 10]$	20	$7.93e^{07}$	$1.03e^{08}$	96.53 %	93.65 %
$IA - 40'\_14$	8	$[15 - 20]$	20	$7.93e^{07}$	$1.03e^{08}$	98.21 %	85.41 %
$IA - 40'\_17$	8	$[25 - 30]$	20	$7.93e^{07}$	$1.03e^{08}$	95.59 %	87.98 %
$IA - 40'\_20$	10	$[5 - 10]$	20	$7.93e^{07}$	$1.03e^{08}$	96.31 %	96.09 %
$IA - 40'\_23$	10	$[15 - 20]$	20	$7.93e^{07}$	$1.03e^{08}$	95.57 %	89.69 %
$IA - 40'\_26$	10	$[25 - 30]$	20	$7.93e^{07}$	$1.03e^{08}$	95.30 %	86.27 %

possible values: 10 %, 20 % and 30 %. These parameters allows to enhance the multi-objective approach of the problem.

Tables 2 and 3 show the set of values given to the input parameters of the instance generator (columns from 1 to 6). As well as, the maximum values of volume and weight, which are possible achieved by each instance. We have generated and tested a total of 54 problem instances, 27 of them correspond to the 1C-20' container, denoted as $1C\_20'\_01$ *to* $1C\_20'\_27$. The rest of instances are denoted as $IA\_40'\_01$ *to* $IA\_40'\_27$ and correspond to the IA-40' container.

4 Experimental Evaluation

To solve the 3DPP, we have applied multi-objective evolutionary algorithms (MOEAs), in particular NSGA-II. MOEAs have shown promise for solving

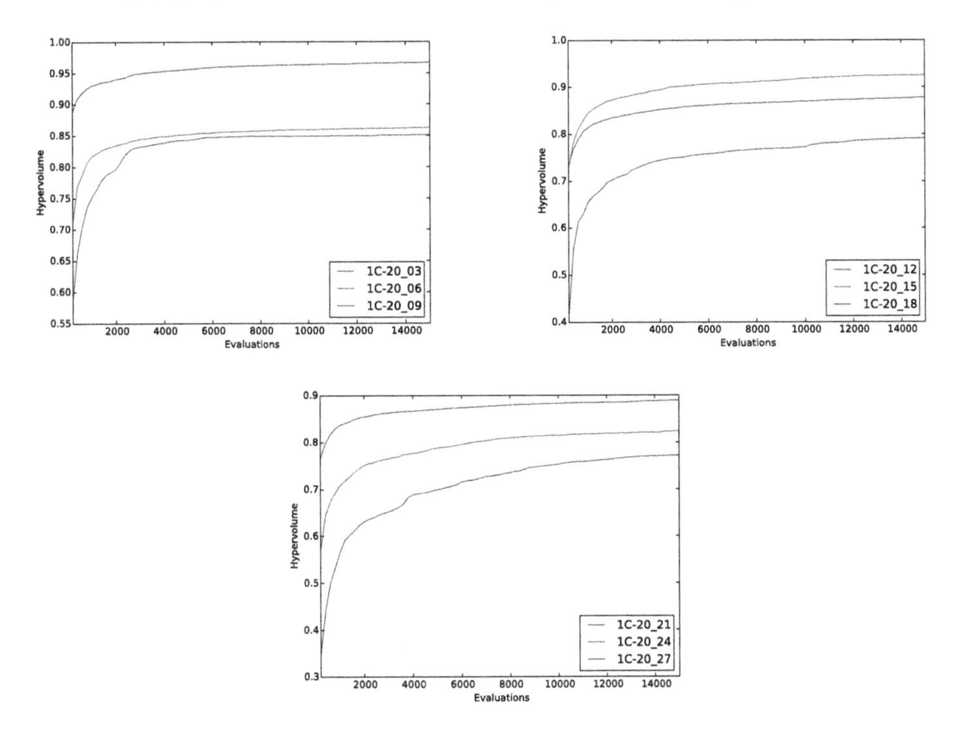

Fig. 3. Hypervolume: different instances for 1C-20' container and WG = 30 %.

cutting and packing problems. We have designed a set of heuristics to evaluate possible solutions for the 3DPP. For this study, we have used two filling heuristics: Multiple-Level Filling Heuristic (MLFH) [14] and Multiple-Level Filling Heuristic in Depth (MLFHD). Last one, it's a modification of the MLFH. Both heuristics are based on the creation and management of pieces levels inside the container. MOEAs and heuristics are provided by *METCO* [22]. METCO is a pluging-based which incorporates a set of multi-objective schemes to tackle MOPs.

The experimental evaluation have been run on a dedicated Debian GNU/Linux operating system of 48 cores, each one consisting of one *AMD Opteron 6164HE*. METCO framework and the problem approach were implemented using C++ and compilated with *gcc 4.7.2*.

Tables 2 and 3 show - for each problem instance - the best possible volume and weight values and the average best volume and weight values obtained by the proposed approach. In this case of study, the results obtained for instances with small-sized boxes improve the rest results obtained for medium and large sized boxes. The analysis shows that small sized boxes are more easily located inside the container. So, it is possible to fill more holes within the container. In the opposite case are the large-sized boxes, which are more complex to pack. With large-sized boxes is much more difficult to fill possible empty small holes inside the container.

Figure 3 shows the hypervolume obtained for some instances with the same weight gain ($WG = 30$) and type of container (1C-20'). In this study, we want to demonstrate that the size of the boxes is a significant aspect when dealing with 3DPP instances. For small sized boxes is more easy to achieve better results. For problem instances with larger boxes, the average volume and weight obtained are not so close to the optimal values, so the quality of solutions decreases when the size of boxes increases. Using the instances show for IA-40' container, the conclusions are the same.

5 Conclusions and Future Work

Most studies proposed on the 3D Packing Problem deal with the mono-objective formulation of the problem. Hence, there are numerous available instances in the literature. However, when dealing with the multi-objective formulation of the problem, this isn't true: there is a clear lack of suitable problem instances. Most studies transform multi-objective formulations into mono-objective approaches, but we have studied the problem considering its multi-objective version. This approach appears to be a interesting line of research, but there is a lack of problem instances. For this reason, we have implemented an instance generator, which could be used to further research in this scope. The set of input parameters allows to create problem instances of varying complexity, which can be mainly classified according to the dimensions and homogeneity of available boxes. In general, the instance generator gives chance to the users to obtain a wide range of problem instances, mainly through the tunning of the defined input parameters.

In order to further analyse the designed instance generator, it would be interesting to generate more problem instances with homogeneous and heterogeneous box types. It would be also very interesting to check whether some extra input parameters will allow to better adapt the variation in box's dimensions and thus in the difficulty of the problem. Another future line of work will focus on the design of a suitable set of instances which can be proposed as a standard test set for the multi-objective 3DPP. It would be also interesting to try to expand the instance generator to other 3D problems.

Acknowledgment. This work was funded by the Spanish Ministry of Science and Technology as part of the 'Plan Nacional de I+D+i' (TIN2011-25448).

References

1. Bischoff, E.E., Janetz, F., Ratcliff, M.S.W.: Loading pallets with non-identical items. Eur. J. Oper. Res. **84**(3), 681–692 (1995)
2. Bischoff, E., Ratcliff, M.: Issues in the development of approaches to container loading. Omega **23**(4), 377–390 (1995)
3. Bortfeldt, A., Gehring, H.: A tabu search algorithm for weakly heterogeneous container loading problems. OR Spectr. **20**(4), 237–250 (1998)
4. Bortfeldt, A., Gehring, H.: A hybrid genetic algorithm for the container loading problem. Eur. J. Oper. Res. **131**(1), 143–161 (2001)

5. Bortfeldt, A., Wäscher, G.: Container loading problems - a state-of-the-art review. In: Working Paper 1, Otto-von-Guericke-Universität Magdeburg, April 2012
6. Burke, E.K., Hyde, M.R., Kendall, G., Woodward, J.: Automating the packing heuristic design process with genetic programming. Evol. Comput. **20**(1), 63–89 (2012)
7. Davies, A., Bischoff, E.E.: Weight distribution considerations in container loading. Eur. J. Oper. Res. **114**(3), 509–527 (1999)
8. Dereli, T., Das, S.G.: A hybrid simulated annealing algorithm for solving multi-objective container loading problems. Appl. Artif. Intell. Int. J. **24**(5), 463–486 (2010)
9. Ding, X., Han, Y., Zhang, X.: A discussion of adaptive genetic algorithm solving container-loading problem. Periodical Ocean Univ. China **34**(5), 844–848 (2004)
10. Eley, M.: Solving container loading problems by block arrangement. Eur. J. Oper. Res. **141**(2), 393–409 (2002)
11. Faina, L.: A global optimization algorithm for the three-dimensional packing problem. Eur. J. Oper. Res. **126**(2), 340–354 (2000)
12. Gehring, H., Bortfeldt, A.: A genetic algorithm for solving the container loading problem. Int. Trans. Oper. Res. **4**(5–6), 401–418 (1997)
13. Gonalves, J.F., Resende, M.G.: A parallel multi-population biased random-key genetic algorithm for a container loading problem. Comput. Oper. Res. **39**(2), 179–190 (2012)
14. González, Y., Miranda, G., León, C.: A multi-level filling heuristic for the multi-objective container loading problem. In: International Joint Conference SOCO 2013-CISIS 2013-ICEUTE 2013 - Salamanca, Spain, Proceedings, pp. 11–20, 11–13 September 2013
15. He, K., Huang, W.: An efficient placement heuristic for three-dimensional rectangular packing. Comput. Oper. Res. **38**, 227–233 (2011)
16. Hifi, M.: Exact algorithms for unconstrained three-dimensional cutting problems: a comparative study. Comput. Oper. Res. **31**, 657–674 (2004)
17. Huang, W., He, K.: A caving degree approach for the single container loading problem. Eur. J. Oper. Res. **196**(1), 93–101 (2009)
18. Ivancic, N., Mathur, K., Mohanty, B.B.: An integer programming based heuristic approach to the three-dimensional packing problem. J. Manuf. Oper. Manage. **2**, 268–289 (1989)
19. Jin, Z., Ohno, K., Du, J.: An efficient approach for the three dimensional container packing problem with practical constraints. Asia Pac. J. Oper. Res. **21**(3), 279–295 (2004)
20. Junqueira, L., Morabito, R., Yamashita, S.D.: Mip-based approaches for the container loading problem with multi-drop constraints. Ann. Oper. Res. **199**(1), 51–75 (2012). http://dx.doi.org/10.1007/s10479-011-0942-z
21. Kang, K., Moon, I., Wang, H.: A hybrid genetic algorithm with a new packing strategy for the three-dimensional bin packing problem. Appl. Math. Comput. **219**(3), 1287–1299 (2012)
22. León, C., Miranda, G., Segura, C.: METCO: a parallel plugin-based framework for multi-objective optimization. Int. J. Artif. Intell. Tools **18**(4), 569–588 (2009)
23. Liu, J., Yue, Y., Dong, Z., Maple, C., Keech, M.: A novel hybrid tabu search approach to container loading. Comput. Oper. Res. **38**, 797–807 (2011)
24. Loh, T.H., Nee, A.Y.C.: A packing algorithm for hexahedral boxes, pp. 115–126 (1992)
25. Moura, A., Oliveira, J.F.: A GRASP approach to the container-loading problem. IEEE Intell. Syst. **20**(4), 50–57 (2005)

26. Parreño, F., Alvarez-Valdes, R., Oliveira, J.F., Tamarit, J.M.: Neighborhood structures for the container loading problem: a VNS implementation. J. Heuristics **16**(1), 1–22 (2010)
27. Pisinger, D.: Heuristics for the container loading problem. Eur. J. Oper. Res. **141**(2), 382–392 (2002)
28. Ren, J., Tian, Y., Sawaragi, T.: A tree search method for the container loading problem with shipment priority. Eur. J. Oper. Res. **214**(3), 526–535 (2011)
29. Scheithauer, G.: Algorithms for the container loading problem. In: Gaul, W., Bachem, A., Habenicht, W., Runge, W., Stahl, W.W. (eds.) Operations Research Proceedings 1991, vol. 1991, pp. 445–452. Springer, Heidelberg (1992)
30. Zhao, X., Bennell, J., Bektas, T., Dowsland, K.: A comparative review of 3d container loading algorithms, April 2014

Solving Repetitive Production Planning Problems. An Approach Based on Activity-oriented Petri Nets

Bozena Skolud[1], Damian Krenczyk[1(✉)], and Reggie Davidrajuh[2]

[1] Faculty of Mechanical Engineering, Silesian University of Technology, Gliwice, Poland
damian.krenczyk@polsl.pl

[2] Faculty of Science and Technology, University of Stavanger, Stavanger, Norway
reggie.davidrajuh@uis.no

Abstract. In the paper, the problem of flow planning in production systems belonging to a class of Cyclic Concurrent Processes Systems is presented. The possibility of using Activity-oriented Petri Nets approach to model the problem as a discrete event system model, and then to perform simulations with the tool known as GPenSIM is shown. For simulation dispatching rules achieved by the analytical method of production order verification based on constraints sequencing methodology and its computer implementation in the system of production orders verification SWZ is used. This paper shows that the results achieved from GPenSIM agrees with that from SWZ. The proposed approach is easy to implement and can be used as an effective production plans verification tool, to identify problems during the implementation of the control rules on production resources in the form of deadlocks, failure to disclose a bottleneck, or problems with production flow synchronization.

Keywords: Production planning and control · Petri nets · Simulation · Flow logic

1 Introduction

In modern production (especially in automotive industry) the tendency to one piece flow organization is observed. However, from the operational point of view, all those operations that should be executed on one machine (assembly station, work stand) are the identical or very similar one to the other one. They use the same manufacturing resources and tooling (alternative but available tools). Each operation characterizes very similar time consumption, and duration of the operations of individual resources cannot exceed the time for a critical resource. This approach is very close to the flow shop organization and can be analyzed similarly but on the other hand, the approach is not flexible. Taking all that into consideration, single-piece flow proposed by producers and presented by Womack [1] is very similar to the flow shop systems and can be analyzed for this type what is the advantage of it. On the other hand, it is not flexible because it is designed to manufacture only very similar products what is the disadvantage.

M. Graña et al. (eds.), *International Joint Conference SOCO'16-CISIS'16-ICEUTE'16*,
Advances in Intelligent Systems and Computing 527, DOI 10.1007/978-3-319-47364-2_38

Production in job shop organization is very difficult to plan, so producers avoid this possibility as much as possible despite the fact it is flexible. It is difficult to plan the production and determine the parameters that characterize it. In most of the cases, producers decide to use dispatching rules (earliest due date, longest processing time, etc.) for control, and planning stage is minimized.

This paper focuses on the problem associated with the planning of the discrete concurrent processes flow, that are executed according to diverse production routes, which can cross one with another one just as it is in a case of job shop [2–5]. This production, however, was boiled down to the cyclic one. Repetitive character let as to concentrate on one cycle in order to calculate parameters characterizing the flow of production only for one cycle. These calculated values are identical for the entire production.

In this paper, the problem of planning the production flow in production systems belonging to a class of Cyclic Concurrent Processes Systems (CCPS) is presented. Discrete concurrent processes are often found in production systems, but the scheduling and timetabling problems associated with this class of systems, are also present in information systems (multi-tasking operating systems) and transport systems (public transport, rail) [3, 6, 7]. With regard to the production systems, this approach makes it possible to manufacture technologically dissimilar products on a shared resource at a high level of resource utilization.

The objective of the approach presented is to provide a combination of two different approaches, which could be considered as complementary. It boils down to show possibilities of Activity-oriented Petri Nets (AOPN) approach to model the problem as a discrete event system model, and then to perform simulations with the tool known as GPenSIM for simulation of given production flow achieved by analytical method of production order verification based on constraints satisfaction methodology and its computer implementation in the system of production orders verification SWZ.

The paper is organized as follows: In Sect. 2, the problem of repetitive production planning is presented. In Sect. 3, a new approach known as Activity-oriented Petri nets is introduced. Section 4 presents an application example, which is modeled with the AOPN approach. The results achieved by this approach are compared with the answer received from SWZ system, in Sect. 5. Finally, Sect. 6 concludes the research with directions for future work.

2 Repetitive Production

The problem of planning the production flow in production systems belonging to the class of Cyclic Concurrent Processes Systems (CCPS) is considered in [6–8]. In the considered class of production systems, cyclic discrete processes are concurrently executed according to exclusive-like mode. The cycle of the system is established by a technological operations processes sequence carried out on resource shared among processes. The order of processes is regulated with local dispatching rules allocated on the resources R_i. The rules are created in the production planning stage, for each production resource.

The cyclic concurrent processes system with dispatching rules is defined as [9]:

$$SC = (M, PP, B, R), \tag{1}$$

where:

$M = \{M_i,\ i = 1,2,\ldots,m\}$ – the set of production resources,

$PP = (P, MP, N,)$ – the structure of the production processes, where:

$P = \{P_j,\ j = 1,2,\ldots n\}$ – the set of production processes,

$MP = \{MP_j,\ j = 1,2,\ldots n\}$ – the set of production processes matrix:

$$MP_j = \begin{bmatrix} mp_{1,1} & mp_{1,2} & \ldots & mp_{1,h} & \ldots & mp_{1,H_j} \\ mp_{2,1} & mp_{2,2} & \ldots & mp_{2,h} & \ldots & mp_{2,H_j} \\ mp_{3,1} & mp_{3,2} & \ldots & mp_{3,h} & \ldots & mp_{3,H_j} \end{bmatrix},$$ *j-th* production process matrix,

where:

h – operation sequence number (according to the order of operations specified by the route),

H_j – the number of operations in the j-th process route,

$mp_{1,h}$ – the resource number, on which the *h-th* operation is performed,

$mp_{2,h}$ – the cycle time of *h-th* operation,

$mp_{3,h}$ – the setup time of *h-th* operation,

$N,$ – the batch size,

$B = \{B_{l,k},\ l = 1,2,\ldots,m;\ k = 1,2,\ldots,m;\ m \neq k\}$ – the set of inter-resource buffers allocated to the neighbouring resources (M_l, M_k),

$R = \{Ri,\ i = 1, 2,\ldots,m)$ – the set of dispatching rules allocated on the resources.

In [3, 6] it has been shown that in the system when processes routes do not create closed loops a deadlock problem does not appear because one of the necessary conditions for deadlock occurrence is not satisfied. Authors of this paper in their previous publications [5–10] presented a concept which is called “synchronised manufacturing”, which guarantees the utilization of resources without disturbances.

Repetitive production means that for every constant period T, the same sequence of operations is repeated for the resources. Period T is determined by a sequence of the access of the operations on the shared resources, which is qualified by a dispatching rule. The dispatching rule Ri controlling the access of the processes to the *i-th* shared machine is the sequence: $Ri = (Pa_1, Pa_2, \ldots, Pa_j, \ldots, Pa_{oi})$ that determines the number of the processes executed on the *i-th* production resource (e.g. machine tool), where: Pa_j – the-a_j-*th* process, $i \in \{1,2,\ldots,m\}$, m – the number of resources, $a_j \in \{1,\ldots,n\}$, n – the number of processes. For illustration, let us consider processes P_1, P_2, and P_3 that compete for the access to machine M_1. The dispatching rule: $R1 = (P_1, P_1, P_3, P_2)$ guarantees the access to the machine tool M_1 twice for process P_1, once for process P_3, and once for process P_2.

2.1 Constraints Sequencing

The constraints used for the synthesis of the system should be applied not randomly but in a precisely determined succession. To determine due time order completion (customer’s demand), the number of elements in one batch should be known.

This information is important only if it is possible to execute at least one element during one cycle. The qualitative functioning of the system means deadlock-free and starvation-free behaviour of the system. Considering these criteria it is possible to conduct the quality validation. This solution may not be the best one. It is very probable that the set of solutions does not contain an optimal one, therefore, the obtained solution is the best solution in a given sense of search.

The conditions of qualitative system functioning determine the subset of admissible process realisations. Basing on this subset the satisfaction problem is formulated. This condition guarantees, the steady-state behaviour of the system and is a base of the analytical determination of the following parameters (Fig. 1):

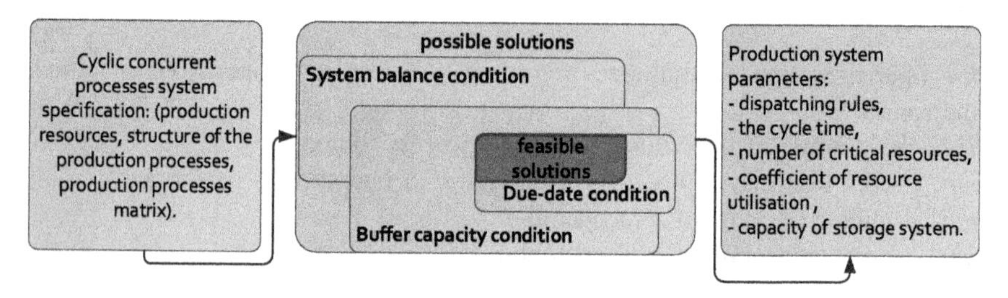

Fig. 1. The conjunction of defined conditions

- Dispatching rules,
- The cycle time (repetitiveness),
- Number of critical resources,
- Coefficient of resource utilisation,
- The capacity of the storage system.

These parameters are the base of the determination of the solution which meets the elements that are definitely quantitative oriented (dedicated) to the producer or to the client.

The System Balance Condition. Considered systems of repetitive production are characterize by a simultaneous realization of processes, when their production routes create a closed loop the deadlock can appear. To avoid the problem of deadlock and starvation the sufficient condition concerning a flow balance and interoperation buffers capacity should be satisfied. The balance condition in the system is satisfied when a number of elements entering the system in one repetitive period should be equal to the number of elements leaving the system in the same period.

The system balance is accomplished in the case when the following equations are satisfied for each process [7, 9]:

$$\chi_1 n_{1,j} = \chi_2 n_{2,j} = \ldots = \chi_i n_{i,j} = \ldots = \chi_m n_{m,j}, \quad (2)$$

where:

χ_i – the repetitiveness of the dispatching rule allocated on the *i-th* machine from the production route,

$n_{i,j}$ – the repetitiveness of the *j-th* process in the rule allocated on the *i-th* machine.

The Condition of the Buffers Space. Simultaneously with balance condition the condition of the buffers space allocation in the system should be satisfied. It means that the capacity of buffer $B_{l,k}$ allocated between machines M_i, and M_k for the *j-th* process should satisfy the following condition:

$$B^j_{1,k} = n_{l,j} \cdot \chi_l, \tag{3}$$

Interoperation buffers capacity condition assures that capacity of the buffer allocated between resources M_l, and M_k should not be less than the number of elements executed according given process during one cycle (see Eq. (2)).

If conditions (2) and (3) as well as condition (4) are satisfied, then there is a set of dispatching rules that enable due time order completion.

$$tz_j \geq \left(to_j + \frac{I_j T}{Q_j} \right), \tag{4}$$

where:

to_j – beginning term of *j-th* order realisation,
tz_j – due date
T – the cycle of the system,
I_j – lot size,
$Q_j = n_{i,j}\chi_i$ – number of elements executed during T.

2.2 System of Order Verification SWZ

The constraints sequencing methodology has been implemented in the system of production orders verification SWZ [8], that assisting in the decision-making processes of the acceptance or rejection of the production orders for production planning. SWZ system enables determining the control procedures assigned to resources in the form of local dispatching rules. The generated rules are designed to ensure the acceptable production execution quality (without deadlocks and starvations) and timely execution of orders. SWZ also generates information about the necessity to reject the production order which does not have a chance to perform in due date with the identified resource constraints.

3 Activity-Oriented Petri Nets

Being a discrete event system, the problem of repetitive production planning can be modelled with Petri nets. Though Petri nets could be used to model repetitive production planning problems, the resulting model could be huge due to the number of production resources involved. Usually, even for a problem with few activities competing for a few resources, the resulting Petri Net model can be very large [11]. Activity-oriented Petri nets (AOPN) is an approach for obtaining compact Petri net models of discrete event systems where resource sharing and resource scheduling dominate [12]. The software tool General purpose Petri net simulator (GPenSIM) is a realization of AOPN on the MATLAB platform; GPenSIM is freely available from the website [13].

3.1 The Two Phases of AOPN Approach

Activity-oriented Petri Nets (AOPN) is a two-phase modeling approach [12]. In the Phase-I, the static Petri Net model is created. In this phase, mainly the activities are considered and they are represented by transitions embedded with places as buffers. The resources are grouped into two groups such as ‘focal’ resources and ‘utility’ resources. Only the focal resources are included in the static Petri Net model; the utility resources will be considered later in the phase-II (the run-time model). Thus, a compact Petri Net model is obtained with only the transitions representing the activities and, if there are any focal resources, they will be represented by places. Using the tool GPenSIM, coding the static Petri net in phase-I will result in the Petri net definition file (PDF).

In the phase-II, the run-time details that are not considered in the phase-I are added to the Petri Net model; e.g. transitions (activities) requesting, using, and releasing of the utility resources are coded in the run-time model. Using the tool GPenSIM, the run-time details in the phase-II will result in the files COMMON_PRE and COMMON_POST. The interested reader is referred to the GPenSIM user manual available from the website [13].

4 Illustrative Example

Practical verification of the proposed systems integration method based on data mapping and data transformation is implemented in the transformation module of SWZ system (which generates an input file for simulation system with a description of the model).

Let’s assume a production system, consisting of four resources M_1–M_4. In the system three concurrent processes P_1, P_2 and P_3 are performed (Fig. 2).

The production routes and times are recorded in the process matrix:

$$MP_1 = \begin{bmatrix} 1 & 3 & 4 \\ 4 & 2 & 3 \\ 0 & 0 & 0 \end{bmatrix}, MP_2 = \begin{bmatrix} 4 & 3 & 1 & 2 \\ 5 & 7 & 5 & 4 \\ 0 & 0 & 0 & 0 \end{bmatrix}, MP_3 = \begin{bmatrix} 2 & 1 \\ 4 & 3 \\ 0 & 0 \end{bmatrix} \quad (5)$$

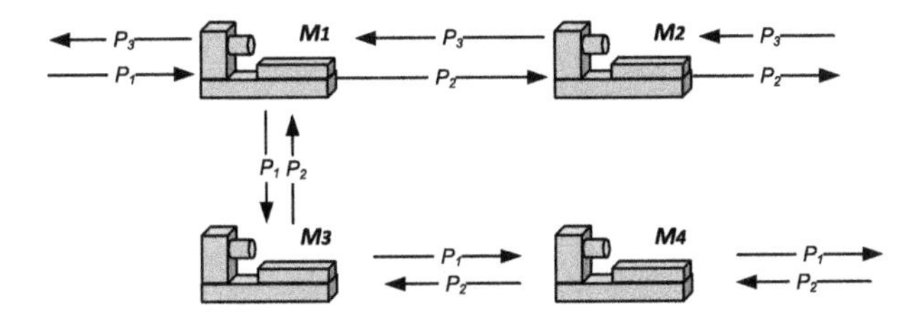

Fig. 2. System of multiassortment production

the first row of the matrix corresponds to the resources over which the route of the process goes; in the second line cycle times on proper resources are given; the third line contains setup times. The dispatching meta-rules containing the sequence of operations executed in the system in starting-up phase, steady state and cease phases have been generated:

```
---Report---SWZ v. 5.0---
The amount of resources in the
system : 4
The amount of processes in the
queue : 3
Processes data:
Process P1: Lot size: 101;    Due
time: 2000;  Operations amount: 3
Process P2: Lot size: 102;    Due
time: 2000;  Operations amount: 4
Process P3: Lot size: 200;    Due
time: 2300;  Operations amount: 2
Rules:
R1={(1,1,2);(1,2,3,3);(2,2,3,3)}
R2={(3,3);(2,3,3);(2,2,2)}
R3={(1,2,2);(1,2);(1,2)}
R4={(2,2,2);(1,2);(1,1)}
Rules realisation times at the
resources :
Rule R1 = 15
Rule R2 = 12
Rule R3 = 9
Rule R4 = 8

System cycle : 15
Realisation times (steady state):
Process P1 = 1485
Process P2 = 1485
Process P3 = 1485

The coefficient of the system
resources utilisation = 0,7333333
```

4.1 The Petri Net Model

It is clear from the description given above that the problem involves 3 processes, having 3, 4, and 2 activities, respectively. In addition, the problem involves 4 resources.

Phase-I of the modeling approach: Let us assume that all the four resources are utility resources, thus need not be considered in the static Petri net model. Hence, the resulting static Petri net model will be as simple as the one shown in the Fig. 3, highlighting only the activities of the 3 processes and the precedence relationship that exist between the activities.

Phase-II of the modeling approach: in this phase, all the run-time details are considered. E.g. the cycle times (expected duration of the activities), the connection between the activities and the utility resources, etc.:

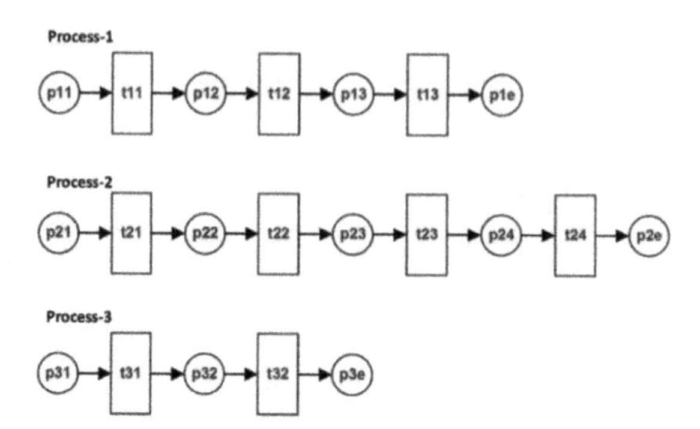

Fig. 3. Static Petri net model of the problem

- The set of resources that can perform the specific activities of a process ($mp_{1,1}$ to $mp_{1,n}$) becomes the resource-allocation-policy (*RAP*).
- The cycle time ($mp_{2,\ j}$) of the process-*i* becomes the firing times of the respective transition *tij*.

These details will be coded in the GPenSIM files such as main simulation file (MSF) and the pre-processor file COMMON_PRE.

4.2 GPenSIM Code for Simulations

Using GPenSIM, the Petri net model of the problem consists of four files:

1. Petri Net Definition File (PDF): this is the coding of the static Petri net shown in the Fig. 3.
2. Main Simulation File (MSF): In this file, the initial dynamics (e.g. initial tokens in places p11, p22, and p33), firing times of transitions, and the availability of the four resources are declared.
3. COMMON_PRE: In this file, the conditions for the enabled transitions to start firing are coded, mainly about reservation of the required resources, and using them if they are allocated. Thus, this file manipulates the *RAP*.
4. COMMON_POST: the post-firing actions of the transitions are coded here, usually releasing the resources after usage.

In addition to these four files, there are three other utility files such as check_RAP and modify_RAP (for checking the resource allocation rules, before firing; these two files are used by COMMON_PRE) and analyseAdditionalGlobalVar (to assign global variables; this file is used by MSF). These three utility files are just to reduce the code (size) of the files COMMON_PRE and MSF.

Due to brevity, complete code for simulations (the seven files mentioned above) are not shown in this paper. However, for reproducibility, the complete code is given on the web page [14]; interested readers are encouraged to visit the web page for downloading the codes and experimenting with them.

4.3 Simulation Results

The simulation are done with following initial input materials: p11 = 101, p21 = 102, and p31 = 200. The Figs. 4 and 5 shows the simulation results. The results obtained in the two approaches are consistent. The simulation performed using modeling of Petri net confirmed the cyclic production flow for the generated rules in accordance with 15 units of time working cycle of the bottleneck (M_1).

```
Final tokens: 101p1e + 102p2e + 200p3e

RESOURCE USAGE SUMMARY:
M1:  Total occasions: 403   Total Time spent: 1514
M2:  Total occasions: 302   Total Time spent: 1208
M3:  Total occasions: 203   Total Time spent: 916
M4:  Total occasions: 203   Total Time spent: 813

*****  LINE EFFICIENCY AND COST CALCULATIONS: *****
  Number of servers:  k = 4
  Total number of server instances:  K = 4
  Completion = 1600
  LT = 6400
  Total time at Stations: 4451
  LE = 69.5469 %
  **
```

Fig. 4. Simulation results showing the summary of the resource usage

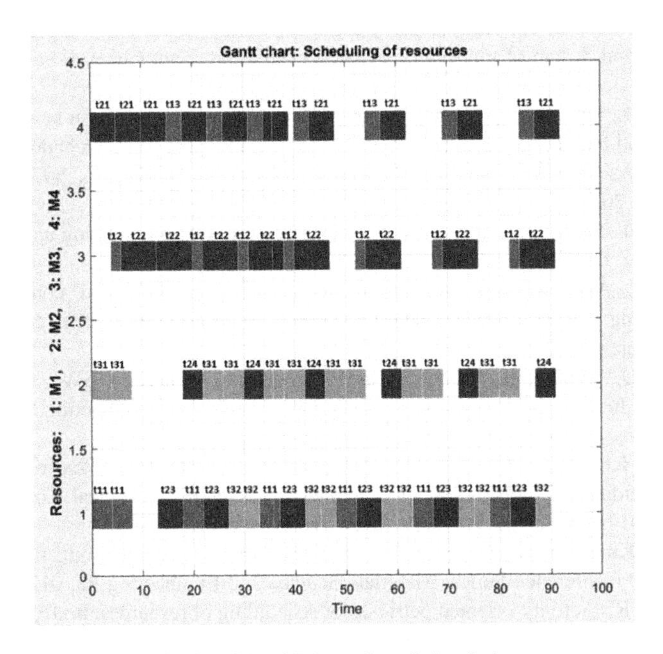

Fig. 5. GPenSIM results of simulation

5 Summary

From the presented analysis one should conclude that elaborated constraints based methodology gives similar results as AOPN. All of the system parameters are exactly the same. The only difference comes from the fact that AOPN is discrete event and does not control time. It causes that theoretically achieved cycle is exactly the same but can be shifted on particular machines. The proposed approach is easy to implement and can be used as an effective verification tool of production plans, to identify possible problems during the implementation of the control rules in the form of deadlocks, failure to disclose a bottleneck or problems with synchronization of production flow. For the next step, the authors propose to check sufficient conditions for deadlock free of starting-up and cease phases realization.

References

1. Womack, J.P., Jones, D.T.: Lean Thinking: Banish Waste and Create Wealth in Your Corporation. Harper Business, New York (2003)
2. Levner, E., Kats, V., de Pablo, D.A.L., Cheng, T.C.E.: Complexity of cyclic scheduling problems: A state-of-the-art survey. Comput. Ind. Eng. **59**(2), 352–361 (2010)
3. Burduk, A.: Artificial neural networks as tools for controlling production systems and ensuring their stability. In: Saeed, K., Chaki, R., Cortesi, A., Wierzchoń, S. (eds.) CISIM 2013. LNCS, vol. 8104, pp. 487–498. Springer, Heidelberg (2013)
4. Heo, S.-K., Lee, K.-H., Lee, H.-K., Lee, I.-B., Park, J.H.: A new algorithm for cyclic scheduling and design of multipurpose batch plants. Ind. Eng. Chem. Res. **42**(4), 836–846 (2013)
5. Krenczyk, D., Skolud, B.: Production preparation and order verification systems integration using method based on data transformation and data mapping. In: Corchado, E., Kurzyński, M., Woźniak, M. (eds.) HAIS 2011, Part II. LNCS, vol. 6679, pp. 397–404. Springer, Heidelberg (2011)
6. Krenczyk, D., Skołud, B.: Transient states of cyclic production planning and control. Appl. Mech. Mater. **657**, 961–965 (2014)
7. Skołud, B.: Approaches to the production flow scheduling and control, intelligent manufacturing for industrial business process. J. Mach. Eng. **8**(2), 5–13 (2008). J. blank;Jędrzejewski (Ed.)
8. Krenczyk, D., Skolud, B.: Computer aided production planning - SWZ system of order verification. In: IOP Conference Series: Materials Science and Engineering, vol. 95, 012135 (2015)
9. Skolud, B., Krenczyk, D.: Rhythmic production planning in the context of flow logic. In: Selected Conference Proceedings IV International Scientific-Technical Conference Manufacturing 2014, pp. 35–43. Poznan University of Technology (2016)
10. Skolud, B., Krenczyk, D., Zemczak, M.: Multi-assortment rhythmic production planning and control. IOP Conference Series: Materials Science and Engineering **95**, 012133 (2015)
11. Davidrajuh, R.: Activity-oriented petri net for scheduling of resources. In: IEEE International Conference on Systems, Man, and Cybernetics (SMC), pp. 1201–1206. IEEE (2012)

12. Davidrajuh, R.: Modeling resource management problems with activity-oriented petri nets. In: Sixth UKSim/AMSS European Symposium on Computer Modeling and Simulation (EMS), pp. 179–184. IEEE (2012)
13. GPenSIM User Manual. http://www.davidrajuh/gpensim/
14. Simulation Code for the repetitive production planning problem. http://www.davidrajuh.net/gpensim/2016-SOCO-Repetitive-PP

The Concept of Ant Colony Algorithm for Scheduling of Flexible Manufacturing Systems

Krzysztof Kalinowski(✉) and Bożena Skołud

Institute of Engineering Processes Automation and Integrated Manufacturing Systems, Faculty of Mechanical Engineering, Silesian University of Technology, Konarskiego 18A, 44-100 Gliwice, Poland
{krzysztof.kalinowski,bozena.skolud}@polsl.pl

Abstract. The paper presents the conception of algorithm for scheduling of manufacturing systems with consideration of flexible resources and production routes. The proposed algorithm is based on ant colony optimisation (ACO) mechanisms. Although ACO metaheuristics do not guarantee finding optimal solutions, and their performance strongly depends on the intensification and the diversification parameters tuning, they are an interesting alternatives in solving NP hard problems. Their effectiveness and comparison with other methods are presented e.g. in [1, 4, 8]. The discussed search space is defined by the graph of operations planning relationships of the set of orders – the directed AND/OR-type graph describing precedence relations between all operations for scheduling. In the structure of the graph the notation 'operation on the node' is used. The presented model supports complex production orders, with hierarchical structures of processes and their execution according to both forward and backward strategies.

Keywords: Ant colony optimisation · Scheduling · And/or graphs

1 Introduction

In most cases production scheduling in real manufacturing systems belongs to NP-complete class of optimisation problems for which there is no known algorithm that can compute an exact solution in polynomial time. Fortunately, finding the optimal solution is usually not necessary - it is sufficient to find good, but not necessary the best, acceptable one that meets all given constraints. In solving this type of problems especially useful are soft computing class methods.

In this paper we focus on the ant colony optimisation (ACO) algorithm as member of swarm intelligence methods. Swarm intelligence (SI) deals with the implementation of rules of collective behaviour of small organisms from biological systems. They are characterized by decentralization in decision-making and self-organization. A review of engineering domain with implementation of ACO where presented in [3]. Some experimental results [1, 16] indicate that ACO algorithms can effectively solve the problems in integrated production planning and scheduling systems. Solutions for precisely defined scheduling problems, focused on e.g. resource constrained project

M. Graña et al. (eds.), *International Joint Conference SOCO'16-CISIS'16-ICEUTE'16*,
Advances in Intelligent Systems and Computing 527, DOI 10.1007/978-3-319-47364-2_39

environment [1, 13], backward scheduling [9], reactive scheduling [12], multiple objectives [5, 16] indicate a high level of similarity in the modelled mechanisms of swarms behaviour.

2 Characteristics of the Production System

Formulation of the problem relates to the typical mechanical engineering manufacturing systems. That kind of system consists of both single and dedicated, and also flexible, multitasking resources which can replace each other in the selected tasks (unrelated parallel). All production orders have defined routes, but they can be realized according to different transitions in the system (routing flexibility) by use multiple machines to perform the same operation in a given task, or use other available technology and machines. Because of parallel resources, there are two sub-tasks to considering: task allocation and scheduling. Production orders, rather non repetitive, can be realised according to forward or backward strategy, some of them are executed in pieces production conditions, some in series. Both machining and assembling (also disassembling) types of operations are allowed. To optimize the production flow a larger number of criteria are used (more than one) and sometimes they are defined by more than one decision maker. Criteria are based on times, dates, costs and others performance measures. The proposed model is built for given decision situation, according to scheduling/rescheduling method described in [20].

3 The Model of Product Flow

On the basis of the above characteristics the model of production flow was elaborated. The developed method of describing complex and multi-variant structures of processes, using and/or graphs, allows for flexible scheduling of operations of final products, as well as their sub-assemblies and parts. The proposed aggregated graph of operations planning relationships in the set of orders (G_{opr}) determines planning precedence constraints between all operations and defines a search space of the scheduling problem. This type of graph is created/modified in each decision situation requiring the creation or correction of a schedule. The procedure of creating this graph was described in [17]. The directed graph G_{opr} has notation ‘operation in the node’ - edges represent the possible transitions between operations during scheduling process. Between the edges in the graph may exist ‘and’ and ‘or’ types of relations. Major impact on the position of an operation in the graph has a scheduling strategy of its order, priority of order and the existence of alternative routes.

The exampled graph G_{opr} shown in Fig. 1 describes a production order consists of 3 processes P_1–P_3. Process P_1 realizes machining operations and has three alternative routes (O_1, O_2, O_4, O_6), (O_1, O_3, O_4, O_6), (O_1, O_5, O_6). Process P_2 consist of 2 disassembling: O_7, O_8 and 2 assembling operations O_9 and O_{10} without alternatives. P_3 has assembling O_{11} and machining O_{12} operation, and it requires the prior execution of processes P_1 and P_2. Edges with ‘and’ relation are marked with arches.

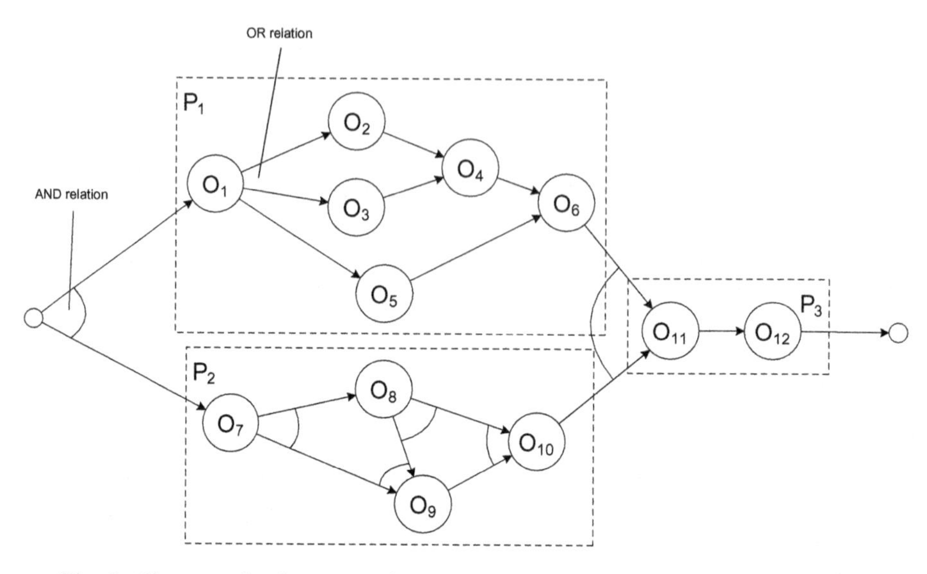

Fig. 1. The example of aggregated graph of operations planning relationships (G_{opr})

4 ACO Based Scheduling Algorithm

As a one way to determine the solution that best meets given set of criteria the method based on ant colony optimisation algorithm is proposed. This kind of metaheuristic belonging to the swarm intelligence group of methods imitates natural behaviour of ants when they collecting food in the anthill environment. The way in which ants find the shortest path between a food source and the anthill can be used to search the graph G_{opr} in order to find the best possible schedule. The following characteristics of ant colony are taken into account:

- pheromone trail of ant and the strength of its impact,
- the impact of pheromone on the ants behaviour,
- the need for finding the shortest (easiest) path to food.

A general flowchart of algorithm based on ant colony behaviour is shown in Fig. 2. The input data for scheduling should contain complete description of the production system and its load. Usually it is described symbolically by three (α, β, γ) where:

α - describes the configuration of the productive system: the size and the relationship between resources according to their functional similarity and routings of processes in adopted set of orders, β - describes hard constraints, which have to be satisfied in an acceptable solution. The collection of such restrictions is associated with both the resources, processes and orders, γ - describes the form of the objective function, taking into account the soft constraints which affects the quality of the schedule. Identifying these parameters enables the complete formulation of the scheduling task. On the base of scheduling task description the aggregated graph of operations planning relationships is built. Creating solutions is based on an iterative generation of solutions for adopted the population of ants. A single solution,

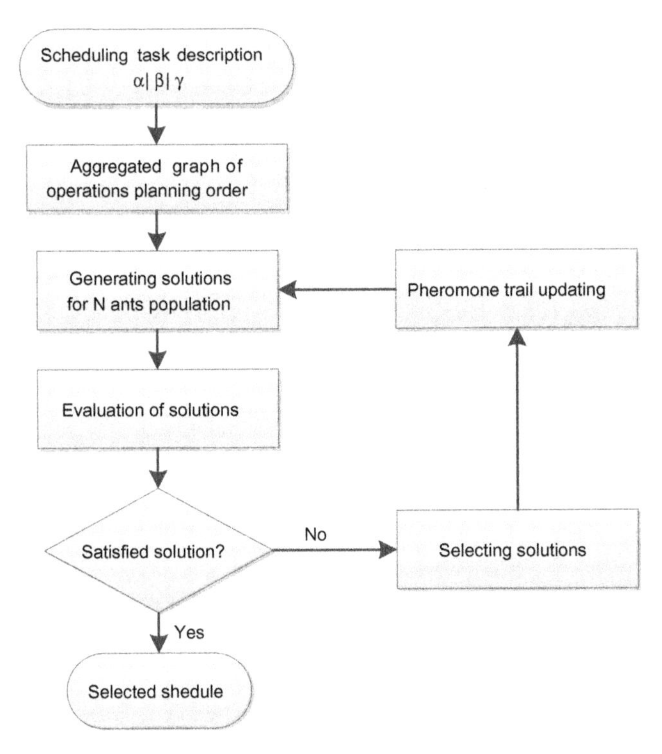

Fig. 2. A general flowchart of scheduling by ACO algorithm

represented by pass of one ant through all required nodes (all nodes excluding those of alternative paths), contains sequence of operations for a complete schedule.

4.1 Graph Searching Procedure

At each solution generation pass, each ant generates a complete solution by selecting the subsequent operations taking into account attractiveness of given path. In Fig. 3 the example of searching is presented. Starting from dummy node O_0 artificial ant can move to O_1 or O_7 node. Assuming that O_1 were selected the next move can be to O_2, O_3, O_5 or O_7. Staying in the node O_5, the O_6 or O_7 can be selected as next, etc.

It should be noted that ants transitions between nodes may but not have to coincide with the edges of the graph representing the relationship between operations.

In Table 1 all the possible transitions (x) between operations in discussed graph are shown. '-'means that the transition is not possible.

Attractiveness of given transition depends on two main factors, the pheromone amount and function of distance (desirability) of transition. Generally, the probability of choosing the next operation for the k-th ant is determined by [1–7]:

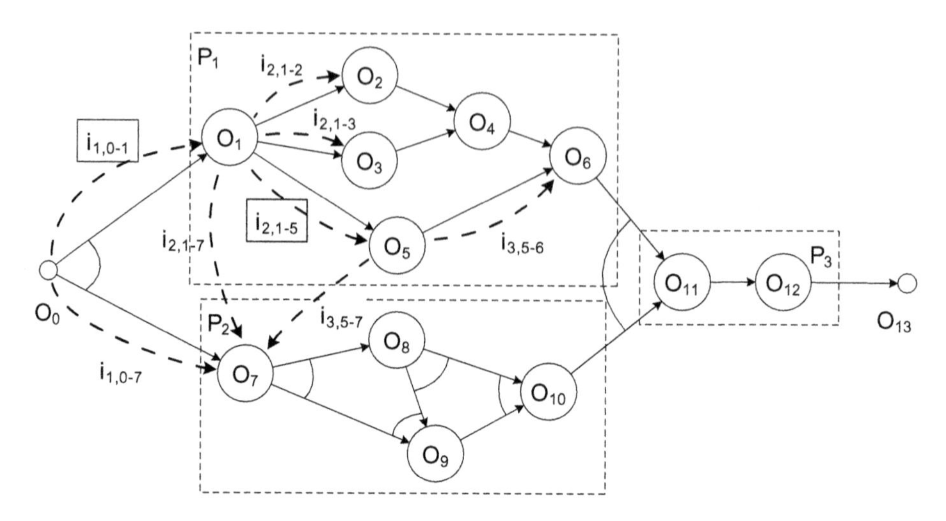

Fig. 3. An example of ant transitions in searching graph by ACO algorithm

Table 1. Matrix of possible transitions during searching the graph from Fig. 1

From\to	O0	O1	O2	O3	O4	O5	O6	O7	O8	O9	O10	O11	O12	O13
O0	–	x	–	–	–	–	–	x	–	–	–	–	–	–
O1	–	–	x	x	–	x	–	x	x	x	x	–	–	–
O2	–	–	–	–	x	–	–	x	x	x	x	–	–	–
O3	–	–	–	–	x	–	–	x	x	x	x	–	–	–
O4	–	–	–	–	–	–	x	x	x	x	x	–	–	–
O5	–	–	–	–	–	–	x	x	x	x	x	–	–	–
O6	–	–	–	–	–	–	–	x	x	x	x	x	–	–
O7	–	x	x	x	x	x	x	–	x	x	–	–	–	–
O8	–	x	x	x	x	x	x	–	–	x	x	–	–	–
O9	–	x	x	x	x	x	x	–	–	–	x	–	–	–
O10	–	x	x	x	x	x	x	–	–	–	–	x	–	–
O11	–	–	–	–	–	–	–	–	–	–	–	–	x	–
O12	–	–	–	–	–	–	–	–	–	–	–	–	–	x
O13	–	–	–	–	–	–	–	–	–	–	–	–	–	–

$$p_k(u,v) = \begin{cases} \frac{[\tau(u,v)]^a[\eta(u,v)]^b}{\sum_{w \in S_k}[\tau(u,w)]^a[\eta(u,w)]^b}, & \text{if } v \in S_k \\ 0 & \text{otherwise} \end{cases} \quad (1)$$

where:

u, v – the id of arc from node u to node v.
τ_{uv} – the current amount of pheromone of given transition $u \rightarrow v$,
η_{uv} – function indicates 'basic' attractiveness of transition $u \rightarrow v$,

a, b – weights controlling the relative importance of the pheromone and distance.
S_k – the set of nodes visited by k-th ant.

The result of distance function can be variable in each decision situation, depending on selected parameters connected with an operation:

$$\eta(u,v) \leftarrow f(e_t d_t + e_s d_s + e_{tb} d_{tb} + e_{te} d_{te} + e_c d_c) \quad (2)$$

where:
d_t – distance based on processing time,
d_s – distance based on setup time,
d_{tb} – distance based calculated beginning time,
d_{te} – distance based calculated end time,
d_c – distance based on costs,
e_x – weights.

4.2 Evaluation and Updating Pheromone Strength

After receiving solutions of all ants obtained best solutions are identified. The evaluation method can use metacriterion issue, by aggregation of partial rates using selected weight based methods e.g. convex combination, or other distance based measures [14, 15]. They may also be useful methods of searching for Pareto optimal solutions, which determine a set of effective solutions without aggregating evaluations of selected criteria. The set of solutions to update of pheromone can contain one or more results.

Another mechanism of changing pheromone strength is called evaporation. The pheromone evaporation factor (ρ) typically decreases strength of existing trail, before adding new to them. In the proposed solution evaporation is worst result based rather time based. The rule of updating pheromone is as follow:

$$\tau(u,v) \leftarrow \tau(u,v) + \Delta\tau_{k1}(u,v) - \rho\Delta\tau_{k2}(u,v) \quad (3)$$

Where $\Delta\tau_{kx}(u,v)$ - indicates the amount of pheromone adding (*k1*) or removing (*k2*) from the trail.

This allows weakening the arcs belonging to a worse solution, and strengthening those used by better rated. The quantity of the pheromone is calculated by:

$$\Delta\tau_{kx}(u,v) = \begin{cases} \frac{Q}{L_{kx}}, & \text{if transition (u,v) is used by kx} \\ 0 & \text{otherwise} \end{cases} \quad (4)$$

where:
Q – positive constant (a constant parameter of the algorithm),
L_{kx} – cost associated with solution by ant kx (as inverse of evaluation).

In the algorithm some control procedures should be also included. They apply to stopping or resetting iterations when obtained schedules not improve quality of solutions.

5 Summary

In the paper the conception of algorithm based on ACO mechanisms for scheduling of flexible manufacturing system is presented. This kind of soft computing methods, classified to widely understood swarm intelligence, is interesting and effective for searching solutions in computationally hard tasks, such as NP-complete problems. Presented scheduling algorithm deals with production orders that can be realised according to different production routes using monolithic approach – the choice of routes are done simultaneously with determining dates of operations executions. Elaborated mechanisms of creating and searching directed and/or type graph enable scheduling both with machining and assembling (disassembling) operations and complex structures of orders. Used notation 'operation in the node' allows direct transformation of multivariant descriptions of processes from CAPP class systems.

In real conditions, according to changes in a production system and environment creating/modifying a scheduling model and searching for a solution should be repeated using data driven and or event driven rescheduling procedures. The software implementation of elaborated algorithm in the KbRS scheduling system and testing (including comparison with other available algorithms and software) using data from the mechanical industry enterprises are planned in further work. Of course - the effectiveness of the ACO-class algorithm strongly depends on the proper selection of values of its parameters – it requires considerable knowledge and experience with their selection depending on the model of the manufacturing system.

References

1. Li, X., Shao, X., Gao, L., Qian, W.: An effective hybrid algorithm for integrated process planning and scheduling. Int. J. Prod. Econ. **126**, 289–298 (2010)
2. Leung, C.W., Wong, T.N., Mak, K.L., Fung, R.Y.K.: Integrated process planning and scheduling by an agent-based ant colony optimization. Comput. Ind. Eng. **59**, 166–180 (2010)
3. Chandra, B., Mohan, R.: Baskaran: a survey: ant colony optimization based recent research and implementation on several engineering domain. Expert Syst. Appl. **39**, 4618–4627 (2012)
4. Rossi, A., Dini, G.: Flexible job-shop scheduling with routing flexibility and separable setup times using ant colony optimisation method. Robot. Comput. Integr. Manuf. **23**, 503–516 (2007)
5. Yagmahan, B., Yenisey, M.M.: A multi-objective ant colony system algorithm for flow shop scheduling problem. Expert Syst. Appl. **37**, 1361–1368 (2010)
6. Blum, Ch., Sampels, M.: An ant colony optimization algorithm for shop scheduling problems. J. Math. Model. Algorithms **3**, 285–308 (2004)
7. Merkle, D., Middendorf, M., Schmeck, H.: Ant colony optimization for resource-constrained project scheduling. IEEE Trans. Evol. Comput. **6**(4), 333–346 (2002)
8. Xing, L.-N., Chen, Y.-W., Wang, P., Zhao, Q.-S., Xiong, J.: A knowledge-based ant colony optimization for flexible job shop scheduling problems. Appl. Soft Comput. **10**, 888–896 (2010)

9. Pereira dos Santos, L., Vieira, G.E.I., Leite, H.V.R., Steiner, M.T.A.: Ant colony optimisation for backward production scheduling. Adv. Artif. Intell. **2012**, Article ID 312132 (2012)
10. Dorigo, M., Maniezzo, V., Colorni, A.: Distributed optimization by ant colonies. In: Proceedings of ECAL91 – European Conference on Artificial Life, pp. 134–142. Elsevier Publishing, (1991)
11. Ponnambalam, S.G., Jawahar, N., Girish, B.S.: An Ant colony optimization algorithm for flexible job shop scheduling problem, New Advanced Technologies, ISBN: 978-953-307-067-4, InTech (2010)
12. Kato, E.R.R., Morandin Jr., O., Fonseca, M.A.S.: Ant colony optimization algorithm for reactive production scheduling problem in the job shop system. In: Proceedings of the 2009 IEEE International Conference on Systems, Man, and Cybernetics San Antonio, TX, USA (2009)
13. Chiang, Ch.-W., Huang, Y.-Q.: Multi-mode resource-constrained project scheduling by ant colony optimization with a dynamic tournament strategy. In: Third International Conference on Innovations in Bio-Inspired Computing and Applications (2012)
14. Diering, M., Dyczkowski, K., Hamrol, A.: New method for assessment of raters agreement based on fuzzy similarity. Adv. Intell. Syst. Comput. **368**, 415–425 (2015)
15. Lei, D.: Multi-objective production scheduling: a survey. Int. J. Adv. Manuf. Technol. **43**, 926–938 (2009)
16. Huang, R.-H., Yang, C.-L.: Overlapping production scheduling planning with multiple objectives - an ant colony approach. Int. J. Prod. Econ. **115**, 163–170 (2008)
17. Kalinowski, K., Zemczak, M.: Preparatory stages of the production scheduling of complex and multivariant products structures. Adv. Intell. Syst. Comput. **368**, 475–483 (2015)
18. Kalinowski, K., Grabowik, C., Kempa, W., Paprocka, I.: The graph representation of multivariant and complex processes for production scheduling. Adv. Mater. Res. **837**, 422–427 (2014)
19. Kalinowski, C. Grabowik, I. Paprocka, W. Kempa: Production scheduling with discrete and renewable additional resources. In: IOP Conference Series; Materials Science and Engineering; vol. 95, pp. 1757–8981 (2015)
20. Kalinowski, K., Grabowik, C., Kempa, W., Paprocka, I.: The procedure of reaction to unexpected events in scheduling of manufacturing systems with discrete production flow. Adv. Mater. Res. **1036**, 1662–8985 (2014)

Multistage Sequencing System for Complex Mixed-Model Assembly Problems

Marcin Zemczak[1(✉)] and Bożena Skołud[2]

[1] Faculty of Mechanical Engineering and Computer Science, Department of Production Engineering, University of Bielsko-Biala, Willowa 2, 43-309 Bielsko-Biala, Poland
mzemczak@ath.bielsko.pl

[2] Faculty of Mechanical Engineering, Institute of Engineering Processes Automation and Integrated Manufacturing Systems, Silesian University of Technology, Konarskiego 18A, 44-100 Gliwice, Poland
bozena.skolud@polsl.pl

Abstract. The paper presents the concept of multistage sequencing system which main task is to aid the processes of production scheduling in modern assembly systems, mainly in automotive industry, but also in other industry branches, where mixed-model production is conducted. Because of NP-hard character of considered scheduling tasks, it is a potential application area of soft computing graph searching techniques. Described sequencing system is a tool for production scheduling optimizing the process in terms defined by the user (e.g. the number of cycle times exceeded on the assembly line stations as well as the average exceeding of the cycle time). The system structure is based on graph searching techniques, and may be implemented in many problems solving methods, including not only production scheduling but also others, e.g. project scheduling. The developed system model is the basis of its implementation in production planning software, creating the sequences of production orders in few different stages.

Keywords: Production scheduling · Sequencing · Mixed-model assembly · Modelling · Graphs

1 Introduction

In order to ensure the quantity and quality of customer service, i.e. keeping deadlines of deliveries or the quality of products produced, new ideas in manufacturing processes management are implemented. Manufacturers strive to eliminate activities that do not add value to their products. This pursue is reflected in the concept of lean manufacturing. An approach consistent with lean is to reduce losses in all spheres of business activity, and thus shortening the production cycle, reducing inventory levels, increase resource utilization, and also reducing capital employed by the enterprise.

The complexity of real production systems and conducted tasks requires a considerable amount of calculations in production planning and control processes, what can be noted especially in scheduling. Those problems are mainly related to:

M. Graña et al. (eds.), *International Joint Conference SOCO'16-CISIS'16-ICEUTE'16*, Advances in Intelligent Systems and Computing 527, DOI 10.1007/978-3-319-47364-2_40

mixed-model, non-repeatable production, complex structures of manufacturing processes or the need for planning as flexibly as possible. Scheduling in real manufacturing systems, in majority belongs to the class of NP hard (NP-complete) problems as there are probably no possibilities for finding optimal solution. Therefore, it is reasonable to implement methods from the soft computing domain.

Increased interest may be noted not only known for a long time and used successfully "slimming" production by applying lean tools, but also in "smart" management systems, according to agile manufacturing, which largely have focus on cooperation and synergies through strategic vision, which would provide the customer appropriately personalized high quality products in time as well as acceptable for its price. Systems based only on lean, are systems that seek to eliminate losses, and agile systems are trying to positively utilize all activities that generate a loss, but cannot be eliminated through the lean tools. In recent times, an approach which combines features of both lean and agile in the literature is described by the term leagile [1, 2].

The search methods, taking into account the rapid product development and shortening product life cycle has led to the development of virtual manufacturing [3], which is the next step in the development of CIM. The essence of the concept of VM is to create a production unit by the selection of its resources with other companies and the integration of these resources with the help of a network connection. Companies that lend their resources, are part of the virtual enterprise, implementing joint production orders. The virtual company is characterized by high integration, interoperability and orientation on the customer-needs [4, 5]. It is now a very good example of the introduction of para-paradigm agile manufacturing production practice in advanced production enterprises. It should be noted that close cooperation between the supplier and the customer reduces costs on both sides, so it is a pure example of a win-win situation to manufacturing partners [6].

2 Modern Assembly Processes Organisation

In the case of modern assembly systems, the fact that assortment produced on them may differ in labour-consumption of specific operations is vital. There are several solutions, which in this case may be applied. One of the techniques to support assembly processes, and also an important element of modern assembly systems is the possibility of concurrent execution of assembly operations (Fig. 1). This is the chance of conducting assembly operations on another workstation, if the time available to perform an operation is too short, by temporarily exceeding the limits of the station by the assembly worker and returning after the completion of assembly operations. After completion of assembly operations on another workstation, the operator after returning to his own assembly station has less time at disposal (its associated with the continuous movement of the transport system) for the execution of oncoming assembly operations.

It should be noted that this kind of solution allows elimination of some of the problems associated with differences in the workload of individual products subjected to assembly operations, but practical use is often limited by the type of assembly operation, and above all the mobility of additional equipment supporting the various assembly operations. Such operations may be e.g. the installation of the dashboard to

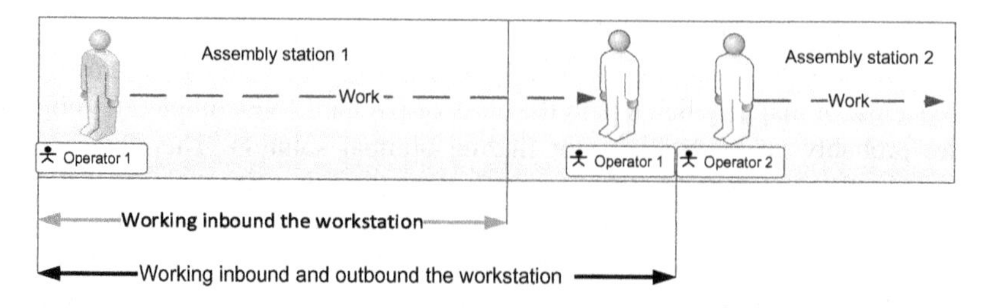

Fig. 1. Excessing the bounds of the assembly station

the car, where the range of possibilities to carry out efficient assembly is limited by the reach of the arm which is transporting the dashboard. Also relevant is the fact that it is permissible only in the case of well-balanced assembly line, in which the sequence is determined dynamically, in order to eliminate interference caused by the excessing of cycle time in assembly stations.

When the MPS (*Master Production Schedule*) is not dynamic, and the program output is fixed, production usually requires an additional assembly worker that in the event of possible cycle time excessing will assist in the execution of assembly operations.

Figure 2 presents the assembly operations conducted on a closed assembly station, where the vertical lines describe the boundaries of the station, with the possibility of utilizing additional operator. Sequence of assembly of models in the form of DACDDB has been established. Operation on the first model, each cycle begins in a reference position, which is the left border of the station. The horizontal solid line shows the start and end of each assembly operation, as well as the length of the station required to perform appropriate operation. The dotted line represents the distance between two successive products. This distance (w) is calculated by multiplying the frequency of appearance of the next product (t_{jA}) and the conveyor speed (v_c). Since w is a constant, dotted lines are the same length. After completion of the assembly operation, an

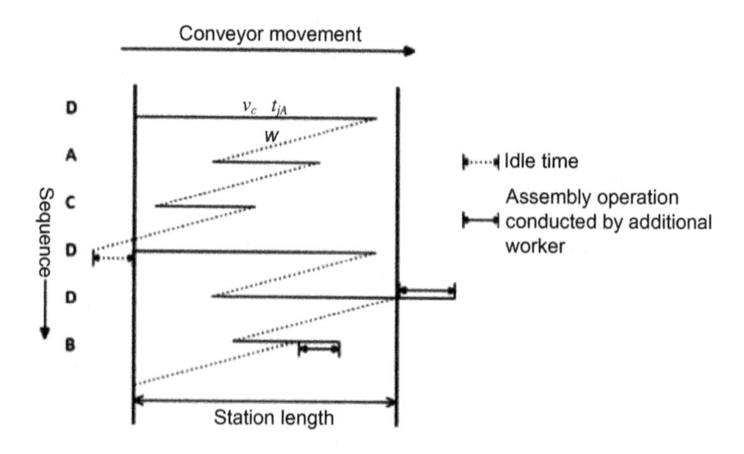

Fig. 2. Operations in closed assembly station with additional worker

employee is moving towards the beginning of the assembly station. If the new model is within the station, the employee can start working on it. Otherwise, the employee must wait until the product arrives at a reference point. Due to the limits of the station, the idle time is inevitable. The dotted line with an arrow indicates that the employee is idle, and its length is equal to a period of inactivity (until a next product appears at the station). For example, a worker is idle before he starts working on the fourth model of the sequence (A). Similarly, the employee may not exceed the right border of the station. Therefore, the worker must move to the left when it comes to the limit. In this case, incomplete operations are carried out by additional operators and are designated by a solid line with arrows. Additional operator for the last product is also a must due to the fact that the employee has to return to the reference point of before the next sequence starts. It should be noted that for the fixed sequence in the above example, removing the possibility of finishing the operation with the use of additional worker may cause in the next cycle the propagation of the interference and instead 2 of six operations executed partly outside the station, it would be five of six operations in the next cycle (Fig. 3).

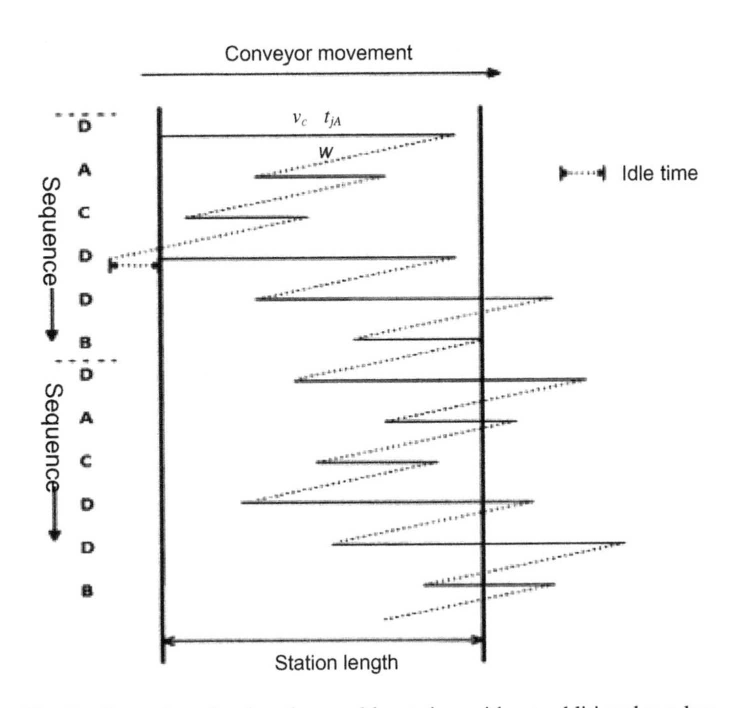

Fig. 3. Operations in closed assembly station without additional worker.

Process management assembly is an essential element of the operation of enterprises, but any intervention and testing of new solutions are usually not welcome, because it interferes with the production system [7, 10]. Therefore, more and more systems for optimizing the assembly prior to the introduction of changes in the physical system, are tested with various kinds of computer simulations.

3 Problems Solving Methods in Scheduling

Methods for solving scheduling problems can be generally divided into exact and approximate methods (Fig. 4).

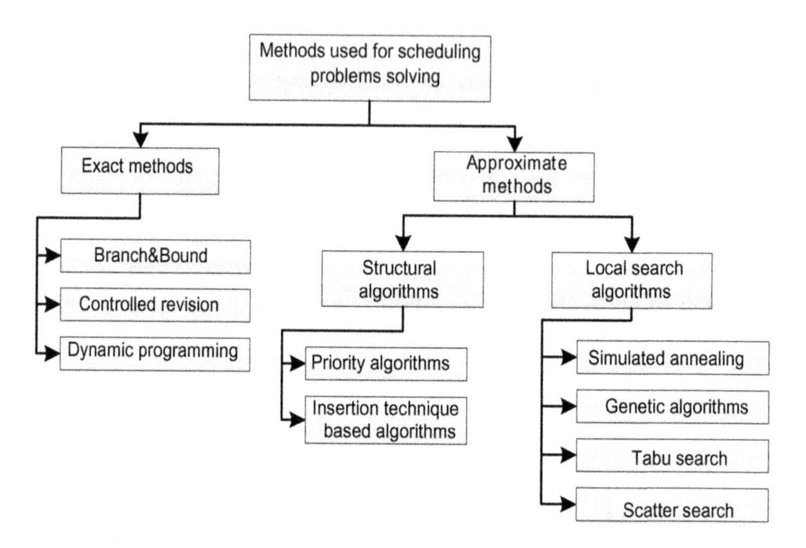

Fig. 4. Division of methods used for scheduling problems solving

The exact methods can be divided into e.g. methods based on the principle of branch and bound, the method of controlled revision, methods based on dynamic programming scheme. Due to the disadvantages and problems connected with their implementation to more complex tasks - the exact algorithms will not be further discussed. Among the approximate algorithms two major groups can be mentioned: structural algorithms that construct a solution to the problem and local search algorithms. Structural algorithms design is significantly dependent on the specifics of the problem. These algorithms can be divided into priority algorithms and algorithms that use the so-called insertions technique. The first rank the relative priorities of the tasks with the use of the rules of priority. The second group of algorithms, before ranking the tasks, creates a set of test solutions by inserting a task into a test position in current solution (constructed in the previous iteration of the algorithm). Among the selected test solutions, the best solution (in terms of objective function values) is chosen, which at the same becomes the base solution in the next iteration of the algorithm. Local search algorithms start operation from initial solution provided by the structural algorithm and work iteratively to improve the solution, using the techniques of reviewing the solution space. Among the most effective algorithms in this group simulated annealing, genetic algorithms, scatter search algorithms and tabu search algorithms can be specified. Presented methods are also implemented with the use of simulation software in order to visualize the solutions [8].

3.1 Scheduling Problems Description

Almost every scheduling problem can be described as a pair (*X*, *F*), where *X* is the set of feasible solutions to the problem (wherein $X \subseteq X^0$, where X^0 is the set of all solutions), *F* is the criterion of optimization (objective function). Solutions space is given explicitly only in a few cases, usually it is a set of independent variables, which can take the values of the specified ranges.

One of the basic concepts of local search algorithms is the movement. It can be described as an activity involving the transition between the two solutions and presented as a function of $v(x)\colon X \rightarrow X$, where $v(x) : x_v \in X, x_v \neq x$. Solution x_v, resulting from movement v is called the neighbouring solution of solution x. For each solution $x \in X$, a set of movements $V(x)$ which generates a neighbourhood $N(x)$ (1):

$$N(x) = \{x_v : \ v \in V(x)\} \tag{1}$$

of solution x can be defined. The definition of a set of movements and neighborhood depends on the analyzed problem and an algorithm used for its solutions.

Another important element is the local optimality of the solution [9]. In the literature, in the context of local search algorithms the term "best solution" found by the algorithm often appears. This concept is the result of a specific shortcut pointing to the solution generating the smallest value of the objective function from all the solutions generated by the algorithm (when the objective function is minimization).

3.2 General Car Scheduling Problem (CSP) Problem Description

In [7] a general definition of instances of the CSP problem is presented. An instance of the problem is defined as a set (V, o, p, q, r), where

- $V = \{v_1, .., v_n\}$ is a set of vehicles to be produced;
- $O = \{o_1, .., o_m\}$ is a set of different options;
- $p : O \rightarrow \infty$ and $q : O \rightarrow \infty$ define the limits capacitive associated with each option $o_i \in O$, this limitation restricts displacement that for any subsequence q_i consecutive cars on the line, no more than five of these options may require o_i;
- $r : V \, x \, O \rightarrow \{0, 1\}$ defines the requirements for options, for example for each car $v_j \in V$ and for each option $o_i \in O$, $r_{ji} = 1$ if the option o_i must be installed in v_j and $r_{ji} = 0$ otherwise.

It should be noted that two different cars of a set V may require the installation of the same configuration options, i.e. the same options are to be installed in them. All vehicles require installation of the same configuration options are grouped in the same class of car. More specifically, they form a k different vehicle classes, so that the set V is subdivided into k subsets $V = V_1 \cup V_2 \cup \ldots V_k$ such that all vehicles in a single subset V_i require the same configuration options.

CSP solution of the problem is to find a set of car set in the sequence *V*, thus defining the order in which they pass along the assembly line. Decision problem is to address the question if find a sequence that satisfies all constraints imposed can be found, and optimization for CSP is to find a sequence that minimizes costs, knowing that the cost function takes into account the non-fulfillment of limitations.

4 Multistage Orders Sequencing Software

For the purpose of sequencing and because of the fact that almost each product (car) is different, data of each order is kept in a shared database. Data itself is created automatically after the order is taken from the client in the place of sale (usually a new record, containing information about the selected version, engine, options, etc.). The system itself is executed in three stages (Fig. 5).

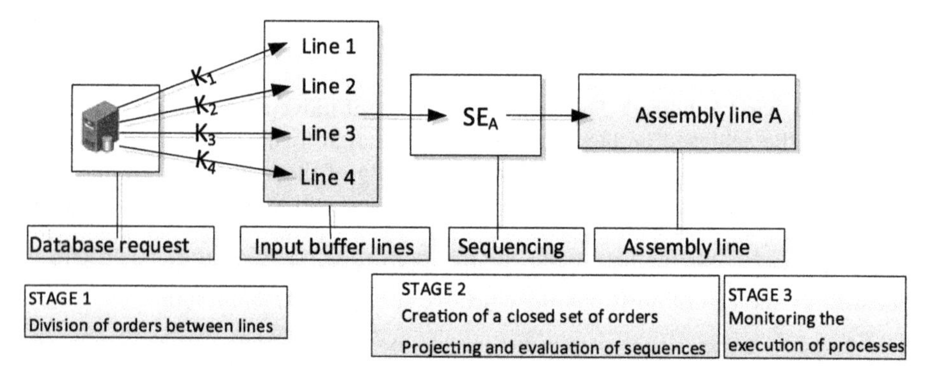

Fig. 5. Main steps performed in orders sequencing system

In the first stage the request for the data is sent, and according to the information obtained from the database orders are assigned to specific sets (K) and then sent to appropriate lines of the input buffer. The idea of the classification is quite simple, according to product labor-consumption matrix. Closed set of orders, with the information about the lines of the buffer they are placed in is being here created.

4.1 Proposed Method

The main goal of the proposed solution is to determine the effective method of creation of task sequence since sequencing according to the basic CSP guidelines is possible only in systems with fixed production plans, and with no interferences in logistics processes. Through the use of computational algorithms, and automatic analysis of the resulting sequence implemented into *Computerized Sequencing System* (CSS), rates of production are able to be checked in a real time. Identifying and applying a set of rules, responsible for obtaining the production orders sequence, provides more efficient planning of the production schedule, and hence better utilization of resources remaining at the disposal of the company. Main parts of the sequencing algorithm are based on the graph search methods, developed in order to adjust the form of the resulting algorithm to the form of a real manufacturing system that the research is based on.

In proposed solution (see Fig. 6) in the form of CSS, two steps of calculations are executed. In the first one, according to selected criteria (e.g. type of product and its version, labour-consumption requirements), orders entering the buffer (e.g. Z1–Z20) are allocated into proper buffer lines, and the report containing information about location of orders is generated. In the second step, sequencing algorithm, based on graph search methods, projects sequences, which are able to be executed and then assessment of each projected sequence based on a chosen criterion is carried out.

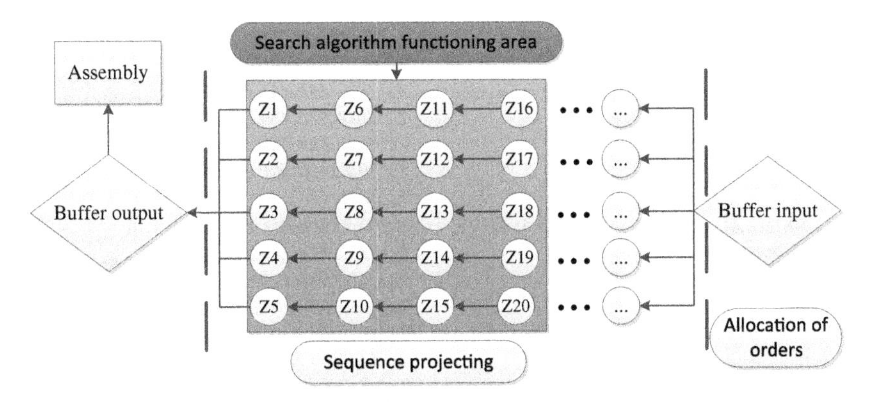

Fig. 6. Search algorithm for sequence projecting functioning area

5 Practical Example

The functioning of a sequencing system containing a combination of graph search methods (with the possibility to implement further algorithms), created for the purpose of sequencing orders in the considered system has been designed. In presented solution the minimization of two basic measures is used: the total time exceeded at the work-stations - min(ExWS) and number of exceedances - min(ExNo). Adapted multi-criteria evaluation method provides the ability to conduct searches in various forms. In Table 1 results of a simple experiment has been presented.

In the first step oncoming orders (*Z*) are divided according to their labour-consumption matrix into four groups (to fill four lines of the buffer on the basis of factors obtained during calculations). After completion of this phase comes the selection of a general method for searching the graph –to search the space various algorithms, based on the breadth-first search or depth-first search schemes may be used. The searching algorithm can also use both of them at the same time, executing the so-called mixed scheme. There are many different algorithms for implementing these strategies, extended by additional procedures for choosing the order of search nodes of the graph, the most promising directions, e.g. a beam search, best-first search, etc. Mentioned strategies differ in the way of passing graph nodes, however a common feature of them is a single execution of the action associated with a node, regardless of the number of passes through the node.

Table 1. CSS report

Orders allocation			Best projected sequence		Min (ExWS)	Min (ExNo)
Z1 (31.68; 2)	Z18 (31; 1)	Z42 (34.4; 4)	1	Z1,Z2,Z10,Z23,Z13,Z35,Z12, Z34,Z5,Z14,Z33,Z11,Z43, Z3,Z42,Z31,Z24,Z41,Z4, Z40,Z36,Z39,Z38,Z30,Z29, Z22,Z32,Z28,Z21,Z2,Z26, Z20,Z9,Z19,Z18,Z17, Z16, Z6,Z25,Z15,	71	15
Z2 (32.59; 2)	Z17 (31; 1)	Z41 (34.4; 4)	2	Z35,Z23,Z10,Z1,Z22,Z21,Z5, Z20,Z2,Z34,Z14,Z33,Z19, Z31,Z43,Z42,Z41,Z30,Z29, Z13,Z18,Z40,Z39,Z17,Z28, Z16,Z12,Z27,Z26,Z32,Z11, Z3,Z9,Z24,Z4,Z36,Z38,Z6, Z25,Z15,	72	18
Z13 (32.59; 2)	Z16 (31; 1)	Z40 (34.4; 4)	3	Z35,Z34,Z1,Z2,Z23,Z13,Z33, Z10,Z22,Z21,Z43,Z20,Z19, Z18,Z5,Z14,Z12,Z31,Z11, Z42,Z17,Z30,Z41,Z3,Z29, Z16,Z24,Z4,Z40,Z36,Z28, Z6,Z27,Z39,Z32,Z38,Z26, Z25,Z15,Z9	78	17
Z12 (32.59; 2)	Z4 (32.36; 2)	Z39 (34.4; 4)				
Z11 (32.59; 2)	Z5 (33.49; 3)	Z32 (34.4; 4)				
Z10 (32.59; 3)	Z14 (33.49; 3)	Z31 (34.4; 3)				
Z3 (31; 2)	Z6 (30.32; 1)	Z30 (34.4; 3)				
Z24 (31; 2)	Z25 (30.32; 1)	Z29 (34.4; 3)				
Z23 (31; 1)	Z15 (30.32; 1)	Z28 (34.4; 3)				
Z22 (31; 1)	Z35 (34.4; 4)	Z27 (34.4; 3)				
Z21 (31; 1)	Z34 (34.4; 4)	Z26 (34.4; 3)				
Z20 (31; 1)	Z33 (34.4; 4)	Z36 (32.13; 2)				
Z19 (31; 1)	Z43 (34.4; 4)	Z38(32.13; 2)				
		Z9 (33.95; 3)				

For presented exemplary data CSS has designed 50 sequences, out of which three best have been presented. The system may stop projecting sequences after obtaining a pre-set stop condition, or projecting sequences may be stopped by the user after achieving certain parameters.

6 Summary

In the paper the solution to a *CSP* problem in the form of *Computerized Sequencing System* has been presented. Through the use of computational algorithms, and automatic analysis of the resulting sequence, rates of production are able to be checked in a real time. Determination of solutions in that multistage process can be accomplished using many different strategies and algorithms. Appropriate sequencing of production orders in the case of mass linear mixed-model production systems is a key factor, since it has a significant impact on efficiency throughout the whole enterprise. Identifying and applying a set of rules and criteria, responsible for obtaining the production orders sequence and its assessment, provides more efficient planning of the production schedule, and hence better utilization of resources that remain at the disposal of the company. The issue in later stages can also be developed in many fields, e.g. the form of the buffer or simultaneous balancing of the assembly line and sequencing of orders. Developed *Computerized Sequencing System* is a very versatile software, and after just few modifications, concerning mainly the structure of the buffer at the assembly department and set of rules in projecting algorithm can be adopted to many other assembly systems.

References

1. Seyedi Seyed, N.: Supply chain management, lean manufacturing, agile production, and their combination. Interdisc. J. Contemp. Res. Bus. **4**(8), 648–653 (2012)
2. Vinodh, S., Aravindraj, S.: Evaluation of leagility in supply chains using fuzzy logic approach. Int. J. Prod. Res. **51**(4), 1186–1195 (2013)
3. Banaszak, Z., Skołud, B., Zaremba, M.B.: Computer-aided prototyping of production flows for a virtual enterprise. J. Intell. Manuf. **14**, 83–106 (2003)
4. Rudnicki, J.: Przedsiębiorstwo agile, Logistyka Produkcji (2011). http://www.log24.pl/artykuly/przedsiebiorstwo-agile,1464
5. Masahiko, O., Iwata, K.: Development of a virtual manufacturing system by integrating product models and factory models. Ann. CIRP **42**(1), 475–478 (1993)
6. Kalinowski, K., Paprocka, I.: Scheduling Schemes Based on Searching the Aggregated Graph of Operations Planning Sequence. Applied Mechanics and Materials, vol. 809/810, pp. 1462–1467. Trans Tech Publications, Zurich (2015)
7. Kalinowski, K., Zemczak, M.: Preparatory stages of the production scheduling of complex and multivariant products structures. In: Herrero, Á., Sedano, J., Baruque, B., Quintián, H., Corchado, E. (eds.) 10th International Conference on Soft Computing Models in Industrial and Environmental Applications. Advances in Intelligent Systems and Computing, vol. 368, pp. 475–483. Springer International Publishing, Switzerland (2015)
8. Allahverdi, A., Al-Anzi, F.S.: A PSO and a Tabu search heuristics for the assembly scheduling problem of the two-stage distributed database application. Comput. Oper. Res. **33**, 1056–1080 (2006)
9. Skolud, B., Krenczyk, D., Zemczak, M.: Multi-assortment rhythmic production planning and control. In: Book Series: IOP Conference Series - Materials Science and Engineering, vol. 95, Article Number: 012133 (2015)
10. Krenczyk, D., Skolud, B.: Transient states of cyclic production planning and control. Appl. Mech. Mater. **657**, 961–965 (2014)

Robustness of Schedules Obtained Using the Tabu Search Algorithm Based on the Average Slack Method

Iwona Paprocka, Aleksander Gwiazda(✉), and Magdalena Bączkowicz

Faculty of Mechanical Engineering, Silesian University of Technology, Konarskiego 18A str., 44-100 Gliwice, Poland
{iwona.paprocka,aleksander.gwiazda, magdalena.baczkowicz}@polsl.pl

Abstract. One of the most important problems consider with the scheduling process is to ensure the needed level of robustness of obtained schedules. One of possible tools that could be used to realize this objective is the Taboo Search Algorithm (TSA). The Average Slack Method (ASM) enables to obtain the best performance of the job shop system. In the paper is presented analysis of two objectives: to achieve the best compromise basic schedule for four efficiency measures as well as to achieve the best compromise reactive schedule. It was investigated of 15 processes executed on 10 machines. It was shown that ASM enables the obtainment of the best performance of the job shop system.

Keywords: Scheduling · Tabu Search Algorithm (TSA) · Average Slack Method (ASM)

1 Introduction

Since the 50 s of the XX century many researchers have interested in scheduling problems, and since then they have made researches to solve this complex problem including different production systems, constrains and criteria using different methods [23]. Often, analysed scheduling problems are NP-hard and a general method which achieves an optimal solution is not known. Therefore solutions which meet given constraints and solutions which are obtained within a predefined period of time are accepted.

In the production practice, it is important to achieve a solution which deals with the uncertainty of the production system [14, 15, 20]. Unexpected disruptions can affect the performance measure and can change the system status [4, 5, 21]. There are many types of disturbances which can generate the differences between a basic schedule and a new realization on the shop floor, namely: machines failures, rush order arrival, priority of job changing, unavailable materials or tools (delay in the arrival or shortage of materials or tools), changing due date of an order (delay or advance), job cancellation, rework or quality problems, operator absenteeism, inserting or deleting of an operation from a process route, new routes of processes, overestimating or underestimating of

M. Graña et al. (eds.), *International Joint Conference SOCO'16-CISIS'16-ICEUTE'16*,
Advances in Intelligent Systems and Computing 527, DOI 10.1007/978-3-319-47364-2_41

operation times, overestimating or underestimating of reliability characteristics (failure-free times, repairing times). Due to the high number of possible disruptions the interest in algorithms which take into account the dynamics of production systems rises. In the following paper the attempt to classify the research related to the disturbances of production systems is undertaken [8]. The disturbances are classified into three groups: disturbances related to resources availability, disturbances related to orders, disturbances related to errors in the estimation of production parameters (Table 1).

Table 1. Research related to the disturbances of production systems

Disturbances related to resources availability		Disturbances related to orders		Disturbances related to errors in the estimation of production parameters	
Machines failures	[2, 3, 11, 12, 16, 18, 25, 27]	Rush order arrival	[19, 25]	Overestimating or underestimating of operation times,	[25]
		Priority of job changing	[9]	Overestimating or underestimating of reliability characteristics (failure-free times, repairing times)	[22]
		Changing due date of an order (delay or advance)			
		Job cancellation	[25]		
Unavailable materials (delay in the arrival or shortage of materials)	[10]	A new job arrival	[17, 25, 27]		
Unavailable tools (delay in the arrival or shortage of tools)	[7]	Rework or quality problems	[19]		
		Inserting or deleting of an operation from a process route	[26]		
Operator absenteeism	[24]	New routes of processes			

The literature proposes two scheduling methods to deal with static and dynamic problems. Deterministic or static (off line) scheduling occurs when a schedule is generated in advance. Dynamic (on line) scheduling refers to situations when a schedule is generated or a control decision is made after a disruption. There are two scheduling strategies for static or deterministic environment of a production system: nominal scheduling relates to the system when all information is deterministic and predictive scheduling relates to the system when some information can be uncertain.

There are two rescheduling strategies for dynamic environment of a production system: dynamic scheduling and reactive scheduling. Rescheduling policies can be triggered by events (event-driven), can be periodic and hybrid [13]. In an event-driven rescheduling policy, reactive schedule is triggered by a single event e.g. after a machine failure occurrence [9]. In the rolling horizon scheduling, reactive schedule is generated at regular intervals e.g. every week or each shift [9]. In a hybrid rescheduling policy schedules are updated periodically and when a special disruption occurs, e.g. a machine failure [28]. Periodic scheduling can generate more stable schedules, under constrain, that the even driven period is less that the constant period.

Managers search for not only high-quality schedules but also schedules which enable to react quickly to unexpected events and reorganize production plans in a cost-effective manner. Searching for an effective algorithm (the algorithm which achieves stable and robust schedules) two predictive-reactive scheduling algorithms are compared: Multi Objective Immune Algorithm (MOIA), and the algorithm based on the Average Slack Method (ASM). The architecture of both algorithm relies on researching the influence of a rescheduling policy over the performance of a production system.

On the basis of the overview of reference publications, the following research points have been identified:

(1) methods of generating basic schedules enabling the obtainment of the best performance of the job shop system when the objective is to minimise makespan, flow time, total tardiness and idle time.
(2) methods of generating basic schedules enabling the obtainment of stable and robust schedules in the case of the disturbance occurrence.

The paper includes an analysis of utilization of the TSA using the ASM approach. Below is presented the algorithm of the TSA.

2 The Architecture of the Tabu Search Algorithm

The Tabu Search Algorithm (TSA) enhances the performance of local search by relaxing basic rules considered this process. In TSA are accepted worsening moves at each step if no improving move is available. Additionally, to the TS algorithm are introduced special prohibitions. Their aim is to discourage the search from coming back to previously analysed solutions. These prohibitions are called tabu. Hence if a potential solution has been previously analysed within a certain short-term period or if it was not compatible with the system of rules, it is marked as "tabu" (forbidden). The algorithm does not consider that possibility next time.

TSA accepts the value of parameters generating the most robust schedules. The scheduling algorithm consists of two parts, i.e. a sequence generator and a sequence evaluator. TSA is applied to effective scanning of the solution space [12]. The ASM is applied to evaluate a reactive schedule using the same criterion as the criteria used for the basic schedule evaluation. However, in this case this criterion is increased by the value of deterioration of the criterion due to a disturbance. The advantage of the AMS is a high correlation between the stability and the average slack value of schedules.

In the tabu-search method, the sequence of jobs is generated and evaluated taking into account the effects of a disturbance [12]. The best neighbouring sequence, achieved in an iteration, is compared with the tabu solution. The tabu solution is the best sequence achieved in all previous iterations. The tabu solution is the input data for the next iteration (Fig. 1). Such functioning of the sub-processes of the algorithm could be repeated continuously to obtain the more improved list of tabu solutions as the base for optimizing decision. To evaluate the obtained schedules are used two criteria: solution robustness *SR* and quality robustness *QR*.

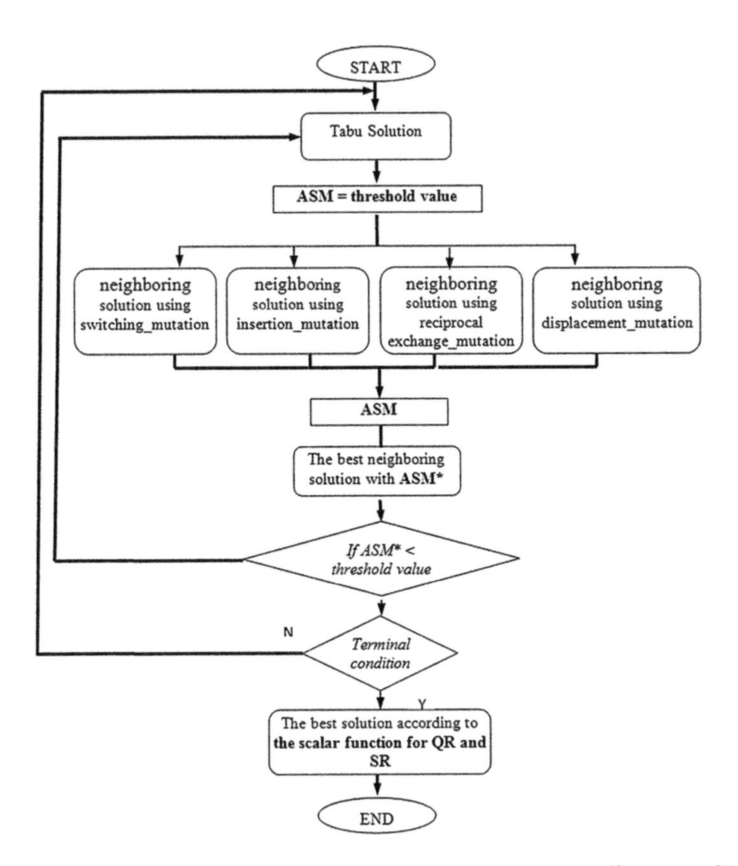

Fig. 1. Architecture of the TSA in the searching process for the trade-off between SR and QR

The best compromise solution is selected from the tabu list in the TSA. The reactive schedules are generated by rescheduling disturbed operation/s to the parallel machine/s that is/are first available. It allows shortening the needed time of rescheduling process.

3 Experimental Design and Job Shop Scheduling Problem Statement

Following the experiments conducted by Abumaizar and Svestka [1], the effect of four experimental factors can be analysed: predictive-reactive scheduling method on a performance measure, size of the predictive-reactive scheduling problem measured by the number of operations for execution after a disturbance, the magnitude (size) of the disruption and flexibility of the production system measured by the routing flexibility (the number of available alternative routes for each job [6]). In this paper the influence of predictive-reactive scheduling methods on performance measures for a job shop system is analysed.

The job shop system can be defined as assigning jobs to machines. Each job consists of a set of operations that have to be processed in a specified sequence. Each operation has to be executed on a given machine. Processing times are known in advance. Release times and due dates are specified, non-preemptive. Moreover in the case of a disturbance, routing flexibility is admissible.

The first objective is to achieve the best compromise basic schedule of 15 processes executed on 10 machines, for four efficiency measures: makespan ($C_{\max} \rightarrow \min$), flow time ($F \rightarrow \min$), total tardiness ($T \rightarrow \min$) and idle time ($I \rightarrow \min$). In order to make comparison possible between the two algorithms, a decision maker defined priorities of criteria. The priorities (weights) of 1st and 3rd criteria equal 0.3, the priorities of 2nd and 4th criteria equal 0.2.

The first machine is the most loaded one. The increased probability of the bottleneck failure is indicated at the time horizon: $[a, b + MTTR]$ where: a = *60* and b = *72* and, MTTR = 6. The failure free time of the bottleneck MTTF equals 66. After a disturbance, each operation v_j can be rescheduled on a machine first available from a set of parallel machines. The second objective is to achieve the best compromise reactive schedule of 15 processes executed on 10 machines, for two criteria: solution robustness *SR* and quality robustness *QR* in the event of bottleneck failure. The quality robustness measures the degradation of the performance (efficiency measure) of the basic schedule due to the disturbance. The solution robustness computes the absolute deviations of starting times of operations in the reactive and basic schedule. In order to make a possible comparison between the two algorithms, a decision maker defined priorities of criteria. Priorities (weights) of SR and QR equal 0.5.

4 Results of Computer Simulations

The objective of the experimental study is to determine which method of predictive-reactive scheduling provides better results considering the following sets of objectives:

- makespan, flow time, total tardiness and idle time for predictive scheduling,
- solution robustness and quality robustness for reactive scheduling, for the job shop system.

In order to achieve the basic schedule for JS problem (10 × 15) using the TSA, three simulations were generated for the input data: terminal condition is the number of iterations $L = 20$, number of clones $C = 4$.

In the first simulation, the best schedule was generated according to the rule of {2 5 8 14 1 0 3 4 7 6 11 12 10 9 13}. The quality of the reactive schedule is FF = 131 with the components of QR = 8 and SR = 254. The quality of this schedule is: $C_{\max} = 118$, $F = 562$, $I = 648$ and $T = 0$ before rescheduling, and $C_{\max} = 117$, $F = 535$, $I = 623$ and $T = 0$ after rescheduling. Remaining solutions are described in Table 2.

Table 2. The best basic schedules achieved by the TSA and $L = 20$

No.	Job shop scheduling problem *(15 × 10)*													
	The priority rule of the basic schedule	The quality of the schedule y					The quality of the reactive schedule y*							
		C_{max}	F	I	T	FFy	C_{max}	F	I	T	FFy*	QR	SR	FFry*
1	2 5 8 14 1 0 3 4 7 6 11 12 10 9 13	118	562	648	0	277,4	117	535	632	0	268,5	8	254	131
2	2 5 8 14 1 3 0 4 6 7 10 11 12 9 13	117	521	638	0	266,9	117	512	632	0	263,9	3	285	144
3	2 5 8 14 1 3 0 4 6 7 11 10 12 9 13	117	517	638	0	266,1	117	518	632	0	265,1	1	279	140

Table 3. The best basic schedules achieved by the TSA and $L = 40$

No.	Job shop scheduling problem *(15 × 10)*													
	The priority rule of the basic schedule	The quality of the schedule y					The quality of the reactive schedule y*							
		C_{max}	F	I	T	FFy	C_{max}	F	I	T	FFy*	QR	SR	FFry*
1	2 5 8 14 1 3 0 4 6 7 10 11 12 9 13	117	521	638	0	266,9	117	512	632	0	263,9	3	285	144
2	2 5 8 14 1 0 4 6 7 10 11 12 3 9 13	117	516	638	0	265,9	117	507	632	0	262,9	3	178	90.5
3	2 5 8 14 1 3 0 4 6 7 10 11 12 9 13	117	521	638	0	266,9	117	512	632	0	263,9	3	285	144

Also, three simulations were generated for $L = 40$ and the remaining input data presented above. In the first simulation, the best schedule was generated according to the rule of {2 5 8 14 1 3 0 4 6 7 10 11 12 9 13}. The quality of the reactive schedule is FF = 144 with the components of QR = 3 and SR = 285. The quality of this schedule is: $C_{\max} = 117$, $F = 521$, $I = 638$ and $T = 0$ before rescheduling, and $C_{\max} = 117$, $F = 512$, $I = 632$ and $T = 0$ after rescheduling. Remaining solutions are described in Table 3.

Using the CSA, the best basic schedule dealing with the uncertainty is generated in the first simulation for $L = 20$ and in the second simulation for $L = 80$.

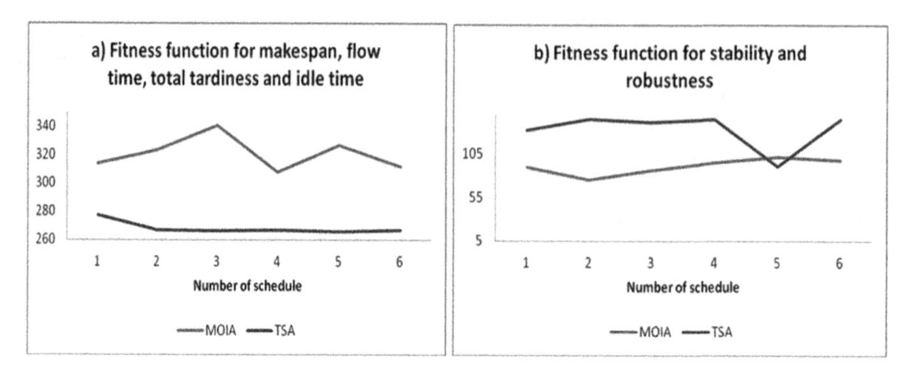

Fig. 2. The fitness function of the basic (a) and reactive schedules (b)

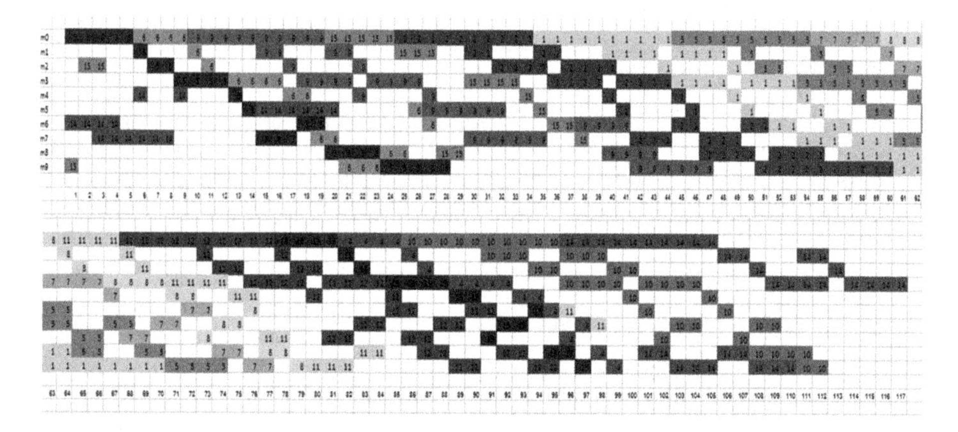

Fig. 3. The best basic schedule, achieved using the TSA, which generates the best reactive schedule in the event of a disturbance.

5 Conclusion

In the paper the effectiveness of the Tabu Search Algorithm (TSA) based on the Average Slack Method (ASM) is presented. The algorithm which achieves stable and robust schedules is searched for. The architecture of this algorithm relies on researching the influence of a rescheduling policy over the performance of a production system. The experimental study was to determine the method of predictive-reactive scheduling that provides enough good results considering such objectives: makespan, flow time, total tardiness and idle time for predictive scheduling, and solution robustness and quality robustness for reactive scheduling.

The Average Slack Method enables the obtainment of the best performance of the job shop system when the objective is to minimise makespan, flow time, total tardiness and idle time. In the Fig. 2 is presented the comparison of the TSA with similarly popular the MOIA (Multi Objective Immune Algorithm) algorithm.

The objective of the future researches is the modification of architecture and evaluation criteria of the TSA to achieve not only good quality schedules for criteria like makespan, flow time, total tardiness and idle time but also stable and robust schedules. The best solution achieved by the TSA is presented in Fig. 3.

It should be emphasized that one of the goals of future studies will also be comparing the range of applicability and other scheduling algorithms.

References

1. Abumaizar, R.J., Svestka, J.A.: Rescheduling job shops under random disruptions. Int. J. Prod. Res. **35**, 2065–2082 (1997)
2. Al-Hinai, N., ElMekkawy, T.Y.: Robust and flexible job shop scheduling with random machine breakdowns using a hybrid genetic algorithm. Int. J. Prod. Econ. **132**, 279–291 (2011)
3. Hamzadayi, A., Yildiz, G.: Event driven strategy based complete rescheduling approaches for dynamic m identical parallel machines scheduling problem with a common server. Comput. Ind. Eng. **91**, 66–84 (2016)
4. Banaś, W., Sękala, A., Foit, K., Gwiazda, A., Hryniewicz, P., Kost, G.: The modular design of robotic workcells in a flexible production line. In: IOP Conference Series: Materials Science and Engineering, vol. 95, p. 012099 (2015)
5. Banaś, W., Sękala, A., Gwiazda, A., Foit, K., Hryniewicz, P., Kost, G.: Determination of the robot location in a workcell of a flexible production line. In: IOP Conference Series: Materials Science and Engineering, vol. 95, p. 012105 (2015)
6. Chen, J., Chung, C.-H.: An examination of flexibility measurements and performance of flexible manufacturing systems. Int. J. Prod. Res. **34**, 379–394 (1996)
7. Cheng, R., Gen, M., Tsujimura, Y.: A tutorial survey of job-shop scheduling problems using genetic algorithms, part II: hybrid genetic search strategies. Comput. Ind. Eng. **36**, 343–346 (1999)
8. Chong, C.S., Sivakumar, A.I., Gay, R.: Simulation-based scheduling for dynamic discrete manufacturing. In: Proceedings of the 2003 Winter Simulation Conference, pp. 1465–1473 (2003)
9. Church, L.K., Uzsoy, R.: Analysis of periodic and event-driven rescheduling policies in dynamic shops. Int. J. Comput. Integr. Manuf. **5**, 153–163 (1992)
10. Duenas, A., Petrovic, D.: An approach to predictive-reactive scheduling of parallel machines subject to disruptions. Ann. Oper. Res. **159**, 65–82 (2008)
11. Pan, E., Liao, W., Xi, L.: A joint model of production scheduling and predictive maintenance for minimizing job tardiness. Int. J. Adv. Manuf. Technol. **60**, 1049–1061 (2012)
12. Goren, S., Sabuncuoglu, I.: Robustness and stability measures for scheduling: single-machine environment. IIE Trans. **40**, 66–83 (2008)
13. Vieira, G.V., Herrmann, J.W., Lin, E.: Rescheduling manufacturing systems: a framework of strategies, policies, and methods. J. Sched. **6**(1), 35–58 (2003)
14. Heng, L., Zhicheng, L., Ling, X.L., Bin, H.: A production rescheduling expert simulation system. Eur. J. Oper. Res. **124**, 283–293 (2000)
15. Jain, A.K., Elmaraghy, H.A.: Production scheduling/rescheduling in flexible manufacturing. Int. J. Prod. Res. **35**, 28–309 (1997)
16. Hasan, S.M.K., Sarker, R., Essam, D.: Genetic algorithm for job-shop scheduling with machine unavailability and breakdowns. Int. J. Prod. Res. **49**(16), 4999–5015 (2011)

17. Zhang, L., Gao, L., Li, X.: A hybrid genetic algorithm and tabu search for multi-objective dynamic job shop scheduling problem. Int. J. Prod. Res. **51**(12), 3516–3531 (2013)
18. Liu, L., Han-yu, G., Yu-geng, X.: Robust and stable scheduling of a single machine with random machine breakdowns. Int. J. Adv. Manuf. Technol. **31**, 645–656 (2007)
19. Mattfeld, D.C., Bierwirth, C.: An efficient genetic algorithm for job shop scheduling with tardiness objectives. Eur. J. Oper. Res. **155**, 616–630 (2004)
20. Matsuura, H., Tsubone, H., Kanezashi, M.: Sequencing, dispatching and switching in a dynamic manufacturing environment. Int. J. Prod. Res. **37**(7), 1671–1688 (1993)
21. Monica, Z.: Optimization of the production process using virtual model of a workspace. In: IOP Conference Series: Materials Science and Engineering, vol. 95, p. 012102 (2015)
22. Paprocka, I., Kempa, W.M., Kalinowski, K., Grabowik, C.: A production scheduling model with maintenance. Adv. Mater. Res. **1036**, 885–890 (2014)
23. Paprocka, I., Kempa, W., Kalinowski, K., et al.: Estimation of overall equipment effectiveness using simulation programme. Mater. Sci. Eng. **95** (2015). Article Number: 012155
24. Paprocka, I., Kempa, W.M., Grabowik, C., Kalinowski, K.: Sensitivity analysis of predictive scheduling algorithms. Adv. Mater. Res. **2014**, 921–926 (1036)
25. Fahmy, S.A., Balakrishnan, S., ElMekkawy, T.Y.: A generic deadlock-free reactive scheduling approach. Int. J. Prod. Res. **47**(20), 5657–5676 (2009)
26. Suresh, V., Chudhari, D.: Dynamic scheduling - a survey of research. Int. J. Prod. Econ. **32** (1), 53–63 (1993)
27. Turkcan, A., Akturk, M.S., Storer, R.H.: Predictive/reactive scheduling with controllable processing times and earliness-tardiness penalties. IIE Trans. **41**, 1080–1095 (2009)
28. Vieira, G.E., Herrmann, J.W., Lin, E.: Predicting the performance of rescheduling strategies for parallel machine systems. J. Manuf. Syst. **19**(4), 256–266 (2000)

Heterogeneous Fleet Vehicle Routing and Scheduling Subject to Mesh-Like Route Layout Constraints

Grzegorz Bocewicz[1], Zbigniew Banaszak[1], and Damian Krenczyk[2](✉)

[1] Department of Computer Science and Management, Koszalin University of Technology, Koszalin, Poland
{bocewicz, banaszak}@ie.tu.koszalin.pl
[2] Faculty of Mechanical Engineering, Silesian University of Technology, Gliwice, Poland
damian.krenczyk@polsl.pl

Abstract. The behavior of a Material Transportation System (MTS) encompassing movement of various transport modes has to be admissible, i.e. collision- and congestion-free, as to guarantee deadlock-free different flows of concurrently transported goods. Since the material flows following possible machining routes serviced by MTS determine its behavior the following questions occur: what kind of MTS structure can guarantee a given behavior, and what admissible behavior can be reachable in a given MTS structure? These questions are typical for vehicle routing problems which are computationally hard. Their formulation within the framework of mesh-like and fractal-like structures enables, however, to get a significant reduction on the size. Such structures enable to evaluate admissible routings and schedules following flow-paths of material transportation in a polynomial time. Considered in the paper production routes followed by MTS are serviced by operations subsequently executed by AGVs and machine tools. The transport operations performed by AGVs are arranged in a streaming closed-loops network where potential conflicts are resolved by priority dispatching rules assigned to shared resources. The main problem boils down to the searching for sufficient conditions guaranteeing MTS cyclic steady state behavior. Implementation of proposed conditions is illustrated through multiple examples.

Keywords: Material Transportation System · Mesh-like structure networks · Declarative modeling · Routing · Scheduling

1 Introduction

A Material Transport System (MTS) configuration is one of the most important aspects of manufacturing facility design. The fundamental questions that need to be answered are what transport technologies are going to be used in a manufacturing environment and how will they be used [1, 5, 6, 8, 10, 16]. The MTS design problem usually focuses either on the technology selection problem or the flow network design problem. In turn,

M. Graña et al. (eds.), *International Joint Conference SOCO'16-CISIS'16-ICEUTE'16*,
Advances in Intelligent Systems and Computing 527, DOI 10.1007/978-3-319-47364-2_42

the flow network design problem boils down to transport modes routing problem, i.e. vehicles routing problem which is a computationally hard [9, 11], and consequent scheduling problem [7, 12]. In a mixed floor and overhead material handling transport system equipped with many different unidirectional AGV and overhead hoist transport modes a guarantee of congestion and deadlocks-free well organized material flow plays a pivotal role. In that context, available transport modes can be seen as means supporting prescheduled material flows and deciding about MTS productivity.

Consequently, assuming the material flows following possible machining routes are serviced by a MTS the following questions arise: what kind of MTS structure can guarantee a given behavior, and what admissible behavior can be reachable in a given MTS structure? These questions are typical for a well known vehicle routing problem [4, 7, 16] which formulation within the framework of mesh-like and fractal-like structures enables to get a significant reduction on the problem size. The main advantages following from the regular structure of MTS layout comes down to the flexibility and robustness being vital to improve the manufacturing flexibility, especially in case of shorter lead times with variable product mix and volumes [5, 13, 17]. Moreover, such structures provide a chance to evaluate variants of admissible routings and schedules following assumed flow-paths of material transportation in a polynomial time. Among numerous reports concerning mesh-like or grid-like as well as fractal-like structures of MTS networks the following ones should be mentioned [5, 13, 17].

In this paper we consider production routes specified by sequences of subsequently executed operations performed on AGVs and machine tools while serviced by MTS facilities. Assuming transport operations are executed by AGVs arranged in a streaming closed-loops network where potential conflicts are resolved by priority dispatching rules assigned to shared resources the main problem boils down to the searching for sufficient conditions guaranteeing MTS cyclic steady state. In this context the contribution can be seen as a continuation of our former work [2].

The paper is structured as follows: the problem statement is provided in Sect. 2; then main results regarding sufficient conditions guaranteeing cyclic steady MTS's state behavior are submitted in Sect. 3. Illustrative example and concluding remarks are submitted in Sects. 4 and 5 respectively.

2 Problem Formulation

Considered MTS layout encompasses a network of AGVs circulating along cyclic routes which mesh-like topology consists of multiple isomorphic substructures. The exemplary network is shown in Fig. 1 (a). Example of such isomorphic substructure encompassing the network of AGVs periodically circulating along cyclic routes while modeled in terms of Concurrent Cyclic Processes Systems (SCCP) [2, 3] is shown in Fig. 1 (b). Distinguished substructures consist of four local cyclic processes ${}^{(i)}P_1$, ${}^{(i)}P_2$, ${}^{(i)}P_3$, ${}^{(i)}P_4$ following operation of four convoys of vehicles (e.g. AGVs) driving over the same path in the same direction. In general case all vehicles from any convoy can be instanced by subsequently numbered entities called sub-processes. For the sake of

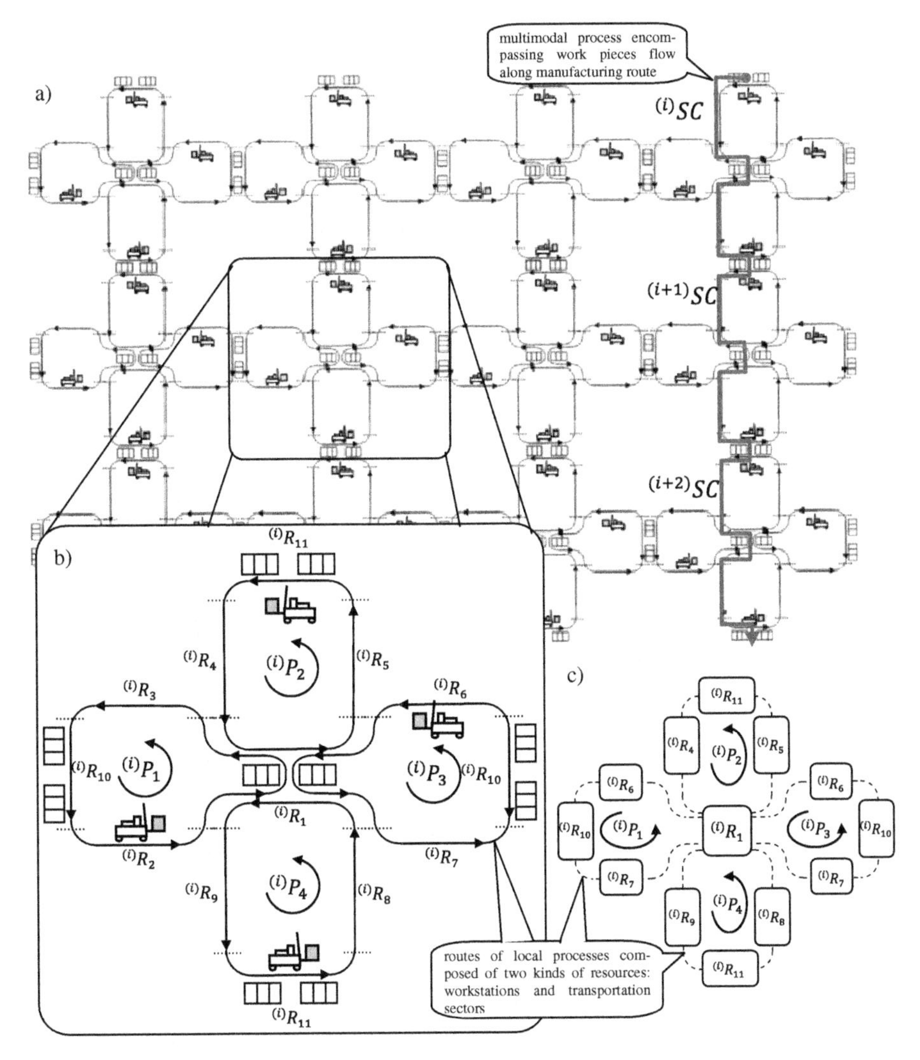

Fig. 1. An example of a mesh-like structure of MTS layout (a), isomorphic substructure (b), graph model of isomorphic structure (c)

simplicity, let us replace convoys of AGVs by unique vehicles traveling along arbitrary assumed cyclic routes.

Component processes are executed along given routes composed of two resources associated with workstations and two resources representing transportation sectors, e.g. for ${}^{(i)}P_2$, $\{{}^{(i)}R_1, {}^{(i)}R_{11}\}$ and $\{{}^{(i)}R_4, {}^{(i)}R_5\}$, respectively – see Fig. 1 (c). Access to the common shared resources ${}^{(i)}R_1$ is synchronized by a mutual exclusion protocol following the given priority dispatching rule ${}^{(i)}\sigma_1 = ({}^{(i)}P_1, {}^{(i)}P_2, {}^{(i)}P_3, {}^{(i)}P_4)$, where ${}^{(i)}\sigma_1$ determines the order in which local processes can access the shared resource

${}^{(i)}R_1$, i.e. ${}^{(i)}P_1$ is allowed to access first, then the processes ${}^{(i)}P_2$, ${}^{(i)}P_3$, and next process ${}^{(i)}P_4$, and then once again ${}^{(i)}P_1$, and so on.

To summarize, the mesh-like structure of layout of considered MTS can be seen as a result of multiple composition of isomorphic substructures encompassing transport processes serviced by AGVs arranged in a streaming closed-loops sub-network. Justification standing behind this postulate is twofold: to guarantee cyclic steady state behavior of MTS and to emphasis what awaited MTS's functionality just follows from its structure. Consequently, assuming a given mesh-like structure of cyclic processes synchronized in their access to the common shared resources the main question concerns of a set of dispatching rules guaranteeing the cyclic steady state behavior of the whole MTS.

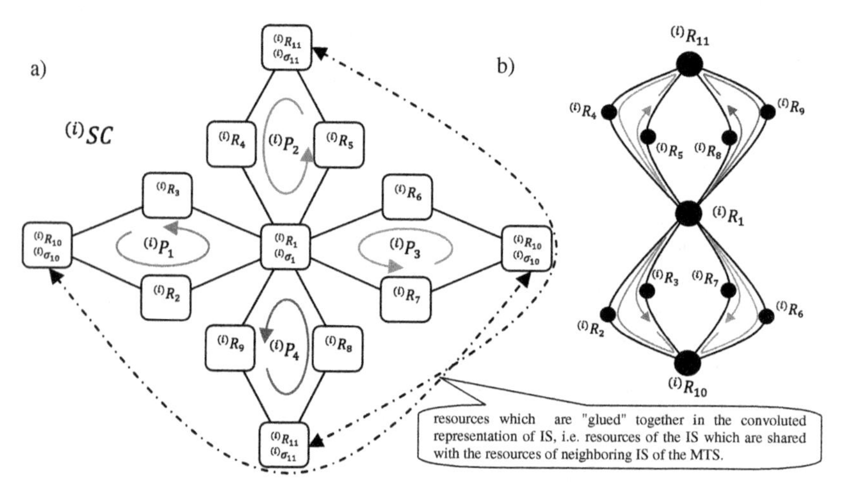

Fig. 2. Graph model of isomorphic substructure (IS) (a), a convoluted representation of IS from Fig. 2 (a) (b)

In opposite however, since the mesh-like structure can be decomposed into a set of isomorphic substructures $\{{}^{(1)}SC, \ldots, {}^{(i)}SC, \ldots, {}^{(lc)}SC\}$ the selection of dispatching rules for each substructure can be carried out independently. So, in case every substructure ${}^{(i)}SC$ has a subset of dispatching rules, then alternative question occurs: Does there exist a set of dispatching rules guaranteeing cyclic steady state of the given isomorphic substructure? The dispatching rules sought should result in a cyclic steady state of so called convoluted representation of an isomorphic substructure. In a convoluted representation some resources are unified, see Fig. 2 (a) (b). Therefore, the dispatching rules assigned to unified substructure resources are sought as to guarantee, together with the rest of rules assigned, its cyclic steady state behavior.

Positive response means that a regular composition of isomorphic substructures encompassing local cyclic steady states implies the cyclic steady state of whole MTS, i.e. guaranteeing congestion and deadlock-free flow of AGVs.

In other words, if for any substructure ${}^{(i)}SC$, there exist its deadlock-free convoluted representation, i.e. encompassing steady state cyclic behavior, then the question boils down to the following one: Does there exist a way of composition of the set of isomorphic elementary substructures $\{{}^{(1)}SC, \ldots, {}^{(i)}SC, \ldots, {}^{(lc)}SC\}$ guaranteeing a cyclic schedule representation of the MTS operation?

Positive solution to the aforementioned allows one to consider cases where final products (goods) following assumed production routes must occur in a certain time windows, i.e. intervals treated as a part of respective production takt times, varying from production order to production order. Assuming production routes are specified by sequences of subsequently executed operations performed on AGVs and machine tools, respectively a new problem can be recognized. Such a new class of problems concerns with finding the best set of routes and schedules for a fleet of AGVs periodically circulating along cyclic routes to serve customers within prescheduled time windows aimed at load/unload operations.

3 Deadlock-Free Avoidance Conditions

Assuming that behavior of each isomorphic substructure ${}^{(i)}SC$ is represented by the cyclic schedule ${}^{(i)}X' = ({}^{(i)}X, {}^{(i)}\alpha)$, where ${}^{(i)}X$ – the set of the initiation moments of local process operations; ${}^{(i)}\alpha$ – the period of local processes executions, the set of relevant dispatching rules ${}^{(i)}\Theta$ of the substructure ${}^{(i)}SC$ can be seen as solution to the following constraint satisfaction problem [2, 3, 14, 15]:

$$PS_i = \left(\left(\left\{{}^{(i)}X', {}^{(i)}\Theta, {}^{(i)}\alpha\right\}, \{D_X, D_\Theta, D_\alpha\}\right), \{C_L, C_M, C_D\}\right) \tag{1}$$

where: ${}^{(i)}X'$, ${}^{(i)}\Theta$, ${}^{(i)}\alpha$ – the decision variables,

${}^{(i)}X'$ – the cyclic schedule of substructure ${}^{(i)}SC$,

${}^{(i)}\Theta$ – the set of priority dispatching rules for ${}^{(i)}SC$,

${}^{(i)}\alpha$ – the period of local processes for ${}^{(i)}SC$,

D_L, D_M, D_D – the constraints C_L and C_M limit possible SCCP behavior [1, 2], C_D – guaranty the smooth (i.e. conflict-free or overlapping-free) execution of the material transport/handling operations competing to the access to the common shared resources.

The schedule ${}^{(i)}X'$ that meets all the constraints from the given set $\{C_L, C_M, C_D\}$ is the solution sought for the problem (1). The constraints C_L, C_M guarantee that in the processes following the substructure ${}^{(i)}SC$ are executed in a cyclic manner [1, 2], however, cannot ensure conflict freeness between the material transport/handling operations of processes competing to the access to the common shared resource in the same convoluted representation of an isomorphic substructure. In order to avoid such situations additional constraints C_D following a principle of so called match-up structures coupling [2] can be implemented.

Sufficient constraints for conflict prevention among processes of the convoluted representation of a given isomorphic substructure are defined in the following way. The two operations $o_{i,j}$, $o_{q,r}$ do not interfere (on the mutually shared resource R_{k_i}) if the operation $o_{i,j}$ begins (at the moment $x_{i,j}$) after the release (with the delay Δt) of the resource by the operation $o_{q,r}$ (at the moment x_{q,r^*} of the subsequent operation initiation) and releases the resource (at the moment x_{i,j^*} of the subsequent operation initiation) before the beginning of the next execution of the operation $o_{q,r}$ (at the moment $x_{q,r}+\alpha_b$). The collision-free execution of the local process operations is possible if the following constraint holds:

$$\begin{aligned}&\left[\left(x_{i,j}\geq x_{q,r^*}+k''\cdot\alpha_b+\Delta t\right)\wedge\left(x_{i,j^*}+k'\cdot\alpha_a+\Delta t\leq x_{q,r}+\alpha_b\right)\right]\\&\vee\left[\left(x_{q,r}\geq x_{i,j^*}+k'\cdot\alpha_a+\Delta t\right)\wedge\left(x_{q,r^*}+k''\cdot\alpha_b+\Delta t\leq x_{i,j}+\alpha_a\right)\right]\end{aligned}\qquad(2)$$

where: $j^*=(j+1)\,MODlr(i)$, $r^*=(r+1)MODlr(q)$, $k'=\begin{cases}0 \text{ when } j+1\leq lr(i)\\1 \text{ when } j+1>lr(i)\end{cases}$, $k''=\begin{cases}0 \text{ when } r+1\leq lr(q)\\1 \text{ when } r+1>lr(q)\end{cases}$.

Satisfying the constraint (2) means that on every mutually shared resource of the convoluted representation the local processes are executed alternately. Consequently, the resulting substructure coupling of any two neighboring isomorphic substructures specified by cyclic schedules $X_a^{'}$, $X_b^{'}$ in the mesh-like network can be determined by the cyclic schedule $X_c^{'}$ being a composition of the schedules $X_a^{'}$, $X_b^{'}$, if: the value of the period of schedule $X_a^{'}$ is the least common multiple of the period of schedule $X_b^{'}$ and such that $m\alpha_a\,MOD\,m\alpha_b=0$; and $\alpha_a\,MOD\,\alpha_b=0$. The derived conditions stand behind of a match-up coupling of isomorphic substructures $\{^{(1)}SC,\ldots,{}^{(i)}SC,\ldots,{}^{(lc)}SC\}$ of the known cyclic behaviors $\{^{(1)}X',\ldots,{}^{(i)}X',\ldots,{}^{(lc)}X'\}$ resulting in a cyclic schedule representation of the MTS operation.

4 Illustrative Example

The assessment of the cyclic behavior (i.e., the existence of the cyclic schedule $X^{'}$) of the mesh-like structure from Fig. 1 (a) can be obtained as a result of evaluating the parameters of its convoluted structure ${}^{(i)}SC$ from Fig. 2 (b).

Assuming that the all transportation times are the same and equal to $t_{i,j}=5$ u.t. (units of time), the relevant problem (1) was implemented and solved in the constraint programming environment OzMozart (CPU Intel Core 2 Duo 3 GHz RAM 4 GB). The first acceptable solution was obtained in less than one second. The problem solution for the substructure from Fig. 2 (b) consists both: the cyclic schedule ${}^{(i)}X'$ (with periods ${}^{(i)}\alpha_1=24$ - see Fig. 3) and the set of dispatching rules ${}^{(i)}\Theta$ shown in Table 1.

The attained schedule is a component of the schedule $X^{'}$ that characterizes the behavior of the whole structure SC. The part of schedule $X^{'}$ being a multiple composition of the schedules ${}^{(i)}X'$ is presented in Fig. 4. The schedule encompasses behavior of a part of the network Fig. 1 (a) determining a multimodal process (MP) execution. Meaning of the MP can be illustrated on example of an FMS equipped with AGVS where the work-pieces pass their origin-destination routes among

Table 1. The dispatching rules of SCCP following ${}^{(i)}SC$ from Fig. 2 (b)

Resources	Dispatching rules
${}^{(i)}R_1$	${}^{(i)}\sigma_1^0 = \left({}^{(i)}P_1^1, {}^{(i)}P_2^1, {}^{(i)}P_3^1, {}^{(i)}P_4^1\right)$
${}^{(i)}R_{10}$	${}^{(i)}\sigma_{10}^0 = \left({}^{(i)}P_1^1, {}^{(i)}P_3^1\right)$
${}^{(i)}R_{11}$	${}^{(i)}\sigma_{11}^0 = \left({}^{(i)}P_2^1, {}^{(i)}P_4^1\right)$

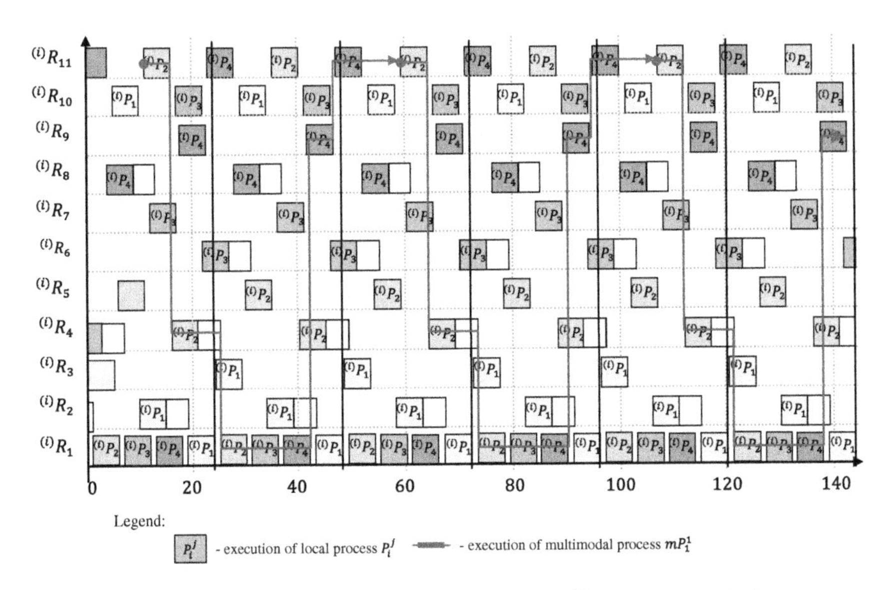

Fig. 3. The Gantt's chart of SCCP following ${}^{(i)}SC$ from Fig. 2 (b)

workstations using local transportation means, i.e. where work-piece flows are treated as multimodal processes. For further reading about multimodal processes treated as processes executed along the routes consisting parts of the routes of local processes refer to [1, 2, 8].

Therefore, the obtained schedule illustrates operation of transportation modes (local processes) and supported by them workpieces flow.

Moreover, it assumes that schedules of locally acting AGVs match-up a given, i.e. already planned, schedules of work-pieces machining. Consequently, assuming a given travel times between subsequent workstations the time windows for servicing relevant load/unload operations can be determined. Also the production cycle and takt time can be estimated analytically, as well. Assuming a takt time (TT) is the amount of time that must elapse between two consecutive work pieces completions and a production cycle (PC) means the period during which the work piece remain in the production process, in the case considered TT = 24, and PC = 126.

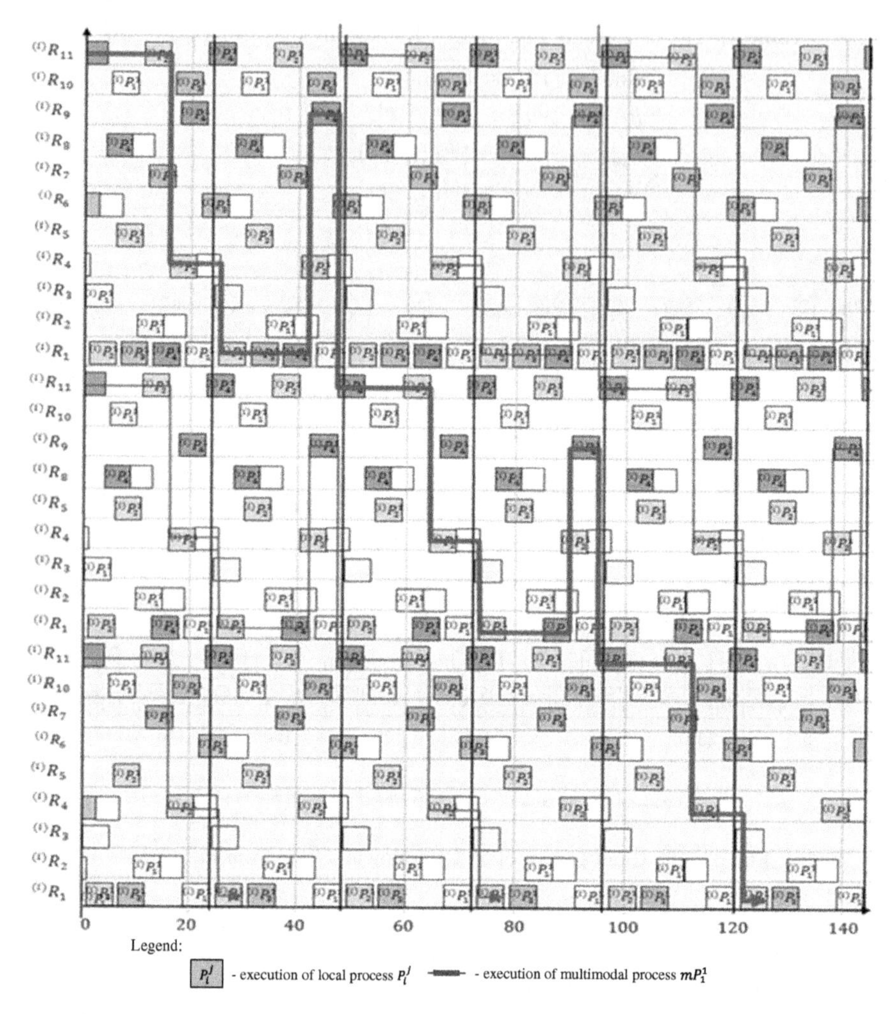

Fig. 4. The Gantt's chart including local processes and the multimodal processes.

Of course, in general case the searching for conditions guaranteeing solvability of the cyclic processes scheduling, i.e. guaranteeing the right match-up of local cyclic acting AGV schedules to a given work-pieces machining schedules, can be stated as well.

5 Concluding Remarks

The paper introduces the concept of a mesh-like multimodal transportation network in which several isomorphic subnetworks interact each other via distinguished subsets of common shared workstations as to provide a variety of demand-responsive work-piece transportation/handling services. In opposite to traditional approach a given network of local cyclic acting AGV services is assumed. In such a regular network, i.e. composed of elementary and structurally isomorphic subnetworks, the work-pieces pass their origin-destination routes among workstations using local AGVs, i.e. AGVs assigned to interacting subnetworks.

Our approach focusing on MTS steady state behavior following assumed the steady state task requirements enables the deadlock-free scheduling and routing of a great number of multimodal transport means. Resulting, so easily obtained schedules (i.e., at cost of polynomial computational complexity) suffer, however, from lack of autonomous access to mutually shared resources. The potential conflicts are resolved by priority dispatching rules assigned to shared resources. The sufficient conditions guaranteeing such rules exist were provided. Our further work will be devoted to improvement of manufacturing flexibility through selection of an appropriate, e.g. fractal-driven, MTS layout configuration.

References

1. Bocewicz, G., Muszyński, W., Banaszak, Z.: Models of multimodal networks and transport processes. Bull. Polish Acad. Sci. Tech. Sci. **63**(3), 635–650 (2015)
2. Bocewicz, G., Nielsen, I., Banaszak, Z.: Automated guided vehicles fleet match-up scheduling with production flow constraints. Eng. Appl. Artif. Intell. **30**, 49–62 (2014)
3. Bocewicz, G., Wójcik, R., Banaszak, Z., Pawlewski, P.: Multimodal processes rescheduling: cyclic steady states space approach. Math. Probl. Eng. **2013**, 1–24 (2013). Article ID 407096
4. Brito, J., Campos, C., Castro, J.P., Martínez, F.J., Melián, B., Moreno, J.A., Moreno, J.M.: Fuzzy vehicle routing problem with time windows. In: Magdalena, L., Ojeda-Aciego, M., Verdegay, J.L. (eds.) Proceedings of IPMU 2008, Torremolinos (Malaga), pp. 1266–1273 (2008)
5. Cheng, C.H., Balakrishnan, J.: Multi-period planning and uncertainty issues in cellular manufacturing: a review and future directions. Eur. J. Oper. Res. **177**, 281–309 (2007)
6. Eguia, I., Lozano, S., Racero, J., Guerrero, F.: Cell design and loading with alternative routing in cellular reconfigurable manufacturing systems. In: 7th IFAC Conference on Manufacturing Modelling, Management, and Control, Russia, pp. 1744–1749 (2013)
7. Fazlollahtabar, H., Saidi-Mehrabad, M.: Methodologies to optimize automated guided vehicle scheduling and routing problems: a review study. J. Intell. Robot. Syst. **77**(3), 525–545 (2015). Springer
8. Friedrich, M.: A multi-modal transport model for integrated planning. In: Proceedings of 8th World Conference on Transport Research, pp. 1–14 (1999)
9. Gambardella, L.M., Taillard, E., Agazzi, G.: A multiple ant colony system for vehicle routing problems with time windows. In: Corne, D., Dorigo, M., Glover, F. (eds.) New Ideas in Optimization, pp. 63–79. McGraw-Hill, London (1999)
10. Khayat, G.E., Langevin, A., Riope, D.: Integrated production and material handling scheduling using mathematical programming and constraint programming. Eur. J. Oper. Res. **175**(3), 1818–1832 (2006)
11. Kumar, S.N., Panneerselvam, R.: A time-dependent vehicle routing problem with time windows for E-commerce supplier site pickups using genetic algorithm. Intell. Inf. Manag. **7**, 181–194 (2015)

12. Pradhananga, R., Taniguchi, E., Yamada, T.: Ant colony system based routing and scheduling for hazardous material transportation. Procedia – Soc. Behav. Sci. **2**(3), 6097–6108 (2010)
13. Silva, A.-L.: Critical analysis of layout concepts: functional layout, cell layout, product layout, modular layout, fractal layout, small factory layout. In: Challenges and Maturity of Production Engineering: Competitiveness of Enterprises, Working Conditions, Environment. Proceedings of the XVI International Conference on Industrial Engineering and Operations Management, São Carlos, Brazil, pp. 1–13 (2010)
14. Sitek, P., Wikarek, J.: A hybrid framework for the modelling and optimisation of decision problems in sustainable supply chain management. Int. J. Prod. Res. **53**(21), 1–18 (2015)
15. Skolud B., Krenczyk D., Zemczak M.: Multi-assortment rhythmic production planning and control. In: Series: IOP Conference Series - Materials Science and Engineering, vol. 95 (2015). Article Number: 012133
16. Wan, Y.-T.: Material Transport System Design in Manufacturing, Ph.D. dissertation, School of Industrial and Systems Engineering, Georgia Institute of Technology, Atlanta, GA (2006)
17. Qiu, L., Hsu, W.-J., Huang, S.-Y., Wang, H.: Scheduling and routing algorithms for AGVs: a survey. Int. J. Prod. Res. **40**(3), 745–760 (2002)

Application of the Hybrid - Multi Objective Immune Algorithm for Obtaining the Robustness of Schedules

Iwona Paprocka, Aleksander Gwiazda(✉), and Magdalena Bączkowicz

Faculty of Mechanical Engineering, Silesian University of Technology, Konarskiego 18A street, 44-100 Gliwice, Poland
{iwona.paprocka,aleksander.gwiazda, magdalena.baczkowicz}@polsl.pl

Abstract. This paper investigates one approach to the no-wait flow shop scheduling problem with the objective to improve the robustness of created schedules. One of the fundamental objectives is obtaining an optimal solution for this type of complex, large-sized problems in reasonable computational time. For this purpose was used a new hybrid multi-objective algorithm based on the features of a biological immune system (IS) and bacterial optimization (BO) to find Pareto optimal solutions. It is proposed the hybrid multi-objective immune algorithm (HMOIA II). Computational results suggest that proposed HMOIA II enables the obtainment of stable and robust schedules in case of the disturbance.

Keywords: Scheduling · Predictive and reactive scheduling · Hybrid - Multi Objective Immune Algorithm (HMOIA II)

1 Introduction

Lately, a great deal of effort has been spent developing methods to generate schedules dealing with the uncertainty [12]. Summing up the researches related to the minimization of the effect of disturbances, there are three research groups concerning the problem of updating production schedules: repairing a schedule which has been disturbed (reactive scheduling), creating a schedule which is robust for a disturbance (predictive scheduling), studies on the influence of rescheduling policies over the performance of a dynamic manufacturing system.

Another classification of methods dealing with uncertainty includes: predictive scheduling, dynamic scheduling and reactive scheduling. The first methods aim in achieving schedules (predictive schedules - PS) which absorb the effects of disturbances [2, 11] e.g.:

- The algorithm which generates a basic schedule and inserts a time buffer prior to a job with a disturbance prediction [11].
- The algorithm based on priority rules: the Least Flexible Job First (LFJ) and the Longest Processing Time (LPT) [7].

M. Graña et al. (eds.), *International Joint Conference SOCO'16-CISIS'16-ICEUTE'16*, Advances in Intelligent Systems and Computing 527, DOI 10.1007/978-3-319-47364-2_43

- The algorithm which generates a basic schedule and inserts a maintenance work prior to a job with the bottleneck disturbance prediction [13, 14].

A reactive schedule (RS) is generated if the effect of a disturbance is excessively large and disruption-affected jobs need to be rescheduled [10, 18]. Methods for reactive scheduling are as follows:

- Right Shift Rescheduling (RSR) relies on shifting ahead in time all directly and indirectly affected operations without changing the defined sequence of jobs on machines [9]. RSR causes the minimal deviation from the sequence of jobs on machines but the major deterioration of the robustness criterion.
- Affected Operations Rescheduling (AOR) relies on shifting ahead in time all directly affected operations without changing the defined sequence of operations of jobs on machines [1]. The AOR achieves more efficient and stable reactive schedules comparing to the schedules achieved by the RSR.
- Rescheduling Affected Job(s) relies on re-inserting the disturbed jobs considering alternative routes without changing the defined sequence of operations of remaining jobs on machines [17].

The method of researching the influence of rescheduling policies over the performance of a production system is for example the Average Slack Method (ASM) [8] which deals with uncertainty by proposing the initial schedule with the best performance in the event of a disruption. The performance of the schedule is computed by adding the initial performance measure of the schedule and the degradation in the performance measure (the slack) due to random disruptions.

Dynamic scheduling aims in dispatching jobs which arrive dynamically over time and are processed on machines and deleted from a production system continuously [3]. In dynamic scheduling a schedule is not created [5, 6]. The literature proposes two types of priority rules to deal with dynamic scheduling: global and local priority rules. The global priority rule relates to a job assignment to a production system e.g. the earliest or latest date of arrival of a job to the system: First Come First Served - FCFS and Last Come First Served - LCFS [14]. The local priority rule relates to the operation assignment to a machine e.g. the operation with the shortest or longest processing time: Shortest Processing Time - SPT, Longest Processing Time - LPT. There are also known various hybrids of priority rules [15]. Dynamic scheduling is also proposed to be applied when the number of disturbances is higher than or equal to the certain predefined value [4].

2 Problem Definition

The job shop system a system of jobs with machines assigned to them. Each job is represented by a set of operations, which must be realized in a specified order. Moreover each operation must be performed on a given machine. It is assumed that processing times are known in advance. It is additionally required that release times and due dates must be specified and non-preemptive. Moreover routing flexibility is admissible in the case of disturbances.

The first objective of the investigations is to obtain the best basic schedule of 15 processes executed on 10 machines, which is a compromise one. This schedule is determined basing on four efficiency measures: makespan ($C_{\max} \rightarrow \min$), flow time ($F \rightarrow \min$), total tardiness ($T \rightarrow \min$) and idle time ($I \rightarrow \min$). It has been determined the weights (priorities) of them. For the 1st and 3rd criteria they are equal to 0.3, for the priorities 2nd and 4th they are equal to 0.2. To analysis are introduced next parameters: processing times, processing routes, setup times, due dates, batch sizes and routing flexibility. They are described in [23].

It was assumed that the first machine is the most loaded one. The probability of the bottleneck failure, which is increased, is: $[a, b + MTTR]$ where: $a = 60$ and $b = 72$ and, MTTR = 6. The MTTF (failure free time of the bottleneck) is equal to 66. Each operation v_j can be rescheduled, if it is needed, on a machine which is first available taking into account a set of parallel machines.

Achieving the best compromise reactive schedule of mentioned 15 processes executed on 10 machines is the second objective of the conducted investigations. It was applied two criteria: solution robustness *SR* and quality robustness *QR* in the event of bottleneck failure. SR represents the difference (absolute deviations) between starting times of operations in the reactive and basic schedules. QR is the efficiency measure of the degradation, due to the disturbance, of the performance of the basic schedule. Weights, which represent priorities of SR and QR, are equal to 0.5.

3 The Architecture of HMOIA II

In the paper, the modification of the Multi Objective Immune Algorithm (MOIA) [20] is presented in order to apply the algorithm for predictive and reactive scheduling problems. The MOIA was previously applied only for nominal scheduling problems [21, 22]. The new Hybrid - Multi Objective Immune Algorithm II (HMOIA II) accepts the value of parameters generating the most robust and stable schedules. The scheduling algorithm consists of three parts, i.e. sequences generation, sequences evaluation and similar sequences reduction.

The immune algorithm is applied to effective scanning of the solution space [16]. The HMOIA II is applied to evaluate schedules using two criteria: solution robustness *SR* and quality robustness *QR*. The quality robustness measures the degradation of the performance (efficiency measures) of the basic schedule due to the disturbance. The solution robustness measures the sum of absolute deviations of the starting times of the operations in the reactive schedule and in the basic schedule. The fitness function adds the weighted values of the quality robustness and solution robustness. Weights of two criteria are equaled.

A pathogen states as the job shop scheduling problem. Thus, a pathogen is described by the fitness function for the scheduling problem.

An antibody chromosome coding procedure according to the idea given by Cheng, Gen, and Tsujimura [19] is applied. The antibody chromosome is a permutation representation of jobs. The chromosome encodes a sequence of genes that represent decimal or binary numbers of jobs accepted for execution in a production system. A gene represents a single job in the scheduling problem. A length of antibody chromosome equals to a number of all jobs executed in the scheduling problem. In the chromosome, a position of gene corresponds to a priority indicator assigned to the job.

To decode a solution representation, jobs are scheduled at the earliest feasible time (taking into account the precedence and resource constraints) according to the order in the antibody chromosome.

In the HMOIA II, the initial population of sequences of jobs is generated, trained and evaluated separately taking into account the effects of disturbance on the makespan C_{max}, total tardiness T, total flow-time F, total idle time of machines I and stability criterion SR.

The training process of antibody population for QR (Fig. 1) starts with antibodies selection to create a matting pool. Parents are randomly matched in couples. The Position-Based Order Crossover operator is applied in the reproduction procedure [24]. In the elite selection procedure, better individual is selected from the two: a parent and an offspring. Next, antibodies undergo Switching mutation operator and also the elite selection procedure is repeated [24]. The training process of antibody population for

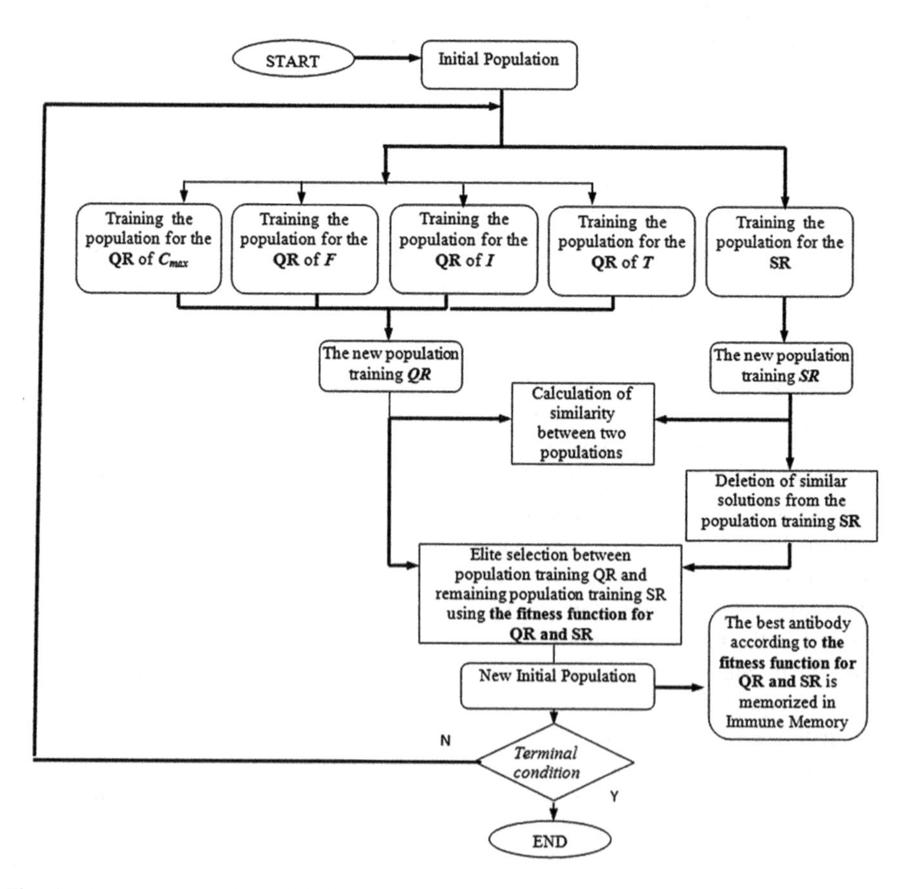

Fig. 1. Architecture of the HMOIA II in the searching process for the trade-off between SR and QR

SR (Fig. 1) differs in the application of the reproduction procedure. A hypermutation operator is applied for generating a new offspring.

The similarity between two populations is calculated: the population of solutions trained and evaluated for the QR and for the SR. In order to prevent a premature convergence of the HMOIA II to a local optima, it is necessary to use *affthres*. The degree of affinity between antibodies from *SCP* is calculated using the Hamming distance. If the degree of affinity between two antibodies is greater than the *affthres*, one antibody stimulates another as both antibodies are similar. An antibody is deleted from the population trained for the SR if it is stimulated by a number of antibodies more than *stimthres*.

A new initial population is generated in the elite selection process using the fitness function to evaluate the solutions. The best sequence which achieves the minimal value of the fitness function in an iteration is memorised (Fig. 1). In HMOIA II, the searching process is stopped after running a predefined number of generations.

The trade-off between SR and QR is measured using the fitness function in the algorithm. The best compromise solution is selected from the immune memory in the HMOIA II.

Results achieved by the HMOIA II are compared to the results achieved by the TSA presented in [8]. The TSA consists of two parts, i.e. a sequence generator and a sequence evaluator. In the TSA, the sequence of jobs is evaluated taking into account the effects of a disturbance [8]. The best neighbouring sequence, achieved in an iteration, is compared with the tabu solution. The tabu solution is the best sequence achieved in all previous iterations. The tabu solution is the input data for the next iteration.

4 Results of Computer Simulations

In the HMOIA II, input parameters are as follows: the size of a sub-population for a single objective scheduling problem, $z = 6$; the size of the initial population for the multi-criteria scheduling problem, $Y = z \cdot O$ (O is the number of objective functions) = $4 \cdot 6$, $y = 1,2,\ldots Y$ (y - an antibody); the number of iterations *termcon* = 0 in the endogenous population. The number of iterations *termcon* = 30 (terminal condition) in the exogenous population; the maximal number of genes undergoing a mutation procedure in a hyper mutation process = 2; the affinity threshold *affthres* (is used to determine if one antibody is similar to another) = 8; the stimulation threshold *stimthres* (is used to define the number of similar solutions which can exist in the exogenous population) = 3.

In order to achieve the basic schedule for JS problem (10×15) using the HMOIA II, three simulations were generated. The experiments were performed on a PC with Inel Pentium CPU B970, 2.3 GHz and 6 GB RAM. The algorithm was coded in Borland C++. The basic schedules undergo disturbance and rescheduling policy to evaluate the deterioration of the quality of basic schedule. The best basic schedules (denoted as y) are selected based on the minimal value of fitness function FFr of their reactive schedules (denoted as y^*). The fitness function consists of two components: SR and QR. In the first simulation, the best schedule was generated according to the rule of {5 9 7 6 0 10 11 14 3 2 8 4 12 1 13}. The quality of the reactive schedule is FF = 88.9 with the components of QR = 0.79 and SR = 177.

The quality of this schedule was also measured using: $C_{\max}$, F, I and T. The quality of the basic schedule is: $C_{\max} = 131$, $F = 539$, $I = 772$ and $T = 41$, and $C_{\max} = 121$, $F = 577$, $I = 672$ and $T = 67$ after rescheduling. Remaining solutions are described in Table 1.

Also, three simulations were generated for *affthres* = 80 and the remaining input data was presented above. In the first simulation, the best basic schedule was generated according to the rule of {2 13 11 14 3 0 4 8 12 1 6 9 5 7 10}. The quality of the reactive schedule is FF = 95 with the components of QR = 2 and SR = 188. The quality of this schedule is also measured using: $C_{\max}$, F, I and T. The quality of the basic schedule is: $C_{\max} = 127$, $F = 501$, $I = 732$ and $T = 71$, and $C_{\max} = 126$, $F = 469$, $I = 722$ and $T = 89$ after rescheduling. Remaining solutions are described in Table 2.

Table 1. The best basic schedules achieved by the MOIA II and *affthres* = 8

No	Job shop scheduling problem *(15 × 10)*													
	The priority rule of the basic schedule	The quality of the schedule y					The quality of the reactive schedule y*							
		C_{max}	F	I	T	FFy	C_{max}	F	I	T	FFy*	QR	SR	FFry*
1	5 9 7 6 0 10 11 14 3 2 8 4 12 1 13	131	539	772	41	313.8	121	577	672	67	306,2	0.79	177	88.9
2	14 12 9 13 3 10 5 4 7 2 11 0 6 8 1	134	448	802	111	323.5	134	457	802	103	322,9	1.89	149	75.45
3	7 1 3 13 6 8 14 2 4 0 10 9 11 12 5	126	604	722	125	340.5	124	589	702	112	329	2.8	168	85.4

Table 2. The best basic schedules achieved by the MOIA II and *affthres* = 80

No	Job shop scheduling problem *(15 × 10)*													
	The priority rule of the basic schedule	The quality of the schedule y					The quality of the reactive schedule y*							
		C_{max}	F	I	T	FFy	C_{max}	F	I	T	FFy*	QR	SR	FFry*
1	2 13 11 14 3 0 4 8 12 1 6 9 5 7 10	127	501	732	77	307,8	126	469	722	89	302,7	2	188	95
2	11 10 2 0 13 8 5 12 4 9 14 1 3 6 7	133	593	792	33	326,8	123	562	692	82	312,3	10.4	192	101.2
3	7 12 5 13 9 14 6 0 3 2 4 11 1 10 8	127	517	732	79	311,6	128	501	742	76	309,8	8.2	186	97.1

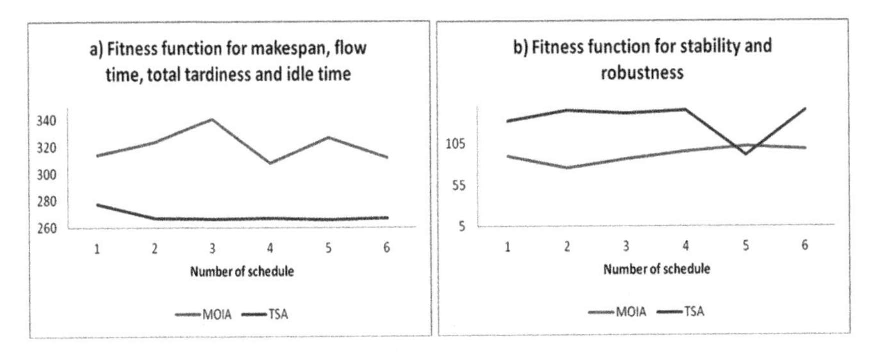

Fig. 2. The fitness function of the basic (a) and reactive schedules (b)

Using the HMOIA II, the best basic schedule dealing with the uncertainty is generated in the second simulation for *affthres* = 8 and in the first simulation for *affthres* = 80.

Reproducibility of solutions is not possible to achieve since only three computer simulations for each *affthres* were run, moreover the terminal condition was equalled to 10 iterations. One can observed the support for reproducibility in the application of the MOIA for batch scheduling problem with dependent setups [20] for example.

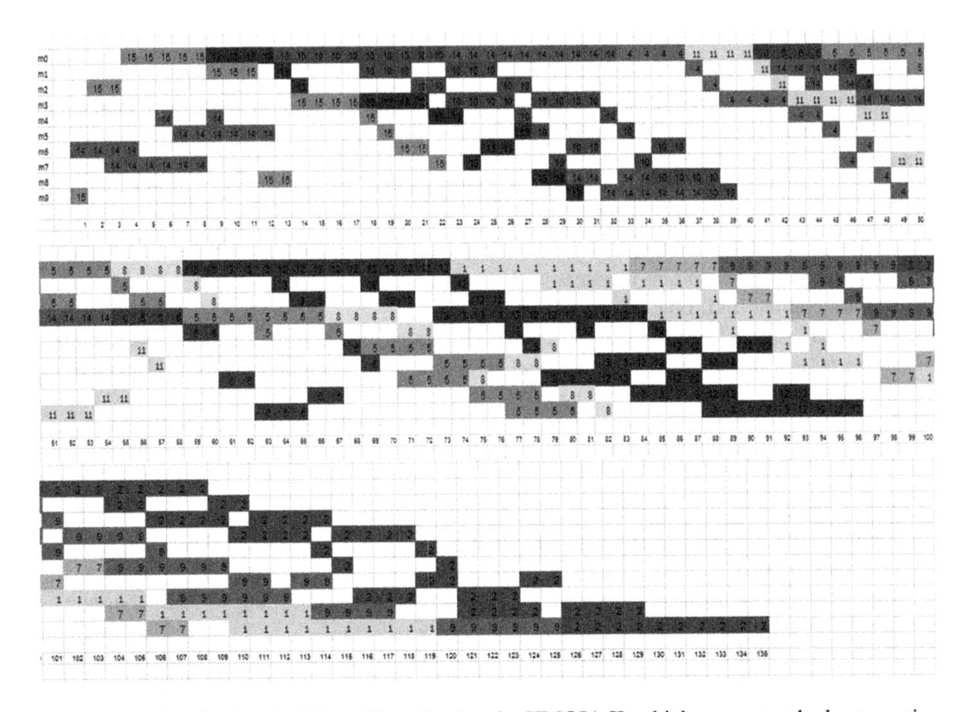

Fig. 3. The best basic schedule, achieved using the HMOIA II, which generates the best reactive schedule in the event of a disturbance.

5 Conclusion

The paper includes the analysis of quality of functioning of the Hybrid - Multi Objective Immune Algorithm II (HMOIA II). The HMOIA II enables to obtain the stable and robust schedules in case of the disturbance what is presented in Fig. 2. The achieved results were compared to the results obtained using a Tabu Search Algorithm presented in [8].

The objective of these quality investigations was to determine the algorithm which allows achieving the robust functioning of the scheduling algorithm. The objectives of particular tests were as follows: makespan, flow time, total tardiness and idle time for predictive scheduling, solution robustness and quality robustness for reactive scheduling.

The objective of the future researches is the modification of architecture and evaluation criteria of the HMOIA II to achieve not only stable and robust schedules but also good quality schedules for such exemplar criteria as makespan, flow time, total tardiness and idle time. The best solution achieved by the HMOIA II is presented in Fig. 3.

The architecture of the algorithm used in the investigations consisted in searching the influence of a rescheduling policy over the scheduling performance of HMOIA II. In the future it is needed to experimentally investigate which parameters of the applied method provides better results considering different objectives of production plans scheduling.

References

1. Abumaizar, R.J., Svestka, J.A.: Rescheduling job shops under random disruptions. Int. J. Prod. Res. **35**, 2065–2082 (1997)
2. Al-Hinai, N., ElMekkawy, T.Y.: Robust and flexible job shop scheduling with random machine breakdowns using a hybrid genetic algorithm. Int. J. Prod. Econ. **132**, 279–291 (2011)
3. Hamzadayi, Alper, Yildiz, Gokalp: Event driven strategy based complete rescheduling approaches for dynamic m identical parallel machines scheduling problem with a common server. Comput. Ind. Eng. **91**, 66–84 (2016)
4. Aytug, H., Lawley, M., McKay, K., Mohan, S., Uzsoy, R.: Executing production schedules in the face of uncertainties: a review and some future directions. Eur. J. Oper. Res. **161**(1), 86–110 (2005)
5. Banaś, W., Sękala, A., Foit, K., Gwiazda, A., Hryniewicz, P., Kost, G.: The modular design of robotic workcells in a flexible production line. IOP Conf. Ser. Mater. Sci. Eng. **95**, 012099 (2015)
6. Banaś, W., Sękala, A., Gwiazda, A., Foit, K., Hryniewicz, P., Kost, G.: Determination of the robot location in a workcell of a flexible production line. IOP Conf. Ser. Mater. Sci. Eng. **95,** 012105 (2015)
7. Duenas, A., Petrovic, D.: An approach to predictive-reactive scheduling of parallel machines subject to disruptions. Ann. Oper. Res. **159**, 65–82 (2008)
8. Goren, S., Sabuncuoglu, I.: Robustness and stability measures for scheduling: single-machine environment. IIE Trans. **40**, 66–83 (2008)
9. Jensen, M.T.: Generating robust and flexible job shop schedules using genetic algorithms. IEEE Trans. Evol. Comput. **7**, 275–288 (2003)

10. Kamrul Hasan, S.M., Sarker, R., Essam, D.: Genetic algorithm for job-shop scheduling with machine unavailability and breakdowns. Int. J. Prod. Res. **49**(16), 4999–5015 (2011)
11. Liu, L., Han-yu, G., Yu-geng, X.: Robust and stable scheduling of a single machine with random machine breakdowns. Int. J. Adv. Manuf. Technol. **31**, 645–656 (2007)
12. Monica, Z.: Optimization of the production process using virtual model of a workspace. IOP Conf. Ser. Mater. Sci. Eng. **95**, 012102 (2015)
13. Paprocka, I., Kempa, W.M., Kalinowski, K., Grabowik, C.: A production scheduling model with maintenance. Adv. Mater. Res. **1036**, 885–890 (2014)
14. Paprocka, I., Kempa, W.M., Kalinowski, K., Grabowik, C.: On Pareto optimal solution for production and maintenance jobs scheduling problem in a job shop and flow shop with an immune algorithm. Adv. Mater. Res. **1036**, 875–880 (2014)
15. Paprocka, I., Kempa, W.M., Grabowik, C., Kalinowski, K.: Sensitivity analysis of predictive scheduling algorithms. Adv. Mater. Res. **1036**, 921–926 (2014)
16. Paprocka, I., Kempa, W.M., Grabowik, C., et al.: Time-series pattern recognition with an immune algorithm. Mater. Sci. Eng. **95** (2015). Article no. 012110
17. Fahmy, Sherif A., Balakrishnan, Subramaniam, ElMekkawy, Tarek Y.: A generic deadlock-free reactive scheduling approach. Int. J. Prod. Res. **47**(20), 5657–5676 (2009)
18. Turkcan, A., Akturk, M.S., Storer, R.H.: Predictive/reactive scheduling with controllable processing times and earliness-tardiness penalties. IIE Trans. **41**, 1080–1095 (2009)
19. Cheng, R., Gen, M., Tsujimura, Y.: A tutorial survey of job-shop scheduling problems using genetic algorithms, part II: hybrid genetic search strategies. Comput. Ind. Eng. **36**, 343–346 (1999)
20. Skołud, B., Wosik, I.: Multi-objective genetic and immune algorithms for batch scheduling problem with dependent setups. In: Recent Developments in Artificial Intelligence Methods, pp. 185–196 (2007)
21. Skołud, B., Wosik, I.: The development of IA with local search approach for multi-objective Job shop scheduling problem. In: Virtual Design and Automation, pp. 235–242. Publishing House of Poznań University of Technology, Poznań (2008)
22. Skołud, B., Wosik, I.: Clonally selection and multi-objective immune algorithms for open job shop scheduling problems. In: 30th International Conference Information Systems, Architecture, and Technology, System Analysis in Decision Aided Problems, Wrocław, pp. 217–230 (2009)
23. Paprocka, I.: On the quality of the basic schedule generation influencing over the performance of predictive and reactive schedules. Adv. Intell. Syst. Comput. (Accepted for publishing)
24. Skołud, B., Wosik, I.: Immune Algorithms in scheduling production tasks (in Polish). Enterp. Manage. (in Polish) Pol. Soc. Prod. Manage. **1**, 47–56 (2008)

Outperforming Genetic Algorithm with a Brute Force Approach Based on Activity-Oriented Petri Nets

Reggie Davidrajuh(✉)

Department of Electrical and Computer Engineering,
University of Stavanger, Stavanger, Norway
Reggie.Davidrajuh@uis.no

Abstract. Scheduling problems are NP-hard, thus have few alternative methods for obtaining solutions. Genetic algorithms have been used to solve scheduling problems; however, the application of genetic algorithms are too expectant, as the steps involved in a genetic algorithm, especially the reproduction step and the selection step, are often time-consuming and computationally expensive. This is because the newly reproduced chromosomes are often redundant or invalid. This paper proposes a brute-force approach for solving scheduling problems, as an alternative to genetic algorithm; the proposed approach is based on Activity-oriented Petri nets (AOPN) and is computationally simple; in addition, the proposed approach also provides the optimal solution as it scans the whole workspace, whereas genetic algorithm does not guarantee optimal solution.

Keywords: Scheduling problems · Genetic algorithms · Activity-oriented Petri nets (AOPN) · GPenSIM

1 Introduction

Genetic algorithms (GA) have been successfully used for solving many problems. GA are also used for solving scheduling problems, as literature study reveals many works on this topic. One of the main steps in genetic algorithms is the reproduction of chromosomes; this is a very important step as newly produced chromosomes are supposed to be positioned closer to the optimal point on the workspace, than the older generation of chromosomes. However, this is not always true, as in scheduling problems, the newly generated pool of chromosomes may contain a large number of invalid (or redundant) chromosomes. Thus, the chromosomes have to be reproduced repeatedly to make the pool large enough to span the workspace, especially around the optimal point. This repetition of reproducing chromosomes until a healthy pool is obtained comes with a cost that can be prohibitive to an extent that even a simple Brute-Force approach that checks every point in the workspace can sometimes outperform GA.

This paper proposes a brute-force approach that involves a Petri net model for scheduling problems. Since scheduling problems involve a large number of resources, the type of Petri net used in this paper is the Activity-oriented Petri nets (AOPN),

M. Graña et al. (eds.), *International Joint Conference SOCO'16-CISIS'16-ICEUTE'16*,
Advances in Intelligent Systems and Computing 527, DOI 10.1007/978-3-319-47364-2_44

which provides compact Petri net models even for a system with a large number of resources. In addition, this paper also introduces a new Petri net simulator known as GPenSIM and show how easily Petri nets can be implemented on a MATLAB platform for simulations. GPenSIM is developed by the author of this paper.

In this paper: Sect. 2 introduces genetic algorithms. Section 3 presents a simple job-shop problem as a scheduling problem. Through this job-shop problem, the idea behind the step of reproduction in GA is explained in Sect. 4. Section 5 presents the brute-force approach that involves a Petri net model of the job-shop problem. Section 6 presents the simulation results from the Petri net model.

2 Genetic Algorithm

A genetic algorithm is a stochastic search algorithm, as it searches the population of chromosomes to find out and select a pool of chromosomes for further reproduction. Randomness plays a central role in genetic algorithms. It is due to the randomness, genetic algorithms do not suffer from the so-called hill-climbing phenomena as in traditional search algorithms [1, 2]. This is because, by randomness, the generated chromosomes should be positioned all over the workspace, inclusive the optimal point.

The population-based genetic algorithms involve several steps (see the Fig. 1):

```
iteration i=1;
 Step-0: Create initial population Pi ;

 repeat
 {   step-1: evaluate fitness of each individual in population Pi;
     step-2: select individuals for reproduction based on fitness;
     step-3: perform crossover & mutation on selected individuals;
     step-4: replace old population with the new generation;

     i++;
 } until terminal condition is met;
```

Fig. 1. The genetic algorithm as a pseudo code (adapted from [3]).

The initial step: the initial step is to prepare the initial population that is ready for the iterations. The iterations consist of the following steps [4]:

1. Evaluation step: evaluation of the fitness of the chromosomes (individuals).
2. Selection step: selection of the individuals for reproduction.
3. Reproduction step: reproduction of offspring from the selected individuals: new chromosomes are made by recombination (crossover) and mutation.

The final step: the final step in the genetic algorithm is the termination of the algorithm when some of the chromosomes achieve the optimal value. Another reason to terminate is that the number of iterations has passed the maximal value, usually a very large value (e.g. ten thousand).

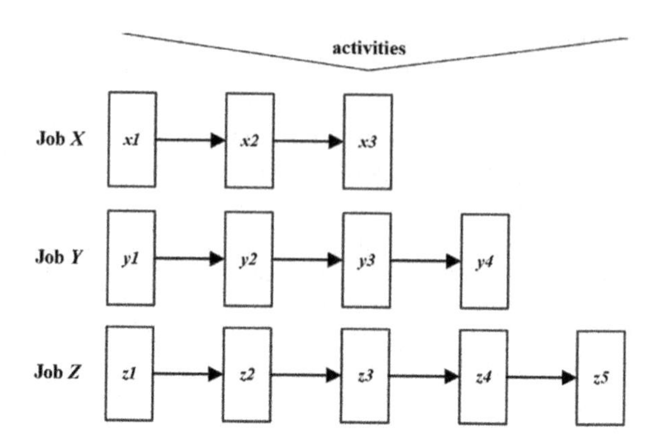

Fig. 2. A JSP consisting of three jobs X, Y, and Z. These jobs have 3, 4, and 5 activities, respectively. The precedence relationship that exists between the activities is only within a job.

3 A Simple Job-Shop Example

A job-shop problem (JSP, also known as job-shop scheduling problem JSSP) consists of a number of jobs. These jobs are made up of a number of activities.

3.1 Characteristics of a Job-Shop Problem

Characteristics of a job-shop problem (see the Fig. 2):

- Each job contains a number of activities.
- Each activity has a single predecessor (not for the first activity of the job) and a single successor (not for the last activity of the job).
- Each activity has its own estimated duration.
- Each activity can only be performed by a specific resource (or a set of resources).

The objective of the JSP is to schedule the set of activities subject to the following constraints:

- *Capacity constraints*: at any time, only a number of resources are available. These resources are *non-reentrant*, meaning only one activity can use a resource at a time.
- *Precedence constraints*: the precedence order that is imposed on activities must be followed.

3.2 A Sample JSP

The example on JSP shown in the Fig. 2 consists of three jobs X, Y, and Z. These three jobs have 3, 4, and 5 activities respectively; the precedence relations between the activities are given below:

X: x1 → x2 → x3
Y: y1 → y2 → y3 → y4
Z: z1 → z2 → z3 → z4 → z5

There are three resources (machines M1, M2, and M3) available to perform the activities. These resources can only perform some specific activities:

- M1: x1, y2, z1
- M2: x3, y1, y4, z3, z5
- M3: x2, y3, z2, z4.

4 The Reproduction Step of GA

This section focuses on the reproduction step of GA, as we see this step as the strength and the weakness of GA. The reproduction step of GA is the strength of GA as the newly generated chromosomes are supposed to edge nearer to the optimal point with each iteration. We say in this paper, that this step is also the weakness, especially when we work with scheduling problems, as the newly generated chromosomes can be invalid, most of the time.

4.1 Chromosome

Let us design the chromosome for the JSP mentioned in the previous section. The JSP has three resources, each of them can only perform specific activities. In addition, activities of a job have a strict linear precedence order, whereas activities from different jobs have no precedence relationship (they can be also executed in parallel). Thus, adhering to the design practice found in the literature, let us design the chromosome with 3 sub-chromosomes, one for each resource, as shown in the Fig. 3.

Chromosome-1:
M1: x1 → y2 → z1
M2: x3 → y1 → z3 → y4 → z5
M3: x2 → y3 → z2 → z4

Chromosome-2:
M1: z1 → x1 → y2
M2: y1 → z3 → y4 → z5 → x3
M3: y3 → z2 → x2 → z4

Fig. 3. Two chromosomes. Each chromosome has three sub-chromosomes, one for each resource; the sub-chromosomes represent the schedule of activities for each resource.

Each sub-chromosome represent dispatch rule for a resource thus, the complete schedule of all activities is defined implicitly by the chromosome. With this encoding, the aim of the iterations is to find the best permutation of the activities in each sub-chromosome, so that the jobs can be finished in the least time.

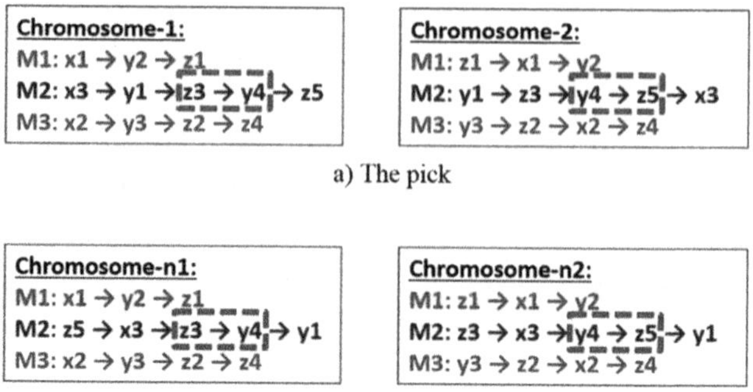

Chromosome-1:
M1: x1 → y2 → z1
M2: x3 → y1 → z3 → y4 → z5
M3: x2 → y3 → z2 → z4

Chromosome-2:
M1: z1 → x1 → y2
M2: y1 → z3 → y4 → z5 → x3
M3: y3 → z2 → x2 → z4

a) The pick

Chromosome-n1:
M1: x1 → y2 → z1
M2: z5 → x3 → z3 → y4 → y1
M3: x2 → y3 → z2 → z4

Chromosome-n2:
M1: z1 → x1 → y2
M2: z3 → x3 → y4 → z5 → y1
M3: y3 → z2 → x2 → z4

b) The new chromosomes after ordered crossover operation

Fig. 4. The ordered crossover operation that exchanges genes between sub-chromosomes M2.

4.2 The Step of Reproduction

Let us assume that we have already selected a pool of chromosomes for reproduction, out of the whole population. As mentioned previously, the reproduction step involves two operations: (1) crossover, and (2) mutation.

The crossover operation: From the pool of selected chromosomes, chromosomes are randomly selected pairwise for an exchange of genes within the pairs. Genes can be exchanged between sub-chromosomes only; e.g. between sub-chromosome M2 of Chromosomes-1 and M2 of Chromosomes-2. In addition, ordered crossover operation is performed instead of the simpler crossover, as the sub-chromosomes cannot possess duplicate entries [5]. An example of the ordered crossover operation is shown in the Fig. 4: in the Fig. 4a, the genes (z3, y4) of sub-chromosome M2 of the chromosome-1 and the genes (y4, z5) of sub-chromosome M2 of the chromosome-2 are chosen as the pick. By the ordered crossover operation, the resulting new chromosomes are shown in the Fig. 4b. Similar crossover operations have to be done between the other two sub-chromosome pairs.

The mutation operation: usually, the mutation operation brings newer genes into the population. However, in the problems like scheduling, bringing newer genes (activities) is not possible as the set of activities are already fixed. Hence, a variation of mutation known as the 'inversion' is performed. By this operation, two genes within a

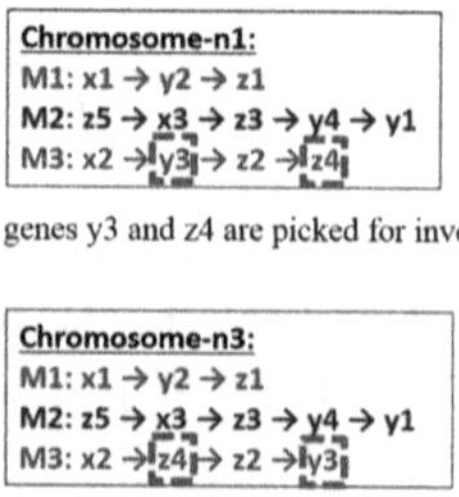

Chromosome-n1:
M1: x1 → y2 → z1
M2: z5 → x3 → z3 → y4 → y1
M3: x2 → y3 → z2 → z4

a) The genes y3 and z4 are picked for inversion.

Chromosome-n3:
M1: x1 → y2 → z1
M2: z5 → x3 → z3 → y4 → y1
M3: x2 → z4 → z2 → y3

b) The new chromosome after the inversion operation.

Fig. 5. The inversion operation that swaps genes within a sub-chromosome

sub-chromosome are simply swapped. A sample inversion operation on the sub-chromosome M3 is shown in the Fig. 5.

4.3 The Redundant Reproduction Step that Generates Invalid Chromosomes

The Fig. 4 shows two new chromosomes, chromosome-n1 and chromosome-n2, generated by the ordered crossover operation. Both of these chromosomes are invalid. This is because, in the sub-chromosome M1 of chromosome-n1, activity y1 is scheduled after y4 and z3 is scheduled after z5; this scheduling breaches the precedence relationship between these activities. Similar violation is found (y1 is scheduled after y4) in M2 of chromosome-n2. The inversion operation shown in the Fig. 5 also introduces noncompliance to the precedence relationship: in the Fig. 5b, the newly generated chromosome-n3 by the inversion operation schedules activity z2 after z4.

Considering the fact that out of the 17280 possible chromosomes ($3! \times 5! \times 4!$), only 2160 chromosomes can be valid ones, adhering to the three precedence order among the three jobs. Further, out of these 2160 valid chromosomes, only 934 are useful chromosomes, as the rest induces deadlock situation and thus never completes. This means the rate for successful generation of a valid and useful chromosome by reproduction is only 0.054 %. This also means, on average, we have to try reproduction step 18 times just to generate a valid and useful chromosome. In a GA application, we may have to generate a pool new chromosomes amounting to hundreds of individuals in each iteration, and usually, ten thousands of iterations are done. This implies the unsuccessful reproduction attempts will amount to (on average) $18 \times 100 \times 10000$. The time and cost involved in these futile attempts are huge.

5 Petri Net-Based Evaluation Instrument

In this section, we propose a brute-force approach that involves a Petri net model, as a better alternative to GA for solving scheduling problems.

5.1 The Brute-Force Approach

The approach consists of the following steps:

- Step-1 Generation: in this step, we will generate a set of all the possible chromosomes. As discussed above, the set will consist of 17280 chromosomes. Thus, the set contains chromosomes representing all the discrete points of the whole workspace, in addition to invalid (redundant) ones.
- Step-2 Filtration: the set of all the possible chromosomes will be fed into the filter. The filter will check the sub-chromosomes M2 and M3, to verify whether the activities belonging to the jobs Y and Z are in order (follow the precedence order). The output of this filter is the set of valid chromosomes. As discussed above, there

will be 2160 chromosomes in the output spanning the entire workspace; each chromosome in this output represents a single discrete point on the workspace.

- Step-3 Evaluation: each valid chromosome will be fed into the Petri net model, in order to compute the job-shop completion time. For the JSP example described above, out of the 2160 valid chromosomes, only 934 chromosomes give finite completion time, and the rest 1226 chromosomes induce deadlock, thus never completes.

The steps involved in the proposed approach are shown in the Fig. 6.

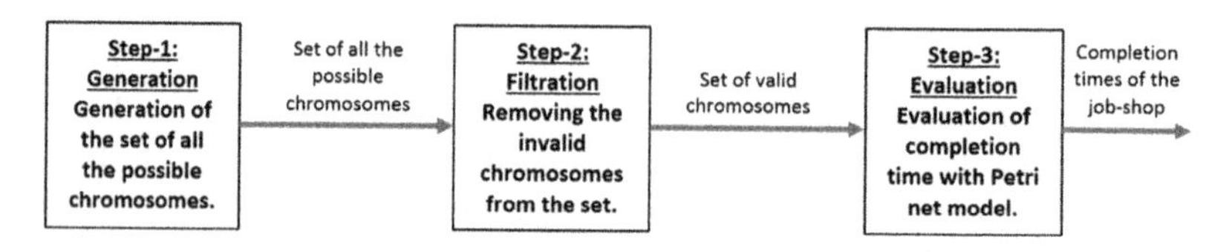

Fig. 6. The three steps of the brute-force approach

5.2 The Petri Net Model

Literature study reveals that Petri nets are not suitable for modeling scheduling problems, as the number of resources involved will result in huge Petri net models. Activity-oriented Petri Nets (AOPN) is a methodology that provides compact Petri net models even for a system with a large number of resources [6, 7]. AOPN is developed by the author of this paper [8]. GPenSIM (general purpose Petri Net simulator) is an implementation of AOPN on the MATLAB platform [9, 10].

AOPN is a 2-phase approach. In the phase-I, only the activities are taken into consideration. In the phase-II, the dynamic details, including the interaction between the activities and the resources, are considered. The static Petri net model obtained by the phase-I is shown in the Fig. 7. For simulations using GPenSIM, the static Petri net model will be coded in the Petri net definition file (PDF). In the phase-II (the run-time phase), all the dynamic details will be coded. The dynamic details such as initial marking (tokens), firing times of the transitions, declaration of the resources are coded into the file known as the main simulation file (MSF). The run-time details such as the activities requesting the resources, using the resources, and then releasing them after use, are coded into the pre-processor and the post-processor files known as the COMMON_PRE and COMMON_POST.

Due to the page limitation, the complete code for simulations (code for the four files) is not given in this paper. However, the complete code is given on the web page [11]; for reproducibility, interested readers are encouraged to visit the web page for downloading the codes and experimenting with them.

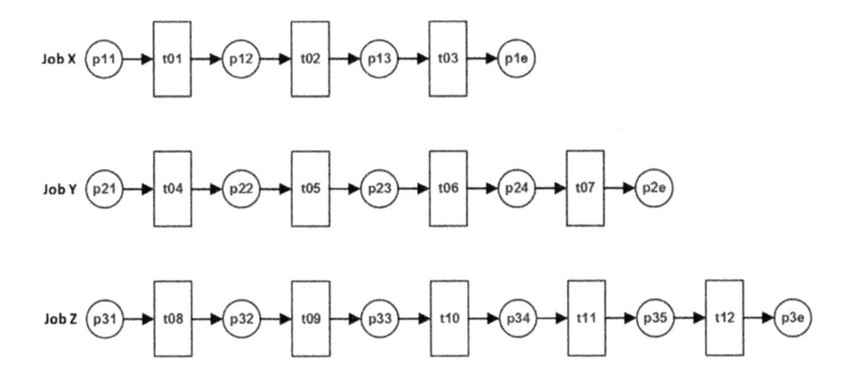

Fig. 7. The static Petri net obtained by the phase-I of the AOPN approach

6 Simulations with GPenSIM

6.1 The Values for Simulation

The duration of the activities (the firing times of the corresponding transitions) in time units (TU) are given below, next to the activities within the brackets.

Job X: x1 (1), x2 (2), x3 (3)
Job Y: y1 (4), y2 (5), y3 (6), y4 (7)
Z: z1 (8), z2 (9), z3 (10), z4 (11), z5 (12)

6.2 The Simulation Results

The summary of the results is shown in the Fig. 8. The results shown in the Fig. 8 indicates that there were 8 optimal points yielding the best completion time of 50 TU. The longest (deadlock-free) completion time is 78 TU, which means each activity was forced to execute one after the other, in a serial manner ($\sum_{i=1}^{12} i = 78$).

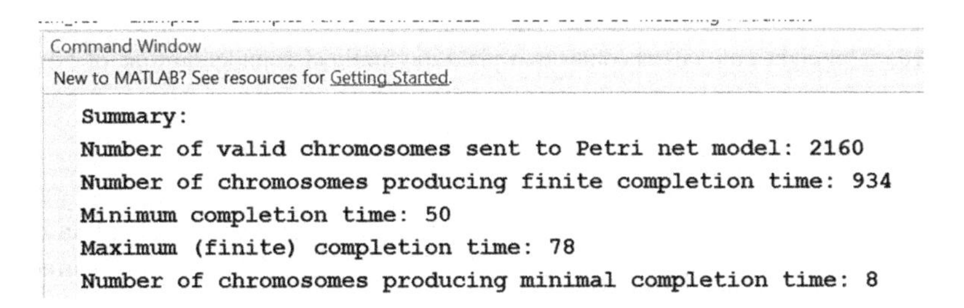

Fig. 8. Summary of the simulation results.

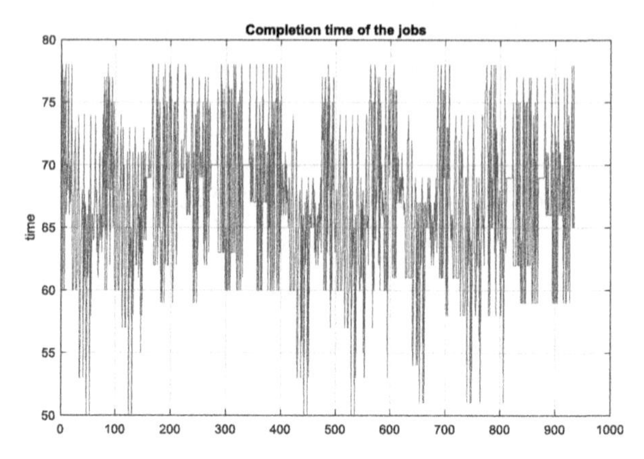

Fig. 9. Only 934 chromosomes yielded a finite completion time of job-shop between 50–78 TU.

The Fig. 9 indicates that there were 934 chromosomes yielding a finite completion time between 50 and 78 TUs. Finally, the eight optimal schedules that yield the minimal completion time of 50 TUs are shown in the Fig. 10.

Chromosome-O1: M1: z1 → y2 → x1 M2: y1 → z3 → y4 → x3 → z5 M3: z2 → y3 → x2 → z4	Chromosome-O2: M1: z1 → y2 → x1 M2: y1 → z3 → y4 → x3 → z5 M3: z2 → x2 → y3 → z4
Chromosome-O3: M1: z1 → y2 → x1 M2: y1 → z3 → x3 → y4 → z5 M3: z2 → y3 → x2 → z4	Chromosome-O4: M1: z1 → y2 → x1 M2: y1 → z3 → x3 → y4 → z5 M3: z2 → x2 → y3 → z4
Chromosome-O5: M1: z1 → x1 → y2 M2: y1 → z3 → y4 → x3 → z5 M3: z2 → y3 → x2 → z4	Chromosome-O6: M1: z1 → x1 → y2 M2: y1 → z3 → y4 → x3 → z5 M3: z2 → x2 → y3 → z4
Chromosome-O7: M1: z1 → x1 → y2 M2: y1 → z3 → x3 → y4 → z5 M3: z2 → y3 → x2 → z4	Chromosome-O8: M1: z1 → x1 → y2 M2: y1 → z3 → x3 → y4 → z5 M3: z2 → x2 → y3 → z4

Fig. 10. The eight optimal schedules that yield the minimal completion time of 50 TUs.

7 Discussion

In a sense, genetic algorithm (GA) scans the entire workspace too, to find an optimal solution. However, in order to reduce time (to perform better than a brute-force approach) GA scans the workspace by generating arbitrary points over it, but in a systematic way, assuming that the generated points are edging closer to the unknown optimal point. Hence, the basic assumption in GA is that by generating fewer but "more fitting" points, the whole workspace can be scanned faster and that some of these generated points will hit on or near the optimal point.

However, a fundamental problem in this basic assumption can be observed by experiments: GA cannot and does not generate "more suitable" points for scanning the workspace, at least not for scheduling problems. This is because most of the generated points by "mutation" and "crossover" are invalid points; it is also important to note that generating points for solving real-life problems (e.g. scheduling) is not cheap as it involves some computations (e.g. "ordered-crossover" instead of simple crossover). Hence, the idea to save time by generating fewer but "more fitting" points is backfired, as most of the "more fitting" points are not fit at all (they are invalid); a large number of points has to be generated and filtered to make one suitable point.

How can a brute-force approach outperform GA? GA generates a large number of points only to see that a few are good ones; a large amount of time wasted in generating the bad ones proposes generating all the points and find the optimal one by brute-force. At least, the brute-force approach will certainly hit the optimal point whereas GA may hit the optimal point by sheer luck!

Further work: The proposal to use a brute-force approach instead of GA has to be tested with scheduling problems with varying sizes. Through these experiments, we can quantify for what size of scheduling problems, it will be better to use the brute-force approach instead of GA.

References

1. Neapolitan, R.: Foundations of Algorithms. Jones & Bartlett, Burlington (2015)
2. Sivanandam, S.N., Deepa, S.N.: Introduction of Genetic Algorithms. Springer, Heidelberg (2008)
3. Mathworks. MATLAB User Manual, Global Optimization Toolbox (2015)
4. Wall, M.B.: A Genetic Algorithm for Resource-Constrained Scheduling. Ph.D. Thesis, MIT (1996)
5. Falkenauer, E., Bouffouix, S.: A genetic algorithm for job shop. In: IEEE International Conference on Robotics and Automation (1991)
6. Davidrajuh, R.: Activity-oriented Petri net for scheduling of resources. In: 2012 IEEE International Conference on Systems, Man, and Cybernetics (SMC). IEEE (2012)
7. Davidrajuh, R.: Modeling resource management problems with activity-oriented Petri nets. In: Sixth UKSim/AMSS European Symposium on Computer Modeling and Simulation (EMS). IEEE (2012)
8. Davidrajuh, R.: Verifying solutions to the dining philosophers problem with activity-oriented Petri nets. In: 4th International Conference on Artificial Intelligence with Applications in Engineering and Technology (ICAIET). IEEE (2014)
9. Davidrajuh, R.: Developing a new Petri net tool for simulation of discrete event systems. In: 2008 Second Asia International Conference on Modelling & Simulation (AMS). IEEE (2008)
10. GPenSIM User Manual. http://www.davidrajuh/gpensim/
11. Simulation Code for the brute-force approach. http://www.davidrajuh.net/gpensim/2016-SOCO-brute-force

Integration of Manufacturing Functions for SME. Holonic-Based Approach

Bozena Skolud, Damian Krenczyk, Krzysztof Kalinowski, Grzegorz Ćwikła(✉), and Cezary Grabowik

Faculty of Mechanical Engineering,
Silesian University of Technology, Gliwice, Poland
{bozena.skolud, damian.krenczyk, krzysztof.kalinowski, grzegorz.cwikla, cezary.grabowik}@polsl.pl

Abstract. Imperfections in the form of losses and delays in production are visible especially in SME dedicated for MTO. Improving the operation of systems at the stage of product design and production organization and management becomes great challenge. The paper proposes new, holonic-based approach to manufacturing systems integration. In this context authors presents results of their achievements in the area of technical and organizational production preparation, and production running. Proposed SME's functional modules (that behave like holons) are generally independent. The concept of the systems integration depending on continuously occurring requirements and conditions involving the creation of a process plan, preparation of schedules and the data acquired from production system is proposed in the paper. As a result of implementation of the presented integration method increases the effectiveness in the integrated areas of decision-making, and moreover, the manufacturing abilities of SME companies.

Keywords: Computer integration · XML data exchange · Computer aided management · Scheduling · Holon · Data acquisition

1 Introduction

Recently, a significant increase in the number of small and medium enterprises (SME) has been observed. In the EU, SMEs constitute 99.8 % of all enterprises in the non-financial business sector, representing almost 67 % of total EU employment, and generated 58 % of the sector's added value. The manufacturing sector is one of the most important among the 5 key sectors in 2014 [1]. To be competitive, manufacturing companies should adapt to changeable conditions imposed by the market. The greater variety of products, the shorter lifecycle of products expressed by a higher dynamics of new products, and the increased customer expectations concerning quality and delivery time are challenges that manufacturing companies have to deal with to remain competitive. Although the optimization of the production process remains a key aspect in the domain of fabrication systems, adaptive production gains more and more field. Flexible manufacturing systems should be able to quickly adapt to new situations like machine breakdown, machine recovery due to physical failure or stock depletion and

M. Graña et al. (eds.), *International Joint Conference SOCO'16-CISIS'16-ICEUTE'16*,
Advances in Intelligent Systems and Computing 527, DOI 10.1007/978-3-319-47364-2_45

also face rush of orders. As a result, a tendency to shift from mass production (make to stock - MTS) and the flow shop production, to production for immediate orders (make to order – MTO) executed in a job shop, is observed.

Imperfections in the form of losses and delays in production, and thus, great opportunities to improve the operation of systems, are visible at the stage of product design and production organization and management. To meet the needs of such enterprises, resulting from the necessity to reduce production costs and shortening the time involved in design, planning and preparation of production, which allow companies to adapt quickly to the requirements of the customers, implementing of innovative and efficient computer systems supporting management are required. Support systems functioning in the sector of SME usually operate in various areas related to the preparation, planning and controlling of the production, but usually they are not integrated. Lack of integration between these systems affects the efficiency in their use and is a potential area where it is possible to increase efficiency, associated with a reduction of costs for SMEs. Today's IT solutions offering more advanced methods of determining schedules are usually created by scientific entities and are usually implemented as a customized software for a specific customer. These solutions, due to the high complexity of the scheduling problem, are based on artificial intelligence methods, heuristic methods, approximate methods, constraint propagation methods and others belonging to the soft computing methods [2, 3].

Product differentiation due to the growing expectations of customers, which in turn becomes the reason for dynamic growth of producers competition and is a defining characteristic of modern manufacturing. For the manufacturers, in the era of a highly competitive, market dynamics and increasing experiences poses a great challenge, which will undoubtedly always boils down to an increase in product offerings. Manufacturers are looking internally for greater efficiencies and to reduce the manufacturing costs, which will not affect on the quality of the products in accordance with customer expectations, leading to "lean" production by eliminating all activities that add no value. Research of new ideas of the production flow organization, in the era of the advanced stage of implementation of Lean Manufacturing concept, forces manufacturers to implement more and more innovative organizational solutions.

Lean philosophy requires the widespread use of tools supporting the management process and the evaluation of changes in production, changes in the configuration of production systems, planning, scheduling, and control at the operational level also for MSP – MTO companies. The response to these needs is the concurrent, multi-assortment production, in which a diversified range of products is manufactured in the same planning period with high flexibility. Therefore, it becomes necessary to support decision-making process in the area of production planning and control. In recent decades, scientific developments in the field of production control have led to new architectures including hierarchical/non-hierarchical, that plays a prominent role in flexible manufacturing. The traditional approach is mainly associated with the initial CIM (Computer Integrated Manufacturing) concept, and usually leads to centralized or hierarchical control structures, in which a supervisor initiates all the activities and the subordinate units respond in order to perform these operations. Due to the complexity of manufacturing problems, the usual practice was to split the global problem into hierarchically dependent functions that operate within smaller areas, such as planning, scheduling, as well as control and monitoring [4].

New approach to manufacturing systems description was inspired by social organisations. Philosophy based on the observation of self-regulation capabilities of social organisation, is called holonic concept. A holon behaviour involves very few characteristics supported by elementary entities named holons [5], co-operating to constitute living organisation called holarchy. Holons are defined as autonomous, co-operating modules, that have a unique identity, yet are made up of subordinate parts. Holarchy is a system of holons that can co-operate in order to achieve a specific objective.

The holon concept can also be applied to industrial environment, such as machine tools, production cells or technological lines. Production system entities, like machines, robots, products, as well as control systems or IT system modules, can be considered as holons, handling data necessary for its actions and communication with other entities. Based on the holon concept, the management process may take into account state of production orders, availability and productivity of machines, activity of workers, flow of material and WIP, etc. In this context authors presents their achievements in the fields of technical and organizational production preparation, as well as production reporting and management.

Software modules developed by authors according to the proposed holons concept can operate independently, but can also be integrated and exchange data in order to allow improved management of production in conditions of constantly changing environment, thanks to on-line updates of production plan (CAPP/CAP) and schedules, based on real-time information feed, assured by the data acquisition system. An expected result of the presented approach is increase of the effectiveness due to improved decision-making and, as a result, better productivity and profitability of SMEs.

2 Data Acquisition in SMEs

Acquisition of data from the production system in the past was not regarded as an important part of the activity of enterprises, but the challenges that companies have to face in contemporary world of globalization and increasing competition made this issue to be regarded as an essential element, allowing for efficient management of the company. This is particularly important in the context of the implementation of concepts like Industry 4.0, Lean Manufacturing and other methods, to optimize the operation of the company, facilitating adaptation to the rapidly changing environment, economic conditions and introduction of modern technologies [6]. Most of these concepts are based on the analysis of real-time data from the production system, however, the issue of systematic acquisition of these data was not an object of too many analysis as far. Recently, attempts were made to employ a comprehensive, systematic approach to the problem of data acquisition and conversion into synthetic information, that can be used to support company management. That type of data acquisition system should provide the following types of data [7]: data on production orders, the state, productivity, location and operation of machinery, data on location and activity of workers, data on the flow and location of materials, products and WIP, and data on the quality.

The work carried out, including an analysis of IT systems used in different parts of companies (e.g. ERP, MES, PPC, SCADA/PLC), hardware solutions, and systems integration practice in companies, allowed the presentation of the methodology supporting development of an integrated, automatically supplying pre-processed information, Manufacturing Information Acquisition System (MIAS) [7]. The proposed methodology is versatile and allow easy design of an efficient data acquisition system for management. The main MIAS functions are: automatic or semi-automatic data acquisition from various data sources, data integration (pre-processing and aggregation), archiving, standardised communication between all elements, data sharing and presentation for various IT systems and managers, and command channel [7].

2.1 Data Sources and Communication in SMEs

Data sources are the foundation of every MIAS, and its availability depends strongly on the level of automation of technological processes in the factory. Data sources can be divided into 3 main groups [8, 9]: automated data sources (data is obtained from control devices and IT systems, that are available only in automated production systems, acquired data is reliable but it has to be reduced and pre-processed to allow effective usage in management), manual data sources (usually the only sources available in non-automated production systems, based on verbal communication and reports, unreliable because of delays, mistakes or forgery), and semi-automatic data sources (based usually on automatic identification systems, like barcodes, RFID or vision systems, intended to collect data from partially-automated or non-automated production systems more reliably than manual sources, assuring verification and quicker access to data, minimizing involvement of employees).

During the design of the data acquisition system, that should operate with minimal involvement of employees, available automated data sources should be used, supported with semi-automatic solutions replacing manual methods of data acquisition.

The following features of SMEs, significant in terms of the ability to create an efficient system of data acquisition, can be specified:

- low availability of automated data sources,
- usually multi-assortment production to order, low repeatability of orders,
- lack of automated means of transport and permanent transport solutions,
- wide variety of materials and WIP flowing through the production system,
- manual data acquisition, verbal or in the form of reports.

The main sources of data in SMES should be automatic identification systems, that can be used both to register the flow of material and WIP, as well as the confirmation of order execution. Generally, the SMEs do not have a problem of anonymity of employees, usually they are just visible all the time, it is possible to visually determine whether carry out their tasks. There is no need for sophisticated RTLSs [10] to locate employees, materials, WIP, products, containers, pallets and means of transport. Sufficient data on the location of these objects should be provided by automatic identification systems, mobile (data collectors) and stationary terminals, installed in workstations and warehouses. Both employees and objects passing through the production system should be equipped with barcode label or RFID tag.

Due to the small amount of data from the control systems there is no need for intensive reduction and initial interpretation of overly detailed data in SMES, so it is not necessary to use sophisticated MES solutions. Data sources can be interfaced by customised Historian database or accessed directly from client applications (visualisation, verification, scheduling, ERP, etc.). Due to the small area of the SMES's production system, it is possible to use standard wired and wireless communications.

3 Systems Data Exchange - System Integration

3.1 CAPP/CAP Systems Integration

It is widely considered, that technical production preparation (TPP) is the totality of actions associated with design of the new product, improving of existing designs, process planning and process planning activation at a workshop level, tools and manufacturing instrumentation design, production planning and scheduling to a production system maintenance. One of the most important functions of the TPP is to satisfy a postulate of the continuous progress of the technical level of products being manufactured. This could have been done by design optimisation, introducing of the newest manufacturing technology, application of the newest software tools that aids design, process plan, production planning and scheduling processes. The base part of TPP ends with emission of design and process plan documentation. TPP can be divided into design (DPP), technological (TPP) and organisational production preparation (OPP). The design production preparation consist of the following subsequent steps: initial studies, elaboration of the design requirements, elaboration of the initial product design, elaboration of the final product design, prototype making, and technical documentation preparation. At the stage of technological production preparation the process plans and manufacturing processes maintenance procedures are being prepared. Whilst the main goal of the organizational production preparation is elaboration of suitable conditions which make full utilization of results achieved at the stages of the DPP and TPP possible.

It is commonly known that the each part of the technical production preparation is supported by suitable IT system. These systems are well fitted to the functions of the particular stage of the TPP. The main drawback of this solution is usually very weak information flow between systems. So, there is still the need of elaboration of an integration tool. As a starting point for a new method of the CAPP/CAP integration, an analysis of definitions used for describing functions and operations which appear in processes of the technical and organizational production preparation and also the simplified functional schemas of the manufacturing process (Figs. 1, 2) were chosen.

The goal function F_C described as a transformation of a blank material into the ready to use part, and the disturbance function Z_i which represents an external disturbance, acts on the considered model. In this system the initial product state S_p constitutes the system input data, it represents the product state before starting of the manufacturing process. As a result of transformation in the system, the initial state S_P is

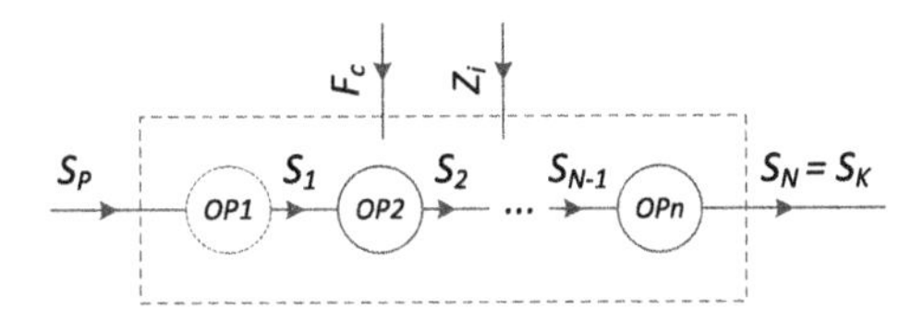

Fig. 1. The simplified functional model of technological process plan: *OP1*, *OP2*, …, *OPn*, – work stations performing particular operations S_1, S_2, …, S_{N-1}, S_N – intermediate product states; F_C – an objective function; Z_i – external disturbance; S_P – the initial state; S_K – the final state

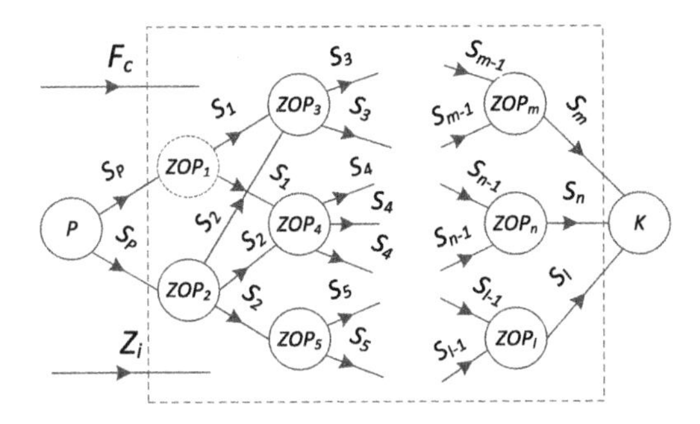

Fig. 2. The simplified functional model of technological process plans: *ZOP1*, *ZOP2*, …, *ZOPn*, – sets of work stations performing particular operations; *P* – the net origin; *K* – net end

transformed into the final product state S_K which responds to the product state after machining. From the manufacturing process nature results that accomplishment of the S_K state is explicitly associated with necessity of getting intermediate states S_1 to S_n.

The integration basis of the CAPP/CAP systems is a method of the multivariant process plan representation. In the proposed solution, the multivariant process is represented by a set of manufacturing features which are responsible for representing specified process plan structure elements (Fig. 2).

The models of the intermediate product states and process plan allows to design various process plans for the specific product. The possibility of proposing process plan variants allows to react elastically on disturbances, which might appear in the production system. It is easy to notice that for the system from the Fig. 1 it is possible to represent the process plan which possess only one route.

This formulation of the process plan as a system is not satisfying from the problem of the integration of CAPP/CAP systems angle. This solution allows to represent different routes. The main difference between these schemas lies in information interchanging represented in the directed graph by vertices and edges. In the modified schema the particular directed graph vertices represent the subsequent workpiece states from S_p to S_k, while graph edges represents the structural elements of the process plan which have to be applied in order to gain the deliberate state.

This seemingly little modification has the great importance for elaboration of a new integration method. So, in case of the unpredictable break appearance in the production system, caused by the sudden disturbance, basis on information on the process state, it is the register of the performed operations and manufacturing cuts, it is possible to draw conclusion about the state which that product gained during the manufacturing process. Taking into account this information, it is possible to make the correct decision and action, which allow to react on this disturbance.

3.2 Data Model for Detailed Scheduling

The functional model of multivariant technological process is a useful form of representation of the possible variants of a process that enables identifying particular stages of a processes realization. This kind of graph, known as a process states graph (G_{sk}), is a directed graph in which nodes describes states of the process, and edges – the possible transitions between states, indicating operations.

A form of the multivariant process representation, alternative to the states graph, is an operations graph (G_{ok}). The graph G_{ok} has a form of directed line graph of the G_{sk} arising by the transformation of the nodes in the arches:

$$G_{ok} = L(G_{sk}), \tag{1}$$

where $L()$ is an unary transformation of a graph into the edge graph.

This form of processes representation is more useful when planning operations. The main advantage of G_{ok} is that it does not exist in the form of multigraph, with possible parallel edges between nodes. Figure 3 shows the transformation $G_{o1} = L(G_{s1})$ of the graph G_{s1} to operation graph G_{o1}.

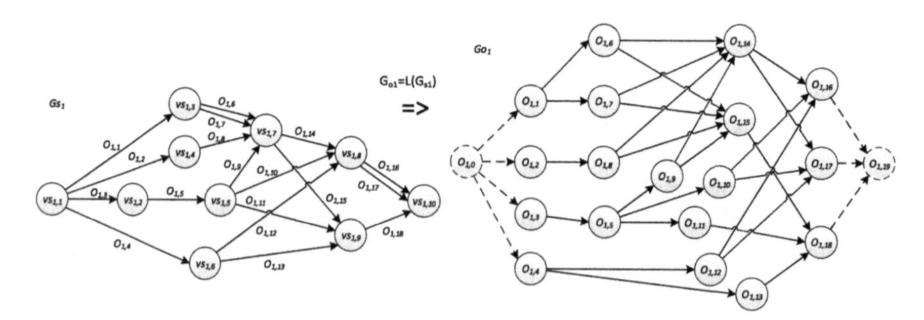

Fig. 3. The transformation $G_{o1} = L(G_{s1})$ of the graph G_{s1} to operation graph G_{o1}.

Determination of the additional dummy start and end node, and dummy activities (in the figure indicated by dashed lines) are required. Merging operation graphs of processes enables creating the aggregated graph of operations planning of the set of orders, which determines relationships between operations of executed production orders, and represents the search space in the given scheduling problem.

The sequence of operations for scheduling of a single production order is mainly determined by the processes structure and the scheduling strategy. Scheduling strategy is usually specified in the input data, according to the accepted model of treating of the customer order. Generally, two basic strategies for scheduling orders: forward and backward, are distinguished. The forward strategy causes that operations of a process are inserted into the schedule in order from first to last, in the same order as given in G_{ok}. In the case of backward strategy, operations are planned from the last operation to the first, so the graph G_{ok} should be transformed (reversed in arcs direction).

Accepted priorities of orders have also major impact on the shape of the aggregated graph. Scheduling orders can be: equivalent - do not have assigned value of priority, or their values are the same, partially arranged - there are groups of orders with the same value of priority, or fully arranged - each order has a unique value of priority. For searching that class of the space many algorithms can be constructed, but all of them should take into account the following, basic problems: a general method for searching the graph, sequencing of parallel processes and operations and selection of variants (branches with alternative relationship).

Selection of a general method for searching the graph – in order to search the space various algorithms, based on the depth-first search or breadth-first search schemes may be used. The searching algorithm can also use both schemes at the same time, representing so-called mixed scheme. There are also many different algorithms for implementing these strategies, extended by additional procedures for choosing the order of search nodes of the graph, the most promising directions, e.g. a best-first search, beam search, etc. These strategies differ in the way of passing through nodes, but a common feature of them is a single execution of the action associated with a node, regardless of the number of passes through the node.

Selection method of determining the order of visiting nodes of the graph (processes and operations on the parallel edges) can be supported by different schedule generation schemes (SGS). A number of different SGS were defined. In general, serial and parallel SGS can be distinguished.

Selection of variants of production flow applies to: variants of technology, variants of processes phases or single operations. Choosing a particular variant is associated with the local optimization, depending on a given situation in a production system. This choice does not guarantee obtaining the best solution (schedule) in a global sense.

3.3 Integration Method

Implementation of the integration process requires the identification of sources and forms of data acquisition from CAPP/PPC/MES systems, the manner of their transformation, methods of storage and transmission, as well as assessment at every step. Achieving this objective will be implemented in three main stages [11].

The first stage is focused on the acquisition of the complete data required in the target system. These data, obtained from planning support systems (actually their representation), will be saved with the use of a neutral format. Obtaining complete information practically requires acquisition of data from various heterogeneous computer systems (in a situation where existing decision-making support systems in the area of planning are not integrated), stored often in different types of databases, therefore it is necessary to develop a definition of structures for data representation in neutral formats, using universal formats, independent of the source data structures, hardware and software platforms. The result of the implementation of this stage are definitions of data structures for the source systems.

In the second stage the processes of data exchange between the source representation of the data obtained in the first stage and the intermediate neutral data model are carried out. In this step data mapping and data transformation methods are used, it is also necessary to develop the definition of the intermediate data model.

In a next step, the data from intermediate model are transformed to the data model of the target system. Therefore, it was necessary to choose the methods of data transformation. The result of the implementation of this phase is complete data for the target system, for a particular condition and the specific structure of the production system, obtained from CAPP/PPC systems stored in a neutral data model (Fig. 4).

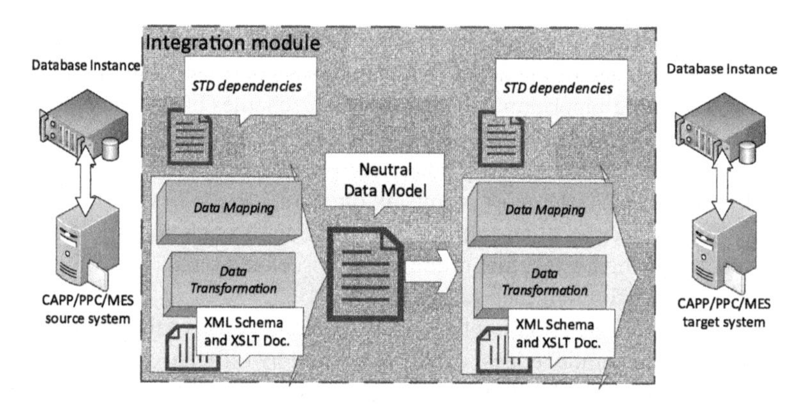

Fig. 4. CAPP/PPC/MES integration

Developed XML Schema, describing the structure of production processes with regard to available resources of the production system, includes a definition of the XML document structure for the data describing the resources that make up the manufacturing system, i.e.: manufacturing resources, inter-operational buffers, input and output buffers for products, which are to be manufactured in the system, etc.

It was decided that the structure of the document will be defined using XML Schema due to the fact that XML Schema is also stored in XML, is more powerful in comparison to the standard DTD, and allows to define additional data restrictions.

The XML Scheme is also used to control the correctness of the document (XML validation) during the transformation process of XML documents in applications. For the purpose of automating the process of transformation Extensible Stylesheet Language Transformations (XSLT) has been used. XSLT is widely used in various software and, like the XML diagrams, the realization of the transformation process based on XSLT document can be executed using the popular processors, like XMLSpy, Sablotron for C++, XSLT, PHP. Data Mapping and Data Transformation are supported by means of XML Path Language (XPath). XPath is a query language for selecting nodes from an XML document, but may be used to compute values (e.g. strings, numbers, or Boolean values) from the content of an XML document.

4 Summary

The paper presets the concept of holonic integration in the field of production preparation and management. The concept is based on idea of social holonic systems, where each module is independent but can cooperate with others, creating complex organization. This type of integration is not hierarchical, system consists of different modules, e.g. production preparation (CAPP), production planning and control (PPC), as well as production data acquisition (MIAS). The integration process requires the identification of sources and forms of data acquisition from CAPP/PPC systems. For that purpose, the methods of defining of data structure, data mapping and data transformation, has been elaborated. This ability to automate the transfer of data between different modules of the company allows designing independent, diverse and highly specialized modules, as well as fully integrate various areas of the company.

References

1. Hope, K. (eds.): Annual Report on European SMEs 2014/2015. SME Performance Review 2014/2015, Final Report, November 2015, European Union (2015)
2. Corchado, E., Sedano, J., Curiel, L., Villar, J.R.: Optimizing the operating conditions in a high precision industrial process using soft computing techniques. Expert Syst. **29**, 276–299 (2012)
3. Sedano, J., Berzosa, A., Villar, J.R., Corchado, E., De La Cal, E.: Optimising operational costs using soft computing techniques. Integr. Comput. Aided Eng. **18**(4), 313–325 (2011)
4. Borangiu, T., Gilbert, P., Ivanescu, N.A., Rosu, A.: An implementing framework for holonic manufacturing control with multiple robot-vision stations. Eng. Appl. Artif. Intell. **22**(4), 505–521 (2009)
5. Bajic, E., Cchaxel, F.: Towards a holon-product oriented management. In: 4th IFAC Workshop on Intelligent Manufacturing Systems IMS 1997, Seoul, Korea, pp. 121–126 (1997)
6. Sekala, A., Kost, G., Dobrzanska-Danikiewicz, A., Banas, W., Foit, K.: The distributed agent-based approach in the e-manufacturing environment. IOP Conf. Ser. Mater. Sci. Eng **95**, article no. 012134 (2015)
7. Ćwikła, G., Krenczyk, D., Kampa, A., Gołda, G.: Application of the MIAS methodology in design of the data acquisition system for wastewater treatment plant. IOP Conf. Ser. Mater. Sci. Eng. **95**, article no. 012153 (2015)
8. Ćwikła, G.: The methodology of development of the Manufacturing Information Acquisition System (MIAS) for production management. Appl. Mech. Mater. **474**, 27–32 (2014)
9. Ćwikła, G.: Methods of manufacturing data acquisition for production management – a review. Adv. Mater. Res. **837**, 618–623 (2014)
10. Deak, G., Curran, K., Condell, J.: A survey of active and passive indoor localisation systems. Comput. Commun. **35**, 1939–1954 (2012)
11. Krenczyk, D., Bocewicz, G.: Data-driven simulation model generation for ERP and DES systems integration. In: Jackowski, K., Burduk, R., Walkowiak, K., Woźniak, M., Yin, H. (eds.) IDEAL 2015. LNCS, vol. 9375, pp. 264–272. Springer, Heidelberg (2015). doi:10.1007/978-3-319-24834-9_32

CISIS 2016: Applications of Intelligent Methods for Security

Feel Me Flow: A Review of Control-Flow Integrity Methods for User and Kernel Space

Irene Díez-Franco(✉) and Igor Santos

DeustoTech, University of Deusto, Bilbao, Spain
{irene.diez,isantos}@deusto.es

Abstract. Attackers have evolved classic *code-injection attacks*, such as those caused by buffer overflows to sophisticated Turing-complete *code-reuse* attacks. Control-Flow Integrity (CFI) is a defence mechanism to eliminate control-flow hijacking attacks caused by common memory errors. CFI relies on static analysis for the creation of a program's control-flow graph (CFG), then at runtime CFI ensures that the program follows the legitimate path. Thereby, when an attacker tries to execute malicious shellcode, CFI detects an unintended path and aborts execution. CFI heavily relies on static analysis for the accurate generation of the control-flow graph, and its security depends on how strictly the CFG is generated and enforced.

This paper reviews the CFI schemes proposed over the last ten years and assesses their security guarantees against advanced exploitation techniques.

Keywords: Control-Flow Integrity · Code-reuse attacks · Operating system security

1 Introduction

Operating systems must ensure that both their own code and the code of their applications remain incorruptible and consequently secure and reliable against attackers. Since code-injection attacks are widely known, adversaries nowadays commonly exploit memory corruption bugs to subvert the *control-flow* of the operating system or the applications that are being executed within it. Rather than focusing on protecting the integrity of code, with complete memory safety or developing safe dialects of C/C++, modern defences try to protect the *control-flow integrity* (CFI) of these systems.

Since CFI [1] was introduced to avoid these problems and issues, different implementations and versions of this techniques have been proposed by the community that try to make it practical while ensuring the completeness of its protection. In addition, new attacks have been proposed that limit the effectiveness of these methods. Due to the raising relevance of CFI methods for the system security community, in this paper we present the first comprehensive literature review and discussion of control-flow integrity defences and the attacks that try to subvert them.

M. Graña et al. (eds.), *International Joint Conference SOCO'16-CISIS'16-ICEUTE'16*, Advances in Intelligent Systems and Computing 527, DOI 10.1007/978-3-319-47364-2_46

2 Control-Flow Integrity

C/C++ code goes hand in hand with memory corruption bugs, which allow an adversary to launch attacks that exploit those memory errors. *Code injection* attacks due to stack-based or heap-based overflows, dangling pointers/use-after-free, and format string vulnerabilities are common, and can be prevented using defences such as write-xor-execute (W⊕E)/Data Execution Prevention (DEP) [4] and stack canaries [16] which are included in nowadays compilers and operating systems. Nevertheless *code-reuse* attacks, like return-into-libc [36], return-oriented programming (ROP) [44], and jump-oriented programming (JOP) [5,12] still can not be fully prevented. Operating systems themselves are not exempt from code-reuse attacks, such as return-to-user (ret2usr) [33], a kernel level variant of return-into-libc, and sigreturn oriented programing (SROP) [6], which exploits the signal handling capabilities of UNIX like systems to deploy gadgets in the same manner that ROP and JOP do with `ret` and `jmp` instructions respectively.

Operating systems deploy statistical defences to protect user space and kernel space against code-reuse attacks; namely address space layout randomisation (ASLR) [47], and Kernel ASLR [22]. However, these defences can be circumvented due to information leakage and just-in-time code-reuse attacks both for user [45] and kernel [29] space.

Taking into account these problems, Abadi et al. introduced *control-flow integrity* (CFI) [1], a defence mechanism to prevent code-reuse attacks, which try to subvert the legitimate execution flow of a program.

CFI works in two phases, firstly, it computes the Control-Flow Graph (CFG) of the program by static analysis, either using its source code or its binary; afterwards, during program execution, CFI enforces that the program follows through the legitimate execution path; otherwise the program is aborted.

In the computation phase, CFI is concerned with *points-to analysis*, the static analysis that deals with the possible values of a pointer, because it affects the precision in which a CFG is generated [8] and consequently, the precision in which the enforcement phase will enforce the legitimate execution path. Taking into account the precision, CFI implementations can be categorised into (i) *flow-sensitive* or *flow-insensitive* and (ii) *context-sensitive* or *context-insensitive.* On the one hand, flow-sensitive algorithms use the control-flow information of a program to determine the possible values of a pointer, whereas flow-insensitive algorithms compute a set of values that are valid for all program inputs [25,26]. On the other hand, context-sensitive algorithms take into account the context when analysing a function, preventing values from propagating to impracticable paths and thus guaranteeing that the context of a call remains independent from other call contexts; in contrast, context-insensitive algorithms allow a function to return to the computed set of all callers [26,52].

In the enforcement phase, CFI may take into account *forward* (e.g. indirect calls or jumps) and *backward* (e.g. return instructions) control-flow transfers. CFI solutions that provide just a forward enforcement of control-flow transfers have been found insecure [10,21,23], whereas solutions that enforce the backward transfers usually rely on a shadow stack, a structure that holds copies of the return addresses present in each of the stack frames of the original stack, causing up to a $\sim 10\,\%$ increase in the program overhead [19], or use the last-branch record registers (LBR) [3,31] which are only available to a subset of CPUs and have a limited storing capacity.

The vast majority of CFI implementations aim to protect the user space, and come in the flavours of compiler extensions, source code or binary code patching frameworks and kernel modules, whereas a small subset intend to secure the operating system deploying kernel modifications or new kernel modules.

2.1 Userland Implementations

The original CFI [1] operates on x86 binaries by machine-code rewriting. For the forward control-flow transfers, the rewriting process includes a ID insertion at each destination, and a ID-check before each source; then at runtime the source ID and the destination ID must coincide. To ensure that a function call returns to the appropriate call site, a backward control-flow transfer, the implementation uses a shadow call stack relying on x86's segmentation capabilities. CFI requires (i) the code to be non-writable, to prevent attackers from rewriting the ID-check, and (ii) the data to be non-executable, to prevent attackers to execute data generated with the expected ID. The first requirement is true in modern OSes, excluding the loading time of dynamic libraries and runtime code-generation, and the second requirement is enforced with W$\oplus$E. This implementation makes the assumption that two destinations are equivalent if they are called from the same source, thus introducing imprecision in the CFG and thereby in the enforcement phase.

MoCFI [20] provides CFI protection on iOS devices' applications running on ARM processors. It addresses the special issues of ARM architecture (e.g. the nonexistence of dedicated return instructions). As the original CFI, it also operates on binaries. The authors generate a CFG of the application and a patchfile containing metadata of the indirect branches and function calls in the application; dynamic libraries used in the application are not protected. In the runtime enforcement phase, the patchfile is used by the MoCFI shared library, generating a patched application which is executed within the CFI policy. MoCFI uses a shadow stack to protect calls and returns. For the forward control-flow transfers however, it cannot protect indirect jumps/calls whose destination cannot be identified on the static analysis; thereby they can target any valid address within the function, or any valid function respectively.

Unlike MoCFI, CCFIR [55] and Bin-CFI [56] are two other binary implementations that include protection for libraries. On the one hand, CCFIR works Windows x86 PE executables, with partial support for libraries. It builds upon Abadi et al.'s approach and incorporates a third new ID-check for returns to sensitive and non-sensitive functions. This implementation suffers from the same imprecision as Abadi et al.'s for forward edges and introduces it in backward edges. On the other hand, Bin-CFI protects stripped Linux x86 binaries including shared libraries. This approach is similar to the original CFI scheme and has lower security guarantees than CCFIR. Recent studies have found both Bin-CFI and CCFIR protections insufficient [21,23], since grouping destinations into equivalence classes is not strong enough to prevent them for being used as ROP/JOP gadgets.

kBouncer [40] is a hardware based Windows toolkit that relies on Intel Nehalem architecture's LBR registers to retrieve the sequence of the latest 16 indirect branch instructions at critical points (e.g. system calls). In total, kBouncer protects the execution of 52 Windows API functions. Similar to kBouncer, ROPecker [14] is a Linux x86 kernel module that utilises the LBR register to prevent code-reuse attacks. Both schemes depend on chain length and gadget length heuristics to prevent such attacks. Nevertheless they can be bypassed by choosing the right sized gadget-chain length [10,21,24].

A recent binary based x86/64 CFI implementation, O-CFI [35], combines code randomisation with CFI checking. O-CFI first computes the permissible destination addresses for each indirect branch, then it transforms the policy that indirect branches must reach to a valid destination into a bounds-checking problem; thereby O-CFI has to check that the destination address exists within min/max address boundaries. These boundaries are protected using code randomisation and then checked making use of Intel's memory protection extensions (MPX) [30]. O-CFI uses a relaxed version of forward and backward control-flow transfer checks and consequently, can be bypassed.

All the previously presented binary level approaches [1,20,35,40,55,56] fail to capture *complete* context sensitivity; whereas just some of them support partial (backward) context sensitivity due to the use of a shadow stack [1,20]. PathArmor [49], is the first binary level scheme to tackle context sensitivity for forward and backward edges. Context-sensitive CFI methods need to keep track of the paths of the executed control-flow transfers, to later on enforce that the execution follows the legitimate path. Instead of using a shadow stack, PathArmor employs LBR registers to emulate a path monitoring mechanism limited by the number of LBR registers (just 16). PathArmor outperforms all previous protection schemes for forward edge transfers. However, shadow stack based approaches are still more reliable for backward edges due to the limitations that current hardware imposes.

Regarding CFI implementations that rely on source-code, Tice et al. [48] present two different forward-edge protection mechanisms integrated in production compilers, Virtual-Table Verification (VTV) and Indirect Function-Call Checks (IFCC) for GCC and LLVM respectively. Stack based attacks have been found effective bypassing VTV/IFCC [15] and the subsequent compilers have been patched. SafeDispatch [32] is a earlier LLVM compiler extension, and like VTV, aims to protect virtual tables (vtables) for `C++` virtual calls; both VTV and SafeDispatch fail to provide full control-flow protection since they focus just on forward edges. Further research has been done with the objective of protecting vtables, resulting in two binary level implementations, VfGuard [42] and VTint [54]; which unfortunately are also limited to partial control-flow protection.

A recent form of control-flow reuse attack, Counterfeit Object-oriented Programming (COOP) [43] can mount Turing-complete attacks using gadgets of C++ virtual functions. COOP is effective against the original CFI, bin-CFI, CCFIR, VTint and partially against IFCC, VfGuard and PathArmor. In contrast, COOP can be prevented at binary level by TypeArmor [50], and at source code level with the compiler extensions SafeDispatch, VTV, VTrust [53] and VTI [7].

Niu and Tan introduced Modular CFI (MCFI) [37], a new scheme which extends CFI with modular compilation. Building upon MCFI the authors present RockJIT [38] which enforces CFI in Just-In-Time compilers; both MCFI and RockJIT induce some imprecision in the edge generation since they apply the same assumption as the original CFI for equivalent targets. Their following contribution, πCFI (per-input CFI) [39], on the contrary, introduces the highest security guarantees for a source code based CFI solution. πCFI differs from all previous CFI implementations in the way it addresses the CFG generation. Conservative CFI implementations utilise static analysis to compute the CFG before the enforcement phase, this analysis is considered hard since it has to take into account *all* the possible input values for the given program; moreover, CFI's security guarantees are strictly bounded to the CFG's precision. Niu and Tan point out than even if a perfect CFG were possible, it would still include unnecessary edges for a given input. Thereby they tackle the CFG generation in the following way; firstly, they generate the conservative CFG for all program inputs (building upon MCFI and RockJIT), then during program execution, given an input, πCFI generates CFG edges on the fly, but just those which comply with the conservative all-input CFG are enforced. This innovative scheme provides less backward edge protection compared to shadow stack approaches, but higher guarantees that other backward edge approaches. Concerning forward edges, πCFI has stronger assurance that original CFI due to the per-input mechanism.

2.2 Kernel-Space Implementations

State-based CFI (SBCFI) [41] is a CFI implementation for Xen and VMware Workstation virtual machine monitors. Unlike CFI enforced in userland, kernel space CFI cannot guarantee that the generated CFG is read only, nor that its data is non-executable since an attacker with access to the kernel space could also have access to page tables, and thus be able to change their properties. Thereby, SBCFI enforces a relaxed CFI by periodically checking the current kernel's CFG against the initial kernel's CFG. This implementation provides light security guarantees since it does not enforce backward edges and the support for forward edges is limited.

Hypersafe [51] is a LLVM framework extension that targets hypervisors. Hypersafe introduces the concepts of *non-bypassable memory lockdown* and *restricted pointer indexing* to introduce CFI on hypervisors. The former method is in charge of guaranteeing the integrity of the hypervisor's code and static data; the later delimits the contents of the targets of the control data (function pointers and return addresses) into a target table, to then rewrite each function pointer or return address to a pointer index to the target table. Using the restricted pointer indexing, Hypersafe can either allow light security guarantees by allowing a function to return to any address entry on the target table, or a more strict scheme, by generating a target table for each function and allowing the function to return to a *subset* of all returns, made specifically for that function. Hypersafe implements backward edge enforcement policies but not as safe as those provided by shadow stack schemes, and for forward edges, in its strict scheme, a policy more accurate than the orignal CFI but less that the most strict user space implementations (PathArmor and πCFI).

kGuard [33] is a GCC compiler extension whose aim is to protect the kernel against ret2usr attacks. kGuard combines CFI with program shepherding. *Program shepherding* [34] is a technique that permits to implement arbitrary restrictions to code origins and control flow transfers. Upon compiling a kernel with kGuard, Control-Flow Assertions (CFA) are introduced before each control-flow transfer. These assertions are comparable to the original CFI checks, but unlike them, CFAs are not checked against a CFG to enforce a valid edge, they just ensure that the target address exists within kernel space instead. This security mechanism cannot withstand ROP/JOP like attacks since it is comparable to just enforcing weak forward control-flow transfers, like the traditional CFI, and also a weak policy for backward transfers.

KCoFI [17] provides CFI for commodity OSs utilising the infrastructure provided by the Secure Virtual Architecture (SVA) [18] virtual machine. This infrastructure is used to handle low level operations regarding the MMU, general I/O, signal dispatch and context switching. KCoFI is built on top of the SVA virtual machine, and thereby requires the OS and applications to instrument to be compiled to the virtual instruction set provided by the SVA architecture. As the original CFI, KCoFI enforces a CFI policy that is not context-sensitive.

3 Discussion

Table 1 summarises the implementations reviewed in this paper, the top part of the table lists userland implementations whereas the bottom part lists kernel space implementations. Regarding userland CFI, on the one hand, binary schemes are more common, nevertheless these schemes are known to be less secure than their source-code based counterparts. On the other hand, some source schemes (VTV, IFCC, SafeDispatch, TypeArmor) tend to focus on just forward edges, and thereby are prone to ROP attacks; while others enforce policies that fall into the *equivalent classes* paradigm which just can partially prevent ROP/JOP. Concerning kernel space, the implementations enforce modified CFI methods due to the peculiarities that protecting a kernel involves. Hypersafe is the strongest implementation followed by KCoFI. The former provides limited context-sensitivity for both edges, whereas the later falls into the equivalent classes paradigm.

In summary, the strongest implementations provide some level of context-sensitivity for both edges. PathArmor utilises the LBR registers for a hardware limited context-sensitivity, πCFI builds upon the modular CFG idea and Hypersafe uses restricted pointer indexing to provide a limited context-sensitivity.

Table 1. Comparison of CFI implementations. *B* stands for binary, *S* source-code, *KM* kernel module, *VMM* virtual machine monitor; *CS* context-sensitive, *EC* equivalent classes, *H* heuristics, ∅ the policy is not enforced. Regarding attacks, *Th.* stands for theoretical attacks.

		Precision		
	Scheme	**Forward**	**Backward**	**Known attacks**
Original CFI	B	EC	CS	COOP
MoCFI	B	EC	CS	Limited JOP
CCFIR	B	EC	EC	[23], COOP
Bin-CFI	B	EC	EC	[21,23], COOP
kBouncer	B	H	H	[10,21,24]
ROPecker	KM	H	H	[10,21,24]
O-CFI	B	EC	EC	Th. ROP/JOP
PathArmor	B	Hardware limited CS	Hardware limited CS	History flush
VTV	S	CS	∅	COOP
IFCC	S	CS	∅	ROP
SafeDispatch	S	CS	∅	ROP
TypeArmor	B	EC	∅	ROP
MCFI	S	EC	EC	Th. ROP/JOP
RockJIT	S	EC	EC	Th. ROP/JOP
πCFI	S	Limited CS	Limited CS	Limited ROP/JOP
SBCFI	VMM	CFG comparison	∅	ROP
Hypersafe	S	Limited CS	Limited CS	Limited ROP/JOP
kGuard	S	Exists in kernel space	Exists in kernel space	ROP/JOP
KCoFI	S	EC	EC	Th. ROP/JOP

Future trends of work are focusing on more precise context-sensitive schemes and addressing the implementation of safer CFI schemes for commodity OSs.

4 Related Work

Data-Flow Integrity. Chen et al. [13] raised awareness of *non-control data* attacks the same year as Abadi et al.'s proposed CFI; one year before, Castro et al. proposed *data-flow integrity* (DFI) [11], a defence mechanism that tries to protect the legitimate data-flow analogously to CFI with control-flow.

Due to the attention that the research community has given to code injection and code-reuse attacks, attackers have directed their efforts into the creation of non-control data attacks, which have been recently found to be Turing-complete [28] and moreover, can be automatically constructed [27].

CFI mechanisms cannot withstand non-control data attacks [9] since they *follow the legitimate execution flow* and thereby pose a major threat for OSs and userland programs. The research community has proposed several approaches to protect userland applications [2,11] and kernel space [18,46] but they have not been proven yet to be practical due to performance issues.

5 Conclusions

The academia has proposed several methods to protect applications and operating systems against code-reuse attacks. These approaches build upon control-flow integrity to prevent code-reuse attacks, but since control-flow integrity's effectiveness is closely bounded with the precision in which its control-flow graph is generated, not all schemes have proven to be effective against code-reuse attacks.

References

1. Abadi, M., Budiu, M., Erlingsson, U., Ligatti, J.: Control-flow integrity: principles, implementations and applications. In: CCS (2005)
2. Akritidis, P., Cadar, C., Raiciu, C., Costa, M., Castro, M.: Preventing memory error exploits with WIT. In: Security & Privacy (2008)
3. AMD: AMD64 Architecture Programmer's Manual: System Programming, vol.2 (2013). http://developer.amd.com/wordpress/media/2012/10/24593_APM_v21.pdf
4. Andersen, S., Abella, V.: Data Execution Prevention. Changes to Functionality in Microsoft Windows XP Service Pack 2, Part 3: Memory Protection Technologies (2004)
5. Bletsch, T., Jiang, X., Freeh, V.W., Liang, Z.: Jump-oriented programming: a new class of code-reuse attack. In: CCS (2011)
6. Bosman, E., Bos, H.: Framing signals-a return to portable shellcode. In: Security & Privacy (2014)
7. Bounov, D., Kıcı, R.G., Lerner, S.: Protecting C++ dynamic dispatch through vtable interleaving. In: NDSS (2016)
8. Burow, N., Carr, S.A., Brunthaler, S., Payer, M., Nash, J., Larsen, P., Franz, M.: Control-flow integrity: precision, security, and performance. arXiv preprint arXiv:1602.04056 (2016)
9. Carlini, N., Barresi, A., Payer, M., Wagner, D., Gross, T.R.: Control-flow bending: on the effectiveness of control-flow integrity. In: USENIX Security (2015)
10. Carlini, N., Wagner, D.: ROP is still dangerous: breaking modern defenses. In: USENIX Security (2014)
11. Castro, M., Costa, M., Harris, T.: Securing software by enforcing data-flow integrity. In: OSDI (2006)
12. Checkoway, S., Davi, L., Dmitrienko, A., Sadeghi, A.R., Shacham, H., Winandy, M.: Return-oriented programming without returns. In: CCS (2010)
13. Chen, S., Xu, J., Sezer, E.C., Gauriar, P., Iyer, R.K.: Non-control-data attacks are realistic threats. In: USENIX Security (2005)
14. Cheng, Y., Zhou, Z., Miao, Y., Ding, X., Deng, H., R.: ROPecker: a generic and practical approach for defending against ROP attack. In: NDSS (2014)
15. Conti, M., Crane, S., Davi, L., Franz, M., Larsen, P., Negro, M., Liebchen, C., Qunaibit, M., Sadeghi, A.R.: Losing control: on the effectiveness of control-flow integrity under stack attacks. In: CCS (2015)
16. Cowan, C., Pu, C., Maier, D., Hinton, H., Walpole, J., Bakke, P., Beattie, S., Grier, A., Wagle, P., Zhang, Q.: StackGuard: automatic adaptive detection and prevention of buffer-overflow attacks. In: USENIX Security (1998)
17. Criswell, J., Dautenhahn, N., Adve, V.: KCoFI: complete control-flow integrity for commodity operating system kernels. In: Security & Privacy (2014)

18. Criswell, J., Lenharth, A., Dhurjati, D., Adve, V.: Secure virtual architecture: a safe execution environment for commodity operating systems. In: ACM SIGOPS Operating Systems Review (2007)
19. Dang, T.H.Y., Maniatis, P., Wagner, D.: The performance cost of shadow stacks and stack canaries. In: ASIACCS (2015)
20. Davi, L., Dmitrienko, A., Egele, M., Fischer, T., Holz, T., Hund, R., Nürnberger, S., Sadeghi, A.R.: MoCFI: a framework to mitigate control-flow attacks on smartphones. In: NDSS (2012)
21. Davi, L., Sadeghi, A.R., Lehmann, D., Monrose, F.: Stitching the gadgets: on the ineffectiveness of coarse-grained control-flow integrity protection. In: USENIX Security (2014)
22. Giuffrida, C., Kuijsten, A., Tanenbaum, A.S.: Enhanced operating system security through efficient and fine-grained address space randomization. In: USENIX Security (2012)
23. Göktaş, E., Athanasopoulos, E., Bos, H., Portokalidis, G.: Out of control: overcoming control-flow integrity. In: Security & Privacy (2014)
24. Göktaş, E., Athanasopoulos, E., Polychronakis, M., Bos, H., Portokalidis, G.: Size does matter: why using gadget-chain length to prevent code-reuse attacks is hard. In: USENIX Security (2014)
25. Hardekopf, B., Lin, C.: Semi-sparse flow-sensitive pointer analysis. In: ACM SIGPLAN Notices (2009)
26. Hind, M.: Pointer analysis: haven't we solved this problem yet?. In: Proceedings of the 2001 ACM SIGPLAN-SIGSOFT Workshop on Program Analysis for Software Tools and Engineering (2001)
27. Hu, H., Chua, Z.L., Adrian, S., Saxena, P., Liang, Z.: Automatic generation of data-oriented exploits. In: USENIX Security (2015)
28. Hu, H., Shinde, S., Adrian, S., Chua, Z.L., Saxena, P., Liang, Z.: Data-oriented programming: on the expressiveness of non-control data attacks. In: Security & Privacy (2016)
29. Hund, R., Willems, C., Holz, T.: Practical timing side channel attacks against kernel space ASLR. In: Security & Privacy (2013)
30. Intel: Intel 64 and IA-32 Architectures Software Developer's Manual: Basic Architecture, vol. 1 (2016). https://www-ssl.intel.com/content/www/us/en/processors/architectures-software-developer-manuals.html
31. Intel: Intel 64 and IA-32 Architectures Software Developer's Manual: System Programming Guide, vol. 3B, Part 2 (2016). https://www-ssl.intel.com/content/www/us/en/architecture-and-technology/64-ia-32-architectures-software-developer-vol-3b-part-2-manual.html
32. Jang, D., Tatlock, Z., Lerner, S.: SafeDispatch: securing C++ virtual calls from memory corruption attacks. In: NDSS (2014)
33. Kemerlis, V.P., Portokalidis, G., Keromytis, A.D.: kGuard: lightweight kernel protection against return-to-user attacks. In: USENIX Security (2012)
34. Kiriansky, V., Bruening, D., Amarasinghe, S.P., et al.: Secure execution via program shepherding. In: USENIX Security (2002)
35. Mohan, V., Larsen, P., Brunthaler, S., Hamlen, K.W., Franz, M.: Opaque control-flow integrity. In: NDSS (2015)
36. Nergal: The advanced return-into-lib(c) exploits: pax case study. Phrack Mag. **58**(4), 54 (2001)
37. Niu, B., Tan, G.: Modular control-flow integrity. In: PLDI (2014)
38. Niu, B., Tan, G.: Rockjit: securing just-in-time compilation using modular control-flow integrity. In: CCS (2014)

39. Niu, B., Tan, G.: Per-input control-flow integrity. In: CCS (2015)
40. Pappas, V., Polychronakis, M., Keromytis, A.D.: Transparent rop exploit mitigation using indirect branch tracing. In: USENIX Security (2013)
41. Petroni Jr., N.L., Hicks, M.: Automated detection of persistent kernel control-flow attacks. In: CCS (2007)
42. Prakash, A., Hu, X., Yin, H.: vfGuard: Strict protection for virtual function calls in COTS C++ binaries. In: NDSS (2015)
43. Schuster, F., Tendyck, T., Liebchen, C., Davi, L., Sadeghi, A.R., Holz, T.: Counterfeit object-oriented programming: on the difficulty of preventing code reuse attacks in C++ applications. In: Security & Privacy (2015)
44. Shacham, H.: The geometry of innocent flesh on the bone: return-into-libc without function calls (on the x86). In: CCS (2007)
45. Snow, K.Z., Monrose, F., Davi, L., Dmitrienko, A., Liebchen, C., Sadeghi, A.R.: Just-in-time code reuse: on the effectiveness of fine-grained address space layout randomization. In: Security & Privacy (2013)
46. Song, C., Lee, B., Lu, K., Harris, W., Kim, T., Lee, W.: Enforcing Kernel security invariants with data flow integrity. In: NDSS (2016)
47. Team, P.: Address space layout randomization (ASLR) (2003). http://pax.grsecurity.net/docs/aslr.txt
48. Tice, C., Roeder, T., Collingbourne, P., Checkoway, S., Erlingsson, Ú., Lozano, L., Pike, G.: Enforcing forward-edge control-flow integrity in GCC & LLVM. In: USENIX Security (2014)
49. van der Veen, V., Andriesse, D., Göktaş, E., Gras, B., Sambuc, L., Slowinska, A., Bos, H., Giuffrida, C.: Practical context-sensitive CFI. In: CCS (2015)
50. van der Veen, V., Göktas, E., Contag, M., Pawlowski, A., Chen, X., Rawat, S., Bos, H., Holz, T., Athanasopoulos, E., Giuffrida, C.: A Tough call: mitigating advanced code-reuse attacks at the binary level. In: Security & Privacy (2016)
51. Wang, Z., Jiang, X.: Hypersafe: a lightweight approach to provide lifetime hypervisor control-flow integrity. In: Security & Privacy (2010)
52. Wilson, R.P., Lam, S., M.: Efficient context-sensitive pointer analysis for C programs. In: PLDI (1995)
53. Zhang, C., Carr, S.A., Li, T., Ding, Y., Song, C., Payer, M., Song, D.: VTrust: regaining trust on virtual calls. In: NDSS (2016)
54. Zhang, C., Song, C., Chen, K.Z., Chen, Z., Song, D.: VTint: protecting virtual function tables' integrity. In: NDSS (2015)
55. Zhang, C., Wei, T., Chen, Z., Duan, L., Szekeres, L., McCamant, S., Song, D., Zou, W.: Practical control flow integrity and randomization for binary executables. In: Security & Privacy (2013)
56. Zhang, M., Sekar, R.: Control flow integrity for COTS binaries. In: USENIX Security (2013)

A Secure Mobile Platform for Intelligent Transportation Systems

Alexandra Rivero-García, Iván Santos-González, and Pino Caballero-Gil(✉)

Universidad de La Laguna, San Cristóbal de La Laguna, Spain
{ariverog,jsantosg,pcaballe}@ull.edu.es

Abstract. The number of vehicles around the world has increased a lot during the last decades and will continue to grow in the next years. This factor is closely related to the increase of air pollution, traffic congestion and road non-safety. Due to this, the automotive industry and research entities have put lots of efforts in exploring the potential of the Intelligent Transportation Systems (ITS). These efforts have been focused on trying to add ITS systems to the new cars of the future. The present work describes a proposal for taking advantage of ITS possibilities using the vehicles that are nowadays on our roads. In particular, the described platform is mainly based on smartphones, but also on sensors and servers. The proposal is based on a hybrid scheme that combines an online mode using the Internet access of the smartphone, with an offline mode using wireless technologies such as WiFi Direct and Bluetooth Low Energy to define various ITS applications related with traffic collisions, violations, jams, signs and lights, and parking management. The described system has been developed for the Android platform, producing promising results.

Keywords: Ad-hoc · VANET · Security · Hybrid · Sensors · Wireless · Android

1 Introduction

Mobile Ad-hoc NETworks (MANETs) [20] are networks based on the principle of constant connectivity among nodes that are in continuous motion. Vehicular Ad-hoc NETworks (VANETs) can be seen as a special type of MANET, where not only mobile nodes, named On-Board Units (OBUs) exist, but also static nodes, named Road-Side Units (RSUs) [1]. These networks have become an important research area during the last years, mainly due to a combination of the quick advance of technology, the proliferation of the Internet of Things, and the promotion of Intelligent Transportation Systems (ITS), including the incipient arrival of autonomous cars. In these networks, the communication among vehicles or OBUs is known as Vehicle-TO-Vehicle (V2V), and the communication between vehicles and fixed infrastructure elements or OBUs and RSUs is known as Vehicle-TO-Infrastructure (V2I).

M. Graña et al. (eds.), *International Joint Conference SOCO'16-CISIS'16-ICEUTE'16*,
Advances in Intelligent Systems and Computing 527, DOI 10.1007/978-3-319-47364-2_47

In the last decade, a great effort in the development of standards and research on VANETs has been made in Europe, US and Japan. The main objective of the different proposed solutions would eventually converge, leading to a common platform that should be already fully developed and available worldwide. However, so far, the wide development of the theoretical IEEE 802.11p standard defined for VANETs has been physically impossible because all proposals need the presence of OBUs and RSUs. This requirement represents an enormous cost and commitment both for institutions, which would have to adapt roads by incorporating RSUs; and for users, who would have to either buy new vehicles or adapt theirs by incorporating OBUs.

This work presents an innovative proposal to implement a VANET using smartphones and cloud servers. Specifically, this system tries to solve the cost problem replacing the OBUs with smartphones, which are able to communicate both online and offline.

This paper is organized as follows. Section 2 provides a brief state of the art. The proposed platform is described in Sect. 3. Section 4 introduces both the online and offline architectures. A brief analysis of the implementation is detailed in Sect. 5. Finally, Sect. 6 closes the paper with a few conclusions and open problems.

2 Related Work

The current state of the art of the research on ITS is based on different initiatives from both academia and industry [9].

One of the most important families of standards for VANET communication is IEEE 1609, also known as Wireless Access in Vehicular Environments (WAVE). It defines the architecture and a standardized set of services and interfaces enable secure V2X wireless communications. This standard defines security services that involve a complex Public-Key Infrastructure (PKI) where the government plays the role of the root Certificate Authority (CA) [11]. The IEEE 1609 standards rely on the amendment to the IEEE 802.11 (based on WiFi) to support V2V and V2I communications [17], the so-called IEEE 802.11p and the equivalent European standard ETSI ITS G5. IEEE 802.11p was used as a basis for the definition of ITS G5 [10], which is now being standardized by the European Telecommunications Standards Institute group for Intelligent Transport Systems.

There are some traffic applications and devices on the market, most of them to try to solve specific problems related to traffic management or even to calculate the fastest route through basic navigation systems using static maps. Thus, there are devices that send data to central servers like TomTom [18]. Waze [19] is a mobile application that allows disseminating events and GPS coordinates to create road maps and traffic density estimation in real time, whose main problem is that it is based on trust and voluntary cooperation of users. Furthermore, Apple Carplay [3] and Android Auto [2] are in-car versions of the two main mobile operating systems, iOS and Android.

Thanks to the high number of kind of vehicles, VANETs will have to be deployed in practice by combining on the one hand both existing and new vehicles, and on the other hand existing infrastructures and technological devices. The use of the Information and Communication Technologies (ICT) has meant a lot of improvements in different technological elements such as mobile phones, sensors and other wireless devices. The strong potential of these technologies has been explored for the Developed ExPerimental Hybrid wIrelesS platform for Intelligent Transportation systems, here called DEPHISIT platform.

3 Proposed Platform

The proposed system consists on the use of a hybrid platform for wireless communications, giving its name to the DEPHISIT system. The main purpose of the system is to offer a group of different applications to improve road safety and the driving experience of users.

This paper presents an innovative solution that represents a new step in the practical development of VANETs, providing technological innovations to the currently available systems. The developed hybrid wireless network works as a VANET that allows receiving information in real time through the LTE connection or WiFi hotspots. Another important aspect of the system is its security, which is strengthened through the use of different encryption techniques. These encryption techniques are especially important to transfer the sensitive information related to the vehicles and to their users. Moreover, the use of the WiFi hotspots allows reducing the cost of use of the system.

Traditionally, this kind of systems has been used only in restricted areas such as ports, airports, military areas, etc., but with the proposal here described it will reach a wider audience. Indeed, the main objective of DEPHISIT is to reach the general public to offer the benefits of this kind of systems. The development of the DEPHISIT platform has included several novel features to enable:

1. Efficient use of private vehicles thanks to tools such as carpooling, and improvement of the experience of public transport users (routing, schedule, delays, etc.).
2. Value-added tools such as parking management, geolocated advertising platform, provision of weather information, optimization of public transportation, etc.
3. Control of traffic situation in real time, especially in cases of large congestions or road works.
4. Measures to reduce accidents through warnings of possible rear-end collisions, serious traffic infringement or dangerous situations like the approaching of emergency vehicles, etc.
5. Efficient use of the vehicle sensors to provide the user with interesting information about aspects of the vehicle or the state of the road.

4 System Architecture

The DEPHISIT platform has two different working modes that are used depending on whether the users have Internet connection or not.

The system architecture is based on the client-server application model. The client is a mobile device used as OBU within the system. Moreover, the FrontEnd web application can be considered as the client part too, but with a different role in comparison to the mobile devices. The server is hosted in the cloud and divided into two parts. The first of them, the Google Cloud Messaging (GCM) server handles all the notifications and is responsible for sending notifications in real time when the receiver clients are alive. On the other hand, the dedicated server stores all the data related to users and system in the DataBase (DB). The use of this dedicated server allows us to manage the real time notifications between different client devices through the Internet connection. It also serves as a gateway for sending notifications between the client and the GCM server. In this case, communications between vehicles are made through the cloud, and DEPHISIT server can notify vehicles about relevant information within their area in real time.

The main architecture is for those users who have access to the Internet, which is nowadays the most common situation in urban areas, where there is good mobile network coverage and/or municipal wireless networks.

The proposed system can also be used without Internet connection. This feature is especially interesting in rural or remote places where the network infrastructures are not deeply deployed and the Internet connection is not available. In this mode, we provide the users with an ad-hoc network that use a direct communication between devices through the WiFi Direct or Bluetooth Low Energy wireless technologies.

If a vehicle wants to report an event, it must send the event through broadcast mode to the rest of the surrounding network, so that the event can be spread across multiple forwarding hops, achieving a rapid deployment in a decentralized way.

When an ad-hoc communication system is created by broadcast, it is possible that a lot of messages would be generated and can get the network overloading. Therefore, this paper uses the idea explained in [6] to avoid overloading the network. The idea is based on 1-hop clustering to reduce the number of VANET communications in dense road traffic scenarios while maintaining the security of communications. So security problems of VANETs are analyses in [12]. In order to certify the authenticity of the users in the ad-hoc network, an authentication scheme has been used [14], based on a Non-Interactive Zero-Knowledge Proof). The new scheme is also used for the establishment of shared keys using the notion under the Diffie-Hellman protocol using Elliptic Curves [4]. The Elliptic Curve Diffie-Hellman (ECDH) Protocol is a variation of the original Diffie-Hellman protocol, which uses the properties of the elliptic curves defined over finite fields.

In this way, the two users agree beforehand on the use of a prime number p, an elliptic curve E defined over Z_p and a point $P \in E$.

Then, the users A and B choose as secret keys two random numbers belonging to Z_p, being these $a \in Z_p$ and $b \in Z_p$.

Later, they obtain their public keys multiplying their secret keys by the point P previously agreed. The next step is the exchange of their public keys in order to compute later the shared key by multiplying their private key by the public key of the other user, obtaining both the same shared key. The strength of this protocol lies on the difficulty to solve the Elliptic Curve Discrete Logarithm Problem. However, since this protocol derives from the original Diffie-Hellman protocol, it is susceptible to suffer the same attacks. On the other hand, to encrypt all communications the SNOW 3G stream cipher is used.

SNOW 3G is the stream cipher algorithm designated in 2006 as basis for the integrity protection and encryption of the UMTS technology. Thanks to the fact that the algorithm satisfies all the requirements imposed by the 3GPP (3rd Generation Partnership Project) with respect to time and memory resources, it was selected for the UMTS Encryption Algorithm 2 (UEA2) and UMTS Integrity Algorithm 2 (UIA2) [13,15].

The SNOW 3G algorithm derives from the SNOW 2 algorithm, and uses 128-bit keys and an initialization vector in order to generate in each iteration 32 bits of keystream. The LFSR used in this algorithm has 16 stages denoted s0, s1, s2, s15 with 32 bits each one. On the other hand, the used FSM (Finite State Machine) is based on three 32-bit records denoted R1, R2 and R3 and uses two Substitution-boxes called S1 and S2.

For large networks with more than 100 nodes, the protocol can be applied with fewer nodes if clusters [5] are used.

On the one hand, different sensors are deployed within the road to act as RSU. These sensors are responsible to obtain different measurements related to the road such as temperature, humidity, flow of vehicles, etc., which are transmitted to the smartphones using the Bluetooth Low Energy API and then are processed in the smartphone using fuzzy logic based on different thresholds to determine whether something has happened in the road. This procedure is applied for example, to determine whether the road is frozen or wet, or whether the traffic flow in the last minutes in some point of the road involves determining that a traffic jam is being created. Then, the corresponding event is announced to the server responsible to warn all nearby vehicles.

On the other hand, a sensor network inside the vehicle allows obtaining different measurements related to the vehicle, such as speed, pressure of tires, gasoline level, oil level, etc., and deducing whether something is on a seat and the corresponding seat belt is being used, including baby chairs, etc. This system uses an OBD2, Arduino and Intel Edison for the different sensors and communication of the obtained measurements through BLE to the mobile device where the data is processed using fuzzy logic to determine what is happening inside the vehicle. In this case, except when a failure in the vehicle happens, the warning is only for the driver of the vehicle and not sent to the server. See Fig. 1.

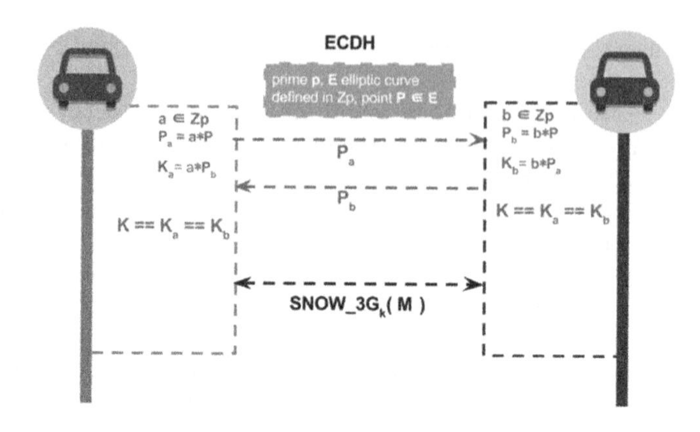

Fig. 1. Communication flow

5 Implementation and Analysis

The proposed system is based on a client-server model. On the one hand the client part has been developed in Android for the 4.0 version or higher. This is an Open Source Project where the Application Programming Interfaces (APIs) like Google Maps v3.0 or Google Cloud Messaging are used. On the other hand, the server was developed using JavaScript technologies through frameworks like 'node.js' and 'express.js'. As a database for all the data centralized on this server, the decision was adopting a No SQL database, specifically MongoDB. Finally, the server was deployed on an instance of Amazon Web Services, specifically under Ubuntu machine with Amazon EC2 account.

The mobile application (see Fig. 2) alerts drivers through sound messages regarding events happening near them, so no need exists to put their attention on the mobile device screen, which at any time displays a map with real-time events.

The deployed application automatically detects various events, such as:

- **Collisions between vehicles:** The application uses internal sensors gyroscopes and accelerometers. In this way the measuring changes on them are calculated and the G-Force is computed on the smartphone. In this proposal 2.7 G-Forces is set as the minimum threshold to consider the existence of a collision because that is the average of G-Forces in a collision between vehicles [7,8].
- **Traffic congestion:** Two different mechanisms interact in this part. On the one hand, the smartphone analyses the speed continuously and it is able to detect variations. When the speed is below a threshold over a certain period of time, the application generates an event of possible congestion. If several vehicles generate the same type of event in the same area and direction, the traffic congestion is considered confirmed. On the other hand, sensors on the road allow analysing the flow of vehicles in different zones.

Fig. 2. Application screenshots

- **Road sign:** The application has preloaded certain road signs in dangerous areas. Through geo-location, the smartphone knows whether it is in the range of these signs and in this case, it recommends the user to act accordingly. Apart of this, some BLE sensors with the information of the road sign are spread on the road.
- **Release of a parking space:** The smartphone detects when the vehicle has been started and is leaving the space where it was parked. Then, it generates an event announcing the release of a parking space.
- **Approach of/to special vehicles:** The application can be used to generate a special vehicle mode, such as: emergency vehicle, heavy vehicle or even cycling mode. In this way, when the application detects the approximation of/to a special vehicle, it warns the user.
- **Traffic lights:** The application can detect the state of a traffic light using light sensors that transmit through BLE the state and the colour of the traffic light. The application tells the driver how to act with respect to the traffic light and in the case the user does not stop at a red traffic light, neighbour vehicles are anonymously warned about the fact that somebody is skipping traffic lights.
- **Baby at a baby chair:** The application detects when a baby is sit or not in a baby chair and whether the seat belt is closed or not. To do this, a weight sensor is used that is configured with the baby weight, and a reed sensor detects whether the seat belt is closed or not. This information is sent through BLE to the driver smartphone to warn in case of danger.
- **Public transport stop:** The application has a pedestrian mode and when this mode is enabled, it allows detecting nearby public transport stops so that when near, the application automatically shows when the next bus, subway, tram or train will arrive. To do this, beacons are used that transmit over BLE in the public transport stop and the mobile application receive them when near.

- **Geolocated advertisements:** Different kinds of advertisements can be defined within the road. When the system detects a nearby advertisement, it sends the advertisement information to the smartphone. The user can filter by categories and decide whether to receive or not each advertisement.
- **Exceeded speed limit:** The application has preloaded speed limits for each road. It checks the vehicle speed continuously and compares it with the corresponding limit so that if the vehicle speed is higher than the speed limit of the road, the system generates an automatic notification of this fault and warns to the driver.

Other events are detected and computed on the server through the constant information being sent from all smartphones that make up the vehicular network. If the online mode is not available, some of these events are detected and computed by the own vehicles through the information received from the vehicles nearby.

The system has two communication modes to events advertising.

- **Zone communication:** This type of communication is used to share information in a zone, independently of the direction of the vehicles. In this way, if a vehicle generates an event of this type, the information is sent to all nearby vehicles. The system collects the last tracking of the last minute of the nearby vehicles and sends the event.
- **Direction communication:** When the system wants to share some information for one road and one way and direction, first of all the system calculates the nearby vehicles as in the previous communication through the haversine formula [16].

$$haversin(\frac{d}{R}) = haversin(\gamma_A - \gamma_B) + cos(\gamma_A)cos(\gamma_B)haversin(\Delta\lambda) \quad (1)$$

where haversin is the haversine function:

$$haversin(\theta) = sin^2(\frac{\theta}{2}) = \frac{1 - cos(\theta)}{2} \quad (2)$$

d is the distance between two points (over the bigger circle of the sphere), R is the sphere radio, in our case the Earths radio, γ_A is the latitude of the point A, γ_B is the latitude of the point B and $\Delta\lambda$ is the difference of the longitudes. Finally, the distance (d) is:

$$d(A,B) = R * arccos(sin(\gamma_A) * sin(\gamma_B) + cos(\gamma_A) * cos(\gamma_B) * cos(\Delta\lambda)) \quad (3)$$

Then system calculates the orientation of each vehicle taking into account the two last tracking points. Taking into account if $\Delta\lambda > 180$ then $\Delta\lambda = (mod 180)$.

$$\Delta\gamma = ln(\frac{tan(\frac{\gamma_B}{2} + \frac{\pi}{4})}{tan(\frac{\gamma_A}{2} + \frac{\pi}{4})}) \quad (4)$$

When the system has all the affected vehicles in a specific route in one way, it sends the information of the event. The system calculate this kind of communication mode to events advertising in real time for each vehicle based on fuzzy logic rules, as in the generation of a new event as in the static notifications.

DEPHISIT packet format is shown in Fig. 3, where each packet sent to notify an event has a size of 29 bytes.

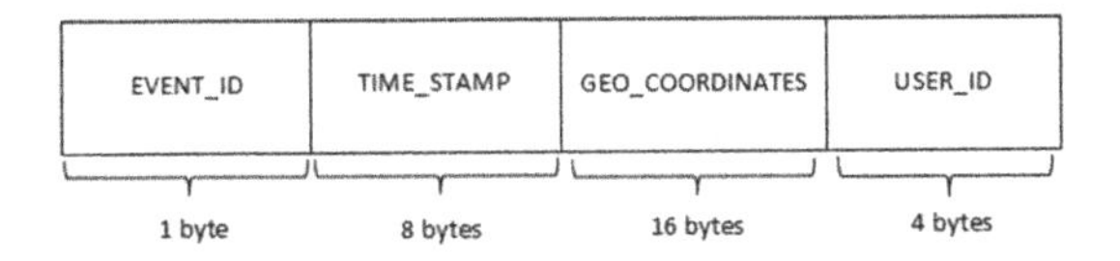

Fig. 3. Event Packet

The package contains the following information:

- EVENT_ID: Event identifier. To discern what type of event it is.
- TIME_STAMP: Date of generation of the event, when it occurred.
- GEO_COORDINATES: Position where the event occurred. Latitude and Longitude.
- USER_ID: User identifier. It identifies the user that detected the event.

Different batteries of tests were used d to check the time processing and a real test with vehicles was done that produced promising results. As a result, the times represented by the averages of all tests in online mode, shown in Table 1, proved that efficiency has been achieved.

Table 1. Average times

Action	Time (in Milliseconds)
Smartphone data event detection and processing	51
Sending event from smartphone to cloud server	116
Cloud server data processing	104
Notification push from cloud server to smartphones	142
Total	**413**

With respect to the offline mode, the communication can be done in two ways: via Bluetooth Low Energy or via WiFi Direct, based on the technology that current smartphones have available. Whenever possible, the channel created by WiFi Direct, due to its higher rate of speed, will be used. In addition to its greater range. Bluetooth Low Energy has a transmission rate of 25 Mbps and WiFi Direct has a transmission rate of 250 Mbps. The maximum range of Bluetooth Low Energy Communication is 60 meters, while WiFi Direct has a range of 200 meters.

6 Conclusions

This work proposes a new solution to help to drive based on ICT tools. It can be considered the first complete system that combines the needs of all kind of vehicles (conventional, emergency vehicles, special vehicles, bicycles, etc.) on all kind of roads (urban, rural, motorways, secondary, national, regional, interregional, local, etc.) in all possible scenarios (either public or private) and under all possible circumstances (bad weather, special roads, tunnels, interaction with ferries/trains, big events, tragedies, etc.). The deployment of a VANET according to the traditional definition, with OBUs and RSUs, has proven not to be trivial due to the complex technical, economic and social requirements that are necessary. Therefore, this paper proposes a practical and low-cost system to generate a VANETs by using only smartphones. The described platform is a hybrid network architecture, which allows V2V communication either directly or through the cloud, and I2V communication based on road sensors. The proposed system has been implemented on the Android platform, with a basic security level including encryption and authentication whenever necessary. Since this is part of work in progress, the next version of this work will include a complete analysis of possible attacks on the platform, and new developed security solutions to protect it against them.

References

1. Alam, M., Ferreira, J., Fonseca, J. (eds.): Intelligent TransportationSystems: Dependable Vehicular Communications for Improved Road Safety. Studies in Systems, Decision and Control, vol. 52. Springer, Heidelberg (2016)
2. Android Auto. www.android.com/auto. Accessed Oct 2015
3. Apple Car Play. www.apple.com/es/ios/carplay. Accessed Oct 2015
4. Batra, M.K., Bhatnagar, P.: Improved Diffie-Hellman Key Exchange Using Elliptic Curve (IDHECC). Scheme for Securing Wireless Sensor Networks Routing Data (2016)
5. Caballero-Gil, C., Caballero-Gil, P., Molina-Gil, J.: Knowledge management using clusters in VANETs-description, simulation and analysis. In: International Conference on Knowledge Management and Information Sharing, pp. 170–175 (2010)
6. Caballero-Gil, C., Caballero-Gil, P., Molina-Gil, J.: Self-organized clustering architecture for vehicular ad hoc networks. Int. J. Distrib. Sensor Netw. **2015**, 5 (2015)
7. Cassidy, J.D., Carroll, L.: Incidence, risk factors and prevention of mild traumatic brain injury: results of the WHO collaborating centre task force on mild traumatic brain injury. J. Rehabil. Med. **36**, 28–60 (2004)
8. Castro, W.H., Schilgen, M., Meyer, S., Weber, M., Peuker, C., Wörtler, K.: Do "whiplash injuries" occur in low-speed rear impacts? Eur. Spine J. **6**(6), 366–375 (1997)
9. ETSI: Intelligent Transport Systems. http://www.etsi.org/technologies-clusters/technologies/intelligent-transport. Accessed Aug 2015
10. Final draft ETSI ES 202 663 V1.1.0 (2009–2011). European Telecommunications Standards Institute. Accessed 16 Apr 2013
11. IEEE 1609 - Family of Standards for Wireless Access in Vehicular Environments (WAVE). U.S. Department of Transportation, 13 April 2013. Accessed 14 Nov 2014

12. Jawandhiya, P.M., et al.: A survey of mobile ad hoc network attacks. Int. J. Eng. Sci. Technol. **2**(9), 4063–4071 (2011)
13. Kitsos, P., Selimis, G., Koufopavlou, O.: High performance ASIC implementation of the SNOW 3G stream cipher. IFIP/IEEE VLSI-SOC, pp. 13–15 (2008)
14. Martin-Fernandez, F., Caballero-Gil, P., Caballero-Gil, C.: Non-interactive authentication and confidential information exchange for mobile environments. In: Herrero, Á., Baruque, B., Sedano, J., Quintián, H., Corchado, E. (eds.) International Joint Conference. Advances in Intelligent Systems and Computing, vol. 369, pp. 261–271. Springer, Heidelberg (2015)
15. Orhanou, G., El Hajji, S., Bentaleb, Y.: SNOW 3G stream cipher operation and complexity study. Contemp. Eng. Sci. Hikari Ltd. **3**, 97–111 (2010)
16. Robusto, C.C.: The cosine-haversine formula. Am. Math. Mon. **64**(1), 38–40 (1957)
17. Status of Project IEEE 802.11 Task Group p: Wireless Access in Vehicular Environments. IEEE. 2004–2010. Accessed 10 August 2011
18. Tomtom: www.tomtom.com. Accessed Oct 2015
19. Waze: www.waze.com. Accessed Oct 2015
20. Yousefi, S., Mousavi, M.S., Fathy, M.: Vehicular ad hoc networks (VANETs): challenges and perspectives. In: 2006 6th International Conference on ITS Telecommunications, pp. 761–766. IEEE, June 2006

Learning Deep Wavelet Networks for Recognition System of Arabic Words

Amira Bouallégue[1(✉)], Salima Hassairi[2], Ridha Ejbali[2], and Mourad Zaied[2]

[1] Higher Institute of Computer Science and Multimedia, University of Gabes, Gabès, Tunisia
amira.bouaallegue@gmail.com
[2] REsearch Group in Intelligent Machines, B.P 1173, Sfax, Tunis, Tunisia
{salima.hassairi.tn,ridha_ejbali,mourad.zaied}@ieee.org

Abstract. In this paper, we propose a new method of learning for speech signal. This technique is based on the deep learning and the wavelet network theories. The goal of our approach is to construct a deep wavelet network (DWN) using a series of Stacked Wavelet Auto-Encoders. The DWN is devoted to the classification of one class compared to other classes of the dataset. The Mel-Frequency Cepstral Coefficients (MFCC) is chosen to select speech features. Finally, the experimental test is performed on a prepared corpus of Arabic words.

Keywords: Deep learning · Wavelet Networks · Arabic words · Speech recognition · Auto-encoder

1 Introduction

Nowadays, signal processing research is developed by several researchers. There are several greatly developed approaches related to speech recognition in the last years. The progress of the speech recognition system is the outcome of different techniques such as the wavelet network (WN). The latter is the combination of the wavelet and the neural networks' theories and consists of a feed-forward neural network. Moreover, the WN uses one hidden layer and its concept spread thanks to the work of Pati [1], Zhang [2] and many other researchers. Among the other methods of recognition systems that are used frequently nowadays is the Deep Learning (DL). The DL has emerged as a new area of machine learning research [3]. Since 2006, DL has been more recently referred to as a representation learning and supervised learning for classification tasks [4,5]. The most used algorithms of DL in the speech recognition field most frequently used by Bengio and Hinton are the Restricted Boltzmann Machines (RBM) and the Deep Belief Network (DBN). The notion of the Auto-encoders was developed by Bengio [1,6] and used for efficient learning coding. This method involves treating each neighboring set of two layers as RBMs so that the pre-training approximates a good solution. Then, it involves using a back-propagation technique to fine-tune

M. Graña et al. (eds.), *International Joint Conference SOCO'16-CISIS'16-ICEUTE'16*,
Advances in Intelligent Systems and Computing 527, DOI 10.1007/978-3-319-47364-2_48

the results [7]. In addition, some researchers working with the speech recognition systems discussed some techniques of feature extraction like the MFCC and the Linear Predictive Coding (LPC). The parameters of the MFCC constitute the nature of speech while they extract the features [8]. So, Our contribution consists of the creation of a new recognition system of Arabic words using some algorithms of the DL and the WN. These algorithms are used in the training and testing phase of our approach. This paper is organized as follows: In Sect. 2, we explain in details the feature extraction based on the MFCC and we describe the idea of the deep learning algorithm of our new approach. The experimental results and discussions are mentioned in Sect. 3. Finally, this work ends with a conclusion.

2 The Proposed Approach

2.1 System Overview

The idea of the new proposed architecture can be summarized in three major phases: Parameterization phase, training phase and recognition phase. Figure 1 illustrates the general architecture of our system and the phases that lead to the creation of the DWN.

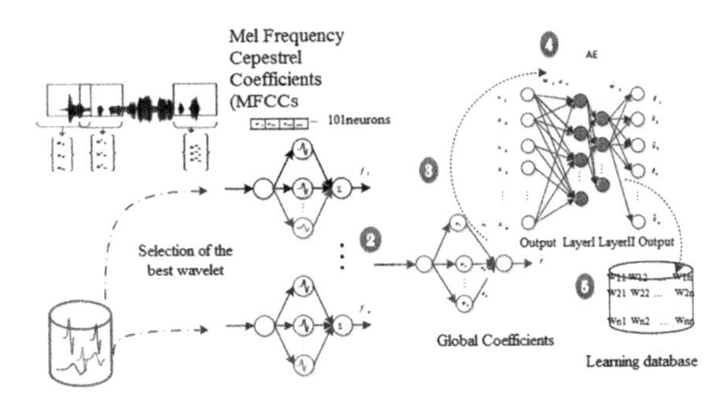

Fig. 1. General architecture of our system.

2.2 Parameterization Phase

Parameterization is the first step in all the process of speech recognition. Principally, parameterization is intended to convert the speech signal into a set of perceptually meaningful features [13]. In that event, there are several signal analysis techniques. Among these, our system is based on MFCCs [9]. The speech signals in our system are processed in a window of 32 ms with an overlap of 10 ms. Figure 2 explains the overall extraction procedure of MFCCs. The MFCC coefficients contain features around Fast Fourier Transform (FFT) and Discrete Cosine Transform (DCT) converted on Mel scale [8]. This method is used mostly to represent a signal in speech recognition because of its robustness. Its advantage is that the coefficients are uncorrelated.

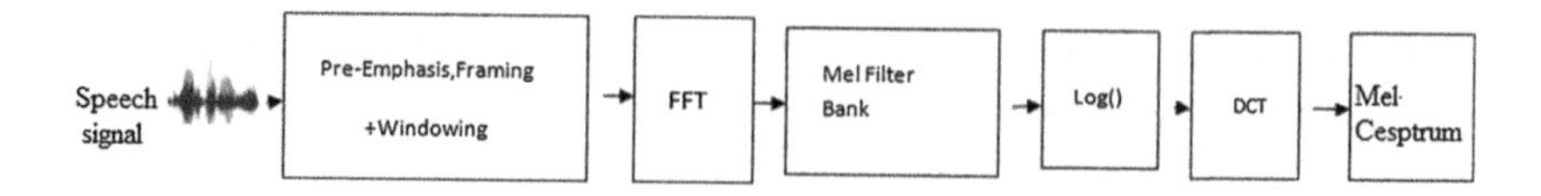

Fig. 2. MFCCs' calculation

2.3 Training Phase

In this part, we introduce the techniques and algorithms that we used for the construction of our deep Wavelet network.

The Best Contributions Algorithm. In stage 1, we present the way to create a wavelet network for each element of MFFC coefficients of all classes of the dataset using the best contribution algorithm [10,11]. Figure 3 learning process consists of learning a 1D wavelet network using the fast wavelet transform (FWT) [12,14].

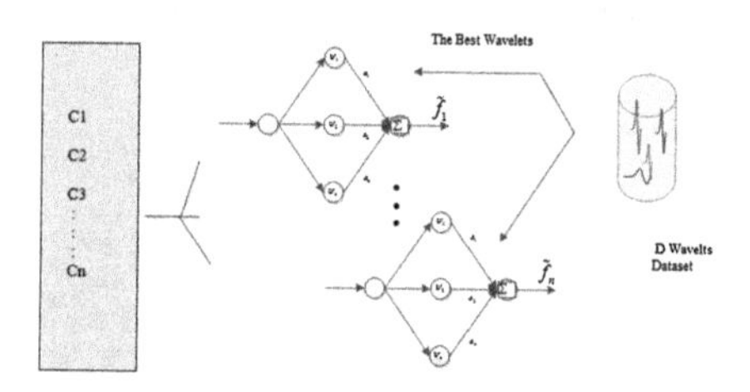

Fig. 3. Wavelet network for a single element of a class

However, We calculate the value of the Peak Signal to Noise Ratio (PSNR) to verify each WN that represents well the input signal. As well, the output of the WN for one signal is illustrated by the Eq. (1):

$$\hat{f}_c = \sum_{i=1}^{n} a_i \psi_i \tag{1}$$

Calculate the Wavelets Scores. In stage 2, we account for the score of wavelets to create a Global wavelet network (GWN) that approximates only the signals of one class in the database (noted Class 1). Then, we divide the wavelet networks obtained previously into two tables [14]. One table contains the wavelet networks of signals of Class 1 and the number of neurons that contain a wavelet can vary for each wavelet network created in step1 as indicated in Fig. 4 and the same applies to the second table that contains all the other signals of all remaining classes of the dataset (noted Class 2).

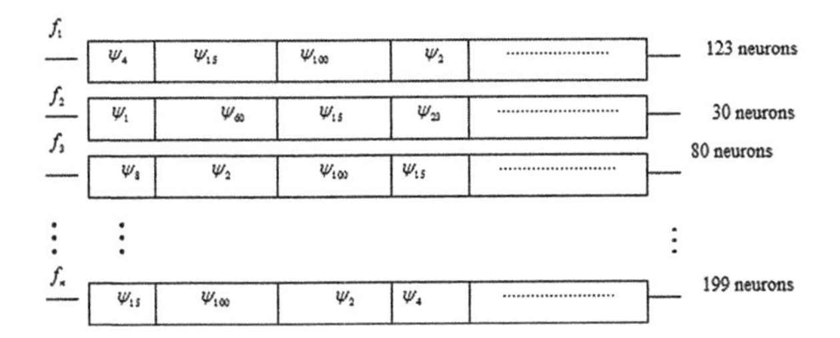

Fig. 4. Tables of Wavelet networks for Class 1

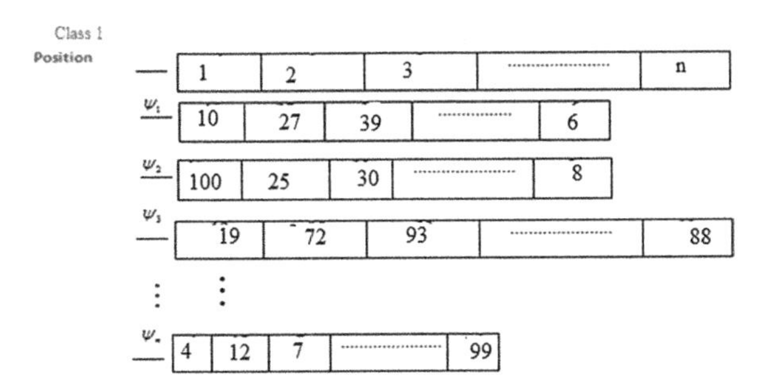

Fig. 5. Tables of the number of apparition for each Wavelet of Class 1 and Class 2

After that, we calculate the number of occurrences of all wavelets in Class 1 in every position and similarly for all wavelets from Class 2. Then, we create a new table that contains the best wavelets to create our GWN. The best wavelet is the wavelet that has the maximum score values in each position (Fig. 5). After the phase of calculating wavelet scores, we create a new GWN for one class as presented in Fig. 6.

Wavelets	ψ_2	ψ_3	....	ψ_n
Global Coefficients	100	72	....	99

Fig. 6. Tables of the best coefficient of each wavelet

To create a GWN, we use the wavelet which has the best coefficient Fig. 7. In this stage, each class of the dataset will be defined with a GWN and to check the efficiency of each network, we calculate the value of the PSNR.

From the GWN to the Wavelet AE. In this stage, we transform our GWN to a Wavelet AE. Figure 8 shows the wavelets that we used for the GWN are used to create the Wavelet AE. In the Wavelet AE, we use the bi-orthogonal wavelets which are generated by the wavelet ψ mother and a dual wavelet $\tilde{\psi}$ [15].

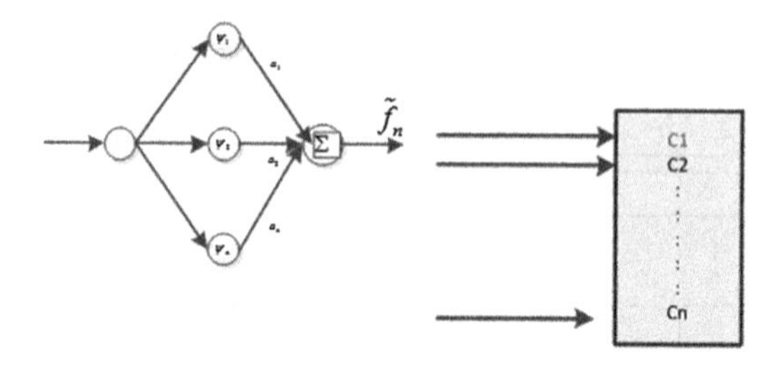

Fig. 7. Global Wavelet network for a class

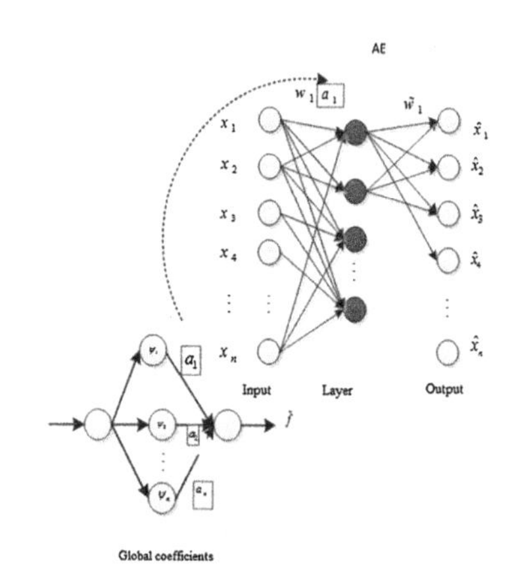

Fig. 8. A wavelet AutoEncoder for a class

So, the result signal of the Wavelet AE is obtained as follows:

$$\begin{aligned} \prec \psi_i, \tilde{\psi}_j \succ &= 0 \quad if \quad i \neq j \\ &= 1 \quad elseif \end{aligned} \tag{2}$$

$$f = x_1, x_2, ...x_n$$
$$\hat{f} = \hat{x}_1, \hat{x}_2, ...\hat{x}_n$$
$$\psi_i = w_{1i}, w_{2i}, w_{3i}, ...w_{ni}$$
$$\tilde{\psi}_i = \tilde{w}_{1i}, \tilde{w}_{2i}, \tilde{w}_{3i}, ...\tilde{w}_{ni}$$

$$then \qquad a_i = \prec f, \psi_i \succ \Rightarrow a_i = \sum_{j=1}^{n} w_{ji} x_j$$

$$and\ finnaly, \qquad \hat{x} = \sum_{j=1}^{n} \tilde{w}_{ji} a_j$$

Creation of a Deep Wavelet Network (DWN). After a series of stacked Wavelet auto-encoders, we create our Deep Wavelet Network (DWN). Figure 9 illustrates the process of creation of a DWN with two hidden layers.

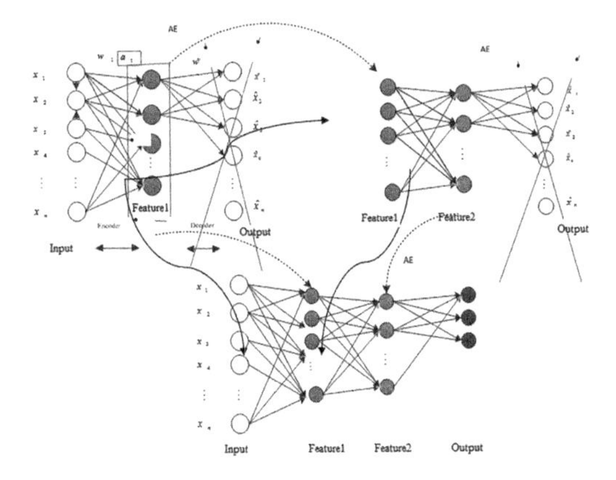

Fig. 9. A deep neural network with two hidden layers

2.4 Recognition Phase

The recognition process (Fig. 10) follows the same procedures of the parameterization and the results of the training phase.

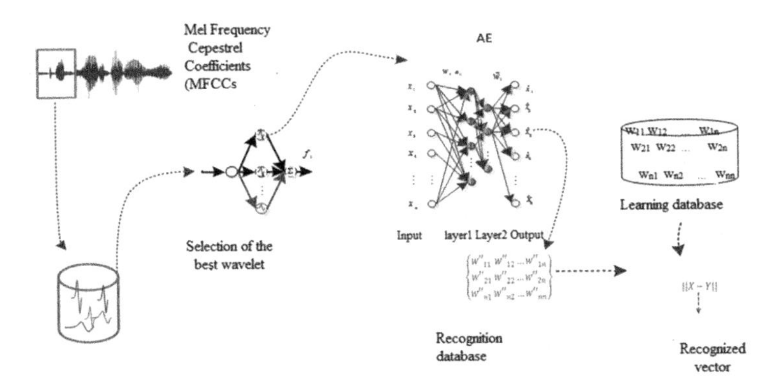

Fig. 10. Recognition phase

So, the recognition algorithm precedes the windowing of every speech signal to extract the acoustic features. Then, it tries to approximate these sets of features with the results of the DWN of the training phase. Furthermore, the approximation parameters are used to identify the pronounced speech. Actually, the result of recognition depends on the similarity between the approximation parameters of the signal to be recognized and each set of parameters in the training base. To calculate the distance between two networks, we adopt the formula of the Euclidian Distance.

$$D = \left\| a_{i=1}^{n} V_i Y_i - a_{j=1}^{n} W_j Y_j \right\| \tag{3}$$

where V_iY_i et W_jY_j are two networks. Moreover, the Euclidian distance between two networks using the same wavelet functions can be expressed as follows:

$$D = (D^t(Y_{i,j})D)^{1/2} \tag{4}$$

with $D = d_1...d_n^t$ et $Y_{i,j} =< Y_i, Y_j >$.
Thus, the distance between two networks with the same wavelet functions is actually the difference between their weights.

3 The Experiment's Results

We have tested our method on a corpus recorded from works of Boudraa in 1998 [16]. We segmented this corpus manually by PRAAT to Arabic words and we chose 18 different words. In our approach, (2/3) of them were used in the training phase and the rest (1/3) in the test phase of the dataset.

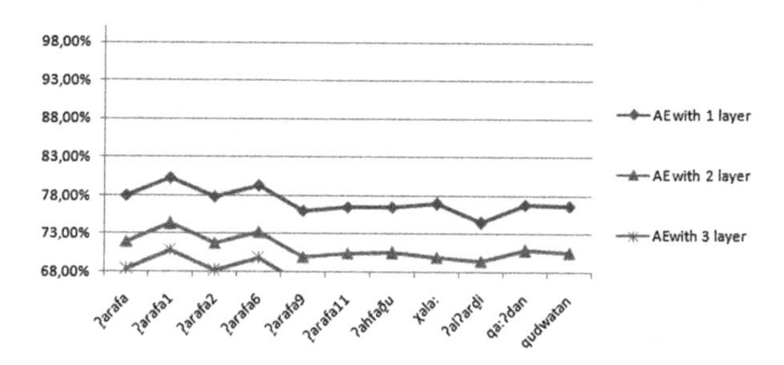

Fig. 11. The values of the PSNR of Class 1 and Class 2 with the Wavelet AE that approximates Class 1

Thus, we notice that the PSNR is good for both classes. Despite that, we rebuilt the global network from class 1. We found that Class 1 and Class 2 have almost the same wavelets. For this reason, we can explain the performance of the signal quality in both classes. In addition, the overall rate obtained by the PSNR of Class 1 is better than that in Class 2 (Fig. 11). In this first experiment, we show that the AE designated and proposed for a class approximates well the signals of the same class.

From Figs. 12 and 13, we show that the Wavelet AE gives a good performance if we compare the original signal to the reconstructed signal, but with increasing numbers of hidden layers, we find that the quality of the reconstructed signals gradually decrease in both classes through the increase of the error and degradation at its level.

The Fig. 14 show the recognition rate given by the DWN compared to the intelligent approach [17] and the WN [18,19]. The method of WN system is a hybrid classifier composed of a neuronal contraption and wavelets as functions

Arabic word	ʔarafa	ʔarafa1	ʔarafa2	ʔarafa6	ʔarafa9	ʔarafa11
AE with 1 layer	77.99 %	80.35%	77.81%	79.36%	76.03%	76.53%
AE with 2 layer	71.97%	74.33%	71.79%	73.34%	70.01%	70.51%
AE with 3 layer	68.44%	70.81%	68.27%	69.81%	66.48%	66.99%

Fig. 12. Table of the values of PSNRs of Class 1 of the AE that approximates Class 1

Arabic word	χala:	ʔalʔarḍi	qa:ʔdan	qudwatan	ʔahfaðu
AE with 1 layer	76.07%	74.50%	76.93%	76.74%	76.59%
AE with 2 layer	70.05%	69.48%	70.91%	70.72%	70.57%
AE with 3 layer	66.52%	65.59%	67.38%	67.19%	67.04%

Fig. 13. Table of The values of PSNRs of Class 2 of the Wavelet AE that approximates Class 1

Arabic word	ʔarafa	χala:	qa:ʔdan	qudwatan	ʔalkabʃa	biθima:riha
WN	84,40%	97,40%	92,50%	93,50%	98,10%	85,40%
Intelligent Approach	92,50%	99,50%	94,50%	92,00%	99,50%	86,00%
Proposed Approach	94,1%	97,81%	95,3%	98,70%	99,6%	87,1%

Arabic word	ba:lana	Biqawlin	ʁabaṭˁa	ka:la	Kunta	Mina	Lahum
WN	87,50%	89,00%	95,00%	83,00%	98,00%	91,20%	89,10%
Intelligent Approach	88,50%	89,95%	96,00%	87,50%	98,00%	97,80%	94,90%
Proposed Approach	94,80%	90.01%	96,6%	97.81%	99,7%	99,89%	95,4%

Arabic word	minkuma:	ṣˁa: ʔiman	ʔalmusa:firu:na	la ða ʔˁathu	zama:nuna:
WN	89,30%	93,70%	83,40%	85,70%	93,50%
Intelligent Approach	85,10%	93,70%	82,50%	84,90%	94,70%
Proposed Approach	87,88%	95,44 %	87,90%	86,4%	96,01%

Fig. 14. Tables of Speech recognition rates using a DWN using MFCC coefficients.

of activation. the purpose of this method is that each acoustic vector will be modeled by WN.

To evaluate the performance of our system, it was compared to other methods in literature. We find, from the tables in Figs. 14 and 15, that our approach provided better recognition rates of the intelligent approach and the WN on Arabic words using MFCC coefficients.

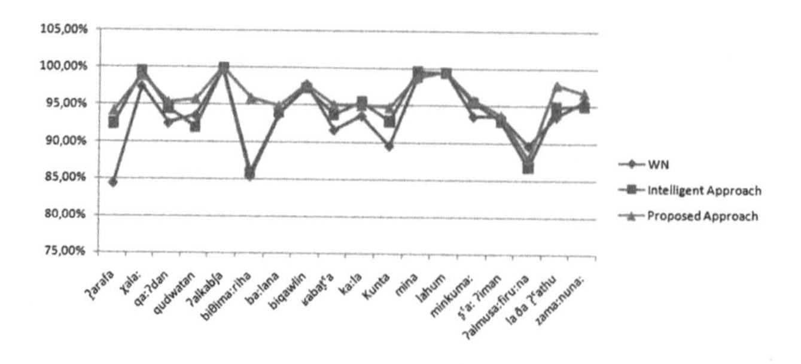

Fig. 15. Comparison of previous works to the novel Approach using MFCC coefficients.

4 Conclusion

This work presents a recognition system of words on three major phases: Parameterization, Training and Recognition. This system leads to the creation of a DWN for each class in the dataset to recognize all the signals of the related class. The architecture of the DWN is obtained after the combination of two theories in the recognition system: the deep learning and the wavelet network. Among the algorithms of the DL used in our approach is the Stacked AE and among the algorithms of the Wavelet Networks are the Best Contribution Algorithms and the Fast Wavelet Transform. The experimental results of our study have demonstrated that the deep wavelet network show enhanced performance with the MFCC to represent acoustic data.

Acknowledgment. The authors would like to acknowledge the financial support of this work by grants from General Direction of Scientific Research (DGRST), Tunisia, under the ARUB program.

References

1. Pati, Y.C., Krishnaprasad, P.S.: Discrete affine wavelet transforms for analysis of feed-forward neural networks. In: Proceedings of the 1990 Conference on Advances in Neural Information Processing Systems, Denver, Colorado, United States, vol. 3, pp. 743–749 (1991)
2. Zhang, Q., Benveniste, A.: Wavelet networks. IEEE Trans. Neural Netw. **3**(6), 889–898 (1992)
3. Hinton, G., Osindero, S., Teh, Y.: A fast learning algorithm for deep belief nets. Neural Comput. **18**, 1527–1554 (2006)
4. Bengio, Y., Courville, A., Vincent, P.: Representation learning: a review and new perspectives. IEEE Trans. Pattern Anal. Mach. Intell. **35**, 1798–1828 (2013)
5. Hinton, G., Deng, L., Dong, Y., Dahl, G.E., Mohamed, A., Jaitly, N., Senior, A., Vanhoucke, V., Nguyen, P., Sainath, T.N., Kingsbury, B.: Deep neural networks for acoustic modeling in speech recognition. In: The Shared Views of Four Research Groups IEEE Signal Processing Magazine, November 2012

6. Liou, C.-Y., Cheng, W.-C., Liou, J.-W., Liou, D.-R.: Autoencoder for words. Neurocomputing **139**, 84–96 (2014)
7. Reducing the Dimensionality of Data with Neural Networks (Science, Hinton and Salakhutdinov), 28 July 2006
8. Shrawankar, U., Thakare, V.: Techniques for feature extraction in speech recognition system: acomparative study. Int. J. Comput. Appl. Eng. Technol. Sci. (2010)
9. Joseph, W.: PICONE: signal modeling techniques in speech recognition. Proc. IEEE **81**(9), 1215–1247 (1993)
10. Zaied, M., Said, S., Jemai, O., ben Amar, C.: A novel approach for face recognition based on fast learning algorithm and wavelet network theory. Int. J. Wavelets Multiresolut. Inf. Process. **9**(6), 923–945 (2011)
11. Jemai, O., Zaied, M., Amar, C.B., Alimi, A.M.: FBWN: an architecture of fast beta wavelet networks for image classification. Int. Joint Conf. Neural Netw. July 2010
12. ElAdel, A., Ejbali, R., Zaied, M., Amar, C.B.: A new system for image retrieval using beta wavelet network fordescriptors extraction and fuzzy decision support. In: International Conference of Soft Computing and Pattern Recognition, pp. 232–236 (2014)
13. LeCun, Y.: Learning invariant feature hierarchies. In: Fusiello, A., Murino, V., Cucchiara, R. (eds.) ECCV 2012. LNCS, vol. 7583, pp. 496–505. Springer, Heidelberg (2012). doi:10.1007/978-3-642-33863-2_51
14. Jemai, O., Ejbali, R., Zaied, M., Amar, C.B.: A speech recognition system based on hybrid wavelet network including a fuzzy decision support system. In: Proceedings of SPIE - The International Society for Optical Engineering, February 2015
15. Hassairi, S., Ejbali, R., Zaied, M.: Supervised image classification using deep convolutional wavelets network. In: 2015 IEEE 27th International Conference on Tools with Artificial Intelligence (2015)
16. Boudraa, M., Boudraa, B.: Twenty list of ten arabic sentences for assessment. ACUSTICA Acta Acoustica **86**(43.71), 870–882 (1998)
17. Ejbali, R., Zaied, M., Ben Amar, C.: Intelligent approach to train wavelet networks for recognition system of arabic words. In: KDIR International Joint Conference on Knowledge Discovery and Information Retrieval, Valencia, Spain, pp. 518–522, 25–28 October 2010
18. Ejbali, R., Zaied, M., Amar, C.B.: Multi-input multi-output beta wavelet network: modeling of acoustic units for speech recognition. Int. J. Adv. Comput. Sci. Appl. 3(3) (2012)
19. Ejbali, R., Zaied, M., Amar, C.B.: Wavelet network for recognition system of Arabic word. Int. J. Speech Technol. **13**(3), 163 (2010)

Intrusion Detection with Neural Networks Based on Knowledge Extraction by Decision Tree

César Guevara(✉), Matilde Santos, and Victoria López

Department of Computer Architecture and Automatic Control,
University Complutense of Madrid, 28040 Madrid, Spain
{cesargue,msantos}@ucm.es, vlopez@fdi.ucm.es

Abstract. Detection of intruders or unauthorized access to computers has always been critical when dealing with information systems, where security, integrity and privacy are key issues. Although more and more sophisticated and efficient detection strategies are being developed and implemented, both hardware and software, there is still the necessity of improving them to completely eradicate illegitimate access. The purpose of this paper is to show how soft computing techniques can be used to identify unauthorized access to computers. Advanced data analysis is first applied to obtain a qualitative approach to the data. Decision tree are used to obtain users' behavior patterns. Neural networks are then chosen as classifiers to identify intrusion detection. The result obtained applying this combination of intelligent techniques on real data is encouraging.

Keywords: Intrusion detection · Pattern recognition · Behavioral profile · Security · Decision tree · Neural networks

1 Introduction

Information networks have become the principal communication channel around the world. Intrusion Detection Systems (IDS) have the function to monitor a network activity for any possible intrusions that could potentially endanger the integrity of communication or information of the network [1]. These systems must be equipped with important features: multiple attack stability, quick turnaround time, and maintenance of network connection [2].

To address this problem, our approach is to obtain behavioral patterns of the authorized users in an automatic way, in order to control the access to those systems. We have applied decision trees for the identification of the user dynamic profiles and then Artificial Neural Networks (ANN) to classify and detect intrusions. This behaviour modeling demands a previous pre-processing and analysis of the available information. These steps provide knowledge of the behavior of the users that will help to test the final results of the detection system [3].

In the literature this problem has been addressed by artificial intelligent techniques but, as far as we know, there are only two close related works. The paper by [4] compares the performance of decision trees (C4.5) and neural network (ANN model),

M. Graña et al. (eds.), *International Joint Conference SOCO'16-CISIS'16-ICEUTE'16*,
Advances in Intelligent Systems and Computing 527, DOI 10.1007/978-3-319-47364-2_49

independently, in detecting network attacks in terms of detection accuracy, detection rate, and false alarm rate. They used KDD Cup 99 data as benchmark. According to the results, decision trees are more effective in intrusion detection than neural networks. The authors of [5] claim that most of the data mining methods applied to IDS, including rule-based expert systems, are not able to successfully identify the attacks which have different patterns from expected ones. Nevertheless, they propose a method based on the combination of decision tree algorithm and Multi-Layer Perceptron (MLP) neural network which is able to identify attacks with high accuracy and reliability. Other works have also applied only neural networks, such as [6, 7], or only decision trees [8].

In this paper we take advantage of both methods, using decision trees to obtain the behavioral pattern of the users and then neural networks to classify the activity of the user into the computer system as normal or intrusion. Moreover, we have tested it on real data with good results.

The paper is organized as follows. In Sect. 2 we describe the pre-processing and analysis of the data and the generation of a multidimensional data model. In Sect. 3, decision trees are applied to represent the behaviour of the users and neural networks to detect anomalous activities. Section 4 shows a simulation tool developed to apply the IDS. Conclusions and future work end the paper.

2 Pre-processing and Data Analysis

The information has been provided by a public institution of the Republic of Ecuador, where different users interact with the system and perform a variety of tasks. We are dealing with 18 different users along three years, 2010, 2011 and 2012. The database is made up of 118.067 records (operations on files) where 116.952 are validated as authorized operations and 1.115 activities have been manually classified as anomalous [3, 11].

The first step involves the selection of the most relevant features of the users' access. Even more, as the information is usually stored in different databases we have first integrated the information and put it together into one repository. This will make easier the tasks of sorting, cleaning and retrieving the information that will be used for the classification. The subsequent analysis and the success of the detection strongly depend on these initial phases.

2.1 Analysis of Variables

The variables should reflect the users' behavior when they are working on the computer system and describe their activities in a detailed manner. In the database there are already several attributes that allow us the definition of a behavioral pattern of a user, such as day of the week, computer used, time, session length, etc. Table 1 shows some of these features as follows: Day of the week, from 1 to 7; hour, from 8 to 20 (every hour); operation, from 1 to 4 (insert, update, delete and find); repository that is acceded, from 1 to 7.

Table 1. Attributes of the database that describes the behaviour of the user.

Attribute	Data type	Description
Logging	DateTime	Day and hour of login
Ip	Varchar	IP address of the workstation
Country	Varchar	Country
Time	DateTime	Day and hour specific task is carried out
Database accessed	Varchar	Table of database that has been acceded
Operation	Varchar	Operation performed in the database
User	Varchar	User performing the activities in the database

Figure 1 shows, as an example, the activity of the user, that is, number of times he has performed each of the operations insert, update, and delete. The same analysis has been carried out with other variables. The behaviour of a user has been recorded: number of times the user performs a specific operation on a database every day of the week, hour of access, and time he spends doing that operation, number of accesses, IP address of the workstation from which operations are performed, etc.

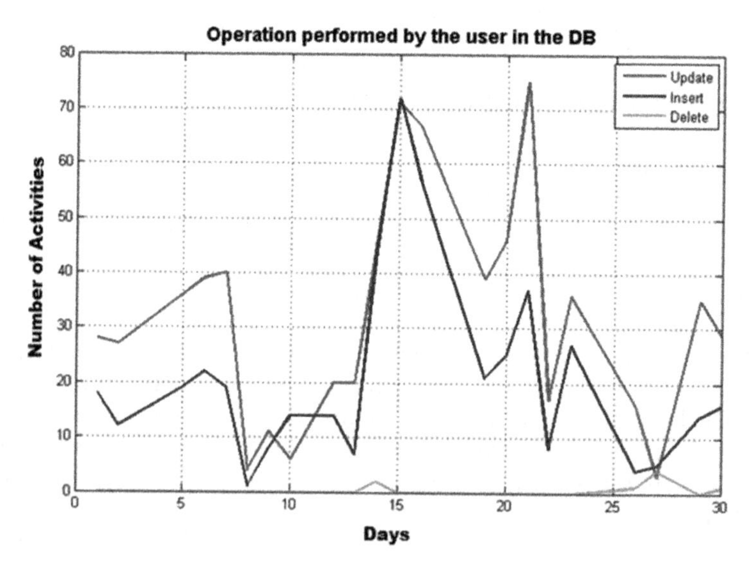

Fig. 1. Operations performed by a user in the DB: Update, Delete and Insert.

2.2 Multidimensional Model of the Data Repository

Several databases have been grouped into one call "tfmdata". This database is multi-dimensional and every dimension describes a set of relevant attributes regarding the access of each user to that information system. The result is what is called OLAP (Online Analytical Processing) [9]. OLAP is used to focus on important features, identify exceptions, or finding iterations.

Four dimensions or variables have been taken into account in the OLAP model: time, user, activity, and workstation (Fig. 2). Different tables are also associated to every user with information regarding the affiliation, department, ranking, position, etc. An activity is defined by a pair: operation and database. Workstation is just the IP and the type of device. The information was decomposed into the variables of an only database to integrate and unify it. This allows us to save space as some attributes had a very large format but little information.

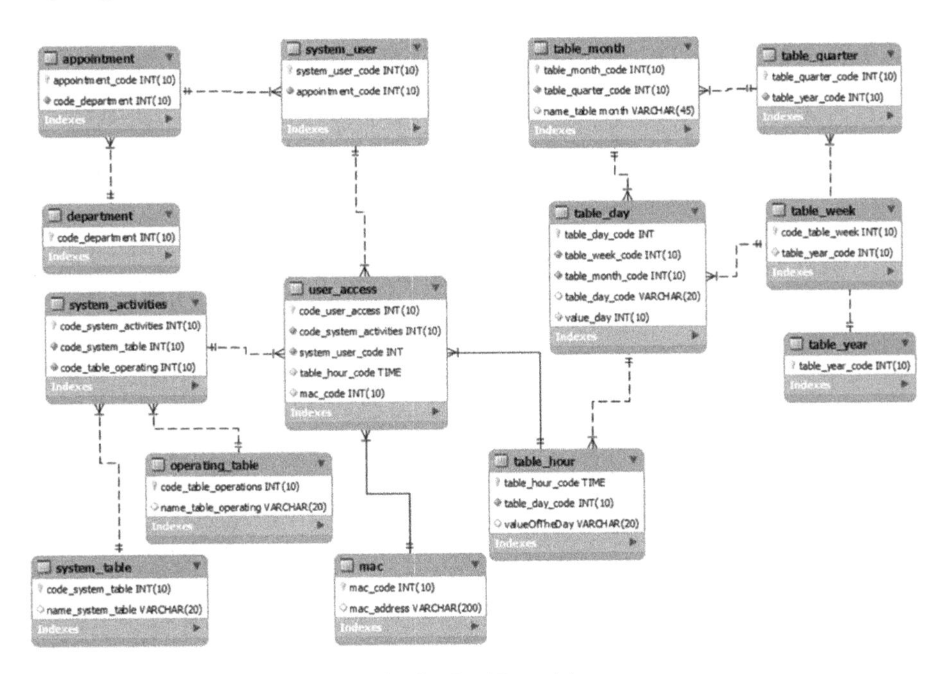

Fig. 2. OLAP model.

2.3 Exploratory Data Analysis

Once all the data are integrated into the OLAP repository, the following step is to analyze the characteristics of the attributes. It may be interesting to obtain some properties such as the maximum, minimum and average values and to represent them graphically. To perform this analysis in an easy, efficient and automated way the tool WEKA and Matlab were used. The result is called mining view and it will be very useful to perform the visual data mining and therefore to get an idea of the possible patterns, the relevant data, etc. This tool scans the data analysis algorithms and gives relevant information of each variable.

Figure 3 shows how the behavior of users #6, #8 and #11 is quite different regarding hours, days, tables, number of operations they perform, etc. For example, at a specific hour of a week day, let say 5 (Friday), each user can be doing a different operation on a different table. Other features are mainly the same every year for all the users, such as the work load. That is why some features that do not give discriminant information have not been considered.

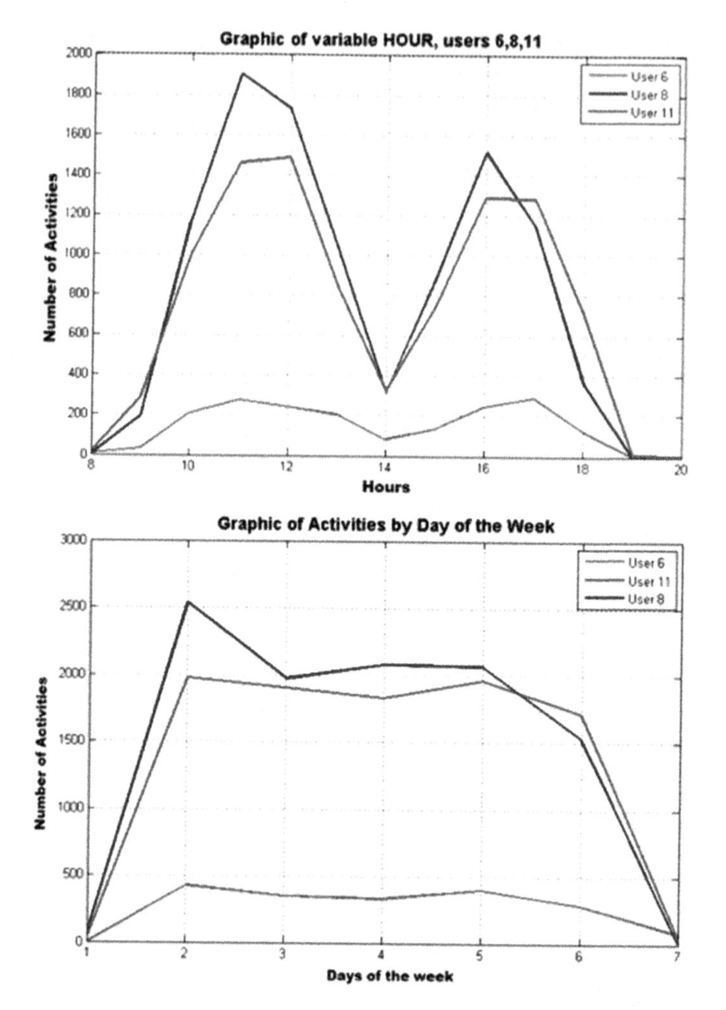

Fig. 3. (a) Variable "hour", (b) Variable "day of the week", for user 6 (green), 11 (red), 8 (blue). (Color figure online)

But the behaviour of each authorized users is not always so clear. Indeed, it is quite difficult to identify whether he/she is performing regular tasks or trying to leak information. Daily tasks can include sending out sensitive information. Even more, the same user can present very different behaviour depending on the tasks he is in charge of at that moment. The same user can show very different behaviour a month or another or even in the same period of time but different years.

3 Application of Intelligent Techniques to Model Users Behaviour

Knowledge extraction from the processed data is the basis to define the behavioral pattern of each user [10]. The user profile has to be unique, dynamic and close to reality. For that reason it is important to work with techniques that keep the models interpretable such as decision trees. The tree represents the expected behaviour of the user. The neural network carries out the final step, the classification. Besides, we are not only determining if the user is an intruder but also the degree of certainty of that prediction. Therefore we have combined these two knowledge-based techniques because the synergy of them allows a more efficient detection. They have been implemented using the software Matlab and the corresponding toolboxes.

3.1 Decision Tree Implementation

Decision trees are useful to express rules that relate multiple attributes and organize the available information in a hierarchical, orderly way [5]. Four decision trees have been generated for periods of one, three, six and twelve months for each user to find which one was more useful for the discrimination. Comparing the trees generated for the different users we realized that selecting the period of three months was enough to obtain a quite different structure for each user and at the same time to keep it simple.

The probability associated to each branch and leaf of the data tree is,

$$P(A) = \frac{valA}{N};\ P(B) = \frac{valB}{N} \tag{1}$$

$$P(T) = P(A1) * P(B2) * P(A2) * P(B2) * \ldots. * P(An) * \mathrm{P}(Bn) \tag{2}$$

Where N is the total number of cases, *valJ* is the number of events in branch *i*, *P(i)* the probability of the event in the *i* node, *P(T)* the final probability at *n*, and *n* the total number of nodes at that point of the tree. As an example, one of the two main branches is shown in Fig. 4 with the likelihood of an action of an authorized user.

This tree represents the behaviors of the users, but it does not describe the behavior that the user does not normally present. That is, if the behavior of the user does not follow any of those branches, how can it be detected if it is an intrusion or an authorized user doing something different than usual? So it needs to be complemented by another technique such as neural networks in order to give a final classification.

3.2 Final Detection of Intrusions by Neural Networks

To identify whether a user is an authorized one or an intruder, a Multi-layer Perceptron (MLP) neural network is applied. After trying several configurations, the final neural network is as follows (Fig. 5). It has four inputs: day, time, operation, and table. These are the most significant variables and the ones that allow us to better discriminate between users, as explained in Sect. 2. The output is a single neuron that represents the

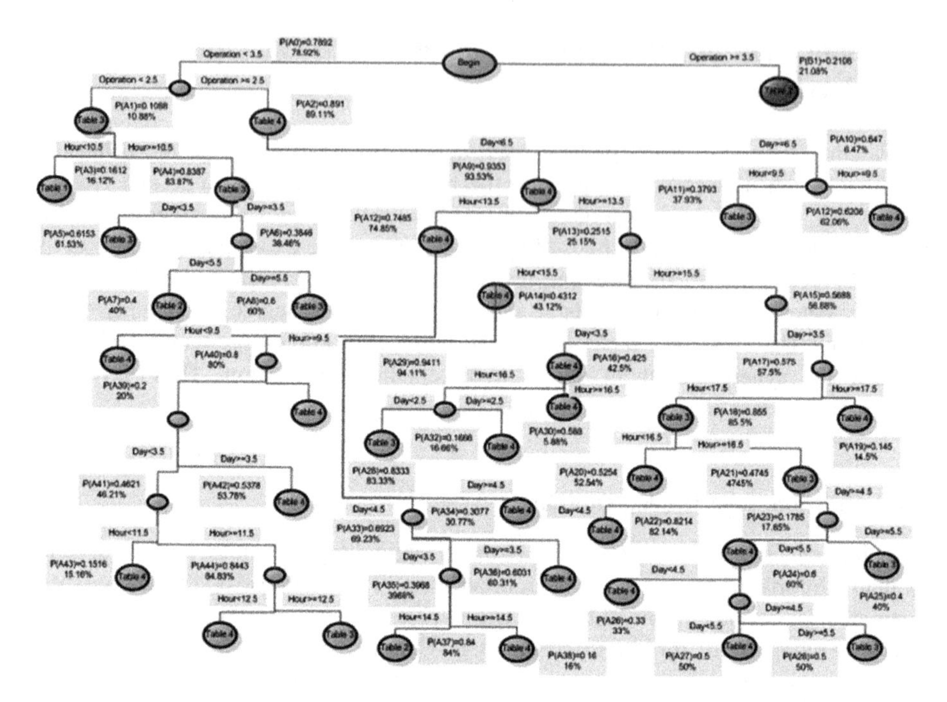

Fig. 4. Right branch of the decision tree, user #6, three months.

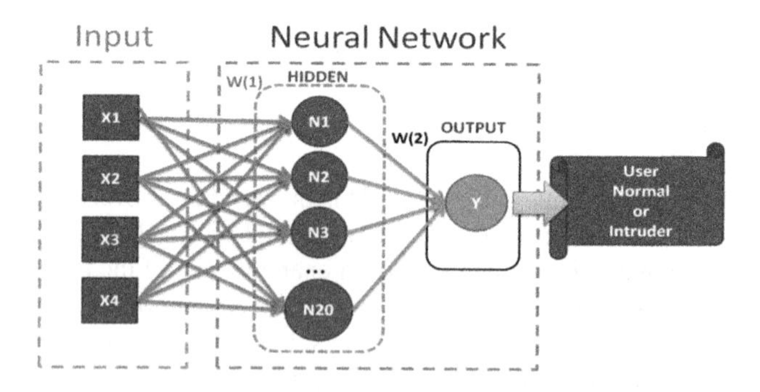

Fig. 5. General structure of the neural network of intrusion detection

user, that is, a number between 0 and 18 (0 stands for intruder and the rest of the numbers identify the user). There is a hidden layer with 20 neurons.

Real data have been taken from the databases. The vast majority corresponds to authorized users. The available samples have been then divided into training and test sets. Learning (performance) is measured in terms of the mean square error. The network is trained until the error is smaller than 0.001.

Once the neural network has learned how to classify the user that is working on the system, in order to validate it we have tested it with the test samples. The results are summarized in Table 2. It shows the percentages of correct and incorrect classification of the 18 users. The classifier has correctly identified 99.92 % of the samples (117.192 out of 118.067), that is, 1860 true positive plus 116.112 true negative samples. There were 35 false positive (normal behaviour detected as intrusion) and 60 false negative (activity identified as normal being intrusion). Besides, Table 2 shows the Attack Detection Rate (ADR) and False Alarm Rate (FAR) of the system.

Table 2. Classification results for Normal and Intrusive activities of 18 users.

True positive	True negative	False positive	False negative	Accuracy	Error	ADR	FAR
1860	116112	35	60	99.92 %	0.08 %	96.88 %	1.85 %

4 Simulation Tool for Intrusions Detection

A simulation tool that implements these techniques for intrusion detection has been developed. The input dataset consists in the activities the user has performed into the computer system, and the output is the identification of the user who logged the system and the probability of being an intrusion. The simulation tool gives three different possible outputs.

- Authorized access: the sequence of activities of the user matches with his behavior pattern, so it seems to be an authorized user and has been rightly identified.
- Unauthorized access: the behavior of the user during that computer session is different from expected. A message is displayed warning it is an intrusion.
- Unrecognized pattern: an X is displayed when the system is not able to clearly identify any of the authorized users, but it could still be an authorized one. In this case the result is a warning of possible intrusion.

A GUI (Graphical User Interface) has been designed to introduce the data and show the final decision (Fig. 6). On the left window, the information about the login of the system is shown (id of the user, day, hour, operation, table, etc.) In this case, it corresponds to user #9, Thursday, 06:00 h, operation 2, on table 3. The decision tree has generated a profile for this user that is shown at the bottom left window (green, hour; light blue, operation; red, table; blue, day). The neural network is then applied to identify which user corresponds to the present activity. The profile of the user that corresponds to the activity is shown at the bottom of the right window.

As it is possible to see, it does not correspond to user #9. Furthermore, the activities' pattern corresponds to user 5. Therefore it is identified as an intrusion.

The application can also detect access of users who do not present the expected behavior but, although it could be an intrusion, it is not clear that it is an intrusion. That means that the user is performing some actions with low probability regarding the corresponding decision tree but his activities somehow match the expected flow of actions. In this case the decision tool gives as result "non-identified user" and sends an alert but it does not block the access.

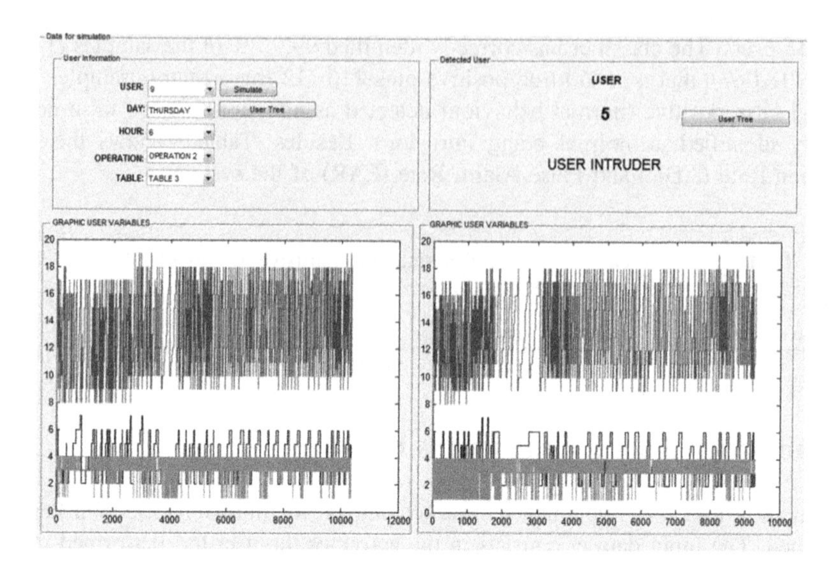

Fig. 6. Simulation tool of the intrusion detection system (Color figure online)

5 Conclusions

Intelligent techniques such as decision trees and neural networks have been proved very efficient when applied to pattern recognition and classification. In this paper we have developed a simulation tool that implements both of them to detect unauthorized accesses to information systems.

The decision tree allowed us to obtain a representation of the behavior of the user close to reality and interpretable. On the other hand, neural networks identify behavioral patterns based on the description given by the decision tree.

Previously, an exploratory analysis of the data was carried out in order to extract the key knowledge and to identify the significant variables that allow the discrimination of different users' behavior.

This decision-making system has been applied to real data provided by a governmental institution of the Republic of Ecuador.

As future work, other artificial techniques will be applied [12]. As it is well known, while many ensemble approaches exist, it remains, however, a difficult task to find a suitable ensemble configuration for a particular dataset [13].

Acknowledgments. This work has been partially supported by the Ministry of Higher Education, Science, Technology and Innovation (SENESCYT) of the Government of the Republic of Ecuador under the scholarship "Convocatoria Abierta 2011 y 2012".

References

1. Haq, N.F., Onik, A.R., Avishek, M., Hridoy, K., Rafni, M., Shah, F.M., Farid, D.M.: Application of machine learning approaches in intrusion detection system: a survey. Int. J. Adv. Res. Artif. Intell. **4**(3), 9–18 (2015)
2. Ahmed, M., Pal, R., Hossain, M.M., Bikas, M.A. N., Hasa, M.K.: A comparative study on the currently existing intrusion detection systems. In: International Association of Computer Science and Information Technology-Spring Conference, 2009, IACSITSC 2009, pp. 151–154. IEEE, April 2009
3. Guevara, C.B., Santos, M., López, M.V.: Negative selection and knuth morris pratt algorithm for anomaly detection. IEEE Lat. Am. Trans. **14**(3), 1473–1479 (2016)
4. Jo, S., Sung, H., Ahn, B.: A comparative study on the performance of intrusion detection using Decision Tree and Artificial Neural Network models. J. Korea Soc. Digit. Ind. Inf. Manag. **11**(4), 33–45 (2015)
5. Esmaily, J., Moradinezhad, R., Ghasemi, J.: Intrusion detection system based on Multi-Layer Perceptron Neural Networks and Decision Tree. In: 2015 7th Conference on Information and Knowledge Technology (IKT), pp. 1–5. IEEE, May 2015
6. Chen, Y., Abraham, A., Yang, B.: Hybrid flexible neural tree based intrusion detection systems. Int. J. Intell. Syst. **22**(4), 337–352 (2007)
7. Liu, G., Yi, Z., Yang, S.: A hierarchical intrusion detection model based on the PCA neural networks. Neurocomputing **70**(7), 1561–1568 (2007)
8. Amudhavel, J., Brindha, V., Anantharaj, B., Karthikeyan, P., Bhuvaneswari, B., Vasanthi, M., Vinodha, D.: A survey on intrusion detection system: state of the art review. Indian J. Sci. Technol. **9**(11), 1–9 (2016)
9. Thomsen, E.: OLAP Solutions: Building Multidimensional Information Systems. John Wiley & Sons, New York (2002)
10. Prakash, P.O., Jaya, A.: Analyzing and predicting user behavior pattern from weblogs. Int. J. Appl. Eng. Res. **11**(9), 6278–6283 (2016)
11. Guevara, C., Santos, M., López, V.: Data leakage detection algorithm based on sequences of activities. In: Proceedings of the 17th International Symposium Research in Attacks, Intrusions and Defenses RAID, vol. 8688, pp. 477–478. Springer, August 2014
12. Santos, M.: An applied approach to intelligent control. Revista Iberoamericana de Automática e Informática Industrial RIAI **8**(4), 283–296 (2011)
13. Aburomman, A.A., Reaz, M.B.I.: A novel SVM-kNN-PSO ensemble method for intrusion detection system. Appl. Soft Comput. **38**, 360–372 (2016)

Using Spritz as a Password-Based Key Derivation Function

Rafael Álvarez(✉) and Antonio Zamora

Department of Computer Science and Artificial Intelligence (DCCIA),
University of Alicante (Campus de San Vicente), Ap. 99, 03080 Alicante, Spain
{ralvarez,zamora}@dccia.ua.es

Abstract. Even if combined with other techniques, passwords are still the main way of authentication in many services and systems. Attackers can usually test many passwords very quickly when using standard hash functions, so specific password hashing algorithms have been designed to slow down brute force attacks.

Spritz is a sponge-based stream cipher intended to be a drop-in replacement for RC4. It is more secure, more complex and more versatile than RC4. Since it is based on a sponge function, it can be employed for other applications like password hashing.

In this paper we build upon Spritz to construct a password hashing algorithm and study its performance and suitability.

Keywords: Password · PBKDF · Cryptography · Spritz · Hash

1 Introduction

In recent times, password hashing strategies have risen to the foreground of authentication and security technologies. Nevertheless, passwords are still the primary means of authentication in most systems and services.

Traditionally, passwords would be stored in the clear in a database or file; this presents many problems since, usually, attackers can obtain this data in multiple ways.

The first step was to use a cryptographic hash function to process the passwords so they would be unrecognizable if the database was compromised; unfortunately, attackers could still discover many passwords because the same password always had the same hash and it was easy to precompute a dictionary of probable passwords hashes and accelerate the process even further.

The next technique involved adding a random binary string (called *salt*) to the password before hashing, therefore avoiding that the same password resulted in the same hash in the database and preventing precomputed dictionaries since the salt had to be taken into consideration.

Recently, attackers have begun exploiting the massively parallel capabilities of modern graphic processing units (GPU) or custom hardware to accelerate

M. Graña et al. (eds.), *International Joint Conference SOCO'16-CISIS'16-ICEUTE'16*,
Advances in Intelligent Systems and Computing 527, DOI 10.1007/978-3-319-47364-2_50

brute force attacks to unprecedented levels, testing all possible passwords of a certain length in very little time.

For this reason, it has become necessary to use password-based key derivation functions (PBKDF) that have customizable parameters to adjust the time and space complexity involved (see [5]) in order to slow down brute force attacks. This is a very active area of research, with many recent proposals (see [2,4,9–11,15,16]) that improve upon the PBKDF2 standard (see [6]).

Spritz (see [13]) is a sponge-based cryptographic function that can be employed as a stream cipher as well as a hash function (among other applications). It is a non-trivial evolution of the well-known RC4 stream cipher (see [12]), incorporating the sponge paradigm over a permutation of N values and extending the internal state of RC4 with more registers. It is meant as a drop-in replacement for RC4.

In this paper, taking advantage of the modular nature of its interface, we modify Spritz to work as a PBKDF with time and space complexity parameters; then analyze its performance regarding both parameters and establish its suitability for this application.

2 Description

2.1 Spritz

Although it has been designed as a replacement for RC4, Spritz (see [13]) is much more complex than RC4. It is based on a sponge-like (see [1]) function design and expands on the internal state of RC4 with additional 8 bit registers: i, j, k, z, and a; besides the S-Box, S.

Due to its sponge interface, Spritz allows many applications beyond that of a traditional stream cipher, including hash functions, authenticated encryption, etc. It has a flexible interface that depends on the specific application and we describe the most relevant methods in the following:

Initialize(n int). Spritz can be configured with an internal state of 2^n elements of n bits. This is particularly useful in the case of a PBKDF since it can be used to modify the amount of memory required by the algorithm.

Absorb(s []byte). This method modifies the internal state as a function of the string of bytes s, which can also be a single byte. The absorbed data can be a key for a stream cipher, a seed for a pseudo random number generator or a whole message for a hash function, among other possibilities. There is also a *AbsorbStop()* method that can be useful to separate different blocks of data to be absorbed, like a password/salt combination.

Drip()(b byte) and Squeeze()(s []byte). These two methods differ in that *Drip()* obtains a single byte of output while *Squeeze()* produces a string of bytes of output (making multiple calls to *Drip()*).

In the case of a PBKDF, the interface is analogous to a hash function, *absorbing* the *password,salt* combination and *squeezing* the output key/hash of the desired length; the main difference is the modifications to Spritz in order to allow parametrization of time/space complexity making a brute force attack more difficult. For further details regarding Spritz, refer to the original paper by Rivest and Schuldt [13].

2.2 Complexity Parameters

Ideally, password hashing functions have two separate parameters: space and time; with space determining the amount of memory the algorithm consumes and time establishing the computational cost of the algorithm. In practice, in most password hashing functions, increasing the space requirements also involves an additional computational cost.

The prototype for the call to the PBKDF function based on Spritz is:

```
SpritzPH(password []byte, salt []byte, m int, t int, size int)
```

where *password* is the user pass-phrase to be processed, *salt* is a random string of sufficient length to deter attacks based on rainbow tables, m is the space complexity parameter (memory), t is the time complexity parameter and *size* is the length of the output in bytes.

Space complexity in Spritz could be implemented by the modulation of the size of the internal state (s-box). We consider the minimum size to be 8 bits, so this value is taken as 2^{m+8} elements of $m+8$ bits (which are internally implemented as 64 bit integers).

Time complexity describes the number of times a certain action is performed in order to increase the computational cost and is taken as 2^t. It can be achieved in several different ways in Spritz, we have chosen three different approaches for further testing:

(A) Drip. Which, basically, calls *Drip()* to obtain 2^t elements ($m+8$ bits long each) from Spritz.

(B) Absorb. In this case, we call *Absorb()* to absorb every byte obtained from *Drip()*. In this way, the internal state is reconfigured more extensively.

(C) Squeeze. This scheme builds on (B), calling *Squeeze()* to obtain longer byte string rather than a single byte which is then absorbed with *Absorb()*. The motivation for this is testing performance benefits (if any) when processing bigger amounts of data per iteration than in the previous scheme.

3 Results

In this section we analyze the computational cost in seconds for each of the three time complexity strategies discussed as we increase the m (space complexity) and t (time complexity) parameters, both independently and simultaneously.

These tests have been implemented in the Go programming language (version 1.6) and conducted on a computer with a 2.6 GHz Intel Core i5 and 16 GB of RAM; passwords and salts have been 32 bytes (256 bit) in length while the output size has been 64 bytes (512 bit); every test has been measured ten times, taking the minimum as the final value.

3.1 Space Complexity

We can see in Fig. 1 the computational cost associated to each different approach (Drip, Absorb, Squeeze) when we increase the space complexity parameter from

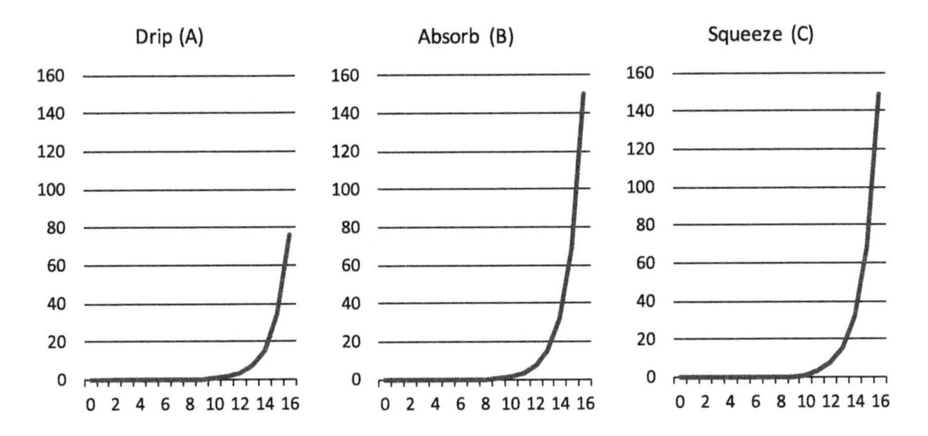

Fig. 1. Computational cost *(in seconds)* as space complexity (m) increases.

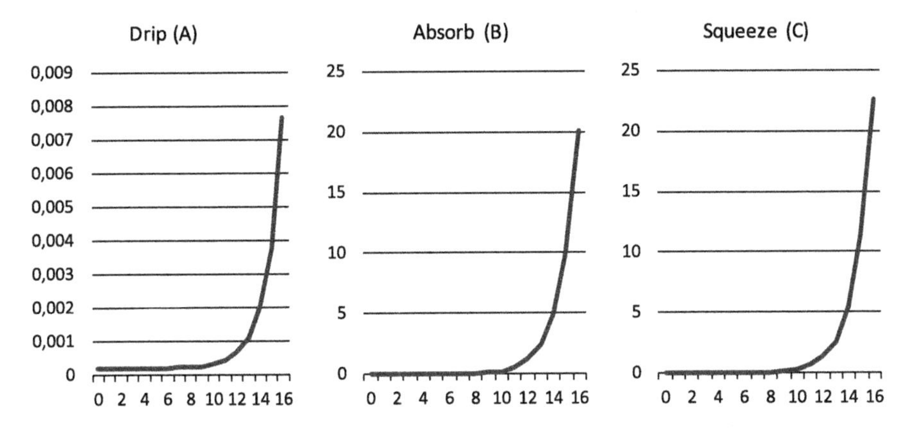

Fig. 2. Computational cost *(in seconds)* as time complexity (t) increases.

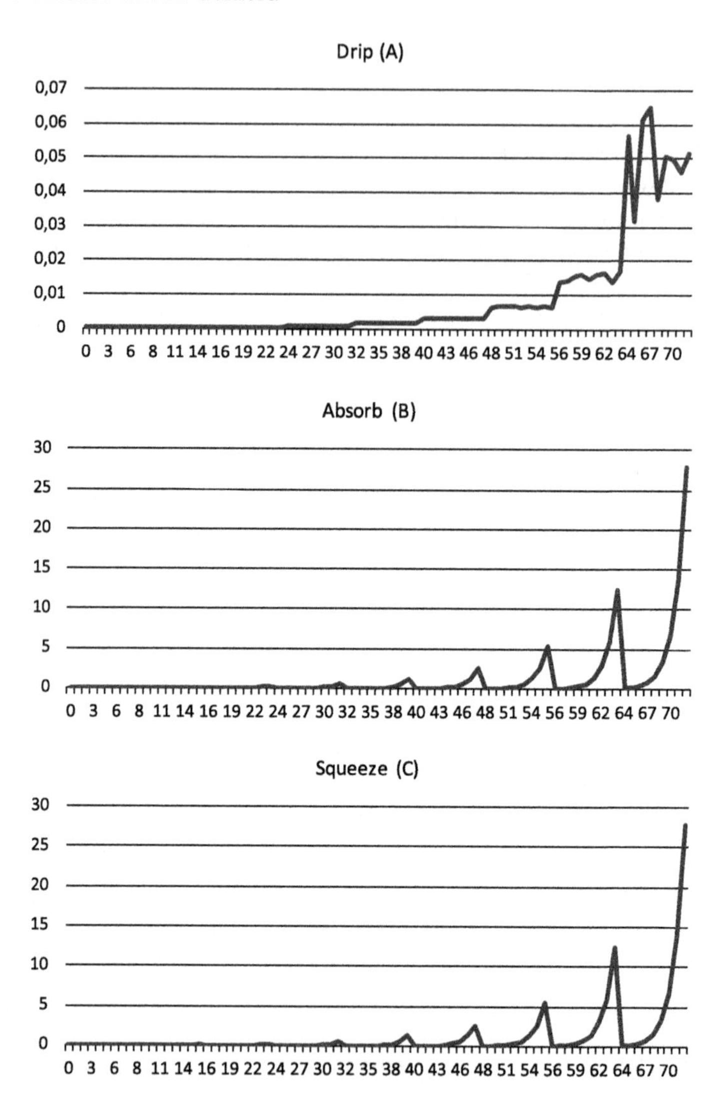

Fig. 3. Computational cost *(in seconds)* as time and space complexity increase simultaneously ($m \times 8 + t$) in a double loop.

0 to 16 (i.e. from 2^8 to 3×2^{24} bytes of memory) while maintaining the time complexity parameter at 0. It can be observed that the computational cost increases exponentially as expected although, after around $m = 8$ or 2×2^{16} bytes of memory, all three variants become very slow.

3.2 Time Complexity

The computational cost associated to the time complexity parameter increasing from 0 to 16 is shown in Fig. 2. As in the case of the space complexity, it grows exponentially as expected. It must be remarked that the *Drip (A)* approach is much faster than the other two, which are very similar to each other. Nevertheless, this performance difference is not as significant in the case of the space complexity.

3.3 Combined Complexity

In the combined complexity study, we measure the computation cost associated to each approach when increasing both parameters (space and time) simultaneously in a double loop from $m = 0, t = 0$ to $m = 8, t = 8$, with the space complexity as the outer loop variable and the time complexity as the inner one; the combined parameter is thus calculated as $m \times 8 + t$, as shown in Fig. 3.

We can observe a clear double loop pattern in the case of approaches *(B)* and *(C)* while the *Drip (A)* approach does not follow the expected pattern, basically only increasing computational cost significantly for the space complexity and showing erratic behavior for the time complexity. For this reason, we believe that the *Drip (A)* approach is not suitable for a PBKDF algorithm based on Spritz; approaches *(B)* and *(C)* are very similar but *(B)* is simpler and has a slightly better performance in some cases (especially in the time complexity analysis) and is, therefore, our recommendation.

Nevertheless, the Absorb *(B)* approach has a space complexity cost profile that limits the maximum amount of memory that could be used within a reasonable amount of time.

4 Conclusions

Originally designed as a drop-in replacement for the well-known RC4 stream cipher, Spritz is a very flexible sponge-based algorithm that can function as other types of cryptographic primitives like hash functions. In this paper, we have analyzed the performance of Spritz as a password hashing or password-based key derivation function. With that aim, we have parametrized the original design with a space complexity parameter that is directly related to the internal state memory footprint and with three different variants of a time complexity parameter to modulate the computational cost per invocation of the algorithm.

After extensive testing, we can conclude that the first approach (*Drip*) is not stable enough and very unbalanced between both complexity parameters, while the other two approaches (*Absorb* and *Squeeze*) are much more stable and follow the expected behavior; both provide very similar results, but *Absorb* would be the preferred candidate for its simplicity and marginally better performance.

Although the presented algorithms are adequate for many uses, they are not well-suited for those situations where a high memory footprint is paramount, since the space complexity profile limits the maximum amount of memory that could be used in practice.

Therefore, as possible future research, it would be interesting to explore further performance modifications that would enhance the amount of memory consumed for a given total computational cost, increasing resistance to brute force attacks mounted on multiple graphical processing units (GPU) or custom/programmable hardware platforms.

Acknowledgments. Research partially supported by the Spanish MINECO and FEDER under Project Grant TEC2014-54110-R.

References

1. Bertoni, G., Daemen, J., Peeters, M., Van Assche, G.: Cryptographic sponge functions (2011). http://sponge.noekeon.org/
2. Biryukov, A., Dinu, D., Khovratovich, D.: Argon2: the memory-hard function for password hashing and other applications. In: Password Hashing Competition Winner (2016). https://github.com/P-H-C/phc-winner-argon2/blob/master/argon2-specs.pdf
3. Fluhrer, S., Mantin, I., Shamir, A.: Weaknesses in the key scheduling algorithm of RC4. In: Vaudenay, S., Youssef, A.M. (eds.) SAC 2001. LNCS, vol. 2259, pp. 1–24. Springer, Heidelberg (2001). doi:10.1007/3-540-45537-X_1
4. Forler, C., Lucks, S., Wenzel, J.: The Catena Password-Scrambling Framework. Version 3.2, Bauhaus-Universitt Weimar (2015). https://www.uni-weimar.de/fileadmin/user/fak/medien/professuren/Mediensicherheit/Research/Publications/catena-v3.2.pdf
5. Hellman, M.E.: A cryptanalytic time-memory trade-off. IEEE Trans. Inf. Theory **26**(4), 401–406 (1980)
6. Kaliski, B.: PKCS #5: Password-Based Cryptography Specification Version 2.0. Internet Engineering Task Force, Network Working Group, Request for Comments (RFC) 2898 (2000). https://tools.ietf.org/html/rfc2898#section-5.2
7. Klein, A.: Attacks on the RC4 stream cipher. Des. Codes Crypt. **48**(3), 269–286 (2008). Springer
8. Paul, G., Maitra, S.: RC4 Stream Cipher and Its Variants. CRC Press, Boca Raton (2012)
9. Percival, C.: Stronger key derivation via sequential memory-hard functions. In: BSDCan - The BSD Conference (2009). http://www.bsdcan.org/2009/schedule/attachments/87_scrypt.pdf
10. Pornin, T.: The MAKWA Password Hashing Function. Version 1.1. Password Hashing Competition finalist (2015). http://www.bolet.org/makwa/makwa-spec-20150422.pdf
11. Provos, N., Mazieres, D.: A Future-adaptable password scheme. In: USENIX Annual Technical Conference, FREENIX track, pp. 81–91 (1999)
12. Rivest, R.L.: The RC4 Encryption Algorithm. RSA Data Security Inc. (1992)
13. Rivest, R.L., Schuldt, J.: Spritz - a spongy RC4-like stream cipher and hash function. In: Presented at CRYPTO 2014 Rump Session (2014). http://people.csail.mit.edu/rivest/pubs/RS14.pdf
14. Sengupta, S., Maitra, S., Paul, G., Sarkar, S.: RC4: (Non-) random words from (non-) random permutations. IACR Cryptology ePrint Archive 2011:448 (2011)
15. Simplicio, M.A., Almeida, L.C., Andrade, E.R., dos Santos, P.C.F., Barreto, P.S.L.M.: Lyra2: Password hashing scheme with improved security against time-memory trade-offs. IACR Cryptology ePrint Archive 2015:136 (2015)

16. Solar Designer: yescrypt - password hashing scalable beyond bcrypt and scrypt. Presented at PHDays 2014. Openwall (2014). http://www.openwall.com/presentations/PHDays2014-Yescrypt/PHDays2014-Yescrypt.pdf
17. Zoltak, B.: Statistical weakness in Spritz against VMPC-R: in search for the RC4 replacement. IACR Cryptology ePrint Archive 2014:985 (2014)

A Multiresolution Approach for Blind Watermarking of 3D Meshes Using Spiral Scanning Method

Ikbel Sayahi(✉), Akram Elkefi, and Chokri Ben Amar

REGIM: REsearch Group on Intelligent Machines,
University of Sfax, National School of Engineers (ENIS),
BP 1173, 3038 Sfax, Tunisia
{phd.ikbe.sayahi,akram.elkefi,chokri.benamar}@ieee.org
http://www.regim.org/members-1/

Abstract. During the last decade, the flow of 3D objects is increasingly used everywhere. This wide range of applications and the necessity to exchange 3D meshes via internet raise major security problems. As a solution, we propose a blind watermarking algorithm for 3D multi-resolution meshes ensuring a good compromise between invisibility, insertion rate and robustness while minimizing the amount of memory used during the execution of our algorithm. To this end, spiral scanning method is applied. It decomposes the mesh into GOTs (a Group Of Triangles). At each time, only one GOT will be loaded into memory to be watermarked. It undergoes a wavelet transform, a modulation then embedding data. Once finished, the memory will be released to upload the next GOT. This process is stopped when the entire mesh is watermarked. Experimental tests showed that the quality of watermarked meshes is kept despite the high insertion rate used and that memory consumption is very reduced (until 24 % of memory reduction). As for the robustness, our algorithm overcomes the most popular attacks in particular compression. A comparison with literature showed that our algorithm gives better results than those recently published.

Keywords: Digital watermarking · 3D multiresolution meshes · Wavelet transform · Spiral scanning

1 Introduction

3D mesh, is a new data type, designed to model complex 3D objects thanks to its dual geometric and combinatorial nature (vertex positions and connectivity). The strengthening of computer graphics and acquisition techniques has generated a variety of areas benefiting from this new data category such as industry, medicine, architecture, 3D games, movies.

On the other side, the information technology revolution, that has affected mainly the telecommunications and networks, has made its own way. The result

M. Graña et al. (eds.), *International Joint Conference SOCO'16-CISIS'16-ICEUTE'16*,
Advances in Intelligent Systems and Computing 527, DOI 10.1007/978-3-319-47364-2_51

being is the emergence of high speed broadband networks allowing the storage and the transfer of digital documents through remote multimedia databases. The 3D models are an example of these documents shared via the net. Sharing 3D meshes between remote users has spawned huge security problems especially that digital copying does not entail any loss of quality [14]. In addition, in contrast to acts of counterfeiting of analog works, the digital reproduction costs are negligible and counterfeiters can act anonymously without leaving a trace of their passage. All these problems leads that legal protection is no longer sufficient to ensure alone the peaceful management of works transmitted to the public [9]. Therefore, the need to use other techniques to strengthen existing legal protections becomes imperative. Digital watermarking [3,7] is then announced as a new technique that aims to limit these "digital abuse" and to preserve the copyrights.

In this context, we propose a new blind watermarking algorithm for 3D multiresolution meshes. Our goal is no longer to have a good compromise between insertion rate, robustness and invisibility only, but also to minimize the hardware resources (memory) used during the execution of our algorithm. This may facilitate the implementation of our prototype as well as the handling of meshes with high definition.

To present our approach, we organized this manuscript as follow: Firstly, a related works section is introduced to present the advancement of research in this area. The next section describes the proposed approach (insertion and extracted steps). To evaluate this approach, given results are discussed is Results and discussion section. This paper ends with a conclusion.

2 Related Works

Since the appearance of the first watermarking algorithm for 3D meshes, a lot of works emerged in order to improvement 3D watermarking field. The common point between all these works is to improve insertion rate, invisibility and robustness. Unfortunately, it is not obvious to correctly extract embedded information due to attacks that may alter and even destroy these data. Therefore, published works used several techniques in multiple domains (spatial, frequency and multiresolution).

The first embedding domain used was the spatial domain. In this case, insertion modifies either the geometric information [11,12,19] or the topological information [1,22]. Working in spatial domain is slightly complex, given that no transformation will be applied on the mesh but results show a failure in visibility criterion.

To deal with the already mentioned problem, several studies have worked in transformed fields such as the frequency domain. The main idea is to calculate the so-called "frequency coefficients" of the host mesh to represent the mesh in the frequency domain. Many tools ensure this new representation such as the Radial Basis Functions [18] or the Manifold harmonics [13].

As was the case for image [8], video [6] and audio [2,4], multi-resolution domain is no longer excluded from watermarking 3D meshes. The used tool is the wavelet transform which allows the generation of multi-resolution coefficients called "wavelet coefficients". These latter are targeted by embedding (approach of Ouled Zaid et al. in [15])

Notwithstanding the diversity of techniques and tools used, published results show that no perfect solution has been proposed until now.

3 Presentation of the Proposed Approach

In this paper, a new blind watermarking algorithm for multi-resolution 3D meshes is proposed. Firstly, our goal is to insert a maximum amount of information while keeping the mesh quality and ensuring robustness against any treatments threatening a correct extraction of the information. These treatments are called 'Attacks'. Secondly, we aim reducing the quantity of memory used during the execution of the watermarking algorithm using spiral scanning method. It consists on decomposing the host mesh into GOTs. For each GOT overloaded in memory, watermarking which can be either an insertion or an extraction step is applied. Once treated, watermarked GOT will be released from memory to upload the next one. This process ends when the entire mesh is treated.

3.1 Insertion Step

Insertion is to embed data, in the form of a binary sequence, into the host mesh while keeping its quality. As it was said, embedding is applied to each GOT saved in memory. As shown in Fig. 1, Hidden data in the host mesh is based on these steps:

Spiral Scanning Acquisition. To treat an object sequentially, it must be split into parts. Only the necessary parts are sent to memory each time. Contrary to 3D meshes, decomposition of image into blocks or video into image looks very

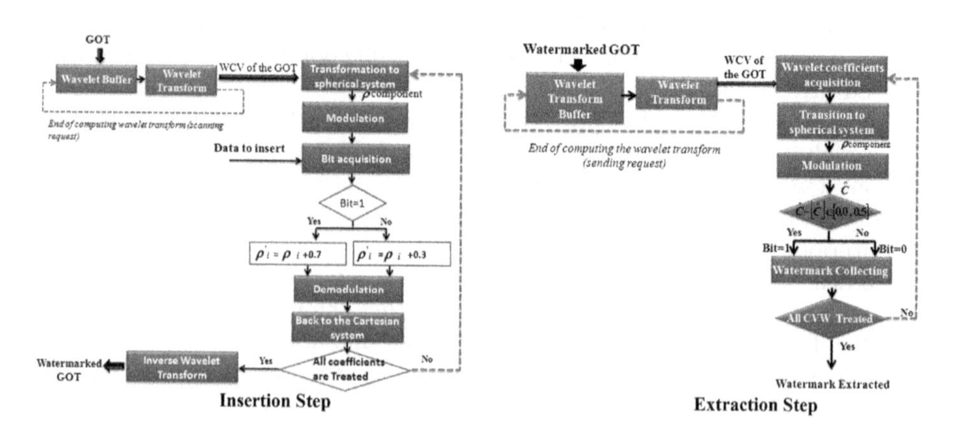

Fig. 1. Overview of our approach: Insertion and Extraction steps.

simple. The difficulties that arises, in the case of 3D objects, is the choice of the starting point and the direction chosen to browse the whole mesh. To overcome these problems, we based our work on spiral scanning method proposed to ensure a progressive compression in [10]. The use of spiral scanning method during compression gives very good results in terms of minimizing the amount of used memory during compression. We will readopt this method for 3D watermarking. Our objective is to ensure that the watermarking process has swept the whole mesh and that there is no untreated part, which is tricky, as the topology of the object is varied

The procedure begins from a low frequency triangle. Then there will be a calculation of the neighborhood to move from this triangle to its neighbor and so forth. Unfortunately, this step is very costly in terms of the number of operations. As a solution, a procedure using the following properties is proposed (Fig. 2):

- Property 1: The neighbors of a central triangle T_c are among the sons of the father of T_c.
- Property 2: Of the sons of the father of a triangle T, the only one that is central is a neighbor. Other neighbors of T are among the neighbors' son of the father of T.
- Property 3: If two triangles T_1 and T_2 are adjacent then one of them is central. Once this is known, it facilitates the search for other neighbors according to property 1.
- Property 4: Of the son of the father of a triangle T, the only one that is central is a neighbor. Other neighbors of T are among the neighbors' son of the father of T.

The acquisition follows then a directed movement so that the scanning of the 3D mesh does not leave untreated portions. To this end, we follow, as a reference, the list $L_0 = \{a, c, b\}$ for acquiring the triangles A_i, B_i and C_i, respectively colored in yellow (near the point a), green (vicinity of the point c) and blue (vicinity of the point b). This represents a complete initialization turn. When detecting a new triangle, we send it to the watermarking buffer and we update the new reference list. The second round will reference the newly created list being $L_1 = \{r_0, r_1, ..., r_i\}$ and so on. The end of transmitting triangles to the watermarking buffer corresponds to a new empty reference list.

Wavelet Transform. The main idea of the multiresolution analysis is to break down a mesh M_i in two parts: a low resolution mesh M_{i-1} grosser and a set of details D_{i-1}: the analysis phase. All these details and meshes of different resolution level are used to reconstruct the original mesh: synthesis phase. Using the formalism of multiresolution analysis, we can write:

$$M_j = M_{j-1} \oplus D_{j-1} \tag{1}$$

D_{j-1} is all the details necessary to rebuild the mesh, M_j refers to the higher resolution from the mesh M_{j-1}, and $\oplus$ is the orthogonal complement operator. To implement a multiresolution analysis, wavelet transform can be applied.

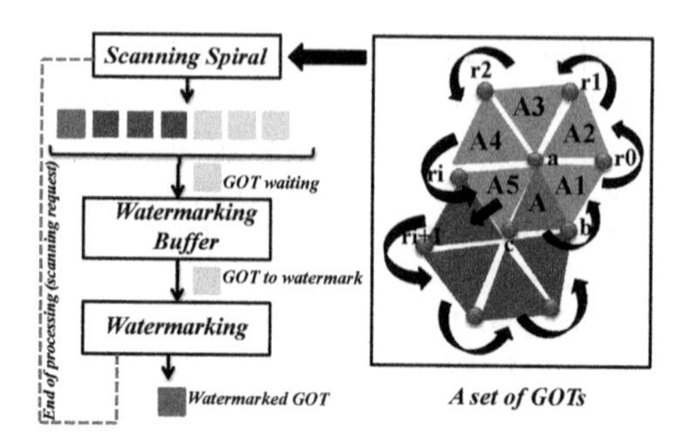

Fig. 2. Spiral scanning method: neighborhood calculation.

Its principle is to decompose the energy of a signal using two basic functions (prediction and update) to obtain a lower resolution mesh and a set of wavelet coefficients. All these coefficients are assembled into a single vector called wavelet coefficient vector (WCV). Especially, this vector will be modified during insertion. As for implementation, a lifting schema, which is a wavelet transform of second generation, is used. The idea is to exploit the spatial and frequency correlation present in the mesh to reduce its entropy. The lifting scheme is divided into three steps:

- Poly-phase transformed: it is a basic operation that divides the signal to two sub-bands.
- Prediction: This step exploits the spatial and frequency correlation present in the signals. During this step the elements of the first component are used to predict the elements of the second component. The difference between the predicted element and the element present in the same second component are the details.
- Update: at this stage the first component elements represent only a sub-sampling of the original signal. To transform these elements to low frequencies, we apply a low pass filter.

3.2 Extraction Step

This step allows the retrieval of the inserted information from the mesh which is not always possible. The extraction step presented is this section is designed to extract all the information correctly. To achieve this goal, spiral scanning method is used again to decompose the mesh into GOTs. For each one, a wavelet transform is applied to get the correspondent WCV. For each coefficient presented in a spherical system, we apply modulation to extract then the inserted bit depending on the results of modulation (see Fig. 1, Extraction part). Once this part of the mesh is processed and the watermark is extracted, the following GOT is loaded into memory to be treated in its turn.

4 Results and Discussion

To achieve experimentation, we used 4 multi-resolution meshes: Venus, Feline, Rabbit and horse. All these meshes are stored in .DAT files.

4.1 Insertion Rate and Invisibility Criteria

To be modified during insertion, the ρ component of each wavelet coefficient undergoes a modulation step (multiplication of wavelet coefficients by a modulation factor empirically determined). This coefficient is related to the strength of our algorithm and the visibility criterion. The higher the value of coefficient is, the more affected the mesh quality becomes. Applied tests show that with a modulation factor equal to 10000, we obtained a good value of PSNR (126.35 db) and MSQE (1.2×10^{-6}). This value will be used in the next experiments. In an attempt to make comparison with recent works, our approach has an insertion rate of about 250,000 bits but with very low values of MSQE and a good PSNR. This compromise represents an improvement over other published results (Table 1).

Table 1. Compromise between insertion rate and invisibility: comparison with literature.

Approach	Insertion rate	MSQE	PSNR
[1]	39707	0, 0	84, 13
[11]	21022	2.7×10^{-5}	–
[15]	10650	0.2×10^{-3}	–
[21]	172974	1.2×10^{-5}	–
[20]	199	3.2×10^{-5}	–
Our approach	250000	$1,2 \times 10^{-6}$	126, 35

4.2 Robustness Criterion

Similarity Transformations. Results of experiments assert that our algorithm is robust against similarity transformation attacks. Indeed, the application of translation, rotation and uniform scaling to watermarked meshes, has not prevented the correct extraction of all inserted data.

Noise Addition. As shown in Table 2, correlation value is acceptable for a noise level down to 10^{-4}. Correct retrieval of information is possible in this range of values. By comparing our results to recently published works, we note that our approach is an outstanding improvement over existing approaches.

Table 2. Correlation depending on noise level

Noise level	10^{-1}	10^{-2}	10^{-3}	10^{-4}	10^{-5}	10^{-6}
C in [16]	0.1	—	0.3	—	0.4	—
C in [17]	0.02	—	0.6	—	0.8	—
C in [5]	—	—	—	0.05	0.3	0.55
Obtained C	0,03	0,12	0,3	0,7	9	1

Table 3. Correlation depending on degree of deformation

dFactor	10^{-5}	10^{-6}	10^{-7}	10^{-8}	10^{-9}	10^{-10}
C in [17]	–	–	–	0.18	0.31	0.43
c in [20]	–	0.3	0.4	0.5	0.8	1
Obtained C	0,02	0,4	0,6	0,78	0,99	1

Table 4. Correlation depending on quantization level

Quantization level	10	12	13	14	15	20
C in [5]	0.3	0.4	0.45	0.6	–	–
Obtained C	0, 08	0, 58	0, 84	0.95	1	1

Smoothing. To study the effect of smoothing during extraction we did many tests. The found results are presented in Table 3. For a dFactor value less than 10^{-7}, we obtained a correlation value near 1. Comparing these results with those of previous published work published [17], we note that with our present approach we have reinforced the robustness against smoothing attack.

Coordinate Quantization. Table 4 shows that we have a good extraction result from a quantization level up to 13. This result became perfect (correlation = 1) with a level up to 14.

Mesh Simplification. Applied tests, whose results are presented in Table 5, show very well that we are able to extract all the inserted information despite the application of simplification from the third iteration. In the first two iterations, correlation value is about 0, 8.

Compression. In order to conclude on the robustness of our algorithm against the compression attack, we applied several tests. We changed each time the compression rate. The result presented in Table 6 shows that for a rate greater than 2, we obtained good correlation values. All of the inserted information was correctly extracted when the compression rate is equal or higher than 2, 5.

Table 5. Correlation depending on simplification degree

Iteration number	1	2	3	4	5	6
C in [16]	–	–	–	0.46	0.31	0.15
C in [11]	–	–	–	0.79	0.68	0.61
C in [15]	–	–	–	0.99	0.97	0.92
C in [5]	–	0.6	0.45	0.25	0.1	0.05
Obtained C	0,8	0,87	0,94	1	1	1

Table 6. Correlation depending on the compression rate

Bit/vertex	0.1	0.5	1	1.5	2	2.5	3
Obtained C	0.04	0.28	0.39	0.58	0.82	0.89	1

4.3 Memory Consumption

In order to minimize the amount of memory used, we decomposed the host mesh into GOTs (spiral scanning method). Each time, we send a GOT to memory to be watermarked. Once treated, it will be deleted to allow loading the next GOT. Our approach reduces remarkably the amount of used memory and allows us to work even with a very small memory space (Fig. 3).

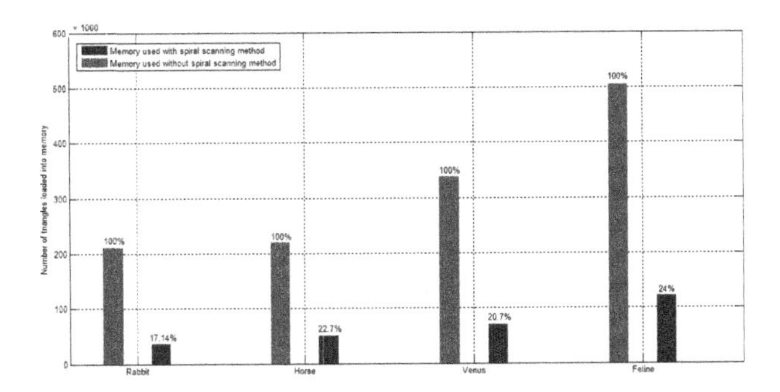

Fig. 3. Memory usage during the execution of our algorithm

Indeed, reducing has reached a value equal to 24 %. This memory space presents the minimum space required for the application of wavelet transform. Although these results are very motivating, improve even more the memory usage (in order to implement the algorithm or handle meshes with high resolution) returns to the improvement of the used wavelet transform.

5 Conclusion

In this paper, we propose a new watermarking approach for 3D meshes which can be summarized as follows: decomposing the mesh into a set of GOTs. At each time t, a GOT will be sent into memory to be watermarked. Treatment includes the application of a wavelet transform in order to generate the vector of wavelet coefficients which is modified during insertion. When this treatment ends, the watermarked GOT is deleted and the next one will be sent into memory. This process is stopped when the entire mesh is watermarked. The particularity of this work is the fact of applying spiral scanning. This method of decomposition allowed a huge gain in memory adopted (reducing memory reached 24 %). Our algorithm can then work even with a very small memory space.

Tests applied in this paper prove that our algorithm preserves mesh quality. It does not cause quality degradation of the mesh despite the large number of bits to be inserted (250000 bits). Previous displayed results, which present a considerable improvement compared to the results of recent works, assert that our algorithm is robust against several attacks such as similarity transformation, Random noise addition, coordinate quantization, smoothing, simplification and compression.

Concerning our future work, we think of changing the method of wavelet transform used to further reduce the amount of memory used. As for the criteria of robustness, we strongly believe in changing techniques used to improve the results already presented, in including mechanisms of tolerance to failures and in studying the robustness of our algorithm against malicious software of last generation.

References

1. Chao, H.L., Min, W.C., Jyun, Y.C., Cheng, W.Y., Wei, Y.H.: A high-capacity distortion-free information hiding algorithm for 3D polygon models. Int. J. Innovative Comput. Inf. Control **9**(3), 1321–1335 (2013)
2. Charfeddine, M., Elarbi, M., Ben Amar, C.: A blind audio watermarking scheme based on neural network and psychoacoustic model with error correcting code in wavelet domain. In: IEEE International Symposium on Communications, Control and Signal Processing (2008)
3. Charfeddine, M., Elarbi, M., Ben-Amar, C.: A new DCT audio watermarking scheme based on preliminary mp3 study application to video watermarking. Int. J. Multimedia Tools Appl. **70**(3), 1–37 (2012)
4. Charfeddine, M., Elarbi, M., Koubaa, M., Ben Amar, C.: DCT based blind audio watermarking scheme. In: IEEE International Conference on Signal Processing and Multimedia Applications (2010)
5. Dae, J.C.: Watermarking scheme of mpeg-4 laser object for mobile device. Int. J. Secur. Appl. **9**(1), 305–312 (2015)
6. Elarbi, M., Ben Amar, C., Nicholas, H.: A dynamic video watermarking scheme in the DWT domain. In: International Conference on Signal Processing and Communications (2007)

7. Elarbi, M., Charfeddine, M., Masmoudi, S., Koubaa, M., Ben Amar, C.: Video watermarking algorithm with BCH error correcting codes hidden in audio channel. In: IEEE Symposium on Computational Intelligence in Cyber Security (2011)
8. Elarbi, M., Koubaa, M., Ben Amar, C.: A wavelet networks approach for image watermarking. Int. J. Comput. Intell. Inf. Secur. **1**(1), 34–43 (2010)
9. Elarbi, M., Koubaa, M., Charfeddine, M., Ben-Amar, C.: A dynamic video watermarking algorithm in fast motion areas in the wavelet domain. Int. J. Multimedia Tools Appl. **55**(3), 579–600 (2011)
10. Elkefi, A.: Compression des maillages 3D multiresolutions de grandes precisions (2011)
11. Hitendra, G., Krishna, K., K., Manish, G., Suneeta, A.: Uniform selection of vertices for watermark embedding in 3-D polygon mesh using IEEE754 floating point representation. In: International Conference on Communication Systems and Network Technologies, pp. 788–792 (2014)
12. Jen-Tse, W., Yi-Ching, C., Shyr-Shen, Y., Chun-Yuan, Y.: Hamming code based watermarking scheme for 3D model verification. In: International Symposium on Computer, Consumer and Control, pp. 1095–1098 (2014)
13. Jinrong, W., Jieqing, F., Yongwei, M.: A robust confirmable watermarking algorithm for 3D mesh based on manifold harmonics analysis. Int. J. Comput. Graph. **28**(11), 1049–1062 (2012)
14. Koubaa, M., Ben-Amar, C., Nicholas, H.: Collusion, mpeg4 compression and frame dropping resistant video watermarking. Int. J. Multimedia Tools Appl. **56**(2), 281–301 (2012)
15. Ouled-Zaid, A., Hachani, M., Puech, W.: Wavelet-based high-capacity watermarking of 3-D irregular meshes. Multimed Tools Appl. **74**(15), 5897–5915 (2015)
16. Roland, H., Li, X., Huimin, Y., Baocang, D.: Applying 3D polygonal mesh watermarking for transmission security protection through sensor networks. Math. Probl. Eng. **2014**(2014), 27–40 (2014)
17. Sayahi, I., Elkefi, A., Koubaa, M., Ben Amar, C.: Robust watermarking algorithm for 3D multiresolution meshes. In: International Conference on Computer Vision Theory and Applications, pp. 150–157 (2015)
18. Xiangjiu, C., Zhanheng, G.: Watermarking algorithm for 3D mesh based on multiscale radial basis functions. Int. J. Parallel Emergent Distrib. Syst. **27**(2), 133–141 (2012)
19. Xiao, Z., Qing, Z.: A DCT-based dual watermarking algorithm for three-dimensional mesh models. In: International Conference on Consumer Electronics, Communications and Networks, pp. 1509–1513 (2012)
20. Ying, Y., Ruggero, P., Holly, R., Ioannis, I.: A 3d steganalytic algorithm and steganalysis-resistant watermarking. IEEE Trans. Vis. Comput. Graph. 1–12 (2016)
21. Yuan, Y.T.: A secret 3D model sharing scheme with reversible data hiding based on space subdivision. 3D Res. **7**(1), 1–14 (2016)
22. Zhiyong, S., Weiqing, L., Jianshou, K., Yuewei, D., Weiqing, T.: Watermarking 3D capd models for topology verification. Comput. Aided Des. **45**(7), 1042–1052 (2013)

Data Is Flowing in the Wind: A Review of Data-Flow Integrity Methods to Overcome Non-Control-Data Attacks

Irene Díez-Franco(✉) and Igor Santos

DeustoTech, University of Deusto, Bilbao, Spain
{irene.diez,isantos}@deusto.es

Abstract. Security researchers have been focusing on developing mitigation and protection mechanisms against *code-injection* and *code-reuse* attacks. Modern defences focus on protecting the legitimate control-flow of a program, nevertheless they cannot withstand a more subtle type of attack, *non-control-data* attacks, since they follow the legitimate control flow, and thus leave no trace. *Data-Flow Integrity* (DFI) is a defence mechanism which aims to protect programs against non-control-data attacks. DFI uses static analysis to compute the data-flow graph of a program, and then, enforce at runtime that the data-flow of the program follows the legitimate path; otherwise the execution is aborted.

In this paper, we review the state of the techniques to generate non-control-data attacks and present the state of DFI methods.

Keywords: Data-flow integrity · Non-control-data attacks · Operating system security

1 Introduction

Programs are formed by control data (e.g. return addresses, function pointers) and non-control data (e.g. variables, constants). Even though non-control data is more abundant, both attackers and defenders have focused their efforts into exploiting or protecting control data. Different attacks have been repeatedly applied to the control flow, such as complex *code-injection* and *code-reuse* attacks. Therefore, memory integrity methods have been proposed in order to secure operating systems and userland programs (e.g. safe dialects of C/C++, secure virtual architectures, and control-flow integrity methods).

Even though non-control-data attacks are not new [8] and their importance has not decreased, modern operating systems and their programs remain vulnerable against these type of attacks. Given the lack of usable methods against these attacks, in this paper we present a review of the most relevant techniques to overcome non-control-data attacks, as well as a overview of the landscape of non-control-data attacks.

M. Graña et al. (eds.), *International Joint Conference SOCO'16-CISIS'16-ICEUTE'16*,
Advances in Intelligent Systems and Computing 527, DOI 10.1007/978-3-319-47364-2_52

2 Non-Control-Data Attacks

The most common memory corruption vulnerabilities, namely *code-injection* and *code-reuse* attacks, have been tackled by the community, creating defences for commodity operating systems and compilers. On the one hand, stack canaries [9] and write-xor-execute (W⊕E)/DEP [3] defence schemes try to prevent *code-injection* attacks resulting from stack, heap or buffer overflows, use-after-free and format string vulnerabilities, whereas Control-Flow Integrity (CFI) [1] and program shepherding [14], along with the statistical defences provided by ASLR [20] and Kernel ASLR [11] concentrate on *code-reuse* attacks arising from classic return-oriented programming (ROP) [15,17], or its newer variants [4,5,7].

Due to the attention given to code-injection and code-reuse attacks, Chen et al. [8] raised awareness of the *pure data* or *non-control-data* attacks, since all the previous approaches [1,3,9,11,14,20] focus just on *control-data*, and thereby cannot endure more subtle non-control-data attacks.

A non-control-data attack differs from a control-data attack because it does not affect the control-flow of a program. Control-data attacks are based on rewriting control data (e.g. return addresses), leaving a trace in the form of an unintended control-flow transfer. This transfer can be detected and prevented at runtime [1].

On the contrary, non-control-data attacks follow legitimate control-flow transfers since they are based on modifying the program's logic or decision-making data. Consequently, they remain invisible to defence techniques which only focus on control data.

2.1 Security-Critical Non-Control Data

Chen et al.'s work identifies the following types of security-critical data that may be subjected to non-control-data attacks:

- **Configuration data**. Many applications, such as web servers, need configuration files in order to define access control policies and file path directives to specify the location of trusted executables. If an attacker was capable of overwriting such configuration data, it would be possible to launch unintended applications (e.g. root shells), and moreover bypass the access controls of the web server.
- **User input data**. A well known practice in software engineering is to distrust user input data, and only after that data has been validated it can be used. If an attacker could change the input data after the validation process, she could execute the program with a malicious input.
- **User identity data**. UIDs and GIDs are stored in memory while authentication routines are executed. If such IDs were tampered with, an attacker could impersonate a user with administration privileges.
- **Decision-making data**. Boolean values (e.g. authenticated or not) are usually used to make decisions in an application, an attacker could change those decision making values to redirect the flow of a program through unintended branches.

```
struct passwd {
  uid_t pw_uid;
  ...
} *pw;
...
int uid = getuid();
pw->pw_uid = uid;
// format string vulnerability
...
void passive(void) {
  ...
  seteuid(0); // set root uid
  ...
  seteuid(pw->pw_uid); // set non-root uid
}
```

Fig. 1. Vulnerable code from the `wu-ftpd` web server.

Moreover, Hu et al. [12] enhanced the previously presented types of security-critical data with the following items:

- **Passwords and private keys**. The disclosure of passwords and private keys could give an attacker full access to a system.
- **Randomised values**. Tags for CFI enforcement, random canary words and randomised addresses are used in many security related mechanisms (e.g. CFI, ASLR, SSP). If an attacker knows the random canaries placed in the stack she can use stack-smashing attacks without alerting the Stack Smashing Protector (SSP).
- **System call parameters**. Tampering with the parameters of security critical system calls (e.g. `execve`, `setuid`) can lead to privilege escalation or unintended program execution.

The following two attacks describe how a vulnerability can be exploited due to a memory error to expose security-critical non-control data using a non-control-data attack.

Figure 1 shows a vulnerable piece of code with a format string vulnerability where the value of `pw->pw_uid` can be rewritten. This vulnerability can be exploited using a non-control-data attack that targets *user identity data*. Concretely, an attacker could overwrite the value with root's UID, to, later on avoid dropping root privileges to normal user privileges in the `passive` function.

The OpenSSL *Heartbleed* vulnerability [21] allowed a remote attacker to expose sensitive data, such as *private keys*, using a non-control-data attack. On the OpenSSL 1.0.1 and 1.0.2β heartbeat request/response protocol, an attacker could request a heartbeat using legitimate payload but with a payload length field larger (up to 65,535 bytes) than the real payload. Then, the heartbeat protocol crafted a response copying the original payload in a buffer allocated of the size indicated by the payload length field. Since the payload length field was not

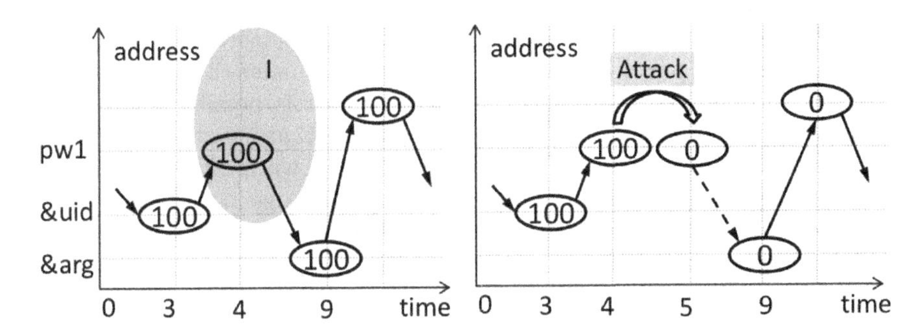

Fig. 2. Original 2D-DFG of the vulnerable code of `wu-ftpd` web server (left) and resulting 2D-DFG after a non-control-data attack (right). `&arg` is the stack address of `setuid`'s argument.

correctly verified against the length of the real payload, a memory leakage was possible.

2.2 Data-Flow Stitching

Hu et al. [12] demonstrated that non-control-data attacks can be automatically constructed using a technique named *data-flow stitching*. This technique is capable of redirecting the data-flow of a program through unintended paths in order to tamper with data or leak sensitive data.

They introduced the concept of *two-dimensional data-flow graph* (2D-DFG) to represent the data dependencies created in a program executed with a concrete input. A 2D-DFG is a directed graph $\mathcal{G} = \{\mathcal{V}, \mathcal{E}\}$ where $\mathcal{V}$ is the set of vertices and $\mathcal{E}$ the set of edges. A vertex v ($v \in \mathcal{V}$) is a variable instance with a value, and it is represented as a point (a, t) in the two dimensions of the 2D-DFG, addresses and time; thereby, a refers to the address or register name of the variable, and t to the execution time when the variable instance is created. A *vertex* $v = (a, t)$ is created when an instruction writes to memory value a at time t; a *data edge* (v', v) is created when the instruction takes v' as the source and v as the destination operands, and finally, an *address edge* (v', v) is created when an instruction uses v' as the address of the operand v.

In order to generate a non-control-data attack, data-flow stitching requires a program with a memory error. The set of memory locations this memory error can write to is called the *influence* I. The new data-flow that wants to be created consists of two vertices, namely source vertex (v_s), and target vertex (v_t); resulting in a data-flow path from v_s to v_t. This new data-flow path would have a new 2D-DFG $\mathcal{G}' = \{\mathcal{V}', \mathcal{E}'\}$, where $\mathcal{V}'$ and $\mathcal{E}'$ are generated by the memory error exploit. The goal of data-flow stitching is to discover a data-flow edge set $\overline{\mathcal{E}}$, where $\overline{\mathcal{E}} = \mathcal{E}' - \mathcal{E}$, to add in the resulting 2D-DFG of the memory error ($\mathcal{G}'$), allowing new data-flow paths from v_s to v_t.

Following the examples given by Chen et al. and Hu et al., Fig. 1 shows a vulnerable piece of code from the `wu-ftpd` web server, and Fig. 2 (left) shows the original 2D-DFG of such program. The format string vulnerability on `wu-ftpd` server can overwrite `pw->pw_uid`, since such vertex is under the influence of the memory error; thereby, a privilege escalation attack is possible using a non-control-data attack generated by data-flow stitching that inserts an edge in the DFG, overwriting `pw->pw_uid` with root's UID, as shown in Fig. 2 (right).

This example requires only the addition of a single edge in the new 2D-DFG, nevertheless data-flow stitching can also generate advanced attacks that need stitching more edges.

2.3 Data-Oriented Programming

Data-oriented programming (DOP) is a technique proposed by Hu et al. [13] to perform computations on a program's memory respecting its legitimate control-flow. DOP's computations are based on non-control-data attacks resulting from memory errors and have been proven to be Turing-complete. DOP is comparable to the computations made using gadgets on code-reuse attacks by ROP [17], JOP [4,7], and sigreturn-oriented programming [5]; however, unlike all the previous approaches, DOP is based on *data-oriented gadgets* that have a small number of differences compared to classic gadgets.

DOP requires the use of (i) *data-oriented gadgets* and (ii) a *gadget dispatcher*. On the one hand, data-oriented gadgets form the virtual instructions (i.e. arithmetic, logical, assignment, load, store, jump and conditional jump) required to simulate a Turing machine. Data-oriented gadgets can simulate these operations using the x86 instruction set the same way ROP and JOP do. In contrast, data-oriented gadgets need to use just memory and not memory or registers to generate its operations. In addition, data-oriented gadgets must follow the legitimate control flow. On the contrary, one of the benefits of data-oriented gadgets is that they can be scattered and consequently, there is no need for them to be executed one after the other.

On the other hand, the gadget dispatcher allows the chaining of the data-oriented gadgets and simulating control operations. This dispatcher allows an attacker to choose the sequence of data-oriented gadgets (e.g. creating loops), resulting in *interactive* and *non-interactive* DOP attacks. Interactive attacks use loops and at every loop iteration, a selector controlled by the memory error selects the sequence of data-oriented gadgets that must be executed. Non-interactive attacks require a single payload where all the data-oriented gadgets must be chained.

3 Data-Flow Integrity

Castro et al. introduced *data-flow integrity* (DFI) [6], a defensive technique that aims to protect programs against non-control-data attacks. DFI targets x86 architecture ensuring that a given program's data stays within the permitted paths. Firstly, DFI generates the data-flow graph (DFG) of the program by static analysis, secondly, it instruments the program introducing data-flow integrity checks, and finally, it enforces at runtime that the data-flow of the program is allowed by the DFG, otherwise the execution is aborted.

DFI relies on *reaching definitions analysis* [2] for the static DFG generation. Reaching definitions analysis is a static analysis technique used by modern compilers to deploy global code optimisation (e.g. dead code elimination), based on data-flow analysis. Data-flow analysis tries to extract information about the flow of data from program execution paths [2], and reaching definitions analysis concretely deals with the definition (i.e. assignment) and use (i.e. read) of variables. Using reaching definitions analysis, DFI can compute a DFG that contains a set of definitions, assigns an identifier to each definition, and maps those identifiers to instructions. In this way, the DFG shows the instructions that assigned a value to each used variable.

DFI uses two different static analyses to generate reaching definitions, (i) a *flow-sensitive intra-procedural* analysis and (ii) a *context-insensitive inter-procedural* analysis. The former is used to compute the reaching definitions of variables that have no definition outside the function in which they are declared, whereas the later computes the reaching definitions of variables with definitions outside the function in which they are declared. This separation is done to increase the performance of the analysis, since flow-insensitive algorithms have less computing overhead.

Once the static DFG has been generated, DFI instruments at runtime the program to check before every variable use that its definition is within the statically generated reaching definitions identifiers. If not, the data-flow integrity property does not hold, and the program must be terminated. Castro et al.'s approach makes use of a *runtime definitions table* (RDT) to keep track of the last definition of each identifier. In order to check if the data-flow integrity holds, the last value of the RDT for a given identifier must be checked against the static DFG.

In order to be effective, DFI itself must remain safe against sabotages, requiring (i) the integrity of the RDT, (ii) the integrity of the code, and (iii) the integrity of DFI's instrumentation. RDT integrity is achieved ensuring that the definitions are within the memory boundaries defined for the RDT, code integrity is accomplished using modern operating systems' W$\oplus$E check on pagination. The integrity of the instrumentation performed by DFI can be ensured relying on DFI (e.g. instrumenting uses and definitions of control-data made by the compiler) or using additional defences, such as CFI [1] and program shepherding [14].

3.1 Kernel Data-Flow Integrity

The operating system is the first line of defence against attacks based on memory corruption on userland applications. However the OS itself is not safe against non-control-data attacks. If an attacker were capable to successfully gain control of the OS, all the defences deployed in userland applications would become futile.

Song et al. [18] utilise DFI in order to enforce kernel security invariants related to access control mechanisms against memory-corruption attacks. They proposed a system named KENALI in order to protect two security invariants,

(i) *complete mediation*, attackers have to be prevented from bypassing access control checks, and (ii) *tamper proof*, the integrity of the data and code of the reference monitors must be maintained.

As to enforce these invariants KENALI uses two techniques: *InferDist* and *ProtectDists*. InferDist is used to distinguish the *distinguished regions*, which are the regions that have essential data for enforcing the security invariants. ProtectDists enforces DFI over these regions and due to invariant (i) complete mediation, CFI must also be enforced.

Furthermore, InferDist uses the kernel CFI mechanisms proposed by Criswell et al. [10] to protect control-data. Regarding non-control data, KENALI enforces that if a security check fails, it will return the `-EACCESS` error code (permission denied). InferDist retrieves these error codes and, through dependency analysis on the conditional variables of the security checks, is able to discover which are the distinguished regions.

Finally, to enforce the DFI over the inference result regions, KENALI distinguishes tree types of data-flow (i) within non-distinguishing regions, (ii) between two different types of regions and (iii) within distinguishing regions. KENALI protects the distinguishing regions using a two-layer scheme. The first layer is a lightweight data-flow isolation mechanism to protect the second type of data-flow (between two different types of regions), and a more heavy DFI enforcement mechanism when the two regions are distinguishing.

4 Related Work

Apart from DFI, *Dynamic Information Flow Tracking* (DIFT) [19] and *Dynamic Taint Analysis* are two techniques that can be applied to prevent non-control-data attacks. DIFT is a hardware mechanism to track malicious information flows. These information flows are controlled by the operating system which denies the usage of suspicious paths for the flow of information through suspicious paths.

DTA [16] is a technique to monitor taint sources of a program while it is executing. DTA is commonly used in malware analysis and vulnerability discovery, and thereby it can be applied to non-control-data attack detection.

However, these techniques are not the focus of this review paper since they have not been applied yet to prevent non-control-data attacks.

5 Discussion and Conclusions

Non-control-data attacks are gaining more importance due to the efforts that have been directed into preventing and mitigating control-data attacks. Despite all the tools and methods deployed to protect commodity operating systems and userland programs against code-injection and code-reuse attacks, there is still a lack of security mechanisms to protect the same operating systems and userland programs against non-control-data attacks. We believe the main challenge of DFI is to overcome the trade off between computing overhead and completeness, because the use of more accurate and thus more slow static analysis techniques in the DFG generation can prevent users from using these tools.

References

1. Abadi, M., Budiu, M., Erlingsson, U., Ligatti, J.: Control-flow integrity: principles, implementations and applications. In: Proceedings of the ACM SIGSAC Conference on Computer and Communications Security (2005)
2. Aho, A.V., Lam, M.S., Sethi, R., Ullman, J.D.: Compilers: Principles, Techniques, and Tools. Addison-Wesley, Reading (2006)
3. Andersen, S., Abella, V.: Data Execution Prevention. Changes to Functionality in Microsoft Windows XP Service Pack 2, Part 3: Memory Protection Technologies (2004)
4. Bletsch, T., Jiang, X., Freeh, V.W., Liang, Z.: Jump-oriented programming: a new class of code-reuse attack. In: Proceedings of the ACM SIGSAC Conference on Computer and Communications Security (2011)
5. Bosman, E., Bos, H.: Framing signals-a return to portable shellcode. In: Proceedings of the IEEE Symposium on Security and Privacy (Oakland) (2014)
6. Castro, M., Costa, M., Harris, T.: Securing software by enforcing data-flow integrity. In: Proceedings of the USENIX Symposium on Operating Systems Design and Implementation (OSDI) (2006)
7. Checkoway, S., Davi, L., Dmitrienko, A., Sadeghi, A.R., Shacham, H., Winandy, M.: Return-oriented programming without returns. In: Proceedings of the ACM SIGSAC Conference on Computer and Communications Security (2010)
8. Chen, S., Xu, J., Sezer, E.C., Gauriar, P., Iyer, R.K.: Non-control-data attacks are realistic threats. In: Proceedings of the USENIX Security Symposium (2005)
9. Cowan, C., Pu, C., Maier, D., Hinton, H., Walpole, J., Bakke, P., Beattie, S., Grier, A., Wagle, P., Zhang, Q.: StackGuard: automatic adaptive detection and prevention of buffer-overflow attacks. In: Proceedings of the USENIX Security Symposium (1998)
10. Criswell, J., Dautenhahn, N., Adve, V.: KCoFI: Complete control-flow integrity for commodity operating system kernels. In: Proceedings of the IEEE Symposium on Security and Privacy (Oakland) (2014)
11. Giuffrida, C., Kuijsten, A., Tanenbaum, A.S.: Enhanced operating system security through efficient and fine-grained address space randomization. In: Proceedings of the USENIX Security Symposium (2012)
12. Hu, H., Chua, Z.L., Adrian, S., Saxena, P., Liang, Z.: Automatic generation of data-oriented exploits. In: Proceedings of the USENIX Security Symposium (2015)
13. Hu, H., Shinde, S., Adrian, S., Chua, Z.L., Saxena, P., Liang, Z.: Data-oriented programming: on the expressiveness of non-control data attacks. In: Proceedings of the IEEE Symposium on Security and Privacy (Oakland) (2016)
14. Kiriansky, V., Bruening, D., Amarasinghe, S.P., et al.: Secure execution via program shepherding. In: Proceedings of the USENIX Security Symposium (2002)
15. Nergal: The advanced return-into-lib(c) exploits: Pax case study. Phrack Magazine **58** (2001)
16. Schwartz, E.J., Avgerinos, T., Brumley, D.: All you ever wanted to know about dynamic taint analysis and forward symbolic execution (but might have been afraid to ask). In: Proceedings of the IEEE Symposium on Security and Privacy (Oakland) (2010)

17. Shacham, H.: The geometry of innocent flesh on the bone: Return-into-libc without function calls (on the x86). In: Proceedings of the ACM SIGSAC Conference on Computer and Communications Security (2007)
18. Song, C., Lee, B., Lu, K., Harris, W., Kim, T., Lee, W.: Enforcing kernel security invariants with data flow integrity. In: Annual Network and Distributed System Security Symposium (NDSS) (2016)
19. Suh, G.E., Lee, J.W., Zhang, D., Devadas, S.: Secure program execution via dynamic information flow tracking. In: Conference on Architectural Support for Programming Languages and Operating Systems (ASPLOS) (2004)
20. PaX Team: Address space layout randomization (ASLR) (2003). http://pax.grsecurity.net/docs/aslr.txt
21. US-CERT: OpenSSL 'Heartbleed' vulnerability (CVE-2014-0160) (2014). https://www.us-cert.gov/ncas/alerts/TA14-098A

CISIS 2016: Infrastructure and Network Security

Towards a Secure Two-Stage Supply Chain Network: A Transportation-Cost Approach

Camelia-M. Pintea[1(✉)], Anisoara Calinescu[2], Petrica C. Pop[1], and Cosmin Sabo[1]

[1] North University Center at Baia-Mare, Technical University Cluj-Napoca, Baia-Mare, Romania
dr.camelia.pintea@ieee.org, petrica.pop@cunbm.utcluj.ro, cosmin_sabo@prime-tech.com

[2] Department of Computer Science, University of Oxford, Oxford OX1 3QD, UK
ani.calinescu@cs.ox.ac.uk

Abstract. The robustness, resilience and security of supply chain transportation is an active research topic, as it directly determines the overall supply chain resilience and security. In this paper, we propose a theoretical model for the transportation problem within a two-stage supply chain network with security constraints called the *Secure Supply Chain Network (SSCN)*. The *SSCN* contains a manufacturer, directly connected to several distribution centres *DC*, which are directly connected to one or more customers *C*. Each direct link between any two elements of *Secure Supply Chain Network* is allocated a transportation cost. Within the proposed model, the manufacturer produces a single product type; each distribution centre has a fixed capacity and a *security rank*. The overall objective of the *Secure Supply Chain Network* is 100 % customer satisfaction whilst fully satisfying the security constraints and minimizing the overall transportation costs. A heuristic solving technique is proposed and discussed.

1 Introduction

Supply Chain Networks as well as other transportations and logistics nowadays are facing security threats. These threats could be natural disasters, such as earthquakes or floods, or human-related threats, such as market uncertainty and computer attacks [28].

As stated by Rice et al. [15]: "The supply network is inherently vulnerable to disruption, and the failure of any one element in it could cause the whole network to fail". Furthermore, Lee et al. [17] highlight the need to create a "secure freight system", across all transportation stages between supply chain partners.

The transport-related problems are combinatorial optimization problems. The researchers use in general heuristic approaches [4,18,20,25], approximation algorithms [2] or hybrid algorithms [8] to solve transportation problems.

It is inherently difficult to model real-world transportation problems: due to the large number of constraints and parameters, often characterised by

M. Graña et al. (eds.), *International Joint Conference SOCO'16-CISIS'16-ICEUTE'16*, Advances in Intelligent Systems and Computing 527, DOI 10.1007/978-3-319-47364-2_53

variability, uncertainty and interdependence, as well as many local optimal or sub-optimal solutions. There are supply chains with the transportation cost directly proportional to the number of units transported [5]. Some transportation problems can be described as fixed-charge transportation problems [1,25].

The problem described in this paper assumes a fixed-charge transportation. The *Secure Supply Chain Network (SSCN)* problem involves two different supply chain configurations. The first supply chain consists of a manufacturer directly linked to all distribution centers. The second chain is from each of the distribution centres to one or more customers. The novelty is introducing secure constraints for distribution centers into the existing two-stage Supply Chain Network with fixed-charge transportation.

The objective function is to minimize the whole transportation cost, from manufacturer to customers, based on several intermediary costs and other parameters, whilst fully satisfying the customers and meeting the security constraints. The manufacturer supplies a single type of product. The objective function takes into account opening fixed costs and transportation costs.

A mathematical model of the problem is given using the *SSCN* parameters and additional security constraints. The applicability of linear algorithms for solving real-world transportation problems is limited due to their inherent complexity. Hybrid models were used to solve these type of problems [19,25]. A related approach is in [6]. Previous work solved this problem using the nearest neighbour in [12], a hybrid approach in [11], a genetic algorithm in [14] and further research with a reverse distribution system in [13].

The paper is organized as follows. Section 2 includes some prerequisites. Section 3 introduced the *SSCN* problem with its formulation and mathematical model. In Sect. 4 is described the new problem *Secure Supply Chain Network* with the security constraints. The last section discusses the strengths and limitations of the *SSCN* problem and concludes with future work suggestions.

2 Prerequisites

The *Supply Chain Network (SCN)* problem is formulated based on a particular *Supply Chain Network* presented in [9,11–13], with fixed-charged capacity.

The *Supply Chain Network* problem considers two stages of a supply chain network: The manufacturer and distribution centres (DCs), and the DCs and customers (C). The first supply chain includes the manufacturer that provides items to distribution centres (DCs); the second supply chain uses these *DC*s to deliver to the customers, according to their demands. An ideal manufacturer, with no capacity limitation in production is considered. Each potential distribution centre has a different capacity to support the customers [9]. The time constraints are not considered. The total cost includes the transportation costs from the manufacturer to potential distribution centres, the opening cost of these DCs, and the transportation cost from the DCs to the customers.

In [10,23,24] was introduced and developed a security constraint model for jobs scheduling. The security modes considered are the *secure* mode, as a conservative approach, the *Risky* mode, the aggressive approach, and the γ-*risky* mode, considering a probability measure of risk. In general the risky or the γ-risky modes are used. The secure mode includes supplementary costs and it allows less flexibility, and therefore it is rarely used.

3 The Secure Supply Chain Network: Problem Formulation and Mathematical Model

The *Secure Supply Chain Network* problem it is a *Supply Chain Network* with security constraints. Each supply chain is endowed with a security constraint set, which includes the security demand and security rank, denoted $SC = \{sd, sr\}$ (see Sect. 4 and Fig. 1).

The objective of the problem is to optimize the transportation cost, from manufacturer to customers, subject to satisfying the security risk constraints of distribution centers.

The mathematical model of the fixed-charged transportation problem with security constraints uses m DCs and n customers. Table 1 shows the input and output data with notations and descriptions; the input data includes besides the SCN input, the security demand (sd) for a DC, the security rank (sr_i) for each DC_i and the security levels values: from very low (vl) to very high (vh).

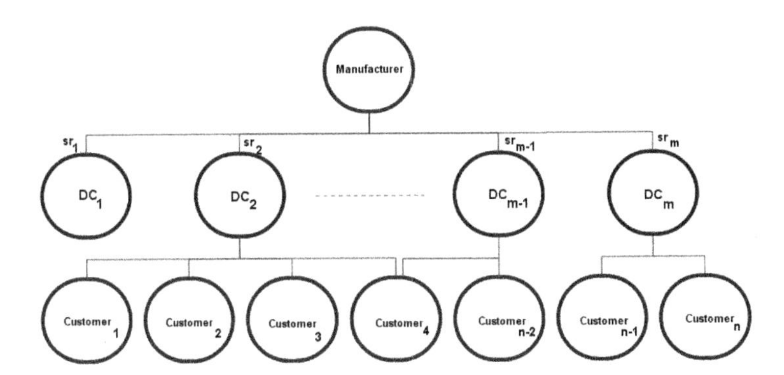

Fig. 1. An example of a two-stage Secure Supply Chain Network (SSCN) with an infinite-capacity manufacturer, m available distribution centres (DCs) and n customers.

The total cost, Z, includes fixed costs, Z_{fc}, and transportation costs, Z_{tc}. The aim is to minimize the total $SSCN$ cost while meeting the security constraints and fully satisfying the customers. Two types of fixed costs are involved: the opening cost f_i for DC_i and the fixed transportation cost from DC_i to customer j, denoted f_{ij}.

The transportation cost per unit includes the transportation cost from the manufacturer to distribution centre i, c_i, and the transportation cost per unit (c_{ij}) from DC_i to a customer j.

Table 1. *Secure Supply Chain Network* parameter description.

Symbol	Description
input	
sd	The security demand for a DC
sr_i	The security rank of DC_i
vl-vh	The security levels values: from very low to very high
f_i	The opening cost for DC_i
f_{ij}	The fixed transportation cost from DC_i to C_j
c_i	The transportation cost from M to DC_i
c_{ij}	The transportation cost from DC_i to C_j
a_i	The capacity of DC_i
d_j	The demands of C_j
x_{ij}	A part of the demand of C_j supplied by DC_i
x_i	A part of the quantity of the M supplied to DC_i
y_i	A binary variable; 1, if DC_i is opened as potential location
y_{ij}	A binary variable; 1, if DC_i is opened for customer C_j
output	
Z	The total cost: the sum of fixed costs (Z_{fc}) and transportation cost (Z_{tc})

Due to the two stages of the supply chain network, the following objectives are specified, in order to minimize the total cost:

- Identify the distribution centre to be opened that satisfies the security requirements, as a function of its security demand, sd, and security rank, sr, as presented in Sect. 4.
- Identify the optimal set of DCs for each customer. The demands of all the customers should be met; a customer could receive products from one or more distribution centres.

The main objective of the problem is to minimize the function Z.

$$Z = Z_{fc} + Z_{tc} \tag{1}$$

$$Z_{tc} = \sum_{i=1}^{m} c_i x_i + \sum_{i=1}^{m}\sum_{j=1}^{n} c_{ij} x_{ij}, \qquad Z_{fc} = \sum_{i=1}^{m} f_i\ y_i + \sum_{i=1}^{m}\sum_{j=1}^{n} f_{ij} y_{ij}, \tag{2}$$

The following Eq. (3) are the constraints for the quantities to be transported.

$$x_{ij} \geq 0,\ \forall i = 1, \dots m,\ \forall j = 1, \dots n, \qquad x_i \leq a_i,\ \forall i = 1, \dots m, \tag{3}$$

where

$$y_i = \begin{cases} 1, \sum_{j=1}^{n} x_{ij} \geq 0 \\ 0, \sum_{j=1}^{h} x_{ij} = 0 \end{cases}, \forall i = 1 \dots m, \quad y_{ij} = \begin{cases} 1, x_{ij} \geq 0,\ \forall i = 1 \dots m, \\ 0, x_{ij} = 0,\ \forall j = 1 \dots n. \end{cases} \tag{4}$$

The secure fixed-charge transportation problem is about to minimize the function Z, Eq. (1), when satisfy the security constraints, Eq. (5). The security objective should be satisfied on both stages of the supply chain network.

In the next section the new constraint (Eq. (5)) of the *SSCN* is introduced.

4 The Secure Supply Chain Network: The Security Constraint

In this section, we propose a security constraint for the *Supply Chain Network* based on the security constraint model for data-intensive jobs running on distributed computing environments [10, 23, 24].

A distribution is assigned to an available unit, for example a distribution centre in our particular problem, only if the condition $sd \leq sr$ is met. Three security modes are considered:

- *Secure* mode: the fully secure mode permits transportation only on the units satisfying the security requirements.
- *Risky* mode: the risky mode takes all possible risks by transporting the products on any available unit, *DC* in our case.
- *γ-risky* mode: use a probability measure of risk, the γ-risky mode, the transportation will be available to the unit taking at the most γ-risk. The *secure* mode is obtained for $\gamma = 0$, and the *risky* mode when $\gamma = 1$.

The current approach of *SSCN* uses the *γ-risky* mode. The security levels are assessed in distributed environments [22] in terms of a qualitative/fuzzy scale with five levels: very high (vh), high (h), medium (med), low (l), very low (vl).

- A transportation is secure if it is made to a completely safe unit, when $sd \leq sr$.
- The risk must be less than 50 % in the security constraint model if a distribution is assigned to a unit with a failure risk (when $sd > sr$).
- The transport is possible if $0 < sd - sr \leq vl$, where vl represents the *very low* security level.
- The transport will be delayed when $vl < sd - sr \leq l$, but should be made before the execution deadline [26]; vl and l are the *very low* and *low* security levels.
- A transport operation is not feasible if $l < sd - sr \leq vh$ based on [22], where l and vh are the *low* and *very high* security levels.

The risk probability, Eq. (5), for the proposed security constraint model is defined as in [10] and represented in Fig. 1. The sr_i is the security risk of the i-th distribution center, DC_i.

$$P(risk) = \begin{cases} 0, & sd - sr_i \leq 0 \\ 1 - e^{-\frac{1}{2} sd - sr_i}, & 0 < sd - sr_i \leq vl \\ 1 - e^{-\frac{3}{2} sd - sr_i}, & vl < sd - sr_i \leq l \\ 1, & l < sd - sr_i \leq vh \end{cases} \quad \forall i = 1, \ldots m. \tag{5}$$

5 Discussions on the Proposed Problem

A hybrid algorithm based on the nearest neighbour searching technique [2], was used in [12] to solve the fixed-charged transportation problem; an improved hybrid algorithm for the problem was introduced in [11]. An efficient reverse distribution system for solving a sustainable *SCN* was proposed in [13]. The Nearest Neighbour transportation model could be applied to minimize the transportation cost in the two-stage supply chain network with security constraints on DCs, based on [11–13]. The model is used when choosing a distribution centre and a connection between DCs and customers.

After the list (L) of selected DCs is specified, based on quantities and security risks, the second supply chain is considered. The reverse technique [13] could be used by starting from the customer demand, and the objective is to use the secure DCs from list L by optimizing the transportation costs, Eq. (1). In real life there could be different threats on the second supply chain, so the risk parameters could be different from those in the first supply chain.

The threats could come from several internal technical issues, manageable factors, or from external factors. Some of the technical problems could be staffing capacity of distribution centres, and the product design and manufacturing [3,21]. External threats could be increases in raw material price, or natural disasters. A natural disaster could lead, for example, to a transportation delay. An external threat over a supply chain could be controllable if there are some previous contingency plans in case of the specific threat [28], such as spare manufacturing capacity and/or transportation links.

The controllable threats are quantified as mathematical parameters, as already described in Sect. 4. Being the most expensive, due to major costs, the secure mode which allows transportation only when $sd \leq sr$, is very rare used. Either the risky mode or the γ-risky mode are used, instead.

For the risky mode, when all possible risks are taken by transporting the products on any available unit, DC or customer, the *SSNC* problem is reduced to the initial *SNC* problem without any secure conditions. As specified in Sect. 4, in the γ-risky mode the transportation will be available to an unit, in our case the distribution centre for both supply chains, taking at most γ-risk.

We consider a numerical example involving four DC_i, $i = \overline{1,4}$. The values for security demand, $sd = 10$, the security risk values for each DC are $sr = \{11, 8, 15, 12\}$ and the levels of security from the very low vl to the very high vh are: $vl = 1$, $l = 2$, $med = 3$, $h = 4$ and $vh = 5$.

Based on Table 2 the distribution will be possible just on DC_1, DC_2 and DC_4, where the security risk is acceptable. The risk probability is computed based on Eq. (5). The current ongoing work is on implementing the proposed algorithm for solving the *SSCN* problem in terms of security constraints. The tests consider different values for the security parameters.

Furthermore could be useful to develop a component-based system, for example modeled as a finite automaton as in [7,27]; could be extended to include time constraints and a security approach using intelligent mobile multiagent systems [16].

Table 2. The influence of security levels of distribution centre when distributing the products from manufacturer to distribution centres, and from distribution centres to customers (based on [22,26]), for security demand $sd = 10$, DC security risk values $sr = \{11, 8, 15, 12\}$ and the levels of security $vl = 1$, $l = 2$, $med = 3$, $h = 4$ and $vh = 5$.

DC	Security level of a DC	Symbolic representation	P(risk)	Risk description for a DC
DC_2	High security level	$sd - sr \leq 0$	0	No risk is involved in the distribution to/from DC
DC_1	Medium security level	$0 < sd - sr \leq 1$	$1 - e^{-16}$	Distribution to/from DC is possible, has no a major risk
DC_4	Low security level	$1 < sd - sr \leq 2$	$1 - e^{-27}$	Delayed distribution to/from DC due to security risks
DC_3	Very low security level	$2 < sd - sr \leq 5$	1	Distribution to/from DC it is not possible

6 Conclusions and Future Work

The security of product distribution is nowadays a challenging problem. The paper proposes a security-based model for a fixed-charged transportation problem within a two-stage supply chain network. The complexity of the initial supply chain problem increases due to the security constraints. Several discussions about the solution for the optimization function based on diverse security levels are presented. Future work will include other constraints, such as time constraints and a dynamic approach of the supply chain problem.

Acknowledgements. The study was conducted under the auspices of the IEEE-CIS Interdisciplinary Emergent Technologies TF.

References

1. Adlakha, V., Kowalski, K.: On the fixed-charge transportation problem. OMEGA Int. J. Manag. Sci. **27**, 381–388 (1999)
2. Arya, S., et al.: An optimal algorithm for approximate nearest neighbor searching in fixed dimensions. J. ACM **45**(6), 891–923 (1998)
3. Calinescu, A., et al.: Applying and assessing two methods for measuring complexity in manufacturing. J. Oper. Res. Soc. **49**(7), 723–733 (1998)
4. Chira, C., Dumitrescu, D., Pintea, C.-M.: Learning sensitive stigmergic agents for solving complex problems. Comput. Inf. **29**(3), 337–356 (2010)
5. Diaby, M.: Successive linear approximation procedure for generalized fixed charge transportation problems. J. Oper. Res. Soc. **42**, 991–1001 (1991)
6. Deaconu, A., Ciurea, E.: A study on the feasibility of the inverse supply and demand problem. In: Proceedings 15th International Conference on Computers, pp. 485–490 (2011)
7. Fanea, A., Motogna, S., Diosan, L.: Automata-based component composition analysis. Studia Universitas Babes-Bolyai, Informatica **50**(1), 13–20 (2006)
8. Pintea, C.-M., Crisan, G.-C., Chira, C.: Hybrid ant models with a transition policy for solving a complex problem. Logic J. IGPL **20**(3), 560–569 (2012)

9. Molla-Alizadeh-Zavardehi, S., et al.: Solving a capacitated fixed-charge transportation problem by artificial immune and genetic algorithms with a Prufer number representation. Expert Sys. Appl. **38**, 10462–10474 (2011)
10. Liu, H., Abraham, A., Snášel, V., McLoone, S.: Swarm scheduling approaches for work-flow applications with security constraints in distributed data-intensive computing environments. Inf. Sci. **192**, 228–243 (2012)
11. Pintea, C.-M., Pop, P.C.: An improved hybrid algorithm for capacitated fixed-charge transportation problem. Logic J. IGPL **23**(3), 369–378 (2015)
12. Pintea, C.-M., Sitar, C.P., Hajdu-Macelaru, M., Petrica, P.: A hybrid classical approach to a fixed-charged transportation problem. In: Corchado, E., Snášel, V., Abraham, A., Woźniak, M., Graña, M., Cho, S.-B. (eds.) HAIS 2012. LNCS (LNAI), vol. 7208, pp. 557–566. Springer, Heidelberg (2012). doi:10.1007/978-3-642-28942-2_50
13. Pop, P.C., et al.: An efficient reverse distribution system for solving sustainable supply chain network design problem. J. Appl. Logic **13**(2), 105–113 (2015)
14. Pop, P.C., et al.: A hybrid based genetic algorithm for solving a capacitated fixed-charged transportation problem. Carpathian J. Math. **32**(2), 225–232 (2016)
15. Rice, J.B., Caniato, F.: Building a secure and resilient supply network. Supply Chain Manag. Rev. **7**(5), 22–30 (2003)
16. Iantovics, B., Crainicu, B.: A distributed security approach for intelligent mobile multiagent systems. Stud. Comput. Intell. **486**, 175–189 (2014)
17. Lee, H.L., Wolfe, M.: Supply chain security without tears. Supply Chain Manag. Rev. **7**(3), 12–20 (2003)
18. Matei, O.: Evolutionary Computation: Principles and Practices. Risoprint, London (2008)
19. Mes, M., et al.: Comparison of agent-based scheduling to look-ahead heuristics for real-time transportation problems. Eur. J. Oper. Res. **181**(1), 59–75 (2007)
20. Nechita, E., et al.: Cooperative ant colonies for vehicle routing problem with time windows. A case study in the distribution of dietary products. WMSCI **5**, 48–52 (2008)
21. Sivadasan, S., et al.: Advances on measuring the operational complexity of supplier–customer systems. Eur. J. Oper. Res. **171**(1), 208–226 (2006)
22. Song, S., Hwang, K., Zhou, R., Kwok, Y.: Trusted P2P transactions with fuzzy reputation aggregation. IEEE Internet Comput. **9**(6), 24–34 (2005)
23. Song, S., Kwok, Y., Hwang, K.: Security-driven heuristics and a fast genetic algorithm for trusted grid job scheduling. Int. Parallel Distrib. Process. IEEE CS **65**, 4–12 (2005)
24. Song, S., Hwang, K., Kwok, Y., et al.: Risk-resilient heuristics and genetic algorithms for security-assured grid job scheduling. IEEE Trans. Comp. **55**(6), 703 (2006)
25. Sun, M., et al.: A tabu search heuristic procedure for the fixed charge transportation problem. Eur. J. Oper. Res. **106**(2), 441–456 (1998)
26. Venugopal, S., Buyya, R.: A deadline and budget constrained scheduling algorithm for escience applications on data grids. In: Hobbs, M., Goscinski, A.M., Zhou, W. (eds.) ICA3PP 2005. LNCS, vol. 3719, pp. 60–72. Springer, Heidelberg (2005). doi:10.1007/11564621_7
27. Vescan, A., Motogna, S.: Overview and architecture of a component modeling tool. Creative Math. Inf. **16**, 159–165 (2007)
28. Li, X., Chandra, C.: Toward a secure supply chain: a system's perspective. Hum. Syst. Manag. **27**(1), 73–86 (2008)

The HTTP Content Segmentation Method Combined with AdaBoost Classifier for Web-Layer Anomaly Detection System

Rafał Kozik(✉) and Michał Choraś

Institute of Telecommunications, UTP University of Science and Technology, ul. Kaliskiego 7, 85-789 Bydgoszcz, Poland
{rafal.kozik,chorasm}@utp.edu.pl

Abstract. In this paper we propose modifications to our machine-learning web-layer anomaly detection system that adapts HTTP content mechanism. Particularly we introduce more effective packet segmentation mechanism, adapt AdaBoost classifier, and present results on more challenging dataset. In this paper we also compared our approach with other techniques and reported the results of our experiments.

1 Introduction

Currently, we can observe that vulnerable web pages are a common element in many high-profile cyber attacks. On the other hand, web applications (as well as web services) have become an inherent element of many supply chains. In example, many web-services work as crucial elements in so called content delivery networks (CDNs) providing data such as weather forecast, software for downloads, news, etc. In result, if one of such elements becomes infected it will impact the whole chain of the connected nodes in a distributed network.

In order to protect the web services against some of the cyber attacks the application layer, one can adapt solutions (called web application firewalls - WAFs), that work as a software module integrated with the web server. In the literature there are also plenty of methods applying more complex schemas for learning normal/anomalous models. However, quite often a challenge for typical WAF software is the fact that HTTP protocol can be used as transport by other protocols (e.g. SOAP, or GWT-RPC). As a result, each protocol uses its inherent request structure to transport user inputs from web browser to web server (see Fig. 1). Typically, to solve this problem it is required to provide additional configuration to the WAF to enable appropriate content type mapping (e.g. from text/x-gwt-rpc to application/x-www-form-urlencoded).

In contrast to our previous work presented in [1], we have proposed more flexible schema for aligning the HTTP packets contents and different approach to data classification. Moreover, we have extended the experiments with the new cyber attack detection tools and the evaluation scenarios.

The paper is structured as follows: the overview of the feature extraction method is presented in Sect. 2. The experimental setup and results are provided in Sect. 3. The conclusions are given thereafter.

M. Graña et al. (eds.), *International Joint Conference SOCO'16-CISIS'16-ICEUTE'16*, Advances in Intelligent Systems and Computing 527, DOI 10.1007/978-3-319-47364-2_54

https://www.google.pl/webhp? sourceid=chrome-instant&ion=1&
espv=2&ie=UTF-8#q=**serialization+ and+deserialization**

https://www.google.pl/maps/place/ **Stacja+PKP+Bydgoszcz+Fordon/**
@53.1439362,18.16896,14z/data=
!4m2!3m1!1s0x4703166afe4d5843: 0x63ecc18530748e8f

http://chart.apis.google.com/ chart?cht=p&chs=500x250&chdl=
first+legend%7Csecond+legend% 7Cthird+legend&chl=first+label%
7Csecond+label%7Cthird+label&
chco=**FF0000|00FFFF|00FF00,6699CC|**
CC33FF|CCCC33&chp=0.436326388889&
chtt=**My+Google+Chart**&chts=000000, 24&chd=**t:5,10,50|25,35,45**

Fig. 1. Example of custom serialization technique used in URL address. Some of the methods use classic "*key* = *value*" separated by "&", while others use customized serialization techniques, e.g. sequence of values separated by "–" or "+".

2 Related Work

There are many solutions (called web application firewalls - WAFs), that work as signature-based filters (e.g. set of rules) over the HTTP traffic. These signatures usually cover common web application attacks (e.g. SQL Injection, Cross Site Scripting, etc.). In example, the ModSecurity [2] plug-in for Apache web server (particularly the "OWASP ModSecurity Core Rule Set – CRS") is intended to provide defensive protection against common application layer attacks. Also the PHPIDS [3] tool is another security measure for web applications. However, in contrast to ModSecurity it can be integrated only with PHP-enabled web servers. The NAXSI [4] is a third party plug-in module for high performance Nginx [5] web server. The NAXSI stands for Nginx Anti XSS and SQL Injection and it provides low-level rules that detect keywords (symptoms) of application layer attacks. In contrast to ModSecurity and PHPIDS, NAXSI learns normal application behaviour instead of attacks patterns.

In the literature there are also methods applying anomaly detection (variety of complex schemas for learning normal/anomalous models). For instance, in [6] authors used χ^2 metric and character distribution approach to detect anomalous HTTP requests. In order to increase the attack detection effectiveness authors used a parser that splits the request into URL address and query string of attributes followed by values. Another approach to HTTP traffic anomaly detection was presented in [7]. Authors applied DFA (Deterministic Finite Automaton) to compare the requests described by the means of tokens. In contrast, in [8] authors have compared different n-grams techniques applied to application layer anomaly detection. In the literature there are also approaches combining

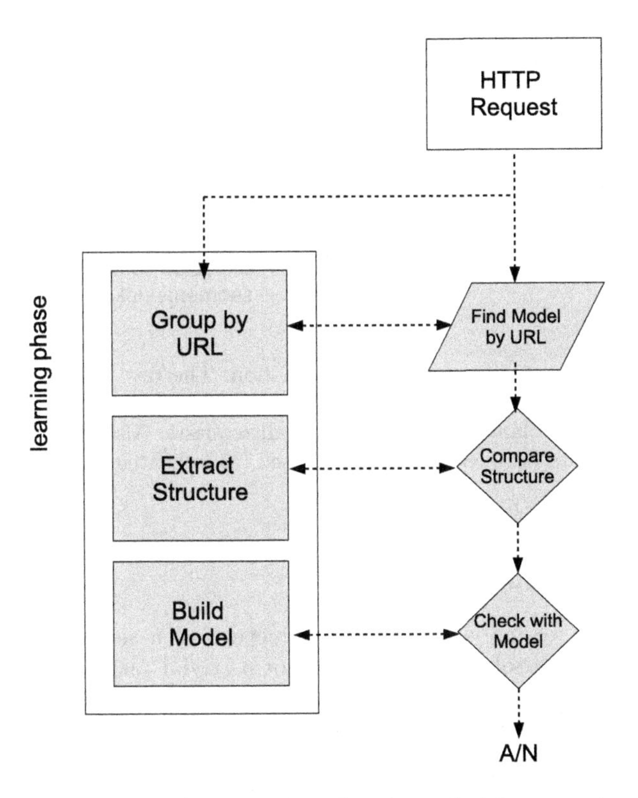

Fig. 2. The proposed algorithm overview. It adapts batch processing approach to machine learning. First, the model is established on learning data, afterwards new data samples are compared with the model in order to detect a cyber attack. The method returns binary result (A/N), where A indicated *anomaly* and N *normal*.

n-grams with Self Organizing Maps [9], Bloom filters [10], and wide variety of different machine-learnt classifiers [11].

3 The Proposed System Overview

There are two phases of the system operation. Before it can be used for attack detection (phase one), the learning procedure has to be applied (phase two). During the learning phase (see Fig.2), the algorithm groups the HTTP request by URL address. The URL address can be accompanied with the arguments (e.g. .../index.php?x=1&y=2). Similar data can be sent in the request content. In this paper, we proposed a technique to extract structure from HTTP requests payloads. Details will be presented in Sect. 3.1. Once the structure is extracted we encode variables and build the model using different variants of machine learning. Details are presented in Sect. 3.2.

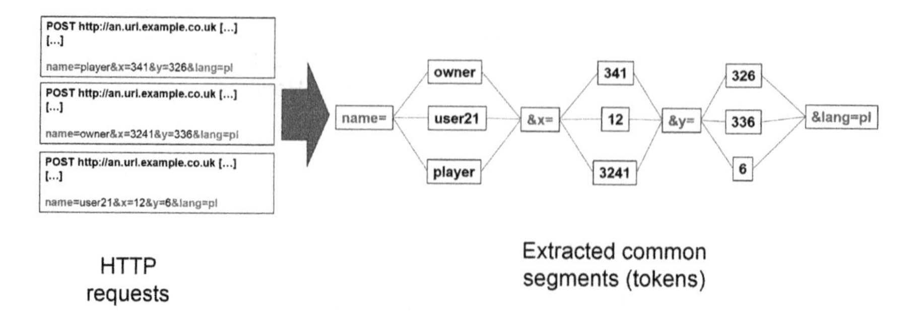

Fig. 3. The overview of idea for tokens extraction. The raw HTTP requests (examples are shown on left side) are analysed to identify the common sequences (&x=,&y=,&name=,&lang=pl) appearing in all requests. Also the left-to-right order is extracted and data between tokens is collected (as indicated on graph shown on the right side).

3.1 Packet Structure Extraction

In our approach, we model the packet structure as a sequence of tokens (see Fig. 3). However, the token extraction is not a trivial task to perform, because single token can appear at different positions in the consecutive HTTP requests. Moreover, the distance between tokens may also vary. Additionally, it may happen that one token is a sub-token of another.

In the bioinformatics science such problem is known as multiple sequence alignment (MSA), where the aligned sequences are represented by DNA chains. Over the last decades, several algorithms have been proposed to solve this problem efficiently. These algorithms can be classified into three categories, namely: exact [12], progressive [13], and iterative [14] methods. The "exact" methods adapt straightforward technique employing dynamic programming to solve the MSA problem. On the other hand, the "progressive" methods perform a heuristic alignment by performing a series of alignments (starting from two closely related sequences). Therefore, the quality of final result is strongly influenced by the initial alignment. To overcome this, the "iterative" methods repeatedly realign initial sequences.

In our approach, we adapted one of the methods that can be classified as a "stochastic iterative" approach. Hereby we use genetic algorithm (GA) to iteratively realign analysed sequences of HTTP payloads in order to extract their structure. However, the MSA method applied directly to raw HTTP payloads may produce poor results. Moreover, the method will be quite slow for long HTTP payload sequences. Therefore, to address these issues we first extract the longest common substrings (LCStr). The LCStr problem is different form LCS problem. For instance for two sequences "ABCD" and "AYCD", the LCS is "ACD", while the LCStr is "CD" (other LCStrs will be "A", "C", and "D"). This approach allows us to identify candidates for tokens and then find their left-to-right alignment

More formally, given n tokens of value v_i, we want to solve the optimisation problem described by Eq. 1.

$$\begin{aligned} \underset{x}{\text{maximise}} \quad & C(x) = \sum_{i=0}^{n} v_i x_i \\ \text{subject to} \quad & \sum_{i=0}^{n} w_i x_i \leq W, \ x_i \in \{0,1\}. \end{aligned} \tag{1}$$

As mentioned before, in order to solve the multiple sequence alignment (MSA) optimisation problem, we adapted genetic algorithms with classical binary chromosome encoding schema and one point crossover. The chromosome in our algorithm represents the candidate solution and it is a string of bits (1 indicates that given token is taken to build the structure of request, while 0 is used to reject the given token). The genetic algorithm works as follows:

1. The population is initialized randomly. The chromosome length is determined by the number of tokens identified during the extraction procedure.
2. The fitness of each chromosome is measured. Individuals are ordered by fitness values.
3. Two chromosomes are selected randomly from the population.
4. Selected chromosomes are subjected to crossover procedure.
5. The procedure is terminated (and individual with best fitness is selected) if maximal number of iteration is exceeded, otherwise it is continued starting from step 2.

Once the tokens are identified, we describe the sequences between tokens using their statistical properties. In this work we have used typical approach for textual data encoding that adapts character distribution histograms. However, instead of classical one byte per bin association, we count the number of characters such that the decimal value in an ASCII table falls into particular ranges. Selected ranges represent different type of symbols like numbers, quotes, letters or special characters. This is significant dimensionality reduction in contrast to sparse 256-bin histograms (one bin per byte character).

3.2 Model Derivation

For the data classification problem we adapted classifiers hybridisation method. Our main goal was to address the problem of data imbalance (usually we have only few examples of attacks while having large volume of data representing normal traffic samples) and to increase the effectiveness of attack detection by adapting an ensemble of one-class classifiers. However, one of the issues of producing the ensemble of classifiers is the diversity problem. According to [15], there are different techniques to improve the diversity, such as data or feature partitioning during learning or boosting or bagging (to exploit local specialisation of given classifiers).

In order to increase the diversity, in our experiments, we investigated hybridisation technique, which combines Reduces Error Pruning Tree (REPTree) [16] classifier with AdaBoost approach. The REPTree is a machine learning technique that uses pruned decision tree. REPTree algorithm generates multiple regression trees in each iteration. Afterwards, it chooses the best one. It uses regression tree adapting variance and information gain (by measuring the entropy). The algorithm prunes the tree using back fitting method.

The reason why we selected the mentioned above classifiers is two-fold. First of all, these classifiers are often used with bagging and boosting techniques, and secondly, these are also efficient and simplistic [16].

4 Experiments

In this paper we use known benchmark dataset for web application cyber attacks and anomalies detection CSIC'10 [17]. However, it does not contain some of the recent web application anomalies. For instance, the original dataset currently contains mainly requests that are complied with "application/x-www-form-urlencoded" encoding standard. On the other hand, the modern web applications are currently using also different types of request content, e.g. "applicaiton/json", "text/x-gwt-rpc", or other proprietary types. Therefore, we decided to enhance it with additional data coming from the monitoring of requests to real web application. We increased the original CSIC'10 dataset by 3.6 %. We added 2.08 % of normal samples (around 1500 requests) and 7.9 % of new attacks (around 2000 of anomalous requests). To gather these samples we set up a GWT-based web application that provides feature-rich GIS (Geographic Information System). In this application we identified several malicious code injection vulnerabilities. We used these vulnerabilities to convey cyber attacks.

In our experiments we have measured different performance characteristics in order to compare adapted in our work classifiers ensembles. For evaluation we adapted classical 10-fold cross validation. The reported results measurements are following $TP\ Rate$ (True Positive Rate indicating the percentage of detected HTTP request labelled as anomalous) and $FP\ Rate$ (False Positive Rate indicating the percentage normal HTTP request labelled (wrongly) as anomalous).

During the experiments we have compared our approach to different detection mechanisms to show that our technique indeed allows as to improve detection rates and to lower the number of false positives.

Because we use character distribution as feature vector, we used the method proposed in [6] as a baseline for results evaluation. In [6] authors used χ^2-test to measure the fitness of the tested sample with so called ICD. The ICD (Idealised Character Distribution) is an averaged character distribution of requests that are considered normal.

All of the experiments have been conducted on the extended CSIC'10+ dataset. In order to explain why the extended CSIC'10+ dataset is more challenging, in Table 1 we have presented the percentage number of detected attacks/anomalies for fixed percentage number of false positives (FPR). It can

Table 1. Effectiveness of our method compared for two different datasets. For all presented methods the percentage of false positives (FPR) is fixed at ~0.007 level

Method	CSIC'10+	CSIC'10 base
Our method	0.92	0.94
BodyParse + ICD	0.88	0.94

be noticed that when we added 8 % of new attacks to CSIC'10 (while creating CSIC'10+), the detection ration (TPR) has decreased by 6 % in case of "Body-Parse + ICD" method and only by 2 % in case of our method.

Table 2. Effectiveness of compared methods.

Type	Method	CSIC'10+	
		TP Rate	FP Rate
Signature-based	PHPIDS	0.204	0.013
	ApacheMod	0.263	0.003
Anomaly-based	Compression-based	0.430	0.000
	ICD	0.332	0.001
	BodyParse + ICD	0.834	0.011
	Our method	**0.915**	**0.007**

Firstly, we have adapted classical HTTP body parameters request segmentation combined with χ^2 metrics of character distribution (in table indicated as BodyParse+ICD). This method assumes that HTTP request is fully compliant with RFC 2616 (Hypertext Transfer Protocol). It means that data sent from web browser to http server is encoded as a sequence of key-values separated by ampersand (e.g. key1=val1&key2=val2). It can be noticed that such method obtains fairly-satisfactory results achieving more than 80 % of attacks detection ration while having 1.1 % of false positives. Moreover, the effectiveness of ICD approach without structure segmentation is even worse (33.2 % of attack detection). In case of compression-based technique, this allows us to achieve 0 % of false positives while detecting 43 % of all anomalies/attacks.

On the contrary, our method (in table indicated as "Our method+ AdaBoost") combined with cost sensitive boosting method for ensemble of classifiers building, allows us to improve the results significantly. As it is shown in Table 2, such approach achieves more than 91 % of attacks detection while having 0.7 % of false positives.

During our experiments we have also evaluated selected signature-based methods. It can be noticed that neither PHPIDS nor ApacheMod perform well when it comes to attacks detection. However, ApacheMod have quite low rate of false alarms (0.34 %). It must be noticed that here we do not count attacks, which signatures are not able to detect, e.g. request is made to files that do not exist on the server. We removed such requests by manual inspection.

5 Conclusions

In this paper we have proposed an algorithm that combines packet structure extraction with statistical content analysis. As for the structure analysis we have adapted stochastic multiple sequence alignment (MSA). For the content analysis we have proposed machine-learnt classifier that engages boosting technique. We have evaluated the proposed method on an extended version of the CISIC'10 dataset in order to assess effectiveness of attacks detection on more recent examples of attacks targeting application layer. As it was shown, our method allows us to achieve satisfactory high ratio of attacks detection while having low number of false positives.

References

1. Kozik, R., Choraś, M., Renk, R., Holubowicz, W.: Patterns extraction method for anomaly detection in HTTP traffic. In: Herrero, A., Baruque, B., Sedano, J., Quintan, H., Corchado, E. (eds.) International Joint Conference CISIS 2015 and ICEUTE 2015, Advances in Intelligent Systems and Computing, pp. 227–236. Springer, Switzerland (2015)
2. ModSecurity project homepage. https://www.modsecurity.org/
3. PHPIDS project homepage. https://github.com/PHPIDS/PHPIDS
4. NAXSI project homepage. https://github.com/nbs-system/naxsi
5. NGINX project homepage. http://nginx.org/en/
6. Kruegel, C., Vigna, G.: Anomaly detection of web-based attacks. In: Proceedings of the 10th ACM Conference on Computer and Communications Security, pp. 251–261 (2003)
7. Ingham, K.L., Somayaji, A., Burge, J., Forrest, S.: Learning DFA representations of HTTP for protecting web applications. Comput. Netw. **51**(5), 1239–1255 (2007)
8. Hadžiosmanović, D., Simionato, L., Bolzoni, D., Zambon, E., Etalle, S.: N-Gram against the machine: on the feasibility of the n-gram network analysis for binary protocols. In: Balzarotti, D., Stolfo, S.J., Cova, M. (eds.) RAID 2012. LNCS, vol. 7462, pp. 354–373. Springer, Heidelberg (2012). doi:10.1007/978-3-642-33338-5_18
9. Bolzoni, D., Zambon, E., Etalle, S., Hartel, PH.: POSEIDON: a 2-tier anomaly-based network intrusion detection system. In: IWIA 2006: Proceedings of 4th IEEE International Workshop on Information Assurance, pp. 144–156 (2006)
10. Wang, K., Parekh, J.J., Stolfo, S.J.: Anagram: a content anomaly detector resistant to mimicry attack. In: Recent Advances in Intrusion Detection, pp. 226–248 (2006)
11. Perdisci, R., Ariu, D., Fogla, P., Giacinto, G., Lee, W.: McPAD: a multiple classifier system for accurate payload-based anomaly detection. Comput. Netw. **53**(6), 864–881 (2009)
12. Sundfeld, D., Melo, A.C.M.A.: MSA-GPU: exact multiple sequence alignment using GPU. In: Setubal, J.C., Almeida, N.F. (eds.) BSB 2013. LNCS, vol. 8213, pp. 47–58. Springer, Heidelberg (2013). doi:10.1007/978-3-319-02624-4_5
13. Higgins, D.G., Sharp, P.M.: Clustal: a package for performing alignment on a microcomputer. Gene **73**, 237–244 (1988)

14. Gotoh, O.: Sequence alignments by iterative refinement as assessed by reference to structural alignments. J. Mol. Biol. **264**(4), 823–838 (1996)
15. Wozniak, M.: Hybrid Classifiers: Methods of Data, Knowledge, and Classifiers Combination. Springer Series in Studies in Computational Intelligence. Springer, Heidelberg (2013)
16. Frank, E.: Data Mining: Practical Machine Learning Tools and Techniques. Data Management Systems, 2nd edn. Morgan Kaufmann, USA (2005)
17. Torrano-Gimnez, C., Prez-Villegas, A., Alvarez, G.: The HTTP dataset CSIC (2010). http://users.aber.ac.uk/pds7/csic_dataset/csic2010http.html

Cluster Forests Based Fuzzy C-Means for Data Clustering

Abdelkarim Ben Ayed(✉), Mohamed Ben Halima, and Adel M. Alimi

REGIM-Lab.: Research Groups in Intelligent Machines,
University of Sfax, ENIS, BP 1173, 3038 Sfax, Tunisia
{abdelkarim.benayed.tn,
Mohamed.benhlima,Adel.alimi}@ieee.org

Abstract. Cluster forests is a novel approach for ensemble clustering based on the aggregation of partial K-means clustering trees. Cluster forests was inspired from random forests algorithm. Cluster forests gives better results than other popular clustering algorithms on most standard benchmarks. In this paper, we propose an improved version of cluster forests using fuzzy C-means clustering. Results shows that the proposed Fuzzy Cluster Forests system gives better clustering results than cluster forests for eight standard clustering benchmarks from UC Irvine Machine Learning Repository.

Keywords: Cluster forest · Clustering · Ensemble clustering · Optimization · Fuzzy logic

1 Introduction

In order to analyze effectively the large quantity of data that we receive across social Medias and other industrial systems, we need to employ data clustering.

Data clustering allows us to identify the different entities of data. The objective of data clustering is to divide data in groups while maximizing the distance between the groups and minimizing the distance between data of the same group.

Several clustering algorithms are being used nowadays [1, 2, 12]; the most used are the hierarchical clustering [1], C-Means and Gaussian mixture model [3]. Hierarchical clustering are simple to use but cannot deal with large data. C-means clustering is more appropriate for large data sets. Gaussian mixture model clustering is more appropriate for data with different group size.

To improve clustering quality, the new clustering trends are using soft computing and ensemble clustering algorithms. Soft computing are based on using fuzzy logic, genetic algorithms and other bio-based algorithms like swarm intelligence, etc. Ensemble clustering algorithms allows using multiple clustering algorithms, generally the same clustering algorithm applied to different subset of data features to improve clustering quality. Cluster Forest [4] is an example of ensemble clustering algorithm and it is inspired from random forest classification algorithm [5]. In [11], the authors present differential equation cluster forests, which is a variant of cluster forests. These algorithms are noise resistant.

M. Graña et al. (eds.), *International Joint Conference SOCO'16-CISIS'16-ICEUTE'16*,
Advances in Intelligent Systems and Computing 527, DOI 10.1007/978-3-319-47364-2_55

In this paper, we propose to use an ensemble-clustering algorithm, which is cluster forest, with fuzzy C-means as the base clustering algorithm.

In the second section, we present cluster forest algorithm. In third section, we present Fuzzy logic clustering. In fourth section, we explain the proposed fuzzy cluster forest algorithm. Finally, we present data sets, results with other works comparison in the fifth section.

2 Cluster Forest Algorithm

Cluster Forest (CF) [4] is an ensemble-clustering algorithm. It has two steps: in the first step, CF makes multiple cluster instances based k-means algorithm on subsets of features, then in a second step, CF aggregates all cluster instances to obtain the final clustering result using spectral clustering algorithm.

2.1 Step1: Growth of Clustering Vectors

The objective of this step is to build a vector of strong features. Clustering with a subset of strong data features allows obtaining better clustering quality compared to clustering with all data features.

To decide that a clustering vector (subset of features) is better than another vector, cluster forests uses a cluster quality measure κ [4].

$$k = \frac{Within\ Cluster\ Distance}{Between\ Cluster\ Distance} \tag{1}$$

After the base clustering algorithm (k-means) is applied to the clustering vector, the cluster quality measure κ is used to compare the clustering vectors. The lower is κ, the better is.

The growth of clustering vectors is composed of two steps: initialization of the clustering vector with m features using feature competitions then growth of the clustering vector by adding iteratively m features that improve the cluster quality measure κ (see Fig. 1) [4]. The values for parameters f (number of features added every time) and d (maximum number of consecutive discard) are usually set respectively to 2 and 3 according to empirical experiments [4].

2.2 Step2: CF Algorithm

This is the real CF algorithm step. It applies k-means to the clustering vector built in step1, and then calculates the co-cluster indicator matrix P.

All these steps are repeated T times (usually 100 times), every instance is considered as a tree growth since CF grows a set of features. The T instances represent a forest of T trees.

To get the final clustering result, all co-cluster indicator matrix P are averaged, regularized then spectral clustering is applied to the result [4, 6] (see Fig. 2).

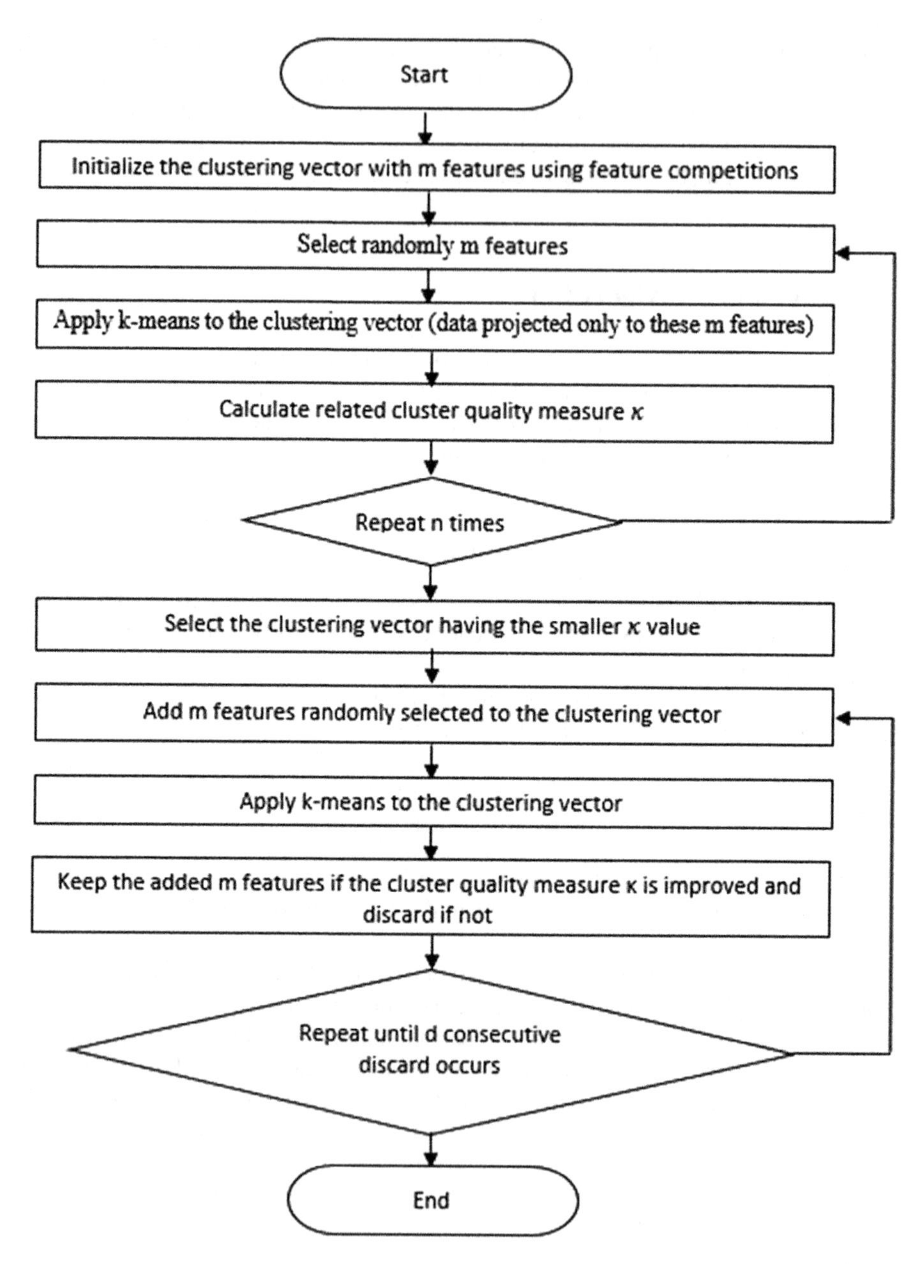

Fig. 1. Growth of clustering vector algorithm

While hard clustering methods assign data points to only one cluster, fuzzy clustering methods, based on fuzzy membership, assign data points to multiple clusters at the same time. To get final hard clustering, the max defuzzification method is used.

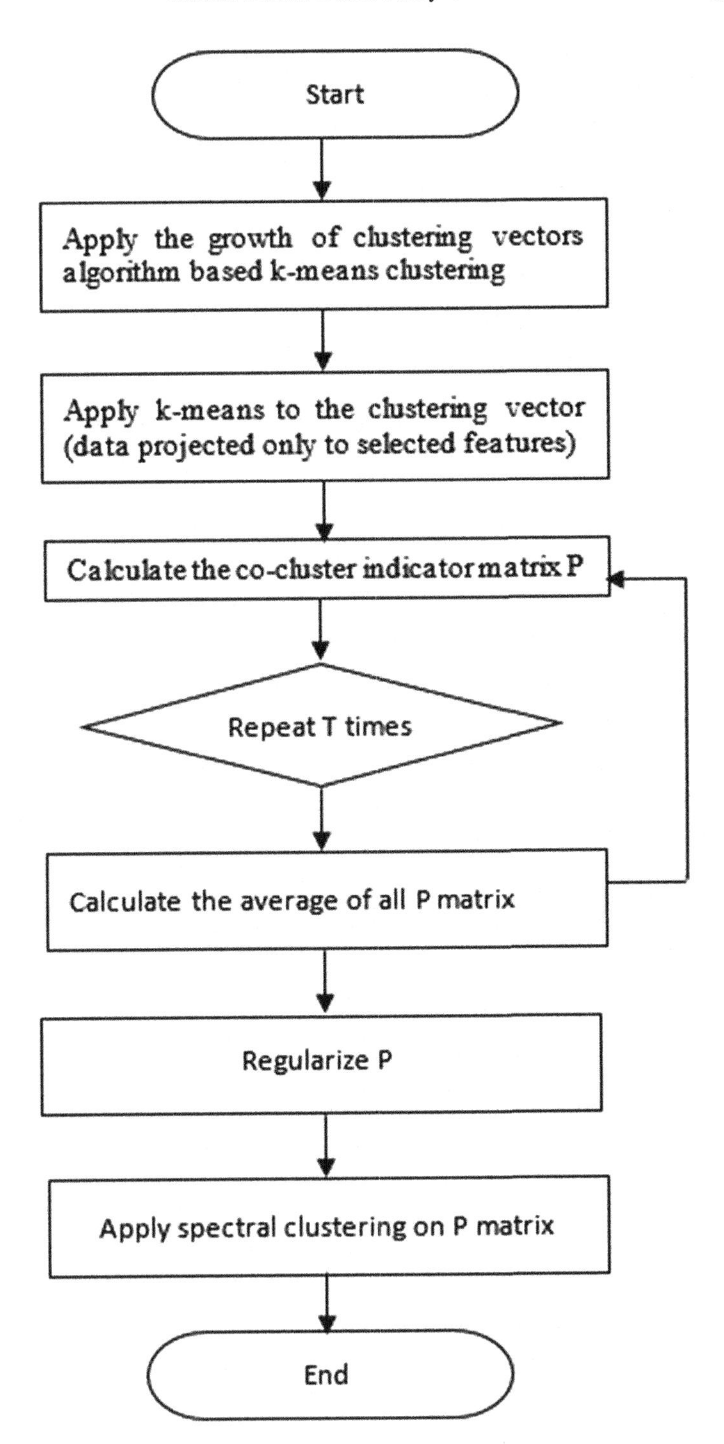

Fig. 2. Cluster forests algorithm

Fuzzy clustering methods are used mainly when data clusters have overlapping boundaries. Most used fuzzy clustering methods are based on objective function like fuzzy c-means and fuzzy Gaussian mixture modals.

3 Fuzzy Cluster Forest

3.1 Fuzzy C-Means Clustering

Dunn [8] was the first to implement fuzzy C-means clustering algorithm in 1973, later Bezdek improved this algorithm in 1981 [7]. For fuzzy C-means, the membership function is represented by matrix (U) having values between 0 and 1 that represent the membership of data points to every cluster, while hard C-means uses only 0 and 1 values for membership function. Fuzzy C-means algorithm is based on an objective function [9, 13].

$$J'(U,V) = \sum_{i=1}^{C} \sum_{j=1}^{N} u_{ij}^{m} ||X_j - V_i||^2 \tag{2}$$

Where

- U: membership matrix
- V: cluster centers matrix
- N: number of data points
- C: number of clusters
- X_j: is the jth measured data point
- v_i: is the center of cluster i

$$V_i = \frac{\sum_{j=1}^{N} u_{ij}^{m} X_j}{\sum_{j=1}^{N} u_{ij}^{m}}, 1 \le i \le C \tag{3}$$

- u_{ij}: $(0 \le u_{ij} \le 1)$ is the membership value of x_j with respect to cluster i.

$$u_{ij} = \frac{1}{\sum_{k=1}^{C} (\| X_j - V_i \| / \| X_j - V_k \|)^{2/(m-1)}} \tag{4}$$

- m: $(m \ge 1)$ represents the parameter of cluster fuzziness, usually this parameter value is set to 2.

Fuzzy C-means clustering algorithms steps are:

1. Chose C cluster centers randomly (matrix V)
2. Compute the membership matrix U
3. Update the cluster centers v_i
4. Compute the objective function J
5. Repeat steps 2 to 4 until convergence $(\| var(J) \le \varepsilon \|)$

6. Use MAX deffuzzification to decide about the belonging of fuzzy cluster memberships.

3.2 Fuzzy Cluster Forest

Cluster forest is an ensemble clustering method using hard C-means (k-means) as the base clustering algorithm. K-means is an effective and simple clustering algorithm, but is not robust for real overlapping data.

In order to improve the clustering quality, mainly for overlapping real data sets, we decide to use fuzzy C-means algorithm in spite of using hard C-means algorithm (see Fig. 3).

4 Results

To validate the accuracy of our developed system, we test it on real word data sets and we compare it with the original cluster forest algorithm and other clustering algorithms.

4.1 Testing Data Sets

We realized our experiments on real world eight UC Irvine data sets [10], see Table 1. These data sets are used as standard benchmarks in several clustering works and include small size to large size data sets. True labels are provided with the eight data sets and are used to evaluate the performance of clustering in order to compare the clustering algorithms.

4.2 Testing Environment

To test our system we use R statistical computing environment version 3.2.2, running on a PC with Windows 10, 8 GB of RAM and Intel core-i7 CPU.

4.3 Evaluation Metrics

To evaluate the performance of clustering algorithms, we use two performance metrics, ρ_r(clustering quality) and ρ_c(clustering accuracy) [4].

The first metric is ρ_r. The idea of this metric is that two pairs of data points that belongs to the same expected cluster must be also in the same found cluster.

$$\rho_r = \frac{Number\ of\ correctly\ clustred\ pairs}{Total\ number\ of\ pairs} \quad (5)$$

The second metric is ρ_c [4]. The idea of this second metric is inspired from the classification accuracy metric.

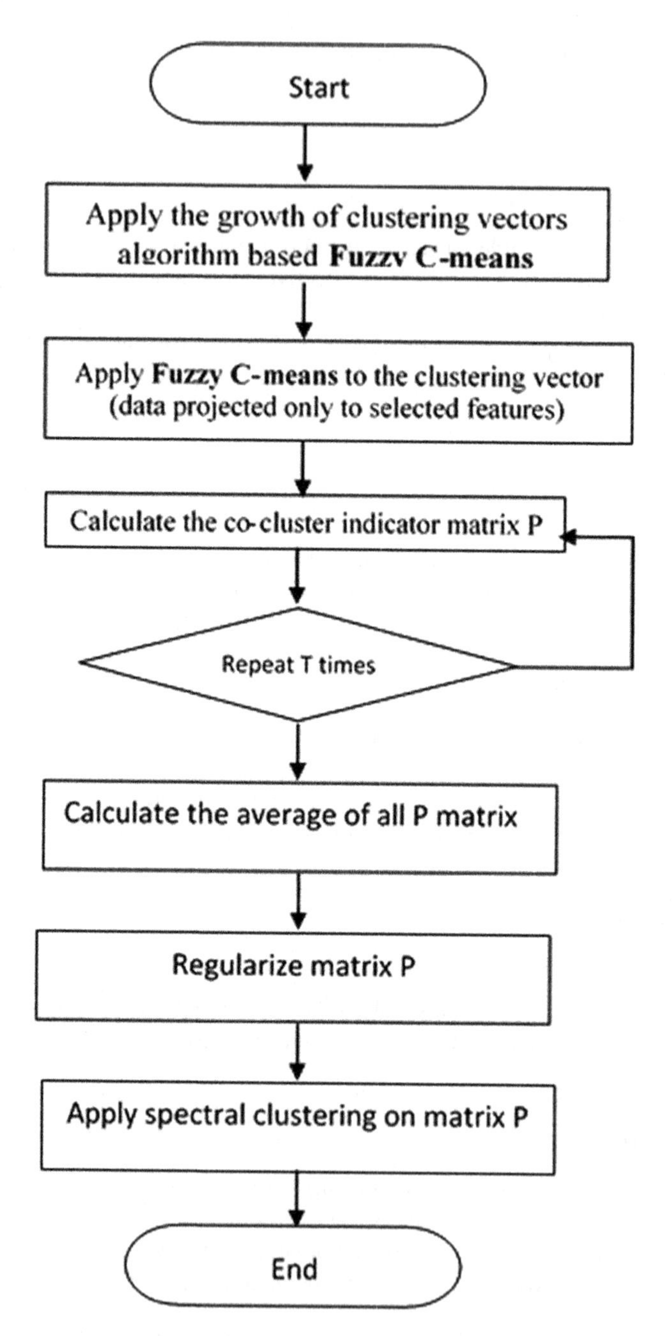

Fig. 3. Fuzzy cluster forests algorithm

$$\rho_c = max_{all\,permutations}\left\{\frac{Number\ of\ correctly\ clustred\ pairs}{Number\ of\ data\ points}\right\} \quad (6)$$

Table 1. Testing data sets

Dataset	Features	Instances	Classes
Soybean	35	47	4
SPECT	22	267	2
ImgSeg	19	2100	7
Heart	13	270	2
Wine	13	178	3
WDBC	30	569	2
Robot	90	164	5
Madelon	500	2000	2

4.4 Experimental Results and Comparisons

We test our clustering algorithm (Fuzzy Cluster Forest), on eight UCI datsets and we compare it to hard C-means, fuzzy C–means and cluster forest algorithms using clustering quality (see Table 2) and clustering accuracy (see Table 3) metrics.

Table 2. Comparison of results for metric ρ_r

Datasets	Hard C-means	Fuzzy C-means	Cluster forest	Fuzzy cluster forest
Soybean	83.81 %	83.16 %	**92.36 %**	92.04 %
SPECT	50.77 %	51.94 %	56.78 %	**57.05 %**
ImgSeg	82.54 %	**83.39 %**	79.71 %	78.84 %
Heart	51.54 %	51.54 %	56.90 %	**60.63 %**
Wine	71.27 %	71.05 %	79.70 %	**91.18 %**
WDBC	75.04 %	75.04 %	79.93 %	**85.95 %**
Robot	51.22 %	54.53 %	63.60 %	**63.53 %**
Madelon	49.98 %	**50.74 %**	50.76 %	50.17 %

Table 3. Comparison of results for metric ρ_c

Datasets	Hard C-means	Fuzzy C-means	Cluster Forest	Fuzzy cluster forest
Soybean	71.49 %	72.34 %	84.43 %	**83.97 %**
SPECT	56.86 %	60.30 %	68.02 %	**68.41 %**
ImgSeg	54.02 %	**54.79 %**	48.24 %	50.20 %
Heart	59.26 %	59.26 %	68.26 %	**72.59 %**
Wine	67.80 %	68.54 %	79.19 %	**93.18 %**
WDBC	85.41 %	85.41 %	88.70 %	**92.38 %**
Robot	36.54 %	**43.94 %**	41.20 %	39.07 %
Madelon	50.32 %	54.73 %	**55.12 %**	52.24 %

For our algorithm parameters, a number of parameters have been fixed manually. For cluster forest parameters, we used 100 trees, the number of chosen features per iteration is fixed to 2 and the number of feature competition iterations to 1. For fuzzy c–means, the fuzzy exponent is fixed to 2 and the maximum number of iterations is fixed to 200. All results are obtained on the average of 10 iterations.

Results shows that fuzzy cluster forest outperforms both hard C-means, fuzzy C–means and cluster forest clustering algorithms.

For our approach fuzzy cluster forests, results are better than classical cluster forests [4] except for the data sets Madelon and ImgSeg due to some irregularities in these data sets.

5 Conclusion

In this paper, we improved the original cluster forest algorithm by replacing the base clustering algorithm (k-means) by fuzzy C-means. Experiments on eight real-world data sets and compared to other clustering algorithms shows that fuzzy cluster forest outperform most other clustering algorithms.

To improve results we are planning to use other soft computing technics and to use parallel computing model like the system described in [13, 14] to reduce computing time.

Acknowledgment. The authors would like to acknowledge the financial support of this work by grants from General Direction of Scientific Research (DGRST), Tunisia, under the ARUB program.

References

1. Berkhin, P.: A survey of clustering data mining techniques. In: Kogan, J., Nicholas, C., Teboulle, M. (eds.) Grouping Multidimensional Data, pp. 25–71. Springer, Heidelberg (2006)
2. Jain, Anil K.: Data clustering: 50 years beyond K-means. Pattern Recogn. Lett. **31**(8), 651–666 (2010)
3. Kalti, K., Mahjoub, M.A.: Image segmentation by gaussian mixture models and modified FCM algorithm. Int. Arab J. Inf. Technol. **11**(1), 11–18 (2014)
4. Donghui, Y., Chen, A., Jordan, M.I.: Cluster forests. Comput. Stat. Data Anal. **66**, 178–192 (2013)
5. Leo, B.: Random forests. Mach. Learn. **45**(1), 5–32 (2001)
6. Andrew, N., Jordan, M.I., Weiss, Y.: On spectral clustering: Analysis and an algorithm. Adv. Neural Inf. Process. Syst. **2**, 849–856 (2002)
7. Zadeh, L.A.: Fuzzy sets. Inf. Control **8**, 338–352 (1965)
8. Ruspini, E.R.: Numerical methods for fuzzy clustering. Inf. Sci. **2**, 319–350 (1970)
9. Begum, S.A., Devi, M.O.: A rough type-2 fuzzy clustering algorithm for MR image segmentation. Int. J. Comput. Appl. **54**(4), 4–11 (2012)
10. Lichman, M.: UCI machine learning repository (2016). http://archive.ics.uci.edu/ml
11. Jan, J., Gajdoš, P., Radecký, M., Snášel, V.: Application of bio-inspired methods within cluster forest algorithm. In: Proceedings of the Second International Afro-European Conference for Industrial Advancement AECIA (2015)
12. Ayed, A.B., Halima, M.B., Alimi, A.M.: Survey on clustering methods: towards fuzzy clustering for big data. In: 6th International Conference of Soft Computing and Pattern Recognition (SoCPaR), pp. 331–336, Tunis, Tunisia (2014)

13. Ayed, A.B., Halima, M.B., Alimi, A.M.: MapReduce based text detection in big data natural scene videos. In: INNS Conference Big Data 2015, BigData'2015, San Francisco, USA, 08–10 August 2015
14. Ayed, A.B., Halima, M.B., Alimi, A.M.: Big data analytics for logistics and transportation. In: Fourth IEEE International Conference on Advanced Logistics and Transport, IEEE ICALT'2015, Valenciennes, France, pp. 311–316, 20–22 May (2015)

Neural Visualization of Android Malware Families

Alejandro González[1], Álvaro Herrero[1(✉)], and Emilio Corchado[2]

[1] Department of Civil Engineering, University of Burgos, Avenida de Cantabria s/n, 09006 Burgos, Spain
agr0095@alu.ubu.es, ahcosio@ubu.es
[2] Departamento de Informática y Automática, Universidad de Salamanca, Plaza de la Merced, s/n, 37008 Salamanca, Spain
escorchado@usal.es

Abstract. Due to the ever increasing amount and severity of attacks aimed at compromising smartphones in general, and Android devices in particular, much effort have been devoted in recent years to deal with such incidents. However, scant attention has been devoted to study the interplay between visualization techniques and Android malware detection. As an initial proposal, neural projection architectures are applied in present work to analyze malware apps data and characterize malware families. By the advanced and intuitive visualization, the proposed solution provides with an overview of the structure of the families dataset and ease the analysis of their internal organization. Dimensionality reduction based on unsupervised neural networks is performed on family information from the Android Malware Genome (Malgenome) dataset.

Keywords: Android malware · Malware families · Artificial neural networks · Exploratory projection pursuit

1 Introduction

Since the first smartphones came onto the market in the late 90 s, sales on that sector have increased constantly until present days. Among all the available operating systems, Google's Android is the most popular mobile platform [1]. The number of Android-run units sold in Q4 2015 worldwide raised to 325.39 million out of 403.12 million units, that is a share of 80.71 %. It is not only the number of devices but also the number of apps; those available at Google Play (Android's official store) constantly increase, up to more than 2.1 million that are available nowadays [2]. With regard to the security issue, Android became the top mobile malware platform as well [3] and it is forecast that the volume of Android malware will spike to 20 million during 2016 when it was 4.26 million at the end of 2014 and 7.10 million in first half of 2015 [4]. This operating system is an appealing target for bad-intentioned apps, mainly because of its open mentality, in contrast to iOS or some other operating systems.

Smartphone security and privacy are nowadays major concerns. In order to address these issues, it is required to understand the malware and its nature. Otherwise, it will not be possible to practically develop an effective solution [5]. According to this idea of

M. Graña et al. (eds.), *International Joint Conference SOCO'16-CISIS'16-ICEUTE'16*, Advances in Intelligent Systems and Computing 527, DOI 10.1007/978-3-319-47364-2_56

gaining deeper knowledge about malware nature, present study is focused on the analysis of Android malware families. To do so, Malgenome (a real-life publicly-available) dataset [6] has been analyzed by means of several neural visualization models. From the samples contained in such dataset, several alarming statistics were found [5], that motivate further research on Android malware. That is the case of the 36.7 % of the collected samples that leverage root-level exploits to fully compromise the security of the whole system or the fact that more than 90 % of the samples turn the compromised phones into a botnet controlled through network or short messages.

To characterize malware families, this study proposes the use of neural models able to visualize a high-dimensionality dataset, further described in Sect. 2. Each individual from the dataset (a malware app) encodes the subset of selected features using a binary representation (details on Sect. 4). These individuals are grouped by families and then visualized trying to identify patterns that exist across dimensional boundaries in the high dimensional dataset by changing the spatial coordinates of family data. The idea is to obtain an intuitive visualization of the malware families to draw conclusions about the structure of the dataset.

Neural visualization techniques have been previously applied to massive security datasets, such as those generated by network traffic [7], SQL code [8], honeynets [9], or HTTP traffic [10]. In present paper, such methods are applied to a new problem, related to the detection of malware.

Up to now, a growing effort has been devoted to detect Android malware [11]. Machine learning [12, 13], has been applied to differentiate between legitimate and malicious Android apps, as well as knowledge discovery [14], and weighted similarity matching of logs [15] among others. Although some visualization techniques have been applied to the detection of malware in general terms [16], few visualization-based proposals for Android malware detection are available at present time. In [17] Pythagoras tree fractal is used to visualize the malware data, being all apps scattered, as leaves in the tree. Authors of [18] proposed graphs for deciding about malware by depicting lists malicious methods, needless permissions and malicious strings. In [19], visualization obtained from biclustering on permission information is described. Behavior-related dendrograms are generated out of malware traces in [20], comprising nodes related to the package name of the application, the Android components that has called the API call and the names of functions and methods invoked by the application. Unlike previous work, Android malware families are visualized by neural models in present paper. Up to the authors knowledge, this is the first time that neural projection models are applied to visualize Android malware.

The rest of this paper is organized as follows: the applied neural methods are described in Sect. 2, the setup of experiments for the Android Malware Genome dataset is described in Sect. 3, together with the results obtained and the conclusions of the study that are stated in Sect. 4.

2 Neural Visualization

This work proposes the application of unsupervised neural models for the visualization of Android malware data. Visualization techniques are considered a viable approach to information seeking, as humans are able to recognize different features and to detect anomalies by means of visual inspection. The underlying operational assumption of the proposed approach is mainly grounded in the ability to render the high-dimensional traffic data in a consistent yet low-dimensional representation. In most cases, security visualization tools have to deal with massive datasets with a high dimensionality, to obtain a low-dimensional space for presentation.

This problem of identifying patterns that exist across dimensional boundaries in high dimensional datasets can be solved by changing the spatial coordinates of data. However, an a priori decision as to which parameters will reveal most patterns requires prior knowledge of unknown patterns.

Projection methods project high-dimensional data points onto a lower dimensional space in order to identify "interesting" directions in terms of any specific index or projection. Having identified the most interesting projections, the data are then projected onto a lower dimensional subspace plotted in two or three dimensions, which makes it possible to examine the structure with the naked eye.

2.1 Principal Component Analysis

Principal Component Analysis (PCA) is a well-known statistical model, introduced in [21], that describes the variation in a set of multivariate data in terms of a set of uncorrelated variables each, of which is a linear combination of the original variables. From a geometrical point of view, this goal mainly consists of a rotation of the axes of the original coordinate system to a new set of orthogonal axes that are ordered in terms of the amount of variance of the original data they account for.

PCA can be performed by means of neural models such as those described in [22] or [23]. It should be noted that even if we are able to characterize the data with a few variables, it does not follow that an interpretation will ensue.

2.2 Maximum Likelihood Hebbian Learning

Maximum Likelihood Hebbian Learning [24] which is based on Exploration Projection Pursuit. The statistical method of EPP was designed for solving the complex problem of identifying structure in high dimensional data by projecting it onto a lower dimensional subspace in which its structure is searched for by eye. To that end, an "index" must be defined to measure the varying degrees of interest associated with each projection. Subsequently, the data is transformed by maximizing the index and the associated interest. From a statistical point of view the most interesting directions are those that are as non-Gaussian as possible.

2.3 Cooperative Maximum Likelihood Hebbian Learning

The Cooperative MLHL (CMLHL) model [25] extends the MLHL model, by adding lateral connections between neurons in the output layer of the model. Considering an N-dimensional input vector (x), and an M-dimensional output vector (y), with W_{ij} being the weight (linking input j to output i), then CMLHL can be expressed as defined in Eqs. 1–4.

1. Feed-forward step:

$$y_i = \sum_{j=1}^{N} W_{ij} x_j, \forall i \quad (1)$$

2. Lateral activation passing:

$$y_i(t+1) = [y_i(t) + \tau(b - Ay)]^+ \quad (2)$$

3. Feedback step:

$$e_j = x_j - \sum_{i=1}^{M} W_{ij} y_i, \forall j \quad (3)$$

4. Weight change:

$$\Delta W_{ij} = \eta . y_i . sign(e_j) |e_j|^{p-1} \quad (4)$$

Where: η is the learning rate, τ is the “strength” of the lateral connections, b the bias parameter, p a parameter related to the energy function and A a symmetric matrix used to modify the response to the data. The effect of this matrix is based on the relation between the distances separating the output neurons.

3 Experiments and Results

As previously mentioned, some neural visualization models (see Sect. 2) have been applied to analyze Android malware. Present section introduces the analyzed dataset as well as the main obtained results.

3.1 Malgenome Dataset

The Malgenome dataset [5], coming from the Android Malware Genome Project [6], has been analysed in present study. It is the first large collection of Android malware (1,260 samples) that was split in malware families (49 different ones). It covered the majority of existing Android malware, collected from the beginning of the project in August 2010.

Data related to many different apps from a variety of Android app repositories were accumulated over more than one year. Additionally, malware apps were thoroughly characterized based on their detailed behavior breakdown, including the installation, activation, and payloads.

Collected malware was split in families, that were obtained by "carefully examining the related security announcements, threat reports, and blog contents from existing mobile antivirus companies and active researchers as exhaustively as possible and diligently requesting malware samples from them or actively crawling from existing official and alternative Android Markets" [5]. The defined families are: ADRD, AnserverBot, Asroot, BaseBridge, BeanBot, BgServ, CoinPirate, Crusewin, DogWars, DroidCoupon, DroidDeluxe, DroidDream, DroidDreamLight, DroidKungFu1, DroidKungFu2, DroidKungFu3, DroidKungFu4, DroidKungFuSapp, DoidKungFuUpdate, Endofday, FakeNetflix, FakePlayer, GamblerSMS, Geinimi, GGTracker, GingerMaster, GoldDream, Gone60, GPSSMSSpy, HippoSMS, Jifake, jSMSHider, Kmin, Lovetrap, NickyBot, Nickyspy, Pjapps, Plankton, RogueLemon, RogueSPPush, SMSReplicator, SndApps, Spitmo, TapSnake, Walkinwat, YZHC, zHash, Zitmo, and Zsone. Samples of 14 of the malware families were obtained from the official Android market, while samples of 44 of the families came from unofficial markets.

The dataset to be analyzed consists of 49 samples (one for each family) and each sample is described by 26 different features derived from a study of each one of the apps. The features are divided into six categories, as can be seen in Table 1.

The features describing each family take the values of 0 (if that feature is not present in that family) or 1 (if the feature is present).

Table 1. Features describing each one of the malware families in the Malgenome dataset.

Category #1: installation		Category #3: privilege escalation	
1	Repackaging	14	exploid
2	Update	15	RATC/zimperlich
3	Drive-by download	16	ginger break
4	Standalone	17	asroot
Category #2: Activation		18	encrypted
5	BOOT	Category #4: remote control	
6	SMS	19	NET
7	NET	20	SMS
8	CALL	Category #5: financial charges	
9	USB	21	phone call
10	PKG	22	SMS
11	BATT	23	block SMS
12	SYS	Category #6: personal information stealing	
13	MAIN	24	SMS
		25	phone number
		26	user account

3.2 Results

For comparison purposes, three different projection models have been applied, whose results are shown below.

PCA Projection. Figure 1 shows the principal component projection, obtained by applying PCA to the previously described data. Figure 1a corresponds to the scatterplot matrix, where the three principal components are shown pairwise; those pairs in the main diagonal of the matrix do not provide with interesting information as the same component is shown in both axes (1-1, 2-2 and 3-3). Figure 1b corresponds to the projection obtained by combining the two principal components.

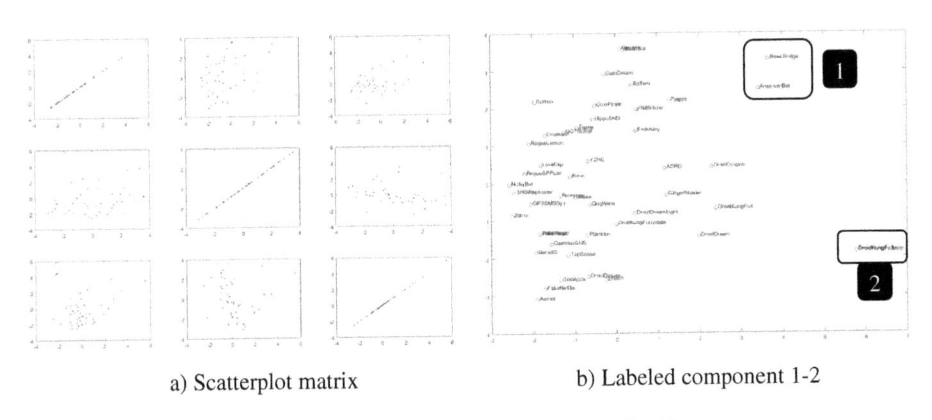

a) Scatterplot matrix b) Labeled component 1-2

Fig. 1. PCA projection of Malgenome families.

In Fig. 1b it can be seen that most of the malware families are grouped in a main group (left side of the figure) while just a few families can be identified away from this cluster (groups 1 and 2). Group 1 gathers two families (BaseBridge and AnserverBot), that are the only two families in the dataset that combine repackaging and update installation. Group 2 gathers four families (DroidKungFu1, DroidKungFu2, DroidKungFu3 and DroidKungFuSapp) that are the only ones in the dataset presenting the encrypted privilege escalation.

Additionally, this projection let us identify that some families are projected at the very same place. By getting back to the data we have realized that these families take the very same values for all the features. This is the case of Walkinwat and FakePlayer on the one hand and for DroidKungFu1, DroidKungFu2, DroidKungFu3 and DroidKungFuSapp on the other hand. It means that, by taking into account the features in the analysed dataset, it will not be possible to distinguish Walkinwat from FakePlayer malware or any of the 4 mentioned variants of DroidKungFu malware.

MLHL Projection. Figure 2 shows the MLHL projection of the analyzed data. As in the case of PCA, Fig. 2a represents the obtained scatterplot matrix and Fig. 2b shows the projection on the two main components. MLHL projection shows the structure of

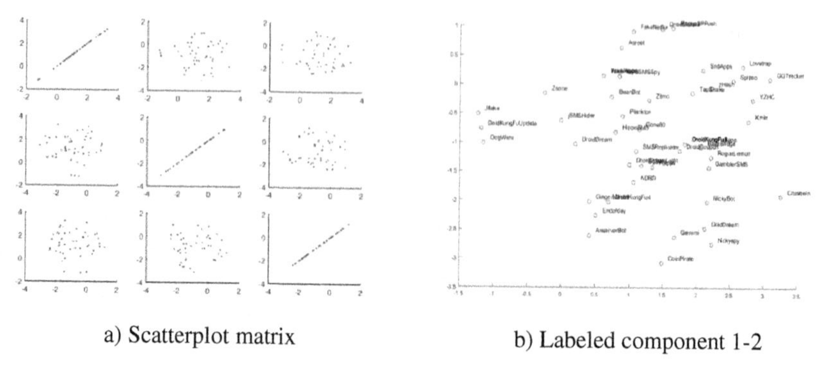

a) Scatterplot matrix b) Labeled component 1-2

Fig. 2. MLHL projection of Malgenome families.

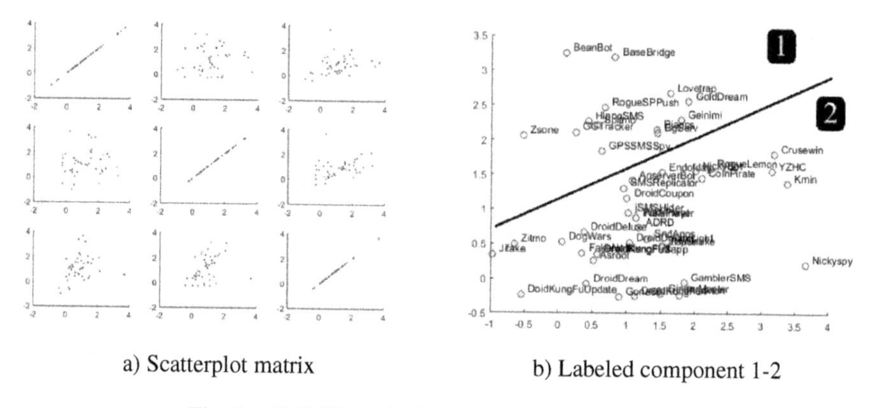

a) Scatterplot matrix b) Labeled component 1-2

Fig. 3. CMLHL projection of Malgenome families.

the data in a way that a kind of ordering can be seen in the dataset. However, as it is more clearly shown in the CMLHL projection (Fig. 3), MLHL is not further described.

The parameter values of the MLHL model for the projections shown in Fig. 2 are: Number of output dimensions: 3. Number of iterations: 100, learning rate: 0. 2872, p: 0.4852.

CMLHL Projection. When applying CMLHL to the analysed dataset, the projection shown in Fig. 3 has been obtained. As in previous figures, Fig. 3a represents the obtained scatterplot matrix and Fig. 3b shows the projection on the two main components. As expected, CMLHL obtained a sparser projection, revealing the structure of the dataset in a clearer way.

The parameter values of the CMLHL model for the projections shown in Fig. 3 are; Number of output dimensions: 3. Number of iterations: 100, learning rate: 0.0406, p: 1.92, τ: 0.44056.

In Fig. 3b it is easy to visually identify at least two main groups of data, labeled as 1 and 2. It means that families in each one of these groups are similar in a certain way. Group 1 gathers all the families with dangerous SMS activity, as SMS activation and

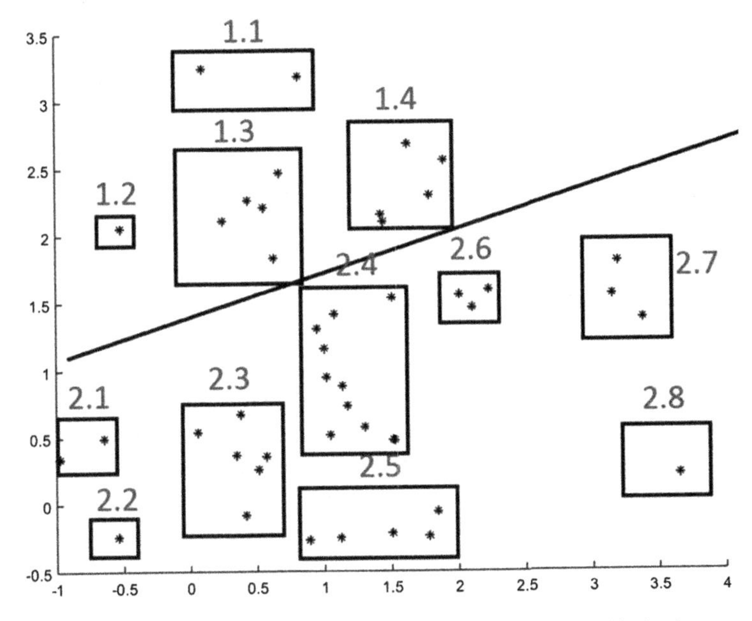

Fig. 4. CMLHL projection of Malgenome families with identified subgroups.

SMS financial charges are present in all the families in Group 1. On the other hand, none of the families in this group present any of the following features: USB or PKG activation, and user-account information stealing. This group is also characterized by the almost complete absence of privilege escalation, as only one of those features (RATC/Zimperlich) is present in only one of the families (BaseBridge). Regarding group 2, none of the families in Group 2 present phone-call financial charges.

Table 2. Families allocation to subgroups defined in CMLHL projection.

Subgroup	Families
1.1	BaseBridge, BeanBot
1.2	Zsone
1.3	GGTracker, GPSSMSSpy, HippoSMS, RogueSPPush, Spitmo
1.4	BgServ, Geinimi, GoldDream, Lovetrap, Pjapps
2.1	Jifake, Zitmo
2.2	DroidKungFuUpdate
2.3	Asroot, DogWars, DroidDeluxe, DroidDream, DroidKungFu1, DroidKungFu2, DroidKungFu3, DroidKungFuSapp, FakeNetflix
2.4	ADRD, AnserverBot, DroidCoupon, DroidDreamLight, Endofday, FakePlayer, jSMSHider, SMSReplicator, SndApps, TapSnake, Walkinwat, zHash
2.5	DroidKungFu4, GamblerSMS, GingerMaster, Gone60, Plankton
2.6	CoinPirate, NickyBot, RogueLemon
2.7	Crusewin, Kmin, YZHC
2.8	Nickyspy

From a deeper analysis of such groups, some subgroups can be distinguished and are identified in Fig. 4. Additionally, the families located in each one of these groups are listed in Table 2.

All the variants of DroidKungFu malware are located in the bottom-left side of the projection (groups 2.2, 2.3, and 2.5). Jifake and Zitmo are gathered in the same subgroup (2.1) as they are the only two families in group 2 presenting the drive-by download installation feature.

4 Conclusions and Future Work

From the projections in Sect. 3, it can be concluded that neural projection models are an interesting proposal to visually analyse the structure of a high-dimensionality dataset in general terms. More specifically, when studying Android malware families, neural projections let us gain deep knowledge about the nature of such apps. Similarities and differences of the studied families are identified thanks to the obtained projections.

After the analysis of the CMLHL projection and the associated allocation of families in groups, it can be said that a coherent ordering is shown, consistent with the seminal characterization of Malgenome dataset [6].

In future work, some other neural visualization models will be applied to the same dataset to better understand the nature of Android malware.

References

1. http://www.statista.com/statistics/266219/global-smartphone-sales-since-1st-quarter-2009-by-operating-system/
2. http://www.appbrain.com/stats/stats-index
3. Micro, T.: The Fine Line: 2016 Trend Micro Security Predictions (2015)
4. http://www.trendmicro.com/vinfo/us/security/news/mobile-safety/mind-the-security-gaps-1h-2015-mobile-threat-landscape
5. Yajin, Z., Xuxian, J.: Dissecting android malware: characterization and evolution. In: 2012 IEEE Symposium on Security and Privacy, pp. 95–109 (2012)
6. http://www.malgenomeproject.org/
7. Corchado, E., Herrero, Á.: Neural visualization of network traffic data for intrusion detection. Appl. Soft Comput. **11**, 2042–2056 (2011)
8. Pinzón, C.I., De Paz, J.F., Herrero, Á., Corchado, E., Bajo, J., Corchado, J.M.: idMAS-SQL: intrusion detection based on MAS to detect and block SQL injection through data mining. Inf. Sci. **231**, 15–31 (2013)
9. Herrero, Á., Zurutuza, U., Corchado, E.: A neural-visualization IDS for honeynet data. Int. J. Neural Syst. **22**, 1–18 (2012)
10. Atienza, D., Herrero, Á., Corchado, E.: Neural analysis of HTTP traffic for web attack detection. In: Herrero, Á., Baruque, B., Sedano, J., Quintián, H., Corchado, E. (eds.) International Joint Conference, vol. 369, pp. 201–212. Springer, New York (2015)
11. Arshad, S., Khan, A., Shah, M.A., Ahmed, M.: Android malware detection & protection: a survey. Int. J. Adv. Comput. Sci. Appl. **7**, 463–475 (2016)

12. Cen, L., Gates, C.S., Si, L., Li, N.: A probabilistic discriminative model for android malware detection with decompiled source code. IEEE Trans. Dependable Secure Comput. **12**, 400–412 (2015)
13. Sanz, B., Santos, I., Laorden, C., Ugarte-Pedrero, X., Nieves, J., Bringas, P.G., Marañón, G. A.: MAMA: manifest analysis for malware detection in android. Cybern. Syst. **44**, 469–488 (2013)
14. Teufl, P., Ferk, M., Fitzek, A., Hein, D., Kraxberger, S., Orthacker, C.: Malware detection by applying knowledge discovery processes to application metadata on the android market (Google Play). Secur. Commun. Netw. **9**, 389–419 (2016)
15. Jang, J.-W., Yun, J., Mohaisen, A., Woo, J., Kim, H.K.: Detecting and classifying method based on similarity matching of android malware behavior with profile. SpringerPlus **5**, 1–23 (2016)
16. Wagner, M., Fischer, F., Luh, R., Haberson, A., Rind, A., Keim, D.A., Aigner, W.: A survey of visualization systems for malware analysis. In: EG Conference on Visualization (EuroVis)-STARs, pp. 105–125 (2015)
17. Paturi, A., Cherukuri, M., Donahue, J., Mukkamala, S.: Mobile malware visual analytics and similarities of attack toolkits (malware gene analysis). In: 2013 International Conference on Collaboration Technologies and Systems (CTS), pp. 149–154 (2013)
18. Park, W., Lee, K.H., Cho, K.S., Ryu, W.: Analyzing and detecting method of android malware via disassembling and visualization. In: 2014 International Conference on Information and Communication Technology Convergence (ICTC), pp. 817–818 (2014)
19. Moonsamy, V., Rong, J., Liu, S.: Mining permission patterns for contrasting clean and malicious android applications. Future Gener. Comput. Syst. **36**, 122–132 (2014)
20. Somarriba, O., Zurutuza, U., Uribeetxeberria, R., Delosières, L., Nadjm-Tehrani, S.: Detection and visualization of android malware behavior. J. Electr. Comput. Eng. **2016**, 17 (2016). doi:10.1155/2016/8034967. Article ID: 8034967
21. Pearson, K.: On lines and planes of closest fit to systems of points in space. Philos. Mag. **2**, 559–572 (1901)
22. Oja, E.: Principal components, minor components, and linear neural networks. Neural Netw. **5**, 927–935 (1992)
23. Fyfe, C.: A neural network for PCA and beyond. Neural Process. Lett. **6**, 33–41 (1997)
24. Corchado, E., MacDonald, D., Fyfe, C.: Maximum and minimum likelihood hebbian learning for exploratory projection pursuit. Data Mining Knowl. Discov. **8**, 203–225 (2004)
25. Corchado, E., Fyfe, C.: Connectionist techniques for the identification and suppression of interfering underlying factors. Int. J. Pattern Recogn. Artif. Intell. **17**, 1447–1466 (2003)

Time Series Data Mining for Network Service Dependency Analysis

Mona Lange(✉) and Ralf Möller

Universität zu Lübeck, Lübeck, Germany
{lange,moeller}@ifis.uni-luebeck.de

Abstract. In data-communication networks, network reliability is of great concern to both network operators and customers. To provide network reliability it is fundamentally important to know the ongoing tasks in a network. A particular task may depend on multiple network services, spanning many network devices. Unfortunately, dependency details are often not documented and are difficult to discover by relying on human expert knowledge. In monitored networks huge amounts of data are available and by applying data mining techniques, we are able to extract information of ongoing network activities. Hence, we aim to automatically learn network dependencies by analyzing network traffic and derive ongoing tasks in data-communication networks. To automatically learn network dependencies, we propose a methodology based on the normalized form of cross correlation, which is a well-established methodology for detecting similar signals in feature matching applications.

1 Introduction

For deriving how susceptible a network is to software vulnerabilities or attacks, it is essential to understand how ongoing network activities could potentially be affected. A network is built with a higher purpose or mission in mind and this mission leads to interactions of network devices and applications causing network dependencies. A monitored infrastructure's missions can be derived through human labor, however missions are subject to frequent change and often knowledge of how an activity links to network devices and applications is not available. So we are challenged to automatically derive these missions as network activity patterns through network service dependency discovery.

In the context of this work, we introduce a novel framework for Mission Oriented Network Analysis (MONA). To motivate our approach, in Sect. 2 we provide a background to other network dependency assessment methodologies and illustrate their limitations. We introduce a network model in Sect. 3 and in Sect. 4 we uncover network dependencies based on network traffic. In a monitored network large amount of unlabeled data, in form of network traffic, are available for knowledge discovery. This allows us to develop a deeper understanding of network activities. For uncovering network dependencies, we propose a methodology based on normalized cross correlation, which is a well-established methodology

M. Graña et al. (eds.), *International Joint Conference SOCO'16-CISIS'16-ICEUTE'16*,
Advances in Intelligent Systems and Computing 527, DOI 10.1007/978-3-319-47364-2_57

for detecting similar signals in feature matching applications. In Sect. 5, we evaluate MONA based on network traffic traces provided by an energy distribution network.

2 Related Work

Network activities in a data-communication network follow from a network having a higher task. Others refer to a network having a higher task as a mission. The concept of missions is sometimes also referred to as mission-centricity in cyber security.

Multiple distinct mission-centric approaches to cyber security have been proposed [1,4,8,10,11]. Albanese et al. [8] use ontologies for integrating available data into a common model. Barreto et al. introduced an impact assessment methodology [4] incorporating vulnerability descriptions [12] and other numerical scores acquired through a BPMN model via human input. Another mission-centric approach is introduced by Jakobson [10], who presents an interdependency model representing an infrastructures operational capacity. Albanese et al. [1] present an interdependency model focusing on cost minimization based on information acquired via human input. Another mission-centric approach is the Cyber Attack Modeling and Impact Assessment Framework [11], which supports computational analysis of a monitored infrastructure and allows impact assessment. An example for a mission-centric approach is the framework for cyber attack modeling and impact assessment [11]. They rely on a mission model for generating attack graphs. All these models do not focus on how to acquire the information used to derive interdependency models. Network dependency analysis allows automatically deriving interdependency models based on network traffic. Recent efforts have explored network-based approaches that treat each host as a black box and passively analyze the network traffic between them. For network administrators that are planning to upgrade or reorganize existing applications a dependency discovery approach named Leslie Graph [2] was designed. The approach aims at identifying complex dependencies between network services and components that may potentially be affected and prevent unexpected consequences. NSDMiner [13] addresses the same problem of network service dependency for network stability and automatic manageability. Sherlock [3] is another approach, which learns an inference graph of network service dependency based on co-occurrence within network traffic. A well-known approach is called Orion [6], which was developed to use spike detection analysis in the delay distribution of flow pairs to infer dependencies. All previously mentioned approach require large amounts of network traffic compared to MONA, and where developed to minimize false negative network service dependencies. To our knowledge, MONA is the first stream-based network service dependency analyzer.

3 IT Network Model

Modeling an IT network requires a basic understanding [7] of the Open Systems Interconnection (OSI) model. For understanding network connectivity, the following layers of the OSI model are of particular interest: data link layer, network layer, transport layer and application layer. We define a network device as a physical device on the network. The data link layer physically links network devices using MAC addresses to identify devices. However, the data link layer only provides point-to-point connectivity. For enabling network connectivity beyond a point-to-point communication, a network layer protocol such as the Internet Protocol (IP) is required. IP addresses are used to identify source and destination of an end-to-end connection. In other words, MAC addresses allow a point-to-point connection, while IP addresses provide an end-to-end connection. Therefore, switches and routers are used to forward packets, i.e. they act as intermediate hosts. Data-communication networks are built with a common higher purpose. This leads to reoccurring interactions between distinct network devices and services, which we call network activity patterns. An example for such a network activity pattern is given in the following example.

Definition 1 (Network Device). *Let MAC and IP be non empty sets of MAC and IP addresses, respectively ($MAC \cap IP = \emptyset$), then*

$$D^{CY} \subseteq \mathcal{P}(MAC) \setminus \{\emptyset\} \times \mathcal{P}(IP) \tag{1}$$

is the set of network devices.

This allows a device to be assigned multiple MAC addresses and IP addresses. Being able to assign multiple MAC addresses to a network device is needed as routers and switches supply multiple point-to-point endpoints. However, switches do not necessarily need to have IP addresses as they work on the data link layer. From this it follows that they are not visible on the network layer. A network captures devices that are communication endpoints and additional intermediate devices, over which endpoints communicate. Network devices that are endpoints can host network services.

Definition 2 (Network Service). *Let S be a set of network services such that a network service $s_i^j \in S$ is hosted by a network device $d_j \in D^{CY}$. Additionally, the network service is associated by a transport protocol $\Psi = \{TCP, UDP\}$ and a port. This allows us to define a relation $SERV$, which links a network device, a transport protocol and a port number to a network service by the follow equation*

$$SERV : D^{CY} \times \Psi \times \mathbb{N} \rightarrow S. \tag{2}$$

To derive all network services hosted by a device d_j, we define a relationship $HOSTS(d_j)$, which returns all network services hosted by d_j. In order to derive the device a service s is hosted by, we write $HOSTS^{-1}(s)$, and associate service s_i with device d_j by writing s_i^j. Given a service $s_i^j \in S$, the corresponding device d_j can be derived by

$$d_j = HOSTS^{-1}(s_i^j). \tag{3}$$

Additionally, for a given IP-address and port, we are able to derive the corresponding network device by

$$DEV : \mathcal{P}(IP) \times \mathbb{N}. \tag{4}$$

This allows us to derive all involved network services for a given IP-address and port pair by $HOSTS(DEV(sIP, sPort)) \rightarrow \mathcal{P}(S)$. Based on network traffic analysis we will detect network services and determine how they communicate in an end-to-end manner with each other. Aside from intermediate devices (e.g. routers and switches), network devices can be categorized into client and server network devices. In the following we will refer to client network devices as clients and server network devices as servers. Clients and servers are able to send requests. Servers additionally provide network services that answer these requests. Generally, the number of clients by far surpasses the number of servers.

Definition 3 (Network Packet). *The basic building block of our approach are network packets exchanged between directly dependent network services. A network packet is exchanged by a source and destination IP address srcIP and dstIP via source and destination port srcPort and dstPort. In addition, a network packet relies on a specific transport layer protocol. In the context of this paper we distinguish the transport layer protocols TCP and UDP. We define a network packet as a 6-tuple*

$$P = (sIP, sPort, dIP, dPort, \psi, t), \tag{5}$$

for source IP addresses sIP*, a source ports* $sPort$*, destination IP addresses* dIP*, destination ports* $dPort$*, a transport protocol* $\Psi = \{UDP, TCP\}$ *and timestamps* t*.*

Requests are often sent through a dynamically assigned port. Dynamically assigned ports are chosen from specifically assigned port ranges [14]. Ephemeral port ranges are available for private, customized or temporary purposes. Although IANA recommends ephemeral port ranges to range from $2^{15}+2^{14}$ to 2^{16}, the range is highly dependent on the operation system. Microsoft assigns ephemeral ports starting as low as 1025 for some windows versions and a lot of Linux kernels have the ephemeral port range start at 32768. We follow the IANA recommended ephemeral port range for clustering purposes.

Definition 4 (Cluster Network Service). *Let* S *be a set of services that are hosted by device* d_j*. All network services communicating through a dynamically assigned port, are grouped by*

$$s_*^j \in S, \tag{6}$$

whereas $*$ *represents a dynamically assigned port and* j *represents the device a network service is hosted on. Known network services have to be linked to ports statically, such that other network services can routinely communicate requests with them. It should also be noted that multiple statically assigned ports could be assigned to the same application.*

Based on Eq. 5, we conduct a network dependency analysis based on packet headers (e.g. IP, UDP and TCP) and timing data in network traffic. Hence, our approach operates on network flows. To identify network flow boundaries, we look into the definition of TCP and UDP flows. A TCP flow starts with a 3-way handshake (SYN, SYN-ACK, ACK) between a client and a server and terminates with a 4-way handshake (FIN, ACK, FIN, ACK) or RST packet exchange. If network services communicate frequently, they may forgo the cost of repetitive TCP handshakes by using KEEPALIVE messages to maintain a connection in idle periods. In comparison the notion of UDP flows is vague, since UDP is a stateless protocol. This is due to the protocol not having well-defined boundaries for the start and end of a conversation between server and client. In the context of this work, we consider a stream of consecutive UDP packets between server and client as a UDP flow, if the time difference between to consecutive packets is below a predefined threshold. In our analysis we exclude all network packet that are necessary for establishing a communication between server and client. So given that additional data is exchanged between network service s_i^j and s_k^l, we term these end-to-end interactions between network services as direct dependencies. The direct dependency $SDEP$ between network services s_i^j and s_k^l is denoted as $SDEP = \{SDEP^{rq} \bigcup SDEP^{rsp}\}$. We distinguish requests and responses exchanged between network services based on Eq. 6. If a network services uses an ephemeral port to send a network packet to a network service on a static port range, we assume it is a request. Thus, an exchanged request $SDEP^{rq}$ is denoted by

$$\begin{aligned} SDEP^{rq} = \{(s_*^j, s_k^l) \mid\ & s_*^j \text{ sends a request to } s_k^l \\ & \text{in the period under consideration,}\} \end{aligned} \tag{7}$$

where k is in the statically assigned port range. Conversely, this means that a network service using its static port range to answer a network service on an ephemeral port is defined as a response. An exchanged response $SDEP^{rsp}$ is written as

$$\begin{aligned} SDEP^{rsp} = \{(s_k^l, s_*^j) \mid\ & s_k^l, \text{ sends a response to } s_*^j \\ & \text{in the period under consideration.}\} \end{aligned} \tag{8}$$

4 Network Service Dependency Discovery

A data-communication network consists of network devices, which interact due to applications via network services. For example an application "email" uses network services $IMAP$ and $POP3$ to access email messages from a remote network device (i.e. host). From this it follows that an application can rely on multiple network services to fulfill a common goal, which is also referred to as mission. Additionally, we note network activities such as accessing email messages lead to network packets being exchanged by directly dependent network service. The purpose of network service dependency discovery is to abstract network packets in order to detect reoccurring communication pattern. Reoccurring communication patterns indicate that the involved network services are dependent. In order to detect reoccurring communication patterns, we first abstract monitored network traffic into communication histograms. This analysis is done online based on continously captured network traffic.

4.1 Communication Histograms

Let us suppose that we are mirroring network traffic from an initial time point t_{min} to a time point t_{max} within an IT network. We are observing network packets $p \in P$, which are defined as a 6-tuple according to Eq. 5. For communicating network services, we build communication histogram with a bin size Δ_t. In the context of work we set Δ_t to 1 second. The number of histogram bins is given by

$$bins = \lfloor \frac{(t_{max} - t_{min})}{\Delta_t} \rfloor, \tag{9}$$

assuming we want to build a communication histogram for network traffic mirrored from time point t_{min} to t_{max} with a bin size Δ_t.

Given that we are monitoring a set of S network services then the data structure for all communication histograms is defined by

$$H : S \times S \times \Psi \rightarrow (\{0, \cdots, bins - 1\} \rightarrow \mathbb{N}_0), \tag{10}$$

where the communication histogram bins $\{0, \cdots, bins - 1\}$ are mapped to $\mathbb{N}_0$. Now for every network packet exchanged between directly dependent network services, assuming it was received during the considered time period, the corresponding bin $I(H(s, s', \psi)$ in the communication histogram is incremented. The corresponding bin in the communication histogram is determined by

$$(t_{min} - t) \mod bins, \tag{11}$$

assuming that the network packet p contains the time stamp t.

4.2 Indirect Dependencies

Network services operate on distributed sets of clients and servers and rely on supporting network services, such as Kerberos, Domain Name System (DNS), and Active Directory. To fulfill a network's mission, network services need to interact. Since, engineers use the divide-and-conquer approach to implement a new task, they are able to reuse network services and do not need to re-implement complex customized ones. This leads to multiple network services interacting for a common high-level task. For the purpose of detecting indirect dependencies, we analyze the communication histograms of directly dependent network services in order to derive re-occurring communication patterns. Detecting re-occurring communication patterns requires clustering direct dependencies. Similarly to previous work, we distinguish two different types of remote-remote dependencies and local-remote dependencies [6]. A local-remote (LR) dependency is an indirect dependency, where a system must issue a request to a remote system in order to complete an outstanding request issued to a local service. A remote-remote (RR) dependency is one in which a system must first contact one host before issuing a request to the desired host.

Definition 5 (Candidates for an Indirect Dependency). *Given a direct dependency $\delta(s_i^j, s_k^l)$, all network services hosted by*

$$HOSTS^{-1}(s_k^l) = d_l \text{ or } HOSTS^{-1}(s_i^j) = d_j \tag{12}$$

are candidates for an indirect dependency.

The communication histograms contain the communication pattern of all involved directly dependent network services.

4.3 Measuring Communication Histogram Similarity

Suppose we have two communication histograms r and $s \in H$, which are candidates for being indirectly dependent. The two communication histograms are time series $r = (r_1, r_2 \cdots, r_{bin})$, consist of *bin* samples.

In statistics, the Pearson product-moment correlation coefficient the linear correlation σ between two variables X and Y by

$$\sigma = \frac{[(X - \mu_X)(Y - \mu_Y)]}{\sigma_X \sigma_Y}, \tag{13}$$

where μ_Y and σ_Y denotes the mean and standard deviation of Y. The Pearson product-moment correlation coefficient is a measure of the linear correlation between two variables X and Y, giving a value between +1 and −1 inclusive, where

1 is total correlation,
0 is no correlation, and
−1 is total anti-correlation.

To apply the Pearson correlation coefficient as a distance measure on a time series [9], we define the Pearson Distance as

$$d_\varrho(r, s) = \frac{\frac{1}{bins}\sum_{t=0}^{bins}(r_t - \mu_r)(s_t - \mu_s)}{\sigma_r \sigma_s}, \tag{14}$$

where μ_r and σ_r are mean and standard deviation of r, such that $-1 \leq d_\varrho \geq 1$.

An indirect dependency implies that the data received by s_k^l is processed on device d_l. Then, data is sent to another network service s_m^l. Due to the processing of data on device d_l, the request might be sent to s_m^l τ_{delay} time steps later. Network latency can lead to communication patterns being shifted due to the time it takes for a network packet to be transferred. Communication patterns are stored in communication histograms and they contain the same pattern, which thus are also shifted due to network latency. Hence, the communication patterns of both direct dependencies would be similar, although shifted by τ_{delay} time steps. In pattern recognition, normalized cross correlation has been proposed to take a shift, such as τ_{delay}, into account.

To overcome the lack of a perfect alignment between two communication networks, we extend the Pearson distance, introduced in Eq. 14, to normalized cross-correlation [5].

$$\varrho_{r,s}(\tau) = \frac{\frac{1}{bins}\sum_{t=0}^{bins}(r_t - \mu_r)(s_{t+\tau} - \mu_s)}{\sigma_r \sigma_s}, \quad (15)$$

The point in time τ_{delay}, where both signals are best aligned can be found by computing

$$\tau_{delay} = \text{argmax}_t \varrho_{r,s}(\tau). \quad (16)$$

If $\varrho_{r,s}(\tau) \geq \theta$, we consider both communication histograms r and s to be correlated and therefore indirectly dependent and shifted by τ_{delay}. Normalized cross-correlation is applied to all indirect dependency candidates and returns a set $ISDEP$, which is derived by applying

$$ISDEP = SDEP \bowtie SDEP \quad (17)$$

on all LR (local-remote) and RR (remote-remote) dependencies.

Algorithm 1. Building communication histograms algorithm

```
1: Input:
2: Network packet P = (sIP, sPort, dIP, dPort, ψ, t),
3: start time t_min, stop time t_max, time step Δ_t
4: Output: Communication histograms H
5:                    ▷ compute number of bins for communication histogram H
6: bins = (t_max − t_min) div Δ_t
                      ▷ all histogram vectors are initialized and filled with zeros
7:                    ▷ fill histogram bins for every network packet in a network flow
8: for all p = (sIP, sPort, dIP, dPort, ψ, t) ∈ P
9:       and t ≥ t_min and t ≤ t_max do
10:     tb = (t − t_min) mod bins
11:     H(SERV(DEV(sIP), sPort),
12:       SERV(DEV(dIP), dPort),
13:                         ψ)[tb]++
14: end for
15: return H
```

5 Experimental Evaluation

The disaster recovery site of an energy distribution network, provided an Italian water and energy distribution company, was available for network traffic analysis. Based on this network, we are able to deploy MONA for online analysis. In addition, we are collect and analyze real-life network traffic with the help of network operators. Real-life network traffic consists of network services frequently to rarely interacting and we are able to test MONA with typical communication

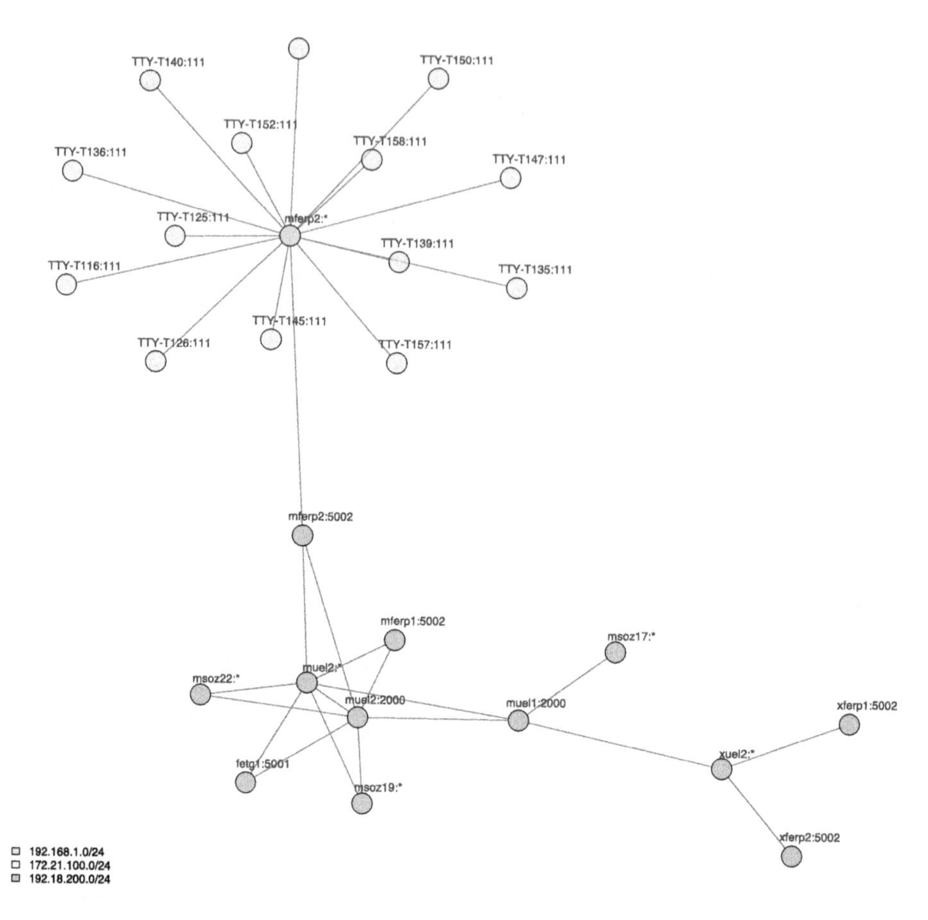

Fig. 1. Network service dependency analysis in an energy distribution network.

patterns occurring in an operational environment. As this network is a real-life network, absolute knowledge of all existing and non existing network dependencies can only be assumed. Data communication networks are dependent on third party software and operators do not have complete knowledge. However, all identified indirect network service dependencies, which are shown in Fig. 1, where classified as true positives.

The node named mferp2 is a communication server for multiple substations, which are identified as TTY-T[116-158]. Also, our evaluation shows, how important Eq. 1 is as mferp2 is one physical device, but is assigned two IP address from two different subnetworks. The communication server mferp2 hosts a network service, defined in Eq. 2. Port numbers that identify the network service are appended to the host names. This network service belongs to an application, which sends requests (see Eq. 7) to all substations in order to be updated with current measurement information. Therefore, a network service dependency joins these substations to the communication server. Mferp2 communicates via muel2 with Human Machine Interface (HMIs) msoz19 and mso22. Another HMI msoz17 wants to access information about the substations TTY-T[116-158]. For this, first muel1 is contacted, who passes the request on to muel2. As an energy distribution network is a critical infrastructure, it needs to be ensured that all communication pathways are always available. Hence, regularly back up servers and alternate communication pathways are tested, even if no information needs to be transmitted. All these network service dependencies where verified by operators as true positives.

6 Conclusion

We have demonstrated a novel network dependency detection approach. During first experiments on online network traffic from the disaster recovery site, we often found new network dependencies that had been previously forgotten by the network operators. Subsequence mining enables deriving a deeper understanding of network activities and leverages well data-communication networks with different numbers of network devices and indirect dependencies.

Acknowledgments. This work has been partially supported by the European Union Seventh Framework Programme (FP7/2007-2013) under grant agreement No. 610416 (PANOPTESEC). The opinions expressed in this paper are those of the authors and do not necessarily reflect the views of the European Commission.

References

1. Albanese, M., Jajodia, S., Jhawar, R., Piuri, V.: Reliable mission deployment in vulnerable distributed systems. In: 2013 43rd Annual IEEE/IFIP Conference on Dependable Systems and Networks Workshop (DSN-W), pp. 1–8. IEEE (2013)
2. Bahl, P., Barham, P., Black, R., Chandra, R., Goldszmidt, M., Isaacs, R., Kandula, S., Li, L., MacCormick, J., Maltz, D.A., et al.: Discovering dependencies for network management. In: ACM SIGCOMM 5th Workshop on Hot Topics in Networks (Hotnets-V), pp. 97–102. ACM (2006)
3. Bahl, P., Chandra, R., Greenberg, A., Kandula, S., Maltz, D.A., Zhang, M.: Towards highly reliable enterprise network services via inference of multi-level dependencies. ACM SIGCOMM Comput. Commun. Rev. **37**, 13–24 (2007)
4. de Barros Barreto, A., Costa, P.C.G., Yano, E.T.: A semantic approach to evaluate the impact of cyber actions on the physical domain (2012)
5. Briechle, K., Hanebeck, U.D.: Template matching using fast normalized cross correlation. In: Aerospace/Defense Sensing, Simulation, and Controls, pp. 95–102. International Society for Optics and Photonics (2001)
6. Chen, X., Zhang, M., Mao, Z.M., Bahl, P.: Automating network application dependency discovery: experiences, limitations, and new solutions. In: USENIX Symposium on Operating Systems Design and Implementation (OSDI), vol. 8, pp. 117–130 (2008)
7. Edwards, J., Bramante, R.: Networking Self-teaching Guide: OSI, TCP/IP, LANs, MANs, WANs, Implementation, Management, and Maintenance. Wiley, New York (2015)

8. Goodall, J.R., D'Amico, A., Kopylec, J.K.: Camus: automatically mapping cyber assets to missions and users. In: Military Communications Conference (MILCOM), pp. 1–7. IEEE (2009)
9. Höppner, F., Klawonn, F.: Compensation of translational displacement in time series clustering using cross correlation. In: Adams, N.M., Robardet, C., Siebes, A., Boulicaut, J.-F. (eds.) IDA 2009. LNCS, vol. 5772, pp. 71–82. Springer, Heidelberg (2009). doi:10.1007/978-3-642-03915-7_7
10. Jakobson, G.: Mission cyber security situation assessment using impact dependency graphs. In: Information Fusion (FUSION), pp. 1–8 (2011)
11. Kotenko, I., Chechulin, A.: A cyber attack modeling and impact assessment framework. In: 2013 5th International Conference on Cyber Conflict (CyCon), pp. 1–24, June 2013
12. MITRE: Common vulnerabilities and exposures (2000). https://cve.mitre.org/
13. Natarajan, A., Ning, P., Liu, Y., Jajodia, S., Hutchinson, S.E.: NSDMiner: automated discovery of network service dependencies. In: IEEE International Conference on Computer Communications (IEEE INFOCOM 2012). IEEE (2012)
14. Touch, J., Kojo, M., Lear, E., Mankin, A., Ono, K., Stiemerling, M., Eggert, L.: Service name and transport protocol port number registry. The Internet Assigned Numbers Authority (IANA) (2013)

Security Analysis of a New Bit-Level Permutation Image Encryption Algorithm

Adrian-Viorel Diaconu[1], Valeriu Ionescu[2(✉)], and Jose Manuel Lopez-Guede[3]

[1] IT&C Department, Lumina – The University of South-East Europe, Bucharest, Romania
[2] Department of Electronics, Computers and Electrical Engineering, University of Pitesti, Pitesti, Romania
[3] Systems and Automatic Control Department, University College of Engineering of Victoria, Basque Country University (UPV/EHU), Vitoria, Spain
manuelcore@yahoo.com

Abstract. A new chaotic, permutation-substitution architecture based, image encryption algorithm has been introduced with a novel inter-intra bit-level permutation based confusion strategy. The proposed image cryptosystem uses Sudoku grids to ensure a high performance diffusion process and key space. This paper presents a fully comprehensive set of security analyses on the newly proposed image cryptosystem, including statistical and differential analysis, local and global information entropy calculation and robustness profile to different types of attacks. Based on the good statistical results, rounded by the theoretical arguments, the conducted study demonstrates that the proposed images' encryption algorithm has a desirable level of security.

Keywords: Inter-intra bit-level permutation · Image encryption · Chaos-based cryptography · Security analysis

1 Introduction

Communication security is an important topic in modern communications because there is much interest in keeping communications private even if a public communication channel is used. Digital imaging security, one aspect of communication security, can be insured through encryption.

In the past two decades plenty of image encryption algorithms have been proposed, most of them relying on a chaotic dynamical system, e.g., [2, 5, 6, 7–10, 15, 16], and ranging from pixel-level operations [2, 8–10, 15] to bit-level operations [17, 18]. The use of dynamic systems is justified by the fact that the core properties of a good encryption system, i.e., confusion and diffusion, can be easily assured by using chaotic maps which, usually, are ergodic and sensitive to the system parameters [2].

In this paper, we use the results from a previous research article [1] where a new chaos-based image encryption scheme is proposed. It is built on the, single round, substitution – permutation (*viz.*, diffusion, *resp.*, and confusion) model and uses a

M. Graña et al. (eds.), *International Joint Conference SOCO'16-CISIS'16-ICEUTE'16*,
Advances in Intelligent Systems and Computing 527, DOI 10.1007/978-3-319-47364-2_58

discrete chaotic dynamic system based Sudoku Grid (slightly modified in comparing with the original proposals [3, 4]). The novelty of the proposed scheme consists of the intra pixel permutation strategy of choice, which improves the statistical performances of the confusion phase.

This paper focuses on the security analysis of the proposed cryptosystem. Section 2 presents the designing stages of the proposed cryptosystem. In Sect. 3, an extensive security analysis of the cryptosystem is performed regarding the following aspects: visual analysis; security assessment by statistical analysis, differential analysis, and key analysis; computational and complexity analysis; simulation results and performance analyses.

2 The Proposed Cryptosystem

The algorithm [1] uses a pseudorandom bit generator (1), called PRNG, in the construction stages of the shuffling and ciphering Sudoku Grid, as well within the intra bit-level permutation step. Performances of the PRNG of choice are fully assessed in [19].

$$y_i = f_1\left(x_i^1, r_1\right) \cdot f_2\left(x_i^2, r_2\right) = \frac{f_1\left(x_i^1, r_1\right) + f_2\left(x_i^2, r_2\right)}{1 - f_1\left(x_i^1, r_1\right) \cdot f_2\left(x_i^2, r_2\right)}, \tag{1}$$

where: x_0^1, x_0^2 and r_1,r_2 are the initial conditions, *resp.*, the control parameters of the f_1, f_2 chaotic maps; x_i^1, x_i^2 are the orbits obtained with recurrences $x_{i+1}^1 = f_1\left(x_i^1, r_1\right), x_{i+1}^2 = f_2\left(x_i^2, r_2\right), \forall i \in N$; one-dimensional chaotic discrete dynamical systems, *i.e.*, f_1 and f_2, are of the form:

$$f_p : [-1, 1] \to [-1, 1], f_p(x_i^p) = \frac{2}{\pi} arctg\left(ctg\left(x_i^p \cdot r_p\right)\right). \tag{2}$$

During performances' testing procedures, f_p s' initial seeding points' values and control parameters' values were chosen with the following values: $x_0^1 = 0.68775492511773$, $r_1 = 5.938725025421$, *resp.*, $x_0^2 = -0.0134623354671, r_2 = 1.257490188615$.

In this case study we will use a 8-bit grayscale image of size $m \times m$ (m = 256), and we represent with I_0 the pixels' values matrix, a Latin Square of the same size as the image to be encrypted (referred to as LS_{256}) and Sudoku Grid (referred to as the SG_{256}) that is derived from the LS_{256} matrix by chaotically resampling its structure, both vertically and horizontally. Ciphering and deciphering stages of the targeted image encryption algorithm are fully described in [1] and summarized as follows.

2.1 Image Encryption Algorithm

Here m_1, m_2, m_3 are row vectors of size m representing the I_0 samples (given by (1)), required to construct LS_{256} first row *resp.*, for its horizontal and vertical resampling and m_4 is representing a random $(m \cdot m) \times 8$ size *uint*8 matrix (where $\forall m_4(k, p) = \overline{1, 8}$ for any $k = \overline{1, m \times m}$ and $p = \overline{1, 8}$) used to compute pixels' intra bit-level permutation.

Image encryption algorithm steps:

(A) Construct LS_{256} matrix:
  (a) initialize LS_{256} matrix as a null matrix;
  (b) sort m_1 ascending and create the linear indices vector m_1';
  (c) construct LS_{256} rows by circular shifting left by 1 the m_1';
(B) Construct SG_{256} matrix derived from LS_{256}:
  (a) copy LS_{256} in SG_{256};
  (b) sort m_2 and m_3 ascending and create the linear indices vectors h_{res} and v_{res};
  (c) Resample/rearrange LS_{256}'s rows as dictated by h_{res};
  (d) Resample/rearrange LS_{256}'s columns as dictated by v_{res};
(C) Shuffle each pixel of the plain-image I_0 into I_{HVPS}:
  (a) reorder the bits for each pixel of I_0 to a location given by the m_4 matrix:

$$I_0(i,j) = Pixel_{bin}(m_4((i-1)\cdot 256 + j, :)) \tag{3}$$

  (b) relocate I_0's rows to a location specified by SG_{256} row;
  (c) relocate I_0's columns to a location specified by SG_{256} column;
(D) Cipher the shuffled image I_{HVPS} into I_{ENC}:
  (a) cipher I_{HVPS} rows:

$$I_{HVPS}(i,:) = I_{HVPS}(i,:) \oplus SG_{256}(i,:); \tag{4}$$

  (b) cipher I_{HVPS} columns:

$$I_{HVPS}(i,:) = I_{HVPS}(i,:) \oplus SG_{256}(i,:); \tag{5}$$

The result is $\mathrm{I_{ENC}}$ with the pixel values matrix of an 8-bit grayscale encrypted image of the size $\mathrm{m} \times \mathrm{m}$, with the correct key $\mathrm{K} = \{\mathrm{x}_0^1, \mathrm{r}_1, \mathrm{x}_0^2, \mathrm{r}_2\}$.

2.2 Image Decryption Algorithm

The original image I_0 is recovered from I_{ENC} as follows:

(A) construct L_{256} matrix and the corresponding SG_{256} matrix,
  (a) replicate encryption algorithm steps (A), *resp.*, (B);
(B) decipher the encrypted image, *i.e.*, I_{ENC} matrix,
  (a) for $i = 1 : m$, decipher I_{ENC}'s rows:

$$I_{ENC}(i,:) = I_{ENC}(i,:) \oplus SG_{256}(i,:); \tag{6}$$

  (b) for $i = 1 : m$, decipher I_{ENC}'s columns:

$$I_{ENC}(:,i) = I_{ENC}(:,i) \oplus SG_{256}(i,:)'; \tag{7}$$

(C) deshuffle the decrypted image, *i.e.*, I_{DEC} matrix,
  (a) for $i = 1 : m$, for $j = 1 : m$, deshuffle I_{DEC}'s columns:

$$I(SG_{256}(j,i),i) = I_{DEC}(j,i); \quad I_{DEC} = I; \tag{8}$$

(b) for $i = 1 : m$, for $j = 1 : m$, deshuffle I_{DEC}'s rows:

$$I(i, SG_{256}(j,i)) = I_{DEC}(i,j); \quad I_{DEC} = I; \tag{9}$$

(c) for $i = 1 : m$, for $j = 1 : m$, compute $\hat{I}_0$'s pixels', *i.e.*, restore pixels' bit levels:

$$\hat{I}_0(i,j) = Pixel_{bin}(m_4((i-1) \cdot 256 + j, :)) \tag{10}$$

Step (C) will modify I_{DEC}, generating a new matrix $\hat{I}_0$, representing the pixel values matrix of the deshuffled image representing the plain-image.

3 Security Analysis of the Cryptosystem

In this section, few statistical estimators are used, according to the used conventional methodology [5, 15], in order to assess the characteristics of the proposed image encryption algorithm. The first step is the visual analysis that detects the presence of similarities between the plain-image and it's scrambled and (or) ciphered versions [5]. By comparing Fig. 1(a) with Fig. 1(b) and (c), one can say that there are no perceptual similarities between the plain-image and its scrambled version, *resp.*, ciphered version (i.e., image's pixels have been rearranged with a random-like appearance).

3.1 Security Assessment by Statistical Analysis

3.1.1 Histogram Analysis

Figure 2 represents histograms of the 'Baboon' plain (a), shuffled (b) and ciphered (c) images. Analyzing them one can notice that shuffled and ciphered images histograms gain a more uniform distribution, a fact which ensures robustness against statistical attacks. To assess histograms goodness of fit, the χ^2 test was used (i.e., with the null hypothesis that the distribution of the values within ciphered image's histogram approaches a uniform one). Screening Table 1 it can be observed that the null hypothesis was accepted for all images subjected to tests.

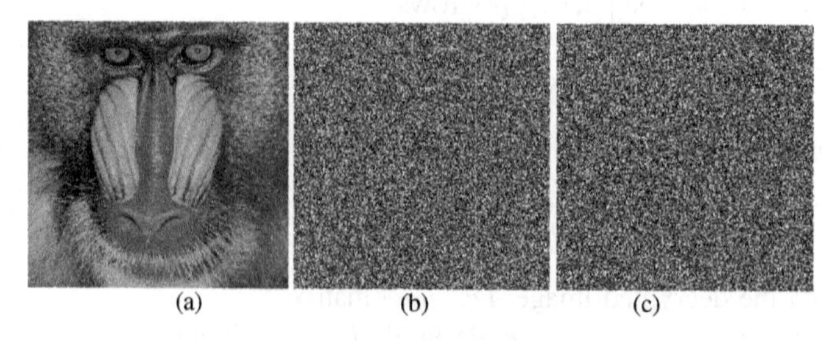

Fig. 1. 'Baboon': (a) testing plain-image, (b) scrambled version of (a), (c) ciphered version of (b)

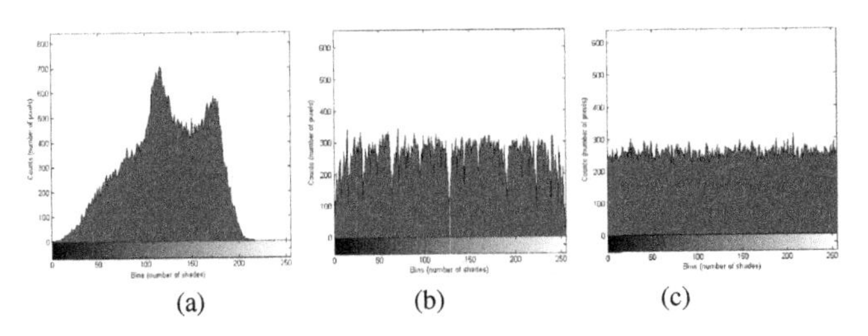

Fig. 2. Histograms: (a) of the 'Baboon' testing plain-image, (b) scrambled version of (a), (c) ciphered version of (b)

Table 1. χ^2 tests' results for different ciphered images

Measure and Decision	Baboon	Lena	Peppers	House	Cameraman
χ^2 p-value	0.7628	0.2846	0.4856	0.9552	0.5920
Decision	H_0 accepted	H_0 accepted	H_0 accepted	H_0 accepted	H_0 accepted

3.1.2 Adjacent Pixels' Correlation Coefficients' Analysis

An efficient and secure image encryption algorithm should be able to reduce considerably the inherent strong correlation of adjacent pixels. For the analysis of APCC, 10.000 pairs of adjacent pixels were randomly selected, for each of the testing directions, the results being summarized in Table 2. Here it can be easily noticed that the correlation coefficients are significantly reduced firstly by the shuffling procedure and then even more by the ciphering procedure of the image encryption algorithm.

3.1.3 Information Entropy Analysis

For this analysis, global and local entropy values were computed, results being summarized in Table 3. Here it can be noticed that both measures have satisfactory values: global entropy values suggest that the image at the output of the image encryption algorithm has equiprobable gray levels; local entropy values, falling within the confidence interval, suggest that the global randomness of the output image is preserved. The conclusion is that the encryption algorithm presented is resistant against entropy attacks.

3.2 Security Assessment by Differential Analysis

Differential analysis – based assessment of an encryption algorithm uses two qualitative indicators, namely NPCR (*i.e.*, number of pixels change rate) and UACI (*i.e.*, unified average changing intensity). NPCR and UACI tests' results are shown in Tables 4 and 5. Analyzing these two tables, and observing the fact that values of both indicators lie within the confidence intervals (*i.e.*, swiftly changes in the plain-image will result in negligible changes to its ciphered version), one can conclude that the proposed encryption algorithm ensures the required strength against any differential cryptanalysis [17, 18].

Table 2. Correlation coefficients of adjacent pairs of pixels

Image	Stage	Testing direction		
		Vertical	Horizontal	Diagonal
Lena	Plain	0.965906071028415	0.936080862763546	0.910862293283444
	Scrambled	0.008975899618625	0.003856171773592	0.002263387143221
	Ciphered	–0.003825257919513	–0.005764482741213	0.002899904924928
Baboon	Plain	0.657323161492229	0.717040906395162	0.629697947164171
	Scrambled	0.014968527105150	–0.020185917128567	0.012658339518317
	Ciphered	0.009959198004718	–0.000857333468853	–0.000531030837520
Peppers	Plain	0.657323161492229	0.717040906395162	0.629697947164171
	Scrambled	0.014968527105150	–0.020185917128567	0.012658339518317
	Ciphered	0.009959198004718	–0.000857333468853	–0.000531030837520
Cameraman	Plain	0.956897902715783	0.953347814808425	0.918201328056903
	Scrambled	–0.005856286261312	–0.007314971573111	0.014811757092666
	Ciphered	–0.003069896371827	–0.010508579725693	0.009211326104522
House	Plain	0.959730495747664	0.933538553901624	0.906094917683073
	Scrambled	–0.000765430673068	0.009417345770564	0.000759311534106
	Ciphered	0.000922496394267	0.000167557017665	0.000177834403900

3.3 Security Assessment for Different Types of Attacks

3.3.1 Additive Noise Attack

The additive noise attack is made by adding random noise to the encrypted image [5]. Two types of noises are considered: salt and pepper noise and speckle noise. In order to test the ability of the proposed image encryption scheme to cope with errors generated by this kind of attacks PSNR measure is used, for which a value near 40 dB or above is desired.

Figure 3(a) and (b) depicts the 'Baboon' recovered image when the encrypted version was attacked with a 'salt and pepper' type of noise (with 0.01 densities), *resp.*, with a 'speckle' type of noise (with 0.01 variance). One can conclude that, with except of speckle noise attack (*i.e.*, where poor performance is attributed to the fact that intrinsic pixels' features, that is, their values, are used within the scrambling phase), the proposed encryption scheme can face, lightly, such harsh attacks. Resulted PSNR values, for different values of noise's densities, are summarized in Table 6.

3.3.2 Cropping Attack

A cropping attack modifies the intercepted encrypted image by deleting one or more of its areas, with or without offset from the center of the image [5]. Figure 3(c) illustrates the decrypted image, resulting from the encrypted version that was attacked by cropping one 1/8 of its area (*i.e.*, at center of the image). To measure the error resistance of proposed encryption scheme against this kind of attacks PSNR measure is also used, with results presented in Table 6. In cases where the ciphered image is attacked by cropping it with 1/16, 1/32 and 1/64 ratio (with the cropped area centered on the image) PSNR values are: 42.55 dB, 48.09 dB and 53.34 dB.

Table 3. Global and local entropy values of the ciphered images

Testing image	Global entropy	Local entropy	Local entropy critical values $k = 30, T_B^{L=256*} = 1936$		
			$h_{left}^{l*0.05} = 7.901901305$ $h_{right}^{l*0.05} = 7.903037329$	$h_{left}^{l*0.01} = 7.901722822$ $h_{right}^{l*0.01} = 7.903215812$	$h_{left}^{l*0.001} = 7.901515698$ $h_{right}^{l*0.001} = 7.903422936$
Lena	7.9968715	7.9019503	Passed	Passed	Passed
Baboon	7.9968534	7.9023835	Passed	Passed	Passed
Peppers	7.9972087	7.9029299	Passed	Passed	Passed
Cameraman	7.9969636	7.9028656	Passed	Passed	Passed
House	7.9971936	7.9021775	Passed	Passed	Passed

Table 4. UACI values for the proposed ciphering scheme

Testing image	UACI value	UACI critical values		
		$UACI_{0.05}^{*-} = 33.3730$ % $UACI_{0.05}^{*+} = 33.5541$ %	$UACI_{0.01}^{*-} = 33.3445$ % $UACI_{0.01}^{*+} = 33.5826$ %	$UACI_{0.001}^{*-} = 33.3115$ % $UACI_{0.001}^{*+} = 33.6156$ %
Lena	33.5521 %	Passed	Passed	Passed
Baboon	33.5091 %	Passed	Passed	Passed
Peppers	33.4703 %	Passed	Passed	Passed
Cameraman	33.5196 %	Passed	Passed	Passed
House	33.4924 %	Passed	Passed	Passed

Table 5. NPCR values for the proposed ciphering scheme

Testing image	NPCR value	NPCR critical values		
		$NPCR^*_{0.05} = 99.5893\ \%$	$NPCR^*_{0.01} = 99.5810\ \%$	$NPCR^*_{0.001} = 99.5717\ \%$
Lena	99.6338 %	Passed	Passed	Passed
Baboon	99.5972 %	Passed	Passed	Passed
Peppers	99.6070 %	Passed	Passed	Passed
Cameraman	99.6023 %	Passed	Passed	Passed
House	99.5957 %	Passed	Passed	Passed

Table 6. Robustness of the proposed encryption scheme against different types of attacks

Attack	'Salt and pepper' noise	'Speckle' noise	Cropping
Parameters	*0.0001 densities*	*0.0001 variance*	
PSNR value [dB]	66.140792284144183	26.736487348855960	
Parameters	*0.001 densities*	*0.001 variance*	
PSNR value [dB]	66.140792284144183	23.197934275836456	
Parameters	*0.01 densities*	*0.01 variance*	
PSNR value [dB]	56.148043078365895	21.113705664627918	
Parameters			*1/8th of the area affected*
PSNR value [dB]			36.764685181302426

3.4 Security Assessment by Key Analysis

3.4.1 Key Sensibility

Another desirable feature, assessed for any newly proposed digital images' encryption algorithms, is the sensibility to the encryption key (*i.e.*, any small changes in the key should lead to significant changes in the shuffled and (or) encrypted, *resp.*, deshuffled and (or) decrypted images). For the proposed cryptosystem, the assessment of key's sensibility was made taking into consideration a $\pm LSB$ variation over one of key's elements. Figures 4 and 5 highlight proposed cryptosystem's sensibility to small changes within the encryption key.

3.4.2 Key Space

A large key space is very important for a digital images' encryption algorithm, as to be able to repel brute-force attacks. For the proposed encryption scheme, the considered key, *i.e.*, $K = \{x_0^1, r_1, x_0^2, r_2\}$, generates a key-space sufficiently large (of at least 10^{67}, taking into consideration the fact that PRNG's seeding points, *resp.*, control parameters have been represented with an accuracy of up to 10^{-19}), as to ensure its immunity to these types of attacks, with respect to the current computers' capacities. Moreover, this key-space can be easily extended by the number of all possible c matrices, derivable from all possible L_{256} matrices (28) [7]. Thus, with a lower bound of $256! \cong 10^{507}$, one can notice that the proposed encryption scheme's key-space is higher than the one of the majority, previously proposed, digital images' encryption algorithms [17].

$$L(n) \geq (n!)^{2 \cdot n} / n^{n^2}. \tag{11}$$

where, n represents Latin Square's order (*i.e.*, dimension).

3.5 Computational and Complexity Analysis

The time complexity of the image encryption algorithm is given by the bit-level operations, i.e., $\Theta(8 \times m \times m)$, the pixel-level operations, i.e., $\Theta(m \times m)$, *resp.*, by the

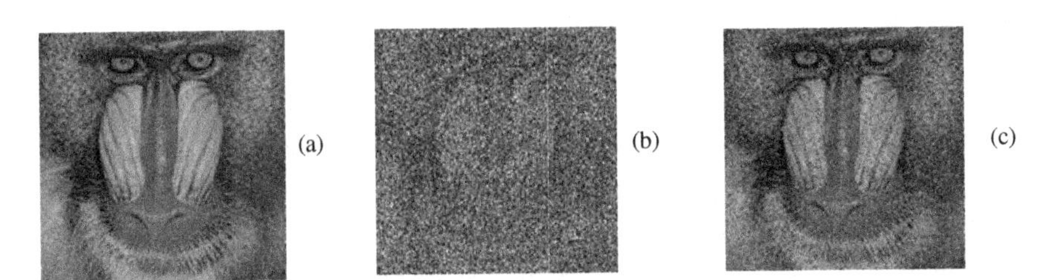

Fig. 3. 'Baboon' recovered image: (a) when the encrypted version was attacked with an 'salt and pepper' type of noise, (b) when the encrypted version was attacked with an 'speckle' type of noise, (c) when the encrypted version was cropped at the center of the image, with an $1/8^{th}$ ratio.

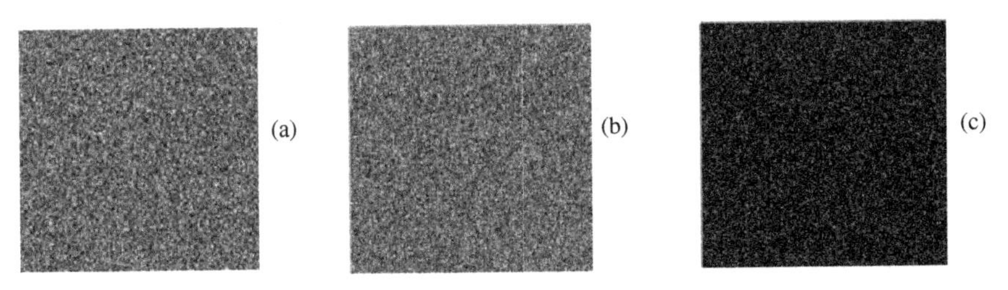

Fig. 4. Proposed cryptosystem's sensitivity to small changes within the encryption key: (a) image encrypted using key K_1, (b) image encrypted using key K_2, and (c) the image representing differences between (a) and (b).

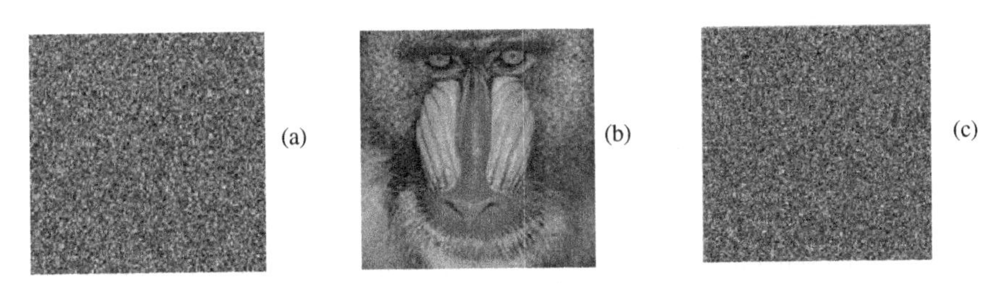

Fig. 5. Proposed cryptosystem's sensitivity to small changes within the encryption key: (a) image encrypted using key K_1, (b) image decrypted using the correct key K_1, and (c) the image decrypted using the wrong key K_2.

diffusion process, i.e., $\Theta(m \times m)$. Thus, the total time complexity of the encryption algorithm is $\Theta(8 \times m \times m)$, comparable or better with ones showcased in [12–14]. In terms of throughput, as shown in Table 7, the processing speed of the proposed image encryption algorithm is comparable with one showcased in [11, 14], better than [6], *resp.*, less than [2, 15]. Analyzing presented testing results that the cryptosystem presented in this paper has the desirable level of security and better or comparable performances, with those reported in literature [8–11], as shown in Table 8.

Table 7. Speed analysis between the proposed and other image encryption algorithms

Criteria	Referenced works						Proposed
	[11]	[15]	[11]	[14]	[8]	15]	
Throughput (MB/s)	0.17	1.52	2.62	2.09	2.98	9.38	1.68

Table 8. Performances' analysis between the proposed and other image encryption algorithms

Criteria		Referenced works						Proposed
		[8]	[20]	21]	[9]	[10]	[11]	
Testing image		Lena, 256 × 256 px, grayscale image						
Global entropy (mean)		7.9984	7.9992	7.997	7.9976	NaN	7.997	7.9968
APCC	H	0.0109	0.0011	0.0055	0.004	NaN	0.0019	-0.0008
	V	0.0139	0.0192	0.0041	–0.0018	NaN	0.0038	0.0099
	D	0.0081	0.0045	0.0002	0.0266	NaN	–0.0019	–0.0005
NPCR		0.9684	0.9979	0.9965	NaN	NaN	0.9965	0.9959
UACI		0.324	0.3335	0.3351	NaN	NaN	0.3348	0.335
Noise attacks	'Salt and pepper'	NaN	NaN	NaN	NaN	NaN	NaN	Success
	'Speckle'	NaN	NaN	NaN	NaN	NaN	NaN	Fail
	Gaussian	NaN	NaN	NaN	Success	Success	NaN	NaN
Cropping attacks		NaN	NaN	NaN	Success	Success	NaN	Success

4 Conclusions

In this paper, a new image cryptosystem based on chaotic map with permutation-substitution architecture [1] was analyzed. For the security analysis, various methods were employed such as key space and key's sensitivity to small changes assessment, differential and statistical analyses etc. The corresponding experimental results show that the proposed image cryptosystem can resist different types of attacks, ensuring a desirable level of security. The achieved time complexity is comparable or better with ones showcased in [12–14]. The proposed algorithm has a highly computational parallelism (*i.e.*, in the inter-pixel permutation stage, all rows could be computed at the same time) and, as a future development, the execution speed should be tested with proper hardware experimental setup.

References

1. Diaconu, A.V., Ionescu, V.M., Iana, G., Lope-Guede, J.M.: A new bit-level permutation image encryption algorithm. In: International Conference on Communications, COMM 2016, Bucharest, Romania, 9–11 June 2016
2. Boriga, R., Dăscălescu, A.C., Priescu, I.: A new hyperchaotic map and its application in an image encryption scheme. Signal Process. Image **29**(8), 887–901 (2014)
3. Wu, Y., Zhou, Y., Noonan, J.P., Agaian, S.: Design of image cipher using Latin squares. Inf. Sci. **264**, 317–339 (2014)
4. Wu, Y., Zhou, Y., Noonan, J.P., Panetta, K., Agaian, S.: Image encryption using the Sudoku matrix. In: Proceedings of SPIE 7708, pp. 77080P-1–77080P-12 (2010)
5. Diaconu, A.-V., Loukhaoukha, K.: An improved secure image encryption algorithm based on Rubik's cube principle and digital chaotic map. Math. Prob. Eng. (2013). doi:10.1155/2013/848392
6. Zhu, Z.-L., Zhang, W., Wong, K.-W., Yu, H.: A chaos-based symmetric image encryption scheme using a bit-level permutation. Inf. Sci. **181**(6), 1171–1186 (2011)
7. Zhou, Y., Bao, L., Chen, C.L.P.: A new 1D chaotic system for image encryption. Signal Process. **97**, 172–182 (2014)
8. Enayatifar, R., Abdullah, A.H., Isnin, I.F.: Chaos-based image encryption using a hybrid genetic algorithm and a DNA sequence. Opt. Lasers Eng. **56**, 83–93 (2014)
9. Sui, L., Duan, K., Liang, J., Zhang, Z., Meng, H.: Asymmetric multiple-image encryption based on coupled logistic maps in fractional Fourier transform domain. Opt. Lasers Eng. **62**, 139–152 (2014)
10. Li, H., Wang, Y., Yan, H., Li, L., Li, Q., Zhao, X.: Double-image encryption by using chaos-based local pixel scrambling technique and gyrator transform. Opt. Lasers Eng. **51**, 1327–1331 (2013)
11. Wang, X., Liu, L., Zhang, Y.: A novel chaotic block image encryption algorithm based on dynamic random growth technique. Opt. Lasers Eng. **66**, 10–18 (2015)
12. Wang, X.Y., Wang, X.M.: A novel block cryptosystem based on the coupled chaotic map lattice. Nonlinear Dyn. **72**, 707–715 (2013)
13. Zhan, Y.-Q., Wang, X.-Y.: A new image encryption algorithm based on non-adjacent coupled map lattices. Appl. Soft Comput. **26**, 10–20 (2015)
14. Zhan, Y.-Q., Wang, X.-Y.: A symmetric image encryption algorithm based on mixed linear-nonlinear coupled map lattice. Inf. Sci. **273**, 329–351 (2014)
15. Dăscălescu, A.C., Boriga, R.: A novel fast chaos-based algorithm for generating random permutations with high shift factor suitable for image scrambling. Nonlinear Dyn. **73**, 307–318 (2013)
16. Li, C., Li, S., Lo, K.-T.: Breaking a modified substitution-diffusion image cipher based on chaotic standard and logistic maps. Commun. Nonlinear Sci. Numer. Simul. **16**, 837–843 (2011)
17. Diaconu, A.-V.: KenKen puzzle-based image encryption algorithm. Proc. Rom. Acad., Ser. A **16**, 313–320 (2015)
18. Diaconu A.-V.: Circular inter-intra pixels bit-level permutation and chaos based image encryption. Inf. Sci. (in press). doi:10.1016/j.ins.2015.10.027
19. Boriga, R., Dascalescu, A.-C., Diaconu A.-V.: Study of a new chaotic dynamical system and its usage in a novel pseudorandom bit generator, Math. Prob. Eng. **2015**, Article. ID 769108 (2013)

20. Wang, X., Luan, D.: A novel image encryption algorithm using a chaos and reversible cellular automata. Commun. Nonlinear Sci. Numer. Simulat. **18**, 3075–3085 (2013)
21. Song, C.-Y., Qiao, Y.-L., Zhang, X.-Z.: An image encryption scheme based on new spatiotemporal chaos. Optik **124**, 3329–3334 (2013)

Characterization of Android Malware Families by a Reduced Set of Static Features

Javier Sedano[1], Camelia Chira[2], Silvia González[1], Álvaro Herrero[3](✉), Emilio Corchado[4], and José Ramón Villar[5]

[1] Instituto Tecnológico de Castilla y León, C/López Bravo 70, Pol. Ind. Villalonquejar, 09001 Burgos, Spain
{javier.sedano, silvia.gonzalez}@itcl.es
[2] Department of Computer Science, University of Cluj-Napoca, Baritiu 26-28, 400027 Cluj-Napoca, Romania
camelia.chira@cs.utcluj.ro
[3] Department of Civil Engineering, University of Burgos, Avenida de Cantabria s/n, 09006 Burgos, Spain
ahcosio@ubu.es
[4] Department of Computer Science and Automation, University of Salamanca, Plaza de la Merced, s/n, 37008 Salamanca, Spain
escorchado@usal.es
[5] Computer Science Department, University of Oviedo, ETSIMO, 33005 Oviedo, Spain
villarjose@uniovi.es

Abstract. Due to the ever increasing amount and severity of attacks aimed at compromising smartphones in general, and Android devices in particular, much effort have been devoted in recent years to deal with such incidents. However, accurate detection of bad-intentioned Android apps still is an open challenge. As a follow-up step in an ongoing research, preset paper explores the selection of features for the characterization of Android-malware families. The idea is to select those features that are most relevant for characterizing malware families. In order to do that, an evolutionary algorithm is proposed to perform feature selection on the Drebin dataset, attaining interesting results on the most informative features for the characterization of representative families of existing Android malware.

Keywords: Feature selection · Genetic algorithm · Android · Malware families

1 Introduction

Bad-intentioned people are taking advantage of the open nature of Android operating system to exploit its vulnerabilities. It is one of the main targets of mobile-malware creators because it is the most widely used mobile operating system [1]. The number of apps available at Android's official store has increased constantly from the very beginning, up to more than 2.1 million [2] that are available nowadays. With regard to the security issue, Android became the top mobile malware platform as well [3] and it is forecast that the volume of Android malware will spike to 20 million during 2016

M. Graña et al. (eds.), *International Joint Conference SOCO'16-CISIS'16-ICEUTE'16*, Advances in Intelligent Systems and Computing 527, DOI 10.1007/978-3-319-47364-2_59

when it was 4.26 million at the end of 2014 and 7.10 million in first half of 2015 [4]. This operating system is an appealing target for bad-intentioned people, reaching unexpected heights, as there are cases where PC malware is now being transfigured as Android malware [5].

To fight against such a problem, it is required to understand the malware and its nature, given that this nature is constantly evolving as it happens with most software. Without understanding the malware, it will not be possible to practically develop an effective solution [6]. Thus, present study is focused on the characterization of Android malware families, trying to reduce the amount of app features needed to distinguish among all of them. To do so, a real-life benchmark dataset [7, 8] has been analyzed by means of several feature selection strategies.

To more easily identify the malware family an app belongs to, authors address this feature selection problem using a genetic algorithm guided by information theory measures. Each individual encodes the subset of selected features using the binary representation. The evolutionary search process is guided by crossover and mutation operators specific to the binary encoding and a fitness function that evaluates the quality of the encoded feature subset. In the current study, this fitness function is defined as the mutual information.

Feature selection methods are normally used to reduce the number of features considered in a classification task by removing irrelevant or noisy features [9, 10]. Filter methods perform feature selection independently from the learning algorithm while wrapper models embed classifiers in the search model [11, 12].Filter methods select features based on some measures that determine their relevance to the target class without any correlation to a learning method.

There are many advantages of feature selection for malware detection; however, little effort has been devoted until now to apply these methods of machine learning to deal with malware features [13]. In [14] just information gain is used to rank the 32 static and dynamic features from a self-generated malware dataset containing 14,794 instances, comprising 30 legitimate apps. Samples of malware come from five different families (GoldDream, PJApps, DroidKungFu2, Snake and Angry Birds Rio Unlocker). To rank the features, four machine learning classifiers (Naïve Bayes, RandomForest, Logistic Regression, and Support Vector Machine) were applied. The top 10 selected features were (in decreasing order of importance): Native_size, Native_shared, Other_shared, Vmpeak, Vmlib, Dalvik_RSS, Rxbytes, VmData, Send_SMS, and CPU_Usage. In a different work [15], 88 dynamic features from 43 apps were collected and then analysed to discriminate between games and tools. The underlying idea of this study was that distinguishing between games and tools would provide a positive indication about the ability of detection algorithms to learn and model the behavior applications and potentially detect malware. To do so, feature selection was applied to identify the 10, 20 and 50 best features, according to Information Gain, Chi Square, and Fisher Score. A similar analysis [16] by same authors proposed a selection from 22,000 static features about 2,285 apps to distinguish between games and tools apps once again. The following classifiers were applied: Decision Tree, Naïve Bayes, Bayesian Networks, PART, Boosted Bayesian Networks, Boosted Decision Tree, Random Forest, and Voting Feature Intervals. The obtained results shown that the combination of Boosted Bayesian Networks and the top 800 features selected using Information Gain yield an accuracy level of 0.918 with a False Positive Rate of 0.172.

Although MRMR has also been previously applied to the detection of malware [17], present study differentiates from previous work as feature selection is now applied from a new perspective, trying to characterize the different Android malware families, to gain deeper knowledge of malware nature. Additionally, to the best of the authors knowledge, this is the very first proposal applying feature selection to an up-to-date and large Android malware dataset in general terms and the Drebin one, more precisely.

More recently, static analysis of Android malware families was already proposed in [18], trying to identify the malware family of malicious apps. The main difference when compared to present work is that family identification relied on apps payload. That is, authors analyzed the Java Bytecode produced when the source code of apps is compiled. It was analyzed through formal methods, being the system behaviour represented as an automaton. With regard to authors previous work [19], an improved evolutionary algorithm for feature selection is now applied, and a more comprehensive and recent dataset is considered in present paper.

The rest of this paper is organized as follows: the proposed evolutionary feature selection algorithm is described in Sect. 2, the experiments for the Drebin dataset are presented in Subsect. 3.1, the results obtained are discussed in Subsect. 3.2 and the conclusions of the study are drawn in Sect. 4.

2 Feature Selection

The big amount of features in the analysed dataset (see Subsect. 3.1), the various feature subsets that may be defined can be extensively evaluated using different methods. The result of these methods can then be aggregated in a ranking scheme. It is proposed to determine an ordered list of selected features using a genetic algorithm based on mutual information as fitness function. The methods described in this section assume a matrix X of N feature values in M samples and an output value y for each sample.

The proposed Genetic Algorithm (GA) encodes in each individual the feature selection by using a binary representation of features. The size of each individual equals the number of features and the value of each position can be 0 or 1, where 1 means that the corresponding feature is selected (the number of features is N). It is proposed to evaluate feature selection results using the mutual information (I).

Defined by means of their probability distribution, the mutual information between two variables has a high value for higher degrees of relevance between the two features. Let $I(X, Y)$ be the mutual information between two features, given by:

$$I(X, Y) = \iint p(x, y) * \log\left(\frac{p(x, y)}{p(x) * p(y)}\right) dxdy \quad (1)$$

The genetic algorithm resulting by using $I(X, Y)$ as the fitness function, named GA-I, is outlined below. N G and t denote the population size, the maximum number of generations and the current generation, respectively.

Algorithm. GA-I Feature Selection

```
Require: X the input variables data set
Require: Y the output vector
P ← a vector of N Individual objects
t ← 0
Generate the initial population P(t): randomly initialize the value of each individual
while t <G do
        Evaluate each individual IND in P(t): calculate I(IND, Y) value
        P(t +1) ← roulette wheel selection from P(t)
        for all individuals IND in P(t + 1) do
                Select mate J from P(t + 1)
                K ←two-point crossover (IND, J)
                if fitness(K) > fitness(IND) then
                        IND ← K
                end if
                L ← mutation(IND)
                if fitness(L) > fitness(IND) then
                        IND ← L
                end if
        end for
        t ← t+1
end while
Return Best Individual in P(t)
```

The GA follows a standard scheme in which roulette wheel selection, two-point crossover and swap mutation are used to guide the search. Each individual is evaluated based on the correlation between the current subset of selected features and the output, given by I.

Furthermore, a second variant of the GA-I algorithm (called *GA-I-W*, where *W* stands for *weighted*) is proposed in order to control the number of features selected in an individual. In order to do that, the fitness function of *GA-I-W* is based on a weighted scheme between the information theory measure and the number of selected features.

Let $k(x)$ be the size of the feature subset encoded in an individual x and w a real parameter between 0 and 1, denoting the weight of each fitness component. The weighted fitness function for an individual x is depicted in Eq. 2.

$$f(x) = w \cdot I(x) + (1 - w) \cdot 1/k(x) \tag{2}$$

The maximization of f would also lead to a minimum number of possible selected features in the individual. It should be noted that the features would only be selected as long as a high value of $I(x)$ still emerges in the current individual. This balance is ensured by the value of the weight parameter w. A 0.5 value w gives the same relevance to both measurements, while higher values of w can be used to give a relative higher importance to the information theory measure value compared to the size of the feature subset.

3 Experimental Study

As previously explained, several approaches for features selection have been applied to the characterization of Android malware. The analysed dataset is described in Subsect. 3.1 and the obtained results are introduced and described in Subsection 3.2.

3.1 Drebin Dataset

The Drebin dataset [7, 8] is a collection of Android apps gathered from the Android official market (Google Play) and from some other un-official sources (alternative markets, websites, forums…) between 2010 and 2012. The gathered apps were analysed through the VirusTotal [20] service, being declared as malicious when more than one of the applied scanners identified the app as an anomalous one. As a result, the dataset contains123,441 benign applications and 5,554 malicious applications (128,995 in total), being one of the largest publicly-available datasets containing legitimate and malicious Android apps. Aps from 179 different families were collected.

Data were extracted from the manifest and the disassembled dex code of the apps, obtained by a linear sweep over the application's content [8]. Every sample in the dataset is associated to an analysed app and the values of the sample represent the given values of app characteristics, such as permissions, intents and API calls.

The following feature sets were extracted from the manifest file of every app [8]:

- Hardware components: contains information about the hardware components requested by the app.
- Requested permissions: contains information about the permission system, the main security mechanism of Android. Permissions declared by the app, and hence requested before installation, are taking into account.
- App components: contains information about the different types of components in the app, each defining different interfaces to the system.
- Filtered intents: contains information about intents (passive data structures exchanged as asynchronous messages for inter-process and intra-process communication).

Additionally, some other feature sets were extracted from the dex information extracted from the apk files of the apps [8]:

- Restricted API calls: contains information about the calls defined in the app to those APIs defined as critical. Although that information must be declared in the manifest file, exactly for being malware, some APIs may be accessed without declaring that in the manifest file (root exploits) and hence the information is double checked with the API calls from the dex code.
- Used permissions: contains information about the permissions that must be granted for the calls identified in previous feature subset. It is once again a way of double-checking the manifest file; the permissions in this case.
- Suspicious API calls: contains information about calls defined in the app to those APIs identified by the authors of the dataset as potentially dangerous. It includes

calls for accessing sensitive data, communicating over the network, sending and receiving SMS messages, execution of external commands, and obfuscation.
- Network addresses: contains information about IP addresses, hostnames and URLs found in the dex code.

The previously defined features sets resulted in an initial set of features for the analysed apps. Each one of these features takes binary values: 0 if the app does not contain such feature and 1 otherwise. To aggregate this information at a family level, feature data were summarized for each family, taking binary values as well: 1 if any app from the family does contain such feature and 0 otherwise.

3.2 Results

The previously described GA-I and GA-I-W algorithms (see Sect. 2) have been applied for the selection of features from the Drebin dataset (see Sect. 3.1). For each one of the algorithms, three sets of experiments have been carried out, with the maximum number of features to be selected taking values of 10, 30 and 50. Two intermediate values of w (0.5 and 0.7) have been selected for comparison purposed. On the other hand, the following values were setting for the parameters of the GA:

- Population size: 100.
- Number of generations: 100.
- Number of runs for the algorithm for each experiment: 50.

The number of features selected by the two different algorithms is 188. As it is too high taking into account the aim of present paper, in order to reduce it, selected features are grouped according to their similar properties. As a result, the following 11 different groups of selected features were generated: activity::, api_call::android/, call::Cipher, call::getWifiState, intent::android.intent.action, permission::android.permission, service_receiver::, url::, real_permission::android.permission, provider::android.appwidget.provider, and feature::android.hardware.screen.landscape. Average values of I() for

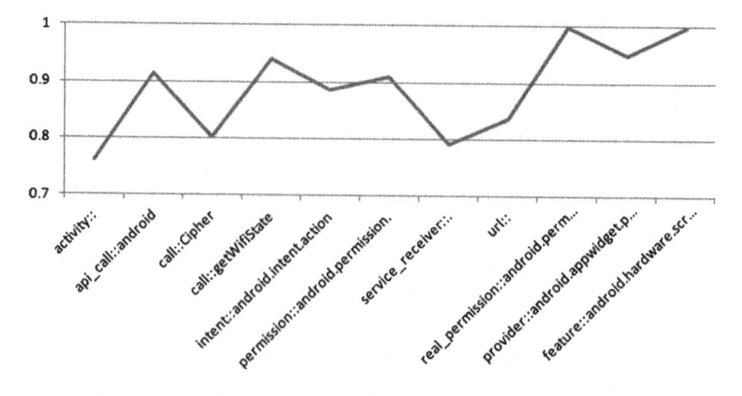

Fig. 1. Average values of *I()* for each group of features.

each one of these groups of features, when running the GA, are shown in Fig. 1. The Y axis shows the values of I and the X axis represents the associated group of features.

To summarize results of the 50 runs for each experiment, percentages were calculated for each group of features and algorithm. In the case of GA-I, percentages are shown in Table 1 while percentages of GA-I-W are shown in Table 2. Additionally, general results are shown in Fig. 2.

Table 1. Percentage of runs in which groups of features are selected by the GA-I algorithm.

Feature	10 Features max	30 Features max	50 Features max
activity::	3,33 %	5,26 %	8,81 %
api_call::android/	0,16 %	0,00 %	0,64 %
call::Cipher	0,21 %	0,21 %	0,21 %
call::getWifiState	0,00 %	0,00 %	0,05 %
intent::android.intent.action	0,00 %	0,21 %	0,48 %
permission::android.permission.	0,11 %	0,21 %	0,64 %
service_receiver::.	1,29 %	2,42 %	3,60 %
url::	3,97 %	7,63 %	13,64 %
real_permission::android.permission.	0,00 %	0,00 %	0,00 %
provider::android.appwidget.provider	0,00 %	0,00 %	0,00 %
feature::android.hardware.screen.landscape	0,05 %	0,00 %	0,00 %

Table 2. Percentage of runs in which groups of features are selected by the GA-I-W algorithm.

	10 Features max		30 Features max		50 Features max	
Feature	W = 0.7	W = 0.5	W = 0.7	W = 0.5	W = 0.7	W = 0.5
activity::	2,58 %	3.23 %	4,30 %	3.63 %	8,16 %	7.78 %
api_call::android/	0,32 %	0.17 %	0,11 %	0.06 %	0,64 %	0.40 %
call::Cipher	0,00 %	0.11 %	0,11 %	0.00 %	0,27 %	0.11 %
call::getWifiState	0,00 %	0.00 %	0,05 %	0.00 %	0,00 %	0.00 %
intent::android.intent.action	0,05 %	0.11 %	0,05 %	0.06 %	0,32 %	0.23 %
permission::android.permission.	0,11 %	0.06 %	0,05 %	0.11 %	0,43 %	0.34 %
service_receiver::.	1,13 %	0.91 %	2,26 %	1.87 %	4,46 %	3.46 %
url::	4,24 %	4.20 %	4,67 %	4.26 %	12,30 %	11.46 %
real_permission::android.permission.	0,00 %	0.00 %	0,00 %	0.00 %	0,11 %	0.00 %
provider::android.appwidget.provider	0,05 %	0.00 %	0,00 %	0.00 %	0,05 %	0.00 %
feature::android.hardware.screen.landscape	0,00 %	0.00 %	0,00 %	0.00 %	0,00 %	0.00 %

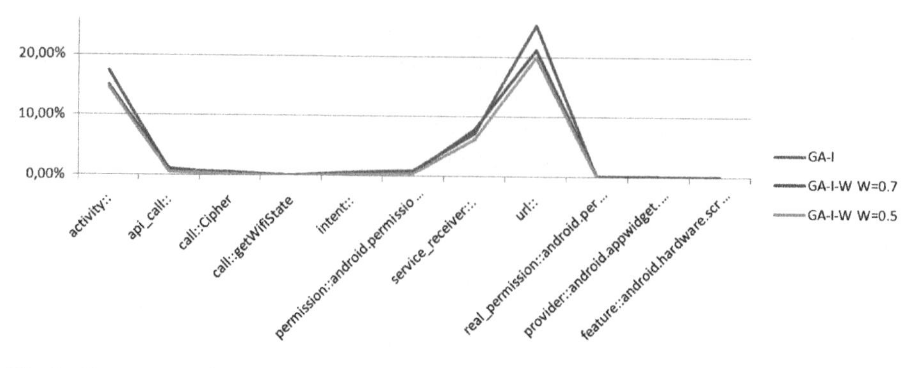

Fig. 2. Percentage of runs in which the group of features are selected by GA-I and GA-I-W algorithms. General results calculated from all the experiments.

As it can be seen from the results shown in tables above, group of features url::, activity:: and service_receiver:: take the highest values of selection by the two algorithms. In general terms, it can be concluded that those groups of features are the most relevant ones for the characterization of Android malware families, according to the results from GA-I and GA-I-W.As url:: gets the highest scores, it means that looking into the URLs that are present in the disassembled core is critical in order to identify the family a malware belongs to. Malicious apps usually establish network connections to retrieve commands or exfiltrate data collected from the device. In a decreasing order of importance, the second group of features is activity::. It means that this kind of app components is the most informative ones. As service components are also important, it can be said that app components are very informative in order to discriminate between malware families.

11 types of features are present in the dataset: activity, api_call, call, feature, intent, network, permission, provider, real_permission, service_receiver, and url. Only one of them (network) is not present (even at a reduced percentage) in the features selected by any of the algorithms. As it can be seen, the group of features that have been selected in most executions are the same for the GA-I and GA-I-W with values of w equal to 0.5 and 0.7.

It should be noted that the GA methods were able to reach the optimum values in the population during the second half of the search process – around generation 30 (see Fig. 3). Each line represents a run of the algorithm, some lines overlap in some executions that were similar - and that is why not every single line of the 50 total lines can be identified. This is due to the low feature quantity limits, which were set to 10, 30, and 50, and led to a size easier to handle (notwithstanding the actual size of the dataset) and enabled to quickly explore many feature subsets.

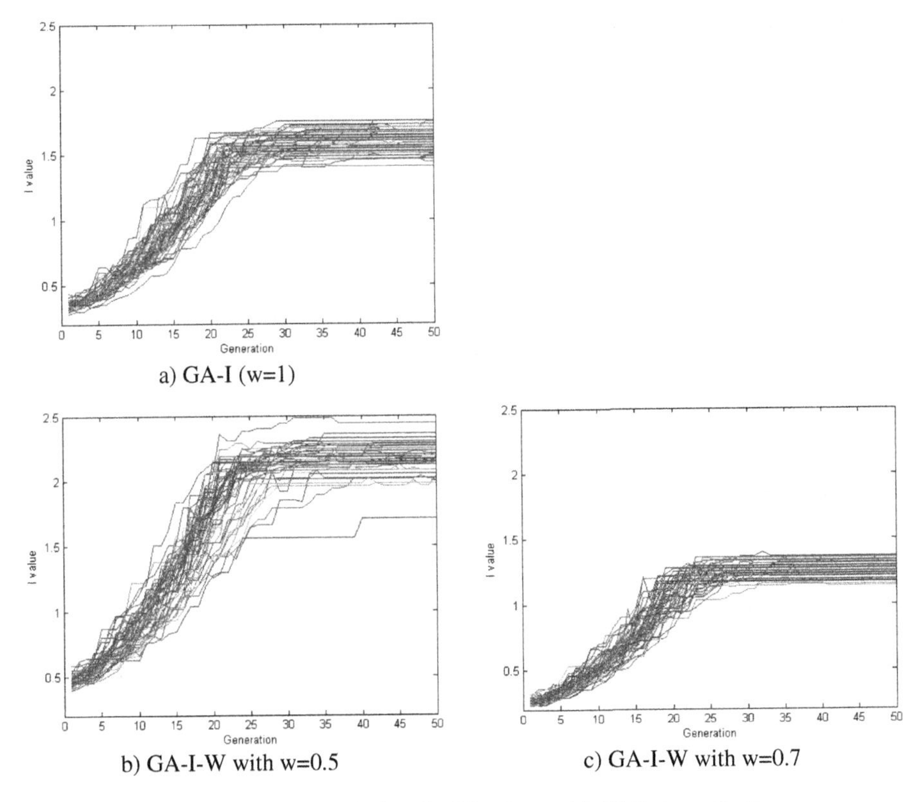

a) GA-I (w=1)

b) GA-I-W with w=0.5

c) GA-I-W with w=0.7

Fig. 3. Fitness evolution for *I()* values with 30 features limit.

4 Conclusions and Future Work

By applying the proposed algorithms of feature selection, Android malware families are characterized. Thanks to such characterization, key features to distinguish from one malware family to the other ones are identified. This is a great contribution for malware detection tools as it is not only important to detect every single intrusion but also to know the family to run appropriate countermeasures.

Experimental results show that the two applied algorithms for feature selection agree on the selection of the 3 main groups of features. By large, url:: and activity:: are identified as the most important features for characterizing malware families as they get the highest percentages in both GA-I and GA-I-W. Feature dealing with URLs in the app code is the most important one to identify the family of a malware sample; it means that it is more important to take into account the external information (where the app is connecting to) rather than the characteristics of the app itself. Consequently, focusing on this piece of information could optimize the analysis of malware. Additionally, from the internal information, activities is the most relevant feature by large.

Future work will focus on proposing further adaptations of feature selection algorithms to ease the characterization of android malware.

Acknowledgments. This research has been partially supported through the project of the Spanish Ministry of Economy and Competitiveness RTC-2014-3059-4. The authors would also like to thank the BIO/BU09/14 and the Spanish Ministry of Science and Innovation PID 560300-2009-11.

References

1. Statista - The Statistics Portal. http://www.statista.com/statistics/266219/global-smartphone-sales-since-1st-quarter-2009-by-operating-system/. Accessed 08 July 2016
2. AppBrain Stats. http://www.appbrain.com/stats/stats-index. Accessed 08 July 2016
3. Micro, T.: The Fine Line: 2016 Trend Micro Security Predictions (2015)
4. Mind the (Security) Gaps: The 1H 2015 Mobile Threat Landscape. http://www.trendmicro.com/vinfo/us/security/news/mobile-safety/mind-the-security-gaps-1h-2015-mobile-threat-landscape. Accessed 08 July 2016
5. F-Secure: Q1 2014 Mobile Threat Report (2015)
6. Yajin, Z., Xuxian, J.: Dissecting android malware: characterization and evolution. In: 2012 IEEE Symposium on Security and Privacy, pp. 95–109 (2012)
7. Spreitzenbarth, M., Echtler, F., Schreck, T., Freling, F.C., Hoffmann, J.: Mobile-sandbox: having a deeper look into android applications. In: 28th International ACM Symposium on Applied Computing (SAC) (2013)
8. Arp, D., Spreitzenbarth, M., Hubner, M., Gascon, H., Rieck, K.: DREBIN: effective and explainable detection of android malware in your pocket. In: 21st Annual Network and Distributed System Security Symposium (2014)
9. Guyon, I., Elisseeff, A.: An introduction to variable and feature selection. J. Mach. Learn. Res. **3**, 1157–1182 (2003)
10. Larrañaga, P., Calvo, B., Santana, R., Bielza, C., Galdiano, J., Inza, I., Lozano, J.A., Armañanzas, R., Santafé, G., Pérez, A.: Machine learning in bioinformatics. Briefings Bioinform. **7**, 86–112 (2006)
11. Ding, C., Peng, H.: Minimum redundancy feature selection from microarray gene expression data. J. Bioinform. Comput. Biol. **3**, 185–205 (2005)
12. Liu, H., Liu, L., Zhang, H.: Ensemble gene selection by grouping for microarray data classification. J. Biomed. Inform. **43**, 81–87 (2010)
13. Feizollah, A., Anuar, N.B., Salleh, R., Wahab, A.W.A.: A review on feature selection in mobile malware detection. Digit. Invest. **13**, 22–37 (2015)
14. Hyo-Sik, H., Mi-Jung, C.: Analysis of android malware detection performance using machine learning classifiers. In: 2013 International Conference on ICT Convergence (2013), pp. 490–495
15. Shabtai, A., Elovici, Y.: Applying behavioral detection on android-based devices. In: Magedanz, T., Li, M., Xia, J., Giannelli, C., Cai, Y. (eds.) Mobilware 2010. LNICST, vol. 48, pp. 235–249. Springer, Heidelberg (2010)
16. Shabtai, A., Fledel, Y., Elovici, Y.: Automated static code analysis for classifying android applications using machine learning. In: 2010 International Conference on Computational Intelligence and Security, pp. 329–333 (2010)
17. Vinod, P., Laxmi, V., Gaur, M.S., Naval, S., Faruki, P.: MCF: multicomponent features for malware analysis. In: 27th International Conference on Advanced Information Networking and Applications Workshops (WAINA), pp. 1076–1081 (2013)

18. Battista, P., Mercaldo, F., Nardone, V., Santone, A., Visaggio, C.: Identification of android malware families with model checking. In: 2nd International Conference on Information Systems Security and Privacy (2016)
19. Sedano, J., Chira, C., González, S., Herrero, Á., Corchado, E., Villar, J.R.: On the selection of key features for android malware characterization. In: Herrero, Á., Baruque, B., Sedano, J., Quintián, H., Corchado, E. (eds.) International Joint Conference. AISC, vol. 369, pp. 167–176. Springer, Heidelberg (2015)
20. Virus Total. https://www.virustotal.com. Accessed 08 July 2016

CISIS 2016: Security in Wireless Networks: Mathematical Algorithms and Models

A Comparison of Computer-Based Technologies Suitable for Cryptographic Attacks

Víctor Gayoso Martínez, Luis Hernández Encinas, Agustin Martín Muñoz(✉), Óscar Martínez-Graullera, and Javier Villazón-Terrazas

Institute of Physical and Information Technologies (ITEFI), Spanish National Research Council (CSIC), Madrid, Spain
{victor.gayoso,luis,agustin}@iec.csic.es
{oscar.martinez.graullera,javier.villazon}@csic.es

Abstract. Developed initially for tasks related to computer graphics, GPUs are increasingly being used for general purpose processing, including scientific and engineering applications. In this contribution, we have analysed the performance of three graphics cards that belong to the parallel computing CUDA platform with two C++ and Java multithreading implementations, using as an example of computation a brute-force attack on KeeLoq, one of the best known remote keyless entry applications. As it was expected, these implementations are not able to break algorithms with 64-bit keys, but the results allow us to provide valuable information regarding the compared capabilities of the tested platforms.

Keywords: Cryptography · CUDA · C++ · Encryption · Java · OpenMP

1 Introduction

In symmetric encryption algorithms, brute-force attacks consist in checking all possible keys until the correct one is found. In the worst case, all the keys from the entire key space are tested, while in average it is necessary to check only half the number of possible keys.

Modern encryption algorithms are designed so that this kind of attack is infeasible, as the search for the key would take millions of years. Legacy algorithms were also designed with that goal in mind, with the difference that their designers could not anticipate the spectacular increase in computing capability that could be employed by organized groups or even individuals in such a task.

For that reason, we have considered to be of interest comparing the computing capability provided by several platforms when using a brute-force approach on legacy algorithms such as KeeLoq, which was designed three decades ago but still can be found in millions of devices. Besides, the simplicity of KeeLoq makes it an ideal candidate for its implementation using CUDA, the GPU-based parallel computing platform created by NVIDIA. In addition to this, we decided to

M. Graña et al. (eds.), *International Joint Conference SOCO'16-CISIS'16-ICEUTE'16*, Advances in Intelligent Systems and Computing 527, DOI 10.1007/978-3-319-47364-2_60

implement other versions of our brute-force application using Java and C++, so we could test their multi-threading capabilities, where in the case of the C++ implementation we have used OpenMP, a well-known multi-threading interface. In this way, we have been able to check the latest improvements in Java regarding its performance and analyse how fast it is compared to a native C++ application. The results allow us to state that these implementations are not able to break algorithms with 64-bit keys, but provide valuable information regarding the compared capabilities of the aforementioned platforms.

The rest of this paper is organized as follows: In Sect. 2, we present a brief overview of the KeeLoq algorithm. Section 3 describes the CUDA, C++/OpenMP, and Java implementations, including portions of the code used in our applications. In Sect. 4, we offer to the readers the experimental results obtained with our implementations. Finally, our conclusions are presented in Sect. 5.

2 KeeLoq Algorithm

KeeLoq is a proprietary block cipher that uses a NLFSR (Non-Linear Feedback Shift Register). Designed in the mid-1980s, KeeLoq was sold by its authors to Microchip Technology Inc. in 1995, and since then it has been used in code hopping encoders and decoders in countless remote keyless entry applications (car remote controllers, garage door openers, etc.) across the world.

KeeLoq operates with 32-bit inputs using a 64-bit secret key, which means there are $2^{64} \approx 1.84 \cdot 10^{19}$ possible keys. Both the encryption and decryption procedures complete 528 rounds, during each of which one bit of the 32-bit state is computed and shifted into the state, as illustrated in Fig. 1. As it can be seen in that figure, the feedback of the NLFSR depends linearly on two register bits, one key bit, and the output of an NLF (Non-Linear Function) that maps five other register bits to a single bit. The key schedule is a simple rotation of the key, i.e.,

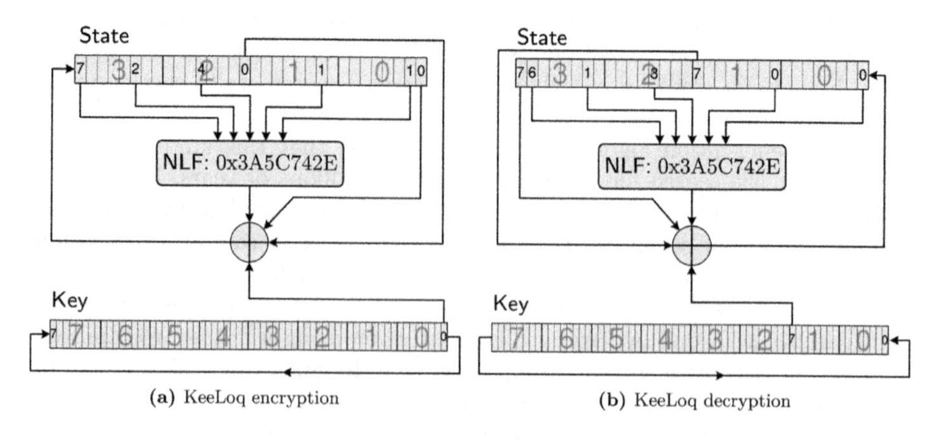

(a) KeeLoq encryption

(b) KeeLoq decryption

Fig. 1. Diagram of the KeeLoq encryption and decryption procedures (source: [1]).

in each clock cycle the key register is rotated to the right for an encryption or to the left for a decryption. Thus, each bit of the key is reused every 64 rounds [2].

3 Programming Platforms

3.1 CUDA

In last years, one of the dominant trends in microprocessor architectures has been the continuous increment of the chip-level parallelism and, as a result of that, multicore CPUs (Central Processing Units) providing 8–16 scalar cores are now commonplace. However, GPUs (Graphics Processor Units) have been at the leading edge of this drive towards increased chip-level parallelism, GPGPU being the term that refers to the use of a GPU card to perform computations in applications traditionally managed by the CPU. Due to their particular hardware architecture, GPUs are able to compute certain types of parallel tasks quicker than multi-core CPUs, which has motivated their usage in scientific and engineering applications [3]. The disadvantage of using GPUs in those scenarios is their higher power consumption compared to that of traditional CPUs [4].

CUDA is the best known GPU-based parallel computing platform and programming model, created by NVIDIA. CUDA is designed to work with C, C++ and Fortran, and with programming frameworks such as OpenACC or OpenCL, though with some limitations. CUDA organizes applications as a sequential host program that may execute parallel programs, referred to as kernels, on a CUDA-capable device. The compute capability specifies characteristics such as the maximum number of resident threads or the amount of shared memory per multiprocessor, which can significantly vary from one version to another (and, consequently, from one graphics card to another) [5].

The code portion displayed in Listing 1.1 contains the main structure of the application, where `num_rounds`, `num_blocks`, and `num_operations_per_block` help to calculate the number of potential keys to be tested, `initial_key` is the first key to be tested by the application, and `decrypted` and `encrypted` are the integer values representing the 32-bit values of the cleartext and the corresponding ciphertext.

```
1  #include <iostream>
2  #include <cuda_runtime.h>
3  #include <stdint.h>
4  #include <ctime>
5  #include <sstream>
6  #include <omp.h>
7  #define bit(x,n)    (((x)>>(n))&1)
8
9  __global__ void keyloqenc(uint32_t *encrypted, uint64_t *mykey,
10                           uint32_t *decrypted)
11 {
12     int index = blockIdx.x*blockDim.x + threadIdx.x, r=0;
13     uint64_t y = *mykey + index;
14     uint32_t x = *decrypted;
15
16     for (r = 0; r < 528; r++)
17         x = (x>>1)^((bit(x,0)^bit(x,16)^(uint32_t)bit(y,r&63)^bit(0x3A5C742E,
18             (bit(x,1)+bit(x,9)*2+bit(x,20)*4+bit(x,26)*8+bit(x,31)*16)))<<31);
19
20     if (x == *encrypted)
```

```
21            *mykey = y;
22    }
23    int main(int argc, char* argv[])
24    {
25        // Variables
26        uint32_t num_rounds = 1, num_blocks = 512;
27        uint32_t num_operations_per_block = 1024;
28        uint32_t decrypted = 0, encrypted = 0;
29        uint64_t initial_key = 0;
30        clock_t begin, end, elapsed_secs;
31
32        uint32_t *dev_encrypted,*dev_decrypted;
33        uint64_t *dev_key, key2;
34
35        // Code for obtaining command-line arguments that change default values
36
37        begin = clock();
38        cudaMalloc((void**)&dev_encrypted, sizeof(uint32_t));
39        cudaMalloc((void**)&dev_decrypted, sizeof(uint32_t));
40        cudaMalloc((void**)&dev_key, sizeof(uint64_t));
41        cudaMemcpy(dev_encrypted,&encrypted, sizeof(uint32_t),
42                    cudaMemcpyHostToDevice);
43        cudaMemcpy(dev_decrypted,&decrypted, sizeof(uint32_t),
44                    cudaMemcpyHostToDevice);
45
46        key = initial_key;
47
48        for (int j = 0; j < num_rounds; j++)
49        {
50            cudaMemcpy(dev_key,&key, sizeof(uint64_t),cudaMemcpyHostToDevice);
51
52            keyloqenc <<<num_operations_per_block, num_blocks>>>
53                        (dev_encrypted, dev_key, dev_decrypted);
54
55            cudaMemcpy(&key2, dev_key, sizeof(uint64_t),cudaMemcpyDeviceToHost);
56
57            if(key2 != key)
58            {
59                // Code for informing the user about the key found
60            }
61
62            key = key + num_blocks*num_operations_per_block;
63        }
64
65        cudaFree(dev_encrypted);
66        cudaFree(dev_decrypted);
67        cudaFree(dev_key);
68        end = clock();
69        elapsed_secs = double(end - begin) / CLOCKS_PER_SEC;
70        std::cout << "Time:" << elapsed_secs << "seconds." << std::endl;
71    }
```

Listing 1.1. Portion of code belonging to the CUDA application.

3.2 C++ and OpenMP

OpenMP (Open Multi-Processing) is an API (Application Programming Interface) that supports shared-memory parallel programming in C, C++, and Fortran on several platforms, including GNU/Linux, OS X, and Windows. The latest stable version is 4.5, released on November 2015.

When using OpenMP, the section of code that is intended to run in parallel is marked with a preprocessor directive that will cause the threads to form before the section is executed. By default, each thread executes the parallelized section of code independently. The runtime environment allocates threads to processors depending on usage, machine load, and other factors.

The code portion included in Listing 1.2 contains the main structure of the application, where `num_cores` is the variable representing the number of concurrent threads, `num_blocks` and `num_operations_per_block` help to calculate the number of potential keys to be tested, `initial_key` is the first key to be tested by the application, and `decrypted` and `encrypted` are the integer values representing the 32-bit values of the cleartext and the ciphertext, respectively.

```
1  #define KeeLoq_NLF       0x3A5C742E
2  #define bit(x,n)         (((x)>>(n))&1)
3  #define f1(x,a,b,c,d,e)
4            (bit(x,a)+bit(x,b)*2+bit(x,c)*4+bit(x,d)*8+bit(x,e)*16)
5
6  uint32_t num_cores = 1, num_blocks = 512;
7  uint32_t num_operations_per_block = 1024;
8  uint32_t r = 0, decrypted = 0, encrypted = 0, result = 0;
9  uint64_t initial_key = 0, key = 0;
10 int index = 0, numthreads = 1;
11 clock_t begin, end, elapsed_secs;
12 int NUMTOTAL = num_blocks*num_operations_per_block;
13
14 key = initial_key;
15 result = decrypted;
16 begin = clock();
17
18 #pragma omp parallel num_threads(num_cores) shared(encrypted)
19         private(i,result,key,r,numthreads)
20 {
21     numthreads=omp_get_num_threads();
22     miclave = initial_key+omp_get_thread_num();
23
24     for(index=omp_get_thread_num();index<NUMTOTAL;index=index+numthreads)
25     {
26         result = decrypted;
27         for (r = 0; r < 528; r++)
28         {
29             result = (result>>1)^((bit(result,0)^bit(result,16)^
30                      (uint32_t)bit(key,r&63)^
31                       bit(KeeLoq_NLF, f1(result,1,9,20,26,31)))<<31);
32         }
33         if(result == encrypted)
34         {
35             // Code to inform the user that the key has been found
36         }
37         key = key+numthreads;
38     }
39
40     end = clock();
41     elapsed_secs = double(end-begin)/CLOCKS_PER_SEC;
42     std::cout << "Time:" << elapsed_secs << "seconds." << std::endl;
43 }
```

Listing 1.2. Portion of code belonging to the C++/OpenMP application.

The `#pragma omp parallel` sentence declares that the code encompassed by it must be executed multi-threaded, as indicated by the `num_threads` reserved word. The variables whose content is shared by all the threads are those indicated by the `shared` identifier, while those that must be managed separately by each thread are indicated by the `private` key word.

3.3 Java

Since its appearance in the mid-1990s, the Java language has experienced a constant growth regarding the number of programmers and commercial deployments, being massively used in web and corporate applications. The latest version, known as Java 8, was launched in 2014.

The code portion displayed in Listing 1.3 contains the structure of the application's main class, where `numCores` is the variable representing the number of concurrent threads, `numBlocks` and `numOperationsPerBlock` help to calculate the number of potential keys to be tested, `initialKey` is the first key to be tested by each thread, and `decrypted` and `encrypted` are the integer values representing the 32-bit values of the cleartext and the ciphertext, respectively.

```
1  int numConcurrentThreads, numTotalThreads, numOperations;
2  int encrypted, decrypted;
3  long firstKey, initialKey, startTime, stopTime;
4  Runnable worker = null;
5  startTime = System.nanoTime();
6  ExecutorService executor = Executors.newFixedThreadPool(numCores);
7
8  for (int i = 0; i < numBlocks; i++){
9      initialKey = firstKey+i*numOperationsPerBlock;
10     worker = new MyRunnableThread(numOperationsPerBlock,initialKey,
11                                   encrypted, decrypted);
12     executor.execute(worker);
13 }
14
15 executor.shutdown();
16 executor.awaitTermination(Long.MAX_VALUE,
17                          java.util.concurrent.TimeUnit.NANOSECONDS);
18 stopTime = System.nanoTime();
19 System.out.println("Time:"+((stopTime-startTime)/1000000)+"seconds.");
```

Listing 1.3. Portion of code belonging to the Java main class.

The `newFixedThreadPool()` method from class `Executor` creates a thread pool that reuses a fixed number of threads (indicated in the application by `numConcurrentThreads`) operating off a shared unbounded queue [6]. At any point, at most `numConcurrentThreads` threads will be active processing tasks, so additional threads added by means of the `execute()` method have to wait in the queue until a thread finishes its task and allows a new thread to replace it.

As soon as the first thread is invoked with the `execute()` method, the code associated to the thread starts to be executed. Once all the threads intended to be executed are added to the thread pool, it is necessary to call the `shutdown()` method, which initiates an orderly shutdown in which previously submitted tasks are executed, but no new tasks will be accepted, and the `awaitTermination()` method, which blocks the main application until all threads have completed their execution after a shutdown request, the timeout occurs or the current thread is interrupted, whichever happens first [7].

Class `MyRunnableThread` includes the code that performs the encryption operation with different keys, and that is executed by each thread, as it can be observed in Listing 1.4.

```
1  public class MyRunnableThread implements Runnable
2  {
3      private int[] iaNFL = {0,1,1,1,0,1,0,0,0,0,1,0,1,1,1,0,
4                             0,0,1,1,1,0,1,0,0,1,0,1,1,1,0,0};
5      long key = 0;
6      int block, result, decrypted, encrypted, operations;
7
8      MyRunnableThread(int operations,long key,int decrypted,int encrypted)
9      {
10         this.clave = clave;
11         this.decrypted = decrypted;
12         this.encrypted = encrypted;
13         this.rounds = rounds;
14     }
15
16     public void run()
17     {
18         int position, value1, value2, value3, value4;
19
20         for(int index=0; index < operations; index++)
21         {
22             block = decrypted;
23             position = 0;
24
25             for(int i=0; i < 528; i++)
26             {
27                 pos = i%64;
28                 value1 = (int)((key >> position) & 0x00000001);
29                 value2 = (block >>0) & 0x00000001;
30                 value3 = (block >>16) & 0x00000001;
31                 value4 = NFLEnc(block) & 0x00000001;
32                 result = (value1^value2^value3^value4) & 0x00000001;
33                 block = (block>>>1)+(result<<31);
34             }
35             if(block == encrypted)
36             {
37                 // Code to inform the user that the key has been found
38             }
39             key++;
40         }
41     }
42
43     public int NFLEnc(int data)
44     {
45         int temp, value;
46         value = value + (((data >> 31) & 0x00000001)<<4);
47         value = value + (((data >> 26) & 0x00000001)<<3);
48         value = value + (((data >> 20) & 0x00000001)<<2);
49         value = value + (((data >> 9) & 0x00000001)<<1);
50         value = value + (((data >> 1) & 0x00000001)<<0);
51         return iaNFL[value];
52     }
53 }
```

Listing 1.4. MyRunnableThread class.

This architecture has the benefit of providing an automatic management of the threads added to the pool, so the programmer does not need to attend them individually, which is an advantage in tasks as the one analysed in this contribution where the processing should stop only if the correct key is found or all the potential keys inside a certain range are tested.

4 Experimental Results

The CUDA tests whose results are presented in this section were completed using the following equipment [8]:

- A GeForce 210 card (compute capability 1.2) with 16 processor cores running at 1.296 GHz and 2 SMs (Streaming Multiprocessors), with a memory bandwidth of 6.40 GB/s, a floating point performance of 39.36 GFLOPS and a texture rate of 4.16 GTexels per second (GT/s), mounted on an Intel Core i7 processor model 3370 at 3.40 GHz.
- A GeForce 555M card (compute capability 2.1) with 144 processor cores running at 1.350 GHz and 2 SMs, with a memory bandwidth of 28.80 GB/s, a floating point performance of 388.8 GFLOPS and a texture rate of 16.20 GT/s, mounted on an Intel Core i7 processor model 2630QM at 2.00 GHz.
- A Tesla C1060 card (compute capability 1.3) with 240 processor cores running at 1.296 GHz and 10 SMs, with a memory bandwidth of 102.4 GB/s, a floating point performance of 622.1 GFLOPS and a texture rate of 48.8 GT/s, mounted on an Intel XEON processor model E5520 at 2.27 GHz.

In all the tests each application has to check the first 503,316,480 possible keys using a given encryption and decryption values. Table 1 includes the running time in seconds of the CUDA implementation for a constant block size of 512. The different grid size values tested are those compatible with the three cards.

Table 1. Running time in seconds using the CUDA implementation for different values of the grid size

	384	512	768	1024	1536	2048	3072	4096
GeForce 210	446.88	446.71	446.52	446.38	446.33	446.24	446.16	446.13
GeForce 555M	100.12	100.21	99.93	100.1	99.90	99.87	99.80	99.85
Tesla C1060	32.92	34.32	32.9	33.181	32.93	32.60	32.46	32.39

Table 2 shows the running time in seconds of the C++/OpenMP and Java implementations when using a different number of concurrent threads. Both implementations have been tested using the Intel Core i7 3370 processor. While the CUDA and C++/OpenMP applications have been compiled with Visual Studio 2010, the Java application has been compiled with NetBeans 8.0 using the JDK (Java Development Kit) version 1.8.0–77.

At it can be observed, even though the second part of Table 2 uses a number of concurrent threads that surpasses the theoretical limit, and as such the C++ implementation does not improve its performance, the Java application allows some improvement when requesting a higher number of concurrent threads. We assume that this is due to optimizations of the Java virtual machine, which apparently manages more efficiently a higher number of threads when communicating with the operating system.

Table 2. Running time in seconds using the C++ and Java multi-threaded implementations for different values of the theoretical number of concurrent threads

	1	2	3	4	5	6	7	8
C++	1057.53	530.821	377.91	288.865	284.278	272.594	263.546	257.524
Java	1651.1638	966.3502	717.64	557.7586	506.0211	456.5317	419.23602	395.47556
	16	32	48	64	92	128	256	512
C++	257.65	257.81	258.35	258.3	258.33	258.94	258.91	259.29
Java	347.44	331.0	327.08	326.17	324.07	323.93	322.44	321.42

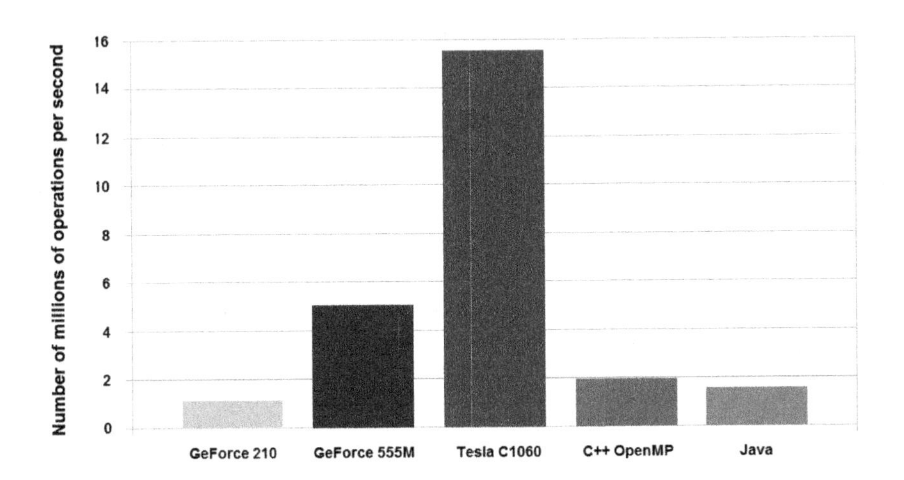

Fig. 2. Performance comparison of the different platforms and implementations

Finally, Fig. 2 shows the highest performance in encryptions per second obtained with the different platforms and implementations.

5 Conclusions

In this contribution we have compared the computer capability of several hardware and software technologies using as an example a cryptographic brute-force attack on the legacy algorithm KeeLoq. More specifically, we have compared the same CUDA application with three different GPUs and both a C++ and Java multi-threading implementations using an Intel Core i7 processor.

The different tests performed with the applications allow us to confirm the suitability of the CUDA platform for intensive workloads of cryptographic computations. The best result obtained with the Tesla device provides a performance eight times better than that of the C++/OpenMP implementation when using the full capacity of the i7 processor. Regarding the comparison between the three CUDA devices, the results are aligned with some of the technical capabilities of the cards such as the memory bandwidth and the texture rate.

Besides, these tests have shown that, with the best configuration in each case, the native C++/OpenMP application provides a performance which is approximately 20 % better than the performance of the interpreted Java code. This confirms the idea that, unless interoperability across several platforms is strongly needed, native applications should be preferred over interpreted applications for tasks of intensive computing, though the difference obtained has been less significant than expected. We consider that this could be explained by the special characteristics of the KeeLoq implementation, which requires a very low amount of memory and, in the case of Java, uses primitive data types and does not require to use resource-consuming classes such as `BigInteger`.

The results show that, as expected, current GPU and CPU technologies cannot still be used for brute-force attacks on 64-bit encryption algorithms (including KeeLoq) without additional data. However, those results clearly demonstrate the superiority of some CUDA cards with respect to advanced CPUs for certain intensive computing tasks. With the current trends in GPU and CPU technologies, this margin could increase in the next years, as for example the latest top-level NVIDIA card released, Tesla P100 [9], could perform between 7 and 10 times better than the Tesla C1060 model used in our tests.

Acknowledgements. This work has been supported by the European Union FEDER funds distributed through Ministerio de Economía y Competitividad (Spain) under the project TIN2014-55325-C2-1-R (ProCriCiS), and through Comunidad de Madrid (Spain) under the project S2013/ICE-3095-CM (CIBERDINE).

References

1. Kasper, T.: Security Analysis of Pervasive Wireless Devices - Physical and Protocol Attacks in Practice Ruhr-University Bochum, Germany (2011)
2. Eisenbarth, T., Kasper, T., Moradi, A., Paar, C., Salmasizadeh, M., Shalmani, M.T.M.: Physical cryptoanalysis of KeeLoq code hopping applications. Cryptology ePrint Archive, Report 2008/058, pp. 1–22 (2008). https://eprint.iacr.org/2008/058.pdf
3. Corp, N.: What is GPU computing? (2016). https://www.nvidia.com/object/what-is-gpu-computing.html
4. Mittal, S., Vetter, J.S.: A survey of methods for analyzing and improving GPU energy efficiency. ACM Computi. Surv. **47**(2), 1–23 (2014)
5. NVIDIA Corporation: Programming Guide (2016). http://docs.nvidia.com/cuda/cuda-c-programming-guide/index.html#compute-capabilities
6. Oracle Corporation: Executors (Java Platform SE 8) (2016). https://docs.oracle.com/javase/8/docs/api/java/util/concurrent/Executors.html
7. Oracle Corporation: Executor Service (Java Platform SE 8) (2016). https://docs.oracle.com/javase/8/docs/api/java/util/concurrent/ExecutorService.html
8. NVIDIA Corporation: CUDA Legacy GPUs (2016). https://developer.nvidia.com/cuda-legacy-gpus
9. NVIDIA Corporation: Tesla P100 (2016). http://www.nvidia.com/object/tesla-p100.html

Cryptanalysis of a Key Authentication Scheme Based on the Chinese Remainder Theorem and Discrete Logarithms

Alberto Peinado(✉)

Departamento de Ingeniería de Comunicaciones,
E.T.S. Ingeniería de Telecomunicación, Universidad de Málaga,
Campus de Teatinos, 29071 Málaga, Spain
apeinado@ic.uma.es

Abstract. In 2015, Kumaraswamy *et al.* have proposed an improvement of the key authentication scheme based on discrete logarithms. That kind of schemes has been widely studied for many years, producing many modifications and improvements designed to overcome the weaknesses detected; most of them leading to key substitution attacks and, in some cases, allowing to recover the user's private key. The improvement proposed by Kumaraswamy *et al.* is based on the Chinese remainder theorem in combination with the discrete logarithm. In this paper, several mathematical inconsistencies are revealed in the definition. Once fixed, a key substitution attack is performed.

Keywords: Cryptanalysis · Key authentication scheme · Discrete logarithm · Chinese remainder theorem

1 Introduction

A major concern in public key cryptosystems is that an intruder could impersonate a legal user by performing the substitution of a public key with a fake key in the public directory, where all the public keys are located, or by intercepting the communications and altering those keys. One of the proposals to solve that problem was originally presented by Horng and Yang [1] in 1996. It consisted on a key authentication scheme (HK scheme) based on discrete logarithms, in which a central trusted server maintained a secure password table to store the identification of each user and the value $f(PWD)$, f being a hash function and PWD his password. The certificate of the corresponding public key was computed as a combination of the user's private key and password. In 1999, Zhan *et al.* [11] showed that the HK scheme suffered from a password-guessing attack, and proposed a new key authentication scheme (ZLYH scheme), also based on discrete logarithms. In 2003, Lee, Hwang and Li [3] proved that the ZLYH scheme did not provide non-repudiation service for the user's public key, and proposed an improvement (LHL scheme) modifying the way of computing the certificates.

M. Graña et al. (eds.), *International Joint Conference SOCO'16-CISIS'16-ICEUTE'16*,
Advances in Intelligent Systems and Computing 527, DOI 10.1007/978-3-319-47364-2_61

Peinado [5] showed in 2004 that the LHL scheme is not secure since the user's private key can be easily recovered from the certificate. A slight modification was suggested in [5] to overcome the vulnerability. Also in 2004, Wu and Lin [8], and Zhang and Kim [12] in 2005, revealed the same vulnerability in the LHL scheme, and proposed improvements that were broken by Yoon and Yoo [10] in 2005, and Li and Liu [4] in 2009, respectively, but no improvements were proposed.

The improvement proposed by Peinado [5] focused the attention in the following years, generating several modifications and improvements [6,7,9]. Recently, in 2015, Kumaraswamy *et al.* have proposed a new improvement of the LHL scheme using the Chinese remainer theorem in order to overcome the weaknesses revealed by Peinado. However, the new improvement also suffers from a vulnerability leading us to perform a key substitution attack.

In Sect. 2, the key authentication scheme presented in [2] is described. Several mathematical inconsistencies are revealed in Sect. 3 and the substitution attack is described in Sect. 4. Finally, the conclusions are presented in Sect. 5.

2 Kumaraswamy *et al.*'s Key Authentication Scheme

Following the general scheme of all of its predecessors, the protocol proposed in [2] considers three phases: setup phase, registration phase and authentication phase. A central trusted server is also considered.

2.1 Setup Phase

Each user U_i of the system has a private key Prv_i and its corresponding public key $Pub_i = g^{Prv_i} \bmod p$, where p is a large prime number and g is a generator in $\mathbb{Z}_p{}^*$. Each user U_i also has a password PWD_i to access the server. The function $f : \mathbb{Z}_p{}^* \longrightarrow \mathbb{Z}_p{}^*$ defined as $f(x) = g^x \bmod p$ is applied to compute $f(PWD_i + r)$, where $r \in \mathbb{Z}_p{}^*$ is selected at random.

The user U_i generates two large random numbers $v_{1,i}, v_{2,i}$ at random and solve for the integer x_i the following two equations using the Chinese remainder theorem.

$$x_i \equiv Prv_i (\bmod v_{1,i}) \tag{1}$$

$$x_i \equiv (PWD_i + r)(\bmod v_{2,i}) \tag{2}$$

Next, $k_{1,i}, k_{2,i}$ are computed using x_i in the following way.

$$k_{1,i} = x_i - Prv_i / v_{1,i} \tag{3}$$

$$k_{2,i} = x_i - (PWD_i + r) / v_{2,i} \tag{4}$$

2.2 Registration Phase

In this phase, the user U_i sends $f(PWD_i + r)$, $R = g^r \bmod p$, Pub_i, $v_{1,i}$, $v_{2,i}$, $f(v_{1,i})$, $f(v_{2,i})$ and $f(x_i)$ to the server secretely. The function f, and the numbers p and g are stored in the public password table into the server. In [2], it is assumed that the password tables cannot be neither modified nor forged.

The server performs the following two verifications.

$$f(PWD_i + r) \stackrel{?}{=} f(PWD_i) \cdot R \tag{5}$$

$$Pub_i \cdot f(v_{1,i})^{k_{1,i}} \stackrel{?}{=} f(PWD_i + r) \cdot f(v_{2,i})^{k_{2,i}} \stackrel{?}{=} f(x_i) \tag{6}$$

If the verification process succeeds, the server assumes that the parameters have been sent by a legal user U_i. The server stores all received values, except R and $f(x_i)$, in a public table.

The certificate C_i of the user's public key is computed as

$$C_i = [f(v_{1,i}) \cdot (PWD_i + r) + x_i \cdot f(v_{2,i})] \bmod (p-1) \tag{7}$$

2.3 Authentication Phase

When someone wants to communicate with a user U_i, the sender gets access to the certificate C_i and the public key Pub_i. The sender also gets all other values from the public password table located at the server. Hence, the sender verifies the following equation.

$$f(C_i) \stackrel{?}{=} f(PWD_i + r)^{f(v_{1,i})} \cdot \left[f(v_{1,i})^{k_{1,i}} \cdot Pub_i\right]^{f(v_{2,i})} \bmod p \tag{8}$$

If the equation holds, the sender accepts the public key Pub_i. Otherwise, it is rejected.

3 Mathematical Inconsistencies

Before describing the main weakness of the protocol, it is important to highlight several mathematical inconsistencies in the original proposal of Kumaraswamy *et al.* [2]. On one hand, although the values $f(PWD_i + r)$ and R are elements in $\mathbb{Z}_p{}^*$, the Eq. (5) is not defined as a modular operation. Hence, the verification cannot be successfully performed. Instead, Eq. (5) must be rewritten as

$$f(PWD_i + r) \stackrel{?}{=} f(PWD_i) \cdot R \mod p \tag{9}$$

Equation (6) suffers from the same problem. Since $f(x_i)$ is an element in $\mathbb{Z}_p{}^*$, Eq. (6) must be rewritten as

$$Pub_i \cdot f(v_{1,i})^{k_{1,i}} \bmod p \stackrel{?}{=} f(PWD_i + r) \cdot f(v_{2,i})^{k_{2,i}} \bmod p \stackrel{?}{=} f(x_i) \tag{10}$$

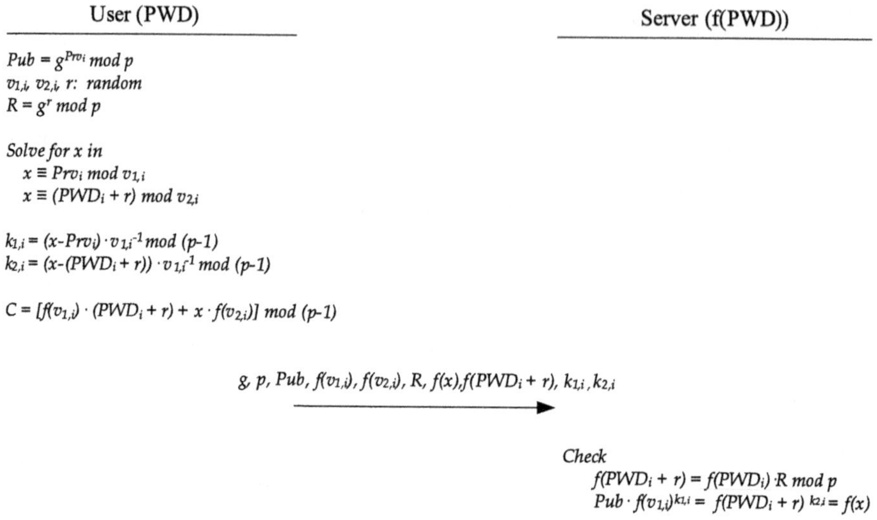

User (PWD) | Server (f(PWD))

$Pub = g^{Prv_i} \bmod p$
$v_{1,i}, v_{2,i}, r$: random
$R = g^r \bmod p$

Solve for x in
$x \equiv Prv_i \bmod v_{1,i}$
$x \equiv (PWD_i + r) \bmod v_{2,i}$

$k_{1,i} = (x - Prv_i) \cdot v_{1,i}^{-1} \bmod (p-1)$
$k_{2,i} = (x - (PWD_i + r)) \cdot v_{1,i}^{-1} \bmod (p-1)$

$C = [f(v_{1,i}) \cdot (PWD_i + r) + x \cdot f(v_{2,i})] \bmod (p-1)$

$g, p, Pub, f(v_{1,i}), f(v_{2,i}), R, f(x), f(PWD_i + r), k_{1,i}, k_{2,i}$ →

Check
$f(PWD_i + r) = f(PWD_i) \cdot R \bmod p$
$Pub \cdot f(v_{1,i})^{k_{1,i}} = f(PWD_i + r)^{k_{2,i}} = f(x)$

Fig. 1. Redefinition of the initial and registration phases

On the other hand, the function f originally defined as $f : \mathbb{Z}_p{}^* \longrightarrow \mathbb{Z}_p{}^*$ must be redefined as $f : \mathbb{Z}_{p-1} \longrightarrow \mathbb{Z}_p{}^*$, since the exponents live in $\mathbb{Z}_{p-1}$. Hence, we have $PWD_i, r \in \mathbb{Z}_{p-1}$.

The verification in Eq. (10) cannot be successfully performed yet using $k_{1,i}$, $k_{2,i}$ as defined in Eq. (3) and (4). As one can observe, $Prv_i/v_{1,i}$ and $(PWD_i + r)/v_{2,i}$ are not, in general, elements in $\mathbb{Z}_p{}^*$. Furthermore, from Eq. (10) we have

$$Pub_i \cdot f(v_{1,i})^{k_{1,i}} \bmod p = g^{Prv_i + v_{1,i} \cdot k_{1,i}} \bmod p \stackrel{?}{=} g^{x_i} \bmod p \quad (11)$$

$$f(PWD_i + r) \cdot f(v_{2,i})^{k_{2,i}} \bmod p = g^{(PWD_i + r) + v_{2,i} \cdot k_{2,i}} \bmod p \stackrel{?}{=} g^{x_i} \bmod p \quad (12)$$

Hence, these equations will hold true when the following relationship are satisfied

$$x_i = Prv_i + v_{1,i} \cdot k_{1,i} \mod (p-1) \quad (13)$$

$$x_i = (PWD_i + r) + v_{2,i} \cdot k_{2,i} \mod (p-1) \quad (14)$$

As a consequence, Eqs. (3) and (4) must be rewritten as

$$k_{1,i} = (x_i - Prv_i) \cdot v_{1,i}{}^{-1} \mod (p-1) \quad (15)$$

$$k_{2,i} = (x_i - (PWD_i + r)) \cdot v_{2,i}{}^{-1} \mod (p-1) \quad (16)$$

Note that $v_{1,i}$ and $v_{2,i}$ must be invertible values. Hence, $v_{1,i}, v_{2,i} \in \mathbb{Z}_{p-1}^*$. Finally, for consistency reasons, the values $f(v_{1,i})$ and $f(v_{2,i})$ in Eqs. (7) and (8) must be previously reduced mod $(p-1)$, since they are defined in $\mathbb{Z}_p{}^*$.

Figure 1 shows the initial and registration phases once the inconsistencies have been fixed. On the other hand, each modification presented in this section keeps satisfying the certificate verification of Eq. (8).

4 Cryptanalysis

The cryptanalysis of the authentication scheme in [2] begins with a discussion about the security model considered in the original definition. As other previous proposals, the security model of the scheme under analysis is not realistic, considering many restrictions on the communication channels and a strong inviolability of the public information. In spite of the security model finally considered, a key substitution attack can be successfully performed, unlike what the authors claimed in [2].

4.1 Security Model

In the current scheme, and most of its predecessors, the security model is based on four security constraints:

1. The key server is trusted by all users.
2. The public password table hosted in the key server cannot be modified or forged by any adversary.
3. There is a secure channel between each user and the key server, by which the user can send messages to the key server in secret and authentic.
4. There is a secure channel between the key server and each sender, by which the key server can send messages to the sender in authentic.

We agree with Shao [6], who performed in 2005 a similar analysis on other improvements of this kind of key authentication schemes, that the last three properties define a poor realistic scenario and they would not be security prerequisites for any cryptosystem. If so, one can deploy a much simpler scheme to distribute public keys, in such a way that once each user U_i generates Prv_i and Pub_i and sends the public key to the trusted key server, the only thing a sender has to do is to ask the key server for the public key he want to use. Because all the channels and storage are secure and trusty, any verification process would be no longer necessary.

It is important to note that the main objective of a key authentication scheme that uses public-key certificates is to avoid repudiation of the public-keys previously employed by the senders, and assure the validity and/or reliability of every certificate. In the next subsection, it is shown a method to generate forged public keys Pub^* with their corresponding certificates C^*, without altering the rest of parameters.

4.2 Key Substitution Attack

The main weakness of Kumaraswamy *et al.*'s scheme resides on the fact that an adversary can produce a fake public key Pub^* and its corresponding certificate C^* using the public parameters hosted in the server. In other words, there exists more than one pair (Pub_i, C_i) satisfying the verification equations Eqs. (9) and (10) for the same values of $f(v_{1,i})$, $f(v_{2,i})$, $k_{1,i}$, $k_{2,i}$ and $f(PWD_i + r)$.

Let us consider the original pair (Pub_i, C_i) generated by a legal user U_i and the values $f(v_{1,i})$, $f(v_{2,i})$, $k_{1,i}$, $k_{2,i}$ and $f(PWD_i{+}r)$ sent to the server and stored in the public table. As all of these parameters are made public, an adversary may compute the following fake pair (Pub^*, C^*) as

$$C^* = C_i + \beta \cdot f(v_{2,i}) \mod (p-1) \tag{17}$$

$$Pub^* = g^{\beta} \cdot Pub_i \mod p \tag{18}$$

When someone wants to check the public key Pub^* and C^* using the verification equation defined in Eq. (8), we have

$$\begin{aligned} f(PWD_i + r)^{f(v_{1,i})} \cdot \left[f(v_{1,i})^{k_{1,i}} \cdot Pub^*\right]^{f(v_{2,i})} & \mod p \\ = g^{(PWD_i+r)f(v_{1,i})} \cdot g^{(v_{1,i}k_{1,i}+\beta+Prv_i)f(v_{2,i})} & \mod p \\ = g^{(PWD_i+r)f(v_{1,i})} \cdot g^{(x-Prv_i+\beta+Prv_i)f(v_{2,i})} & \mod p \\ = g^{(PWD_i+r)f(v_{1,i})+x\cdot f(v_{2,i})} \cdot g^{\beta\cdot f(v_{2,i})} & \mod p \\ = g^{C+\beta\cdot f(v_{2,i}} \mod p = g^{C^*} \mod p = f(C^*) & \end{aligned} \tag{19}$$

Note that Eqs. (17) and (18) are satisfied for every element $\beta \in \mathbb{Z}_{p-1}$. Hence the adversary can select any value for β and then generate the fake pair (Pub^*, C^*).

This weakness allows the attacker to perform a key substitution attack. Since the attacker cannot generate the private key Prv^* corresponding to the fake public key Pub^*, the immediate effect of this weakness is a denial of service, if a man in the middle attack is performed to deliver fake values when honest users request information to the server. However, this weakness could also be employed by a dishonest user to perform a repudiation attack, because in that case the private key Prv^* can be computed as

$$Prv^* = Prv + \beta \mod (p-1) \tag{20}$$

5 Conclusions

It has been shown that the new scheme proposed by Kumaraswamy *et al.* in [2] suffers from an important weakness. This allows an attacker to generate as many fake pairs of public key and certificate as he wants, in order to perform a Denial of Service attack. Furthermore, it has also been proven that a dishonest user could generate fake private and public keys breaking the non-repudiation service required by any key authentication scheme.

Acknowledments. This work has been supported by MINECO-Spain under project "Cryptographic protocolos for cybersecurity: identification, authentication and information protection(ProCriCiS)" (TIN2014-55325-C2-1-R) and by Universidad de Málaga, Campus de Excelencia Andalucía Tech.

References

1. Horng, G., Yang, C.S.: Key authentication scheme for cryptosystems based on discrete logarithms. Comput. Commun. **19**, 848–850 (1996)
2. Kumaraswamy, P., Rao, C.V.G., Janaki, V., Prashanth, K.: A new uthentication scheme for cryptosystems based on discrete logarithms. J. Innov. Comput. Sci. Eng. **5**, 42–47 (2015)
3. Lee, C.C., Hwang, M.S., Li, L.H.: A new key authentication scheme based on discrete logarithms. Appl. Math. Comput. **139**, 343–349 (2003)
4. Li, J., Liu, J.: Remarks on zhang-kim's key authentication scheme. Int. J. Netw. Secur. **8**, 199–200 (2009)
5. Peinado, A.: Cryptanalysis of LHL-key authentication scheme. Appl. Math. Comput. **152**, 721–724 (2004)
6. Shao, A.: A new key authentication scheme for cryptosystems based on discrete logarithms. Appl. Math. Comput. **167**, 143–152 (2005)
7. Sun, D.Z., Cao, Z.F.: Improved public key authentication scheme for non-repudiation. Appl. Math. Comput. **168**, 927–932 (2005)
8. Wu, T., Lin, H.Y.: Robust key authentication scheme resistant to public key substitution attacks. Appl. Math. Comput. **157**, 825–833 (2004)
9. Yoon, E.J., Ryu, E.K., Yoo, K.Y.: Cryptanalysis and further improvement of peinado's improved LHL-key authentication scheme. Appl. Math. Comput. **168**, 788–794 (2005)
10. Yoon, E.J., Yoo, K.Y.: On the security of Wu-Lin's robust key authentication scheme. Appl. Math. Comput. **169**, 1–7 (2005)
11. Zhan, B., Li, Z., Yang, Y., Hu, Z.: On the security of HY-Key authentication scheme. Comput. Commun. **22**, 739–741 (1999)
12. Zhang, F.G., Kim, K.: Cryptanalysis of Lee-Hwang-Li's key authentication scheme. Appl. Math. Comput. **161**, 101–107 (2005)

A SCIRS Model for Malware Propagation in Wireless Networks

Angel Martín del Rey(✉), José Diamantino Hernández Guillén, and Gerardo Rodríguez Sánchez

Department of Applied Mathematics, University of Salamanca, Salamanca, Spain
{delrey,diaman,gerardo}@usal.es

Abstract. The main goal of this work is to propose a novel mathematical model to simulate malware spreading in wireless networks considering carrier devices (those devices that malware has reached but it is not able to carry out its malicious purposes for some reasons: incompatibility of the host's operative system with the operative system targeted by the malware, etc.) Specifically, it is a SCIRS model (Susceptible-Carrier-Infectious-Recovered-Susceptible) where reinfection and vaccination are considered. The dynamic of this model is studied determining the stability of the steady states and the basic reproductive number. The most important control strategies are determined taking into account the explicit expression of the basic reproductive number.

Keywords: Wireless networks · Mobile malware · Spreading · Mathematical modeling

1 Introduction

The proliferation and development of both mobile devices (smartphones, tablets, wereables devices, etc.) and wireless networks have been crucial factors in the technological progress occurred in the last years. As a consequence the notion of Internet of Everything will become reality in the next future. This fascinating scenario in which people and objects are connected every time and everywhere exhibits serious threats related to existence of malware and its propagation.

The fight against malware is mainly focused in the design of efficient algorithms to successfully detect and remove it. Nevertheless it is also interesting to design mathematical models to simulate malware spreading in a wireless network. The importance of these models lies both in predicting the behavior of malware and in evaluating control measures.

Mathematical models to simulate malware spreading on wireless networks are based of Mathematical Epidemiology and their dynamic is usually defined by means of a system of differential equations. Several proposals have been appeared in the literature (see, for example [1,4]), and all of them are compartmental, that is, the devices are classified into different classes taking into account their state in regards to the malware: susceptible (non-infected devices), exposed (infected

M. Graña et al. (eds.), *International Joint Conference SOCO'16-CISIS'16-ICEUTE'16*, Advances in Intelligent Systems and Computing 527, DOI 10.1007/978-3-319-47364-2_62

devices in which the malicious code is not active: it is not performed it payload and it is not able to propagate), infectious (infected devices in which the malware is active), recovered devices, isolated devices, quarantined devices, etc. [3].

To our knowledge, none of these models has considered the compartment of carrier devices. This class is formed by those devices that malware has reached but it is not able to carry out its malicious purposes, and it is due to several reasons: incompatibility of the malware with the operative system of the host (such is the case of OS X-based devices when an epidemic of Android malware occurs), etc.

The main goal of this work is to fill this gap and to propose a novel mathematical model considering carrier devices. Specifically, a SCIRS (Susceptible-Carrier-Infectious-Recovered-Susceptible) model is introduced in this paper. The dynamic of this model is studied determining the stability of the steady states and the basic reproductive number. Taking into account the explicit expression of the basic reproductive number, some control strategies are determined.

The rest of the paper is organized as follows: In Sect. 2 the description of the SCIRS model for mobile malware propagation is shown; the determination of control measures from the basic reproductive number are introduced in Sect. 3, and finally the conclusions and further work are presented in Sect. 4.

2 The SCIRS Model for Mobile Malware Propagation

2.1 Description of the Model

The model proposed in this work to simulate malware spreading in a wireless sensor network is a compartmental model where devices are divided into four classes: susceptibles (S), carriers (C), infectious (I) and recovered (R). Specifically it is a SCIRS model with temporal immunity (which is acquired by security measures) and vaccination (installation of anti-virus software). The dynamic of the model is shown in Fig. 1.

Specifically, the model can be described as follows:

(1) Susceptible devices are infected (and become infectious or carriers) according to the rate a and the mass-action law. Set δ the fraction of susceptible devices with the same operative system that this targeted by the malware, then they

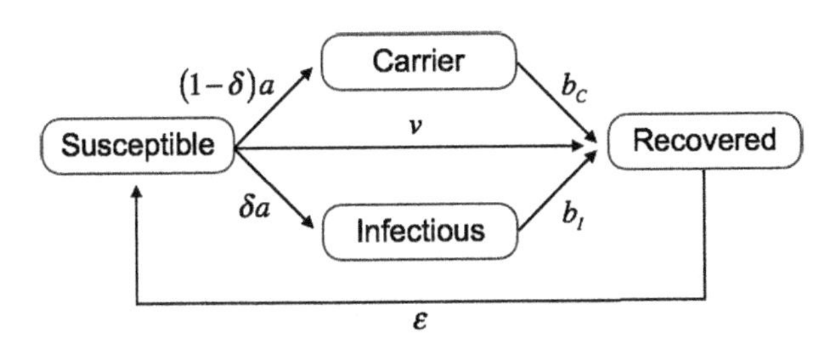

Fig. 1. Dynamic of the SCIRS model

become infectious when the malware reaches them; on the other hand, susceptible devices based on a different operative system become carriers when the malware infect them. Moreover, a susceptible device can acquire temporal immunity (and reaches the recovered states) according to "vaccination" rate v when anti-virus software is installed.

(2) Infected devices (both infectious and carriers) acquire temporal immunity if malware is successfully detected and removed according to the recovered rate b.

(3) Finally, recovered devices become susceptible again taking into account the rate ϵ, which represents the loss of temporal immunity.

Finally, as the scenario where the malware is spreading is a wireless network, it is supposed that the contact topology can be represented by means of a complete graph and, consequently, the dynamic can be modeled using a system of ordinary differential equations whose variables are the number of susceptible, carrier, infectious and recovered devices.

2.2 The Equations that Govern the Dynamic of the Model

Suppose that $S(t), C(t), I(t)$ and $R(t)$ stand for the number of susceptible, carrier, infectious and recovered mobile devices at time t, respectively, then, taking into account Sect. 2.1, the dynamic of the model is governed by means of the following system of ordinary differential equations:

$$S'(t) = -a \cdot S(t) \cdot I(t) - v \cdot S(t) + \epsilon \cdot R(t), \tag{1}$$

$$C'(t) = a \cdot (1 - \delta) S(t) \cdot I(t) - b_C \cdot C(t), \tag{2}$$

$$I'(t) = a \cdot \delta \cdot S(t) I(t) - b_I \cdot I(t), \tag{3}$$

$$R'(t) = b_C \cdot C(t) + b_I \cdot I(t) + v \cdot S(t) - \epsilon \cdot R(t), \tag{4}$$

with the following initial conditions:

$$S(0) = S_0, C(0) = C_0, I(0) = I_0, R(0) = N - S_0 - C_0 - I_0, \tag{5}$$

$$S(t) \geq 0, C(t) \geq 0, I(t) \geq 0, R(t) \geq 0, \tag{6}$$

and $N = S(t) + C(t) + I(t) + R(t)$ for every t. Furthermore, in Table 1 the coefficients involved in the model are shown.

The Eq. (1) states that the variation of the number of susceptible mobile devices is equal to the difference between the recovery devices that have lost the immunity $\epsilon \cdot R(t)$, and the susceptible devices that have lost such status (that is, the susceptible devices infected, $a \cdot S(t) \cdot I(t)$, and the susceptible devices that are "vaccinated", $v \cdot S(t)$).

The Eq. (2) shows that the variation of the number of carrier devices at every step of time is equal to the difference between the new susceptible devices (whose operating systems is different from the targeted one) that have been infected, $a \cdot (1 - \delta) \cdot S(t) \cdot I(t)$), and those carrier devices that have been recovered once the malware has been detected and successfully deleted: $b_C \cdot C(t)$.

Table 1. Coefficients of the SCIRS model

Coefficient	Description	Range
a	Transmission coefficient	[0,1]
v	Vaccination coefficient	[0,1]
ϵ	Loss of immunity coefficient	[0,1]
δ	Fraction of mobile devices based on the targeted OS	[0,1]
b_C	Recovered coefficient for carrier devices	[0,1], $b_C \ll b_I$
b_I	Recovered coefficient for infectious devices	[0,1], $b_I \gg b_C$

The evolution of the number of infectious devices is given in Eq. (3). Note that the variation of this type is given by the different between the new susceptible devices (with the same operating system as the targeted by the malware) that have been infected, $a \cdot \delta \cdot S(t) I(t)$, and the infectious devices that have been recovered: $b_I \cdot I(t)$.

Finally, Eq. (4) shows that the variation of the number of recovered devices is the difference between the new recovered devices (both carriers and infectious), $b_C \cdot C(t) + b_I \cdot I(t)$ plus the "vaccinated" susceptible devices: $v \cdot S(t)$, and the devices that lose the immunity: $\epsilon \cdot R(t)$.

2.3 Steady States of the System

The steady states of the model appears when the variables remain constant over time, that is, when the system reaches an equilibrium state and the number of susceptible, carriers, infectious and recovered mobile devices does not change. These states are the solutions of the nonlinear system $0 = S'(t) = C'(t) = I'(t) = R'(t)$. A simple computation shows that there are two solutions: the disease-free equilibrium point

$$E_0^* = (S_0^*, C_0^*, I_0^*, R_0^*) = \left(\frac{\epsilon N}{\epsilon + v}, 0, 0, \frac{vN}{\epsilon + v}\right), \tag{7}$$

and the endemic equilibrium point $E_1^* = (S_1^*, C_1^*, I_1^*, R_1^*)$, where:

$$S_1^* = \frac{b_I}{a\delta}, \tag{8}$$

$$C_1^* = \frac{b_I(1-\delta)}{b_C\delta} I_1^*, \tag{9}$$

$$I_1^* = \frac{b_C(aN\delta\epsilon - b_I v - b_I\epsilon)}{a(b_I b_C + b_I\epsilon + \delta\epsilon(b_C - b_I))}, \tag{10}$$

$$R_1^* = \frac{b_I(b_I b_C - b_I v - a b_C N\delta + v\delta(b_I - b_C))}{\delta b_C(b_I v + b_I\epsilon - aN\delta\epsilon)} I_1^*, \tag{11}$$

Note that the endemic equilibrium point exists, that is, it makes sense in the case of malware spreading (all variables must be non negative), when the total number of devices exceeds a certain threshold value:

$$N > \left\{ \frac{b_I (v + \epsilon)}{a\delta\epsilon}, \frac{b_I (b_C - v) + v\delta (b_I - b_C)}{a\delta b_C} \right\}. \tag{12}$$

As a consequence, it is shown that a malware outbreak can evolve to two different scenarios:

1. The number of infected devices does not increase and the malware outbreak dies out (disease-free state). In this case, the final number of susceptible devices decreases to S_0^* whereas the number of recovered devices increases to R_0^* and there are not infectious or carrier devices ($C_0^* = I_0^* = 0$).
2. A growth of the number of infected devices occurs reaching the endemic steady state. In this case, the final number of infected devices is I_1^*. Moreover, all susceptible devices will be infected if $b_I = 0$.

2.4 The Basic Reproductive Number

The basic reproductive number, R_0, is the most important parameter in the study of malware propagation since its numerical value will indicate whether or not a malware outbreak become epidemic (the number of infected computers will grow). As is shown in the following subsections, if $R_0 > 1$ the number of infected mobile devices increases (and the outbreak becomes epidemic) whereas if $R_0 \leq 1$ the malware will not spread. Furthermore, the analysis of this threshold parameter yields to the determination of efficient control measures.

It is possible to compute the R_0 from the system of ordinary differential Eqs. (1)–(4) and the knowledge of the disease-free steady state by simply applying the *next generation method* [2]. Consequently, following this method a simple computation shows that the basic reproductive number associated to our model is:

$$R_0 = \frac{a\delta\epsilon N}{b_I (\epsilon + v)}. \tag{13}$$

Note that this threshold parameter depends on all coefficients of the systems with the exception of the recovered coefficient for carriers, b_P.

2.5 Local and Global Stability of the Steady States

Using the Theory of Stability for systems of ordinary differential equations (see [5]) the following results holds (some proofs are shown in the Appendix):

Theorem 1. *The disease-free steady state* $E_0^* = \left(\frac{\epsilon N}{\epsilon+v}, 0, 0, \frac{vN}{\epsilon+v}\right)$ *is locally and globally asymptotically stable if and only if* $R_0 \leq 1$.

Theorem 2. *The endemic equilibrium* $E_1^* = (S_1^*, C_1^*, I_1^*, R_1^*)$ *defined by (8)–(11) is locally and globally asymptotically stable if* $R_0 > 1$.

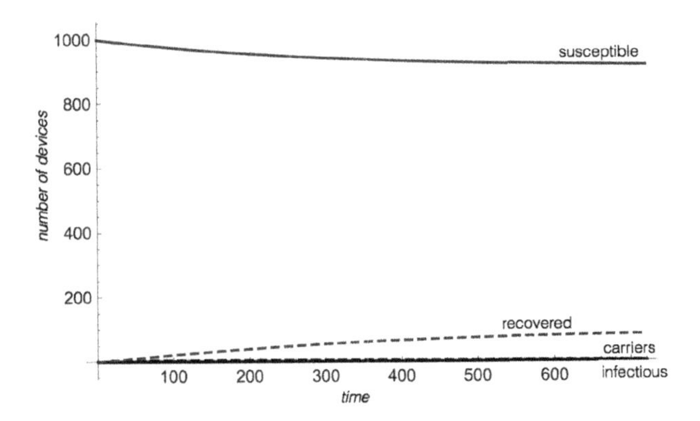

Fig. 2. Evolution of the different compartments when $R_0 \leq 1$ (disease-free steady state).

As a consequence, if $R_0 \leq 1$ the malware outbreak dies out (the system reaches the disease-free steady state), whereas if $R_0 > 1$ the malware outbreak becomes epidemic (the system reaches the endemic steady state). These results are illustrated with the following simulations:

Disease-Free Steady State. Suppose that there are 1001 devices in the network such that initially all devices are susceptible with the exception of only one, that is infectious: $S(0) = 1000, I(0) = 1, C(0) = R(0) = 0$. Moreover, set $a = 0.00009$, $v = 0.0002$, $\epsilon = 0.002$, $b_C = 0.001$, $b_I = 0.02$ and $\delta = 0.2$. In Fig. 2 the evolution of the number of susceptible, carrier, infectious and recovered devices is shown. The time is measured in hours and the simulation period comprises the first 720 h after the onset of the first infectious device.

Note that in this case $R_0 \approx 0.819 \leq 1$ and consequently the number of infected computers does not increase. Moreover, the system reaches the following disease-free steady state:

$$E_0^* = (S_0^*, C_0^*, I_0^*, R_0^*) = (910, 0, 0, 91). \tag{14}$$

Endemic Steady State. On the other hand, if we set $a = 0.0002$, $v = 0.0002$, $\epsilon = 0.002$, $b_C = 0.005$, $b_I = 0.01$ and $\delta = 0.2$, then $R_0 \approx 3.64 > 1$ and consequently the outbreak becomes epidemic. This behavior is shown in Fig. 3. In this case (and for the sake of clarity) the simulation period represents the first 2880 h after the appearance of the first infectious devices. Furthermore, the endemic steady state is given by the following values:

$$E_1^* = (S_1^*, C_1^*, I_1^*, R_1^*) \approx (250, 21, 171, 559). \tag{15}$$

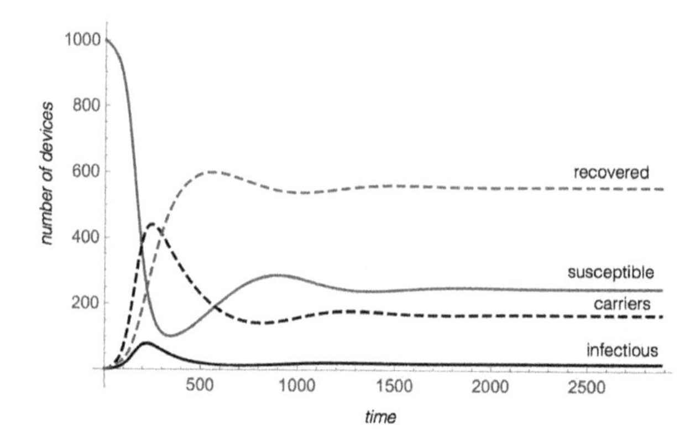

Fig. 3. Evolution of the different compartments of the model when $R_0 > 1$ (endemic steady state).

3 Determining Control Measures

As was previously mentioned, the basic reproductive number, R_0, plays an important role in the control of an epidemic. In this sense, and in order to prevent that a malware outbreak becomes an epidemic process, it is mandatory to reduce R_0 as necessary. In our case, this threshold parameter depends on all parameters of the system (the transmission coefficient a, the total number of mobile devices N, the vaccination coefficient v, the loss of immunity coefficient ϵ, the fraction of the mobile devices based on the targeted operative system δ, and the recovered coefficient from infectious devices b_I) with the exception of the recovered coefficient from carrier devices b_C. Then $R_0 = R_0\left(a, N, \delta, \epsilon, b_I, v\right)$.

Note that

$$\frac{\partial R_0}{\partial a} = \frac{N\delta\epsilon}{b_I\left(v+\epsilon\right)} \geq 0, \quad \frac{\partial R_0}{\partial N} = \frac{a\delta\epsilon}{b_I\left(v+\epsilon\right)} \geq 0, \tag{16}$$

$$\frac{\partial R_0}{\partial \delta} = \frac{aN\epsilon}{b_I\left(v+\epsilon\right)} \geq 0, \quad \frac{\partial R_0}{\partial \epsilon} = \frac{aN\delta v}{b_I\left(v+\epsilon\right)^2} \geq 0, \tag{17}$$

then the function R_0 increases as the coefficients a, N, δ and ϵ increases. As a consequence to reduce the value of R_0 it is necessary to reduce the numeric value of a, N, δ and/or ϵ.

On the other hand, as

$$\frac{\partial R_0}{\partial b_I} = -\frac{aN\delta\epsilon}{b_I^2\left(v+\epsilon\right)} \leq 0, \quad \frac{\partial R_0}{\partial v} = -\frac{aN\delta\epsilon}{b_I\left(v+\epsilon\right)^2} \leq 0, \tag{18}$$

then R_0 decreases as b_I and/or v increases.

In short, from the analysis of the basic reproductive number, the following control measures are obtained to control the malware outbreak:

(1) Reducing the total number of mobile devices on the network N or the number of devices running under the targeted operative system δ by means of, for example, isolation.
(2) Reducing the transmission coefficient a by reducing the number of effective contacts between devices or extreme caution when opening suspicious messages.
(3) Reducing the loss of immunity coefficient ϵ by using efficient anti-virus software.
(4) Increasing the recovery rate b_I by improving the performance of antivirus software.
(5) Increasing the vaccination coefficient v by sensitizing users to install security countermeasures.

4 Conclusions and Further Work

In this work, a novel mathematical model to simulate malware spreading over wireless networks has been introduced. Its dynamic is defined by means of a system of ordinary differential equations in which the compartment of carrier devices is considered.

From the analysis of the basic reproductive number associated to the model, the main efficient security countermeasures are presented.

Further work aimed at improving this model by considering the propagation of malware over complex networks where different graphs defining the topology can be considered. Note that in this work, it is suppose that the topology of connections are based on a complete graph (all devices are in contact with all devices).

Acknowledgments. This work has been supported by Ministerio de Economía y Competitividad (Spain) and the European Union through FEDER funds under grants TIN2014-55325-C2-2-R, and MTM2015-69138-REDT. J.D. Hernández Guillén thanks Ministerio de Educación, Cultura y Deporte (Spain) for his departmental grant.

Appendix: Proofs of Theorems

Proof of Theorem 1

The system of differential equations governing the dynamic of the model (1)–(4) can be reduced to the following system taking into account that the total number of devices remains constant ($N = S(t) + C(t) + I(t) + R(t)$):

$$S'(t) = -a \cdot S(t) \cdot I(t) - v \cdot S(t) + \epsilon \cdot R(t), \tag{19}$$

$$C'(t) = a \cdot (1 - \delta) S(t) \cdot I(t) - b_C \cdot C(t), \tag{20}$$

$$I'(t) = a \cdot \delta \cdot S(t) I(t) - b_I \cdot I(t), \tag{21}$$

The Jacobian matrix associated to this system of ordinary differential Eqs. (19)–(21) in the disease-free steady state is given by:

$$J\left(E_0^*\right)=\begin{pmatrix} -v & 0 & -\frac{aN\epsilon}{v+\epsilon} \\ 0 & -b_C & \frac{aN\left(1-\delta\right)\epsilon}{v+\epsilon} \\ 0 & 0 & -b_I+\frac{aN\delta\epsilon}{v+\epsilon} \end{pmatrix}, \tag{22}$$

so that a simple calculus shows that its eigenvalues are:

$$\lambda_1=-b_C,\quad \lambda_2=-v,\quad \lambda_3=\frac{aN\delta\epsilon-b_I\left(v+\epsilon\right)}{v+\epsilon}=b_I\left(R_0-1\right). \tag{23}$$

Consequently $\mathrm{Re}\left(\lambda_1\right)=-b_C<0$ and $\mathrm{Re}\left(\lambda_2\right)=-v<0$. Moreover, $\mathrm{Re}\left(\lambda_3\right)=b_I\left(R_0-1\right)<0$ iff $R_0\leq 1$, thus finishing. On the other hand, the global stability of the disease-free steady state is also satisfied when $R_0\leq 1$; this result is easily obtained.

Proof of Theorem 2

The Jacobian matrix of the system (19)–(21) in the endemic equilibrium steady state E_1^* is:

$$J\left(E_1^*\right)=\begin{pmatrix} -v-\frac{b_C(-b_Iv-b_I\epsilon+aN\delta\epsilon)}{b_Ib_C+\delta\epsilon b_C+b_I\epsilon-b_I\delta\epsilon} & 0 & -\frac{b_I}{\delta} \\ \frac{b_C(1-\delta)(-b_Iv-b_I\epsilon+aN\delta\epsilon)}{b_Ib_C+\delta\epsilon b_C+b_I\epsilon-b_I\delta\epsilon} & -b_C & \frac{b_I(1-\delta)}{\delta} \\ \frac{b_C\delta(-b_Iv-b_I\epsilon+aN\delta\epsilon)}{b_Ib_C+\delta\epsilon b_C+b_I\epsilon-b_I\delta\epsilon} & 0 & 0 \end{pmatrix}, \tag{24}$$

such that the explicit expression of its characteristic polynomial is the following:

$$\begin{aligned} p\left(\lambda\right) &= p_0\lambda^3+p_1\lambda^2+p_2\lambda+p_3 \\ &= \lambda^3+\lambda^2\frac{b_C\delta\epsilon(aN+b_C+v)+b_I\left(b_C^2-\delta\epsilon(b_C+v)+v\epsilon\right)}{b_I(b_C-\delta\epsilon+\epsilon)+b_C\delta\epsilon} \\ &\quad +\lambda\frac{b_C\left(b_I\epsilon(a\delta n-b_C-\delta v+v)+b_C\delta\epsilon(aN+v)-b_I^2(v+\epsilon)\right)}{b_I(b_C-\delta\epsilon+\epsilon)+b_C\delta\epsilon} \\ &\quad -\frac{b_Ib_C^2(b_I(v+\epsilon)-a\delta n\epsilon)}{b_I(b_C-\delta\epsilon+\epsilon)+b_C\delta\epsilon}. \end{aligned} \tag{25}$$

Note that $p_0>0$, $p_1>0$ if $R_0>\delta$, $p_2>0$ if $R_0>1$, and $p_3>0$ iff $R_0>1$; as a consequence, the coefficients of the characteristic polynomial of $J\left(E_1^*\right)$ are positive if $R_0 > 1$. Now, by applying the Routh-Kurwitz stability criterion, the real part of the eigenvalues of $p(\lambda)$ will be negative when the following conditions hold:

$$\Delta_1 = p_1 > 0, \Delta_2 = \begin{vmatrix} p_1 & p_3 \\ 1 & p_2 \end{vmatrix} > 0, \Delta_3 = \begin{vmatrix} p_1 & p_3 & 0 \\ 1 & p_2 & 0 \\ 0 & p_1 & p_3 \end{vmatrix} = p_3 \Delta_2 > 0. \tag{26}$$

Note that $p_1 > 0$ if $R_0 > 1$ as was mentioned previously. On the other hand as

$$\Delta_2 = \frac{b_C \Omega \Theta}{b_C \delta \epsilon + b_I \left(b_P + \epsilon \left(1 - \delta\right)\right)^2}, \tag{27}$$

where

$$\Omega = b_I b_C v R_0 + b_C v \delta \epsilon + b_I b_C \epsilon + b_I \epsilon v \left(1 - \delta\right), \tag{28}$$

$$\begin{aligned} \Theta = {} & b_i b_C^2 + b_i b_C v R_0 + b_C^2 \delta \epsilon + b_C v \delta \epsilon + b_i b_C \epsilon \left(R_0 - \delta\right) \\ & + b_i v \epsilon \left(1 - \delta\right) + b_I^2 \left(v + \epsilon\right) \left(R_0 - 1\right), \end{aligned} \tag{29}$$

then $\Delta_2 > 0$ if $R_0 > 1$ since $b_C > 0$, $b_C \delta \epsilon + b_I \left(b_P + \epsilon \left(1 - \delta\right)\right)^2 > 0$, $\Omega > 0$, and $\Theta > 0$ if $R_0 > 1$. Finally, taking into account the last results, it is easy to check that $\Delta_3 = p_3 \Delta_2 > 0$ when $R_0 > 1$. Consequently, the endemic steady state is locally asymptotically stable. The global stability follows from simple but tedious computations.

References

1. Abazari, F., Analoui, M., Takabi, H.: Effect of anti-malware software on infectious nodes in cloud environment. Comput. Secur. **58**, 139–148 (2016)
2. Diekmann, O., Heesterbeek, J.A.P.: Mathematical Epidemiology of Infectious Diseases. Wiley, Chichester (2000)
3. Karyotis, V., Khouzani, M.H.R.: Malware Diffusion Models for Modern Complex Networks. Morgan Kaufmann, Cambridge (2016)
4. Liu, W., Liu, C., Liu, X., Cui, S., Huang, X.: Modeling the spread of malware with the influence of heterogeneous immunization. Appl. Math. Model. **40**, 3141–3152 (2016)
5. Merkin, D.R.: Introduction to the Theory of Stability. Texts in Applied Mathematics, vol. 24. Springer, New York (1997)

Malware Propagation Models in Wireless Sensor Networks: A Review

Araceli Queiruga-Dios[1(✉)], Ascensión Hernández Encinas[1], Jesus Martín-Vaquero[1], and Luis Hernández Encinas[2]

[1] Department of Applied Mathematics, University of Salamanca, Salamanca, Spain
{queirugadios,ascen,jesmarva}@usal.es
[2] Institute of Physical and Information Technologies, Spanish National Research Council, Madrid, Spain
luis@iec.csic.es

Abstract. Mathematical models to study to simulate the spread of malware are widely studied today. Malware spreading in Wireless Sensor Networks (WSNs) has special relevance as these networks consist on hundreds or even thousands of autonomous devices (sensors) able to monitor and to communicate with one another. Malware attacks on WSNs have become a critical challenge because sensors generally have weak defense capabilities, that is why the malware propagation in WSNs is relevant for security community. In this paper, some of the most important and recent global mathematical models to describe malware spreading in such networks are presented.

Keywords: Malware · Epidemic spreading · Wireless sensor networks · Global models

1 Introduction

The development of wireless sensor networks started in the military field, around 1980 [3]. A Wireless Sensor Network (WSN) is a set of networked microsensors with a large number of nodes spatially distributed in not predetermined positions and with no specific design, in a position closed to the phenomenon being measured or inside it. Each node is a device called a sensor, which is able to self-organize and whose aim is monitoring and controlling physical phenomena in health areas, defense, surveillance, or environment; i.e., measuring variables such as temperature, humidity, sound, vibration, pressure, contaminants, etc. They can also process the collected data, store and send it to other devices [1]. The microsensors devices may be in the air, under water, on the ground, in vehicles, inside buildings, and also on bodies, and they have the unique feature of the cooperative effort between the sensor nodes [9,34].

In recent years the popularity of the spread of malware in WSNs is increasing and there are many research results [4,10,12]. The security in WSNs is of great importance as the data collected by the sensor nodes could be highly sensitive [4], and furthermore, most of the sensor devices could operate in a hostile

M. Graña et al. (eds.), *International Joint Conference SOCO'16-CISIS'16-ICEUTE'16*, Advances in Intelligent Systems and Computing 527, DOI 10.1007/978-3-319-47364-2_63

environment. WSNs are networks in danger of being attacked by intrusions, eavesdropped, or invaded by any kind of malware to interfere with their normal operations [23,25], or even to destroy them [37]; in fact, the malware spreading caused more damages in WSNs than on the Internet [11].

Some results from numerical simulations showed that the process of malware spreading is very sensitive regarding the high density, the power consumption (limited energy), the small communication range, and the topology of nodes, and also their sleep and work interleaving schedule policy. According to that, the process of malicious software spreading in a WSN has three features, which do not occur on other networks [27]:

(1) When malware exists on a host on a network like the Internet, it tries to infect other hosts by randomly scanning other hosts' IP addresses, whereas malware in a WSN node can spread to its neighbors, and these can directly communicate with the node.
(2) Due to the sleep and work intervals, the malware on a working node can spread to neighbor nodes that are working, but the sleeping neighbors of that working node do not become infected. Moreover, while a node is sleeping, any malware on that node cannot infect other nodes.
(3) When the energy of the nodes is exhausted, more and more nodes become dead nodes that cannot be infected anymore and that will not participate in the process of spreading malware in a WSN. Malware on a dead node immediately disappears from the network.

To study the malware propagation in WSNs, researchers based their first works on the traditional malware spreading on the Internet. The epidemical models have been also extended to this case [10]. Most of the mathematical models dealing with the dynamic of malware spreading in WSNs use systems of ordinary differential equations (SODE) (see, for example, [7,26]). Especially interesting is the model proposed in [41] based on a system of delayed ODE where all sensors of the network are considered identical.

The aim of this work is to present the most recent global mathematical models to describe malware spreading in WSNs. A critical analysis with some possible improvements of the existing models will also be included. Apart from those continuous models, there are also some individual-based models where each node is considered as an individual and the whole WSN as an evolving system of autonomous interacting entities [13]. The main examples of this paradigm are cellular automata [33] and agent-based models [24]. We will see that models based on individuals are an open field of possibilities to model the malicious code spreading in WSNs.

This paper is organized as follows: In Sect. 2 the description and features of WSNs are detailed; in Sect. 3 the global models proposed in the last few years are described, and a critical analysis is presented in Sect. 4. Finally the conclusions are presented in Sect. 5.

2 Wireless Sensor Networks

The features of the WSNs are different depending on whether they are used in industry networks or in a standard network. In the first case, as all sensors are vital to the operation of a plant, a failed node must be replaced. On the other hand, in standard networks, individual nodes can lose power or be destroyed, even though the network will continue working as a whole. WSNs have the self-restored ability, i.e., if a node fails, the network finds new ways to guide the data packets. Thus, the network will survive as a whole. Sensor nodes usually spend much time in sleep mode because of the low power consumption.

Sensor nodes are usually scattered in a sensor field with a specific topology. The network topology describes the physical network distribution. This is how the devices are connected to achieve optimum performance. There are different topologies as in any other network: Tree, star, cluster tree, or mesh [38]. Besides these classic net topologies, in industry the wireless nodes in star topology are communicated through a gateway device, that acts as a bridge with a wired network. There are also routers that connect with the gateway [39].

Each of the sensor nodes have the ability to collect data and send it back to end users. This is done through the sink node (also known as base station) by a multihop architecture without structure. The sink node is a more sophisticated node with better energy, communication, and computing capabilities, which can communicate with the task manager node via Internet or by satellite. The protocol stack used by the sink and all sensor nodes combines the power with the proper routing, integrating data with network protocols, communicating the power efficiently through the wireless medium and promoting cooperative efforts of sensor nodes. The protocol stack consists of application, transport, network, data link, physical, power management plane, and task management plane layers [1]. A routing protocol in the network layer is the responsible for deciding what departure route and which input packet should be transmitted. Due to the constraints of WSNs, routing protocols specifically developed for wired networks or wireless networks such as MANET, are not always suitable for WSN [38].

Routing protocols in WSNs can be usually classified into proactive or reactive, depending on how the route is determined. Proactive determine the route before it is needed and modifies routes when network topology changes. Whereas proactive routing protocols invoke a route on demand. There are many other criteria to classify routing protocols. In terms of the structure of the network, three subcategories can be distinguished: Flat, hierarchical, and location-based routing protocols [38]. In terms of protocol operations there are five subcategories, which are based in: Queries, negotiations, multiway, quality of service, and based on consistencies. These categories and subcategories are not mutually exclusive. In location-based protocols, the main idea is to use the advantages of the locations of wireless sensor nodes for routing data. The address of each node is determined based on their physical location that can be determined using the Global Positioning System (GPS) or another positioning technique. The distance between neighbors can be calculated depending on the signal strength. The two most common location-based routing protocols are based on the Geographic Adaptive Fidelity (GAF), and Geographic and Energy-Aware Routing (GEAR).

3 Global Mathematical Models for Malware Spreading in Wireless Sensor Networks

Most of the mathematical models suggested to model the malware spreading in wireless sensor networks are global models based on epidemic theory [2]. The global models study the dynamics of the complete system, the evolution of the set of nodes devices as a whole, providing the global evolution of the system [19], and without considering the local interactions of the sensor nodes.

In general, the classic global epidemic model considers a population of $N(t)$ identical sensor nodes that are uniformly and randomly distributed, and are divided into compartments: Susceptible (healthy) sensor nodes: $S(t)$; Infected sensor nodes: $I(t)$; and Recovered (immunized) sensor nodes: $R(t)$. This is the case of the first SIR (Susceptible-Infective-Recovered) model formulated by the epidemic model [22], that can be solved exactly on a wide variety of networks, and is defined by a nonlinear SODEs originally proposed by Kermack and McKendrick (see [15]). They introduced the threshold number R_0, also known as stability or reproductive number, to determine when a disease becomes epidemic that occurs when $R_0 > 1$.

WSNs models consider a ripple based propagation of a broadcast protocol which grows with time and from a central infected node. The model proposed in [8] approximate this observation by considering nodes on the periphery of the infected circular region trying to infect their susceptible neighboring nodes outside this circle (once infected, are compromised and cannot be recovered). These susceptible neighbors are situated in a circular strip of width equivalent to a node's communication radius R_c, outside the infected circle. The model consider $I(t)$ and $S(t)$ the sub-population functions, adding $I'(t)$ as the number of infected nodes that lie in the circular strip of thickness R_c from the circumference. Solving the differential equations, the result is that initially only one node was compromised. In particular, in [5] authors have used the random graph model of epidemic theory to simulate the spread of node compromise in a WSN. However, the model does not capture the temporal effects of an epidemic [7], it is focused on capturing the final outcome of the infection but fails in the analysis of the temporal dynamics of the compromise propagation.

In the SIR model all the susceptible sensor nodes are assumed to be working forever, it does not take into account the sleep and work interleaving schedule. A modified SIR model, called SIR with Maintenance (SIR-M), was proposed in [29] for malware spreading in WSNs. This model describes the dynamics of the virus spreading from a single node to the entire network. The spreading process, which is sensitive to the network topology and the energy consumption, starts when an infected sensor node spreads the malware (through a normal operation of a broadcast protocol) to its neighboring susceptible nodes (located inside its signal transmission range), and these recently infected node repeat the process. The SIR-M model introduces a maintenance mechanism to improve the network's anti-malware capability, and to a decrease the number of infected nodes. During maintenance mode the susceptible and recovery nodes pass the check and go to sleep, while the infected nodes take some time for treatment.

Depending on the period of maintenance, a fraction of the maintained infective nodes, will become recovery nodes. The remainder of the nodes will remain in the group of infective nodes. When many nodes become infected, the network will not operate normally, resulting what is called a network failure. A failure state is achieved when the number of infected nodes is greater than a threshold value.

When nodes communicate with each other, they consume their individual energy and become dead. The iSIRS model proposed in [31] is a non-linear dynamic feedback differential system, which supposes an improvement of the SIR model considering the concept of dead state of nodes in WSNs. In the iSIRS model four sets of nodes (statical nodes) are considered: Susceptible, Infectious, and Recovered sets, as detailed above, and also a Dead set, $D(t)$. Due to the energy consumption of nodes, a susceptible node, an infectious node or a recovered node could become a dead node. The iSIRS model did not effectively describe the process of malware propagation, specially in large scale WSNs, as it did not consider the sleep and work interleaving schedule policy which is generally used to schedule sensor nodes to prolong the lifetime of a WSN. To overcome this disadvantage of the iSIRS model, the same authors have proposed in [32] a expanded iSIRS (EiSIRS) model to precisely describe the process of malware propagation in WSNs. In EiSIRS model, at any instant t, in addition to the Susceptible, Infectious, and Recovered working node sets: $S(t), I(t), R(t)$, respectively, the following sleeping node sets are considered: $S'(t), I'(t), R'(t)$, and also the Dead node set, $D(t)$, as before. This model consideres that: (1) all the malware reside in nodes I or I'; (2) At the initial instant $t = 0$, it verifies: $I'(0) = R(0) = R'(0) = S'(0) = D(0) = 0$, $S(0) > 0$ and $I(0) > 0$. (3) In a unit time the state of each node is one of the seven states. A node moves from its current state to another with the SIRS mechanism of malware propagation and considering the sleep and work interleaving schedule policy for nodes. (4) A node in S can become a node in I, D or S'; a node in I can become a node in D, R or I'; a node in R can become a node in S, D or R'; a node in S' can become a node in S; a node in I' can become a node in I; and a node in R' can become a node in R.

A more recent model, also based in SIR epidemic model was proposed by Feng et al. in [10]. In their improved SIRS model, susceptible sensors nodes are infected when they reach the malware, and the infected are recovered when malware is detected and removed. On the other hand, some recovered devices become susceptible again when they lose the immunity that they had because of the antivirus. Authors consider communication radius, energy consumption, and distributed density of nodes in the WSN. They achieved that decreasing the value of communication radius or reducing distributed density of nodes are effective methods to prevent malware spread in WSNs. Furthermore, they have proved that $R_0 = 0$ is the threshold value whether worms are eliminated. Moreover, if $R_0 \leq 1$ malware can be eliminated, and if $R_0 > 1$, malware will exist consistently, and the endemic equilibrium is reached. These considerations have also been stablished by Mishra and Keshri [21] in their

Susceptible-Exposed-Infective-Recovered-Susceptible model with a vaccination compartment model (SEIRS-V), where a new node state called Exposed was defined. This compartmental epidemic model supposed an improvement of the previously SEIRS model proposed by the same authors two years before [20] for malware propagation on the Internet. In SEIRS-V model new sensor nodes can be included in the network, and not working sensor nodes (due to malware attack or hardware/software problems) can be excluded. Furthermore, all the sensor nodes are considered susceptible towards the possible malware spreading. The Exposed compartment includes the sensor nodes with the symptoms of attack, i.e., before fully infectious (the usual speed of transmission of data becomes slow). SEIRS-V model also uses a maintenance mechanism in the sleep node state to improve the network's antivirus capability. As the sensor nodes need some time to clean the malware in a WSN (with antivirus software), and the recovered and the vaccinated sensor nodes have a temporary immunity period after they may be infected again, a delayed is added to the SEIRS-V model [40]. When stability conditions are satisfied, authors get a critical value τ_0 of the delay: For a lower value the system is stable and for a higher one, the system is unstable.

Other frequently used models are the SIS (Susceptible-Infected-Susceptible) and the SI (Susceptible-Infected) models, which do not have the recovered subset $R(t)$, as they do not assume recovered state. In SIS model, the infected nodes fall back into the susceptible subset $S(t)$ after their infectivity duration. Based on the classical SI model, and taking into account the network topology, a topologically-aware worm propagation model (TWPM) was proposed by Khayam and Radha [16]. The TWPM considers the N sensor nodes of the WSN equipped with omnidirectional antennas which have a maximum transmission range. The sensor nodes are placed on a rectangular grid divided into segments. Each segment can receive traffic from its neighbor sensors (the eight segments surrounding the central one). Since nodes are uniformly distributed inside a segment, infectious contacts are received by each segment. Similarly, infected nodes in a segment will infect the rest of neighbors. The TWPM describes both the spatial and temporal dynamics of the spread of malware in [17]. In this case authors applied signal processing techniques for modeling dynamic spatial-temporal propagation of worms in a WSN (with uniformly distributed nodes). The physical binding characteristics of data, and also the network protocols and transport are integrated into the proposed propagation model, that is focused on the dynamics of unknown worms dissemination. As was mentioned in [6], although the proposed TWPM presents a closed form solution for computing the infected fraction of the WSN, it does not consider the simultaneous effects of any recovery process on the malware spreading. Moreover, it is difficult to use the model to represent different broadcast protocols and study their epidemic characteristics against each other.

Continuing with SI models, in [28] a SI with maintenance model is defined, with similar characteristics to the SIR-M from the same authors. This new model describes the sensor node that could perform system maintenance before going to sleep, which would improve the antiviral ability of the network without increasing hardware cost or charge for signaling. In this model each sensor node installs an antivirus program, which is automatically triggered in the sleep mode, and could begin to restore infected nodes on a regular basis. The model describes the spatial-temporal dynamic features of the virus spreading and is suitable for all types of networks, such as wireless networks, social networks and computer networks. Due to the maintenance mechanism, the number of infective nodes will be controlled to a certain value and cannot be increased anymore.

The existing models do not considered the relation between the virus spreading and the medium access control mechanism (MAC). The novel SI model proposed in [36] considered the dynamic behavior of viruses in the WSN with a MAC mechanism that can reduce the number of infected nodes in the networks.

Most of the models have been defined for WSN flat structures. Xiao-Ping and Yu-Rong [35], established a malware propagation model based on cluster structure of Geographic Adaptive Fidelity. The simulation analysis showed that the GAF network cluster architecture could inhibit the spread of malware, but the model only verified that the network topology could inhibit the spread of malware, without any defense mechanism. From the standpoint of inhibiting the spread of viruses, [12] proposed to regionalize the network and added nodes detection in the regional area (unless the broadcast routing protocol is used).

In [11] authors proposed a model to control worm propagation using the spatial correlation parameters. The same year, in [14] monitoring nodes are added to the WSN to establish the model of virus spreading, which describes that packets with virus can trigger the monitoring node to broadcast the antivirus packages over the network and thus stop the virus spreading. In [30] virus propagation is studied in the small world of WSN with tree-based structures and the threshold of the outbreak of a virus on the network is also discussed.

In [18], the wireless sensor network is considered as a hierarchical tree-based small world, where the viruses or malware are called sensor worms which attack the network to propagate the epidemic until all susceptible nodes are infected. Moreover, these authors consider a percolation threshold of Cayley tree equals 1 when there is no shortcut in the network, in such a way that the malware propagation stops when the infection probability is smaller than the percolation threshold. The malware easily attacks the network from side-to-side while the infection probability is larger than the percolation threshold. Yang, Zhu, and Cao [37] have also used the sensor worm concept and defined a model for a sensor worm attack as a SI model.

4 Critical Analysis of the Existing Models

Wireless sensor networks have specific characteristics, that make them different from other networks, such as computer networks, medical networks, or social networks. A SIR and SI maintenance models are described in [29] and [28], respectively, for malware spreading in WSNs, but these models do not take into account the constraints of a WSN. The same occurs with [6,10,17], they do not include the constraints of WSNs. Furthermore, SIS models do not consider the situation when hosts may die out because they are infected by malware [31]. Neither does a SIS model consider the situation in which a host may be immune to the same type of malware cleaned from this host. So, these SIS models cannot properly describe the process of malware propagation on WSNs.

The model proposed in [31] shows that the process of malware propagation is sensitive to the network topology and the energy consumption of nodes in WSNs. Moreover, this model should take into account the sleeping and working interleaving schedule policy. In [32] the sleep and work interleaving schedule policy for sensor nodes are supported, and it can also describe the process of multi-worm propagation in WSNs. Simulation results show that the process of worm propagation in WSNs is sensitive to the energy consumption of nodes and the sleep and work interleaving schedule policy for nodes.

In the SIR model, all the susceptible hosts are assumed to be working forever. However, this assumption does not hold in WSNs due to the limited energy of nodes and the sleep and work interleaving schedule policy used in large scale WSNs. Wang and Li derived an iSIR model describing the process of worm propagation with energy consumption of nodes in WSNs [31]. Numerical simulations are performed to observe the effects of the network topology and energy consumption of nodes on worm spread in WSNs. However, the authors have not performed mathematical analysis based on this model [10].

5 Conclusions

In this review, we have examined the current state of the global models proposed to model the malware spreading in WSNs. Due to the characteristics of WSNs, such as frequent topology change, high density of nodes, limited energy of nodes, smaller communication range of nodes, and the sleep and work interleaving schedule policy for nodes, the mechanism of worm propagation in WSNs is significantly distinct with that of worm propagation on the Internet and other networks. Some of the models proposed for WSNs only improved the existing models of malware propagation on the Internet by limiting the range of worm propagation, without considering the above important characteristics of malware propagation in a WSN. More recently proposed models already take account of the specific features of WSNs.

The SIR-based proposed models assume that all individuals have the same number of contacts, and that all contacts transmit the disease with the same probability. Among the open issues related to the modeling of WSNs, as there is no proposal (up to date) including the agent-based model, the individual behavior should be considered. Another open issue is to study protocol models considering the nodes being mobile.

Acknowledgments. This work has been supported by Ministerio de Economía y Competitividad (Spain) and the European Union through FEDER funds under grants TIN2014-55325-C2-1-R and TIN2014-55325-C2-2-R.

References

1. Akyildiz, I.F., Su, W., Sankarasubramaniam, Y., Cayirci, E.: Wireless sensor networks: a survey. Comput. Netw. **38**(4), 393–422 (2002)
2. Anderson, R.M., May, R.M., Anderson, B.: Infectious Diseases of Humans: Dynamics and Control, vol. 28. Oxford University Press, Oxford (1992)
3. Chong, C.Y., Kumar, S.P.: Sensor networks: evolution, opportunities, and challenges. Proc. IEEE **91**(8), 1247–1256 (2003)
4. Conti, M.: Secure Wireless Sensor Networks: Threats and Solutions. Advances in Information Security, vol. 65. Springer, New York (2015)
5. De, P., Liu, Y., Das, S.K.: Modeling node compromise spread in wireless sensor networks using epidemic theory. In: Proceedings of World Wireless Mobile Multimedia Networks, pp. 237–243 (2006)
6. De, P., Liu, Y., Das, S.K.: An epidemic theoretic framework for evaluating broadcast protocols in wireless sensor networks. In: Mobile Adhoc Sensor Systems, pp. 1–9 (2007)
7. De, P., Das, S.K.: Epidemic Models, Algorithms, and Protocols in Wireless Sensor and Ad Hoc Networks. Wiley, New York (2008)
8. De, P., Liu, Y., Das, S.K.: An epidemic theoretic framework for vulnerability analysis of broadcast protocols in wireless sensor networks. IEEE. Trans. Mobile Comput. **8**(3), 413–425 (2009)
9. Fadel, E., Gungor, V.C., Nassef, L., Akkari, N., Malik, M.G.A., Almasri, S., Akyildiz, I.F.: A survey on wireless sensor networks for smart grid. Comput. Commun. **71**, 22–33 (2015)
10. Feng, L., Song, L., Zhao, Q., Wang, H.: Modeling and stability analysis of worm propagation in wireless sensor network. Math. Probl. Eng. 8 (2015). Article ID: 129598
11. Guo, W., Zhai, L., Guo, L., Shi, J.: Worm propagation control based on spatial correlation in wireless sensor network. In: Wang, H., Zou, L., Huang, G., He, J., Pang, C., Zhang, H.L., Zhao, D., Yi, Z. (eds.) APWeb 2012. LNCS, vol. 7234, pp. 68–77. Springer, Heidelberg (2012). doi:10.1007/978-3-642-29426-6_10
12. Hu, J., Song, Y.: The model of malware propagation in wireless sensor networks with regional detection mechanism. Commun. Comput. Inf. Sci. **501**, 651–662 (2015)
13. Jorgensen, S.E., Fath, B.D.: Individual-based models. Dev. Env. Model. **23**, 291–308 (2011)
14. Kechen, Z., Hong, Z., Kun, Z.C.: Simulation-based analysis of worm propagation in wireless sensor networks. In: IEEE Conference on Multimedia Information Networking and Security, pp. 847–851 (2012)
15. Kermack, W.O., McKendrick, A.G.: Contributions to the mathematical theory of epidemics, part I. Proc. Roy. Soc. A. **115**(772), 700–721 (1927)
16. Khayam, S.A., Radha, H.C.: A topologically-aware worm propagation model for wireless sensor networks. IEEE Conference on Distributed Computing Systems Workshops, pp. 210–216 (2005)
17. Khayam, S.A., Radha, H.: Using signal processing techniques to model worm propagation over wireless sensor networks. IEEE Sig. Process. Mag. **23**(2), 164–169 (2006)
18. Li, Q., Zhang, B., Cui, L., Fan, Z., Athanasios, V.V.: Epidemics on small worlds of tree-based wireless sensor networks. J. Syst. Sci. Complex. **27**(6), 1095–1120 (2014)

19. Martín del Rey, A.: Mathematical modeling of the propagation of malware: a review. Secur. Commun. Netw. **8**(15), 2561–2579 (2015)
20. Mishra, B.K., Pandey, S.K.: Dynamic model of worms with vertical transmission in computer network. Appl. Math. Comput. **217**(21), 8438–8446 (2011)
21. Mishra, B.K., Keshri, N.: Mathematical model on the transmission of worms in wireless sensor network. Appl. Math. Modell. **37**, 4103–4111 (2013)
22. Newman, M.E.J.: Spread of epidemic disease on networks. Phys. Rev. E. **66**(1), 016128 (2002)
23. Perrig, A., Stankovic, J., Wagner, D.: Security in wireless sensor networks. Commun. ACM **47**(6), 53–57 (2004)
24. Railsback, S.F., Grimm, V.: Agent-Based and Individual-Based Modeling. Princeton University Press, Princeton (2012)
25. Sen, J.: A survey on wireless sensor network security. Int. J. Commun. Netw. Inf. Secur. **1**, 55–78 (2009)
26. Shen, S., Li, H., Han, R., et al.: Differential game-based strategies for preventing malware propagation in wireless sensor networks. IEEE Trans. Inf. Forensic Secur. **9**(11), 1962–1973 (2014)
27. Shengjun, W., Junhua, C.: Modeling the spread of worm epidemics in wireless sensor networks. In: 5th International Conference on Networking and Mobile Computing in Wireless Communications, pp. 1–4 (2009)
28. Tang, S.J.: A modified SI epidemic model for combating virus spread in wireless sensor networks. Int. J. Wireless Inf. Netw. **18**(4), 319–326 (2011)
29. Tang, S., Mark, B.L.: Analysis of virus spread in wireless sensor networks: an epidemic model. In: IEEE International Workshop on Design of Reliable Communication Networks, pp. 86–91 (2009)
30. Vasilakos, A.V.J.: Dynamics in small worlds of tree topologies of wireless sensor networks. J. Syst. Eng. Electr. **3**, 001 (2012)
31. Wang, X., Li, Y.: An improved SIR model for analyzing the dynamics of worm propagation in wireless sensor networks. Chin. J. Electron. **18**, 8–12 (2009)
32. Wang, X., Li, Q., Li, Y.: Eisirs: a formal model to analyze the dynamics of worm propagation in wireless sensor networks. J. Comb. Optim. **20**, 47–62 (2010)
33. Wolfram, S.: A New Kind of Science. Wolfram Media, Champaign (2002)
34. Wu, M., Tan, L., Xiong, N.: Data prediction, compression and recovery in clustered wireless sensor networks for environmental monitoring applications. Inf. Sci. **239**, 800–818 (2016)
35. Xiao-Ping, S., Yu-Rong, S.J.: A malware propagation model in wireless sensor networks with cluster structure of GAF. J. Telecommun. Sci. **27**(8), 33–38 (2011)
36. Ya-Qi, W., Xiao-Yuan, Y.J.: Virus spreading in wireless sensor networks with a medium access control mechanism. Chin. Phys. B **22**(4), 040206 (2013)
37. Yang, Y., Zhu, S., Cao, G.: Improving sensor network immunity under worm attacks: a software diversity approach. In: Proceedings of ACM international symposium on Mobile Ad Hoc Networking and Computing, pp. 149–158 (2008)
38. Yang, S.H.: Wireless Sensor Networks. Principles, Design and Applications. Springer, London (2014)
39. Yick, J., Mukherjee, B., Ghosai, D.: Wireless sensor network survey. Comput. Netw. **52**(12), 2292–2330 (2009)
40. Zhang, Z., Si, F.: Dynamics of a delayed SEIRS-V model on the transmission of worms in a wireless sensor network. Adv. Diff. Equat. **2014**(1), 1–15 (2014)
41. Zhu, L., Zhao, H.: Dynamical analysis and optimal control for a malware propagation model in an information network. Neurocomputing **149**, 1370–1386 (2015)

A Study on the Performance of Secure Elliptic Curves for Cryptographic Purposes

Raúl Durán Díaz[1], Victor Gayoso Martínez[2](✉), Luis Hernández Encinas[2], and Agustin Martín Muñoz[2]

[1] Department of Automatics, University of Alcalá, Madrid, Spain
raul.duran@uah.es

[2] Institute of Physical and Information Technologies (ITEFI), Spanish National Research Council (CSIC), Madrid, Spain
{victor.gayoso,luis,agustin}@iec.csic.es

Abstract. Elliptic Curve Cryptography (ECC) is a branch of public-key cryptography based on the arithmetic of elliptic curves. In the short life of ECC, most standards have proposed curves defined over prime finite fields satisfying the curve equation in the short Weierstrass form. However, some researchers have started to propose as a more secure alternative the use of Edwards and Montgomery elliptic curves, which could have an impact in current ECC deployments. This contribution evaluates the performance of the three types of elliptic curves using some of the examples provided by the initiative SafeCurves and a Java implementation developed by the authors, which allows us to offer some conclusions about this topic.

Keywords: Edwards curves · Elliptic curve cryptography · Java · Montgomery curves · Point arithmetic · Weierstrass curves

1 Introducción

In 1987, Neal Koblitz [1] and Victor Miller [2] independently suggested using elliptic curves defined over finite fields for implementing different cryptosystems. This branch of public-key cryptography is typically known as ECC (Elliptic Curve Cryptography), and its security is based on the difficulty of solving the ECDLP (Elliptic Curve Discrete Logarithm Problem).

One of the most important aspects when working with elliptic curves is their selection mechanism. Even though some standards include several sample curves or even the description of the procedures for generating them (for example, X9.63 [3], IEEE 1363 [4] or NIST FIPS 186-4 [5]), in most cases the information contained in those standards has important limitations, such as the lack of clarity in the selection procedure regarding the seeds and prime numbers involved or the insufficient explanation for some of the requirements taken into account.

In this scenario, at the beginning of the last decade a working group called ECC Brainpool focused on this topic and elaborated a first set of recommendations in 2005 [6]. Five years later, the Brainpool specification was revised and

M. Graña et al. (eds.), *International Joint Conference SOCO'16-CISIS'16-ICEUTE'16*,
Advances in Intelligent Systems and Computing 527, DOI 10.1007/978-3-319-47364-2_64

published as an RFC (Request for Comments) [7]. The Brainpool initiative was considered as the first international effort with the goal of producing a truly transparent curve generation procedure, and the curves suggested in its specification were initially considered to be secure without any hint of doubt.

Some time after that, researchers Daniel Bernstein and Tanja Lange published an analysis in which they reviewed the existing elliptic curve generation mechanisms, including the one devised by Brainpool. In their site SafeCurves [8], they compared not only the strength of the curve parameters and the soundness of what they called "ECC security" (basically the strength against rho attacks and transfers of the ECDLP to other fields where the DLP is easier to solve, the class number associated to the trace of the curve, and the rigidity of the definition of the curve parameters), but also what they termed "ECDLP security" (a concept in which they included the resistance to attacks based on the Montgomery ladder, the strength of the associated twisted curves, the completeness of the addition formulas, and the indistinguishability of elliptic curve points from random binary strings). The main result of that analysis was that all the schemes included in the standards overlooked some aspects of the ECDLP security and, for that reason, required to increase the complexity of the implementations in such a way that it opened the door to side channel attacks [8].

As a solution, Bernstein and Lange decided to propose new curves different to those provided by previous specifications. Going one step further, they evaluated 20 curves obtained from different sources (two ad-hoc curves used as examples of faulty designs, one SEC 2 curve, one ANSSI curve, two Brainpool curves, three NIST curves, five Montgomery curves, and six Edwards curves), showing that the only curves that were able to fulfil all their security requirements were the Edwards and Montgomery curves.

However, from the point of view of availability, both Montgomery and Edwards curves have not been popular choices so far, and in that respect *traditional* curves are the dominant options both in hardware and software implementations. In addition to that, the extra security offered by Edwards and Montgomery curves could affect the performance of the point operations which are the core of the scalar multiplication operation (the product of a point of the elliptic curve by an integer, an operation needed in any protocol involving elliptic curves).

Based on the work by Bernstein and Lange, this contribution addresses the issue of the performance of the Edwards and Montgomery curves proposed at SafeCurves compared to the Weierstrass curves suggested in the standards. In order to do that, Sect. 2 provides a brief review of the most important concepts regarding elliptic curves and the three types of curves defined over finite fields considered in our analysis. Section 3 contains the algorithms that minimize the number of finite field operations in the point addition and doubling procedures, which allows us to compare their theoretic complexity to the practical results obtained after implementing those operations in a Java application. Section 4 shows the performance results obtained with the aforementioned implementation. Finally, in Sect. 5 we offer to the interested readers our conclusions on this topic.

2 Elliptic Curves

2.1 Definition

An elliptic curve defined over a field $\mathbb{F}$ is a cubic, non-singular curve whose points $(x, y) \in \mathbb{F} \times \mathbb{F}$ verify the following equation, known as the Weierstrass equation:

$$E : y^2 + a_1xy + a_3y = x^3 + a_2x^2 + a_4x + a_6,$$

where $a_1, a_2, a_3, a_4, a_6 \in \mathbb{F}$ and $\Delta \neq 0$, where Δ is the discriminant of E that can be computed as follows [9]:

$$\begin{aligned}
\Delta &= -d_2^2d_8 - 8d_4^3 - 27d_6^2 + 9d_2d_4d_6, \\
d_2 &= a_1^2 + 4a_2, \\
d_4 &= 2a_4 + a_1a_3, \\
d_6 &= a_3^2 + 4a_6, \\
d_8 &= a_1^2a_6 + 4a_2a_6 - a_1a_3a_4 + a_2a_3^2 - a_4^2.
\end{aligned}$$

An elliptic curve point is singular if and only if the partial derivatives of the curve equation are null at that point. The curve is said to be singular if it possesses at least a singular point, while it is non-singular if it does not have any such points.

The non-homogeneous Weierstrass equation can also be expressed in the following homogeneous form [10]:

$$Y^2Z + a_1XYZ + a_3YZ^2 = X^3 + a_2X^2Z + a_4XZ^2 + a_6Z^3.$$

This equation defines a curve which includes a special point called the point at infinity, which is typically represented as $\mathcal{O} = [0 : 1 : 0]$ and that has no correspondence with any point of the non-homogeneous form. However, this point is very important as it works as the identity element of the addition operation when working with Weierstrass and Montgomery elliptic curves.

2.2 Elliptic Curves Over Finite Fields

Most cryptosystems defined over elliptic curves use only one of the following finite fields $\mathbb{F}_q$ with $q = p^m$ elements: prime fields $\mathbb{F}_p$ (where p is an odd prime number and $m = 1$) and binary fields $\mathbb{F}_{2^m}$ (where m can be any positive integer). However, due to a combination of licence issues and security concerns [11], prime fields have been favoured in the latest specifications at the expense of binary fields (see, for example, Brainpool [7], NSA Suite B [12] or BSI TR-03111 [13]). Following that criterion, in what follows we have accordingly focused our study on prime fields.

The peculiarities of prime fields allow to simplify the general Weierstrass equation and to obtain in the process what is called the short Weierstrass form represented as $y^2 = x^3 + ax + b$, where $4a^3 + 27b^2 \not\equiv 0 \pmod{p}$.

As in the case of the general Weierstrass equation, the identity element of the short Weierstrass form is the point at infinity $\mathcal{O}$, while the opposite element of a point $P = (x, y)$ is the point $-P = (x, -y)$. Adding two points $P_1 = (x_1, y_1)$ and $P_2 = (x_2, y_2)$ such that $P_1 \neq \pm P_2$ produces a point $P_3 = (x_3, y_3)$ whose coordinates can be computed as follows [14]:

$$x_3 = \frac{(y_2 - y_1)^2}{(x_2 - x_1)^2} - x_1 - x_2,$$

$$y_3 = \frac{(2x_1 + x_2)(y_2 - y_1)}{x_2 - x_1} - \frac{(y_2 - y_1)^3}{(x_2 - x_1)^3} - y_1.$$

In comparison, when $P_1 = P_2$ it is necessary to use and alternative addition formula, so in this case the point $P_3 = 2P_1$ obtained through the doubling operation has the following coordinates [14]:

$$x_3 = \frac{(3x_1^2 + a)^2}{(2y_1)^2} - 2x_1,$$

$$y_3 = \frac{(3x_1)(3x_1^2 + a)}{2y_1} - \frac{(3x_1^2 + a)^3}{(2y_1)^3} - y_1.$$

Edwards curves where introduced in [15] and are defined according to the equation $x^2 + y^2 = c^2(1 + dx^2y^2)$, where $cd(1 - dc^4) \not\equiv 0 \pmod p$. An Edwards curve is said to be *complete* if d is not a quadratic residue module p.

The identity element in Edwards curves is the point $(0, c)$ while, taking into consideration a point $P = (x, y)$, its opposite element is $-P = (-x, y)$. The result of adding any two points $P_1 = (x_1, y_1)$ and $P_2 = (x_2, y_2)$ is the point $P_3 = (x_3, y_3)$ whose coordinates can be computed as follows [14]:

$$x_3 = \frac{x_1y_2 + y_1x_2}{c(1 + dx_1x_2y_1y_2)},$$

$$y_3 = \frac{y_1y_2 - x_1x_2}{c(1 - dx_1x_2y_1y_2)}.$$

Readers should note that, in the case of Edwards curves, the equation for both adding two points P_1 and P_2 such that $P_1 \neq \pm P_2$ and doubling a point are exactly the same. Moreover, it is not necessary to implement any logic for detecting if the points to be added are such that $P_2 = -P_1$, as the Edwards addition equations also take into account that circumstance.

Finally, Montgomery curves conform to the equation $By^2 = x^3 + Ax^2 + x$, where $B(A^2 - 4) \not\equiv 0 \pmod p$. As in the case of Weierstrass curves, the identity element in Montgomery curves is the point at infinity $\mathcal{O}$, while the opposite element of $P = (x, y)$ is the point $-P = (x, -y)$. The addition of two points $P_1 = (x_1, y_1)$ and $P_2 = (x_2, y_2)$ such that $P_1 \neq \pm P_2$ is the point $P_3 = (x_3, y_3)$ with the following coordinates [14]:

$$x_3 = \frac{B(y_2 - y_1)^2}{(x_2 - x_1)^2} - A - x_1 - x_2,$$

$$y_3 = \frac{(2x_1 + x_2 + A)(y_2 - y_1)}{x_2 - x_1} - \frac{B(y_2 - y_1)^3}{(x_2 - x_1)^3} - y_1.$$

Unlike Edwards curves, it is necessary to use different equations for the doubling operation, so in this case the coordinates of the point $P_3 = 2P_1$ can be computed as follows [14]:

$$x_3 = \frac{B(3x_1^2 + 2ax_1 + 1)^2}{(2By_1)^2} - A - 2x_1,$$

$$y_3 = \frac{(3x_1 + A)(3x_1^2 + 2Ax_1 + 1)}{2By_1} - \frac{B(3x_1^2 + 2Ax_1 + 1)^3}{(2By_1)^3} - y_1.$$

2.3 Transforming Formulas

While Edwards curves can be always expressed in the Montgomery and short Weierstrass forms, the converse is not always possible. In order to transform an Edwards elliptic curve into a Montgomery elliptic curve as displayed in Sect. 2.2, it is necessary to use the following equivalence formulas [8]:

$$A = B - 2, \quad B = \frac{4}{1 - dc^4}.$$

In this way, a curve point (x_E, y_E) which belongs to an Edwards curve can be converted to a point (x_M, y_M) of the associated Montgomery curve, where the equations for obtaining (x_M, y_M) are as follows:

$$x_M = \frac{c + y_E}{c - y_E}, \quad y_M = c\frac{x_M}{x_E}.$$

Besides, in order to transform a Montgomery elliptic curve into the short Weierstrass form as represented in Sect. 2.2, it is necessary to use the following equivalences [8]:

$$a = \frac{3 - A^2}{3B^2}, \quad b = \frac{2A^3 - 9A}{27B^3}$$

In this specific case, a curve point (x_M, y_M) belonging to a Montgomery curve can be converted to a point (x_W, y_W) of the associated short Weierstrass curve, where the transforming equations are the following ones:

$$x_W = \frac{x_M + \frac{A}{3}}{B}, \quad y_W = \frac{y_M}{B}.$$

3 Implementation

As it is well known, there are two possible approaches when implementing mathematics using modern programming languages: either minimize the amount of memory used by the implementation, or minimize the number of operations. Given that the goal of this contribution is to compare the performance of the different curve implementations, the approach taken has been to reduce as much as possible the number of finite field operations, which can be of the following types: addition/subtraction, multiplication, squaring, and multiplicative inversion (for the sake of simplicity, in what follows we will refer to both additions and subtractions simply as additions).

Depending on the processor, operating system, and programming language, the exact relationship between the computing cost of those operations may vary, and for instance at [14] there are several examples with different equivalences between the inversion, multiplication and squaring costs, though it is commonly accepted that the addition cost is negligible compared to that of the other field operations.

As our implementation was designed to work on personal computers, the amount of memory used was not a concern; however, if the goal had been to implement the curve arithmetic in smart cards, it would have been necessary to adopt the opposite approach, as in that case the amount of memory for card applications is certainly limited and reusing variables would have been an avoidable need.

Algorithm 1 shows the details of the point addition and point doubling operations in Weierstrass curves. In the case of point addition, the elements involved are $P_1 = (x_1, y_1)$ and $P_2 = (x_2, y_2)$ such that $P_1 \neq \pm P_2$. This point addition implementation needs 6 additions, 2 multiplications, 1 squaring, and 1 inversion in $\mathbb{F}_p$, while the point doubling implementation requires 8 additions, 2 multiplications, 2 squarings, and 1 inversion in $\mathbb{F}_p$.

Algorithm 1. Point addition (left) and point doubling (right) in Weierstrass curves

1: $t_1 \leftarrow y_2 - y_1 \pmod p$	1: $t_1 \leftarrow (x_1)^2 \pmod p$
2: $t_2 \leftarrow x_2 - x_1 \pmod p$	2: $t_2 \leftarrow t_1 + t_1 \pmod p$
3: $t_2 \leftarrow t_2^{-1} \pmod p$	3: $t_2 \leftarrow t_2 + t_1 \pmod p$
4: $\lambda \leftarrow t_1 \cdot t_2 \pmod p$	4: $t_2 \leftarrow t_2 + a \pmod p$
5: $x_3 \leftarrow \lambda^2 \pmod p$	5: $t_3 \leftarrow y_1 + y_1 \pmod p$
6: $x_3 \leftarrow x_3 - x_1 \pmod p$	6: $t_3 \leftarrow t_3^{-1} \pmod p$
7: $x_3 \leftarrow x_3 - x_2 \pmod p$	7: $\lambda \leftarrow t_2 \cdot t_3 \pmod p$
8: $y_3 \leftarrow x_1 - x_3 \pmod p$	8: $x_3 \leftarrow \lambda^2 \pmod p$
9: $y_3 \leftarrow y_3 \cdot \lambda \pmod p$	9: $x_3 \leftarrow x_3 - x_1 \pmod p$
10: $y_3 \leftarrow y_3 - y_1 \pmod p$	10: $x_3 \leftarrow x_3 - x_1 \pmod p$
11: **return** (x_3, y_3)	11: $y_3 \leftarrow x_1 - x_3 \pmod p$
	12: $y_3 \leftarrow y_3 \cdot \lambda \pmod p$
	13: $y_3 \leftarrow y_3 - y_1 \pmod p$
	14: **return** (x_3, y_3)

Algorithm 2 shows the details of the point addition in Edwards curves, where as the reader may recall no restriction applies to the points $P_1 = (x_1, y_1)$ and $P_2 = (x_2, y_2)$. This implementation requires 3 additions, 9 multiplications, and 2 inversions in $\mathbb{F}_p$.

Finally, Algorithm 3 shows the details of the point addition and point doubling operations in Montgomery curves. Similarly to Weierstrass curves, in the case of point addition the elements involved are $P_1 = (x_1, y_1)$ and $P_2 = (x_2, y_2)$

Algorithm 2. Point addition and doubling in Edwards curves

1: $t_1 \leftarrow x_1 \cdot y_2 \pmod p$
2: $t_2 \leftarrow y_1 \cdot x_2 \pmod p$
3: $t_2 \leftarrow t_1 + t_2 \pmod p$
4: $t_3 \leftarrow x_1 \cdot x_2 \pmod p$
5: $t_4 \leftarrow y_1 \cdot y_2 \pmod p$
6: $t_5 \leftarrow t_3 \cdot t_4 \pmod p$
7: $t_5 \leftarrow t_5 \cdot d \pmod p$
8: $t_5 \leftarrow t_5 \cdot c \pmod p$
9: $t_6 \leftarrow t_5 + c \pmod p$
10: $t_6 \leftarrow t_6^{-1} \pmod p$
11: $x_3 \leftarrow t_2 \cdot t_6 \pmod p$
12: $t_7 \leftarrow t_4 - t_3 \pmod p$
13: $t_8 \leftarrow c - t_5 \pmod p$
14: $t_8 \leftarrow t_8^{-1} \pmod p$
15: $y_3 \leftarrow t_7 \cdot t_8 \pmod p$
16: **return** (x_3, y_3)

Table 1. Comparison of the number of operations needed

	Edwards				Montgomery				Weierstrass			
	Add.	Mul.	Sqr.	Inv.	Add.	Mul.	Sqr.	Inv.	Add.	Mul.	Sqr.	Inv.
Addition	3	9	0	2	10	9	1	1	6	2	1	1
Doubling	3	9	0	2	12	9	2	1	8	2	2	1

such that $P_1 \neq \pm P_2$. This point addition implementation for Montgomery curves needs 10 additions, 9 multiplications, 1 squaring, and 1 inversion in $\mathbb{F}_p$, while the point doubling implementation requires 12 additions, 9 multiplications, 2 squarings, and 1 inversion in $\mathbb{F}_p$.

Table 1 summarizes the operations needed for both the addition and doubling operations with the three types of curves considered in this work.

4 Experimental Results

The tests whose results are shown in this section were performed using an Intel Core i7 processor model 2630QM at 2.00 GHz, while the Java application was compiled with the Java Development Kit version 1.8.0_66.

We have used the six Edwards curves displayed in SafeCurves as starting point, using the formulas included in this contribution for obtaining the associate Montgomery and short Weierstrass curves. Those curves are the ones identified as E-221 (221 bits), Curve1174 (251 bits), E-382 (382 bits), Curve41417 (414 bits), Ed448-Goldilocks (448 bits), and E-521 (521 bits).

Starting with each base point offered in SafeCurves for the Edwards curves, we have used that point (and the corresponding points of the Montgomery and Weierstrass curves) inside a loop that computes one million addition operations and another million doubling operations. Table 2 shows the running time in seconds of those tests.

Algorithm 3. Point addition (left) and point doubling (right) in Montgomery curves

1: $t_1 \leftarrow y_2 - y_1 \pmod p$	1: $t_1 \leftarrow (x_1)^2 \pmod p$
2: $t_2 \leftarrow (t_1)^2 \pmod p$	2: $t_2 \leftarrow t_1 + t_1 \pmod p$
3: $t_2 \leftarrow t_2 \cdot B \pmod p$	3: $t_2 \leftarrow t_1 + t_2 \pmod p$
4: $t_3 \leftarrow x_2 - x_1 \pmod p$	4: $t_3 \leftarrow x_1 + x_1 \pmod p$
5: $t_3 \leftarrow t_3^{-1} \pmod p$	5: $t_4 \leftarrow A \cdot t_3 \pmod p$
6: $x_3 \leftarrow t_2 \cdot t_3 \pmod p$	6: $t_5 \leftarrow t_2 + t_4 \pmod p$
7: $x_3 \leftarrow x_3 \cdot t_3 \pmod p$	7: $t_5 \leftarrow 1 + t_5 \pmod p$
8: $x_3 \leftarrow x_3 - A \pmod p$	8: $t_6 \leftarrow (t_5)^2 \pmod p$
9: $x_3 \leftarrow x_3 - x_1 \pmod p$	9: $t_6 \leftarrow t_6 \cdot B \pmod p$
10: $x_3 \leftarrow x_3 - x_2 \pmod p$	10: $t_7 \leftarrow B + B \pmod p$
11: $t_4 \leftarrow x_1 + x_1 \pmod p$	11: $t_7 \leftarrow t_7 \cdot y_1 \pmod p$
12: $t_4 \leftarrow t_4 + x_2 \pmod p$	12: $t_7 \leftarrow (t_7)^{-1} \pmod p$
13: $t_4 \leftarrow t_4 + A \pmod p$	13: $t_8 \leftarrow t_6 \cdot t_7 \pmod p$
14: $t_4 \leftarrow t_1 \cdot t_4 \pmod p$	14: $t_8 \leftarrow t_7 \cdot t_8 \pmod p$
15: $t_4 \leftarrow t_3 \cdot t_4 \pmod p$	15: $x_3 \leftarrow t_8 - A \pmod p$
16: $t_5 \leftarrow t_1 \cdot t_2 \pmod p$	16: $x_3 \leftarrow x_3 - t_3 \pmod p$
17: $t_5 \leftarrow t_3 \cdot t_5 \pmod p$	17: $t_9 \leftarrow t_3 + x_1 \pmod p$
18: $t_5 \leftarrow t_3 \cdot t_5 \pmod p$	18: $t_9 \leftarrow t_9 + A \pmod p$
19: $t_5 \leftarrow t_3 \cdot t_5 \pmod p$	19: $t_9 \leftarrow t_5 \cdot t_9 \pmod p$
20: $y_3 \leftarrow t_4 - t_5 \pmod p$	20: $t_9 \leftarrow t_7 \cdot t_9 \pmod p$
21: $y_3 \leftarrow y_3 - y_1 \pmod p$	21: $t_{10} \leftarrow t_5 \cdot t_8 \pmod p$
22: **return** (x_3, y_3)	22: $t_{10} \leftarrow t_7 \cdot t_{10} \pmod p$
	23: $y_3 \leftarrow t_9 - t_{10} \pmod p$
	24: $y_3 \leftarrow y_3 - y_1 \pmod p$
	25: **return** (x_3, y_3)

Table 2. Running time in seconds for 10^6 addition/doubling operations

	Edwards		Montgomery		Weierstrass	
	Addition	Doubling	Addition	Doubling	Addition	Doubling
E-222	81.5329	84.2961	57.4369	59.4419	41.6957	44.9321
Curve1174	102.8423	102.8315	70.1401	72.3692	50.6845	54.3952
E-382	175.2261	182.7206	129.0495	129.0018	89.7586	96.1143
Curve41417	193.9862	202.3829	142.7497	142.2198	99.3412	106.2448
Ed448	216.1019	225.5697	158.5627	156.1755	110.7912	117.9702
E-521	284.5245	297.9542	212.0656	209.4322	145.9185	155.5230

5 Conclusions

As it was expected, given the number and type of operations needed for the addition and doubling operations in each curve, the best results have been obtained with curves of the short Weierstrass type, and the worst results have been produced by the Edwards curves. More specifically, the running time with Edwards curves is almost twice the time needed when using Weierstrass curves, which have the simpler definition for the operations tested.

If we consider negligible the running time for additions, and join the squaring operations with the multiplications, then the relationship obtained between multiplications and inversions is almost 1:100, which means than an inversion is, from the computationally point of view, as costly as 100 multiplications. This value matches the estimation of [14], so we have been able to confirm that data in our testing environment.

Along different tests we have always obtained a small but noticeable difference in the results for both operations in Edwards curves. Given that exactly the same algorithm is used in both cases, the difference can only be explained by the specific points used in each test, which are not the same. Nevertheless, that result deserves a more-in-depth investigation in the next phase of our research, together with an extension of its scope so it takes into consideration alternative coordinate systems (projective, Jacobian, etc.).

Acknowledgements. This work has been supported by the European Union FEDER funds distributed through Ministerio de Economía y Competitividad (Spain) under the project TIN2014-55325-C2-1-R (ProCriCiS), and through Comunidad de Madrid (Spain) under the project S2013/ICE-3095-CM (CIBERDINE).

References

1. Koblitz, N.: Elliptic curve cryptosytems. Math. Comput. **48**(177), 203–209 (1987)
2. Miller, V.S.: Use of elliptic curves in cryptography. In: Williams, H.C. (ed.) CRYPTO 1985. LNCS, vol. 218, pp. 417–426. Springer, Heidelberg (1986). doi:10.1007/3-540-39799-X_31
3. American National Standards Institute: Public Key Cryptography for the Financial Services Industry: Key Agreement and Key Transport Using Elliptic Curve Cryptography. ANSI X9.63 (2001)
4. IEEE: Standard specifications for public key cryptography. Institute of Electrical and Electronics Engineers, IEEE 1363 (2000)
5. National Institute of Standard and Technology: Digital Signature Standard (DSS). NIST FIPS 186-4 (2009). http://nvlpubs.nist.gov/nistpubs/FIPS/NIST.FIPS.186-4.pdf
6. Brainpool: ECC Brainpool standard curves and curve generation, version 1.0 (2005). http://www.ecc-brainpool.org/download/Domain-parameters.pdf
7. Lochter, M., Merkle, J.: Elliptic Curve Cryptography (ECC) Brainpool standard curves and curve generation. Request for comments (RFC 5639), Internet Engineering Task Force (2010)
8. Bernstein, D.J., Lange, T.: SafeCurves (2014). http://safecurves.cr.yp.to/
9. Menezes, A.J.: Elliptic Curve Public Key Cryptosystems. Kluwer Academic Publishers, Boston (1993)
10. Cohen, H., Frey, G.: Handbook of Elliptic and Hyperelliptic Curve Cryptography. Discrete Mathematics and its Applications. Chapman & Hall/CRC, Boca Raton (2006)
11. Bernstein, D.J.: Curve25519: new Diffie-Hellman speed records. In: Yung, M., Dodis, Y., Kiayias, A., Malkin, T. (eds.) PKC 2006. LNCS, vol. 3958, pp. 207–228. Springer, Heidelberg (2006). doi:10.1007/11745853_14

12. National Security Agency: NSA Suite B cryptography (2009). http://www.nsa.gov/ia/programs/suiteb_cryptography/index.shtml
13. Bundesamt für Sicherheit in der Informationstechnik: Elliptic curve cryptography. BSI TR-03111 version 2.0. (2012). https://www.bsi.bund.de/SharedDocs/Downloads/EN/BSI/Publications/TechGuidelines/TR03111/BSI-TR-03111_pdf.pdf?__blob=publicationFile
14. Bernstein, D.J., Lange, T.: Explicit-formulas database (2016). https://hyperelliptic.org/EFD/
15. Edwards, H.M.: A normal form for elliptic curves. Bull. Am. Math. Soc. **44**, 393–422 (2007)

A SEIS Model for Propagation of Random Jamming Attacks in Wireless Sensor Networks

Miguel López, Alberto Peinado(✉), and Andrés Ortiz

E.T.S.Ingeniería de Telecomunicación, Dept. Ingeniería de Comunicaciones, Universidad de Málaga, Andalucía Tech, Campus de Teatinos, 29071 Málaga, Spain
m.lopez@uma.es, {apeinado,aortiz}@ic.uma.es

Abstract. This paper describes the utilization of epidemiological models, usually employed for malware propagation, to study the effects of random jamming attacks, which can affect the physical and MAC/link layers of all nodes in a wireless sensor network, regardless of the complexity and computing power of the devices. The random jamming term considers both the more classical approach of interfering signals, focusing on the physical level of the systems, and the cybersecurity approach that includes the attacks generated in upper layers, mainly in the MAC/link layer, producing the same effect on the communication channel. We propose, as a preliminary modelling task, the epidemiological mathematical model Susceptible–Exposed–Infected–Susceptible (SEIS), and analyze the basic reproductive number, the infection rate, the average incubation time and the average infection time.

Keywords: Cyber security · Jamming attacks · Epidemiological models · Wireless sensor networks

1 Introduction

The epidemiologic theory employs mathematical models to study and analyse the propagation of diseases. The current models were first proposed in 1927 by Kermack and McKendrick to describe epidemics in India [1]. The close relationship between the behaviour of biological and computer infections brought the researchers to apply these models in 1991 to predict the computer virus propagation [2]. Since then, the models have been constantly updated to describe malware propagation on different kind of networks, including wireless sensor networks (WSN) [3–10].

WSN consist of a large number of small nodes deployed to control or monitor critical infrastructure, industrial processes, environments and other applications based on the collection of data in real time. The operation of these networks is characterized by cooperation between nodes to create wireless communication paths, providing better performance when traditional networks are impossible to deploy or very expensive. However, multicast nature of wireless technology and aggressive environments are factors that significantly increase the probability of executing various attacks on these networks [11]. On the other hand, the nodes in WSN are usually constrained devices in power supply, computation capability and memory storage. This fact can be considered

M. Graña et al. (eds.), *International Joint Conference SOCO'16-CISIS'16-ICEUTE'16*,
Advances in Intelligent Systems and Computing 527, DOI 10.1007/978-3-319-47364-2_65

a natural barrier against certain types of attacks, although some of them have proven their effectiveness such as those based on code injection or memory corruption vulnerabilities [12–14].

In any case, there are some attacks not based on malware infection that can be launched by apparently legal nodes to damage the network, removing functionalities and degrading its performance, resulting in a complete denial of service attack (DoS). A corrupted node may interfere intentionally in the normal operation of the network wireless medium, at the physical or MAC level, saturating the channel by injecting and continuously transmitting data packets (with or without sense), causing abnormalities and errors in transmission and reception of data. This would cause in the affected nodes an extra effort in data transmission reducing their energy, and therefore the overall lifetime of the network [15]. This type of DoS attack falls into the general category of jamming, (physical or link jamming) and could have devastating consequences even in the presence of a small number of attackers nodes [16].

The three major wireless standards that operate in the 2.4 GHz ISM (Industrial, Scientific and Medical) band, and have the technical performance and maturity required to be implemented in real WSN applications are IEEE 802.15.1, IEEE 802.15.4 and IEEE 802.11. The IEEE 802.15.1 standard (and their new release 4.0 known as Bluetooth Low Energy) has been designed primarily for consumer electronic devices with very short communication range and low-cost applications and presents a limited deployment as WSN. Industrial and critical infrastructures applications are based on 802.11 networks, mostly organized around an Access Point (AP) or Base Station (BS). The IEEE 802.15.4 is the preferred wireless technology for real- time applications and generic WSN.

Considering the physical layer, standards IEEE 802.15.1 and 802.11 are more resistant to interference due to their multichannel scheme and modulation techniques. IEEE 802.15.4 specification suffers also from attacks generated at MAC layer, producing a DoS effect, as it is reported in [17]. Hence, this standard is, at the same time, the most implemented in WSN and the least resistant against any type of jamming. The IETF has published in 2008 the IEEE 802.15.4e amended with multichannel scheme and new MAC procedures. However, most of devices currently in use is not updated [18].

In this paper we present a dynamic study of the spread of the effects of jamming attacks in a WSN based on the epidemiological mathematical model Susceptible–Exposed–Infected–Susceptible (SEIS), determining whether the attack fades or persists over time. In this model, it will therefore be of particular interest the study of the basic reproductive number, which represents the transition phase of non-equilibrium process of spread of a disease. This parameter constitutes an epidemic threshold: if the parameter is above the threshold, the infection will spread and become a persistent epidemic; otherwise, the parameter is below the threshold, the disease will die out. For practical reasons and due to the expansion of IEEE 802.14.4 networks, this model is intended to be applied in that type of WSN, although the model, at this preliminary stage, is applicable to any other WSN, since the three technologies suffer from weaknesses at MAC level allowing to successfully perform a data-link jamming.

The rest of the paper is organized as follows. In Sect. 2 introduces the background of the epidemiological theory in which the proposed model is based. Section 3 gives the analysis of the jamming attack under an epidemiological approach. In Sect. 4 we evaluate the model and give the results of our simulations; presenting the conclusions in Sect. 5.

2 Fundamentals of Epidemiological Theory

Modern epidemiological theory tries to predict the propagation of a disease by calculating a threshold value to determine when the disease becomes epidemic, considering a particular population [19]. In this way, epidemiological theory is used to analyze a disease and to establish strategies and plans for prevention and control, such as vaccination or quarantine.

Most epidemiological models derived from the mathematical development of Kermack–McKendrick [1], in which the population is divided into different compartments or classes taking into account the characteristics of the disease. Diseases that confer immunity have a different compartmental model from diseases without immunity. The most basic compartmental models are SIR, SIS and SEIS (S:Susceptible, I: Infected, R:Recovered, E:Exposed). SIR is the simplest model. In the SIS model, a susceptible individual gets the disease and then, after an incubation period (*i.e.* the time when the disease persists), the individual becomes susceptible again passing from one compartment to another continuously. The SEIS model takes into account a period of latency or incubation of the disease before the individual is definitely considered in the group of infected. Considering $S(t)$, $I(t)$ and $R(t)$ the number of individuals in Susceptible, Infected and Recovered compartments, respectively, at time t, and defining β as the infection rate and γ as the period of recovery of infected individuals represented by $1/t_d$, where t_d is the expected duration of the disease, the SIR model can be represented by the following relationships between groups or compartments:

$$S(t) \xrightarrow{\beta} I(t) \xrightarrow{\gamma} R(t) \tag{1}$$

In epidemiology, one of the most important objectives is to determine if the infection dies out in the population or it becomes epidemics. To do that, the basic reproductive number R_0 is defined as a threshold, in such a way that represents the average number of secondary infections that occur when the first infected individual (patient zero) is introduced into a population of N individuals fully susceptible [20]. Hence, if $R_0 < 1$ the infection dies out while if $R_0 > 1$ we may have a case of an epidemic disease. Assuming that the first infected individual causes $\beta \cdot N$ contacts per unit time generating new infections, and that the average infection period is γ, then the basic reproductive number is $R_0 = \beta \cdot N/\gamma$. This models are called autonomous invariant systems since the infection rate β is assumed constant, *i.e.* does not change with time, so it depends on each disease; while the recovery rate γ is the inverse of the infectious period and depends only on the population of infected individuals who have at all times [20], when infectious diseases are considered.

3 Epidemiological Model for Jamming Propagation

Here we describe the epidemiological model, which will analyse the infection of jamming attack from a single initial infected node, the *zero patient* which, in this case, will be the interfering node that generate the attack, and that further will produce the effects on the other nodes through the communication channel occupancy. Therefore, for modelling purposes, the effect produced by the jamming attack will be considered as the infectious disease, since it has the potential to affect progressively to other network nodes. The aim of this research is to study the influence of different rates β, γ and ε, where ε is the average incubation time, in the basic reproductive number of the system, that marks the difference between an epidemic process that spread the attack throughout the network, or a process in which the attack dies out. The following sections describe the considered WSN model, the type of jamming to the study and the proposed epidemiological model.

3.1 Network Model

We consider a WSN consisting on N identical sensor nodes randomly distributed, following a uniform distribution in a region of area A, and a Aggregation Node (AN) or data collector. Accordingly, $\rho = N/A$ is the average density. We also consider that all nodes are equipped with omnidirectional antenna which a maximum transmission range r_0. Therefore, the probability that k nodes are within the communication range of a particular node is given by:

$$p(k) = \binom{N-1}{k} p^k (1-p)^{N-1-k} \tag{2}$$

Assuming a uniform distribution of nodes, we have that

$$p = \frac{\pi r_0^2 \rho}{N} = \frac{\pi r_0^2}{A} \tag{3}$$

where p is the probability of existence of a link at the physical level, that is, that at least two nodes are within their communication range, which can be assumed as a probability of disease transmission.

3.2 Propagation Model for Jamming Attacks

Broadly speaking, jamming is a type of attack which aim interfere intentionally in the normal operation of the network wireless medium, at the physical level and access level, saturating the channel by injecting and continuously transmitting data packets (with or without sense), causing abnormalities and errors in transmission and reception of data. From a more traditional point of view, jamming is considered as an interruption of the communication channel by transmitting interfering signals; this is clearly closely related to attacks in the physical layer of the communications system, and therefore has no ability to generate needless frames. However, knowledge of the communications protocol used in the WSN can deploy attacks from upper layers to achieve the same effect by sending valid frames, occupying the channel and thwarting the normal communication process. According to the taxonomy recently proposed by Lichtman et al. [21], such attacks falls into the category of cyber attacks usually related to the DoS, while other authors named it as link layer jamming [17].

As discussed before in both cases, physical or MAC layer, we can find several strategies with different levels of efficiency, to carry out such attacks [22]. The simplest is the continuous emission of a signal to saturate or interfere a wireless channel (constant Jammer), so that legitimate traffic is completely blocked. However, there are techniques of signal modulation, which have proven resistant to this type of interference in the case of attacks at physical level. As for the flood attacks from the MAC level, while effective are easily detectable. Moreover, the strategy known as reactive jamming [23] has demonstrated to be more efficient in WSN environments. In this case, the interfering node is silent when there is no activity on the channel, and starts generating data packets as soon as a transmission is detected. Another strategy more efficient than the constant jamming is the random jamming. In this case, the attacker seeks to randomly inject signals or packets in the channel. Specifically, the attacker activates its radio interface during a time t_j and goes to sleep mode, resuming the attack again after a time t_s. Values for t_j and t_s can be random or fixed values.

This study focuses on the modeling of the random jamming originated at physical or MAC layer. Thus, it is considered the existence of a node into the network (patient zero) that is compromised or modified to perform jamming attacks (acquires the disease) and then initiates it as described above. The effects of the attack (infection) will spread, first, through the nearest neighbor nodes, showing symptoms such as the increase in the number of packets forwarded between nodes, in the lost packets and, in the consumption of resources due to the collisions produced. These initially infected nodes will produce the same effect in their neighbors. It should be noted that in this model nodes that are able to overcome the stage of infection of the disease will became susceptible again.

3.3 Epidemiological Analysis of the Spread of Attack Jamming

The proposed model considers a stationary and uniformly distributed random population of N individuals (sensor nodes), with a density ρ, that never develop immunity to the disease. There will be nodes that despite being exposed to attack a certain time can be kept operating normally, not having symptoms and not being able to infect others. This process can be described with the epidemiological SEIS model since thereby obtain a more accurate scheme of attack propagation, by including in the group of infected only those nodes that the attack effectively induce directly or indirectly malfunction that can spread throughout the network. Based on this premise, let us define $S(t)$, $E(t)$ and $I(t)$ as the number of individuals susceptible, exposed and infected at time t, where $S(t) + E(t) + I(t) = N$. In addition, take β, as the rate of infection or contagion (the probability that an individual sick through contact with an infected), ε as the average incubation or latency time of the disease, and γ as the average time duration of infection (for a single individual), equivalent to the time in which the attacker injects packets randomly. In this model, the flow to move from one group to another can be described as follows:

$$S(t) \xrightarrow{\beta} E(t) \xrightarrow{\varepsilon} I(t) \xrightarrow{\gamma} S(t)$$

Hence, taking again the homogeneous contact between individuals as reference and assuming an infection rate given by the product $\lambda = p{\cdot}\beta$, the equations describing the model are:

$$\begin{aligned} \frac{dS(t)}{dt} &= -\lambda \cdot S(t) \cdot I(t) + \gamma \cdot I(t) \\ \frac{dE(t)}{dt} &= \lambda \cdot S(t) \cdot I(t) - \varepsilon \cdot E(t) \\ \frac{dI(t)}{dt} &= \varepsilon \cdot E(t) - \gamma \cdot I(t) \end{aligned}$$

This system of differential equations offers a view of the rate of change in the proportion of individuals in each group and allow us to determine the point where the spread of the attack becomes an epidemic process. In principle, we are unable to solve this system analytically but we learn a great deal about the behaviour of its solutions by taking a qualitative approach. To begin, we remark that the model makes sense only so long as the population shares of individuals remain non-negative. Then, we seek for the solutions of the differential equation system by locating the steady states or equilibrium points, also called fixed points, which are constant solutions that satisfy the condition $dS(t)/dt = 0$, $dE(t)/dt = 0$ and $dI(t)/dt = 0$. Considering that $S(t) + E(t) + I(t) = N$, we can find possible solutions $I_0(t) = 0$; or $I^*(t) > 0$ with $S^*(t) = \gamma/\lambda$. In the first case, we have a solution of the equation system that gives us an attack-free equilibrium point $P_0 = (N, 0, 0)$. In the second case, $I^*(t) > 0$, we obtain the equilibrium point

$$P^* = \left(\frac{\gamma}{\lambda}, \frac{\gamma(N \cdot \lambda - \gamma)}{\lambda(\gamma + \varepsilon)}, \frac{\varepsilon(N \cdot \lambda - \gamma)}{\lambda(\gamma + \varepsilon)}\right)$$

which represents the endemic equilibrium.

On the other hand, in order to obtaining the basic reproductive number, we need to solve this system eliminating dt and integrating both sides of equations, finding the solution curves in the phase plane [24]. Since R_0 represents the average number of secondary infections that occur when an infectious individual is introduced into the population of susceptible nodes, it can be shown that for $R_0 = \lambda{\cdot}\varepsilon{\cdot}N/\gamma$, the dynamics of the proposed system is completely determined by this threshold parameter, so if $R_0 < 1$, the point P^0 is asymptotically stable and therefore the only equilibrium point of the system, and indicates the effects of jamming attack dies-out with time; whereas if $R_0 > 1$ the equilibrium point P^* is asymptotically stable, which may be looking at a case in which the jamming attack will spread through the network. The rigorous mathematical analysis of local and global stability of these points can be performed using well known functions as Lyapunov, or applying linearization techniques and *Jacobian* operators [25, 26], which is out of the scope of this paper.

4 Simulations

In this paper, we used MATLAB to simulate the proposed epidemiological model. We assume an area $A = 75 \times 75$ m, assuming a uniform distribution of $N = 150$ nodes. In addition we establish a minimum radius r_0 of 25 m, to ensure that all nodes have at least one neighbour within its coverage, thus obtain the following expression of the basic reproductive number:

$$R_0 = \frac{\lambda \varepsilon N}{\gamma} = \frac{\pi r_0^2 \beta}{A \gamma} N$$

This allows us to determine the threshold value that makes the attack becomes epidemic, depending on the infection rate β, the average incubation period of the disease ε and the average duration of time infection γ. In order to observe the influence of each parameter on the propagation of the attack, we have performed numerical simulation using for β, γ and ε the values derived from reports or studies commonly used in the simulation of such networks. We will assume that the infection starts with an attacker node $I(0) = 1$, so that $S(0) = N–I(0)$ and $E(0) = 0$. Figure 1a illustrates the propagation in a WSN with $\beta = 0.03$, an average latency of the disease $\varepsilon = 1/8$, and a period of infection $\gamma = 1/10$, yielding $R_0 < 1$ ($R_0 = 0.19635$), where the number of infected nodes stabilizes at relatively low value. Figure 1.b shows the WSN with $R_0 > 1$ ($R_0 = 1.9635$) producing the infection of a large number of nodes. Similarly, Fig. 1.c shows the influence of ε where an increment in a factor of 10 produces a higher infection than that obtained for the transmission rate β. Figure 1.d shows that increasing the average disease time latency causes the number of nodes which pass to the infected group increases, reducing the number of exposed, and favouring the spread of jamming attack.

It is important to note that if we know the values of β, γ, ε, the equation of R_0 allows us to adjust the maximum transmission range r_0 to guarantee $R_0 < 1$, and therefore that the WSN does not fall under the effect of an epidemic, as it is shown in Fig. 2. That critical radio can be obtained as:

$$R_0 = \frac{\pi r_0^2 \beta \varepsilon}{A \gamma} N = 1 \Rightarrow r_c = \sqrt{\frac{A \gamma}{\pi \beta \varepsilon N}} \tag{4}$$

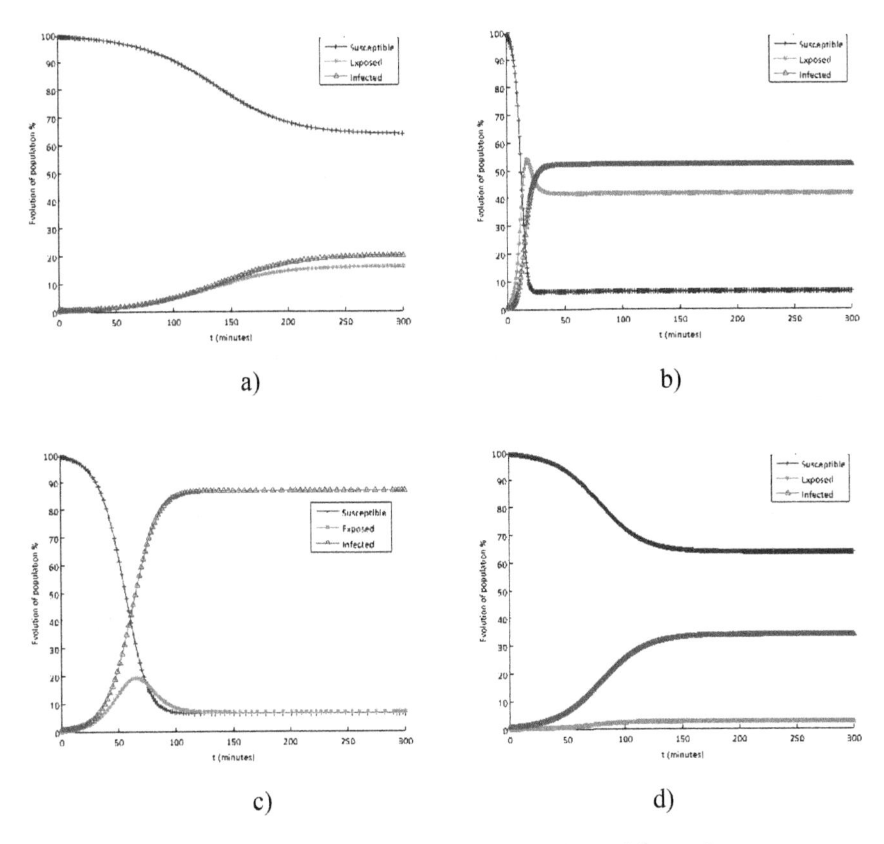

Fig. 1. Analysis of WSN for different values of β, γ and ε

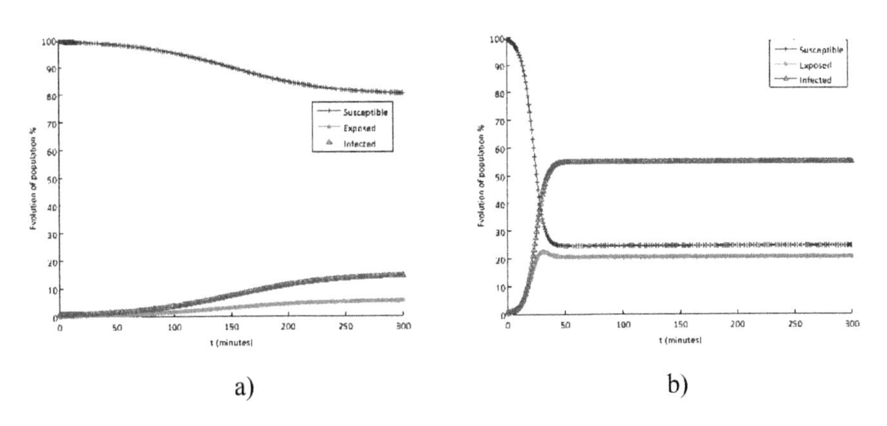

Fig. 2. Analysis of WSN for different values of r_0

5 Conclusion

This paper has shown a first approach to the study of jamming propagation in their random version, by applying the SEIS epidemiological model, analysing the system balance and stability. The values obtained provide us with a threshold given by the basic reproductive number R_0, which indicates the higher the value, the more difficult it will contain the jamming attack. This analysis provides an important tool for predicting the attack effects and modelling possible countermeasures. For example, basic epidemiological models consider that the proportion of the population needs to be vaccinated in order to prevent the sustained spread of a particular disease is given for $1-1/R_0$, thus knowing the value R_0 associated with a particular type of jamming, we could develop a security mechanism that can detect and act, for example by dynamic routing nodes, or even by disabling certain nodes, in order to mitigate the effects of the attack.

Acknowledgements. This work has been supported by the project "Protocolos criptográficos para la ciberseguridad: identificación, autenticación y protección de la información (ProCriCiS)" (TIN2014-55325-C2-1-R) of the Ministry of Economy and Competitiveness and FEDER funds.

References

1. Kermack, W.O., McKendrick, A.G.: Contributions to the mathematical theory of epidemics, part I. Proc. Roy Soc Edin A **115**, 700–721 (1927)
2. Kephart, J.O., White, S.R.: Directed-graph epidemiological models of computer viruses. In: Proceedings of IEEE Symposium on Security and Privacy, pp. 343–359 (1991)
3. Azni, A.H., Ahmad, R., Mohamad Noh, Z.A.: Correlated node behavior in wireless ad hoc networks: an epidemic model. In: Proceedings of 7th International Conference for Internet Technology and Secured Transactions, pp. 403–410 (2012)
4. De, P., Liu, Y., Das, S.K.: Deployment-aware modeling of node compromise spread in wireless sensor networks using epidemic theory. ACM Trans. Sensor Networks **3**, 1–33 (2009)
5. Keshri, N., Mishra, B.K.: Optimal control model for attack of worms in wireless sensor network. Int. J. Grid Distrib. Comput. **7**, 251–272 (2014)
6. Mishra, B.K., Keshri, N.: Mathematical model on the transmission of worms in wireless sensor network. Appl. Math. Model. **37**, 4103–4111 (2013)
7. Mishra, B.K., Srivastava, S.K.: A quarantine model on the spreading behavior of worms in wireless sensor network. Trans. IoT Cloud Comput. **2**, 1–12 (2014)
8. Tang, S.: A modified epidemic model for virus spread control in wireless sensor networks. Int. J. Wirel. Inf. Networks **18**, 319–326 (2011)
9. Xiaoming, W., Yingshu, L.: An improved SIR model for analyzing the dynamics of worm propagation in wireless sensor networks. Chin. J. Electron. **18**, 8–12 (2009)
10. Martín del Rey, A.: Mathematical modeling of the propagation of malware: a review. Secur. Commun. Networks **8**(15), 2561–2579 (2015)
11. Chowdhury, M., Kader, M.F.: Asaduzzaman: security issues in wireless sensor networks: a survey. Int. J. Future Gener. Commun. Networking **6**, 97–116 (2013)

12. Francillon, A., Castelluccia, C.: Code injection attacks on harvard-architecture devices. In: Proceedings of the 15th ACM conference on Computer and communications security, pp. 15–26. ACM (2008)
13. Habibi, J., Gupta, A., Carlsony, S., Panicker, A., Bertino, E.: Mavr: code reuse stealthy attacks and mitigation on unmanned aerial vehicles. In: Distributed Computing Systems (ICDCS), pp. 642–652. IEEE (2015)
14. Braden, K., Crane, S., Davi, L., Franz, M., Larsen, P., Liebchen, C., Sadeghi, A.R.: Leakage-resilient layout randomization for mobile devices. In: Network and Distributed Systems Security Symposium (NDSS) (2016)
15. Modares, H., Moravejosharieh, A., Salleh, R., Lloret, J.: Security overview of wireless sensor network. Life Sci. J. **10**, 1627–1632 (2013)
16. Znaidi, W., Minier, M., Babau, J.P.: An ontology for attacks in wireless sensor networks. Unité de recherche INRIA Rhône-Alpes, Rapport de recherche N° 6704, pp 1–13 (2008)
17. Sokullu, R., Korkmazy, I., Dagdevirenz, O., Mitsevax, A., Prasad, N.R.: An investigation on IEEE 802. 15. 4 MAC layer attacks. In: Proceedings of The 10th International Symposium on Wireless Personal Multimedia Communications (WPMC) (2007)
18. Guglielmo, D., Brienza, S., Anastasi, G.: IEEE 802.15.4e: A survey. Comput. Commun. **88**, 1–24 (2016)
19. Hethecote, H.W.: The mathematics of infectious diseases. SIAM Rev. **42**, 599–653 (2000)
20. Brauer, F., van den Driessche, P., Wu, J.: Mathematical Epidemiology. Springer, Heidelberg (2008)
21. Lichtman, M., Poston, J.D., Amuru, S., Shahriar, C., Clancy, T.C., Buehrer, R.M., Reed, J. H.: A communications jamming taxonomy. IEEE Secur. Priv. **14**, 47–54 (2016)
22. Wei, Y., van Hoesel, L.L., Doumen, J., Hartel, P., Havinga, P.: Energy efficient link layer jamming attacks against wireless sensor network MAC protocols. In: SANS 2005 (2005)
23. Mohammadi, S., Jadidoleslamy, V.: A comparison of link layer attacks on wireless sensor networks. Int. J. Appl. Graph Theory Wireless ad hoc Networks Sensor Networks **3**, 35–56 (2011)
24. Zhu, L., Zhao, H.: Dynamical analysis and optimal control for a malware propagation model in an information network. Neurocomputing **149**, 1370–1386 (2015)
25. Hirsch, M.W., Smale, S., Devaney, R.L.: Differential equations, dynamical systems, and an introduction to chaos. In: 3rd Ed., Academic Press (2012)
26. Perko, L.: Differential Equations and Dynamical Systems, 3rd edn. Springer, Heidelberg (2010)

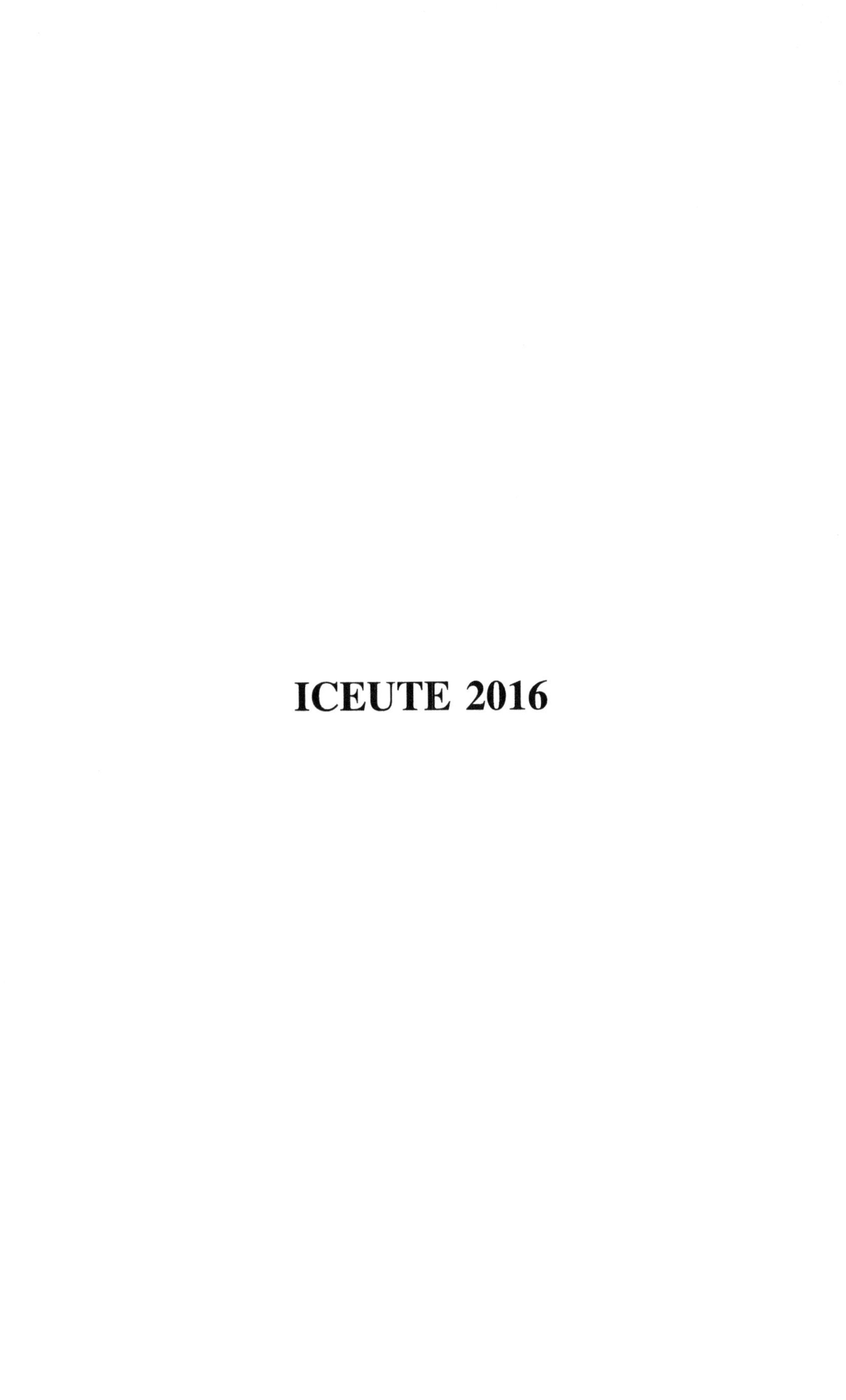

ICEUTE 2016

Educational Big Data Mining: How to Enhance Virtual Learning Environments

Pietro Ducange[1(✉)], Riccardo Pecori[1], Luigi Sarti[1,2], and Massimo Vecchio[1]

[1] SMART Engineering Solutions and Technologies (SMARTEST) Research Centre, eCampus University, Via Isimbardi 10, 22060 Novedrate (CO), Italy
pietro.ducange@uniecampus.it

[2] Istituto per le Tecnologie Didattiche del Consiglio Nazionale delle Ricerche, Via de Marini 6, 16149 Genova (GE), Italy

Abstract. The growing development of virtual learning platforms is boosting a new type of Big Data and of Big Data Stream, those ones that can be labeled as *e-learning Big Data*. These data, coming from different sources of Virtual Learning Environments, such as communications between students and instructors as well as pupils tests, require accurate analysis and mining techniques in order to retrieve from them fruitful insights. This paper analyzes the main features of current e-learning systems, pointing out their sources of data and the huge amount of information that may be retrieved from them. Moreover, we assess the concept of educational Big Data, suggesting a logical and functional layered model that can turn to be very useful in real life.

Keywords: Big Data · Educational data mining · e-learning · Virtual Learning Environment

1 Introduction

E-learning, literally electronic learning, refers to a set of virtualized distance learning techniques and technologies that exploit electronic communication mechanisms and use their functions as a support in both teaching and learning procedures. Education institutions have been increasing more and more the usage of Virtual Learning Environments (VLEs), mainly fostered by the development of the so called *Society 2.0*, a neologism to define the interconnected world where we currently live, characterized by the bandwidth evolution in telecommunications technologies and their availability at very low costs anywhere and anytime. Mobile phones, tablets, laptops but also traditional desktop computers, and every other technological instrument able to communicate, allow users to handle a huge amount of educational data, such as videos, tests, homework, comments, tags, and statistics. These are instruments that both current and future students, the so-called "millenials" and "digital natives", extensively use day by day; therefore their exploitation for learning purposes is straightforward and eased by their inherent willingness of sharing information. Anyway, besides

M. Graña et al. (eds.), *International Joint Conference SOCO'16-CISIS'16-ICEUTE'16*, Advances in Intelligent Systems and Computing 527, DOI 10.1007/978-3-319-47364-2_66

the improvements in technological devices, the always growing demand for alternative ways of learning is driven also by the necessities and functions required by non-traditional students such as full-time or part-time working people, with no chances to attend traditional lectures, and impaired or injured students, forced at home. Moreover, there may be other potential learners with special educational needs or learning disabilities and students living abroad. The advantages of e-learning systems are multifaceted: flexibility, ease of usage, costs reduction, cooperation and collaboration among students and between instructors and pupils [1], repeatability and consistency control, friendly environment, presence of common areas of knowledge, availability of on-line tutoring services [2], progress monitoring, community and group creation and management; anyway the main pro is the possibility for a potentially infinite number of students to attend lessons and classes.

As the learning process becomes more and more computerized, a great amount of data is going to be generated and stored during the years. These pieces of information can be considered, to all intents and purposes, Big Data as they are surely characterized by some of their main dimensions: volume, variety, velocity, veracity, value and so forth [3]. As a consequence Big Data mining techniques and instruments are becoming fundamental in order to usefully employ the information generated by VLEs and to enrich the learning model of academic institutions through the intrinsic value of this information. This has resulted in the need and penetration of Big Data technologies and tools into education, to process the large amount of data involved and to promote various improvements such as shortening response time, detecting early drop outs [4], scaffolding, and so on. This process gave rise to the so-called "Educational Data Mining" (EDM) [5], allowing for the analysis and visualization of data, for creating feedback supporting instructors, for recommendations and forecasts of students' performances, for group creation, for courseware constructing and students' behavior detection, and so forth.

However, the main issues of distance learning environments and data mining techniques regard the requirement for *increasing scalability*, both at infrastructure and mining algorithm levels, *real-time responsiveness*, for quick responses towards students' requirements and needs, and *adaptivity*, as periods of workload peaks may rapidly follow more "quiet" times. Other issues concern costs control, resource management, as well as security and privacy of data [6].

Considering the recent literature, authors of [7] suggest the integration of cloud computing platforms for distance learning environments together with EDM techniques, whereas the contribution in [8] provides a survey of Big Data technologies in learning systems concluding that the major investigated topic of current researches is "Performance Prediction using Data Mining". Regarding security aspects, the contribution in [9] proposes an innovative trustworthiness-based methodology, similar to the one in [10], for increasing data security in Computer-Supported Collaborative Learning environments.

Moreover, always in [9], the authors highlight the important difference between EDM, which mainly regards the development, the research, and the application of automated methods to detect patterns in large sets of educational data, and learning analytics, which simply concerns collection, measurement and analysis of data about learners and instructors (professors, lecturers, tutors). Some traditional mining techniques have been already applied to educational data [11], but EDM is still an emerging research field. In this paper we try to highlight the fruitful usage of some possible data mining techniques to a layered e-learning scenario.

The rest of the paper is composed by Sect. 2 regarding VLE data sources, Sect. 3 concerning Big Data technologies, Sect. 4 suggesting effective deployments of Big Data and Big Stream technologies for enhancing distance learning environments and, finally, by Sect. 5 that summarizes the work with some conclusions and outlines future perspectives.

2 Data Sources in Virtual Learning Environments

In this section, we discuss the main sources of data that are usually produced and managed by VLE platforms of distance learning institutions. Avgeriou et al. [12] describe a general architecture of a VLE system, identifying four layers, as shown in Fig. 1, that can be considered macro-sources of data:

- Application-specific subsystems and modules that are specific of the VLE and are not meant to be shared or reused across different applications:
  - user profile management, including students portfolios and grades, system usage statistics, working groups, assignments and so on;
  - course management: creation, customization, administration and monitoring of courses;
  - educational resource delivery: tracking of learning material usage, assessment results;

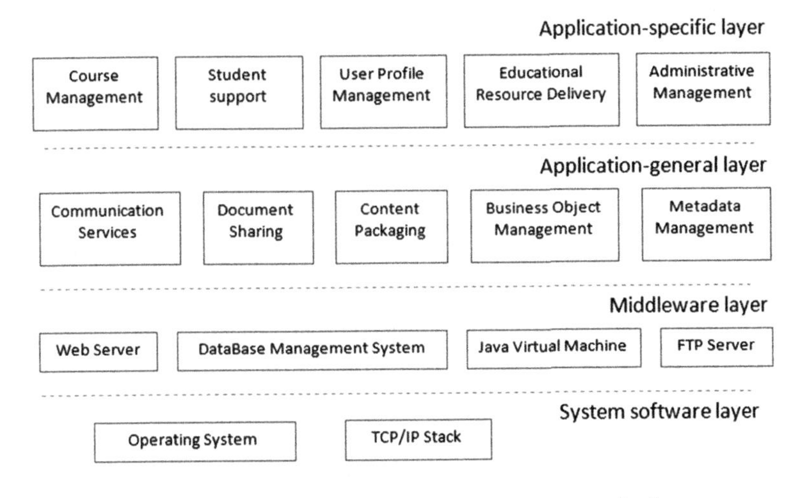

Fig. 1. A VLE layered model (adapted from [12]).

  - administrative management: registration, authentication, student records and logs, access rights, views, etc.;
  - student support: private and shared annotation spaces, bookmarks, notepads, statistics, recommendations, help-desk.
- Application-general subsystems and modules that are not specific of the VLE and whose services can be reused by a number of applications:
  - communication services: email, chat, asynchronous discussion forums and blogs, audio and video conferencing, synchronous facilities such as shared whiteboards and shared desktops;
  - document sharing: file versioning;
  - content packaging: SCORM (Shareable Content Object Reference Model) containers [13];
  - business object management: connection with the database, persistent object factory;
  - meta-data management: user tagging, search engine.
- Middleware reusable building blocks that offer utility services, platform independence and the like: JVM, APIs, DBMS, Web server and system administrator tool-kits.
- System software like operating systems, TCP/IP stack, etc.

As specific example of VLE, we provide a brief description of the operational platform of the eCampus University[1], the Italian distance learning university to which the authors are affiliated.

The platform features a broad range of functions, some of which are particularly interesting from the educational data sources perspective:

- *Lessons*: the learning resources are articulated into lessons and study sessions made of slides, digital documents and videos. All the interactions between the learner and the resources are tracked, in conformity with the SCORM Runtime Environment specifications.
- *Tests*: formative and summative evaluation of the learner's progresses can be performed online; assessments can be carried out online and the results are recorded in the learners' profile.
- *Monitoring tools*: besides lesson and test delivery, all learners' online activities are tracked, including login and logout to the platform, document up- and down-load, contributions to asynchronous (forums) and synchronous (chat) communication and sharing channels.
- *Student records and e-portfolios*: exam results, essays and other student-produced material.
- *e-tivities*: concept maps, wikis and other collaborative and cooperative instruments.

In conclusion, we can state that VLEs produce a large amount of data that can be handled by both predictive and descriptive models: such models may provide suggestions, recommendations and indications to all the actors involved in a distance learning process. Clearly, this huge amount of data calls for specific techniques to be handled for educational data mining purposes.

[1] www.uniecampus.it.

3 Gathering, Storing and Managing Big Data

We talk about Big Data whenever the normal application of current technologies does not enable users to obtain cost and quality effective answers, to data-driven questions, in an acceptable time interval. Indeed, handling the total amount of data coming from VLEs may be a hard and ineffective task if distance learning institutions continue to employ the classical technologies and paradigms for programming and storage. For this reason, more suitable technologies must be adopted for dealing with the main features, namely the velocity, the variety and the volume [3], which characterize Big Data.

As regards the programming paradigms, MapReduce [14] is one of the most important: it was defined and patented by Google. MapReduce is mainly based on the Map and Reduce Tasks, which work on data organized in a distributed file system: input data are mapped to key values that are combined and grouped together, according to similar characteristics, and finally reduced to a minimum set of output values [14]. Apache Hadoop [15] represents one of the most popular frameworks implementing the MapReduce paradigm: its success is due to its performances, open source nature, installation facilities and its distributed file system. In particular, the Hadoop Distributed File System (HDFS) spreads multiple copies of the data chunks into different machines and allows parallel processing based on the MapReduce paradigm. Since Hadoop may be inefficient in some situations, such as repeating the same searches, the Apache Spark framework may be used as an alternative [16]. Spark is a cluster computing framework and allows for reusing a working set of data across multiple parallel operations. On the other hand, Spark cannot be used standalone but it requires an external cluster manager and a third-party distributed storage system, like HDFS for example. As regards the storage issue, MongoDB [17], which is a document-based database system storing data in a JSON-like fashion, may be suitable for organizing Big Data in a distributed environment. It is a NoSQL database that exploits the key-value paradigm for fast queries that can be distributed across clusters of computers.

When a large amount of data arrives with a very high speed, such as in a VLE, we are dealing with streams of information. In this case, Hadoop, Spark and MongoDB are surely not enough for mining useful knowledge from VLEs. Apache Storm [18] or Apache Samza [19] are modern frameworks that may be more suitable for handling data streams that are quickly produced, change very fast and need a real-time analysis. Storm allows to quickly analyze big streams of data. It is based on a distributed real-time computation system, which employs a master-slave approach: the architecture includes both a complex event processor and a distributed computation framework. Samza [19] is a framework processing stream messages as they arrive, one at a time: streams are divided into partitions that are in reality an ordered sequence of read-only messages.

4 Big Data Mining for Enhancing VLEs

As said in Sects. 2 and 3, the great amount of users of educational systems generates a huge quantity of educational data that can be in turn stored in opportune databases for future mining or log purposes (Static Big Data) or analyzed on-the-fly (Big Data Streams). Big Data interpretation and analysis are becoming fundamental as, also in the educational scenario, the variability of the usage of the different data sources, such as blogs, chats, forums, videos, and the like, can generate high peaks of application and underutilized periods that must be properly managed to foster positive outcomes. This is witnessed also by the work in [20] that highlights the importance, in an educational framework, of recent learning data rather than the whole past history.

In this section, we give a glimpse on how data mining approaches, applied to e-learning big data, may provide useful advantages at the different levels of actors involved in a distance learning process. In the following, we take inspiration from the discussions on learning analytics provided in [21] and refer to the model in Fig. 2 to describe the benefits at different layers. As shown in Fig. 2, both static and stream data can be collected and pre-filtered integrating in the VLEs the technologies discussed in Sect. 3. First, such data may be used for extracting useful knowledge in the specific e-learning domain: both descriptive (for clustering and pattern mining purposes) and predictive models (for classification and prediction issues) can be generated. Then, these models may be used for giving suggestions, indications and advices to all the actors involved in a VLE through data mining techniques. In this perspective, a number of data mining libraries, namely Mahout, MLib and SAMOA, are already available for Hadoop, Spark, and Storm/Samza, respectively [22]. In the following, we discuss how the extracted knowledge may be exploited at the various levels of an e-learning framework.

At a *macro-level*, cross-institutional data mining models may provide useful indications to policy makers on current institutional practices and processes. This may concern the evaluation of the national learning system as a whole but it may also encompass the ranking of educational institutions on the basis of

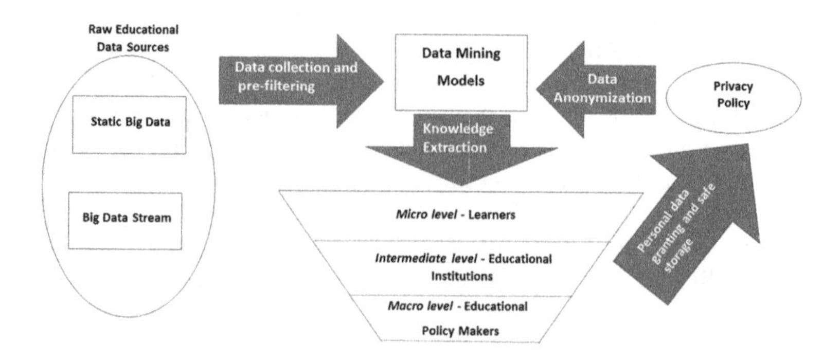

Fig. 2. Big Data mining applied to e-learning environments.

certain metrics such as the number of drop outs, the number of new enrolments, etc., with the purpose of funding best performing institutions through public grants, enlarge their chance to enroll students and employ professors, and so on.

At an *intermediate level*, predictive and descriptive models may help faculty members and stakeholders to better manage both students and instructors' activities and performances, as well as planning and scheduling courseware. Regarding learners, by means of clustering algorithms, instructors can visualize different groups of students and create various student profiles. This can be achieved on the basis of students' periodic feedbacks, surveys and official performances. Moreover, classical information such as previous backgrounds, typology of student (worker or full-time), can be exploited for building student clusters and profiles. In this way, instructors can adapt different teaching and learning strategies on the needs of the students, create personalized learning paths and so on. Performance prediction models can also foster new teaching methodologies in order to enhance both the quality and the outcomes of the courses: collaborative and cooperative instruments such as wikis, concept maps, webquests, virtual desks and so on, may be deployed to make needing students more involved in the learning process and so promote their knowledge according to their specific profile (worker, disabled, elderly people, etc.). Moreover, attrition risk prediction can enable lowering of early drop outs, make easier the understanding of the motivation of withdrawls and make possible to promptly enact proper retaining countermeasures. These can be based on peer-to-peer team working, students' passion-driven education methodologies and other educational strategies that entail not only traditional cognitive techniques but also constructivist and connectivist ones. On the other hand, data mining techniques can be deployed to monitor instructors performances and the effectiveness of their scaffolding role. For example, their ability to follow the students' learning needs may be inferred through the so called text mining and sentiment analysis of blogs, posts, forums, and the like. Conversely their compliance to the management guidelines may be assessed through other aggregated indicators to ease the process, on one side, of granting permanent contracts or productivity awards, or, on the other side, to make provision for penalties or layoffs. Finally recommender systems may boost discussions among instructors deploying similar teaching strategies, in order to collectively exchange and enhance ideas and good practices of educating and explaining subjects [23].

At a *micro level*, data mining models support learners in various ways. First of all, the personalization of the learning path can be made according to the students or groups of students' own needs and necessities. For instance, a map can be built on the basis of individual learning patterns; this can be a measure of the cognitive and social presence [24] of each learner in forums and blogs and it can provide indications on the evolution of individual and group learning processes. Another important aspect is the identification of critical skills or deficits that could be opportunely improved and boosted through particular and ad-hoc learning techniques. Also in this case, recommender systems could be opportunely re-adapted to the educational scenario in order to suggest to the students the courses, the learning groups and the instruments (webinars, forums, virtual office hours, etc.) that best fit their own skills, enhancing the aforementioned constructivist and connectivist views of education. This can be achieved both considering the past experiences of a certain student and all previous interactions of students with a similar profile. Finally, predictive models can result very useful in suggesting future educational and working routes. For example, they can highlight which graduate or post graduate course to attend in the near future or which are the student's main winning soft skills and aptitudes in order to have success in the labour market.

An important aspect, on the right of the model proposed in Fig. 2, concerns privacy and security of educational data. These are concepts intimately connected to trustworthiness, a fundamental prerequisite of any successful e-learning environment [6], and they can be analyzed from both the students and the institutions point of view, between which an opportune balance should be stricken. The students should be encouraged to grant access to their personal data as institutions and instructors, respecting privacy and safe storage regulations, may take advantage of this private information to create more and more trained models of the students themselves, in a multifaceted way encompassing both professional and personal life data. Data mining techniques may guarantee a higher level of privacy and security, both integrating advanced encryption techniques and considering aggregated data when dealing with sensitive information, and this in turn may foster students' willingness to share more and more anonymized data, which are precious sources for tailoring opportune learning strategies, and to perform collaborative and cooperative tasks together with peers.

Finally it should be stressed the importance of choosing the most relevant and significant metrics in the data mining techniques deployed and the significance of their qualitative understanding. This sheds light on the interpretation of the data mining results that should be performed by human beings through an epistemological approach that goes beyond the traditional empirical and positivist one. It is fundamental to pay attention to the whole learning process of the learners considering them as precious human beings and not as simple enrolment numbers.

In conclusion, we suggest to distance learning institutions to re-design their VLEs, including the aforementioned technologies for handling Big Data and building data mining models. Moreover, it is also fundamental to guarantee in the staff of such institutions the presence of qualified figures, such as system administrators and data scientists, trained for working with e-learning Big Data.

5 Conclusions and Future Directions

In this paper, we have analyzed the main sources of information in Virtual Learning Environments and summarized the main benefits that Big Data mining techniques can provide in such learning scenarios. We discussed a layered model that encompasses also security and privacy features. In this perspective, we have shed light on the different levels where Big Data mining can take place and on the respective benefits. We have also stressed the significance of going beyond pure numerical data in order to have an epistemological interpretation of the new knowledge derived from educational data and the importance for the students to trust instructors and institutions in order to gather as many data as possible to opportunely train mining techniques.

A possible future direction of investigation may regard the application of mining techniques from the e-learning scenario to the m-learning one, where mobility must be accurately taken into consideration. In this scenario, an important characteristic to be considered is surely the ability to analyze and get results in real time. Therefore the Big Data Stream mining techniques may conquer more and more a major part in an educational framework that could be based more and more on informal learning strategies endowed with brief summaries of contents. Finally, another possible future perspective may regard the application of Big Data mining procedures to the e-learning activities and platforms of specific social networks, like the one recently launched by Linkedin: the legacy Lynda[2].

References

1. Schneider, B., Blikstein, P.: Unraveling students interaction around a tangible interface using multimodal learning analytics. J. Educ. Data Min. **7**(3), 89–116 (2015)
2. Clement, B., Roy, D., Oudeyer, P.Y., Lopes, M.: Multi-armed bandits for intelligent tutoring systems. J. Educ. Data Min. **7**(2), 20–48 (2015)
3. Vidal Alonso, O.A.: Big data and elearning: a binomial to the future of the knowledge society. Int. J. Interact. Multimedia Artif. Intell. **3**(6), 29–33 (2016)
4. Knowles, J.E.: Of needles and haystacks: building an accurate statewide dropout early warning system in Wisconsin. J. Educ. Data Min. **7**(3), 18–67 (2015)
5. Hegazi, M.O., Abugroon, M.A.: The state of the art on educational data mining in higher education. Int. J. Comput. Trends Techn. **31**(1), 46–56 (2016)
6. Ivanova, M., Grosseck, G., Holotescu, C.: Researching data privacy models in eLearning. In: 2015 International Conference on Information Technology Based Higher Education and Training, pp. 1–6. IEEE, June 2015
7. Fernández, A., Peralta, D., Benítez, J.M., Herrera, F.: E-learning and educational data mining in cloud computing: an overview. Int. J. Learn. Technol. **9**(1), 25–52 (2014)
8. Sin, K., Muthu, L.: Application of big data in education data mining and learning analytics-a literature review. ICTACT J. Soft Comput. Spec. Issue Soft Comput. Models Big Data **5**(4), 1035–1049 (2015)
9. Miguel, J., Caballe, S., Xhafa, F.: A knowledge management process to enhance trustworthiness-based security in on-line learning teams. In: 2015 International Conference on Intelligent Networking and Collaborative Systems, pp. 272–279 (2015)
10. Pecori, R., Veltri, L.: Trust-based routing for kademlia in a sybil scenario. In: 22nd International Conference on Software, Telecommunications and Computer Networks (SoftCOM), pp. 279–283. IEEE (2014)

[2] http://www.lynda.com/.

11. Nithya, P., Umamaheswari, B., Umadevi, A.: A survey on educational data mining in field of education. Int. J. Adv. Res. Comput. Eng. Technol. **5**, 69–78 (2016)
12. Avgeriou, P., Retalis, S., Skordalakis, M.: An architecture for open learning management systems. In: Manolopoulos, Y., Evripidou, S., Kakas, A.C. (eds.) PCI 2001. LNCS, vol. 2563, pp. 183–200. Springer, Heidelberg (2003). doi:10.1007/3-540-38076-0_13
13. Bohl, O., Scheuhase, J., Sengler, R., Winand, U.: The sharable content object reference model (scorm) - a critical review. In: Proceedings of IEEE 2002 International Conference on Computers in Education, pp. 950–951 (2002)
14. Dean, J., Ghemawat, S.: Mapreduce: simplified data processing on large clusters. Commun. ACM **51**(1), 107–113 (2008)
15. White, T.: Hadoop: The Definitive Guide, 4th edn. O'Reilly Media, Sebastopol (2015)
16. Zaharia, M., Chowdhury, M., Franklin, M.J., Shenker, S., Stoica, I.: Spark: cluster computing with working sets. In: Proceedings of the 2nd USENIX Conference on Hot Topics in Cloud Computing, p. 10 (2010)
17. Chodorow, K.: MongoDB: The Definitive Guide. O'Reilly Media, Sebastopol (2013)
18. Evans, R.: Apache storm, a hands on tutorial. In: 2015 IEEE International Conference on Cloud Engineering, p. 2 (2015)
19. Feng, T., Zhuang, Z., Pan, Y., Ramachandra, H.: A memory capacity model for high performing data-filtering applications in Samza framework. In: 2015 IEEE International Conference on Big Data, pp. 2600–2605 (2015)
20. Galyardt, A., Goldin, I.: Move your lamp post: recent data reflects learner knowledge better than older data. J. Educ. Data Min. **7**(2), 83–108 (2015)
21. Shum, B.: Learning analytics policy brief. UNESCO Institute for Information Technology in Education (2012)
22. Landset, S., Khoshgoftaar, T.M., Richter, A.N., Hasanin, T.: A survey of open source tools for machine learning with big data in the hadoop ecosystem. J. Big Data **2**(1), 1–36 (2015)
23. Drachsler, H., Verbert, K., Santos, O.C., Manouselis, N.: Panorama of recommender systems to support learning. In: Ricci, F., Rokach, L., Shapira, B. (eds.) Recommender Systems Handbook, pp. 421–451. Springer, New York (2015)
24. Pozzi, F., Manca, S., Persico, D., Sarti, L.: A general framework for tracking and analysing learning processes in computer-supported collaborative learning environments. Innovations Educ. Teach. Int. **44**(2), 169–179 (2007)

Application of the PBL Methodology at the B.Sc. in Industrial Electronics and Automation Engineering

Isidro Calvo[1(✉)], Jeronimo Quesada[2], Itziar Cabanes[3], and Oscar Barambones[1]

[1] Department of Systems Engineering and Automatic Control, University College of Engineering of Vitoria-Gasteiz, (UPV/EHU), Vitoria-Gasteiz, Spain
{isidro.calvo,oscar.barambones}@ehu.eus
[2] Department of Electronics Technology, University College of Engineering of Vitoria-Gasteiz, (UPV/EHU), Vitoria-Gasteiz, Spain
jeronimo.quesada@ehu.eus
[3] Department of Systems Engineering and Automatic Control, Faculty of Engineering of Bilbao, (UPV/EHU), Bilbao, Spain
itziar.cabanes@ehu.eus

Abstract. This work describes one application of the PBL (Project Based Learning) methodology at the curriculum of the "Industrial Informatics" module of the B.Sc. Degree in Industrial Electronics and Automation, taught at the University College of Engineering of Vitoria-Gasteiz. The choice of this methodology is based on the project orientation expected for the future engineers. Authors intended to reproduce, at reduced scale, the problematic of working in multidisciplinary teams with strict completion times. This approach forced students to get involved in the learning process while carrying out the tasks. During the process, students detected the learning needs by themselves in order to accomplish the project, and had to learn how to apply, in a proactive and autonomous way, the necessary techniques during the project implementation.

Instructors proposed implementing a controller for a SCARA robot, which is a typical configuration found in industrial environments. Students take the "Industrial Informatics" module (first semester) previously to the "Robotics" module (second semester) of the third year, so the authors proposed a simplified configuration for the SCARA robot with only two degrees of freedom (2DoF). The scale model of the robot was built with the LEGO Mindstorms NXT kit, which provides an interesting flexibility/price compromise. This approach forced students to apply concepts acquired in previous modules, as well as skills and techniques that will be of use in the future.

Keywords: Collaborative learning · Active methodologies · Course design and curriculae · Supervising and managing student projects · Project based learning (PBL) · Robotics · Lego mindstorms

M. Graña et al. (eds.), *International Joint Conference SOCO'16-CISIS'16-ICEUTE'16*, Advances in Intelligent Systems and Computing 527, DOI 10.1007/978-3-319-47364-2_67

1 Introduction

This work describes the application of the Program Based Learning (PBL) methodology at the "Industrial Informatics" subject during the first semester of the third year at the B.Sc. Degree in Industrial Electronics and Automatics taught in the University College of Engineering of Vitoria-Gasteiz (UPV/EHU). The authors aimed at solving some of the problems identified by [6] at similar subjects, namely: (1) Lack of motivation of the students; (2) Poor attendance to the classroom after the first sessions and (3) Leaving the required tasks for the last weeks. The application of active methodologies at the "Industrial Informatics" subject continued previous experiences by the same authors [5, 19]. In our opinion the Project Based Learning methodology (PBL) allows adopting fundamental competences [2, 4, 14] that future engineers will be required in their future professional life. As a matter of example, engineers will have to develop engineering projects in multidisciplinary teams while satisfying strict requirements and deadlines. By using PBL authors intended to reproduce, at small scale, the same situations that the students will face in their future profession. This approach forces students to get involved in the educative process. E.g. in order to develop the proposed project they must detect their learning needs, and learn how to apply some techniques in a proactive and autonomous way.

PBL is a constructivist educational approach that organizes curriculum and instruction around carefully crafted "ill-structured" problems [3]. Guided by teachers acting as cognitive coaches, students develop critical thinking, problem solving and collaborative thinking as they identify problems, formulate hypotheses, conduct data searches, perform experiments, formulate solutions and determine the best "fit" of solutions to the conditions of the problem. PBL enables students to embrace complexity and enhance their capacity for creative and responsible real-world problem solving [15]. PBL methodology has been applied at different educative levels from elementary school to university education. For example, [16] describe its application to a whole institution. Some university networks use them coordinately [13]. Application examples in the engineering degrees may be found at [8, 11]. The following works describe successful implementations in: (1) Electronics [1, 12], (2) Electrical Engineering [10]; (3) Power Systems [9], (4) Mechatronics [7]; or (5) Robotics [17].

The authors proposed programming a robot controller in the "Industrial Informatics" subject, since robotics is a motivating discipline for students due to its increasing presence in today's society [17]. In addition, programming a robot is a transversal task that requires the application of concepts learnt in several disciplines, promoting global thinking.

The layout of the current work is as follows: Sect. 2 presents a brief overview of the subject. Subsection 2.3, describes the PBL activity. Section 3 presents some results of the first year of implementation. In Sect. 4, the article draws some conclusions.

2 Description of the "Industrial Informatics" Subject

2.1 Learning Objectives

"Industrial Informatics" is aimed at deepening into the computer concepts used in the modern industrial informatics applications, including industrial communication issues, which with the adoption of the Industry 4.0 and Industrial Internet of Things paradigms are gaining higher relevance. This learning objective is described under TEEO0I10 (*Applied knowledge about industrial informatics and communications*) of the B.Sc. in Electronic Informatics and Automation Engineering. However, other learning objectives found at the verified compliance ANECA report for the degree [18] must be also achieved in this subject (See Table 1). Note that C10, C12, C13 and C14 are transverse learning objectives, to be achieved along several subjects.

Table 1. Learning objectives of the "Industrial Informatics" subject according to [17]

Learning Objective	Description
C3	Knowledge about basic and technological subjects, including new learning new methods and theories aimed at achieving versatile students adaptable to new situations
C4	Ability to solve problems with initiative, decision making, creativity, critical reasoning and knowledge transmission skills
C5	Knowledge to carry out measurements, complex calculations, valuations, expert's reports, task plans and similar works
C10 (*Trans*)	Ability to work in multilingual and multidisciplinary environments
C12 (*Trans*)	Taking responsible and organized attitude at work, favorable at lifelong learning
C13 (*Trans*)	Applying the scientific methodology strategies: Qualitative and quantitative analysis of the situation and problems, consider hypothesis and solutions according to the industrial engineering methodology, esp. industrial electronics
C14 (*Trans*)	Working efficiently in groups, integrating skills and knowledge to make decisions in the scope of industrial engineering, esp. industrial electronics

2.2 Syllabus

This subsection describes the syllabus of the subject. Some topics are only taught in the classroom, whereas others are only developed by means of the PBL methodology, mainly in the laboratory. The PBL project involves mainly *Advanced programming in C* (Topic 3), and *Design of Embedded Systems* (Topic 4). The Lego NXT Mindstorms brick was used at this task, since it *is* an embedded system. It may be programmed in several programming languages but the authors chose NXC since it offers C-like syntax and is available under public license. Due to time restrictions some topics were mainly taught in the classroom, but the acquired knowledge was reinforced through the PBL experience. This was the case of the *Introduction* (Topic 1), which introduced concepts related to computer control, and *Operating Systems* (Topic 2), since during the project the students used the Application Programmer's Interface (API), provided by the firmware which abstracts the hardware from the programming environment. Lastly, in this experience, communication concepts, the core of *Introduction to Computer Networks* and *Industrial Communications* (Topics 5 and 6) were not introduced in the PBL due to lack of time. Following, there is a brief description of the topics of the "Industrial Informatics" subject:

0. **Presentation** ***(Classroom)***: Introduction to the subject. The syllabus, bibliography, objectives and the assessment system are described and explained.
1. **Introduction** ***(Classroom)***: Description of different control based applications. The role of the computer in different kinds of industrial systems. Differences between centralized control and distributed control.
2. **Operating Systems** ***(Classroom)***: Major objectives of the operating systems. Types of operating systems. Components of an operating system. Role of the kernel. Task scheduler.
3. **Advanced programming in C** ***(PBL)***: C programming: Variables. Control flow statements. Modularity and functions. Complex data structs. Use of library functions. Use of the operating system API. Decomposition of a complex problem into modules (*top-down* design). Unitary tests. Building complex software out of modules (*bottom-up* design). Introduction to concurrency.
4. **Programming embedded systems** ***(PBL)***: Difficulties programming embedded systems. Execution cycle at the embedded systems: (1) data acquisition; (2) execution of the control algorithm; (3) actuation
5. **Introduction to computer networks** ***(Classroom)***: Organization of the communication into layers. Description of the OSI reference model. Description of the TCP/IP stack.
6. **Industrial communications** ***(Classroom)***: Data communications in industrial environments. Automation pyramid. Fieldbuses. Computer networks in industrial environments.

2.3 Assessment

Table 2 summarizes the assessment followed at this subject. The PBL activity accounts for 45 % of the total qualification. However some collaborative activities developed in the classroom (10 %) are also closely related to the PBL activity. Another 45 % of the total qualification is obtained by means of individual exams. The C programming exam is carried out at the middle of the term, after the finalization of the first subproject, while the final exam covers the rest of the topics of the subject, operating systems, communications, etc.

The PBL activity was divided into two subprojects that sought different objectives. Students were divided randomly in teams of three people. During the development of both subprojects students must provide several deliverables with short deadlines which are collected by means of Moodle, even though a combination of tools is under consideration [19].

Table 2. Summary of the assesment

Activity type	Percentage
Collaborative activities in the classroom	10
Individual C programming exam	20
First subproject	10
Second subproject	25
Project presentation	10
Final exam during the exam period	25

2.4 First Subproject

The major objectives of the first subproject were: (1) Consolidate C programming concepts; (2) Learn the basics about robotics (classification of robots, forward and inverse kinematics, trajectory planning, etc.); (3) Facing complex projects that require the application of the *divide and conquer* technique; (4) Make students familiarize with Integrated Development Environments (IDE) in order to create applications that will be executed in the same computer where they were created; (5) Learn programming concepts out of scope in the second subproject such as working with files; (6) Adopt modular programming approaches, and reuse the modules (blocks of C code) in the second subproject.

This subproject involved writing a modular C program that computes the movements that a hypothetical polar robot should make in order to achieve a specific trajectory. More specifically, the arithmetic spiral or spiral of Archimedes was chosen (see Fig. 1), since it requires moving both motors simultaneously.

Students should search information in the Internet in order to know how to create this trajectory with the robot motors. Later, they should create a C program which provides a text file as output with the intermediate positions of the motors. This file should be visualized (e.g. with Excel or Matlab) by the students in order to check the

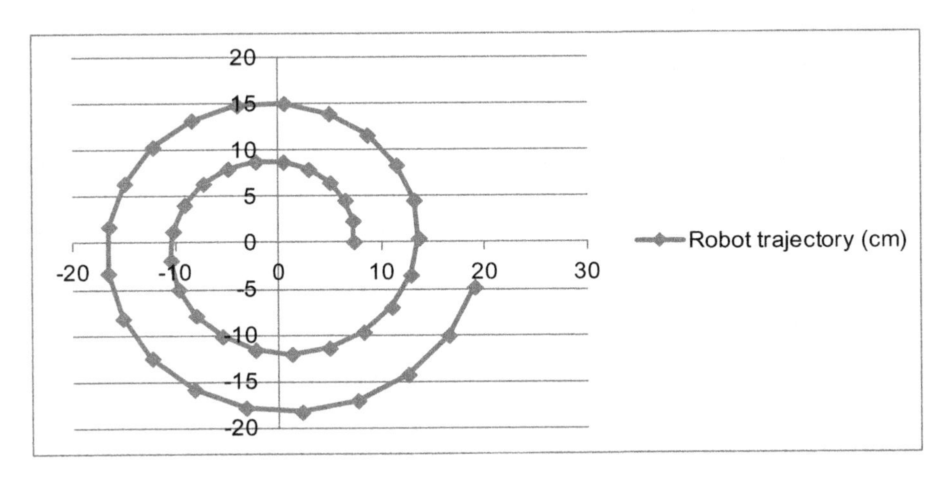

Fig. 1. Sample trajectory obtained with the required simulation C program

success of this subproject (Fig. 1). Even though this subproject was a challenge for the students all groups were very motivated and achieved to solve it in the deadline.

In order to guarantee C programming skills, students had to pass one individual exam in which their programming abilities were tested by means of a computer exam. Passing this exam was a requirement to start the second subproject.

2.5 Second Subproject

This subproject is the core of the PBL activity. The proposed activity was to create a program that allows a specific mechanical LEGO configuration (see Fig. 2) to draw an n-sided regular polygon.

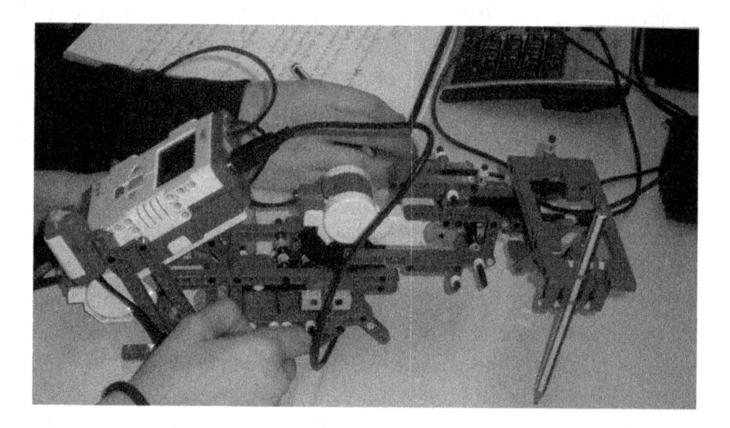

Fig. 2. Detail of the LEGO robot and CPU

For this subproject a planar (or 2° SCARA) robot configuration was used, since it is relatively easy to build its mechanics. However, the trajectory equations are a bit more complex than the equations for the polar robot proposed in the first subproject. Even though in this case the students were provided a base mechanical configuration, they had to design the robot end effector to hold best the pencil, which proved to be a crucial and interesting task.

In this subproject students had to deal with several challenges: (1) Programming an embedded platform; (2) using cross-programming and debugging; (3) accessing inputs and outputs; (4) execute periodically the cycle: Read inputs-Execute the program-Write outputs; (5) solving the problems of non-perfect mechanical systems, such as mechanical vibrations, looseness, etc.; (6) dealing with concurrence.

Also, students had to be very concerned about time management. Actually, this was one key issue. Students should be able to divide the subproject in small tasks that were assigned independently.

Finally, the use of the LEGO platform was very helpful: it allowed the visualization of their programming and incremented the stress on the students since they had to do something that *had to work*. Actually, students got really involved at this stage (see Fig. 3).

Fig. 3. Working atmosphere during the second subproject

2.6 Final Presentation

When the project was finished, students were required to present the work to the rest of the teams. The layout for the presentation was provided as detailed:

1. Project aims
2. Task division and time schedule (Start data – Finish data for every task)
3. Aims completion level
4. Final robot configuration
   (a) Views (Photos): Plan-Elevation-Section + Isometric projection
   (b) Fundamental variables: length of arm 1, length of arm 2, relations between gears
5. Major difficulties found during the PBL experience
6. Video of the robot (less than 2 min)
7. Conclusions
8. Comments and questions

Presentations should be shorter than 10–15 min leaving up to 5 min for questions. The use of the video proved to be very interesting since students had to get familiar with this kind of technologies (cameras, video edition, etc.).

3 Results

This article explains the implementation experience carried out in the academic year 2012/13. Even though the subject was designed for around twenty students, the first year there were only eight. The major reason for such a low number of students was that the B.Sc. in Industrial Electronics and Automation Engineering was a new degree and these were the first students that were passing all the courses successfully. This fact made much easier the implementation of the PBL experience, since the group was quite homogeneous. This year all students finished successfully the PBL project. Typically, the design of the subject produces more drop outs than fails. Anyway, the following years, with numbers of students around 20 or even 40 (in 2015/16), the academic results have been also very positive with very few drop outs. Figure 4 shows the academic results for the 2012/13 academic year for activity type for the eight students involved.

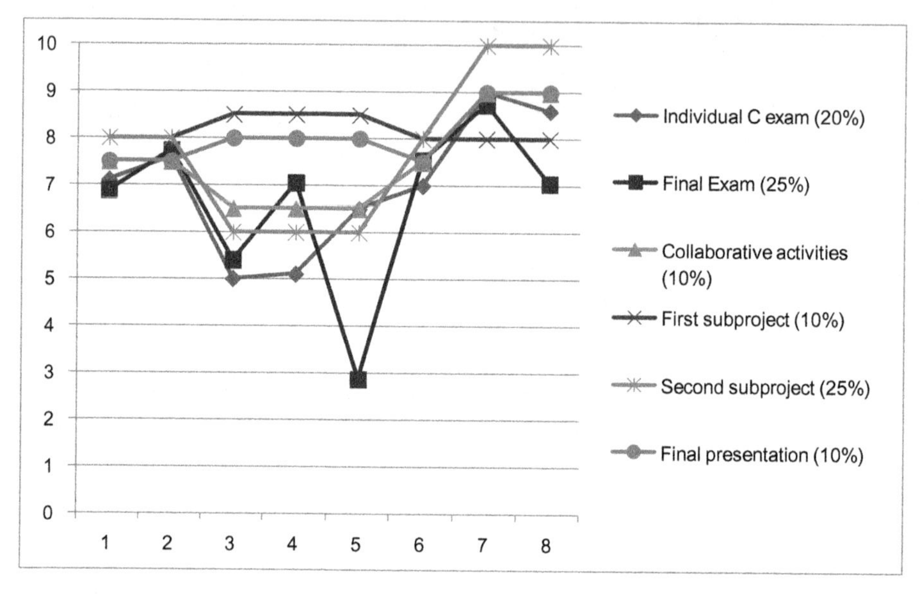

Fig. 4. Academic results for the 2012/13 academic year

4 Conclusions

The PBL approach presented in this work was quite ambitious since and several problems arise such as: (1) the initial schedule was very tight and students had to provide the intermediate deliverables with short deadlines; (2) both subprojects imposed a considerable work effort for both the students and the teacher; (3) some students may drop out from the continuous assessment, which complicates the work of the teacher since a task of the same complexity should be proposed; (4) communication issues were not included in the PBL due to lack of time.

Fortunately, the application of the PBL methodology as described in this work to a small group of students simplified its implementation. In general terms, the instructors are satisfied with the obtained results. Students were also satisfied with the experience, according to the results of the surveys addressed to them and informal discussion. The same methodology has been applied in the following four academic years (from 2012/13 to 2015/16) to groups from 20 to 40 students with similar results.

By using this approach the students acquired the required learning objectives (Table 1): They gained new knowledge about basic and technological subjects (C3); they used a proactive approach and had to transmit knowledge (C4); they had to complete a project and hand over several project related deliverables (C5); the proposed project is intrinsically multidisciplinary (e.g. it combines Maths, Physics, Computer Science, Programming, Control Engineering, and Robotics), and they also had to find and use some resources, mainly written in English (C10); Students had to manage their time responsively in order to optimize their work (C12); They also had to apply the scientific methodology strategies (C13); and finally, they definitely had to work in teams with people chosen randomly.

The authors also consider that the quality/price ratio of the Lego Mindstorms kit is interesting. Even though it is not easy to make mechanical structures with the elements provided, students had to solve a myriad of problems related to these non-perfect structures using their abilities and skills. This is a typical situation in engineering companies where it is not always possible to find a cheap and optimal solution. Unfortunately, our students are too used to think that there is a unique valid solution and that the others are wrong, and they will try to get the "right" solution by any means.

References

1. Aziz, S.M., Sicard, E., Ben Dhia, S.: Effective teaching of the physical design of integrated circuits using educational tools. IEEE Trans. Educ. **53**(4), 517–531 (2010)
2. Barron, B.: Doing with understanding: lessons from research on problem- and project-based learning. J. Learn. Sci. **7**(3&4), 271–311 (1998)
3. Barrows, H.S.: The Tutorial Process. Southern Illinois University School of Medicine, Springfield (1988)
4. Boss, S., Krauss, J.: Reinventing Project-Based Learning: Your Field Guide to Real-World Projects in the Digital Age. International Society for Technology in Education, Eugene (2007). ISBN 978-1-56484-238-1
5. Calvo, I., Lopez-Guede, J.M., Zulueta, E.: Aplicando la metodología project based learning en la docencia de ingeniería técnica en informática de Gestión. Revista de Formación e Innovación Educativa Universitaria **3**(4), 166–181 (2010). [In Spanish]
6. Curty, M., Comesaña, P., Marquez, O.W.: Experiencias metodológicas en la titulación de Ingeniería de telecomunicación: utilización de una plataforma de teleenseñanza en el proceso de evaluación continua. Revista de Formación e Innovación Educativa Universitaria **3**(2), 77–87 (2010). [In Spanish]
7. Habash, R., Suurtamm, C.: Engaging high school and engineering students: a multifaceted outreach program based on a mechatronics platform. IEEE Trans. Educ. **53**(1), 136–143 (2010)
8. Hadim, H.A.; Esche, S.K.: Enhancing the engineering curriculum through project-based learning. In: 2002 32nd Annual Frontiers in Education, FIE 2002, vol. 2, pp. F3F-1–F3F-6 (2002)
9. Hosseinzadeh, N., Hesamzadeh, M.R. (2009): A course in power system analysis based on project based learning methodology. In: 2009 Power & Energy Society General Meeting, PES 2009, pp. 1–6. IEEE
10. Hosseinzadeh, N., Hesamzadeh, M.R., Senini, S.: A curriculum for electrical power engineering based on project based learning philosophy. In: 2009 IEEE International Conference on Industrial Technology ICIT 2009, pp. 1–5 (2009)

11. Latorre Dardé, R. (2007): Diseño de actividades de aprendizaje activo en la asignatura procesos industriales de ingeniero industrial. In: Actas del 15° Congreso de Innovación Educativa en las Enseñanzas Técnicas Valladolid, España, 18–20 de julio 2007 [In Spanish]
12. Macias-Guarasa, J., Montero, J.M., San-Segundo, R., Araujo, A., Nieto-Taladriz, O.: A project-based learning approach to design electronic systems curricula. IEEE Trans. Educ. **49**(3), 389–397 (2006)
13. Ponsa, P., Amante, B., Roman, J.A., Oliver, S., Diaz, M., Vives, J.: Higher education challenges: introduction of active methodologies in engineering curricula. Int. J. Eng. Educ. **25**(4), 799–813 (2009)
14. Pucher, R., Mense, A., Wahl, H.: How to motivate students in project based learning. In: 2002 6th IEEE Africon Conference in Africa, AFRICON, vol. 1, pp. 443–446 (2002)
15. Ram, P., Ram, A., Sprague, C.: From student learner to professional learner: training for lifelong learning through online PBL. In: International Conference on Problem-Based Learning, Lahti, Finland, June 2005. http://www.cc.gatech.edu/faculty/ashwin/papers/er-05-03.pdf
16. Steedman, M., Smith, K., Keleher, P., Martin, F.: Successful cross-campus management of first year engineering courses. In: 36th Annual Frontiers in Education Conference, pp. 14–19 (2006)
17. Solis, J., Takanishi, A.: Practical issues on robotic education and challenges towards roboethics education. In: 2009 18th IEEE International Symposium on Robot and Human Interactive Communication, RO-MAN 2009, pp. 561–565 (2009)
18. UPV/EHU, ANECA, B.Sc. in Industrial Electronics and Automation Engineering Degree, Verified compliance ANECA report, April 2016. https://gestion-servicios.ehu.es/plantillas/Ingenier%EDa%20Electronica%20Industrial%20y%20Autom%E1tica.pdf [Only in Spanish]
19. Quesada, J., Calvo, I., Sancho, J., Sainz, J.A., Sanchez, J., Gil-Garcia, J.M., Sebastian, R., Castro, M.: Combining moodle and redmine as e-learning tools in project based learning of industrial electronics. In: 2013 7th IEEE International Conference on e-Learning in Industrial Electronics (ICELIE), pp. 86–91 (2013)

Coordination and Cooperative Learning in Engineering Studies

Karmele Artano-Pérez(✉), Aitor Bastarrika-Izagirre, Ruperta Delgado-Tercero, Pilar Martínez-Blanco, and Amaia Mesanza-Moraza

Department of Mining and Metallurgical Engineering and Materials Science, Faculty of Engineering of Vitoria-Gasteiz (EUIVG), University of the Basque Country UPV/EHU, C/Nieves Cano, 12, 01006 Vitoria-Gasteiz, Spain
{karmele.artano, aitor.bastarrika, ruper.delgado, mp.martinez, amaia.mesanza}@ehu.eus

Abstract. This project describes the coordination undertaken in the Bachelor's degree in Geomatic and Surveying Engineering in the University of the Basque Country (UPV/EHU) after the integration in the European Higher Education Area, explaining how cooperative learning can positively contribute to the teaching-learning process.

Keywords: Competency based learning · Cooperative project · Active methodologies · Coordination

1 Introduction

In the 2010–2011 academic year, the transition began from the qualification of "Ingeniero Técnico en Topografía" (Technical Engineer in Topography – ITT for its Spanish acronym) to the "Grado en Ingeniería en Geomática y Topografía" – (Degree in Geomatic and Surveying Engineering - IGT for its Spanish acronym) in the Faculty of Engineering of Vitoria-Gasteiz (EUIVG). The new Degree certification allows for the regulated exercise of the profession of ITT (Orden CIN/353/2009).

The transition is regulated by the RD 1393/2007, the Orden CIN/353/2009 and the UPV/EHU Resolution of the 20th of December of 2010. The main characteristics of the new certification are as follows:

- The measure of the subjects is done in ECTS (European Credit Transfer System) credits.
- The duration is 4 years, one year longer than the old certification (ITT), with a total weight of 240 credits.
- The subjects are compulsory or optional (whereas in the old qualification they could be core subjects, university compulsory or optional). Compulsory subjects are grouped in three modules: Basic module (FB) (60 credits, 25 %), a common module of the topographic field (CRT) (60 credits, 34 %), and the specific technologies module (TE) (48 credits, 20 %). Besides, students must develop an end of degree project (EDP, 5 %).

M. Graña et al. (eds.), *International Joint Conference SOCO'16-CISIS'16-ICEUTE'16*, Advances in Intelligent Systems and Computing 527, DOI 10.1007/978-3-319-47364-2_68

- Subjects are quarterly (whereas in the old plan they could also be annual subjects). Besides, it is specified that a student cannot take more than 30 credits per quarter. For this reason, most subjects must amount to 6 credits, even though some of them are actually 9 credit subjects. The EDP goes from having 4, 5 in the old plan to 12 credits, and each optional subjects (16 %) have a weight of 4, 5 credits.
- The last two terms (4th year) have practically no compulsory teaching so that the students can: opt for foreign placements within the ERASMUS programme (hence facilitating the mobility compromise which the principle of Bolonia promotes), developing company placements or taking the necessary optional subjects (39 credits), therefore completing the required 240 credits.

The European Union project for the creation of the European Higher Education Area (EHEA) has meant the acceptance of a change within the internal management of all the processes which intervene in the development of the studies in the IGT degree, both at organizational (Sect. 2) and methodological level (Sect. 3), being both closely intertwined. The coordination of the degree studies allows for this interconnection, and establishes and guarantees the necessary quality parameters for continuous development in the delivery of the studies and the obtained results.

2 Coordination in ihe IGT Degree

The task of coordinating the IGT degree is organized in three levels. In the first level the vertical coordination of the degree is carried out, via a lecturing team (Degree Commission) comprised of the quality subdirector, the degree coordinator, the course coordinators, two lecturers who provide teaching in the degree and a student. In the second level a course or horizontal coordination is established, through four lecturing teams of the four courses of the degree, comprised of the course coordinator, the subject coordinators for each course and the course´s delegate student. In the third level the subject coordination is undertaken, in lecturing teams trained up by lecturers who provide training on that subject, and the lecturer responsible for the subject.

2.1 Degree Coordination Process

The IGT degree coordination is developed following different lines of action, defining for each of them the improvement actions in order to achieve its objectives. The lines of action can in turn be distributed in four groups: organization of studies, methodology, monitoring and promotion. Each of these is briefly explained below.

- ORGANISATION OF STUDIES. It is constituted by those lines of action related to the general organisational aspects of the qualification.
  - TEACHING GUIDES: The objective is for all the teaching guides for all the subjects of the degree to have the optimum level of information for the student and for its publication, being at the same time homogeneous between the degree subjects.

  - TIMELINES OR EVENT PROGRAMS: The drawing up of the event programs for each course of the qualification, agreed during the course commissions, intends to distribute in a homogeneous way the work of the students throughout the period of study.
  - TEACHING VECTORS: The adequacy for each subject´s teaching vector is analysed, and, if necessary, the relevant modifications are requested.
- METHODOLOGY. In this section the lines of action related to the teaching methodologies and evaluation systems are grouped.
  - TEACHING /LEARNING METHODOLOGIES (T/L): Information is gathered annually about the teaching methodologies in the different subjects within the degree, driving the adoption of new teaching methodologies, active learning methodologies and specially the cooperative learning methodology.
  - ASSESSMENT SYSTEMS: Data related to the assessment systems for all subjects is gathered, boosting the implementation of continuous evaluation methodologies.
- MONITORING. These lines of action have the objective of establishing monitoring or control of the development of the study plan, as well as the academic results obtained.
  - OVERLAPS AND DEFICIENCIES OF SPECIFIC CONTENTS: The analysis of the specific competencies of the compulsory subjects of the IGT degree has the objective of detecting duplicities in content /specific competencies and possible deficiencies. It is carried out internally in the degree and course commissions, as well as externally, during the Director Conferences of those universities who offer the IGT degree at state level. This has the aim of tailoring the degree to the needs of today´s society.
  - WORK AND EVALUATON OF THE TRANSVERSAL COMPETENCIES: In this line, a general analysis of the distribution of the transversal competencies (TC) is carried out, and the distribution between courses and subjects, promoting the creation of unified tools for the work on the transversal competencies (TC) and the rubrics for their evaluation.
  - ACADEMIC RESULTS (AR) INDICATOR ANALYSIS: The analysis of each course´s academic results indicators and for the whole qualification falls within a process of evaluation of the quality of the teaching of the degree, which allows for the establishment of improvements related to the implementation of new teaching methodologies, timetable optimisation and evaluation systems. Ultimately, it constitutes the key process in the design of the lines of action in the degree coordination.
- PROMOTION.
  - GEOMATICS: The actions of promotion of the Degree are developed both at local level, within the UPV/EHU, and specifically within the EUIVG, and at state level, together with all the universities which provide the IGT Degree.

2.2 Transversal Competencies in the IGT Degree

The European Union Project for the creation of the EHEA has meant the acceptance of a change in the university teaching methodology, with the Competency Based Learning (CBL) system being adopted. This system is focused on the student, in his or her capacity and responsibility, as well as in the development of his or her autonomy. Until Bolonia, the university system was based fundamentally on the transmission of knowledge to the student. Now, as well as obtaining the specific competencies (SC) related to the university qualification chosen by the student, the system must guarantee the acquisition of dexterity and skills specific to the profession to which the studies are focused, and which allow the graduate to be efficient and effective in his or her professional and social environment. Through the application of the CBL, it is required to work not only the SC of the qualification, but also the TC, being the latest the ones which prepare people for their insertion in the complex educational, occupational and social systems, multidimensional and with knowledge which changes continuously [1].

In the case of the UPV/EHU IGT Degree, the adaptation to the EHEA has meant the creation of the need for coordinated work and evaluation of the TC, which in the previous qualification (ITT) was achieved in a non-global way and in the individual context of each subject. Being aware of the need of a coordinated environment for the work on the TC, in the case of the UPV/EHU IGT degree, the difficultly was that in the verification request statement of the official certificate [2], the qualification TCs are not specified. This is why a meticulous analysis of the qualification SCs has been necessary. From this, two of them, where transversality has been considered associated, have been selected:

C4. Ability for decision-making, leadership, human resources management and inter-disciplinary team management, related to spatial information.

C7. Management and implementation of research projects, development and innovation in the engineering field.

Starting from these two competencies, which, although they are considered specific, could be related to the TC, and with the criterion of trying to clearly identify the most important TCs, the following have been defined for this qualification [3]: TC1: Teamwork; TC2: Written communication; TC3: Oral communication; TC4: Decision-making; TC5: Innovation; and, TC6: Quality. The analysis of each of these competencies has allowed for the identification of different levels of proficiency for each of them, establishing levels from one to three, according to the degree of depth concerned. After the fact, it has been established in which course the different levels (Table 1) would be worked on, and the corresponding indicators for each of them have been detailed, developing simple evaluation rubrics to measure the satisfaction level (bad – B, regular – R, and good – G), achieved in the levels of each of the TCs.

The transmission of the design of the work and evaluation of the TCs to the study plan of the IGT degree qualification requires the coordination of the lecturers who provide training of the different subjects in the qualification. This coordination is achieved through the qualification commission, which has been responsible for the proposals of the qualification´s TCs, the proficiency and distribution levels and the definition of the indicators for its evaluation. This is transferred to the course

Table 1. TC Proficiency levels and distribution by course in the IGT Degree (UPV/EHU)

4-EDP	Level 2	Level 3	Level 2	Level 2	Level 1	Level 2
4th course	Level 2	Level 2	Level 1	Level 2		Level 2
3rd course	Level 1	Level 2	Level 1	Level 1		Level 2
2nd course	Level 1	Level 2	Level 1	Level 1		Level 2
1st course		Level 1				Level 1
	TC1	TC2	TC3	TC4	TC5	TC6

commissions, where it is established in which subjects and to what level the accorded TCs will be worked on, establishing in which of them it will be evaluated and obtaining consensus on the working tool.

2.3 Coordination of the 2nd Course of the IGT Degree

In this section, the coordination undertaken on the 2nd course of the IGT Degree is explained, given that this is the course with the highest percentage of subjects related to the common module of the topography field (CRT), which in turn corresponds with the block of highest percentage of ECTS credits (T:theory; P:practice).

The IGT Degree studies in Vitoria-Gasteiz in the second course are organized in two terms of 30 credits each, through 7 subjects of 6 ECTS credits, and 2 subjects of 9 ECTS credits. These last ones are set out in the second term (Table 2). In this course most of the subjects belong to the CRT module, except Geomorphology and Economics and Business Management, which belong to the FB module, and Satellite Geodesy and Mathematic Cartography which belongs to the TE module.

Table 2. Subject distribution per term in the 2nd course.

First term			Second term		
Subject	Credits		Subject	Credits	
	T	P		T	P
Geomorphology (G)	4, 5	1, 5	Economics and Business Management (EBM)	4, 5	1, 5
Digital image processing (DIP)	3, 0	3, 0	Photogrammetry (P)	6, 0	3, 0
Cartographic design and production (CDP)	3, 0	3, 0	Geographic Information Systems (GIS)	1, 5	4, 5
Geometric Geodesy (GG)	3, 0	3, 0	Spacial Geodesy & Mathematic Cartography (SGMC)	6, 0	3, 0
Topographic methods (TM)	3, 0	3, 0			

Aside from the SC, the TC, already discussed in Sect. 2.2 must be achieved in the Bolonia framework. Thus, in the 2011–12 course, the second course commission established the most appropriate subjects for its evaluation (Table 3), creating a common rubric for the four courses of the qualification, thus allowing for the evaluation of the TC in a homogeneous way throughout the qualification.

Table 3. Transversal competencies for evaluation in the second course.

Subject	TC	Level
Economics and Business Management (EBM)	CT1	Level 1
Geomorphology (G)	CT2	Level 2
Photogrammetry (P)	CT3	Level 1
Geographic Information Systems (GIS)	CT4	Level 1
Satellite Geodesy and Mathematic Cartography (SGMC)	CT5	Level 2

In order for the student body to achieve the total of the SC and TC required in each course, during the exam period of the previous term, the timeframe of the following term´s subjects is agreed in the corresponding course commission. This way, the first term´s event calendar is agreed in June, and the second in the first few weeks in January. During its creation the tasks to be developed by the students was uniformly distributed throughout the fifteen weeks of the term, avoiding work overload for students in specific weeks.

In the second course, it can be said that the timeframes respond to the outline in Table 4, where it can be appreciated that in the first three weeks they have no tasks to develop, being the last weeks (14 and 15) the ones with the biggest workload. In view of this situation, during the last two academic years we have tried to avoid these concentrations of workload and they have been lightened, reaching the conclusion that they can hardly be avoided, noting that the calendar helps to achieve the TC and SC more effectively. Besides, it has been necessary to analyze the overlaps and deficiencies between subjects, specially within the same course (horizontal coordination).

Finally, it must be indicated that the coordination for this course has managed to achieve that in a great part of the subjects, active methodologies are applied, leading to the development of a teaching method of learning based in a cooperative project with encompasses the subjects of TM, GG, SGMC, GIS, DIP and P, as explained in the following section.

3 Cooperative Learning

Nowadays, one of the labor market´s biggest demands is that the graduates and future professionals possess communication and teamwork skills, as well as a wide understanding of social, environmental or economic issues. Achieving and dominating all of these competencies is a task which is acquired all throughout the learning itinerary of the student, as has been evidenced in Sect. 2.2.

Table 4. Event calendar by subjects and terms

WEEK	First Term		Second Term	
	Task	Subject	Task	Subject
4	Report Submission	DIP	Control	SGMC
5	Control	G	Practical Exam	GIS
	Practical exam	CDP		
6			Field work submission	P
7	Report Submission	G		
8	Report Submission	TM	Practical submission	SGMC
	Report Submission	DIP	Practical exam	GIS
9	Control	G	Presentation	P
	Exam	CDP		
10	Report Submission	TM		
	Control	G		
11	Report Submission	TM	Presentation	P
12	Control	GG	Practical exam	GIS
13	Exam	DIP	Presentation	EBM
14	Report Submission	TM	Presentation	GIS
			Presentation	EBM
15	Practical Exam	GG	Presentation	EBM
	Exam	CDP	Practical exam and integration	GIS

Between the lecturers, it is very frequent to carry out teamwork activities which can be developed both in a collaborative or cooperative manner. Even though according to the consulted bibliography both can be considered extremely similar, there are some differences depending on the desired objective, the structures and the lecturer´s role [6].

Thus, the students work in a collaborative manner when one of the members of the team is in charge of carrying out a specific task, unifying everybody´s individual tasks into a final project in a second phase. By contrast, in a cooperative project all the members of the team develop all of the assigned tasks together [4].

Not all the teamwork activities can be considered cooperative learning, given that for this to be true all of the following elements must be fulfilled [7]: positive interdependence, team and individual responsibility in the achievement of the objectives, personal interaction, attitude and interpersonal skills development, as well as teamwork, monitoring and periodical group reflection about the work performed. For this learning the lecturer´s role is essential, becoming a moderator, coordinator and mediator, rigorously planning in advance a series of well-defined and measurable activities which allow for the achievement of such learning.

3.1 The Cooperative Project in the Second Course of the IGT Degree

The EUIVG, with the appearance of the new degrees, promoted from the start a vertical coordination between the courses and a horizontal coordination between the subjects within the same course (Sect. 2). As a result of this latest coordination, and prior to the start of the 2013–2014 course, the idea arose to set into motion a cooperative project (CP) which encompassed 6 out of the 9 subjects in the second course of the degree and which involved 5 lecturers.

The objective set to the students was that at the end of the course they had to be capable of incorporating their obtained cartography in a geographic information system, either via topographic methods or via photogrammetric techniques, in a determined and defined area of the university campus. This way, in order to achieve their objective, the students would acquire all the necessary skills of the different involved subjects over the two terms of the course. For this, it was necessary for the lecturing team to design and plan a series of closely linked activities to be developed (Table 5), and defining an itinerary to guide the student in the achievement of the objective set. This meant a lot of meeting hours for the involved lecturers, a lot of coordination work, and an extensive knowledge, not only of each lecturer's own subject, but also of the rest of the lecturers', so that each of the practices in every subject could be adequately defined within the context of CP.

Table 5. Task to be developed in each subject

Task	Subject	Weeks
1 – Analysis of support photography characteristics	DIP	6,7,8
2 – Projection, observation and XX of the topographic network	TM	6,7
3 – Topographic surveys	TM	8,9,10
4 – Analysis of photogrammetric flights	GG	16,17
5 – Planning for photogrammetric flights	P	17
6 – Planning for photogrammetric support	P	17
7 – UTM calculation of the topographic network	SGMC	18
8 – Ground point measuring with GPS-RTK	SGMC	19,20
9 – Re-observation of the topographic network by GPS-RTK	SGMC	21,22
10 – Projection, observation and calculation of the geodesic-topographic network using GNSS	SGMC	23-26
11 –GIS integration in the observed networks	GIS	27
12 – Photogrammetric support verification	P	28
13 – Photogrammetric restitution	P	29
14 – GIS integration in the restitution/topographic survey	GIS	30

The evaluation of the acquired knowledge by the students during the CP was carried out independently for each subject, using the rubrics established by the degree qualification commission for the evaluation of the TC, and the rubrics designed by each lecturer for the evaluation of the SC.

Table 6. Porcent of student rating regarding difficulty and interest

Subject	Difficulty level					Initial interest					Final interest				
	1	2	3	4	5	1	2	3	4	5	1	2	3	4	5
P	0	0	83	17	0	0	0	67	33	0	0	0	33	67	0
SGMC	0	0	0	43	57	0	43	36	21	0	0	29	57	14	0
GG	0	0	0	43	57	3	22	41	28	0	3	16	51	22	0
TM	1	1	34	46	17	0	6	60	30	4	0	4	60	31	5
GIS	0	0	80	20	0	0	10	40	50	0	0	10	20	50	20
DIP	0	0	88	12	0	0	0	88	12	0	0	0	56	44	0

Table 7. Student rating related to lecturing planning

Subject	Planning	Methodology	Development
P	4, 0	4, 1	4, 2
SGMC	3, 5	3, 3	3, 4
GG	3, 5	3, 4	3, 6
TM	4, 2	3, 9	3, 9
GIS	4, 0	3, 8	3, 7
DIP	3, 9	4, 1	4, 3
Average	**3, 9**	**3, 8**	**3, 9**

4 Conclusions

The adaptation of the university qualifications to the requirements of the EHEA means substantial changes to the organization and development of the study plan of the Degree, as well as the T/L methodologies. The coordinated work of lecturers and the management team is the key aspect which allows for the continuous improvement of the degree. The work carried out by the course and qualification coordinators in the different lines of action make it possible to design the improvement actions which will be undertaken in the different lecturer teams.

The Degree Commission works on those aspects related to the organization of the studies and does the monitoring so as to obtain an "alive" degree, adapted to today´s society needs. Also, the abilities and skills that the student will attain, specially through the TC, is a quality that companies require in their future engineers. That´s why the lecturer has to adapt to the methodologic changes necessary for that aim; the role of the lecturer changes for the students and also in relation to the other lecturers of the qualification, their active participation in the lecturing teams being necessary.

The University of the Basque Country UPV/EHU, specifically the vice rectorate of innovation and degree studies, issues a questionnaire to the students at the end of every term, in order to know their opinion regarding the quality teaching of the lecturers. In these questionnaires they answer 40 questions of a different nature, and from the results very interesting conclusions can be drawn. Thus, one of the aspects which is covered is the perception of the student about the difficulty of the subject and their degree of interest before and after taking that subject.

The results for the 2013–2014 academic years have been gathered for the 6 subjects which conform the Cooperative Project (Table 6), and in all of them it can be appreciated that the interest that the students show once the subject is finished is always higher than the interest shown at the beginning. So their expectations are always enhanced, even for those subjects they have classed as difficult or very difficult.

Other aspects which are covered relate to the planning of the teaching. In other words, ensuring this teaching is focused on the development of the competencies; the teaching methodology to comment on the modalities of the teaching-learning process, the resources and the proposed activities. Also, the teaching development on the lecturer-student relationship in any of its different facets. These results are shown in Table 7, where it can be appreciated that the general rating from the students is quite positive, achieving an average on these three items of 3, 9 out of 5, 0. The lowest ratings are always related to those subjects which the student perceives as most difficult, although they always surpass the 3, 5 points on average anyway.

Besides, we must point out that both the active methodologies and the collaborative learning in this new design from the EHEA mean a challenge for university education, given that they intend for the student to achieve a higher success at the time of understanding and internalising the knowledge, especially the specific skills.

References

1. Morin, E.: Los Siete Saberes Necesarios Para la Educación del Futuro. UNESCO, Paris (1999)
2. http://www.ingeniaritza-gasteiz.ehu.es/p232-content/eu/contenidos/informacion/ingtop_intranet/eu_intranet/adjuntos/plan%20estudio%20geomatica.pdf
3. Villa, A., Poblete, M.: Aprendizaje basado en competencias. Una propuesta para la evaluación de las competencias genéricas. Universidad de Deusto, Bilbao (2007)
4. Los proyectos colaborativos y cooperativos en internet. https://sites.google.com/site/deinfantil/home/consideraciones-de-david-moursund-1. Consulta 14 April 2016
5. Basilotta Gómez-Pablos, V.; Herrada Valverde, G.: Aprendizaje a través de proyectos colaborativos con TIC. Análisis de dos experiencias en el contexto educativo". Revista electrónica de tecnología educativa, 44 (2013)
6. Guerra Azócar, M.: Aprendizaje cooperativo y colaborativo, dos metodologías útiles para desarrollar habilidades socioafectivas y cognitivas en la sociedad del conocimiento. http://www.monografias.com/trabajos66/aprendizaje-colaborativo/aprendizaje-colaborativo.shtml. Consulta 14 April 2016
7. Ramos Hernanz, J.A., Puelles Pérez, E., Arrugaeta Gil, J.J., Sancho Saiz, J., Zubimendi Herranz, J.L., Ruiz Ojeda, M.P.: Aplicación del aprendizaje cooperativo en diferentes asignaturas de ingeniería. IX Jornadas de redes de investigación en docencia universitaria. Alicante (2011)

Welcome Program for First Year Students at the Faculty of Engineering of Vitoria-Gasteiz. Soft Skills

Estíbaliz Apiñaniz-Fernandez de Larrinoa(✉), Javier Sancho-Saiz, Amaia Mesanza-Moraza, Ruperta Delgado-Tercero, I. Tazo-Herrán, J.A. Ramos-Hernanz, J.I. Ochoa de Eribe-Vázquez, J.M. Lopez-Guede, E. Zulueta-Guerrero, and J. Díaz de Argandoña-González

Faculty of Engineering of Vitoria-Gasteiz, University of the Basque Country UPV/EHU, C/Nieves Cano, 12, 01006 Vitoria-Gasteiz, Spain
{estibaliz.apinaniz,javier.sancho}@ehu.eus

Abstract. It is well known that there are high dropout rates in engineering faculties. There are plenty of reasons that can explain these bad results, among which we can mention group integration problems, lack of motivation or difficulty to adapt to the university study methodology. In this work we propose a Four-day Welcome Program for first-year students to help these new students to overcome some difficulties they will have to face during their time at the Engineering School. Our objective is to improve the results of the first year students by motivating them from the first day.

Keywords: Dropout rates · Return rates · Skills

1 Introduction

It is well known that there are high dropout rates in engineering faculties [1]. There are several reasons that can explain these bad results such as: integration problems due to the difficulty some students find in new social groups; some of the students do not really know what engineering is, or what they will learn during those 4 years and what they would be able to do once they have finished. In addition, the profile of engineering students is generally the one of creative people, of people who want to learn new things and find applications for what they are doing and they do not find what they were searching for during the first year; there is a clear lack of motivation; they also find difficulties to adapt themselves to new learning methods; engineering degrees require almost full time occupation. In conclusion, the new studies do not meet the students' expectations.

In order to deal with these issues a number of initiatives have been carried in the Faculty of Engineering of Vitoria-Gasteiz. In [2] a first experience with a remote laboratory for a basic course of control engineering is described. Several years later, once the Bologna framework is implanted, more initiatives have taken place, i.e., in [3] a description of an educational innovation project in the field of Industrial Informatics

M. Graña et al. (eds.), *International Joint Conference SOCO'16-CISIS'16-ICEUTE'16*,
Advances in Intelligent Systems and Computing 527, DOI 10.1007/978-3-319-47364-2_69

can be found, while in [4–6] some successful efforts on the scope of computer science are explained. Anyway, more generic descriptions of some experiences implemented at the Faculty of Engineering of Vitoria-Gasteiz are given in [7–11].

In order to help the students to face their studies with motivation we offered them a 4-day Welcome-Program. This short course takes place before the official beginning of the academic year. All first year students are invited to take part in it and they can get one ECTS credit if, once finished, they write an essay on what they have learned. This Welcome Program took place for the first time in the 2014/15 academic year, so there have already been editions. In each edition an average of 150 students attended the course. During the 12 h of the course the students were able to see different subjects: presentation of the faculty and the degree; working as an engineer; personal development program; library and digital sources, transversal skills and study techniques.

The structure of the paper is as follows: Sect. 2 introduces the main goal of the education innovation project. The third section describes the structure and the methodology of the welcome program, while Sect. 4 gives results based on a students' survey. Finally, Sect. 5 presents the main conclusions of the paper.

2 Goal of the Education Innovation Project

The Faculty of Engineering of Vitoria-Gasteiz is a faculty of the University if the Basque Country. It offers 5 Bachelors degree programs and many complementary courses related to engineering.

In the Faculty they work 109 teachers and 24 people of administration and services. Currently there are 920 students in 5 degrees:

Mechanical Engineering, Industrial Automation and Electronic Engineering, Industrial Chemical Engineering, Computer Engineering Management and Information Systems and Geomatics and Surveying Engineering.

The results of the dropout rate for the 2013–14 year were:

- Industrial Automation and Electronic Engineering: 45.71 %
- Computer Engineering Management and Information Systems: 38.78
- Mechanical Engineering: 36.26 %
- Industrial Chemical Engineering: 36.84 %
- Geomatics and Surveying Engineering: 24.0 %

We have already mentioned several reasons for these bad results: the studies in the first year do not meet their expectations due to the difficulty of the subjects or because they find it difficult to get used to their new situation, different classmates, different teaching methodologies. In several cases they do not know what they can do as engineers once they have finished, so they lack motivation.

Taking into account the latter reasons and with a clear goal, to improve the results of our first year students, we decided to offer the students a Welcome program. After offering it for two years, the University of the Basque Country gave us a Educational Innovation Project entitled: Specialized Educational Innovation Group in welcoming the new students, their tutoring and development of soft skills in engineering.

By means of this course we would like to improve the return rates of the first year students and to decrease the dropout rate in all the degrees we offer. We find it also very important that our students to feel well in their groups, therefore, we would like to improve the adaptation of the newcomers by letting them meet each other before the official course starts. During the last years, we have seen that there is a low participation of the students in the activities of the faculty, therefore with this course we would also like to encourage our students to participate in daily life of the faculty.

In addition, we would like to offer the students a life experience on soft skills that are essential in their formation as engineers. We also find also very important to provide the students working tools and habits that can help them in their studies. It is very important to motivate the students by means of giving them the opportunity to listen to young graduates, who are working in different areas so that the new students know what an engineer can do and what he would like to do.

Besides, with this course we also would like to make our students get soft skills that can make them stand out from the rest of engineering students.

3 Structure and Methodology of the Course

Taking into account all the skills we wanted the student to work on, the course was divided into 4 different modules of 3 h each (1 day/module):

- Presentation of the Faculty + Working as an engineer
- Personal Development Project + Library
- Study Techniques
- Oral presentations + creativity + problem solving + leadership + team work

After these modules the students were asked to fill in a final enquiry that is very important to know if the course meets their expectations and fulfil our goals.

The 150 students were divided into three groups; Table 1 shows the schedule of one of the groups.

Table 1. Schedule of the course for a group of 50 students.

<table>
<tr><th></th><th>MONDAY</th><th>TUESDAY</th><th>WEDNESDAY</th><th>THURSDAY</th></tr>
<tr><td>9.30 - 10.30</td><td>Presentation</td><td>Personal Development Project (PDP)</td><td rowspan="3">Team work/ Leadership/ Creativity/ Oral presentations/ Problem solving</td><td>Study Techniques</td></tr>
<tr><td>10.30-11.30</td><td rowspan="2">Working as an engineer</td><td rowspan="2">Library</td><td rowspan="2">Study Techniques/ Final enquiry</td></tr>
<tr><td>11.30-12.30</td></tr>
</table>

Taking into account the goals we have already mentioned, we wanted the students to take actively part in the courses. In order to get this objective each module had Workshop format where most of the time the students had to work in teams.

The presentation was made dividing the students in groups of 50 people taking into account their degree. After a short introduction, the students were divided into 3–4 people teams and they had to answer short questions on the university and faculty. In order to do that, they could use internet and some other tools that had been shortly described in the introduction and that they would have to use during all their stay at the university. Afterwards, the group was divided in two groups of 25, and they were introduced to their tutors. The tutors are teachers that help the students during their four years in the faculty. To finish with the first day, they were gathered again in the main hall and some young engineers of the faculty explained to them the work they were doing. They also explained the new students what is important to finish the studies successfully and which skills are needed to find a job.

As far as the second module is concerned, one part took place in the library where the personal from the library explained the students how to use different information sources (hardcopies or digitals ones). In addition, an expert professor taught the students via active methodologies how to build a Personal Development Project. In this part students were requested to think about their objectives in their studies and life and how they thought they could fulfil them. They were taught about the importance of not only technical knowledge but also the soft and social skills for working in engineering: teamwork, manage and resolve conflicts, and about the importance of Life Long Learning.

The third module was presented by four teachers of the Faculty. They gave a short presentation on team work, creativity, problem resolution and oral presentations, using videos and games. Then, the 50 student group was divided into 4 groups (12 people) and each group had to develop a short guided project. This 12 people group was divided in 4 subgroups to do the project. Once finished, the subgroups belonging to different projects gathered together and explained each other what they had been working on. The students valuated this last oral presentation by taking into account some items that had been given previously by the teachers.

The Study Techniques module was prepared by an external enterprise Norgara, which are specialized psychologists. They help the student to realize what they lack when studying and how they can overcome those difficulties.

4 Assesment of the Students and Results

The students of the University College of Engineering of Vitoria-Gasteiz found this Welcome Program very interesting (the average score on the survey was 3.6 out of 5) and the average opinion of the course was very satisfactory. In Table 2 we can see the opinion of the students for each module.

1. Presentation

In general the students found it interesting, but they would have liked to see the different rooms and areas of the faculty that were not shown. They were pleased to know from the very beginning the tutors that would help them during their four years.

2. Working as an engineer

Table 2. Opinion of the students for each module, they could rate the module from 1–5

AVERAGE VALUE

	Presentation	Working as an Engineer	Personal Development Project	Library	Team work/ Leadership/ Creativity/ Oral presentations	Study techniques
AVERAGE	3,67	3,70	3,45	3,08	3,99	3,71

The possibility of speaking to former students that are successfully working was very well rated. Some of the student thought that they spoke too much on what they were currently doing and they would have liked to know how they faced their studies and hear some suggestions about how to be successful with them.

3. Personal Development Project

The students liked this module, they found it interesting, well explained and entertaining.

4. Library

The student found the explanation too long.

5. Team work/Leadership/Creativity/Oral presentations/Problem solving

The students enjoyed this module, they could visit different laboratories and they did some interesting projects. They rated very positively the possibility of working in groups where they could meet other students.

6. Study techniques

Most of the students found it interesting and entertaining. However, some others thought 3 h was too long and that they should have had a rest.

Only two editions are not enough to compare the first year results of the students and we have to take into account that there might be some other reasons for the changed of the results in the last two years, such as retirement of some of the first year subject teachers or changes in the teaching methodology. Anyway, the results have improved (see Table 3), and in our opinion some of the goals have been achieved. The

Table 3. Dropout rates for the last two years

DEGREE	DROPOUT RATE (%)	
	2013-14	2014-15
Industrial Automation and Electronic Engineering	45,71	34,18
Computer Engineering Management and Information Systems	38,78	31,88
Mechanical Engineering	36,26	29,27
Industrial Chemical Engineering	36,84	26,56
Geomatics and Surveying Engineering	24,00	17,44

students meet other students and work with them from the very first day, they gain knowledge as far as the engineering career is concerned, they are encouraged to build a personal development project that would be very useful for their studies and life, and they are taught to use the tools the university and faculty offer.

5 Conclusion

The students of the University College of Engineering of Vitoria-Gasteiz found this Welcome Program very interesting as it can be concluded from the inquiries. In addition, the results of the first year students have improved in the last two years, when this Welcome Program has been implemented. As we said before, only two editions are not enough to compare the first year results of the students. Anyway, the results have improved (dropout rate) and in our opinion some of the goals have been achieve.

In addition, the results obtained with the Four-day Welcome Program are consistent with the results of other experiences that have introduced cooperative learning techniques [12–14].

In conclusion, we find this course very profitable for the students and the faculty and we will try to offer it in the future.

Acknowledgements. The authors would like to thank the UPV/EHU for the financial support inside the Educational Innovation Projects 2016.

References

1. Felder, R.M., Brent, R.: Understanding student differences. J. Eng. Educ. **94**(1), 57–72 (2005)
2. Calvo, I., Zulueta, E., Oterino, F., Lopez-Guede, J.M.: A remote laboratory for a basic course on control engineering. Int. J. Online Eng. [iJOE] **5**, 8–13 (2009)

3. Lopez-Guede, J.M., Graña, M., Larrañaga, J.M., Oterino, F.: Educational innovation project in the field of industrial informatics. Procedia Soc. Behav. Sci. **141**, 20–24 (2014)
4. Oterino-Echavarri, F., Lopez-Guede, J.M., Zulueta, E., Graña, M.: Educational innovation: interaction and relationship inside a sub-module. Procedia Soc. Behav. Sci. **186**, 395–400 (2015)
5. Lopez-Guede, J.M., Soto, I., Moreno Fdez de Leceta, A., Larrañaga, J.M.: Educational innovation in the computer architecture area. Procedia Soc. Behav. Sci. **186**, 388–394 (2015)
6. Lopez-Guede, J.M.: Experiencia docente mediante la metodología de aprendizaje basado en problemas. IKASTORRATZA e-Revista de Didáctica **14**, 72–85 (2015)
7. Lopez-Guede, J.M., Graña, M., Oterino, F., Larrañaga, J.M.: Retrospective vision of a long term innovative experience. Procedia Soc. Behav. Sci. **141**, 15–19 (2014)
8. Lopez-Guede, J.M., Graña, M., Larrañaga Lesaca, J.M.: Innovation in engineering: evolution and perception of teaching innovation initiatives at the faculty of engineering of vitoria. J. Altern. Perspect. Soc. Sci. In press
9. Sancho, J., Ochoa de Eribe, J.I., Etxebarría, I., Calvo, I.: Píldoras formativas sobre metodologías docentes en la Escuela Universitaria de Ingeniería de Vitoria-Gasteiz. IKASTORRATZA e-Revista de Didáctica **14**, 1–3 (2015)
10. Sancho, J., Olalde, K.: Panorama global de aplicación de metodologías activas en estudios universitarios en ingeniería. IKASTORRATZA e-Revista de Didáctica **14**, 4–22 (2015)
11. Quesada, J., Sancho, J., Sainz, J.A., Sánchez, J., Gil-García, J.M., Sebastián, R., Castro, M.: Combining moodle and redmine as e-learning tools in project based learning of industrial electronics. In: IEEE Xplore Digital Library, pp. 86–91 (2013)
12. Ramos Hernanz, J.A., Puelles Pérez, E., Arrugaeta Gil, J.J., Sancho Saiz, J., Zubimendi Herranz, J.L., Ruiz Ojeda, M.P.: Aplicación del aprendizaje cooperativo en diferentes asignaturas de ingeniería. IX Jornadas de Redes de Investigación en Docencia Universitaria. Alicante (2011)
13. Lobato, C., Apodaca, P.M., Barandiarán, M.C.: San José, M.J., Sancho, J., Zubimendi, J.L.: Development of the competences of teamwork through cooperative learning at the university. Int. J. Inf. Oper. Manage. **3**, 224–240 (2010)
14. Lobato, C., Apodaca, P., Barandiarán, M., San José, M.J., Sancho, J., Zubimendi, J.L.: Cooperative learning at the university and dimensions of the competences of teamwork. In: Proceedings of the IASK International Conference Teaching and Learning (2009)

Revisiting the Simulated Annealing Algorithm from a Teaching Perspective

Paulo B. de Moura Oliveira[1(✉)], Eduardo J. Solteiro Pires[1], and Paulo Novais[2]

[1] Department of Engineering, School of Sciences and Technology, INESC TEC – INESC Technology and Science, 5001–801 Vila Real, Portugal
oliveira@utad.pt
[2] Centro ALGORITMI/Departamento de Informática, Universidade do Minho, Braga, Portugal

Abstract. Hill climbing and simulated annealing are two fundamental search techniques integrating most artificial intelligence and machine learning courses curricula. These techniques serve as introduction to stochastic and probabilistic based metaheuristics. Simulated annealing can be considered a hill-climbing variant with a probabilistic decision. While simulated annealing is conceptually a simple algorithm, in practice it can be difficult to parameterize. In order to promote a good simulated annealing algorithm perception by students, a simulation experiment is reported here. Key implementation issues are addressed, both for minimization and maximization problems. Simulation results are presented.

Keywords: Simulated annealing · Meta-heuristics · Artificial intelligence education

1 Introduction

Hill-climbing algorithms and simulated annealing are two randomly based algorithms which are mandatory in any course curricula involving search techniques, such as Artificial Intelligence [1]. Indeed, hill-climbing concepts are incorporated in most of the existing random-based search and optimization techniques. Simulated annealing (SA) is a technique which not only accepts up-hill climbing movements, but also accepts down-hill movements (assuming a maximization case), according to a Boltzmann probability distribution law. Since its proposal [2, 3], SA has been used to solve many problems in different scientific areas such as the cases reported in [4–7]. Moreover, many variations and refinement of the basic SA algorithm have been proposed over the years, e.g. [8–13].

SA is conceptually a simple algorithm. However, when it comes to its practical implementation, the adequate selection of their heuristic parameters can be a difficult task. Similar to any other metaheuristic, the parameters selection is problem dependent. When teaching the SA algorithm there are some specific issues which are crucial for promoting its clear understanding and perception by students, such as the ones stated by the following questions:

M. Graña et al. (eds.), *International Joint Conference SOCO'16-CISIS'16-ICEUTE'16*,
Advances in Intelligent Systems and Computing 527, DOI 10.1007/978-3-319-47364-2_70

Which starting/ending temperatures should be used?
What cooling schedule should be used?
What probability function should be used?

In this paper a teaching experiment is proposed using simple continuous function optimization examples, which aims to provide answers to these and other questions. The remaining of the paper is organized as follows. Section 2 reviews the simulated annealing fundamental issues. Section 3 presents the proposed teaching simulation experiment with illustrative results. Finally, Sect. 4 presents some concluding remarks and outlines further work.

2 Simulated Annealing: Fundamental Issues

The SA algorithm is based on a thermodynamic analogy with temperature annealing of metals, in order to minimize an energy value to avoid material ruptures. Thus, the analogy relates an energy value, $E(t)$ to the current solution $x(t)$ objective function value and *E_new* to the new trial solution, $x(t + 1)$ objective function value. In every algorithm iteration a difference between the new energy value and the current value is evaluated as follows:

$$\Delta E(t) = E\_new - E(t) \quad (1)$$

Students often switch the order in which difference (1) is evaluated without reflecting this change in the SA algorithm, leading to completely different results. A SA algorithm for a minimization problem is presented in Fig. 1(a) (for maximization see the commented steps), where *T* represents the temperature, *Tmax* the maximum temperature, *Tit* the number of local search cycle iterations per temperature value, *Tnew* the new decreased temperature and p a probability value evaluated with a law to be defined. Figure 1(b) illustrates that in the minimization case, the goal is to move downhill, randomly accepting some uphill moves depending on a probabilistic decision. In the maximization case (see Fig. 1(c)) the goal is to move uphill, randomly accepting some downhill moves depending on a probabilistic decision. The probability function can be represented by:

$$p(t) = \frac{1}{1 + e^{\frac{(E\_new - E(t))}{T}}} = \frac{1}{1 + e^{\frac{\Delta E(t)}{T}}} \quad (2)$$

Following the analogy with the thermodynamic cooling process, the temperature is decreased using a predefined schedule or by using a function. Considering as an illustrative minimization example, the temperature decreased in the interval [90,0], with a decrement of 10 every Tit = 10 iterations; Assuming an uphill movement expressed by a constant positive energy variation, ΔE = 20, the probability variation in 200 iterations obtained with expression (2) is represented in Fig. 2(a).

However, often expression (2) is approximated by expression (3), and the corresponding probability variation in the same conditions as the ones used in Fig. 2(a),

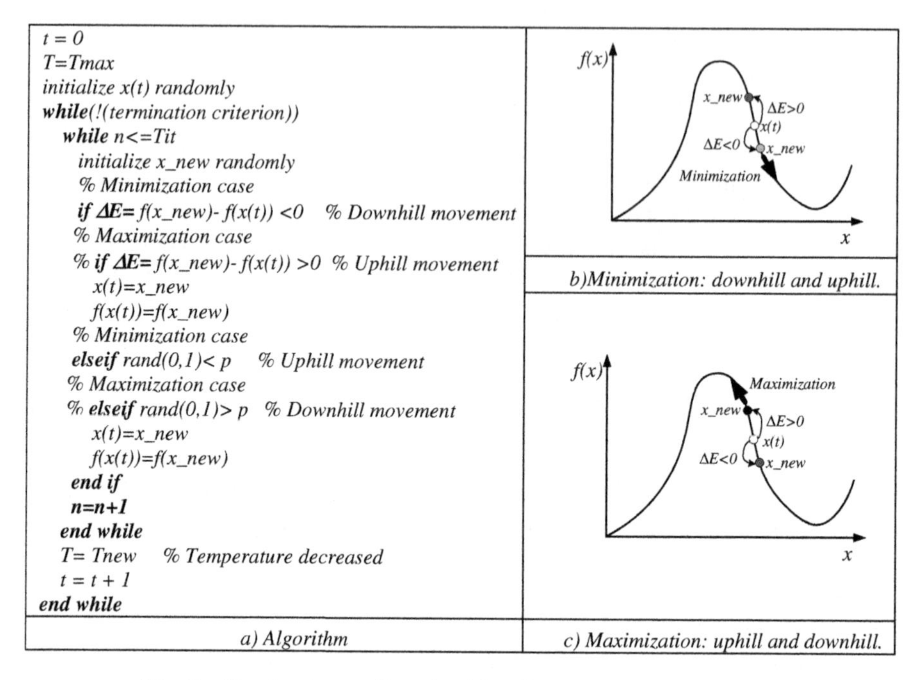

t = 0
T=Tmax
initialize x(t) randomly
while(!(termination criterion))
while n<=Tit
initialize x_new randomly
% Minimization case
if ΔE= f(x_new)- f(x(t)) <0 % Downhill movement
% Maximization case
% if ΔE= f(x_new)- f(x(t)) >0 % Uphill movement
x(t)=x_new
f(x(t))=f(x_new)
% Minimization case
elseif rand(0,1)< p % Uphill movement
% Maximization case
% elseif rand(0,1)> p % Downhill movement
x(t)=x_new
f(x(t))=f(x_new)
end if
n=n+1
end while
T= Tnew % Temperature decreased
t = t + 1
end while
a) Algorithm
f(x)
x_new
ΔE>0
x(t)
ΔE<0
x_new
Minimization
x
b)Minimization: downhill and uphill.
f(x)
Maximization
x_new
ΔE>0
x(t)
ΔE<0
x_new
x
c) Maximization: uphill and downhill.

Fig. 1. Simulated annealing algorithm for minimization problems.

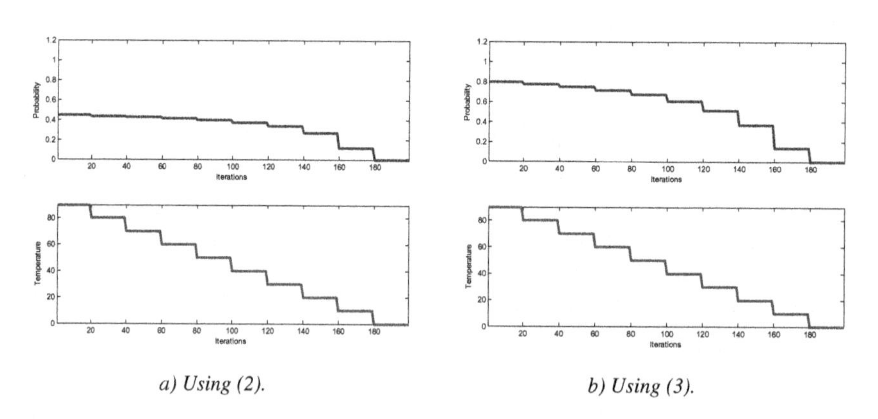

a) Using (2). *b) Using (3).*

Fig. 2. Minimization: probability variation with a constant uphill variation, $\Delta E = 20$.

results in the plots presented in Fig. 2(b). As it can been seen by comparing these two figures, there are significant differences, particularly in the starting probability which is higher in Fig. 2(b). However, if the simulations presented in Fig. 2 are repeated with $\Delta E = 2$, the results are quite different, as presented in Fig. 3. Indeed, in the case of using Eq. (3) for small delta energy amplitude values, the search will have a high probability to accept uphill movements for most of the search with an abrupt change when the temperature reaches zero, at iteration 180.

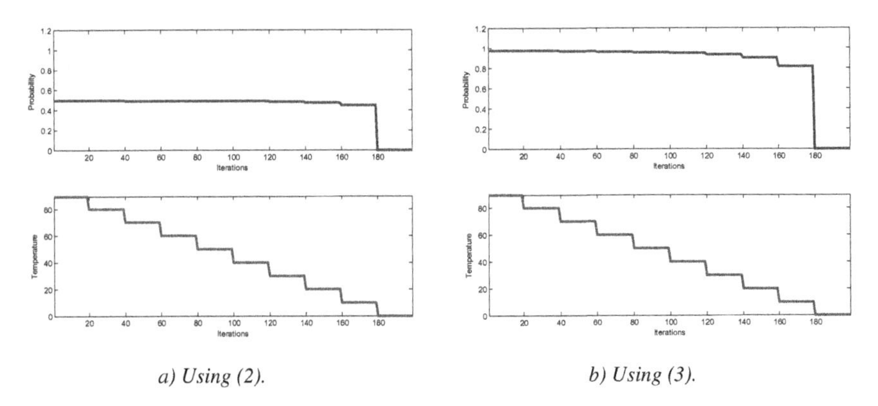

a) Using (2). *b) Using (3).*

Fig. 3. Minimization: probability variation with a constant uphill variation, $\Delta E = 2$.

$$p(t) = e^{\frac{-\Delta E(t)}{T}} \tag{3}$$

The results presented in Figs. 2 and 3 can be significantly different if instead of using a stepwise decreasing schedule for the temperature variable, the following commonly function [17] is used:

$$T_{new} = \alpha T \tag{4}$$

where $0 < \alpha < 1$ is a decreasing factor which has to be adequately selected, as it influences significantly the temperature scheduling profile. Using the same simulation conditions as the ones used to obtain the results presented in Fig. 3, updating the temperature using (4) with $\alpha = 0.94$, outcomes in Fig. 4. The examples presented so far, clearly indicate that testing objective function amplitude variation should be performed prior to the optimization, by simulating the resulting probability function profiles, in order to help in the SA parameters setting. For learning purposes it is also interesting to present the SA algorithm for maximization problems. As it can be observed the probabilistic decision stated in the SA algorithm depicted in Fig. 1(a) (see commented section) is based on uniformly randomly generated value in the interval [0,1] being higher than the probability value, p. The reason for the sign inversion relative to the minimization algorithm, is related with the probabilistic function profile used (note that *if rand(0,1) < (1-p)* could also be used). Figure 5 presents a maximization example, showing the probability variation obtained using (2) with constant $\Delta E = -2$, which represents a downhill movement. As it can be seen, initially downhill movements are accepted if the random value is larger than 0.5 and the gradually the acceptances are reduced to zero as the probability tends to one.

Concerning the initial temperature (defined as *Tmax* in the algorithms presented), in some studies [14–18] indications and formulas have been proposed to evaluate this value. A common indication relates the selection of a high initial temperature with longer computation times. An expression to estimate the initial temperature value [17, 18] is:

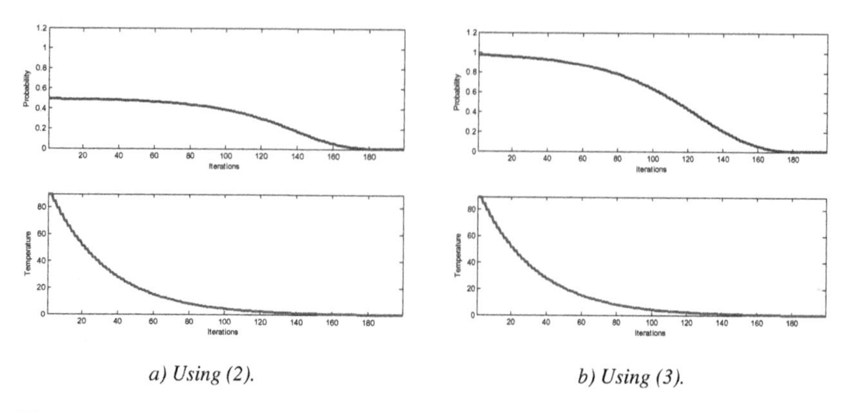

a) Using (2). *b) Using (3).*

Fig. 4. Minimization: probability variation with constant *ΔE = 2*, and *Tnew = 0.94T.*

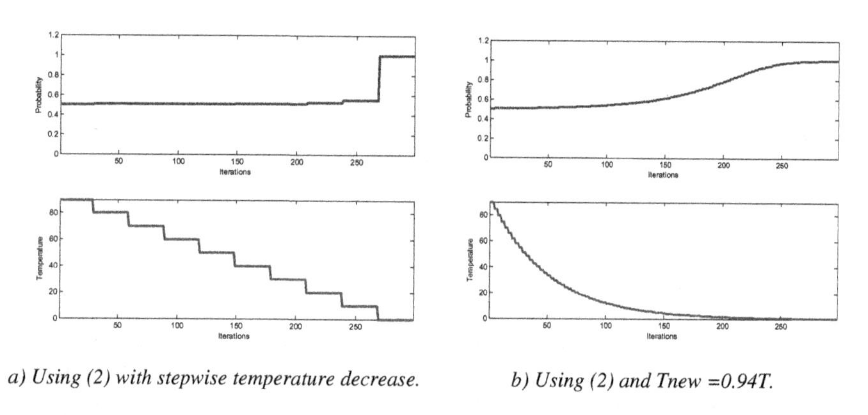

a) Using (2) with stepwise temperature decrease. *b) Using (2) and Tnew =0.94T.*

Fig. 5. Maximization: probability variation with constant variation, ΔE = −2.

$$T_{\max} = \frac{\overline{\Delta E}}{\ln(p_i)} \tag{5}$$

where: $\overline{\Delta E}$ represents an average of some uphill movements (in the minimization case) of the objective function values, and pi the initial probability. The initial temperature value along with the probability law defines the initial probability. As the search should be generalist in the beginning and specialist in the end, negative movements should be allowed at least with 0.5 of probability. However, higher values are often used [15], which when very near to 1 turn the search into a random walk. Here, students are urged to plot the probability function considering several energies deltas in order to verify the starting and ending probability values, for the specific adopted temperature cooling schedule.

3 Simulation Experiment

The first function to be deployed in the assignment proposal is represented by:

$$f_1(x) = 4\sin(5\pi x + 0.5)^6 \exp(\log_2(x - 0.8)^2), \quad 0 \le x \le 1.6 \tag{6}$$

with the corresponding search space illustrated in Fig. 6(a). As it can be seen from this figure, the function has a global maximum located at $x = 0.066$ corresponding to function maximum of 1.633. It has also several local minima and flat intermediate region (plateau). Thus, this function is adequate to illustrate typical search traps. The experiment requires students to implement the SA to maximize function (6), addressing the following topics:

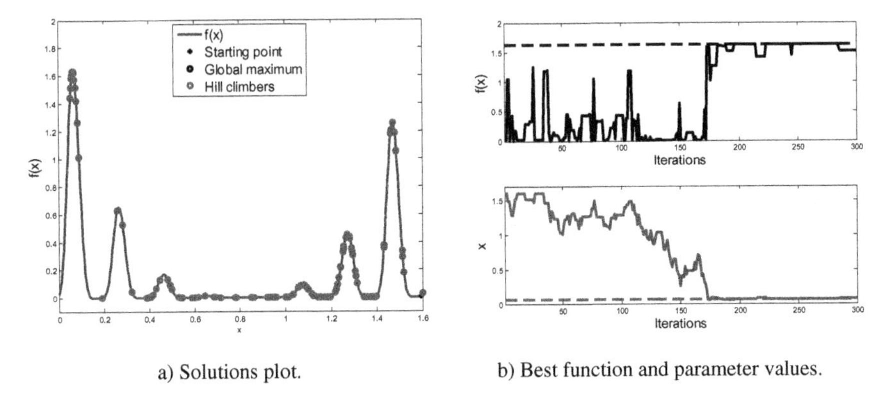

a) Solutions plot.

b) Best function and parameter values.

Fig. 6. Simulating annealing successful run.

The function used to generate the acceptance probability;
The initial and final values for the temperature;
The cooling rate;
The number of repetitions per temperature value;
Neighborhood radius selected, etc.

The results obtained for a single run of a SA algorithm are presented in Figs. 6 and 7. The conditions used to perform this test are the following: number of total iteration is 300. The temperature is decreased 100 times (termination criterion) using $Tnew = 0.93T$, and for each temperature 3 repetitions are executed ($Tit = 3$). The initial temperature is $Tmax = 10$, the probability law used is represented by (2) and new solutions are generated using a radius of 0.16 (one tenth of the x parameter range). The results presented in Fig. 7, present part of the temperature scheduling, the probability simulation using a constant negative delta ($\Delta E = -2$), using (2), the probability for all evaluated energy deltas (negative and positive) in this simulation. Figure 7, also plots the downhill movements acceptances (66) signed by the horizontal points at 1, and uphill movements (60) signed by the horizontal points at 0.

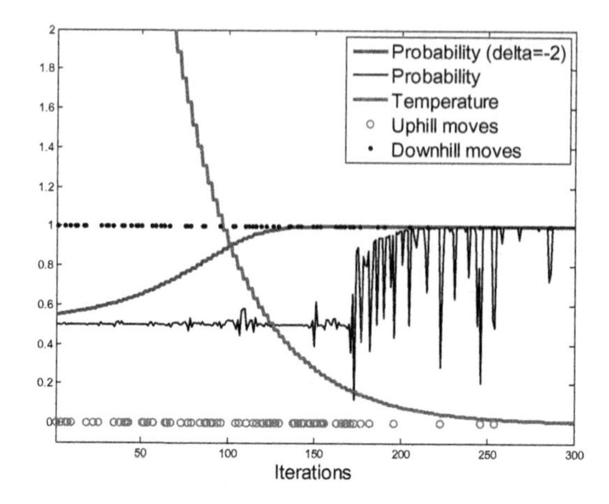

Fig. 7. Probability, downhill and uphill movements for test of Fig. 6.

From these results it is possible to conclude that from the total number of 300 performed functions evaluations in 174, there were no uphill or downhill movements. This value is mostly related with the search part, when the temperature decreases and the rejection probability of downhill moves is 1.

Figures 8 and 9 illustrate an unsuccessful trial run. As it can be observed from these figures the search converged to a local maximum. The second function to be optimized in this experiment is defined by:

$$f_2(x,y) = 0.5 + \frac{\sin(sqrt((x^2+y^2)^2 - 0.5)}{1+0.001(x^2+y^2)^2}, \quad -2.048 \leq x, y \leq +2.048 \tag{7}$$

with the minimum located at point (0,0) as illustrated in Fig. 10(a).

For this function the SA, depending on the starting point and the search radius adopted, the search can be trapped in one of the four corner values (minima). The

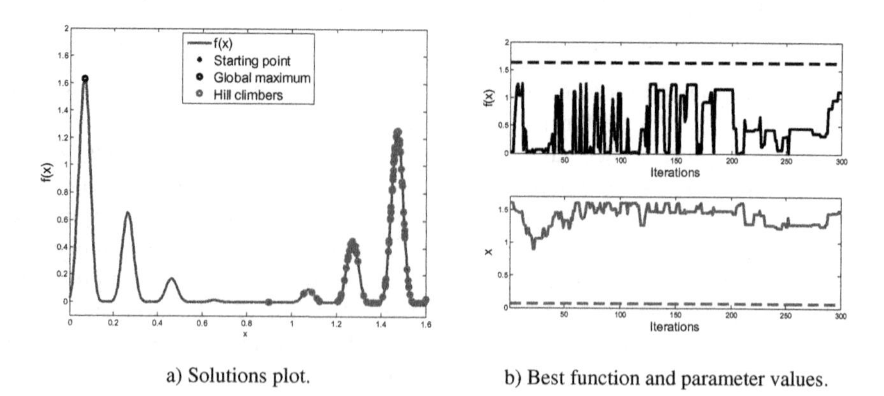

a) Solutions plot.

b) Best function and parameter values.

Fig. 8. Simulating annealing unsuccessful run.

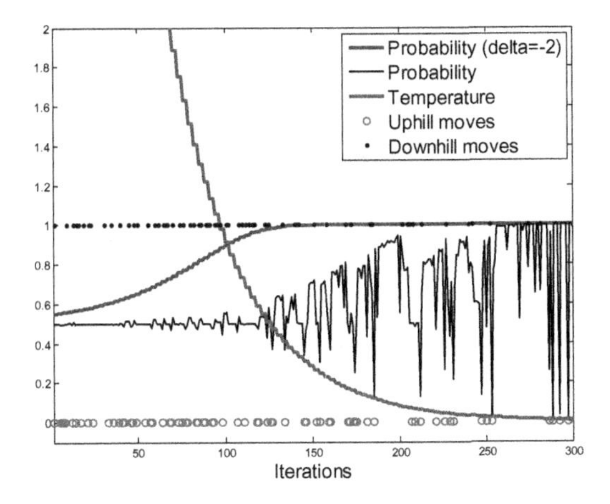

Fig. 9. Probability, downhill and uphill movements for test of Fig. 8.

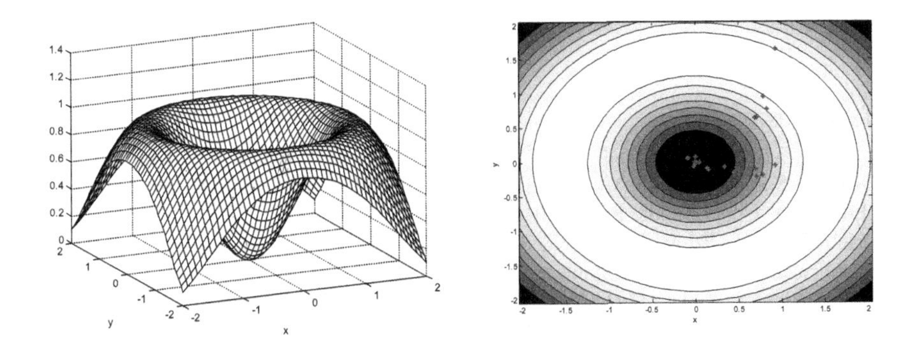

Fig. 10. Function (7) SA trial.

conditions used to perform this test are the following: number of total iteration is 1000. The temperature is decreased 100 times (termination criterion) using Tnew = 0.93T, and for each temperature value 10 repetitions are executed (Tit = 10). The initial temperature is Tmax = 50, the probability law used is represented by (2) and new solutions are generated using a radius of 0.1. A successful test trial for minimizing function (7) is presented in Fig. 10(b). As it can be seen from Fig. 10(b), the SA annealing starts from a position near the point (1,1.7) and converges for the optimum solution at the point (0,0). Function (7) can be further used by students to percept the SA algorithm limitations, and experiment the influence of the amplitude of the search radius used to randomly generate neighbors solutions.

This simulation has been used by course students enrolling in the artificial intelligence course in UTAD University, in the academic year of 2015–2016 (1st semester), as well as in the course of Intelligent Systems of UMinho University (2st semester 2015–2016). Positive feedback was received from students in classes and via a pedagogical enquiry, about the usefulness of the proposed simulation experiment to their learning of SA. Due to the time issues regarding the processing of the pedagogical data, the results could not be presented in this paper.

4 Conclusion

In this paper the simulated annealing algorithms was revisited from a teaching perspective. Fundamental issues concerning this algorithm were stated and a simulation teaching experiment proposed. The global objective of the experiment is to promote student perception of key issues related to simulated annealing search. The experiment allows to test, explore and learn how to adjust simulated annealing parameters, promoting students to perceive the key heuristic parameter selection. A formal learning inquiry data should be processed in the nearby future to further evaluate student's perception of the proposed experiment. While in this paper, the proposed experiment was tested with the simulated annealing algorithm, it can also be used to test other classical nature and biological inspired metaheuristics, such as genetic algorithm and particle swarm optimization.

References

1. Russel, S., Norvig, P.: Artificial Intelligence: A Modern Approach, 3rd edn. Pearson Education, Upper Saddle River (2014)
2. Metropolis, N., Rosenbluth, A.W., Rosenbluth, M.N., Teller, A.H., Teller, E.: Equation of state calculations by fast computing machines. J. Chem. Phys. **21**(6), 1087–1092 (1953)
3. Kirkpatrick, S., Gelatt, C.D., Vecchi, M.P.: Optimization by simulated annealing. Science **220**, 671–680 (1983)
4. Nandhini, M., Kanmani, S.: A survey of simulated annealing methodology for university course timetabling. Int. J. Recent Trends Eng. **1**(2), 177–178 (2009)
5. Wang, C., Mua, D., Zhao, F., Sutherland, J.W.: A parallel simulated annealing method for the vehicle routing problem with simultaneous pickup–delivery and time windows. Comput. Ind. Eng. **83**, 111–122 (2015)
6. Wang, S., Zuo, X., Liu, X., Zhao, X., Li, J.: Solving dynamic double row layout problem via combining simulated annealing and mathematical programming. Appl. Soft Comput. **37**, 303–310 (2015)
7. Behnck, L.P., Doering, D., Pereira, C.P., Rettberg, A.: A modified simulated annealing algorithm for SUAVs path planning. IFAC-PapersOnLine **48**(10), 63–68 (2015)
8. Ingber, L.: Very fast simulated re-annealing. Mathl. Comput. Model. **2**(8), 967–973 (1989)
9. Ingber, L.: Practice versus theory. Mathl. Comput. Model. **18**(11), 29–57 (1993)
10. Mirhosseini, S.H., Yarmohamadi, H., Kabudian, J.: MiGSA: a new simulated annealing algorithm with mixture distribution as generating function. In: 4th International Conference on Computer and Knowledge Engineering, pp. 455–461. IEEE (2014)
11. Debudaj-Grabysz, A., Czech, Z.J.: Theoretical and practical issues of parallel simulated annealing. In: Wyrzykowski, R., Dongarra, J., Karczewski, K., Wasniewski, J. (eds.) PPAM 2007. LNCS, vol. 4967, pp. 189–198. Springer, Heidelberg (2008)
12. Misevičius, A.: A modified simulated annealing algorithm for the quadratic assignment problem. Informatica **14**(4), 497–514 (2003)

13. Ali, M.M., Törn, A., Viitanen, S.: A direct search variant of the simulated annealing algorithm for optimization involving continuous variables. Comput. Oper. Res. **29**, 87–102 (2002)
14. Park, M.-W., Kim, Y.-D.: A systematic procedure for setting parameter in simulated annealing algorithms. Comput. Ops. Res. **25**(3), 207–217 (1998). Elsevier
15. Ameur, W.B.: Computing the initial temperature of simulated annealing. Comput. Optim. Appl. **29**, 369–385 (2004). Kluwer Academic Publishers
16. Shakouri, H.G., Shojaee, K., Behnam, M.T.: Investigation on the choice of the initial temperature in the simulated annealing: a Mushy State SA for TSP. In: 17th IEEE Mediterranean Conference on Control & Automation, Thessaloniki, Greece, pp. 1050–1055, 24–26 June 2009
17. Nourani, Y., Andresen, B.: A comparison of simulated annealing cooling strategies. J. Phys. A: Math. Gen. **31**, 8373–8385 (1998)
18. Lee, C.-Y., Lee, D.: Determination of initial temperature in fast simulate annealing. Comput. Optim. Appl **58**, 503–522 (2014)

Minecraft as a Tool in the Teaching-Learning Process of the Fundamental Elements of Circulation in Architecture

Maria Do Carmo López Méndez[1], Angélica González Arrieta[1], Marián Queiruga Dios[2], Ascensión Hernández Encinas[3], and Araceli Queiruga-Dios[3](✉)

[1] Department of Computer Science and Control, University of Salamanca, Salamanca, Spain
{mariadocarmolopez,angelica}@usal.es
[2] Pontifical University of Salamanca, Salamanca, Spain
maqueirugadi.chs@upsa.es
[3] Department of Applied Mathematics, University of Salamanca, Salamanca, Spain
{ascen,queirugadios}@usal.es

Abstract. In this paper, we make a pedagogical proposal to study the basic elements of circulation in architecture undergraduate degrees, using the Minecraft game as a tool. The theoretical basis for the proposal are Vygotsky's sociointeractional theory and Ausubel's theory of meaningful learning. We find ourselves reflecting on Information and Communication Technologies (ICT), gamification and video games in education. We outline some basic elements about circulation in buildings, its types, functionality, and accessibility, and the creativity needed to solve circulation problems in architecture. We introduce the Minecraft game, its characteristics, elements and use it as educational tool. We conclude that video games, specifically Minecraft, are of high interest in education as they develop skills for problem solving, collaborative work, research motivation and proactivity.

Keywords: Education · Sociointeractionism · Meaningful learning · ICT · Gamification · Video games · Circulation in architecture · Minecraft

1 Introduction

Digital technological advances are changing the most diverse fields of human knowledge. Communication and economy are maybe the fields most affected by these advances. Gines Roca [6] points out that we are experiencing a unique situation in history because it is the first time that a single new technology has changed the production systems and the knowledge transmission systems simultaneously. We live at a time when we can search, create and publish information simultaneously, in real time and ubiquitously. For the first time in human history, we share information and knowledge arbitrarily. Production systems have

M. Graña et al. (eds.), *International Joint Conference SOCO'16-CISIS'16-ICEUTE'16*, Advances in Intelligent Systems and Computing 527, DOI 10.1007/978-3-319-47364-2_71

changed so deeply in the beginning of this century that universities should adapt to changes.

Communications and economy are fields of the human knowledge strongly affected by technological advances, especially those taken place in the last thirty years. Education, hence, have also been affected, since school, at all its levels, essentially prepare children and young adults to live in that changing world. Production processes, communications, and technology change at a different speed from education. We are learning from educational failures in the past. Today, knowledge is increasingly fluid, we have to think in network, and it is also necessary to think globally. No one is a good engineer if one doesn't understand some fundamentals on economics, communication or biology.

We propose an educational activity based on some aspects from Vygotsky's sociointeractionist theory and from the theory of meaningful learning of Ausubel. It also considers some specific topics: ICT incorporated into education, gamification and circulation spaces in building projects in architecture. This paper is both a reflection on some of the impacts of new Information and Communication technologies in the teaching-learning processes and the importance of educational games within these processes. We have chosen a game that we consider important under several aspects to make an educational proposal. We made the proposal for an undergraduate degree because we understand the need to break the misconception that games are only for children. Games are excellent learning tools for all ages, essentially because in all ages we like to play. Just remember the iconic image of old men playing chess at the park. Games are for all ages because, whatever age we are, we like challenges.

Another objective, wider but equally important, is to promote a reflection on the positive impact that digital games can produce as teaching and learning strategies. Game elements, extensively studied in Gamification, can be productive at school, as long as skills and competencies development possibilities are clear and the teacher's role as mediator in the teaching-learning process is effective. This mediation especially refers to intervening in game variants structuring, monitoring progress, assess the activity's general aspects and each student's individual development. This paper also aims to be a multidisciplinary experience, with the understanding that knowledge operates as a network.

2 Educational Theories and Gamification

We propose to make a simple journey on sociointeracionist learning theory, specifically what Vygotsky's theory ([3,15]) says about the importance of the mechanisms of social interaction for the acquisition of higher skills; of the symbolic; of language and games on the development of children. Lev Vygotsky was a psychologist, precursor of Soviet neuropsychology, and also one of the most important theorists of psychology of human development. He defined some concepts that brought significant progress in the pedagogy of the twentieth century. The essence of his theory is that human development only occurs through social interaction, where cognitive skills are increasingly acquired. According to

Vygotsky [17], the activities performed on a shared basis ensure the internalization of local sociocultural thinking structures. These structures are incorporated, becoming constituent of the learner. Vygotsky made an extensive study on the importance of social mediation in learning.

We also review the concept of zone of proximal development which is essentially an area arising from processes of human learning. This idea is an awakening of various internal development processes that only happens when we interact with others. That interaction is placed in a historic and cultural environment, which is structured as well as structuring. These processes, when they take place, they become learning and open other areas of learning. Therefore, understanding that the interactions between individuals are essential for human development and that learning is a process that precedes this development, the role of schools and educational activities in general is huge.

Continuing on the subject of learning processes, we emphasize the concept of meaningful learning from Ausubel (see [1,2]), a psychologist and constructivist pedagogue who synthesized what he told: "If I had to reduce all of educational psychology to just one principle, I would say this: The most important single factor influencing learning is what the learner already knows. Find out this and teach him accordingly" [8]. We searched within his theory, which adds to Vygotsky's findings, theoretical support for an educational proposal based on a digital game. Both Vygotsky's psychogenic development studies and Ausubel's learning theory are important basis for educational technology.

In meaningful learning theory the teaching-learning process is focused on the student and, as Vygotsky proposed, it has a psychological perception, since it focuses in the internal processes that people make in order to learn. But Ausubel had a more pedagogical concern, as he especially addressed educational processes, from the necessary elements to make it possible, the conditions that guarantee the acquisition of knowledge, and the assimilation and retention of new contents until the evaluation [9].

Furthermore, meaningful learning theory is according to the teaching-learning processes that actually produce changes in the learner. To make possible these changes, there must be an acquisition of new knowledge with personal meaning and a logic within the learner's cognitive structures. That is, new knowledge interacts with the concepts and cognitive structures that the student already has. From that interaction the student builds its own, unique knowledge. Ausubel had great interest in studying the environments and elements of school learning, believing that school was able to make cognitive changes that would become truly meaningful to the student. The main contributions of Ausubel's studies for education are the importance given to the active participation; the motivation, necessary to create a subjective willingness to learn; the understanding as a fundamental part of the teaching-learning process; and the learning-to-learn dynamic as an essential educational need in schools.

For our educational proposal we extract from Vygotsky's constructivist theory the importance of social interactions for the development of internal structures and from Ausubel's constructivist theory the importance of meaning to learning. From both theories we emphasize the learner's active role to acquire new knowledge, which is the act of learning itself.

Ten years ago, in education debates we use to talk about digital divide: the difference between digital immigrants, who were educated before the technology revolution and think in a traditional way, and digital natives, who used to live with computers, video games and mobile phones from childhood, developing a new way of thinking [10]. The digital divide was the major concern, the educational confrontation between a generation that learned and taught without digital technology and the generation that does not understand the world without it. Today we can say that all of our young generations who have access to new technologies are digital students. For them the world is interactive, knowledge is multimode, entertainment is online, learning is multifocal. They are multi-task; processing several things at the same time in different ways, in different places, and who can learn faster about several subjects that we never imagined it would happen.

New ICT have not only changed the way we do things, they changed the way we think about things, the way we interact with things and with each other. The digital revolution, which replaced the mechanical, analogical and electronic technology by digital technology, which includes computers, mobile phones, video games and the internet, is not only a change of resources and an explosion of innovative resources, it is also a new form of individual and social growth. The development of these technologies is the consequence of a largely cultural revolution but, more than that, it is a discontinuity resulting in digital natives thinking and processing information in a new way [10].

This paper presents some functionalities that games can bring into education. We started from the possible concepts of game and we have chosen the one that we considered is the best: "a game is what happens within the magic circle" [16], that is, when we play we enter a new world, which has its own rules and enable new learning environments. It is through games that children learn basic things like talking and socializing.

From the business strategy called gamification: "...the use of games elements and games design techniques in non-game contexts" [5], we took some affirmations on how some aspects of games can be used as an educational strategy for developing skills and competencies.

The digital technology allows a huge amount of sensory stimuli on screens and produce each year an unimaginable number of innovations in video games. To illustrate this, in 2014, the video game industry had been roughly double that of the film industry with nearly ten million sales. This is a cultural phenomenon that tells us how much it impacts young people.

Perhaps one of the greatest contributions of ICT in the learning process is to provide tools of intellectual autonomy. Therefore, the digital technology in learning environments cannot be interpreted only as innovative educational resources, but as a fact that changes the concept of learning and ways of learning and teaching.

3 Minecraft as an Educational Tool for Architecture Students

The circulation subject in architecture undergraduate degree includes basic concepts such as the functionality, the integration, the accessibility, and the types of circulation [4]. It also contains the internalization of circulation mechanisms; the need for reality approximation in architectural projects; the necessary development of creativity and circulation solving skills in architecture.

Circulation is an important matter in architectural projects as it directly affects rooms disposition. We use Ching's concept of circulation: the thread that brings together the indoor and outdoor spaces of a building and is marked by fluidity [4]. To understand the possibilities of educational application in the proposed activity, we describe the elements of circulation: approach and access to buildings, routes layout, route-space relations, and space shape [4].

Circulation mechanisms enable and organize circulation routes and communication between spaces [12]. Circulation in buildings, since the last century, is no longer a space subordinate to the rooms and becomes a space that has its own logic, a logic that seeks freedom and fluidity. These features can be understood through the construction dynamics offered by Minecraft and that's why the game is an interesting educational tool in architecture degree.

Minecraft is a construction game created by Markus Persson, founder of Mojang AB. This game allows the player total freedom to move around the virtual world and change the elements of that world, with a strategy called open world [14]. Once purchased, the game can be played online or via downloadable installation. Minecraft has a free trial limited version.

In Minecraf, a player, through a character, individually or collaborating with other players, creates and breaks blocks and builds an infinite number of structures, which becomes his own three dimensional virtual world. To obtain different material blocks, the player has to mine them with the appropriate tools. With these blocks he creates structures and constructs his world, and including farms, fights and makes tools [7]. The player learns that it's necessary to get and then manage resources, and to use them in the best possible way. The game is a mix of adventure, survival and exploration. The player also learns that he needs to replenish his resources, as these, as in real life, are not inexhaustible. Perhaps, the best definition for Minecraft players is virtual architects that solve problems in architecture.

The Minecraft Edu application is an educational version of Minecraft. It came from a partnership between educators, programmers and Mojang AB. According to Sáez López and Domínguez Garrido [13], Minecraft Edu is an "open virtual world in which no plot or story is proposed, leaving total freedom of exploration to the subject". In this game, students and teacher have an avatar, which is the representation of their own character. Using a local web connection they are all connected. The teachers can, according to their goals, point out what should be developed. Minecraft Edu has a design that allows the teacher's control over the students' activities. From June 2016, Microsoft launches a new Minecraft Education Edition. Sáez López and Domínguez Garrido stressed that "most of the users consider that Minecraft Edu enhances creativity, develops the capacity for wonder, it is fun, it uses historical buildings-oriented contents in an effective way, it also and provides interaction advantages associated to microblogs."

The educational proposal is to create challenges for Architecture degree students involving the topic of circulation, the interior design, and the construction of buildings in the virtual world of Minecraft. The challenges could be launched before or after the theoretical study of circulation, and could include three possible general situations: internal circulation of a building, circulation between buildings, and access circulation. As in any educational proposal, the previous knowledge of students and the specific objectives or skills to be developed should be taken into account.

The aim of this proposal is to develop creativity in designing the internal circulation of buildings and to solve problems in that field, under the idea that the rooms are different functional spaces that are connected by circulation spaces. The trainer would propose each student to build a house with a certain number of rooms and they should make three versions of the same house with different circulation solutions. According to the students' foreknowledge and to the criteria for their construction and evaluation, the buildings would be more or less complex. The circulation could be seen as spaces subordinate to the rooms, or with the same importance as they allow fluidity to housing. It is imperative that students think about what circulation choices should they do and why they made one choice or another, to listen to the perception that their classmates and their teacher have of their choices. Another possibility is to work in teams, sharing experiences and criticisms. Other possibilities are: (a) to create a house in which circulation spaces are integrated in the different areas; (b) to cause a specific sensation, like enhancing the perspective; (c) to create buildings with more linear or more spiral circulations; (d) to design horizontal or vertical circulations; (e) to design circulations that work through the filtration mechanisms (where the basic element are the passing rooms [13]); (f) to generate circulations using the pipeline system (which determines the physical separation between rooms and circulation such as galleries, corridors, stairs, etc. [13]). In fact, there are many different versions of the strategy and they are easily adaptable.

It is important to consider that Minecraft's possibilities for basic studies in circulation are very interesting, but the game may not be as useful in more advanced levels in architecture courses. Minecraft's matrix world $1 \times 1 \times 1$ fails in the reproduction of proportions of architectural elements of particular importance, such as the walls and coverings width.

The key point of using Minecraft as an educational tool in architecture degree is the possibility of building not only virtual projects, but meaningful learning, which, as we have seen, comes from active experiences. Building with Minecraft offers the future architect the experience of making by themselves all the paths. The game allows to perceive different aspects, nuances and details similar to the people who will live in real life buildings. It is a good teaching strategy when we realize that playing, creating, experimenting and making mistakes improve the learning process. Furthermore, to play games increase the students motivation as they have clearly defined the goals, goals, and expectations [11], and they can improve their own project playing efficiently.

4 Conclusions

We started this study talking about the need for educational systems to change their perspective on teaching-learning processes. Joint efforts need to be made to bring all the possibilities that ICT offer in this critical area of human development closer to students.

If we consider that social environment and historical context structure our learning [15], we must accept that the schools and universities can not have the same proposals and educational strategies as in the last century. A digital generation needs educational institutions that understand the fluidity of this century, that master and use current technology, that invest in an also fluid and adaptive learning.

Universities and educational centers can no longer be repositories of information, they should be environments for meaningful learning, always considering that people have different times and ways of learning, different levels of intelligence, different ways of interacting with knowledge and countless study strategies. It is also important to understand that what used to be learned in school is no longer what's needed to be learned now. The educational needs of each generation are unique. We have chosen a game as an educational proposal.

Today, teaching and learning are concepts that come together and mixed. Social mediation in teaching-learning is important; to educate is to share; the distances between lecturers, professors and students, rational and emotional, real and virtual, teaching and learning is being reduced.

We prepare young generations to understand the digital world where they are growing. We have the immediate challenge of changing our perspective of what is to educate. It is no longer possible to add ICT in education as new tools to do old things. We need to do new things in new ways.

We conclude with the proposal of using the game Minecraft as a tool in the teaching-learning process of the fundamental elements of circulation in architecture. We don't make a technical presentation of the game's possibilities for the activity, that, we hope, may be built as a future work by a multidisciplinary team.

Acknowledgments. This work has been partially supported by the University of Salamanca, grants ID2015/0252 and ID2015/0248.

References

1. Ausubel, D.P.: The Psychology of Meaningful Verbal Learning. Grune and Stratton, New York (1963)
2. Ausubel, D.P., Novak, J.D., Hanesian, H.: Educational Psychology: A Cognitive View. Holt, Rinehart and Winston, New York (1968)

3. Chaiklin, S.: The zone of proximal development in Vygotsky's analysis of learning and instruction. Vygotsky's Educ. Theor. Cult. Context **1**, 39–64 (2003)
4. Ching, F.D.: Forma, espacio y orden. Gustavo Gilli, Barcelona (1998)
5. Deterding, S., Dixon, D., Khaled, R., Nacke, L.: From game design elements to gamefulness: defining gamification. In: Proceedings of International Academic MindTrek Conference, pp. 9–15 (2011)
6. Fumero, A., Roca, G.: Web 2.0. Fundación Orange, Madrid (2007)
7. Barrio, F.G., Barrio, M.G.: Aprender jugando. Mundos inmersivos abiertos como espacios de aprendizaje de los y las jóvenes. Revista de estudios de juventud **101**, 123–137 (2013)
8. Novak, J.D., Gowin, D.B.: Learning How to Learn. Cambridge University Press, Cambridge (1984)
9. Palmero, M.L.R.: La Teoría del Aprendizaje Significativo en la Perspectiva de la Psicología Cognitiva, Octaedro (2008)
10. Prensky, M.: Digital natives, digital immigrants. Horiz. **9**(5), 1–6 (2001)
11. Dios, M.Q., Encinas, A.H., Dios, A.Q., Queiruga, D.: Motivational programme for undergraduate students. J. Teach. Educ. **2**(2), 131–137 (2013)
12. Sáez, J.: Circulación, fluidez y libertad. Análisis **81**, 87–115 (2012)
13. Sáez-López, J.M., Domínguez-Garrido, M.C.: Pegagogical integration of the application minecraft edu in elementary school: a case study. Píxel-Bit **45**, 95–110 (2014)
14. Short, D.: Teaching scientific concepts using a virtual world-minecraft. Teach. Sci. J. Aust. Sci. Teach. Assoc. **58**(3), 55 (2012)
15. Vygotsky, L.S.: The Collected Works of LS Vygotsky: Problems of the Theory and History of Psychology, vol. 3. Springer, New York (1997)
16. Werbach, K., Hunter, D.: For the Win: How Game Thinking Can Revolutionize Your Business. Wharton Digital Press, Philadelphia (2012)
17. Wertsch, J.V.: Vygotsky and The Social Formation of Mind. Harvard University Press, Cambridge (1988)

Skills Development of Professional Ethics in Engineering Degrees in the European Higher Education Area

Lidia Sanchez[1(✉)], Javier Alfonso-Cendón[1], Hilde Pérez[1], Héctor Quintián[2], and Emilio Corchado[2]

[1] Department of Mechanical, Computer and Aerospace Engineering, University of Leon, Campus de Vegazana s/n, 24071 Leon, Spain
lidia.sanchez@unileon.es

[2] Departamento de Informática y Automática, Universidad de Salamanca, Salamanca, Spain

Abstract. In this paper, an experience to approach the competence about ethical aspects of the profession is presented. Following an existing methodology, several cases are presented to the students in order to determine if people involved have had a professional or ethical behaviour. Codes of professional ethics or conduct have been also discussed with the students. The experience has been successful since students have actively participated and valued the methodology positively. This solves the lack of prior training in these ethical aspects.

Keywords: Skills acquisition · Competence development · Professional ethics · Professional conduct

1 Introduction and Rationale

Ethical behaviour is a fundamental part of the exercise of any profession. Exercising a profession has always been considered to entail not only knowing how to apply the theoretical and practical knowledge acquired, but also how to do this with integrity. This idea, universally accepted in professions related to medicine or law, also applies in areas such as engineering [1, 4]. However, although studies have been conducted in this field, few have focused on engineering degrees, and this used to be reflected in the curricula; these detailed what engineers should know, but did not specify how they should exercise their profession.

Today, competencies related to social responsibility and ethics have been included in the new undergraduate and master's degrees. These days, the learning outcomes required to obtain a degree not only specify the areas of knowledge that should be acquired, the methods involved and how to solve problems (the "know how"); they also include competencies related to "knowing how to be" and "knowing how to act". For example, the learning outcomes for the university master's degree in computer engineering are described as follows: *Students should be capable of integrating knowledge and tackling the complexity of forming opinions based on information that,*

M. Graña et al. (eds.), *International Joint Conference SOCO'16-CISIS'16-ICEUTE'16*,
Advances in Intelligent Systems and Computing 527, DOI 10.1007/978-3-319-47364-2_72

being incomplete or limited, requires reflection on the social and ethical responsibilities entailed in the application of their knowledge and opinions.

This cross-curricular competence also appears in the undergraduate computer engineering degree: *Student should possess the ability to locate and interpret relevant data (usually within their field of study) in order to form opinions that include a reflection on issues of a social, scientific or ethical nature.*

Thus, having identified the need to inculcate this competence in our future graduates, the next step is to decide how to teach and assess it. In addition, its importance must be impressed upon our students, although this may appear run counter to what happens in society given the high number of cases of unethical behaviour reported in the news.

Here, we propose a method for assessing competencies related to professional ethics. We applied a programme developed in [2] to students taking a course in computer forensics and auditing, given in the 2nd year of the master's degree in computer engineering, which included the above-mentioned competencies.

Section 2 describes the programme and objectives of this work. In Sect. 3, the followed methodology is explained. Section 4 gives details of the implementation and in Sect. 5 the qualitative and quantitative results are discussed. Finally, conclusions of this work are gathered in Sect. 6.

2 Description of the Programme and Objectives

The programme introduces the concepts of professional ethics and ethical behaviour in the exercise of the profession. We defined the following objectives:

- Students should learn how to reflect on what they consider ethical or unethical in different situations.
- Students should acquire the capacity for critical thinking, and should possess the capacity to assess whether the behaviour in a given situation has been professional.
- Students should learn how to argue their opinions.
- Students should respect other opinions different to their own.

3 Methodology

The programme was based on the presentation of a series of cases published in [5] that posed ethical dilemmas. We used those that were employed in the programme reported in [2]. In [5], a procedure is presented that encourages the reader to reflect on ethics and ethical behaviour. In other words, it is not knowledge that is pursued, but rather self-assessment as an educational experience: through this procedure, the educational experience is the result of painstaking adult thought. By presenting a series of cases, students are prompted to think about ethical concepts. The ultimate goal is for participants to consider the experience worthwhile. The article cited proposes a method to achieve said educational experience:

1. Read an introduction about whether there is a need for ethical considerations in the field of computer science.
2. Read the ACM (Association for Computing Machinery) code of professional conduct.
3. Read a case, discuss it, reach a conclusion and take notes.
4. Consider the case in more detail and try to analyse it from different points of view. This latter step makes it possible to determine whether the analysis has given rise to other thoughts and if the initial opinion is maintained or not.
5. Opinions rather than answers are given about the cases; however, ethical principles are suggested for application to each case [3], and the conclusions obtained can be compared with those in the report.
6. Go back to step 3 and analyse the following case.

First, students had to decide whether the people concerned in the case presented had behaved ethically or unethically, or if the case was unrelated to ethics [1]. Some of the selected cases were as follows:

- Case II.5. A computer engineer who accepts a grant to design a program that is impossible to achieve.
- Case III.5. A programmer who develops software based on an existing program.
- Case V.7. President of a software company that is advertising a program he/she knows to have bugs.
- Case VI.6. Safety manager who monitors his/her colleagues email.

Afterwards, the ethical principles that a computer engineering professional should abide by were discussed and students were shown several codes of ethics such as those of the ISACF (Information Systems Audit and Control Foundation) and the British Computer Society. In brief, students are told that professional success is not only linked to their skills or dedication, but also to their abidance by a code of professional conduct.

In particular, special reference is made to when an information systems audit is performed. Some of the aspects highlighted include:

- The need not to put the personal interests of the auditor before those of the auditee.
- The responsibility inherent in their work, which will influence decision-making in the audited organisation.
- Respect for company policy, even if this differs from policies in the rest of the sector.
- Discretion with respect to the disclosure of data.
- The moral integrity to not take advantage of the knowledge acquired for use against the auditee or third parties.

To conclude, students completed the questionnaires again, responding to the same cases, in order to determine whether the comments and professional ethics analysed in each case in class had influenced their opinions.

4 Timing

The computer forensics and auditing course is taught on Fridays from 16:00 to 20:00, and this programme was implemented on 27 November 2015 as follows:

- 16:00–16:30 Completion of the questionnaire.
- 16:30–19:00 Introduction to professional ethics and analysis of different codes of conduct.
- 19:00–19:30 Debate on various topics related to professional ethics.
- 17:30–20:00 Completion of the questionnaire.

It was not possible to observe the established deadlines since this was a master's degree course and some students were unable to attend class. These therefore carried out the activities online via the Moodle platform.

5 Qualitative and Quantitative Results

The results obtained indicate that students are aware of the importance of professional ethics. Class debate, discussion of behaviours and high levels of participation all demonstrated their interest in this matter. The students also highlighted the lack of reflection on this topic in other degree courses, and all of them stated that this had been the first time that various situations were analysed in order to reflect on how they would act.

This programme provides a taste of possible scenarios that students may encounter in their professional lives, and they were grateful for an opportunity, free from the obligation or liability of a real situation, to reflect on the different aspects involved: the principles of quality, caution, professional behaviour, capacity, their own criteria, trust, discretion, economics, respect for the profession, independence, moral integrity, legality, free competition, non-interference, non-discrimination, responsibility, confidentiality, truthfulness, etc.

Some responses were extremely interesting; for example, almost all students considered unethical to use a company's email system for personal use (93.8 %), or to use another person's data (86.7 %). Only an 11.7 % thinks it is ethical to sell software with bugs. Moreover, most of them also considered unethical to evade responsibility for it (88.2 %), which indicates a high degree of responsibility, an unquestionably positive quality in our future professionals.

Other questions elicited the same disparity of opinion as among the experts, for example the ethics of a situation in which a programmer copies his/her best programs for his/her friends, but these include links to commercial programs which are then unintentionally included in the copy. Results also show how the debate of the topic affects their opinions. So, for this scenario, where friends of a programmer use copied programs that he has given to them, including the commercial portions, the majority of them thought that it was an unethical behaviour (almost 60 %). However, after discussion and commenting expert opinions, they changed their mind to a 42 % thinking that it was not unethical and a 30 % who answered that there was not an ethical issue.

In other cases they totally agree with experts' opinion from the beginning as occurred in scenario IV.7, where a instructor refuses students' request to discontinue the experiment in which the students were taught in two different ways but with the same evaluation procedure, being one way much better than the other. Identically to the experts, an 88.2 % of the students labelled this behaviour as unethical. A 70.6 % also considered that it is not ethical to use students as subjects of this kind of experiments.

Finally, there were certain scenarios where they disagree with the experts' opinion as happened in scenario III.5: a programmer produces new software built on an existing system. Although an 88 % of the experts considered that this was an ethical behaviour, the students do not think so, showing a high variability in their opinion. Half of them changed their opinion from the initial analysis and the percentages where similar, around 33 % each. This can be caused by the lack of awareness of law but it is important to point out because it can be a way of innovation and for that reason it should be encouraged for the good of the society, as experts think. If students consider this unethical, their achievements can be less than their capabilities.

6 Conclusions

In this paper, an experience to deal with a competence that handles the ethical aspects of the profession has been presented. Several cases have been discussed with the students and ethical codes have been analysed as the ACM (Association for Computing Machinery) code of professional conduct and the ISACA Code of Professional Ethics. Although students know that this is a valuable aspect in their future jobs, they have realized there is a lack in the development of this competence during their previous studies. Students have actively participated in the experience and have expressed their interest in this kind of activities, opining that they are more conscious about how a professional might behave and what is expected from them.

Future works will be oriented to develop this experience in undergraduate degrees in order to improve the sense of ethic and the responsibility of future professionals.

Acknowledgements. This work would have not been possible without the full cooperation of the students and faculty from the University of León who kindly participated in this experience.

References

1. Barroso Asenjo, P., Gonzalez Arencibia, M.: Education on informatics ethics: a challenge to social development. In: ETHICOMP (2010)
2. Dolado, J.J.: An experience in software measurement in academic settings. In: International Conference on Software Process Improvement, INSPIRE, Bilbao, pp. 107–113 (1996)
3. Parker, D.B., Swope, S., Baker, B.N.: Ethical conflicts in Information and Computer Science. In: Technology and Business QED Information Sciences (1990)
4. Singer, J., Vinson, N.G.: Ethical issues in empirical studies of Software Engineering. IEEE Trans. Softw. Eng. **28**(12), 1171–1180 (2002)
5. Weiss, E.: The XXII self-assessment: the ethics of computing. Commun. ACM (ACM) **33** (11), 110–132 (1990)

Expert System for Evaluating Teachers in E-Learning Systems

Bogdan Walek(✉) and Radim Farana

Department of Informatics and Computers,
University of Ostrava, 30. Dubna 22, 701 03 Ostrava, Czech Republic
{bogdan.walek, radim.farana}@osu.cz

Abstract. This paper deals with a fuzzy expert system for evaluating teachers in e-learning. The special fuzzy expert system for evaluating teacher using satisfaction questionnaire with evaluative linguistic expressions and separate knowledge bases of expert system is proposed. Proposed expert system was created using Linguistic Fuzzy Logic Controller. In the proposed fuzzy expert system the theory of Natural Fuzzy Logic is applied.

Keywords: e-learning system · Evaluation · Fuzzy expert system · Fuzzy logic · Questionnaire · Student · Teacher

1 Introduction

Nowadays, there are many e-elarning systems and distance learning using e-learning systems is very popular. The basis of e-learning is the availability of such a system that allows access to online courses and materials to anyone and any time [1]. E-learning offers several advantages such as the creation and organization of the course according to the specific needs and requirements of the target users (students) [2].

Courses in e-learning systems may (and should) contain different types of information and interactive elements. The major interactive elements and types of information that can draw a student in the learning process are the following [3]:

- Audio files
- Video files
- Interactive images
- Fill-in the right answers using interactive elements
- Interactive review questions, etc.

Main role of teachers in e-learning systems is to lead students to the best possible understanding of e-learning courses and to obtain the best possible knowledge and skills. The main tasks of teachers in e-learning systems are:

- communicate with students
- create and evaluate e-learning tests
- evaluate student's knowledge and skills after finishing e-learning course.

M. Graña et al. (eds.), *International Joint Conference SOCO'16-CISIS'16-ICEUTE'16*,
Advances in Intelligent Systems and Computing 527, DOI 10.1007/978-3-319-47364-2_73

This paper focuses on proposing of an expert system that will appropriately evaluate satisfaction surveys (questionnaires) and will propose specific activities that can be improved in e-learning courses. Some procedures and ideas published in this paper are based on paper [3].

2 Problem Formulation

In an e-learning course there are usually sub-tests or final-tests, which verify the student's acquired knowledge. After the evaluation of the test, the user learns whether they have passed the test or not. After filling in every test the user is full of impressions, which may include the following:

- The materials of the course were well or poorly constructed
- The quality of e-learning test or e-learning course is good or bad
- The level of communication with teacher was very low
- The level of quality of teacher is very high

To record these impressions and findings of the user's satisfaction rate with the course the surveys are prepared – mainly in the form of a questionnaire. Currently, there are several kinds of satisfaction surveys – one of them will be mentioned in this paper [3] (Table 1).

Table 1. Satisfaction questionnaire

Question	Answer1	Answer2	Answer3	AnswerN
Q1	A1-1	A1-2	A1-3	A1-N
Q2	A2-1	A2-2	A2-3	A2-N
Q3	A3-1	A3-2	A3-3	A3-N
QN	AN-1	AN-2	AN-3	AN-N

An example of a survey like this might be the following satisfaction questionnaire, for short only with 3 questions here:

1. Do you think that teacher in this e-learning course was suitably prepared?

   yes rather yes rather no not at all

2. Do you think that communicate tools in e-learning system were at a sufficient level?

   yes rather yes rather no not at all

3. Were you satisfied with the content of this the e-learning course?

   yes rather yes rather no not at all

From this questionnaire it can be stated that such a classically built questionnaire will not suffice to capture user's experience in e-learning system and their evaluation of teacher and e-learning course. Such a questionnaire does not contain some important questions and answers that have an impact on user's behavior and does not represent user's ideas and suggestions adequately, which should be taken into consideration.

3 Problem Solution

For these reasons, an expert system which will evaluate the questions in the questionnaire together with the proposal of activities that should be taken into account when creating the next course is proposed. The expert system also generates questionnaires containing important issues for teacher evaluation. Suggestions of activities which are the output of the expert system, together with their evaluation, are then visualized to the teacher (creator of the course) and thus easily understandable.

Nowadays, there are several intelligent systems for e-learning described in [4–7]. The proposed expert system uses concepts and parts of these systems.

The proposed expert system is based on the general model for a decision making process presented in [8]. A general model for a decision making process was also used in papers [9–12] (Fig. 1).

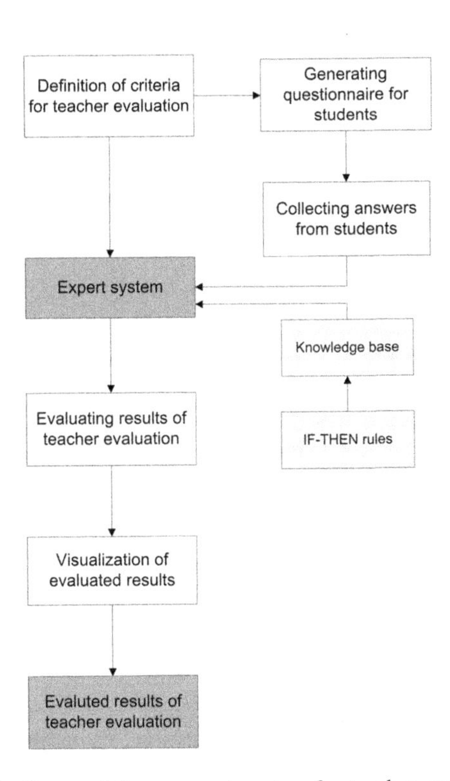

Fig. 1. Proposed fuzzy expert system for teacher evaluation

3.1 Definition of Criteria for Teacher Evaluation

In first step, it is very important to define appropriate criteria for evaluating the teacher in e-learning system. For each course and teacher specific criteria have to be defined. Based on these criteria a special questionnaire will be generated and specific rules of knowledge base in expert system will be created. Here are the possible criteria for the teacher evaluation:

- Level of teacher preparedness and motivation to study
- Level of teacher communication with students
- Level of course quality (course prepared by teacher)
- Level of test quality (test prepared by teacher)
- Difficulty of tests

3.2 Generating Questionnaire for Students

In the next step, a satisfaction questionnaire for students is generated. The satisfaction questionnaire is created based on the criteria for teacher evaluation which were defined in the previous step. The questionnaire is generated from XML file which consists of specific criteria and their values for the teacher evaluation. The questionnaire is created as a web form which is filled in by students after finishing e-learning course.

The structure of a sample XML file with criteria and their values is shown below [3]:

```
<?xml version="1.0" encoding="UTF-8" ?>
<questionnaire>
<question number="1" value="question1">
  <answer>value1</answer>
  <answer>value2</answer>
  <answer>value3</answer>
</question>
<question number="2" value="question2">
  <answer>value1</answer>
  <answer>value2</answer>
  <answer>value3</answer>
</question>
<question number="3" value="question3">
  <answer>value1</answer>
  <answer>value2</answer>
  <answer>value3</answer>
</question>
</questionnaire>
```

where,

question – represents the specific question (criterion) of the satisfaction questionnaire

answer – represents the answer for the question.

3.3 Collecting Answers from Students

In this step, after completion of the satisfaction questionnaire by the users, the results of the filled-in satisfaction questionnaire are stored in a database. The results are stored for a subsequent evaluation which will be processed by the expert system. Each filled-in questionnaire will be evaluated by the expert system and the results will be visualized and shown to the teacher (creator of e-learning course).

3.4 Creating an Expert System

Next, the expert system with the knowledge base is created. The expert system is used for the teacher evaluation. The knowledge base consists of IF-THEN rules which evaluate the selected criteria for teacher evaluation. For each criterion for teacher evaluation the separated knowledge base is evaluated. For example, for criterion called "Level of teacher's communication with students" the specific knowledge base will be created (based on questions of questionnaire related to this criterion). Evaluation of criterion is then provided by expert system. So, if there are four criteria defined in first step, then four knowledge bases will be defined in this step.

The method of creating knowledge base is visually shown in the Fig. 2:

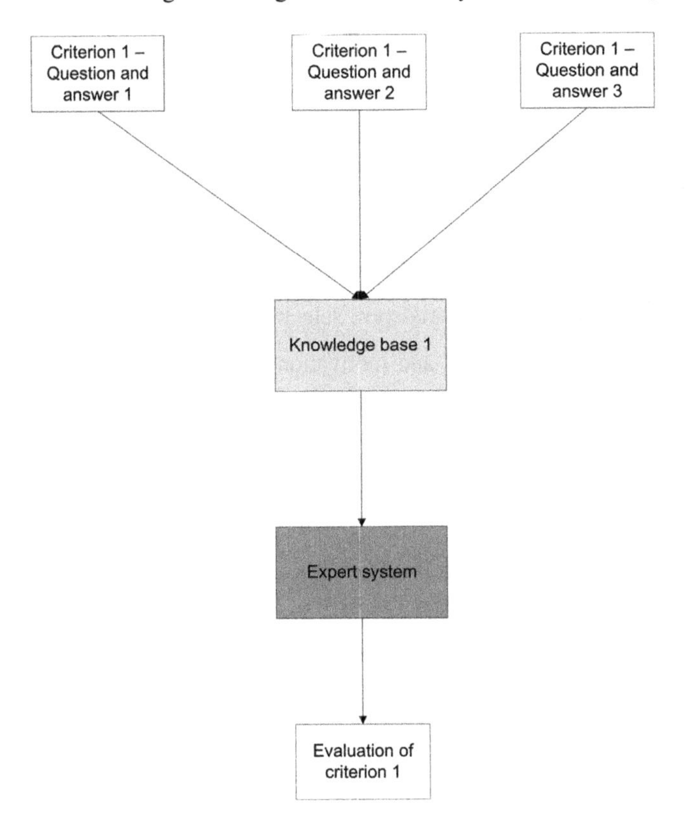

Fig. 2. Method of creating knowledge base

3.5 Evaluating Results of Teacher Evaluation

In this step, the results from the satisfaction questionnaire are evaluated by the expert system. First, the areas for this evaluation are created. These areas are created based on the criteria for evaluation and they summarize the results of the satisfaction questionnaire. Here are a few areas which can be defined for the course evaluation:

- Teacher preparedness
- Teacher communication
- Course quality
- Test quality, etc.

The expert system evaluates the degree of satisfaction with a specific area. A high degree of the area means that users are very satisfied with this area.

3.6 Visualization of Evaluated Results

Finally, the evaluated results of the selected criteria for teacher evaluation are visualized and shown to the teacher (author of e-learning course). Based on these results, the teacher can improve or modify the e-learning course.

4 Verification

For verification, we chose e-learning course with 10 students. The verification is divided into a few steps and explained in the following subchapters.

4.1 Definition of Criteria for Teacher Evaluation

For teacher evaluation, these criteria were selected:

- Level of teacher preparedness and motivation to study
- Level of teacher communication with students
- Course quality
- Test quality

4.2 Generating Questionnaire for Students

In this step, the XML file with the criteria and their values was generated. For each criterion, the linguistic variables were used:

- Very low
- Low
- Medium
- High
- Very high

Part of XML file is shown below:

```
<?xml version=″1.0″ encoding=″UTF-8″ ?>
<questionnaire>
<question number=″1″ text=″What was the level of
preparedness of the teacher during the course?″>
  <answer>very low</answer>
  <answer>low</answer>
  <answer>medium</answer>
  <answer>high</answer>
  <answer>very high</answer>
</question>
</questionnaire>
```

Based on the satisfaction questionnaire, the web form for the students was created. The web form is shown in figure Fig. 3:

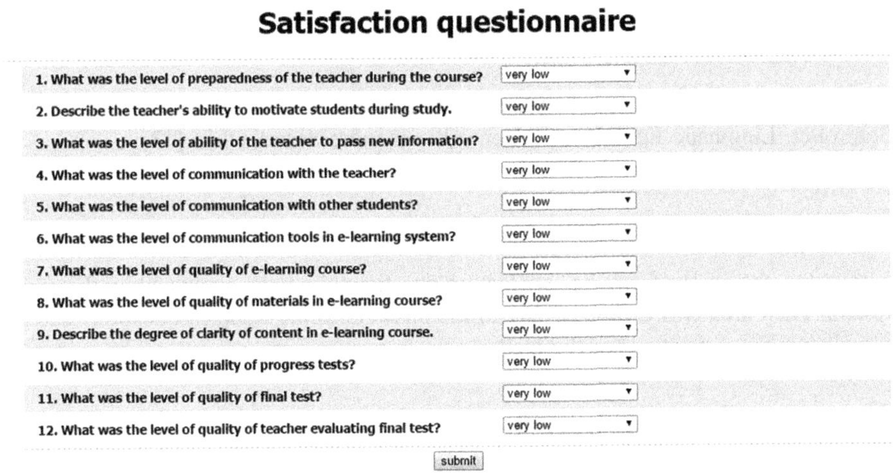

Satisfaction questionnaire

1. What was the level of preparedness of the teacher during the course?	very low
2. Describe the teacher's ability to motivate students during study.	very low
3. What was the level of ability of the teacher to pass new information?	very low
4. What was the level of communication with the teacher?	very low
5. What was the level of communication with other students?	very low
6. What was the level of communication tools in e-learning system?	very low
7. What was the level of quality of e-learning course?	very low
8. What was the level of quality of materials in e-learning course?	very low
9. Describe the degree of clarity of content in e-learning course.	very low
10. What was the level of quality of progress tests?	very low
11. What was the level of quality of final test?	very low
12. What was the level of quality of teacher evaluating final test?	very low

submit

Fig. 3. Satisfaction questionnaire for students

4.3 Collecting Answers from Students

In this step, the results from the students were collected and the filled-in questionnaires were stored in a database.

4.4 Creating an Expert System

The knowledge base of the expert system for each criterion consists of IF-THEN rules. The selected criteria (areas) are:

- Level of teacher preparedness and motivation to study (PREPAREDNESS)

- Level of teacher communication with students (COMMUNICATION)
- Course quality (COURSE)
- Test quality (TEST)

Each knowledge base has its own set of IF-THEN rules. Here are a few examples of IF-THEN rules for the area PREPAREDNESS:

```
IF (LEVEL_PREPAREDNESS IS HIGH) AND
(LEVEL_MOTIVATE_STUDENTS IS HIGH) AND
(LEVEL_TRANSMIT_NEW_INFORMATION IS MEDIUM) THEN
PREPAREDNESS IS HIGH

IF (LEVEL_PREPAREDNESS IS MEDIUM) AND
(LEVEL_MOTIVATE_STUDENTS IS HIGH) AND
(LEVEL_TRANSMIT_NEW_INFORMATION IS MEDIUM) THEN
PREPAREDNESS IS MEDIUM

IF (LEVEL_PREPAREDNESS IS VERY LOW) AND
(LEVEL_MOTIVATE_STUDENTS IS LOW) AND
(LEVEL_TRANSMIT_NEW_INFORMATION IS LOW) THEN
PREPAREDNESS IS LOW
```

The creation of the expert system knowledge base was performed in the LFL Controller. Linguistic Fuzzy Logic Controller is more described in [13].

4.5 Evaluating Results of Teacher Evaluation

In this step, the results from the satisfaction questionnaire were evaluated by the expert system. Each area was evaluated due to the answers to the filled-in questions and the knowledge base specified for the area.

4.6 Visualization of Evaluated Results

The evaluated results of the teacher evaluation were visualized and shown to the teacher. The results are shown in the figure Fig. 4:

From the evaluated results, these conclusions were made:

- For some students, the level of teacher preparedness and level of communication with students is low, for other students these criteria are medium
- For some students, the quality of course and the quality of test is medium, for other students these criteria are high or very high

From these conclusions some suggestions were proposed to the teacher (creator of e-learning course) for improving the course:

- Prepare carefully e-learning courses
- Import the level of communication with students
- Improve the clarity of the content and the structure of the course.

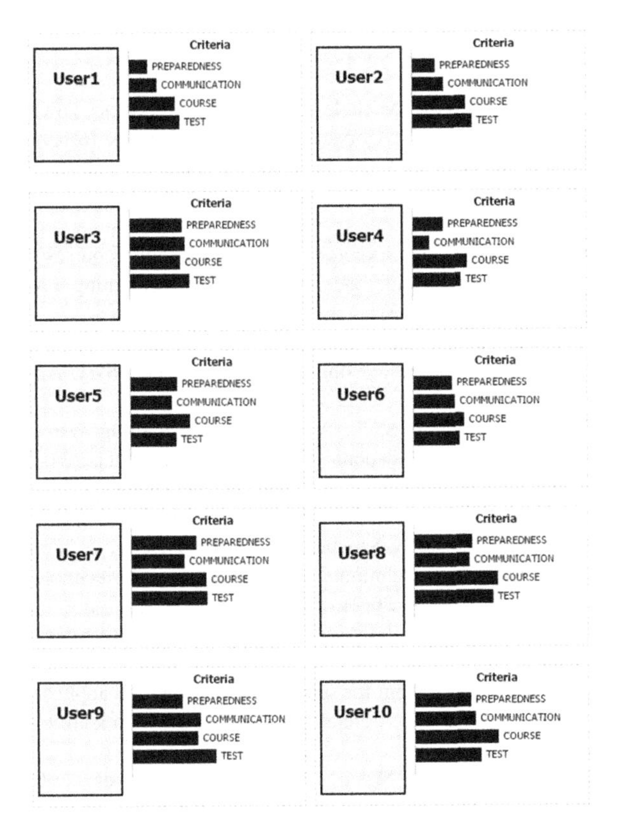

Fig. 4. Evaluated results of teacher evaluation

5 Conclusion

In this paper, the expert system for evaluating teachers in e-learning systems with generating improved type of a satisfaction questionnaire was proposed. The expert system evaluates selected areas of teacher evaluation based on a filled satisfaction questionnaire from students, defined criteria for the evaluation and separate knowledge bases. Some suggestions for the teacher of the e-learning course were proposed.

In future work it is appropriate to implement and verify the expert system in other e-learning systems.

Acknowledgment. This work was supported by the project "LQ1602 IT4Innovations excellence in science" and during the completion of a Student Grant SGS02/UVAFM/2016 with student participation, supported by the Czech Ministry of Education, Youth and Sports.

References

1. De Nicola, Antonio, Missikoff, Michele, Schiappelli, Frederica: Towards an ontological support for elearning courses. In: Corsaro, Angelo, Meersman, Robert, Tari, Zahir (eds.) OTM-WS 2004. LNCS, vol. 3292, pp. 773–777. Springer, Heidelberg (2004)
2. Gruber, T.R.: A translation approach to portable ontologies. Knowl. Acquisition **55**(2), 199–220 (1993)
3. Walek, B., Bartoš, J., Klimeš, C.: Expert system for evaluating courses in distance learning systems. In: 3rd International Conference on Innovative Computing Technology, INTECH 2013, London, pp. 246–251 (2013)
4. Huang, M., Huang, H., Chen, M.: Constructing a personalized e-learning system based on genetic algorithm and case-based reasoning approach. Expert Syst. Appl. **33**(3), 551–564 (2007)
5. Tzouveli, P., Mylonas, P., Kollias, S.: An intelligent e-learning system based on learner profiling and learning resources adaptation. Comput. Educ. **51**(1), 224–238 (2008)
6. Woolf, B.P.: Building Intelligent Interactive Tutors: Student-centered Strategies for Revolutionizing e-Learning. Elsevier Inc., Burlington (2009), ch. 1, 6, 7, 9
7. Gladun, A., Rogushina, J., Garcia-Sanchez, F., Martínez-Béjar, R., Fernández-Breis, J.T.: An application of intelligent techniques and semantic web technologies in e-learning environments. Expert Syst. Appl. **36**(2), 1922–1931 (2009)
8. Klimeš, C.: Model of adaptation under indeterminacy. Kybernetika **47**(3), 355–368 (2011). Prague
9. Walek, B., Bartoš, J.: Expert system for selection of suitable job applicants. In: Proceedings of the 11th International FLINS Conference, pp. 68–73. World Scientific Publishing Co. Pte, Ltd., Singapore (2014)
10. Walek, B., Bartoš, J., Smolka, P., Masár, J., Procházka, J., Klimeš, C.: Creating component model of information system under uncertainty. In: 18th International Conference on Soft Computing Mendel 2012, Brno, pp. 221–226 (2012)
11. Walek, B., Bartoš, J., Klimeš, C.: Process-oriented component modeling under uncertainty. In: AWER Procedia Information Technology and Computer Science, Istanbul, vol. 1, pp. 940–945 (2013)
12. Walek, B.: Fuzzy tool for customer satisfaction analysis in CRM systems. In: 36th International Conference on Telecommunications and Signal Processing, TSP 2013, Rome, pp. 11–14 (2013)
13. Habiballa, H., Novák, V., Dvořák, A., Pavliska, V.: Using software package LFLC 2000. In: 2nd International Conference Aplimat 2003, Bratislava, pp. 355–358 (2003)

The Quadrotor Workshop in Science Week. Spread of Technical and Scientific Applications in Society

Julian Estevez(✉)

Computational Intelligence Group,
University of the Basque Country (UPV/EHU), San Sebastian, Spain
julian.estevez@ehu.eus

Abstract. Along the Science Week in San Sebastian, a workshop about quadrotors was presented where basic concepts of these machines and physical laws in which they are based were explained. This activity was framed in a workshop for high school pupils and families.

Keywords: Physics · Quadrotors · Interaction · Experiment

1 Introduction

The use of science by society is crucial for the provision of scientific advice to policy makers for the benefit of society. Thus, one of the biggest education challenge reveals to be the achievement of a wide science spread for people, so that they become more informed about the problems and issues affecting society [2,3].

Formal education recognizes that school curricula must go further from traditional classes. Museums and other informal contexts meet the conditions to enhance learning process, which is difficult to get in a formal teaching process in class. These conditions involve fun, social interaction, real problems observation, novelty, etc. [4]. Science Week is included in those informal events, and in the frame of this event, we prepared a workshop about the social penetration of drones.

In this article, we introduce the social utility of quadrotors and physical laws in which they are based on. Finally, the development of the workshop in San Sebastian Science Week is described.

2 Which is the Utility of a Quadrotor?

A quadrotor is a small machine that flies thanks to three or more blades (Fig. 1), and it is radio operated.

In last years, multirrotors proliferation has increased through different applications, both for hobbyists and for industry applications. Quadrotors present

M. Graña et al. (eds.), *International Joint Conference SOCO'16-CISIS'16-ICEUTE'16*,
Advances in Intelligent Systems and Computing 527, DOI 10.1007/978-3-319-47364-2_74

Fig. 1. General view of a quadrotor

some advantages among helicopters: smaller size, more agile, they can take off and land in a smaller space. Besides, quadrotors can incorporate different technologies, such as autonomous navigation, artificial vision, and they are much cheaper than helicopters. Moreover, quadrotors do not risk passengers' life. The quadrotor in this article was built with pieces bought in Hobbyking.

3 Concepts and Physical Laws to Keep in Mind

Every multirrotor flight is based on two physical phenomena: angular momentum conservation, and flight theory. In the workshop, reasons for aircrafts to fly and Bernoulli principle were explained. These theories are not strictly applied to quadrotors [1,5,6], although they are very useful as a first approach to flight theory for young pupils.

Why do aircrafts have wings?

The aircraft can fly thanks to two effects, based on the wing. First, when a curved solid goes through the air or any other medium with a minimum speed, that fluid sticks to the surface of the wing and creates a pressure on it. This effect is called Coanda effect in Fluid Mechanics (Fig. 2).

Due to the difference between the curvature downside and upside of the wing, higher pressure is created downside. Bernoulli principle is the responsible of this effect and it states that on the surface where the air flows at high speed, the pressure is low. On the other hand, when speed is lower, pressure becomes higher (Fig. 3).

In order to understand this effect, let us think in a tube that goes from a wide to narrow section, with flowing water along it. The flow of water along the tube must be identical, as the water volume entering must coincide with the water volume coming out. Thus, when the tube gets narrower, in order to keep

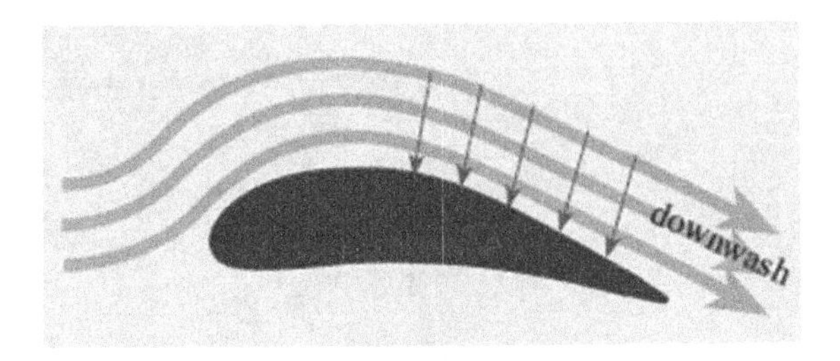

Fig. 2. Coanda effect

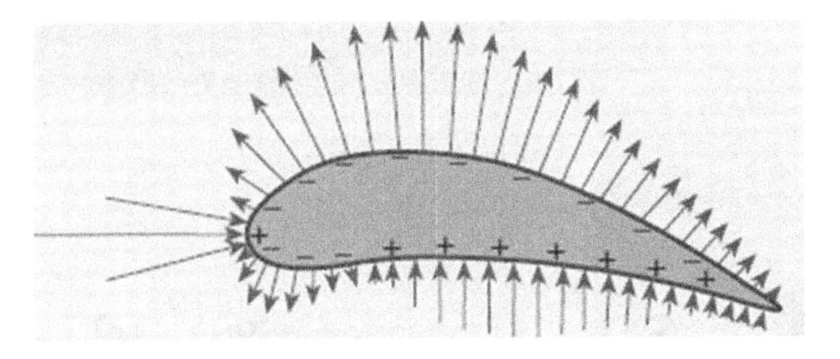

Fig. 3. Pressure distribution on a wing

a constant flow, the water speed increases and relating to Bernoulli principle, pressure decreases.

Coming back to the wing, Bernoulli principle increases pressure downside, as mentioned. And now comes the key: it is the difference of pressures on the two surfaces which creates a vertical force that lifts the plane in the air.

Gyroscopic movement: angular moment in a rotating body

Every spinning rotor behaves like a gyroscope. A gyroscope is an example of a mechanical device with a changing rotation axis. Figure 4 shows a gyroscope consisting in a little free wheel that can rotate along its axis.

It is important to remember the basic concepts of the conservation of angular momentum. Figure 5a shows a person sat on a spinning wheel, holding a rotating bicycle wheel. At the beginning, the chair stays still and the wheel spins very fast along its axis, which is now horizontal. The angular momentum of the system ($L_{sis}i$) coincides with the initial angular momentum of the wheel ($L_{r,e}i$). Thus, the angular momentum does not have any vertical component.

Now, if the person tilts the wheel upwards (Fig. 5b), the angular momentum of the wheel points upwards and the chair starts spinning clockwise along the axis of the platform.

The movement of the chair is provoked by the strength that the person makes to point the angular momentum of the wheel upwards. Due to the vectorial product, an angular momentum upwards requires of horizontal forces. Moreover, due

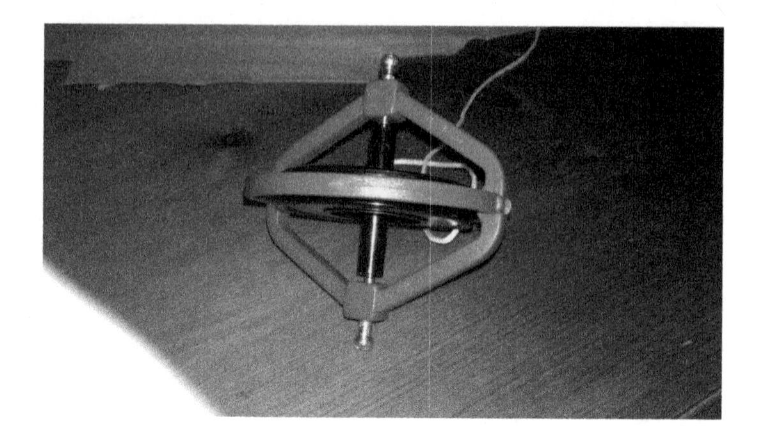

Fig. 4. Gyroscope

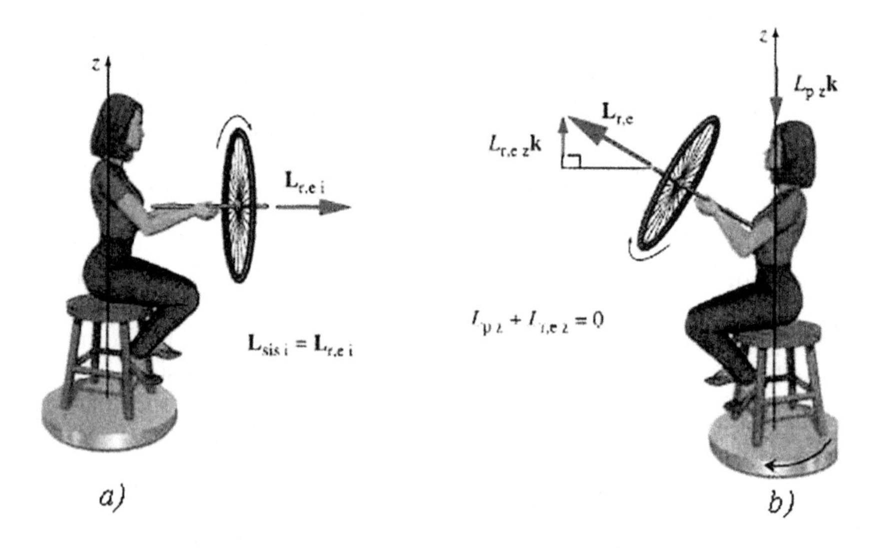

Fig. 5. Angular momentum conservation

to the action-reaction principle, the wheel produces a vertical strength downwards on the person, caused by the angular momentum vertical component of the rotating wheel.

As the total angular moment is kept constant (in absence of external forces), we can conclude that the angular momentum downwards $L_{pz}k$ is equivalent to the one that the spinning wheel generates in the vertical axis. It is precisely this strength on the person which provokes the chair to spin.

To summarize, the spinning of the chair is due to the conservation of the angular momentum, which depends on the angular speed and inertia moment. If any component of the angular speed is altered (such as the tilt of the wheel), the system reacts to keep the angular momentum constant.

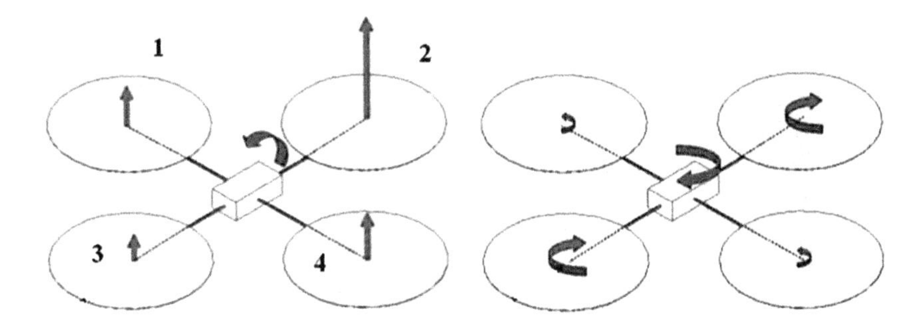

Fig. 6. Roll and yaw angles induced

This is the same phenomenon happening in the quadrotor. When its rotors rotate at high speed, it creates an angular momentum perpendicular to the rotating plane, depending on the module of the angular speed and the inertia moment of the rotor. If all the angular speeds of the rotors are identical, the angular momentum turns to 0 due to the different shape of the blades. However, if not, quadrotor can rotate along different axis (Fig. 6), inducing roll, pitch and yaw rotations.

4 The Workshop of the Quadrotor

It was developed during the Science Week of San Sebastian in 2015, in two days. Maximum capacity was of 35 people. First day, public was formed by high school students, and on the second, families and children attended. Two sessions only differed in the depth of the explanations.

The workshop started storytelling the origins of the quadrotor and stressing that once more, every scientific and technical advance came to solve a particular problem. We explained how Paul Wallich, a physicist, had to go with his child to the bus-stop everyday in Vermont (USA) [8]. There, winters are extremely cold, and Mr Wallich decided that some days are worth to stay at home, even the distance to the bus stop is just 400 m. Therefore, Mr Wallich decided that it would be convenient to build a small "wasp" that could follow and record his child. He then built a drone with real-time image retransmission to Mr Wallich computer. These are some of the advantages of being a physician dad.

Next in the workshop, some videos showing the applications of quadrotors were shown, insisting on the pros and cons of these drones against helicopters.

The second activity consisted in flying a quadrotor among students, so that they could appreciate its movements (Fig. 7).

The third activity started with a series of questions aiming to enhance the curiosity and reasoning capacity of the attendants. Questions were about the two previous activities, and we can classify those questions in two: (a) *Why do aircrafts fly?* (b) *What makes the quadrotor so agile?*

Fig. 7. Drone flight

Related to the first question, we showed a video that explains the basic ideas of Coanda and Bernoulli (YouTube)[1]. Explanation for high school students was a bit deeper because their school curricula includes basic ideas about fluid mechanics. For the second question, we started manipulating the toy gyroscope (Fig. 4). Next, another video about the angular conservation principle was showed (YouTube)[2], and its influence on the quadrotor was explained.

Participants' opinions in two workshops were very positive. High school students valued the new attracting phenomena that learned, as well as the possibility to see and touch the "mathematical abstract concepts". In the case of families and children, they valued the possibility to update their knowledge about technological applications which are in fashion nowadays.

5 Conclusions

The objective of the present activity consisted on the presentation of a quadrotor and transmit to attendants the basic concepts so that they can acquire a critic opinion about these machines, their advantages and their social impact in coming years. Another basic aspect consisted in emphasizing that engineering permits to create machines that work with a combination of electricity, materials, physics

[1] https://www.youtube.com/watch?v=AvLwqRCbGKY.
[2] https://www.youtube.com/watch?v=5cRb0xvPJ2M.

and mechanics, and trying to deny the simplistic idea that most of technology reduces to physical laws [7].

It is essential to highlight that the workshop does not pursue to teach concepts, as Bernoulli's principle. This is the reason why none of these concepts were evaluated, and the evaluation of the activity was based on the questionnaire about the organization of the event, which focuses on the visitants' perceptions and experiences. The workshop had two objectives. First, making people aware about new technical applications based on a logical speech where scientific concepts are explained. Second, getting high school students observe the practical sense of the scientific concepts that they had learned at class.

The great success of the event in the two workshop sessions shows that people are interested in this kind of activities. As a future line work, we will include extra gadgets and toys that permit understand other physical laws.

References

1. Babinsky, H.: How do wings work? Phys. Educ. **38**(6), 497 (2003)
2. Bybee, R.W.: Achieving technological literacy: a national imperative. Technol. Teach. **60**(1), 23–28 (2000)
3. Casal, J.D., Diaz, J.: Sacudiendo el aula: una experiencia sísmica de colaboración entre profesores y divulgadores. Alambique: Didáctica de las ciencias experimentales **72**, 84–91 (2012)
4. Domínguez-Sales, C., Guisasola, J.: Diseño de visitas guiadas para manipular y pensar sobre la ciencia del mundo clásico grecolatino. el taller logos et physis de sagunto. Revista Eureka sobre Enseñanza y Divulgación de las Ciencias **7**(2) (2010)
5. Liebl, M.: Investigating flight with a toy helicopter. Phys. Teach. **48**(7), 458–460 (2010)
6. Silva, J., Soares, A.A.: Understanding wing lift. Phys. Edu. **45**(3), 249 (2010)
7. Valdés, P., Valdés, R., Guisasola, J., Santos, T.: Implicaciones de las relaciones ciencia-tecnología en la educación científica. Revista Iberoamericana de Educación, **28**(1) (2002)
8. Wallich, P.: Arducopter parenting. IEEE Spectr. **49**(12), 26–28 (2012)

Erasmus Innovative European Studies

Jose Manuel Lopez-Guede[1(✉)], Erol Kurt[2], Necmi Altin[2], Manuel Graña[3], and Valeriu Ionescu[4]

[1] Faculty of Engineering of Vitoria, Department of Systems Engineering and Automatic Control, Basque Country University (UPV/EHU), Nieves Cano 12, 01006 Vitoria, Spain
jm.lopez@ehu.es
[2] Faculty of Technology, Department of Electrical and Electronics Engineering, Gazi University, 06500 Teknikokullar, Ankara, Turkey
[3] Faculty of Informatics, Department of Computer Science and Artificial Intelligence, Basque Country Univesrsity (UPV/EHU), Paseo Manuel de Lardizabal 1, 20018 San Sebastian, Spain
[4] Faculty of Electronics, Communications and Computers, Department of Electronics, Computers and Electrical Engineering, University of Pitesti, Targu din Vale 1, 110040 Pitesti, Romania

Abstract. This paper introduces an ongoing Erasmus+ project granted in the KA2 "Cooperation for innovation and the exchange of good practices" call. The project was applied by the Gazi University (Turkey) and its duration is three years. It engages five universities and involves B.Sc., M.Sc. and Ph.D. students in the scope of renewable energies from different points of view, focusing on topics following the main interests of the partners.

1 Introduction

Active Learning is a wide paradigm that groups several methods, and it is based on the responsibility and involvement of the students in its own learning process [1,2]. One of these methods is named Cooperative Learning, that is a paradigm in which the learning activities are planned looking for the positive interdependence between the participants of such learning [3,4]. The Erasmus+ projects [5] are a well known way of engaging students of different countries, and offers an appropriate platform to use active learning, and more specifically, cooperative learning strategies to profit from the different skills and competencies obtained in the educational systems of each of the participating countries. An especially suitable framework is the call related to Key Action 2 "Cooperation for innovation and the exchange of good practices", which is very appropriate since the design of the projects focuses on the obtention of joint results. In this paper authors describe a project applied by Gazi University focused on the above referenced characteristics, paying attention to some of the main relevant parts of a project.

The remainder of the paper is organized as follows. Section 2 introduces the project giving a brief description about it. One of the most relevant parts, i.e.,

M. Graña et al. (eds.), *International Joint Conference SOCO'16-CISIS'16-ICEUTE'16*, Advances in Intelligent Systems and Computing 527, DOI 10.1007/978-3-319-47364-2_75

the partners of the project are described in detail in Sect. 3, while Sect. 4 gives a list of topics that are going to be explained by the academiciand of the Basque Country University. A comprehensive list of the expected results is given in Sect. 5, and finally, Sect. 6 explain our main conclusions.

2 Description of the Project

Clean Energy Research is a key task for the EU Agenda due to the energy Independence of the European Community. The European and Anatolian plates are not rich of fossil fuels such as coal, gas and oil. Even the carbon release from the fossil fuels prevents the European Community to use these traditional energy resources and motivate them to find out and improve alternative and renewable energy resources. In this manner, the efficiency of the renewable energy devices such as solar, wind, hydro-electric and harvesters play an important role to obtain maximal power output from the nature. In addition, new trends and technological ideas on these devices should be introduced to the students in the undergraduate and graduate levels (i.e. B.Sc., M.Sc. and Ph.D.). The priority of this project is to handle this task in order to improve the educational aspects of different institutions in Europe. Initially, a curriculum study will be performed among the partners to present the best solutions of the state-of-art.

3 Description of the Partners

In this section we are going to give a brief description of the history of each partner of the project and the actual capabilities to face it successfully.

3.1 Gazi University

Gazi University the biggest fourth university among the Turkish Higher Education Institutions. It has all branches of faculties, vocational schools, institutes, etc. The university has the third engineering education rank among all Turkish universities. It has 77,000 students and 3,500 academicians. The university has been founded as a Educational High School around 1926 by the founder of Turkish republic Kemal Ataturk. Later it was transformed into an Educational Institute around 1970s and became a university in 1982. From the beginning of that date, it has been one of the historical and powerful educational institutions in the country. Gazi University has not only educated the students and made the research activities for years, but also played a very important role educating the academicians of other universities in the country. Since the M.Sc. and Ph.D. opportunities and related research facilities are very powerful, many other universities send their students to get these degrees from Gazi University. Therefore, Gazi University has been declared as the "founder of universities" throughout the country.

Presently, the university has 21 faculties, 4 higher vocational schools, 1 Turkish Music Conservatory, 11 vocational schools, 48 research centers, and 7 institutes. It has especially high educational and research ranks for Education, Medicine and Engineering. The university educates 1,500 foreign students in addition to the Turkish students and has a wide European and US networks among the other universities. Especially, the foreign students come from Middle Asia, Africa, Middle East Caucasian and European countries. In the graduate institutes, at least 5,000 master and doctoral students are educated.

The university has the European quality systems and Erasmus certificates. It has several campuses, some of them are situated in the central Ankara while others in the closed towns of Ankara. According to Scopus Research Searches, the university has important effect on Medicine, Electrical and Electronics engineering and energy fields.

3.2 The University of the Basque Country

The University of the Basque Country (UPV/EHU) is a teaching and research Institution officially founded In 1985. The UPV/EHU is the Spanish university offering the highest number of degrees, one third of these degrees having the quality mention from the Spanish Ministry of Education. The UPV/EHU has been recently recognized as an International Excellence Research Campus by the Spanish Ministry of Science and Innovation. According to the ranking of Shanghai, UPV/EHU is one of the leading universities in Europe for its teaching quality, its commitment to continuous training, and its research, development and innovation excellence. UPV/EHU holds agreements with over 400 international universities. Since the first Research Framework Programmes, the UPV/EHU has been very active and has participated in many collaborative projects and Marie Curie actions. With regards to the 7th Framework Programme, up to the date of submission the project explained in this paper, the University of the Basque Country participates in 94 projects, coordinating 22 of them and is the beneficiary 4 ERC Grants, accounting with more than 24.6 million euro of FP7 financing.

The project is being carried out with the support of the Vicerectorship for Teaching Quality and Innovation, which is structured in four main areas:

- Quality Cathedra: Contributes to the knowledge, implementation and improvement of quality management in all areas of the organization of the university, helping to achieve the highest standards of excellence.
- Institutional Evaluation Service: A service of the university which aims to guide and promote the process of evaluation, verification and accreditation as well as those related to improving the quality of higher education. It also works with reference quality agencies in developing their programs in the university.
- Faculty Evaluation Service: A service dedicated to promote, design, develop, advise, facilitate and train faculty evaluation process with the desire to contribute to the improvement of teaching quality.

- Educational Advisory Service: It is a service which manages courses and training according to the needs of the faculty. It puts in place processes to gather information about which are the formation necessities.

All these areas converge towards a methodology named IKD-Ikasketa Kooperatibo eta Dinamikoa in Basque (Dynamic and Cooperative Teaching-Learning in English), characterized by the following principles:

- Active Education: IKD invites students to become the architects of their own learning and an active element in the governance of the university. To get this, it encourages learning through active methodologies, ensures continuous and formative evaluation, articulates the acknowledgment of its previous experience (academic, professional, vital and cultural), and promotes mobility programs (Erasmus, SENECA) and cooperation.
- Territorial and social development: The IKD model development requires an ongoing process through which the university is committed to its social environment and community, with public vocation and economic and social sustainability criteria, promoting values of equality and inclusion. It also takes into consideration peculiar characteristics of each of the three provinces where sits the university, to contribute to their empowerment and to extract from them their formative potential. A curricula development responsible with the social environment is done through internships, collaboration with social initiatives, social networks, the relationship with companies and mobility programs that promote international experience and cooperation of our students.
- Institutional Development: IKD curricula development drives institutional policies that promote cooperation between the agents involved in teaching, in an environment of confidence and dynamism. It promotes programs that encourage institutional structuring through the figures of the course or module coordinator, quality commissions and promoting teaching teams, which are key elements in this new teaching culture. Other institutional actions such as offering different types of education (part-time attendance, semi-face, non-face), significant and sustainable use of information and communication technologies (ICTs), institutional regulations concerning assessment, infrastructure design of educational institutions and public spaces (IKDguneak-IKDplaces), the extension of hours of use of space, should be considered from a perspective that encourages IKD culture.
- Professional development: First, the continuous training of the people involved in teaching activities (faculty and support staff to teaching), in order to promote adequate professional development. Training programs (ERAGIN, BEHATU, FOPU) project to support educational innovation (PIE) and assessment tools for teaching (DOCENTIAZ), among others, are actions that support the construction of IKD.

3.3 University of Pitesti

The University of Pitesti is a public educational and research institution from Romania. The University has approximately 500 hired staff and approximately

10,000 students. Its organization is composed of several faculties of different educational fields: Engineering, Social Science, Economics, Law, Science, etc. The university has the Faculty of Electronics, Communications and Computers, which contains the Department of Electronics, Computers and Electrical Engineering. In this department work 30 people specialized in Electrical Engineering, Electronics and Software Engineering. Most of them hold PhDs degrees in their fields of specialization.

The University of Pitesti offers large educational possibilities to the young people from Romania and from other regions in the world. The priorities of the University of Pitesti are focused on the development of a high quality scientific research activity and on training young people as future high specialists able to find a proper job in the Romanian and European labor market. Among the priorities, it is possible to mention the large international collaboration that the university developed through a series of partnerships, projects and programs financed by the European Community.

Its target is to develop in Pitesti a business oriented university, a university deeply rooted in the everyday reality, a university that strongly interferes with the social-economic sector by offering its assistance finding the correct solutions to the numerous problems this sector is facing at present.

The University of Pitesti has various faculties, among which is the Electronic, Communications and Computers faculty. In this faculty there is an Electric Engineering Department which provides courses in the field of Renewable Energy. Also the staff of this department is actively involved in research projects with national and international partners. Among the interest fields, the teachers of Electric Engineering Department are concerned of Renewable Energy and ways of developing this area in Romania as well as other parts of the world. The University of Pitesti is a strong regional Higher Educational Institution that has collaborations with several national and multinational companies that are located in Arges county and Romania. Among these there are automotive, energy, auxiliary industry companies.

The University of Pitesti is closely related with research and educational institutions, i.e., National Research Institute of Electrotechnics, National Research Institute of Chemistry, Nuclear Research Institute, Polytechnic University of Bucharest, University of Targovtste, University of Plolesti, etc. Also being an educational institution, the University of Pitesti has very dose collaboration relation with the Pre-university Educational Regional Department of Arges county and various technological high schools of its area. Through the International Relations Department the University of Pitesti, and the Faculty of Electronics, Communications and Computers, it holds International agreements with several European Universities spread in various countries as Turkey, Spain, Poland, France and Norway.

3.4 Klaipeda University

It is situated in a territory with a population of 650,000 with prospective industrial and business potential as well as a rapidly developing marine metropolis and region that is famous for exclusive cultural heritage, tourism, recreation and resort facilities. KU was established in 1991. Its mission is to develop the university as a modem marine Center of research, arts, and studies in the Baltic Sea Region educating highly qualified specialists.

Present strategic priorities of the University are: development of rational and sustainable academic structure, formation of the Marine Valley (integrated research, study, and business center with its infrastructure comprised of research, education, and social facilities provided by highly qualified human resources). Currently, the prospective area of University Campus is comprised of 24 ha is being supplied by investments. The Business and Technology Incubator, the building of research laboratories, and student dormitories are being erected in the nearest future. Academic and research activities have been organized at seven faculties. Environmental, life, health, social, and technology sciences as well as humanities and arts have been taught there. Today around 6,000 full-time, part-time, and unclassified students are attending lectures at Bachelor, Master, PhD, vocational training, and other study programs.

Research and experimental development is organized at departmental level in faculties as well in two centers of research excellence: the Marine Science and Technology Center and the Institute of Baltic Sea Region History and Archaeology. Around 100 R&D projects are pursued annually at KU. They significantly contribute to the research, scholarly production of the University as well as to its financial sustainability.

University expands its internationalization by taking part in the main academic networks across the Baltic Sea Region and Europe; it has tied agreement-based cooperation with more than 40 foreign universities. More than 190 partner universities in Europe are Erasmus Mobility partners of KU. There have been launched joint study programs in MA and PhD collaborating with Lithuanian and abroad universities.

Klaipeda University has been awarded the ISO 9001, ISO 14001, OHSAS 18001, SA 8000 Quality Management Systems certification by TUV Uolektis. The quality of research and education provided by the university are recognized internationally as well as the status of environmentally friendly and socially responsible university. These standards bring teaching, care, social, and environmental benefits together (for example, the certificate proving the standard SA 8000:2008 for the following fields: organization and execution of studies, research performance, knowledge and technology transfer, projects management).

3.5 University of Perugia

The University of Perugia was founded in 1308. In that year, Pope Clement V issued a bull entitled Super sperula, which granted the Studium of the city, i.e., the authority to engage in higher education. The bull made Perugia a leggere generaliter, giving its degree courses universal validity and recognition. Formal Imperial recognition of the University was conveyed in 1355, when Emperor Charles I granted Perugia the permanent right to have a university and to award degrees to students from all nations. In the 14th century, the university offered degrees in two fields: Law and General Arts. Today, research, education and consulting activities in the various disciplines are organized in 16 departments, with about 23,500 students, 1,100 professors and researchers and 1,000 staff members.

4 Contributions of the Basque Country University

In this section we provide a brief list of the topics that are going to be handled by the lecturers of the Basque Country University. These topics are integrated in the curricula that has been designed by all partners of the project. The topics are the following:

1. Active and passive flow control devices for wind turbines
2. Wind turbine control design
3. Reinforcement learning for speed variable wind turbine power generation control
4. Application of Adaptive Backtracking Search Algorithm for pitch control of a mini wind generator
5. Modeling and control of photo voltaic systems
6. Consumption based Energy Model Transitions: Energy Democracy
7. Composites for renewable energies
8. Implementation of flexible control systems
9. Geo spatial Analysis of Renewable Energy.

5 Expected Results

In this section we describe the main expected outputs of the project, which can be summarized as follows:

1. Online Education and Training Material (e-learning): web-site preparation and update will be performed. The web-site for the project will be open for the awareness, announcement and education on the renewable energy issues world-widely.
2. Experience gained by the project partners in the management and undertaking of transnational partnerships.
3. Exchange of ideas and good practises: Student poster presentations on renewable energy researches for the exchange of ideas and a good academic practises will be prepared for a scientific presentation after several technical transnational meetings.
4. Exchange of ideas and good practises: Innovative laboratory researches with the students and teaching staff in each working group. Cooperative practises will be performed in several technical transnational meetings.
5. Transnational sharing of experience and best practice: Short term mobilities of students (undergraduate, M.Sc. and Ph.D.) for the workshop courses and laboratory practises in each partner country. The students will take part in a special theme in order to develop intellectual project outputs. Apart from their formal education in their institutions, workshops will give a good opportunity to practise on the renewable energy topics.

6. Transnational sharing of experience and best practice: Short term mobilities of teaching staff (undergraduate, M.Sc. and Ph.D.) for formal education in partner countries. The teaching staff will contribute to the annual workshops in other partner universities and explain their methodology in detail in their laboratories.
7. Cooperation processes and methodologies: The formation of working groups abroad Europe for energy education methodology. I will be done directly in researches and trainings (technical transnational meetings, workshops, conferences with many intellectual outcomes) within the working groups including partnering institutions with students and teaching staff.
8. Certification system: The certificates will be given to the students after the compilation of the teaching/learning activities such as the workshops, conferences, laboratory practises and certificates of gratitude will also be given to the teaching staff.

6 Conclusions

In this paper we have introduced a project that is being developed in the context of the Erasmus+ framework of the European Union. We have started the paper with a rough description of the scope and the main objectives of the project. We have described in a more detailed way all the partners of the project, paying attention to their specificities, giving a deeper insight into the Basque Country University. So, we have added a full list of the topics which are going to be handled by their academicians during the development of the project, which cover different knowledge areas. Finally, we have gathered a comprehensive list of the expected results at the end of the project.

Acknowledgments. This work and the described project were supported by the European Union through the Erasmus + Programme call of 2015, project number 2015-1-TR01-KA203-021342.

References

1. Bonwell, C., Eison, J.: Active learning: creating excitement in the classroom aehe-eric higher education report no. 1 (1991)
2. Felder, R.M., Brent, R.: Cooperative learning in technical courses: procedures, pitfalls, and payoffs (1994)
3. Felder, R.M., Brent, R.: Effective strategies for cooperative learning. J. Cooperation Collab. Coll. Teach. **10**(2), 69–75 (2001)
4. Felder, R.M., Brent, R.: Active learning: an introduction. ASQ High. Educ. Brief **2**(4), 1–5 (2009)
5. Comission, E.: Erasmus programme

Study of Huffman Coding Performance in Linux and Windows 10 IoT for Different Frameworks

Alexandru-Cătălin Petrini and Valeriu-Manuel Ionescu(✉)

University of Pitesti, Pitesti, Arges, Romania
manuelcore@yahoo.com

Abstract. In undergraduate classes it is important to understand theoretical aspects in relation with practical laboratory applications. Rasp-berry Pi is a single board computer that is inexpensive and can be used to showcase the intended behavior. Huffman encoding is used in lossless data compression and is a good example for educational applications that target any operating system and can be easily implemented in Java and C#. This paper presents the implementation of Huffman algorithm on multiple frameworks (Oracle Java, OpenJDK Java, .NET Framework, Mono Framework, .Net Core Framework) in order to test its performance for Linux and Windows IoT.

Keywords: Huffman · Algorithm · Performance · Windows iot · NET core · Frameworks · Raspberry pi

1 Introduction

In Computer Science data compression is defined as encoding given data sets such that they are stored in computer memory using less data bits than before [1]. The purpose is to use less resources and a faster, more efficient data transfer rate.

Huffman coding is an encoding technique that offers lossless data compression. Huffman Algorithm is highly used in most encountered compression formats (BZIP2, GZIP, PKZIP, JPEG, PNG), offering a way to avoid newer patented coding algorithms.

Raspberry Pi is a platform used in many homemade applications because of its price and connectivity options. The Raspberry Pi can be connected to a network and accessed remotely like shown in Fig. 1. As most mobile devices are ARM based, at undergraduate level it is important to prepare the students for this type of hardware, in order to detect the strengths and weaknesses of such platforms compared to × 86 systems. The fact that there are multiple operating systems which run on the ARM based Raspberry Pi is also a plus in an educational environment: Linux based OS with Raspbian, Ubuntu MATE, PiNet and Fedora; Windows 10 IoT; Risc OS; FreeBSD. For a student classroom it is very easy and simple to remotely connect to such systems and test their programs written in different frameworks (Java, .NET, etc.) on different operating systems.

Windows IoT was introduced by Microsoft in 2015 [2] and presents an alternative to Linux as an operating system for Raspberry Pi [3]. It is important to test the

M. Graña et al. (eds.), *International Joint Conference SOCO'16-CISIS'16-ICEUTE'16*,
Advances in Intelligent Systems and Computing 527, DOI 10.1007/978-3-319-47364-2_76

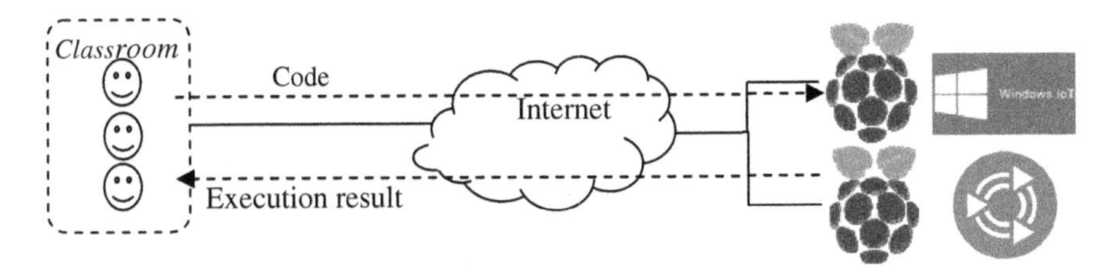

Fig. 1. Students in the classroom and at home can access the remote systems and run their code

performance of this new operating system and compare it with the already proven Linux OS. Microsoft supports the .NET Core Framework along with Windows IoT therefore it is also necessary to consider the performances of this framework.

This paper presents the implementation of Huffman Algorithm in multiple frameworks (Java Oracle, OpenJDK Java, .NET Framework, .NET Core). The tests use two Raspberry Pi installations, one with Ubuntu MATE and the other with Windows 10 IoT. The code was run on these systems and the execution results were saved and compared in order to draw conclusions regarding the code portability and execution performance. This type of educational applications help students understand the advantages of common languages that run on multiple platforms and its disadvantage when trying to obtain maximum performance. Students must understand that there is a complementary relation between code portability and execution speed. The paper is structured as follows. In Sect. 2 the Huffman algorithm is presented. Section 3 introduces the frameworks, the operating systems and the hardware platform used for testing. Section 4 presents the modifications of the code in order to make it compatible with the tested frameworks. In Sect. 5, the test results are analyzed and Sect. 6 presents the conclusions and feature research.

2 Huffman Algorithm

Huffman coding is a technique used by many software compression tools. It belongs to the group of algorithms that use codes (sequences of bits) of variable length to replace individual symbols (like characters). The length of Huffman codes (words) is related to the frequency of the characters in the text to be coded. Symbols that have a high appearance rate are assigned smaller words than the more rare ones, thus reducing space.

Huffman words are optimal prefix codes - meaning that no other code can be the prefix of another. Text coding is obtained by concatenating the words assigned to each text character. Huffman coding produces lossless data compression, where coding and decoding are unambiguous due to prefix codes.

Example: Assume 100,000 characters composed of symbols from the alphabet a; b; c; d; e; f. These symbols (that compose a text) must be saved in an efficient way. If we use a fix code like the one in Table 1a, the resulting file would be 300,000 bits. In case of ASCII code this would be 1,600,000 bits. Considering the alphabet in Table 1a and the character frequency in Table 1b, with the variable length prefix codes from Table 1c, the coding would need 224,000 bits, which means 25 % less used space.

Table 1. Values associated to the alphabet [4]

(a) Alphabet with three bit fixed length codes

a	b	c	d	e	f
000	001	010	011	100	101

(b) Alphabet with associated frequencies

a	b	c	d	e	f
45	43	12	16	9	5

(c) Alphabet with variable length codes

a	b	c	d	e	f
0	101	100	111	1101	1100

(d) Alphabet with Huffman codes

a	b	c	d	e	f
0	100	101	110	1110	1111

2.1 Mathematical Description of Huffman Algorithm

First step is the calculus of every character appearance. In this kind of situations, one can use standard character appearance frequencies specific to each language.

Let $C = \{c_1, c_2, \ldots c_n\}$ and $f_1, f_2, \ldots, f_n$ be their associated frequencies. Let l_i be the length of the string of bits that codes each symbol. Then the total length of the binary representation would be:

$$L = \sum_{i=1}^{n} l_i * f_i (1)$$

For file compression, one must minimize this expression in a bottom up manner:

- At first, consider a partition of the set $C = \{\{c_1, f_1\}, \{c_2, f_2\}, \ldots, \{c_n, f_n\}\}$ given by a forest of single leaved trees.
- The final tree is obtained after n-1 unifications.

The unification of two trees A_1 and A_2 consists of obtaining a new tree A, where the left sub tree is A_1 and the right subtree is A_2 and the frequency of the root represents the sum of the frequencies of the two trees.

Every unification is made between the two trees that have the lowest frequency values in the forest. In the end there will be only one tree in the forest. The task is to compress a text file formed of characters in the alphabet $\{a, b, c, d, e, f\}$ with the associated frequencies $\{45, 13, 12, 16, 9, 5\}$.

The Huffman tree construction starts with building the forest of single leaf trees, presented in Fig. 2.

The Huffman tree construction steps:

Step 1: Unification of e and f trees, with the lowest frequencies (Fig. 2b) [4].

Step 2: Unification of b and c trees that now have the lowest frequencies (Fig. 2c).

Step 3: Unification of d and the tree that has a frequency value of 14 (Fig. 2d).

Step 4: Unifying trees with frequency values 25 and 30 (Fig. 2e).

Step 5: Unifying remaining two trees gives the Huffman tree associated to the characters in the alphabet (Fig. 2f). The resulting codes are presented in Table 1d [4].

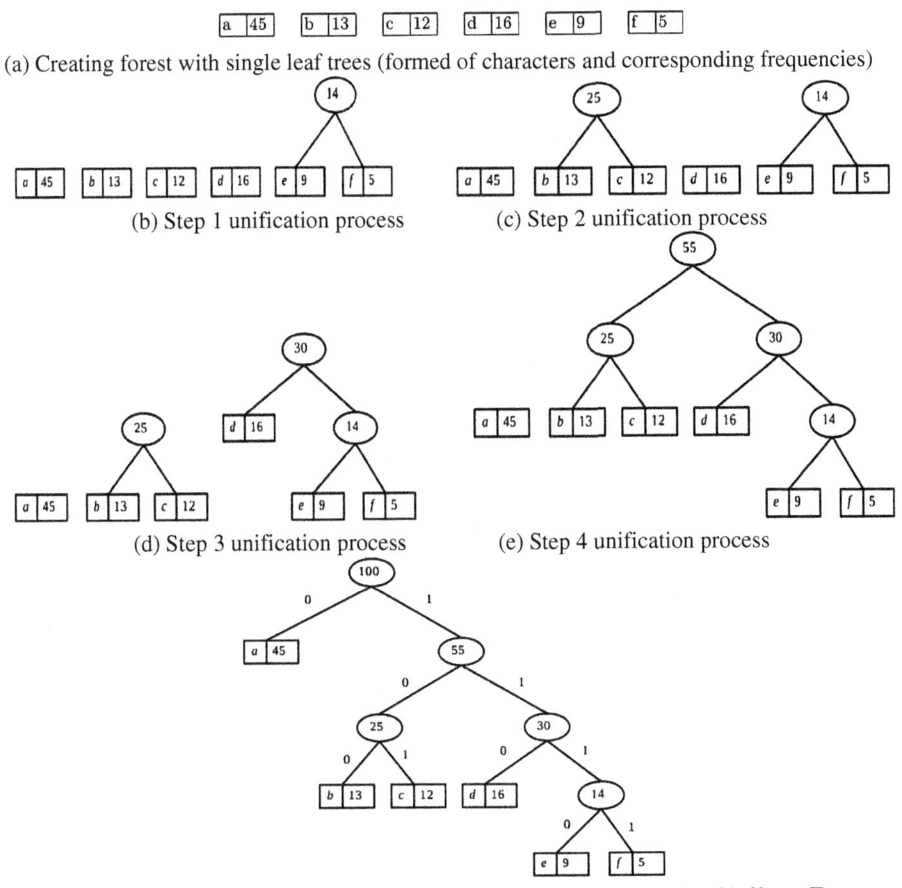

(a) Creating forest with single leaf trees (formed of characters and corresponding frequencies)

(b) Step 1 unification process

(c) Step 2 unification process

(d) Step 3 unification process

(e) Step 4 unification process

(f) Step 5 - unifying remaining two trees and obtaining the resulting Huffman Tree

Fig. 2. Steps in obtaining the Huffman Tree [4]

2.2 Pseudo Code Description of Huffman Tree Construction

Step 1: Initialization

- each character is associated a tree with a single node;
- one organize the characters in a stack that holds the trees;
- the stack represents the forest and the trees are ordered ascending.

Step 2: Repeat for n-1 times:

1. remove the first two elements from the stack, respectively X and Y;
2. unify X and Y trees:
   - create a new node Z - root of the new tree;
   - Z:left: = X;
   - Z:right: = Y;

   - Z:frequency: = X:frequency + Y:frequency;

3. insert Z in the stack so that the stack remains ordered.

Step 3: The remaining node in the stack is the root of the Huffman Tree. Codes are generated by traversing the tree in preorder [4].

3 Hardware and Software Platforms

We test the performance of Huffman Algorithm on the hardware platform Raspberry Pi 2 and the assumption that led to the results presented in this paper was that Java was supported on the recently released Windows IoT for Raspberry Pi (ARM processor architecture).

Our search for getting Java on this OS led us in contact with Azul Systems which provides a commercialized build of OpenJDK named Zulu, as a result of their partnership with Microsoft. They are also building an embedded version for Windows IoT on ARM, but only succeeded in releasing a beta for the Linux based systems. Azul Systems contacted us and promised they would get in touch again, when a version of Zulu Embedded for Windows IoT ARM will be available.

Currently, they provide Zulu Embedded for Windows IoT on $\times$ 86 architecture.

Given the mentioned situation, we opted to port the code to C#, as in Linux one can use Mono, and in Windows IoT, .NET Framework. Further research showed that .NET Core - the open-source project launched by Microsoft, is planned to replace the actual . NET Framework [5].

At this moment, the main work in .NET Core is focused on ASP .NET 5 which offers a way to run same code on Linux and Windows IoT for ARM based Raspberry Pi 2. However, there is no complete equivalency between the C# solutions presented, as Mono and .NET Framework are different implementations of the same standards (published by Microsoft), and .NET Core is based on stable Mono for running applications on Linux, whilst the runtime CoreCLR for Windows IoT is in beta development stage.

The operating systems decided upon are: Ubuntu Mate 15.10 and Windows 10 IoT Core. We also considered testing our implementations for headed and headless runs of the operating systems.

4 Huffman Software Application

The application used for performance testing is based on the Java application described in the university undergraduate paper [6]

The code was modified as follows [7]:

1. Removal of elements that could obstruct performance of the algorithm:
   - graphical interface - we removed the GUI application, but we were forced to use a very simple, one button interface in the case of the native IoT C# version.

- the script model of the application - the original application was composed of two applications itself: the main one - with the algorithm -designed to be used as a console command, and the graphical front-end, that would use the before mentioned and do some estimate calculus on the results. Therefore, we kept the command-line program and hardcoded the tests in Main() method (or function for C#).
- minimizing static memory access - by writing data into memory streams instead of sending them directly on the MicrosSD card. Upon algorithm completion, memory streams are flushed onto the card. Access of static memory is made only outside of the algorithm sections

2. Time of execution was measured only for algorithm sections
3. The algorithm flow is almost identical from platform to platform. There is a small difference between Java and C# versions on the number of rows the time measuring takes place.

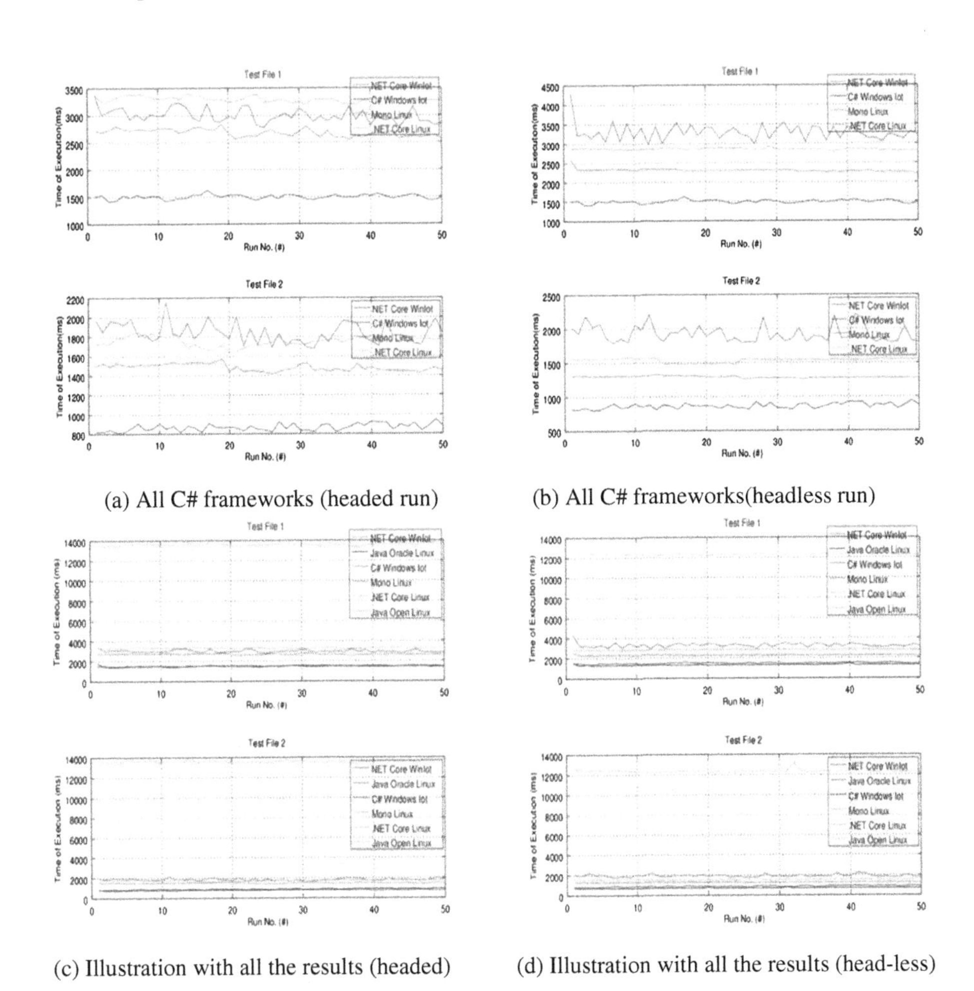

(a) All C# frameworks (headed run)

(b) All C# frameworks(headless run)

(c) Illustration with all the results (headed)

(d) Illustration with all the results (head-less)

Fig. 3. Overview on all the results [7]

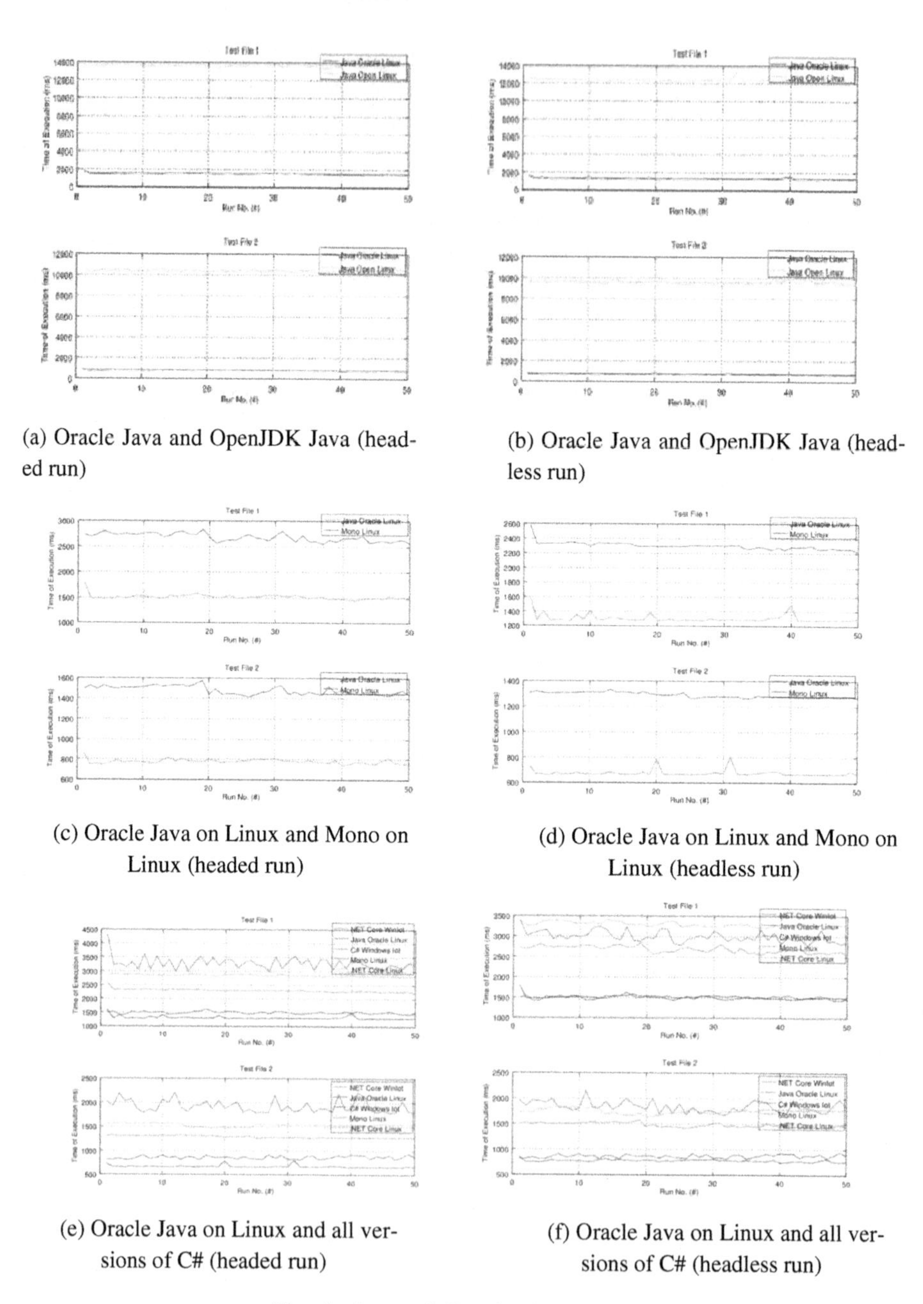

(a) Oracle Java and OpenJDK Java (headed run)

(b) Oracle Java and OpenJDK Java (headless run)

(c) Oracle Java on Linux and Mono on Linux (headed run)

(d) Oracle Java on Linux and Mono on Linux (headless run)

(e) Oracle Java on Linux and all versions of C# (headed run)

(f) Oracle Java on Linux and all versions of C# (headless run)

Fig. 4. Java and C# related results [7]

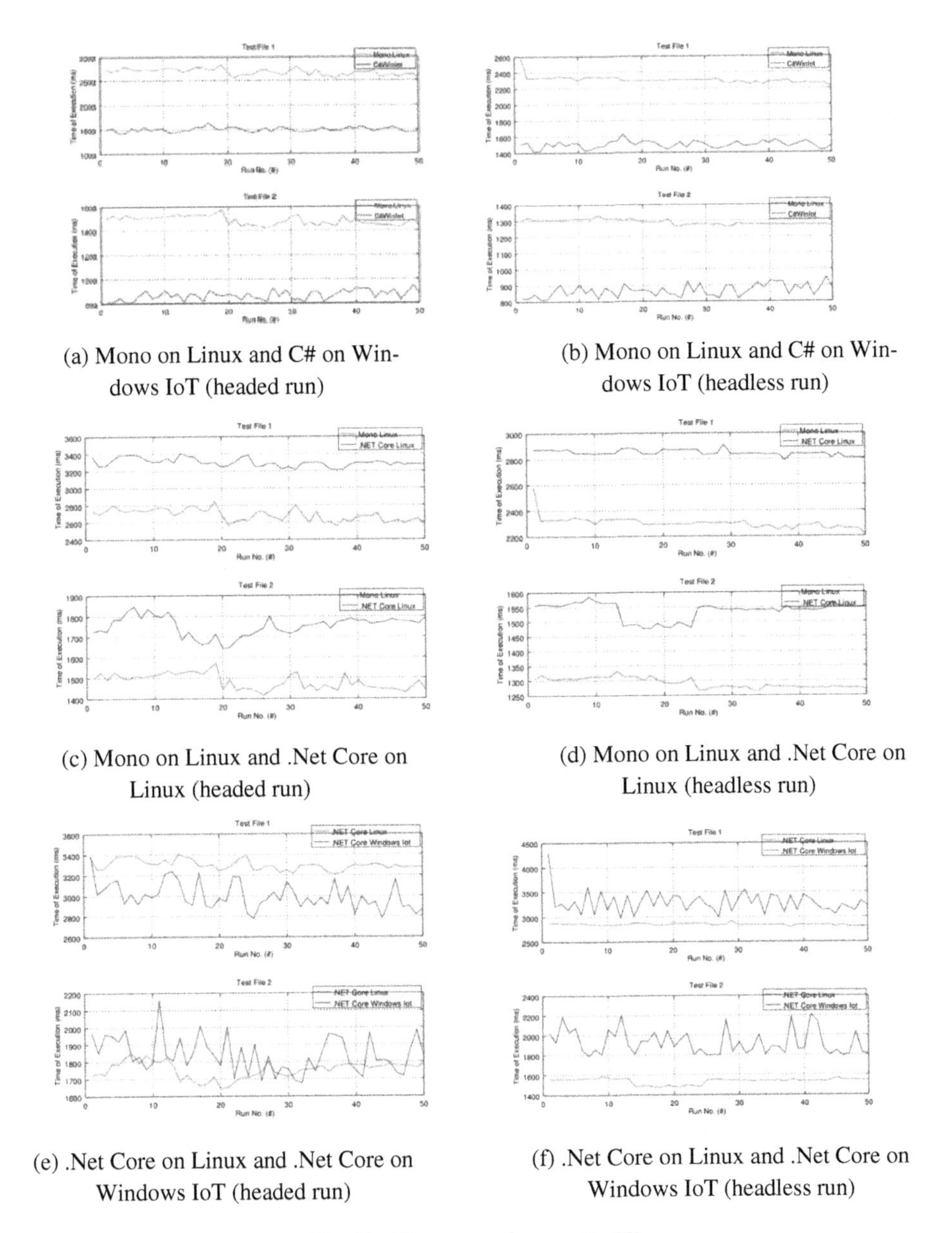

(a) Mono on Linux and C# on Windows IoT (headed run)

(b) Mono on Linux and C# on Windows IoT (headless run)

(c) Mono on Linux and .Net Core on Linux (headed run)

(d) Mono on Linux and .Net Core on Linux (headless run)

(e) .Net Core on Linux and .Net Core on Windows IoT (headed run)

(f) .Net Core on Linux and .Net Core on Windows IoT (headless run)

Fig. 5. C# comparative results [7]

4. The flow of the algorithm itself is based on portability and is identical for C# implementations and the construction uses equivalent semantic structures from Java to C#.
5. Necessary code adaptations were made for writing to the SD card on Windows IoT, keeping in mind that the execution of the algorithm is sequential.

5 Implementation Results

In the following representations, there are only time results for C# Windows IoT in headed run. Some important remarks must be made:

- There is a big discrepancy between Oracle Java and OpenJDK Java (Fig. 4a, b). This is because OpenJDK is an open-source rewritten version of Java that is optimized in a different manner. Figure 3c, d suggests that OpenJDK has the poorest performance of all the frameworks.
- Oracle Java performs closely the same with native IoT C#. The best results are obtained by these two frameworks. One remark that could certainly arise from here is that Windows Iot is very well optimized with Microsoft's IoT C#.
- The open-source Mono and .Net Core have poor results compared to proprietary . NET (Figs. 3a, b and 5a, b).
- In general, headless results are better than headed, and this is normal, because the graphical interface paradigm is usually a great stress to the system.
- .Net Core for Windows IoT shows great instability comparing with .Net Core for Linux (Fig. 5e, f), but this can be explained by the fact that the Linux version is based on Mono, which is a long run project - therefore stable. .Net Core for both operating systms shows more stability in headless tests.
- Mono (Linux) has better results than .Net Core Linux (Fig. 5c, d).
- From all tested frameworks on both operating systems, Oracle Java is the best choice for Linux, and Microsoft's proprietary C# has the best performance of all other cases that involve C# programming language (Fig. 3a, b).

6 Conclusions

This paper presented the implementation of the Huffman coding algorithm and the results with different operating systems and some corresponding frameworks.

The results show that the fastest framework is Oracle Java on Linux. Comparing platforms for .NET execution shows that Microsoft's C# for Windows IoT has the best performance of all the other C# alternatives.

This paper presented an example of using the Raspberry Pi to evaluate and compare code execution times across multiple operating systems, using multiple programming frameworks. The same can be extended to other algorithms and operating systems, making this setup an invaluable environment for programming, computer networks, operating systems and other related courses. Raspberry Pi continues to evolve while keeping the compatibility with earlier versions and the low price tag, therefore upgrading the setup in the future should also be inexpensive to do.

Accessing the platforms remotely frees classroom space and making platforms accessible from home allows the students to run their code for homework. Also, because the Raspberry Pi cost is low (such as Raspberry Pi Zero), it makes easy the transition to home projects for students and encourages the usage of multiple operating systems to explore their advantages. Future research on this topic includes testing the application in other languages and operating systems available for the Raspberry Pi.

References

1. Mahdi, O.A., Mohammed, M.A., Mohamed, A.J.M.: Implementing a novel approach an convert audio compression to text coding via hybrid technique. IJCSI Int. J. Comput. Sci. Issues **9**(6), 53–59 (2012)
2. Microsoft: Release Notes - Windows Iot. https://ms-iot.github.io. Accessed 10 Feb 2016
3. Raspberry Pi: Raspberry Pi Downloads - Software For The Raspberry Pi. Accessed 10 Feb 2016
4. Hashemian, R.: Direct Huffman coding and decoding using the table of code-lengths. In: Proceedings of the Information Technology: Coding and Computing [Computers and Communications], ITCC 2003, pp. 237–241 (2003)
5. Anderson, T.: Why Microsoft's .NET core is the future of its development platform (2015). http://www.theregister.co.uk/2015/11/20/microsoft_net_core_development_platform_fork. Accessed 10 Feb 2016
6. Petrini, A.-C.: Graphical Application For Text File Compression Through Huffman Method. University of Pitesti. Print (2014)
7. Petrini, A.-C.: Programming Frameworks Performance For ARM Systems on Windows IoT And Linux. University of Pitesti. Print (2016)

Virtualization Laboratory for Computer Networks at Undergraduate Level

Valeriu Manuel Ionescu(✉) and Alexandru-Cătălin Petrini

University of Pitesti, FECC, Pitesti, Arges, Romania
manuelcore@yahoo.com

Abstract. A networking hands-on laboratory involves connecting cables, configuring network equipment and testing the resulted network and its traffic. Virtualization, which emulates hardware with software, is used today to better use the hardware resources and offers flexibility when it comes to network connectivity design. This paper presents the challenges related to the use of type 1 and type 2 virtualization solutions in a university laboratory, focusing on the solutions Oracle VirtualBox and Microsoft Hyper-V.

Keywords: Virtualization · Education · Hypervisor · Oracle VirtualBox · Microsoft Hyper-V

1 Introduction

Virtualization represents the creation of a software version of a real system that can be used independently from the underlying hardware structure. Hardware virtualization (and in particular network hardware virtualization) is just one of the virtualization aspects where the real characteristics of the hardware platform are abstracted and are presented to the user as software options. The virtualization brings advantages such as: cost reduction through power consumption; improving system management speed; system security and design flexibility.

The most important components of a virtualization system are: the host system (the hardware where the virtualization system is running), the hypervisor (the manager that handles how the Virtual Machines -VM will share the host resources), the virtual machines (with a guest operating system) and the data store (where the virtual machine are stored as files).

Virtual machines and services are usually accessed and managed through a computer network. As the changes to a virtual machine are easy to make, the underlying network structure should be similarly flexible therefore the components of this network should themselves be virtualized.

There are several virtualization levels. In full virtualization, there is almost a complete model of the underlying physical system resources that allows any and all installed software to run without modification. In partial virtualization, not everything in the target environment becomes simulated. Not all software programs installed on the guest operating system can run unmodified. Finally, there is paravirtualization,

M. Graña et al. (eds.), *International Joint Conference SOCO'16-CISIS'16-ICEUTE'16*,
Advances in Intelligent Systems and Computing 527, DOI 10.1007/978-3-319-47364-2_77

essentially a way to improve performance by having a software interface working between the virtual machine and the underlying physical hardware system itself [1].

Virtual machines have particular needs when it comes to directing the information flow and the CPUs have extensions that help with this process. Intel (Intel VT - IVT) and AMD (AMD-V) have both released hardware virtualization technology to facilitate the running of hardware virtualization on computers. This technology has been designed to boost the power of the hypervisor. Although these two pieces of hardware virtualization technology are separate and independent, they perform broadly the same function. As increasing numbers of modern computers are sold with incorporated hardware virtualization technology, more powerful CPUs will continually be developed to allow for better creations of simulated environments.

The interaction between the virtual networking environment and the real network is defined in the Edge Virtual Bridging IEEE standards. There are two main standards that relate to network connectivity of a virtual machine: Virtual Ethernet Bridge and Virtual Ethernet Port Aggregator.

Virtual Ethernet Bridge (VEB) where physical machines acting as hosts provide the environment for virtual machine communication based on switching connectivity. There are advantages of executing all inter-VM communications inside the VEB [2] such as the lack of switching loops (and therefore the lack of STP presence), network traffic isolation, no MAC learning, etc. that in general increase the security of the network created. This concept is used in most hypervisor solutions today such as VMware ESX or Microsoft Hyper-V. The disadvantages are mainly related to the interaction with specialized devices and interfaces, the limited feature set when compared to real switches and the possibility to actually circumvent in a badly configured VM some of the security aspects.

The second is the Virtual Ethernet Port Aggregator (VEPA) [3] where the hypervisor virtual network forwards the frames always from the VMs to an external switch (and never between VMs) and allows it to handle problems like creating a loop free environment and security. The advantages are the avail ability of monitoring tools but the disadvantages relate to introducing a new hardware complexity layer.

Figure 1 presents the two types of network connectivity between virtual machines that can easily be found and tested in this proposed laboratory structure as the students will run the VMs both on a single hypervisor on the local machine and on different hypervisors found on different machines connected via a real network infrastructure [4].

Figure 1a shows the situation where there are multiple hypervisor systems present in the local network. This could be a real situation where a business has acquired/created multiple virtualization solutions that are difficult to interact (for example Hyper-V and VMWare solutions). If communication of service chaining is necessary between the VMs situated on different Hypervisors, the external infrastructure must be used and configured. Figure 1b shows the situation where the VMs are all on the same hypervisor and the inter VM communication can be handled internally by the VEB.

From the network connectivity perspective, network hardware virtualization allows the translation of network resources (such as routers, switches, firewalls, etc.) in software resources that can be reused by multiple virtual machines in a secure fashion.

Network hardware virtualization has a complex design, therefore in order to simplify this aspect, various GUI tools were created to help users visualize or edit the

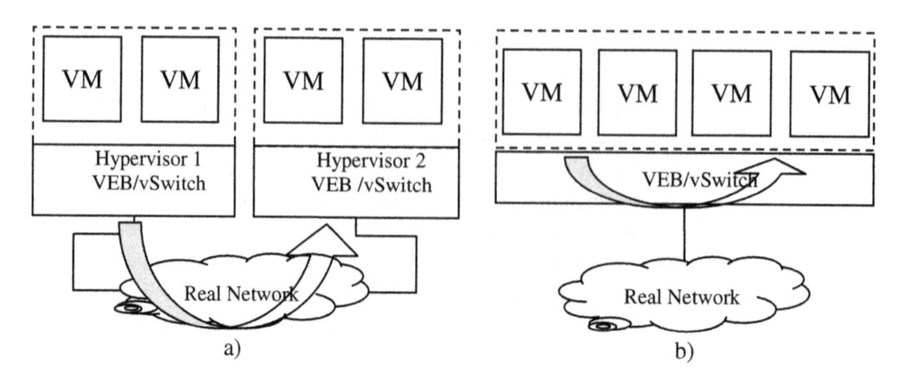

Fig. 1. Using hardware connectivity (a) and virtual connectivity (b) is necessary depending on the VM machine location

network configuration. Different hypervisors have different complexities for the network connection editors. Listed in Fig. 2 are the GUI editors for Hyper-V, VirtualBox and VMWare. The editor network complexity varies and so are the connectivity options. All solutions however offer access to advanced configuration options in the command line.

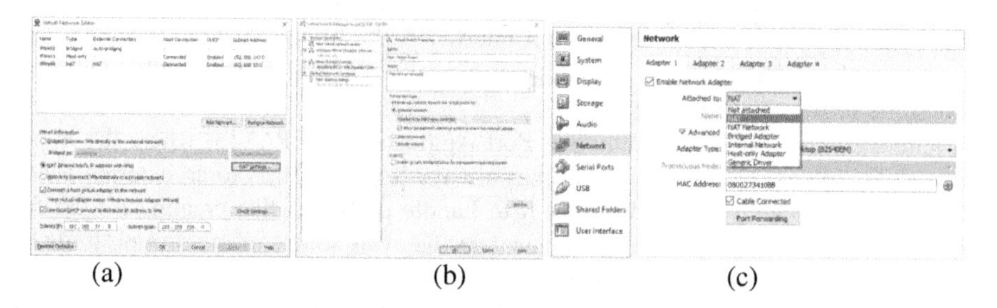

Fig. 2. GUI editors for virtual networks as found in VMWare (a) Microsoft Hyper-V (b) and VirtualBox (c)

Using virtualization in education is not a new concept [5, 6]. While a few years ago using virtualization was limited due to costs, today the advantages and the available computing power make the implementation available for any university laboratory.

This paper presents case studies that students can use in order to understand and test service chaining in a virtualization laboratory based on the Hyper-V and VirtualBox. By introducing this type of laboratory, the students are able to understand concepts that related to network virtualization and concepts that can help them understand later (usually in a master's course) the notions related to Software Defined Networks – SDN (especially the infrastructure layer implementation) and cloud infrastructure implementation.

The paper is structured as follows: Sect. 2 presents the testing methodology, Sect. 3 presents the structure of the network virtualization laboratory proposed for students.

Section 4 presents and discusses test results. Finally, Conclusions show ideas that can be used in the implementation of future virtualization laboratories.

2 Laboratory Setup

Tests were made on a desktop units equipped with Intel Core i7-4970 K CPU (4 GHZ), 32 GB RAM memory, Realtek Ethernet Controller and different storage media:

- 240 GB SSD drive for Windows 10 with Hyper-V and VirtualBox 5;
- 300 GB 7200 RPM HDD for Hyper-V 2012 and for Windows Server 2016 Technical Preview 5.

2.1 Hyper-V Setup and Virtual Machines Configuration

In Hyper-V 2012 case, after installing on the desktop unit, it was necessary to configure network and firewall rules via the two console windows, for the client. A Windows 10 machine was used as client. The client is in the same network with the server, and must be logged in with the same username as the server in order to be able to manage the system. The network does not have Active Directory. To avoid creating a new user on the server, we activated the existing (but disabled) Administrator account. Furthermore, it was necessary to install the necessary tools (Hyper-V manager and PowerShell) on the client. Firewall must be configured via command line in PowerShell to enable managing the Hyper-V server from the client. The Hyper-V manager was used to configure the server and create the two virtual machines.

In Windows 10 and Windows Server 2016 the user must install Hyper-V role to enable local virtualization server. To be able to create virtual machines it is necessary to have virtualization extensions for the local machine. Each virtual machine must have two network cards for testing purposes.

2.2 VirtualBox Setup and Virtual Machine Configuration

VirtualBox installation is similar to any standard Windows application. The only issue that was observed is that in order to take advantage of the 64 bit characteristics of the processor, one must disable Hyper-V role, if already enabled. Windows 10 and Windows Server 2016 ask for a system restart. After VirtualBox is installed the Hyper-V role can be enabled back without a problem.

To better use VirtualBox features, it is recommended to install VirtualBox Guest Aditions extension. We proceeded then to create two virtual machines with the same characteristics and install Ubuntu 14.04 x86_amd64 on both of them. One was used as a router and the other as a client and service chaining was performed.

2.3 Testing

On Hyper-V we tested the performance of Generation 1 and 2 machines by giving them server and client roles, and inter-changing these roles. We use the stress [7] command line utility.

Once we configure routing capabilities of the router system, on the client we run the command:

ping www.upit.ro

and record ping timeout values.

On the server we used the following command in roder to stress the router system:

stress --cpu < *value* > *--timeout* < *value* >

for ex.:

stress --cpu 4 *--timeout* 20*s*

and observe the changes in ping timeout values on the client operating system.

For comparing results on different *–cpu* values, we save *ping* command output in text files:

ping www.upit.ro > <*filename*>

for ex.:

ping www.upit.ro > *ping_results.txt*

The ping timeout values are read with Matlab/Octave commands and used in graphics.

3 Laboratory Structure

3.1 Installing a VirtualBox System

This is a very short part of the laboratory where the type 2 hypervisor is installed. A reboot may be required as the necessary drivers are installed. The only problems found at this stage are possible incompatibilities of VirtualBox with the machine and the operating system (Several machines run Windows XP). They were solved by installing earlier versions of the software.

3.2 Installing a Hyper-V Server 2012 R2 System

This is a rather lengthy part of the laboratory and should be done in parallel with the VirtualBox installation as Hyper-V needs a dedicated machine to be installed on.

The estimated time for a complete install and configuration in a non Active Directory laboratory is one hour and a half.

The solution also needs an active Windows 8+ machine in order to manage the remote hypervisor. An alternative is to install directly evaluation versions of Windows Server/Hyper-V Server 2012 or 2016 that include the GUI tools to manage the hypervisor.

In Hyper-V there are three types of virtual switches: External, Internal, and Private. A fourth switch, the NAT switch, is only included in Windows 10 Fall Update/1511 (and Windows Server 2016 TP4) and can be created via the new New-NetNat command [8] in PowerShell. The new NAT switch does not include a DHCP server for automatic network configuration. The 2016 versions are also necessary to test the New-NetNat command and the NAT switch.

3.3 Connecting a Virtual Machine to the Network in Network Address Translation (NAT) Mode

In this test we have used VirtualBox Hypervisor to create this type of network. The operating system was Ubuntu Linux 14 x64 (Fig. 3).

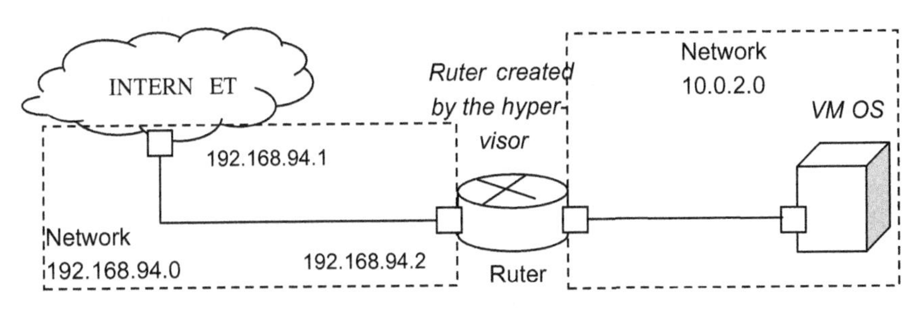

Fig. 3. Network setup for the NAT mode VM

Estimated time for these tests is one hour.

For VirtualBox the time needed to create the configuration is short and it is a good way to start the laboratory as the configuration is minimal and the included DHCP server allows fast auto configuration. Tests that were be performed:

- identifying the fact that the hypervisor acts like a router with NAT (for example via a simple traceroute command or by observing the TTL value);
- monitor the external network with Wireshark to see that the packets exchanges between networks are not accessible from the external;
- configure port forwarding;

Supplementary tests that include observing the increased latency due to the included router and the latency variation with host CPU load (because of the hypervisor usage or other virtual machine usage).

3.4 Connecting a Virtual Machine to the Network in Bridged Networking Mode

In this mode, both VirtualBox and Hyper-V hypervisors were tested. If the external network is statically configured and there was no DHCP server present the students have to manually configure the network connection and immediately observed the particularities of this configuration as several IP conflicts were detected. If there is a DHCP server present in the external network the virtual machines will obtain the configuration from this machine (Fig. 4).

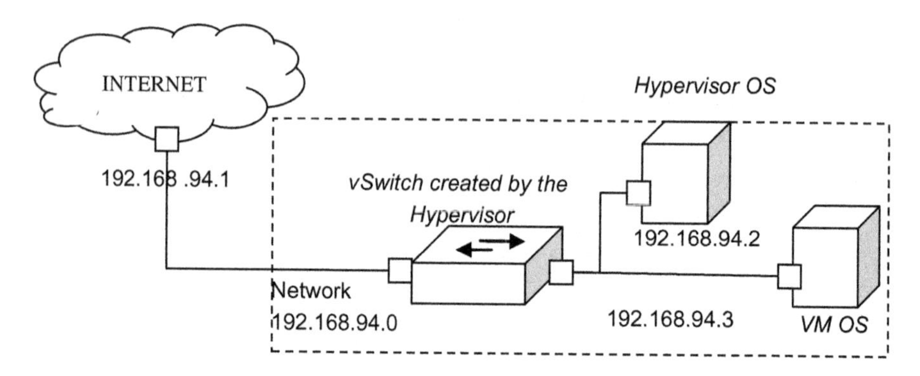

Fig. 4. Network setup for the NAT mode VM

The estimated laboratory time for these tests is one hour. Tests that were be performed:

- observing the conflicts with other real and virtual hosts in the network;
- observing the TTL value and traceroute command results;
- observing the broadcast network traffic of the virtual machine;
- using the VM OS as a router for real and virtual machines running on the same system or on remote systems.

4 Laboratory Testing Results

Under excessive processor use, students have tested the routing performance, network latency and packet loss. The results are presented in Fig. 5.

Results showed that:

- Hyper-V on Windows 10 desktop server provides the best stability in our tests (Fig. 5). This may be due to the fact that the operating system was installed on an SSD drive. VirtualBox results are better than the other two operating systems install on hard drives.
- Generation 1 type virtual machines are more stable when used on desktop systems, regarding standard deviation analysis (Fig. 5c). On Hyper-V 2012, Generation 2 is more stable in first two tests and proves otherwise in 1000 Hogs Test.
- VirtualBox has good results and denote high stability.
- Windows Server 2016 with Hyper-V role installed and used as a desktop virtualization server has poor performances because of it is high load of software dedicated

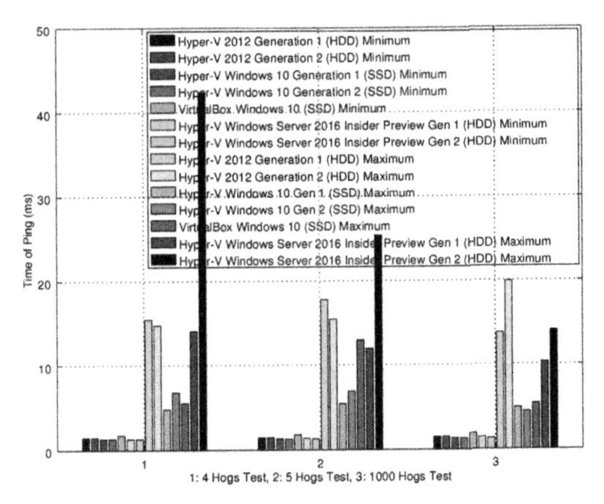

(a) Minimum and Maximum Values

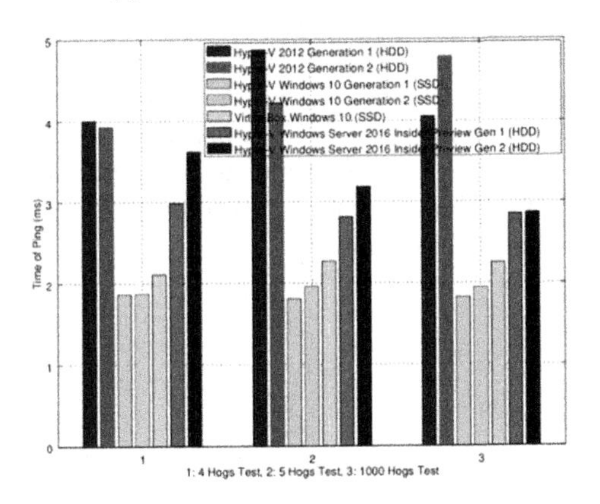

(b) Average Values

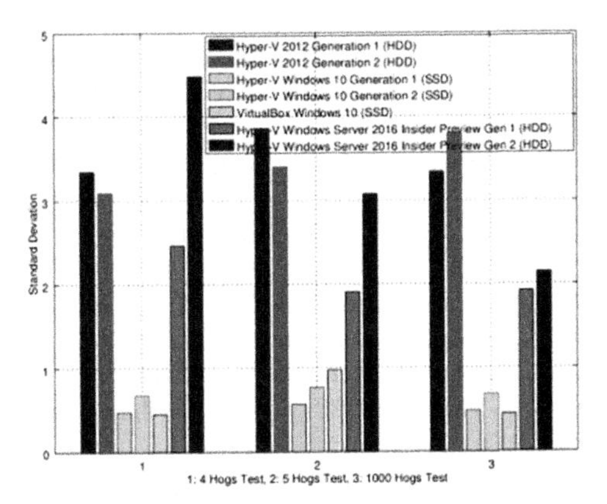

(c) Standard Deviation Values

Fig. 5. Statistical analysis of the results

to server type functions. One other cause may be that it is still in development stage, therefore unstable.

Surprisingly, the ping response time analysis of the results indicates that the network traffic is more stable at high processor loads.

5 Conclusions

This paper presents case studies that students can use in order to understand and test service chaining, virtual switch configuration and virtual network stability in case of stress.

The two hypervisors that were tested were Microsoft Hyper-V (in Server 2012 R2 and Windows 10) and Virtual Box (running on Windows 10). The challenges related to a hand on approach were presented. Applications for this laboratory were presented. This laboratory, due to its hands on approach is a practical way for students to learn basic virtual networking concepts and practice virtualization server administration.

By introducing this type of laboratory at undergraduate level, the students are able to understand concepts that relate to network virtualization and concepts that can help them understand later (usually in a master's course) the notions related to Software Defined Networks – SDN (especially the infrastructure layer implementation) and cloud infrastructure implementation.

References

1. VMware. Understanding Full Virtualization, Paravirtualization, and Hardware Assist (2015). https://www.vmware.com/files/pdf/VMware_paravirtualization.pdf. Accessed 9 Oct 2015
2. VMware. VMware Virtual Networking Concept (2015). https://www.vmware.com/files/pdf/virtual_networking_concepts.pdf. Accessed 9 Oct 2015
3. Hudson, C., Congdon, P.: Edge Virtual Bridging with VEB and VEPA (2009). http://www.ieee802.org/1/files/public/docs2009/new-hudson-vepa_seminar-20090514d.pdf. Accessed 9 Oct 2015
4. Greenberg, A., Maltz, D.A.: What Goes Into a Data Center? (2009). http://goo.gl/U49tND. Accessed 9 Oct 2015
5. Felde, N., Lindinger, T., Reiser, H.: Virtualizing an IT Lab for Higher Education Teaching, Tagungsband zum 1. GI/ITG KuVS Fachgespräch "Virtualisierung", pp. 97–104 (2008)
6. IBM Corporation. Virtualization in Education, IBM Global Education White Paper (2007)
7. Waterland, A.: stress(1): impose load on/stress test systems - Linux man page (2015). https://people.seas.harvard.edu/~apw/stress/. Accessed 9 Oct 2015
8. Microsoft. PowerShell: New-VMSwitch (2015). https://technet.microsoft.com/en-us/library/hh848455.aspx. Accessed 9 Oct 2015

Using the Phone's Light Sensor to Detect the TV Video Stream

Valeriu Manuel Ionescu(✉), Cosmin Stirbu, and Florentina Magda Enescu

University of Pitesti, FECC, Pitesti, Arges, Romania
manuelcore@yahoo.com

Abstract. Current smart devices (phones, tablets, etc.) have integrated light sensors to adjust the screen's brightness to the ambient light. The light sensors have become more sensitive and are even able to read the RGB light components. In Android, this information can be accessed without special access rights for the application. An application can use the information from the light sensor to detect the ambient light variations and relay this information to a server where it can be used to determine the video information being displayed. This paper details the data flow and tests the implementation for a single video flow on multiple light sensors.

Keywords: Light sensor · Video detection · Android · TV channel

1 Introduction

The current generation of smart devices has a large number of sensors (such as: light sensors, accelerometer, pressure, proximity, etc.) that are used to improve the user experience by analyzing the environment and making the smart device react accordingly.

Sensor information can produce sensitive information for the user privacy. Many companies collect the user data by integrating spyware/data collection modules inside applications and transfer it via internet connection to databases where it can further be analyzed and even associated with the user (via cookies, location or other tracking methods) in order to produce detailed information about a user's behavior.

For example, it is detailed that inaudible information can be used to communicate between smart devices in order to connect them and use this information for tracking the user's actions across multiple devices [1, 2].

User tracking, used by legitimate applications but also by spyware/privacy intrusion, is possible because the user usually installs a similar set of applications on multiple devices and the information collected by a similar application running on all devices (or different applications that run the same data collection software in the background) can be combined on a single server to further detail the user's information and behavior.

There are however obstacles in this process, because the applications need permissions from the user to access the smart device's sensors, connectivity options and

M. Graña et al. (eds.), *International Joint Conference SOCO'16-CISIS'16-ICEUTE'16*,
Advances in Intelligent Systems and Computing 527, DOI 10.1007/978-3-319-47364-2_78

folders. For example there are applications that ask permission to the user's microphone even if the application does not need that information to function properly. This would be regarded as a highly suspect request by the user, and that application may be uninstalled because of this. This access request may pass for uses that do not analyze the application permissions closely, but many people may be reluctant to give access to multimedia content, contact list, video or audio sensor data. This is especially true for Android 6.0 (API level 23) that asks for user's permissions one at a time [3], while the application is running and not at the install of the application.

Light sensors were introduced in smart devices in order to adjust the screen brightness in order for the user to comfortably read the display. Applications that access the light sensor do not require these rights and are installed with general rights that eliminate the user's reticence to accept these applications. The use of light sensors as a source of information is restricted, however, by a multitude of factors:

- there may be no light sensor: many of the low end phones do not have a light sensor and relay on manual brightness adjustment. However, as time passes and the price of these sensors continues to fall, more devices will have this type of sensors integrated;
- there are many light sensor types: while the number of light sensor types is large, the framework to access them is the same and it is fairly easy to determine if the sensor has the characteristics that are necessary to be used in the detection process;
- power consumption: using the light sensor has a certain power consumption (ranging in the 0.15–0.75 W) but it is small and the energy savings that are made to the display consumption eliminate this potential issue;
- light sensor accuracy: some light sensors are set to a limited number of states (such as 8 states) even if the light sensor itself is able to produce many more states and cannot be used to accurately determine the ambient light variations. This is indeed a problem for a light sensing application, but the results can be analyzed and excluded if minimal characteristics are not met;
- light sensor information cannot be polled; a change in light level will generate an event that will be handled by the application. This behavior will produce variations that are luminance driven and not time driven and will be taken into account when comparing video data information with light sensor data, for example by interpolating the results based on the event timestamp.

This paper will analyze the possibility of using the light sensor information on order to determine if the user is watching a known video sequence, by monitoring the light intensity variations. The light sensor information is gathered and compared with a known video sequence using a correlation function. The results will show the validity of this assumption.

The application of these results can be extended to monitoring known video streams originating from the TV channels. This can be used for example to monitor which channels does a user often watch in order to determine the impact factor of a certain station (market share) and the advertising spots that it is viewing. Other papers have investigated the use of the light sensor information in order to gain more knowledge about the environment [4] or even to help breaking the PIN number [5] but this is the first time the light sensor was used for this type of action.

Multiple light sensors were tested in order to determine the validity of the results. Similarly multiple different video sequences, of similar length, were used and tests were made to determine if the light sensor information is sufficient to pair the sensor data with the correct video sequence.

This paper targets the Android platform and both RGB and luminance only sensors, but can be extended to other platforms as well (Windows and iOS). The number of users that can be targeted by these applications is large and the information obtained can be used standalone or paired with additional sensor information in order to further refine the user monitoring process.

The paper is structured as follows. In the chapter Light sensor information, the tested sensors are presented, compared and the methods used to obtain and store the light sensor data are detailed. In Sect. 2, the solution chosen to have reference values to make the compare is presented. In the Sect. 3 it is presented the way the video source was processed to obtain a reference data set. Section 4 outlines the conditions for testing. In Sect. 5, the results obtained from comparing light sensor information with the reference data are presented and discussed. The final chapter presents the future developments for the ideas presented in this paper.

2 Light Sensor Information

In general a light sensor captures the light intensity and adjusts the screen luminosity in order for the screen brightness to adjust to the ambient light, but also for the reduction in power consumption because the display (a major consumption part of a smart device) would dim accordingly.

A light sensor reports this value in lux that can read by a smart device application. The number of states that a light sensor can report varies with the sensor type. In Table 1 several light sensors are presented and their characteristics, as interrogated by and Android application via the methods of the Sensor class. The `getMaximumRange()` method can read the maximum range of measurement and `getPower()` method can read a sensor's power requirements.

As seen in Table 1, there are variations in sensor power consumption and maximum range. Better light sensors usually have higher power consumption such as TMD4093 RGB Sensor from Galaxy S6. Also some phone manufacturers have restricted the sensor states in order to get fewer events from the light sensor. For example, the light sensor of Allview X2 Soul (RPR410 Light Sensor) will only generate events at: 0, 60, 90, 225, 640, 1280, 2600, 6400 lux with no intermediary values. This will produce that does not vary its brightness too often, keeping the screen at a low level in low light conditions.

High end devices have RGB light sensors (such as Galaxy Smartphones) that read the light components separately and the phone uses this data to fine tune image representation according to the ambient light [6].

The light sensor is available from the early versions of the Android platform (starting with Android 1.5, API Level 3), therefore the applications can target a wide range of API levels (with some modifications related to the way the data is collected).

Reading the light sensors data is documented in the Android developers' reference only for the luminance information, and shows that the light sensor reports the current illumination in SI lux units [7]. The reporting mode is On-change, meaning that the events are generated if the measured values have changed. Similar behavior is for the proximity, step counter and heart rate sensors.

Table 1. Characteristics read from the tested light sensors

Phone type	getName()	getVendor()	getPower()	getMaximumRange()
Galaxy S5	TMG399X RGB Sensor	AMS	0.75	60000
HTC One E9	CM32181 Light sensor	Capella Microsystems, Inc	0.15	10240
Galaxy S4	CM3323 RGB Sensor	Capella Microsystems, Inc	0.75	60000
Allview X2 Soul	RPR410 Light Sensor	ROHM	0.13	10240
Galaxy S4 mini	GP2A Light Sensor	SHARP	0.75	65000
Galaxy Note 3	TMD27723 Light Sensor	Taos	0.75	32768
Sony M4	CM36286 Light sensor	Capella	0.75	3277
Galaxy S6	TMD4093 RGB Sensor	AMS, Inc.	0.75	60000
Huawei P8 Lite	Light sensor	TAOS_TMD27723	0.75	10000
Allview Viper I	LIGHT	MTK	0.13	10240
Motorola X Style	CT1011 Ambient Light	TAOS	0.25	65535

It is detailed in the Android developer's documentation [7] that the light sensor (`Sensor.TYPE_LIGHT`) returns a single value. However, the `event.values.length` returned the value 3, therefore we could read 3 components: `event.values[0]`, `event.values[1]`, `event.values[2]`. It is noted that on many devices that did not have RGB sensors both of the: `event.values[1]`, `event.values[2]` were 0 all the time, even if the length was reported as 3. The application has filtered these situations by tentatively reading the data and eliminating the constant 0 components.

The `event.values[1]` is the most sensitive of the 3 values and is highly sensible to blue light and has an infrared filter, while sensors `event.values[0]` and `event.values[2]` do not have this filter. The minimum values are: 0 for `event.values[0]` and `event.values[2]` and 1596 for `event.values[1]` (with the value 4.2949655E9 being returned sometimes for `event.values[1]` when near its lowest value). Figure 1 presents the sensor component's reaction to different colors and direct sunlight for the Galaxy S6 Smartphone.

The tests showed that the most sensitive sensor component (especially in low light) is `event.values[1]`. The least sensitive value is `event.values[2]` both in low and high lighting conditions.

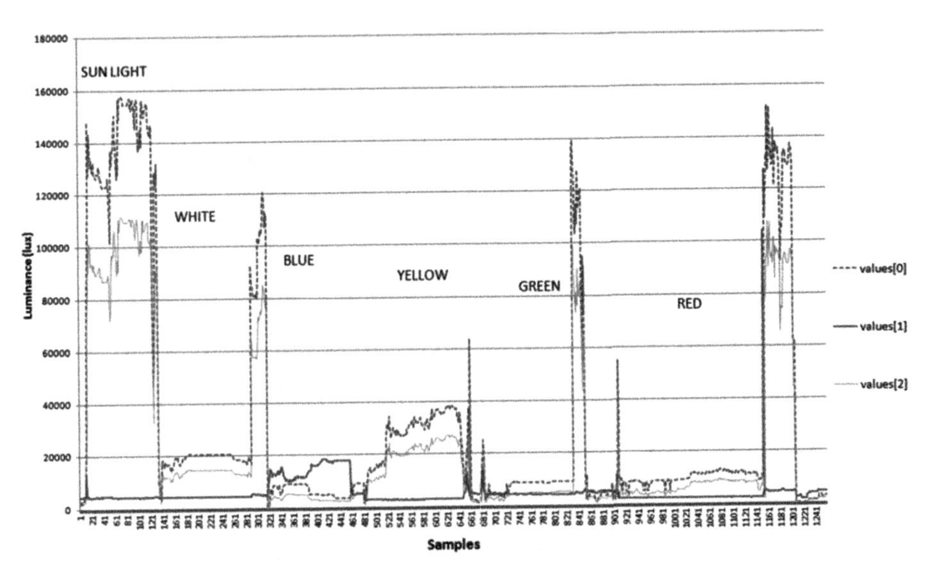

Fig. 1. Test that shows the reaction of the 3 components of the light sensor to different light filters and to direct sunlight

3 Video Source Processing

In order to obtain relevant results, the video source was processed using the DotImaging.IO [8] library to extract luminance information from the video data, using the photometric luminance formula (1) in order to get the best chance to match the light sensor captured luminance.

$$\text{Luminance} = (0.2126 * \text{R} + 0.7152 * \text{G} + 0.0722 * \text{B}) \quad (1)$$

Separately, all the RGB information was extracted in order to be compared individually with the RGB information from the light sensors that are capable to extract separate color information.

The first step was to generate a reference data set that will be used to compare the data recorded in various conditions. First we have tried to generate a synthetic video information data set by processing the video luminance information and RGB components directly with the above formulas for luminance, then modify this to approach the behavior of a light sensor (for example having a minimum latency of 3 ms in light variance for the light level, having a margin within which the luminance is perceived as the same and accentuating the variation in high intensity areas as many light sensors are tweaked to react little or not to low light intensity variations and very sensitive to react to large luminance variations because they try to keep the power consumption down in low light and make the screen visible in bright light environments). The light sensor information and video information were interpolated to obtain the same number of samples and then compared with the correlation function in Matlab as we have targeted a fixed video timescale. The results are presented in Fig. 2.

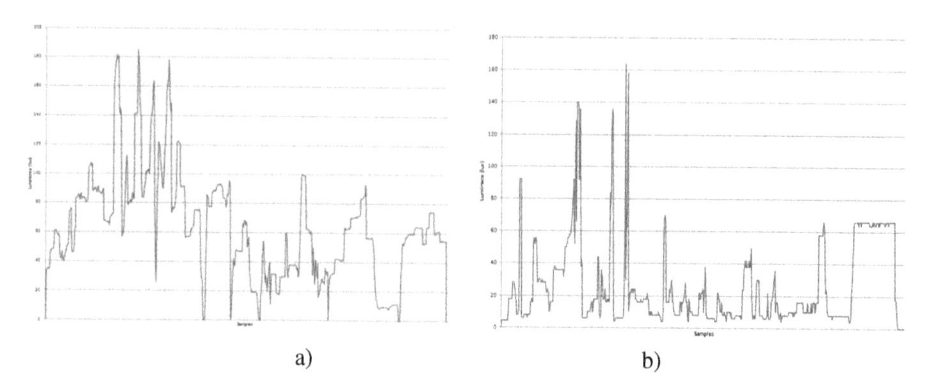

Fig. 2. The processed source video (a) compared to the light sensor information obtained from Galaxy S6 as a reference signal (b)

The results showed that this information is not sufficient (we have obtained a best correlation value of 0.3098 between the values for Fig. 2(a) and (b), the values in Fig. 2(b) were captured in very good conditions as described below) to verify a good similarity between the source video and the light sensor information.

In order to quickly improve these results, we have used as a reference signal the light sensor of Galaxy S6 that gives all 3 light sensor components, recorded in perfect darkness conditions, with the sensor pointed directly at the middle of the 80 cm TV screen at a distance of 0.5 m.

This would allow the entire TV screen to influence the light sensor, not only a single area of the TV. In all the tests below this was considered the reference signal, to which the other signals were compared.

4 Testing Methodology

The targeted case study is where a user watches the TV in the evening and places the phone face facing up on a table and the light sensor is not covered. This is the case as most users frequently check the phone's screen and do not want to miss notifications. The test room had white walls and ceiling.

We have tested the shape of luminosity data during various types of TV shows. A special situation was detected for video advertisements. As it is known for TV audio information [9, 10] an advertisement is created to attract attention and this is reflected in higher perceived audio output. During the video tests the same behavior was detected for video information regarding the luminosity variations. Figure 3 presents the light sensor information recorded during advertisements and after, when a movie was being shown. The luminosity variations are larger when the advertisements are presented. Larger luminosity variations are easier to catch by the light sensor.

Because the sequence of advertisement messages is usually different for each TV channel and each advertisement pause, they could be used as "markers" that should especially be tested when comparing light senor with video information. Helpful

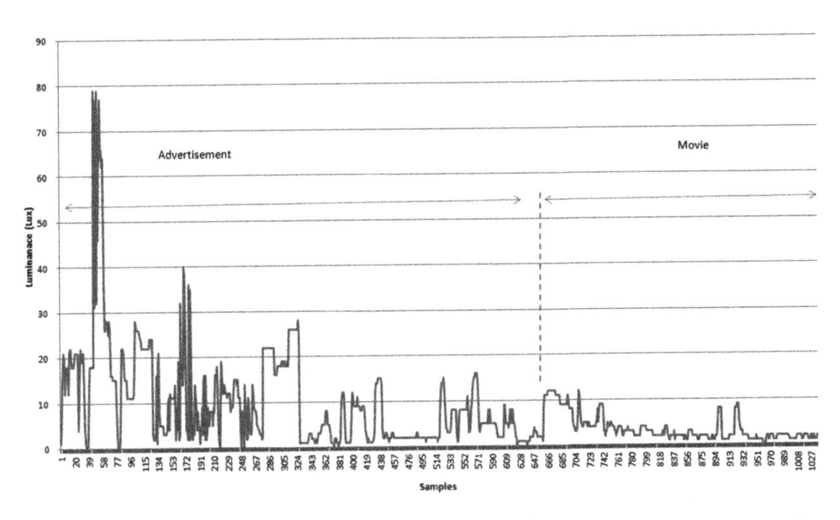

Fig. 3. Luminosity variations are more sudden during advertisement compared to a movie or TV show and make it easier to detect the video stream being played

information would be the timestamp for the recorded light sensor data, but this depends on how well the smart device's clock is synchronized.

In conclusion, advertisements generate a significant light variation and are the best place to detect a light variation pattern in the video stream.

The final consideration on the test methodology is that the tests presented in this paper target a fixed length video stream that will be detected. We have used the correlation function from Matlab in order to compare the light sensor information received from different sensors with the reference of light sensor data.

The signals received from the light sensor are strings of numeric values paired with a timestamp that are stored in text files. Because different sensors react differently to the light (latency and resolution) they will generate a different number of events (samples) for a certain time span. The timestamp recorded for each event allows the server to determine the time length for a certain captured sequence. In order to compare the sequences it was necessary to interpolate the values using the interp1 Matlab function, then analyze if the two sequences data are correlated [11]. The results will range between −1 and 1, with zero meaning no correlation, 1 max correlation and −1 being the maximum negative correlation that means you can make one vector from the other using a negative scale factor.

5 Test Results

First tests targeted the way the distance from the TV influences the correlation value. Tests were made with Galaxy S6 and `event.values[0]` was measured. Tests were made where the Smartphone was placed closer and further from the TV set. The data was correlated with the reference data set (0.5 m). The datasets are presented in Fig. 4.

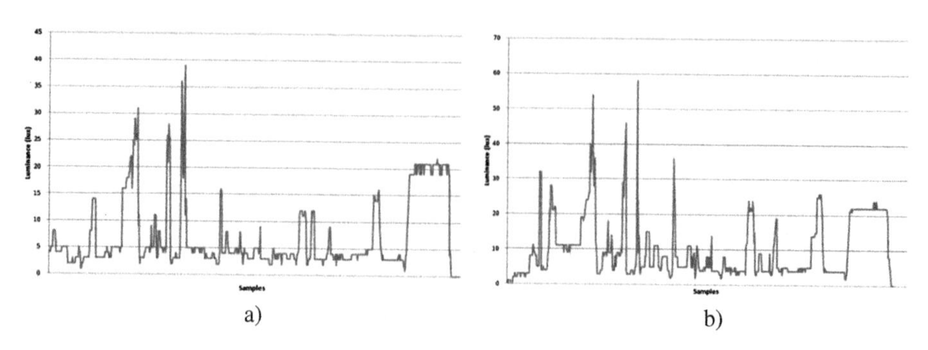

Fig. 4. Data from the light sensor when the Smartphone was away (a) and closer to the TV set (b)

Table 2. Correlation results for different distances of the light sensor to the TV set for Galaxy S6

Distance from TV	1 m	2 m
Correlation with reference	0.4754	0.3080

Table 3. Correlation values for phones with different sensors

Phone	Galaxy S5	HTC M9	Sony M4	Galaxy S4	P8 Lite	Allview X2 Soul	Galaxy S4 mini	Moto X Style
Correlation	0.71588	0.0141	0.2534	0.7746	0.2285	0.0862	0.6426	0.2922

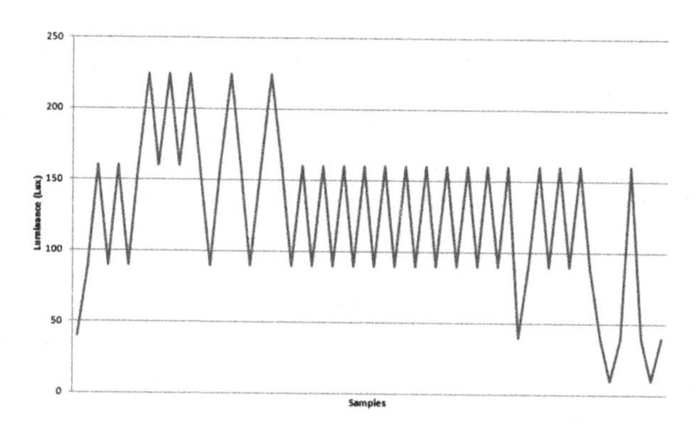

Fig. 5. Light sensor data set for the same video from HTC M9 shows a reduced transition count making it completely different form the reference data

The correlation results for different distances of the light sensor to the TV set for Galaxy S6. As seen in Table 2, the correlation is higher for the phone placed closer to the TV.

The correlation is still significant therefore this information is usable for the intended purpose. Next we have compared the information for the `event.values [0]`. The light sensors that had a reduced transition count show data that is very

Table 4. Correlation values for event.values[1] in phones that have RGB light sensor

Phone	Galaxy S5	Galaxy S4
Correlation	0.5009	0.4938

different form the reference data set. The results are presented in Table 3. Figure 5 shows that the information from HTC M9 light sensor shows a reduced transition count therefore it is completely different form the reference data and the correlation results confirm this. In a different test, the `event.values[1]` were then verified for correlation. This is the most sensitive sensor and it is highly sensitive to blue light. The number of tested phones was reduced to Galaxy S5 and S4. The correlation was constantly significant in this case. The results are presented in Table 4.

6 Conclusions

This paper proposed the idea of using light sensor information in order to obtain information about the video stream that the user is watching. This data can be used as standalone information or can be used in conjunction with other types of sensor and information gathering techniques in order to monitor the user's activity (for example in television market share research) and behavior.

Tests were made using multiple smart phones with various types of light sensors. Both luminance and RGB light sensors were tested. A reference data set was recorded, based on the Galaxy S6 light sensor. Light sensor data was captured from Android using the undocumented `Sensor.TYPE_LIGHT` values along with a timestamp and processed in Matlab in order to check if it can be correlated with the reference data.

The results have showed that a good correlation can be made especially for high end light sensors (RGB sensors). When the smart device is from the screen, the in-formation captured by the light sensor is harder to correlate with the reference signal but it still shows correlation. On the other hand, when the light sensor is less expensive or the transitions are restricted in order to have lower power consumption, the registered transitions only have a slight or no correlation with the reference signal. Signals captured from different video streams show no correlation.

The tests made at this stage targeted short and known length video streams and focused on how capture the sensor information. In the future, we intend to use a better method for comparing the data that would be suitable to detect the video sequence in a long, live, stream compared to the fixed video sequences that we have used for this paper.

References

1. Federal Trade Commission. Comments for November 2015 Workshop on Cross-Device Tracking (2015). https://cdt.org/files/2015/10/10.16.15-CDT-Cross-Device-Comments.pdf. Accessed 9 Oct 2015
2. Hanspach, M., Goetz, M.: On covert acoustical mesh networks in air. J. Commun. **8**(11), 758–767 (2013)

3. Google Inc. Compatibility Definition. Android 6.0, 16 October (2015). http://goo.gl/eD03sq. Accessed 9 Oct 2015
4. Mertz, C., Koppal, S.J., Sia, S., Narasimhan, S.: A low-power structured light sensor for outdoor scene reconstruction and dominant material identification. In: 2012 IEEE Computer Society Conference on Computer Vision and Pattern Recognition Workshops, Providence, RI, pp. 15–22 (2012)
5. Spreitzer, R.: PIN skimming: exploiting the ambient-light sensor in mobile devices. In: Proceedings of the 4th ACM Workshop on Security and Privacy in Smartphones & Mobile Devices, SPSM 2014, Scottsdale, AZ, USA, pp. 51–62 (2014)
6. ams AG. TMG3993, Gesture, Color, ALS, and Proximity Sensor Module with mobeam, ams Datasheet (2015). http://goo.gl/j3Z4qb. Accessed 9 Oct 2015
7. Android Open Source Project. Sensors Overview (2016). http://developer.android.com/guide/topics/sensors/sensors_overview.html. Accessed 9 Oct 2015
8. Jurić, D.: Introducing Portable Imaging IO Library for C# (2015). https://github.com/dajuric/dot-imaging/. Accessed 9 Oct 2015
9. Couling, J.: TV Loudness: time for a new approach (2003). http://www.dolby.com/in/en/professional/broadcast/products/aes-tv-loudness-john-couling.pdf. Accessed 9 Oct 2015
10. FCC. USA gov. Loud Commercials (2011). https://goo.gl/l2s26r. Accessed 9 Oct 2015
11. Raghavender Rao, Y., Prathapani, N., Nagabhooshanam, E.: Application of normalized cross correlation to image registration. IJRET Int. J. Res. Eng. Technol. **3**(5), 12–16 (2014)

Comparing Google Cloud and Microsoft Azure Platforms for Undergraduate Laboratory Use

Valeriu Manuel Ionescu[1(✉)] and Jose Manuel Lopez-Guede[2]

[1] FECC, University of Pitesti, Pitesti, Arges, Romania
manuelcore@yahoo.com
[2] Systems and Automatic Control Department,
University College of Engineering of Vitoria,
Basque Country University (UPV/EHU), Vitoria, Spain

Abstract. Knowing the advantages of public cloud services is important for undergraduate students that will work in the IT domain. The study of cloud technology should cover not only the theoretical aspects but should also give hands on access and experience with the management interface. Two important players on this market offer access to services as a free trial that allows testing and light cloud platform usage: Microsoft Azure and Google Cloud. This paper compares these free services and their usability in an undergraduate laboratory at the Computer Science specialization and proposes a laboratory structure that should cover this process of investigation.

Keywords: Microsoft Azure · Google cloud platform · Compare · Educational

1 Introduction

There is a constant push in recent years from big companies such as Microsoft, Google, Amazon to attract internet users and businesses to their cloud services, but the strategy for these companies varies in regard to the free access offering. Some promote free software based on their solution, while others give free but limited access to their solutions. It could be argued that the best way to proceed in interacting with cloud technology in an undergraduate environment is to create a private cloud with free technologies such as Open Stack or CloudStack, however as most commercial applications will target public cloud platforms from Amazon, Microsoft or Google, having hands-on experience with these platforms at undergraduate level can give an advantage to the students. It is therefore important to compare the solutions that offer free access to the commercial level user interface and their usability in and educational environment (laboratory class), in order to test operating systems and services for undergraduate students.

The idea of Cloud Computing was introduced with the first implementation of the internet when a globally connected world with applications that could run across the globe seemed only a short distance away [1]. However the real start was made by Amazon Web Services in 2002, which provided a suite of cloud-based services to the internet users. In 2009 Google and others started to offer browser-based enterprise applications, though services such as Google Apps. Microsoft is the latest to join the

M. Graña et al. (eds.), *International Joint Conference SOCO'16-CISIS'16-ICEUTE'16*,
Advances in Intelligent Systems and Computing 527, DOI 10.1007/978-3-319-47364-2_79

cloud offerings and is promoting many cross platform services that seamlessly interact with their Azure cloud platform. It is clear that cloud computing can bring enormous benefits for IT users. The top Cloud Providers in this moment are: Amazon, Microsoft, Google and Rackspace as seen in Fig. 1.

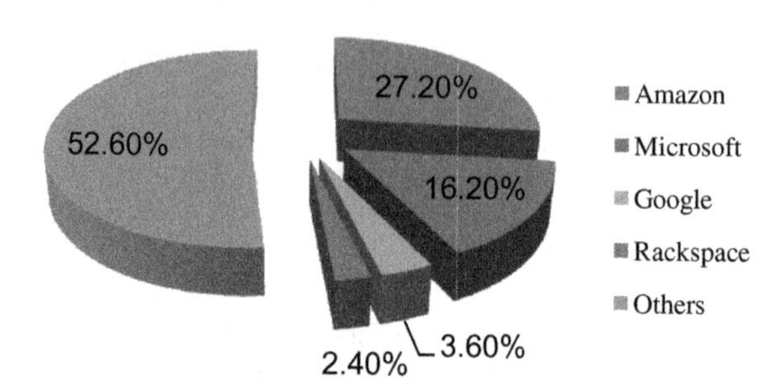

Fig. 1. Amazon is the estimated market leader when it comes to Cloud storage in 2015, followed by Microsoft [2].

This paper presents the results of using Microsoft Azure and Google Cloud platform at undergraduate level students in laboratory classes of Operating Systems and Computer Networks in order to introduce the Cloud concept. Cost control in a university laboratory is an important issue because the expenses must be planned ahead. Many cloud providers offer educational access to the cloud services [3–5] but in some cases the offer can be limited geographically or needs an authorized institution Id. For the cases that do not meet the requirements, the free access is the option left. For example Amazon free cloud services have the policy that starts automatically billing the credit card when the user ends the free offering, presenting the problem of unintentional costs in the classroom: "When your free usage expires or if your application use exceeds the free usage tiers, you simply pay standard, pay-as-you-go service rates" [6].

The classes where the laboratory was presented were the 3rd and 4th year students in Computer Science that were familiar with basic operating and distributed system courses. The purpose was to: test the usability of the free offerings in a student laboratory, with limitations in time usage and service accessibility; demonstrate cloud services and use them to interact with test applications; test their ease of use, limitations and make comparative analysis.

The paper has the following structure: In the chapter *Google Cloud,* the characteristics of this platform will be presented and similarly the chapter *Microsoft Azure* will present the other cloud services platform. The chapter *Comparing Google Cloud and Microsoft Azure trial offers* compares various aspects of these platforms and the final chapter *Laboratory structure and observations*, observations will be presented regarding the laboratory development and student observations related to the platform utilization. The final chapter *Conclusions* presents the future developments for the ideas presented in this paper.

2 Google Cloud

Google Cloud is a platform provided by Google that is growing fast and is hosted on the same support infrastructure that Google is using for the final users, such as Google Search and YouTube.

The platform offers a set of modular services based on Cloud and developing instruments, for example: hosting and computation, cloud storage, websites development, Virtual Machine (VM), translate Application Programming Interface and predictions. The free offer to this platform consists in access to the services but it is limited to 300$ offered for any Google Cloud Platform services over 60 days. A credit card is necessary for user confirmation in order to avoid non human users. When the trial ends, the account will be paused and an option will be available to upgrade to a paid account. The user will not be charged during or after the free trial ends.

Google Cloud Platform offers global coverage, low cost, low latency, and application availability for the customers. They continue to expand Cloud Platform locations over time. Currently, there are 15 Google Cloud Datacenters [7], and they will add two new regions in 2016: US Western, Oregon and East Asia, Tokyo, Japan. Table 1 presents a speed test (download and latency) to Google Cloud various services to Romania.

In regard to the virtual machine creation, Google supports importing raw device images, Amazon Machine Images and VirtualBox Images, and a collection of operating systems, including Red Hat Enterprise Linux, SUSE and Windows Server: Debian GNU/Linux, CentOS, CoreOS, openSUSE, Ubuntu, Red Hat Enterprise Linux, SUSE Linux Enterprise Server, Windows Server 2008/2012.

Steps necessary to create a Google Cloud VM:

- Access www.cloud.google.com
- Log into your account/sign-up
- Access: My Console
- From the dashboard, select: Compute Engine
- Select: Virtual Machines
- Create a new instance
- Select the desired configuration;
- Finish the process by pressing: Create.

Table 1. Google Cloud performance test

Service	Downlink(MB/s)	Latency(MB/s)
CDN	55.2	22
Storage (eu)	59.29	157.5
Storage (us-east1)	50.94	161
Storage (us-central1)	46.64	161
Storage (us-east3)	39.31	319.5
Storage (us-central2)	37.68	164

3 Microsoft Azure

Microsoft Azure is a Cloud development platform (released in 2010 as Windows Azure and renamed to Microsoft Azure in 2014) created by Microsoft for building, deploying and managing applications and services through a global network of data. The platform provides IaaS (Infrastructure as a Service, VMs, servers, storage, load balancers, etc.), PaaS (Platform as a Service, cloud services allowing customers to develop, run and manage applications) and SaaS (Software as a Service) services and supports many different programming languages, tools and frameworks, including Microsoft specific (Visual Studio) and third-party software and systems.

Microsoft has recently proven itself as one of the fastest expanding Cloud providers in the industry with a large number of datacenters [8] that allows providing regional content and good global service load balancing. The free offering from Microsoft presented in Table 2 gives 200$ for a period of 30 days to the user and, as with the free Google Cloud trial, when it ends the user will be warned that it will have to switch to a paying account.

Microsoft makes it easy to create a virtual machine using a custom image. The easiest way is to create a virtual hard disk file and import it into Azure. Although you can build VHD-based images from scratch, System Center Virtual Machine Manager can help with the image creation process.

Microsoft built its Azure public cloud on top of Windows Server and Hyper-V. It is easy to migrate VMs between local data centers and Azure. The process isn't seamless, but is relatively easy once connectivity is established between Azure and a local network. The operating systems available for the virtual machines include: CentOS; Datastax Enterprise; Debian GNU/Linux; Docken on Ubuntu Server; Hortonworks Data Platform; MapR Distribution Including Hadoop in Azure; OpenSUSE; Red Hat Enterprise Linux; Service Fabric Cluster; SLES; SQL Server 2016; Ubuntu Server; Windows Server (up to 2016, while Google Cloud is limited to Windows Server 2008/2012 offerings).

Steps to create a VM on Microsoft Azure platform:

- Log into your Microsoft account
- Access: www.portal.azure.com
- Choose: New, then select: Virtual Machine
- Select the operating system for the Virtual Machine
- Configure your virtual machine from the cascade of windows
- Start the virtual machine.

Table 2. Facilities offered by Microsoft Azure (trial version)

Number of hours	Unlimited
Websites	10
Databases	1 GB SQL instances
Storage	20 GB
Number of storage transactions	1.000.000

Azure is available in 24 regions around the world, and has announced plans for 8 additional regions. Geographic expansion is a priority for Azure because it enables the customers to achieve higher performance by accessing close datacenters and it supports their requirements and preferences regarding data location. The closest Microsoft Azure datacenters to Romania are West Europe (Netherlands) and North Europe (Ireland). Another interesting aspect of Microsoft Azure is that after you finish the free trial they do not restrict your access to that account and they let you explore what is new and how to use the platform better until you create a full account.

4 Comparing Google Cloud and Microsoft Azure Trial Offers

The first aspect compared regards the extent of the free offering. While the money received at sign-up for each platform differs (Google Cloud: 300$/60 days, Microsoft Azure: 200$/30 days), the prices for the services should also be considered. As one of the easiest things to create on the cloud platform is a Virtual Machine (VM), we compared the top configurations VMs that we can get with the money for the free trial and the results are presented in Fig. 2 (a) and (b). It is visible that both the pricing and the top specification differ. If the user decides to use all the money for the VM, the Google Cloud presents a better offer. If we compare the price of the same VM configuration on each platform we obtain the results presented in Fig. 2 (c) that highlight the Google Cloud advantage.

The second aspect investigated was the latency for different platform services. This aspect influences the type of application that can be deployed (for example latency sensitive and jitter sensitive applications may be better suited for a specific platform). For this test we have used the site https://cloudharmony.com/ that is able to connect to various platform components. Table 3 presents the latency to these datacenters from Romania.

Table 3. Comparing latency of Microsoft Azure and Google Cloud closest to Romania

Component	Google Cloud latency (ms)	Microsoft Azure latency (ms)
DNS - Domain Name System	51	75
VM/Compute Engine	126	56
Storage	70.5	61
Websites/Google CDN	22	49.5

Both platform offer good latency but overall, Microsoft Azure has a lower latency comparing to Google Cloud. The tests also showed that Microsoft Azure's latency variation is quite low, and doesn't fluctuate as much as Google Cloud's latency (good jitter). The latency and jitter results were tested also for the applications that the students have placed in the virtual machines and confirm these results.

The Bandwidth was investigated next. The results are presented in Table 4 and target only the Europe datacenters of both providers.

Table 4. Comparing download and upload speed for Google Could and Microsoft Azure

Platform	Download	Upload
Google Cloud	13.33 Mbps	4.664 Mbps
Microsoft Azure	12.38 Mbps	2.824 Mbps

The results show that the speeds are comparable for both upload and download, with a slight advantage for Google Cloud. The download is faster for the virtual machines that were tested indicating that they are better suited to provide content (such as a web site) then to be the target of content upload.

For a university laboratory these tests show to the students that all types of applications can be developed, from low latency applications to file upload/download solutions and even inter-VM communication. Other factors that contribute to the Quality of Service (even if for the free services there is no Service Level Agreement) and can be investigated are service availability (providers keep sites dedicated to downtime), jitter (latency variation) and packet loss.

Limitations such as 8 core running at the same time, crypto currency mining, CPU time/day, bandwidth or Web Sockets per instance are left high enough to make the implementation of simple application servers unrestricted. Also it is possible to run Google App Engine web application with low traffic and capacity requirements even after the free trial period.

5 Laboratory Structure and Observations

Prior to the laboratory where the cloud provider services were tested, the students were presented with two courses that had introduced the virtualization subject and the structures used in virtualization: cluster, grid, cloud. The students have also had prior laboratory hands-on experience with virtual machines in the Operating Systems and Computer Network classes, in the form of type 2 Hypervisor solution: VirtualBox.

The laboratory started with a short introduction of the two platforms and the limitations of the platforms that will be used. The targets were to: inspect the platform functions and their ease of use; deploy a web site on the virtual machine and access its content by using a browser; place a simple server application on the cloud platform and to test its connectivity to a local client application. The operating system that was installed was Ubuntu Linux as it is available on all platforms.

The first observation was that the Microsoft Azure interface has a very short learning curve and is very intuitive by virtually presenting all the information in a single screen. The user starts with the main menu where you can find the most common services that you can use or explore on this platform. If the user wants to install a different option that is not present in this menu it is possible to click on the "New button" and to access a second menu with all Microsoft Azure applications. To simplify even more this approach, there is a search bar present. This proved for the students to be an advantage when they first explored the platform.

On the other hand the Google Cloud interface has adopted the same minimalist design and most options are placed in menus that are out of sight most of the time in

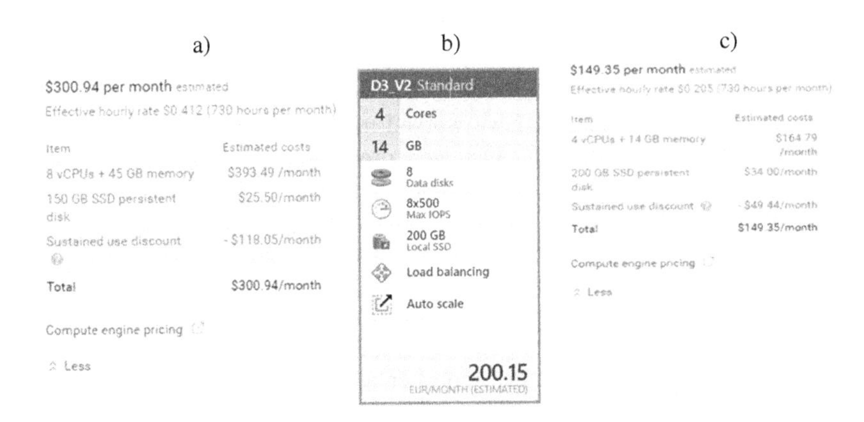

Fig. 2. Comparing VM's that can be created for the freely offered money in a) Google Cloud (top configuration ~ 300$) b) Microsoft Azure (configuration for ~ 200$) c) Google Cloud (for the same money as Microsoft Azure ~ 200$)

order to give the user more screen space for information selection. Google Cloud places on the first page all the information about the user account and in the left corner a hidden menu. When the user opens that menu it can access all applications, sorted and presented by categories. For most actions on the Google Cloud Platform, it is necessary to install the API Manager, so this option is placed it the second place after Home. This positioning proved to be a slowdown in the beginning but after the student learned the site structure, it proved to be an acceptable design choice.

Regarding the platform performance, the network latency was observed and the down-load and upload were monitored. The results are visible in Tables 3 and 4.

The next action was to perform a web site's deployment. Visual Studio was be used to deploy it on both the Google Cloud platform, and Microsoft Azure, but the latter integrates better by presenting a solution to deploy directly to Microsoft Azure platform [8]. If this option is chosen, a dialog box will pop up, where the user has to log into the Microsoft account, then options are presented to select the last publish details, as site name, Region and Service plan. After the website has been deployed on Microsoft Azure it is possible to verify the success in the Visual Studio console and in the Microsoft Azure console. This simple option inserted into the publishing flow subtly helps to keep the users inside a Microsoft software ecosystem.

6 Conclusions

This paper preselected comparatively the Google Cloud and Microsoft Azure platforms. A full four hour laboratory was used to make the students use both platforms and complete successfully the proposed tasks.

Both platforms proved not too difficult to use so a first time user can work on them without too much problems. Microsoft Azure proved to have the lease steep learning curve. Students were able to understand how services that are installed "in the cloud" are actually functioning and they were able to publish their own web site. Another simple application that can be created during this laboratory is a Google App Engine application that is deployed in the cloud and is accessed via the browser.

The level where the students have worked most of the time is IaaS that is very closely related to the operations that they were used to make in the virtual machines on the local computer and ensured a comprehensive transition.

The similarities between the two cloud platforms showed that both can be used individually for laboratory applications because they offer similar basic services and the differences are mainly in the naming conventions that the two providers use.

Complex services such as application virtualization (SaaS) or DNS load balancing were not tested, but can be tested in the future, the only limit being the time necessary to make the applications and test the implementation.

The two platforms proved to have differences and reflected each company's approach to the application implementation and went deeper then the user interface. The services and their pricing are different but each platform can present advantages that can influence a user to choose it.

References

1. Srinivasan, S.: Security, Trust, and Regulatory Aspects of Cloud Computing in Business Environments (2014). ISBN: 9781466657885
2. Bocchi, E., Drago, I., Mellia, M.: Personal cloud storage: usage, performance and impact of terminals. In: 2015 IEEE 4th International Conference on Cloud Networking (CloudNet), pp. 106–111 (2015)
3. Google Inc. Education Grants for Computer Science (2016). https://cloud.google.com/edu/
4. Microsoft Corporation. Try Microsoft Azure (2016). https://www.microsoftazurepass.com/azureu
5. Amazon Web Services, Inc. AWS Educate - Teach Tomorrow's Cloud Workforce Today (2016). https://aws.amazon.com/education/awseducate/
6. Amazon Web Services Inc. AWS Free Usage Tier FAQ (2016). https://aws.amazon.com/free/faqs/
7. Kaufmann, A., Dolan, K.: Price Comparison: Google Cloud Platform vs. Amazon Web Services, Whitepaper (2015). https://cloud.google.com/files/esg-whitepaper.pdf
8. Collier, M., Shahan, R.: Microsoft Azure Essentials - Fundamentals of Azure. Microsoft Press (2015). ISBN: 978-0-7356-9722-5

Author Index

Printed in the United States
By Bookmasters